Selected Titles in This Series

First International Congress of Chinese Mathematicians

AMS/IP
Studies in
Advanced
Mathematics

Volume 20

First International Congress of Chinese Mathematicians

Proceedings of ICCM98,
December 12–16, 1998
Morningside Center of Mathematics
Chinese Academy of Sciences
Beijing, China

Lo Yang and S.-T. Yau, Editors

American Mathematical Society · International Press

Shing-Tung Yau, General Editor

2000 *Mathematics Subject Classification.* Primary 05–XX, 08–XX, 11–XX, 14–XX, 22–XX, 35–XX, 37–XX, 80–XX.

Library of Congress Cataloging-in-Publication Data

International Congress of Chinese Mathematicians (1st : 1998 : Beijing, China)
First International Congress of Chinese Mathematicians : proceedings of ICCM98, December 12–16, 1998, Morningside Center of Mathematics, Chinese Academy of Sciences, Beijing, China / Lo Yang and S.-T. Yau.
p. cm. — (AMS/IP studies in advanced mathematics, ISSN 1089-3288 ; v. 20)
Includes bibliographical references.
ISBN 0-8218-2652-2 (alk. paper)
1. Mathematics—Congresses. I. Lo, Yang, 1939– II. Yau, Shing-Tung. III. Title. IV. Series.

QA1 .I7283 1998
510—dc21 00-066350

∞ The paper used in this book is acid-free and falls within the guidelines
established to ensure permanence and durability.
Visit the AMS home page at URL: `http://www.ams.org/`
Visit the International Press home page at URL: `http://www.intlpress.com/`

10 9 8 7 6 5 4 3 2 1 06 05 04 03 02 01

Contents

Preface

S.-T. Yau
Director of the Morningside Institute

The International Congress of Chinese Mathematicians was held in Beijing, beginning December 12, 1998. This was an historic event. It was the first occasion where the majority of Chinese mathematicians from all over the world gathered together to present their research. The results and the presentations were spectacular. The Morningside Foundation, as was led by Mr. Ronnie Chan, was very generous in providing the largest part of the funding of the conference including both the Morningside awards and the Morningside lectures. The Morningside lectures were given by Borcherds, Coates, Graham, and Stroock. The idea of the Morningside lectures was to invite distinguished mathematicians to give lectures which may stimulate research in areas of mathematics that are not strongly represented in China. We want to thank the executive committee which oversaw the conference and the scientific committee which made the selections. The communications committee, as was led by Huai-Dong Cao, was important in coordinating the communications between Chinese mathematicians all over the world. The main support for the conference came from the Morningside Institute and the Mathematics Institute. Of course, both institutes are part of the Chinese Academy of Sciences. President Lu of the Academy has been most supportive of both institutes.

The idea of establishing the Morningside Institute was initiated by President Lu and strongly endorsed by the Morningside Foundation, as led by Mr. Ronnie Chan. The Morningside Foundation has provided a large sum as a donation to support the Morningside Institute—both for its building and for operation expenses.

The Morningside Institute is an open institute that helps promote research among young scholars that come from all over China and many other countries. Professor Yang Lo and his staff from the Mathematics Institute of the Academic Sinica has been running the institute in a most efficient way. There are six programs for each year. These programs were set by a scientific committee under my chairmanship. Many prominent mathematicians have made outstanding contributions during these programs. S.W. Zhang, in particular, has been coming for several months each year to run the program on number theory and arithmetic geometry. I appreciate the efforts of these mathematicians.

The support of President Lu for this conference was indispensible and the majority of the funding was provided by the Morninside Foundation including all funds for the Morningside Lectures and the Morningside Medals. Ms. Happy Harum at the Morningside Foundation has done a tremendous and important job during the construction of the Morningside Institute building, in the preparation of this conference, and with the design of the Morningside Medals.

The conference will be held in different places every three years. In 2001, it will be held in Taiwan. We expect to have programs similar to the one held in 1998, including Morningside awards and Morningside lectures.

It is our sincere hope that friendship and cooperation among Chinese mathematicians will be promoted by this conference.

Inauguration Speech

Ronnie Chan
Director of the Morningside Foundation

The wealth of a nation, to a good degree, is a function of how advanced are its sciences, which in turn depends on the rigor of research. It is thus not surprising that countries with the highest standards in this regard are among the richest. Intensive analyses and relentless debates are hallmarks of their educational institutions.

China is an ancient civilization with a superb history of literary studies and the arts, a source of immense joy and pride to its people. However, although the country had enjoyed periods when sciences flourished, it had fallen behind the West in the past few centuries. To catch up, a tradition of rigor in research and a healthy respect for related institutions must be established. Intellectual honesty must exist both in individual research and in the scientific community as a whole including its peer review system.

Capital is also a very realistic necessity. Let no one sneer at it, just as no one should sneer at the pursuit of knowledge. We should applaud those who willingly donate to scientific endeavors, just as we honor the dedicated scientists. Install a plaque in gratitude to the generous, commemorate a building or an institute in their name—but go no further! To lightly confer honorary titles merely for a donation is a denigration of the awarding organization and an insult to the academicians working within. It will only stunt the development of the desired traditions. Donors must also refrain from meddling in academic freedom, and donations with certain strings attached should be rejected. We must preserve intellectual integrity and the sanctity of research.

It is with this understanding that Morningside has made possible the establishment of the new Center for Mathematics at the Chinese Academy of Sciences (CAS), and has endowed the Morningside Medals of Mathematics to honor world class mathematicians of Chinese descent. Proceedings in this volume are results of the symposium inaugurating these new institutions. Since science in itself is borderless, I am pleased to see many of the world's brightest minds converging on Beijing to interact with their local counterparts. The newly completed Mathematics Building at CAS will provide a comfortable, and hopefully nurturing environment for its users to push the envelope of knowledge.

All these could not have happened without the leadership and dedication of my dear friend Professor Shing-Tung Yau, now at Harvard. His love for his mother country and admiration for his adopted one are both contagious. My thanks also go to Professor Lu Yongxiang, president of CAS and Professor Yang Lo who, besides being a distinguished mathematician, has done so much to make the above happen. Our common goal is to help develop world class Chinese Mathematicians who will contribute to that branch of knowledge which is considered by many to be the Mother of all sciences.

第一屆全世界華裔數學家大會獻辭

丘成桐

公元一九九八年冬，華裔數學家，會于京師，講學修睦也。 聚三地之精英，集九州之豪士。 四方學者，同根同心。 獻明月之章，傳不朽之作，猗歟乎盛哉！

中土數學，源于九章，盛于陳華。 劉徽註術，祖沖割圓，陳氏作類，華氏堆壘。 此先人之智慧，今世之光華也。 當歐戰初萌，國難方殷，孫楊拔賢士於津滬，姜熊傳心法於清華。 雲南講學，江表立所。薪傳至今，已歷三世。 門生故舊，遍於天下。 俟陳氏去國，續領風騷，竟存一脈於海外。 而華氏返京，繼絕存亡，終創大業於國內。 斯時也，華熊子弟，得數論精妙，奠複變根基，蘇馮門下，發方程微義，開計算先河。 郁郁其文，無愧於先，彬彬大業，有傳於後矣。 然後十年動亂，九州震盪，上下受刑，天地為悲。 父子殊途，師友異路。學忠馬列，文必頌聖。 長者不敢有謀，學者不敢有志矣。 幸得將軍擒凶，國乃太平。 小平開國，世傳中興。百姓樂業，萬國來華。 惟國士不比外賓，商賈有逾學人。 於是家家望子放洋， 戶戶經商期富。文革宿怨未消，而國初銳氣消磨矣。 晚近遊子思歸， 學士東來，能者屢見于台灣，業績不斷于香江。 兩地科研，有聲於外國; 海疆數學，不亞於中土矣。 歐美諸君，承先賢之餘蔭，得大師之薰陶，共馳域外，各取明珠。 無奈華夏雖眾，長城未修， 天地雖寬，瑕疵難容。 終究德不如歐美，力不逮乎日蘇。 根之腐矣，枝葉不榮，葉之枯矣，根莖何養？ 今我同胞，其無相煎，如足如手，其無相負，如師如友。 咨爾賢俊，其能養士，而敦而敬，其能養氣， 而剛而正。 出則博文，入則約禮，究天地之造化，爭日月之光華，庶可立德立言，不朽後世矣。其辭曰：

仲冬嘉會，華廈初築。聚我精英，言歡修睦。集我同志，切磋演繹。思入風雲，玄想無極。

真理同探，永世其傳。何以為歡，必有歌弦。何以為慶，必有德言。盛筵必再，以待千年。

晨興數學所奠基典禮講辭

丘成桐

在這個世紀結束，新世紀來臨的時候，數學家盼望的不是萬兩黃金，也不是千年霸業。畢竟這些都會成為灰燼。我們追求的是永恆的真理，我們熱愛的是理論和方程。它比黃金還要珍貴和真實，因為它是大自然表達自己的唯一方法；它比詩章還要華美動人，因為當真理赤裸裸逞現時，所有頌詞都變得渺小；它可以富國強兵，因為它是所有應用科學的泉源；它可以安邦定國，因為它可以規劃現代社會的經絡。

希望大家能拼著赤子的熱誠，在科學院和陳氏兄弟的幫助下，不分界域，同心協力在中國建立一個有世界水平的數學中心。

ICCM98 Beijing Conference

Lo Yang
Director of the Institute of Mathematics, the Chinese Academy of Sciences
Deputy Director of the Morningside Institute, the Chinese Academy of Sciences

The ICCM98 was held December 12–16, 1998, at the Morningside Center of Mathematics, the Chinese Academy of Sciences in Beijing. The conference was initiated by Prof. Shing-Tung Yau of Harvard University, Fields Medalist, member of the Academy of Sciences, USA, and foreign member of the Academy of Sciences, China. More than 400 mathematicians from overseas, Hong Kong, Macau, Taiwan, and Mainland China participated in the conference. Among them were many outstanding young mathematicians from China who went abroad to study in the eighties; many well-established overseas Chinese mathematicians; more than 30 mathematicians from Taiwan; and more than 40 mathematicians from Hong Kong and Macau; and more than 200 mathematicians from the mainland.

Several distinguished guests attended the conference: R. Graham (former President of the American Mathematical Society), J.-P. Bourguignon (President of the European Mathematical Society and Director of Institut des Hautes Études), J. Jost (Director of the Max-Planck Mathematics Institute), M. Taylor (President of the London Mathematical Society), Seng Luan Lee (President of the Mathematical Society of Singapore), T. Sunada (Treasurer of the Japan Mathematical Society) and Kun Soo Chang (President of the Korean Mathematical Society).

The conference also invited four distinguished mathematicians to deliver Morningside Mathematics Lectures. The lecturers were R. Borcherds of Cambridge University, J. Coates of Cambridge University, R. Graham of Lucent Technologies, and D. Stroock of MIT.

On the first day of the conference, the morning of December 12, the first Morningside Medals Award ceremony was held in the People's Great Hall. The following distinguished guests were invited to the ceremony: Chen Siwei, Vice Chairman of the Standing Committee of the National People's Congress; Zhu Guang-Ya, Vice Chairman of the Chinese People's Political Consultative Conference; Lu Yong-Xiang, President of the Chinese Academy of Sciences; Ronnie Chan, Chairman of the Board of Trustees of the Morningside Group; Zhang Yu-Tai, the first General Secretary of the Association of Science and Technology, China; Wang Nai-Yan, Deputy Director of the National Science Foundation, China; and Duan Rui-Chun, Deputy Secretary of the Ministry of Science and Technology, China.

The Morningside Medals were established, through the Morningside Center of Mathematics, to encourage outstanding young mathematicians of Chinese descent in their pursuit of mathematical truth. The Morningside Medals Committee had seven members: L. Carleson, J. Coates, R. Graham, M. Kashiwara, P. Lax,

D. Stroock, and Shing-Tung Yau (chairman). At the ceremony it was announced that the two Gold Medal recipients were Chang-Shou Lin of National Chung-Cheng University and Shouwu Zhang of Columbia University; four Silver Medal recipients were Raymond Chan of Chinese University of Hong Kong, Chong-Qing Chen of Nanjing University, Kefeng Liu of Stanford University, and Tong Yang of City University of Hong Kong.

In the afternoon of December 12, the four Morningside Mathematics Lectures were delivered in the Conference Hall of Science and Technology. Before the Morningside Lectures, Mr. Ronnie Chan of the Morningside Group, Hong Kong, delivered a speech on scientific development and was well received by the audience. He expressed his wish to give full support to the advancement of mathematics within China.

In addition, in the early morning of December 12, the newly finished Morningside Mathematics Building was dedicated by Mr. Lu Yong-Xiang, the President of the Chinese Academy of Sciences; Mr. Ronnie Chan, Chairman of the Board of Trustees of the Morningside Group; and Prof. Shing-Tung Yau of Harvard University. The Morningside Mathematics Building was funded partly by the Morningside Group and partly by the Chinese Academy of Sciences.

In the evening of December 12, a concert was held in the Hall of Science and Technology.

From December 13 to December 16, 1998, there were 13 one-hour plenary lectures and 55 invited 45-minute lectures, as well as more than 200 contributed talks in various areas of mathematics.

It was a very successful conference. All the participants highly enjoyed the conference which reached a very high level academically.

It was decided that the next ICCM meeting will be held in December 2001, in Taipei.

Congratulation Speeches

AMS/IP Studies in Advanced Mathematics
Volume 20, 2001

Jean-Pierre Bourguignon
Director,
Institut des Hautes Études Scientifiques,
Bures-sur-Yvette, France,
and
President,
European Mathematical Society

Very honourable hosts,
Ladies and Gentlemen,
Dear colleagues and friends,

To honour young people having shown exceptional talent is always a welcomed event. I am very honoured to be given the opportunity to congratulate the laureates of the Morningside Awards and the Morningside Centre for its initiative, both as director of the Institut des Hautes Études Scientifiques and as president of the European Mathematical Society, that established similar awards at the European level.

Professor Jacques-Louis Lions, president of the *Académie des Sciences de Paris*, but also chairman of the Prize Committee for the European Prizes to be given at the opening of the Third European Congress of Mathematics to be held in Barcelona (Spain) in 2000, asked me to present to you his sincere regrets for being unable to attend this ceremony. He would have liked to congratule the laureates himself.

The world around, young people bring enthousiasm to their endeavours. Research institutes, such as the Morningside Centre of the Academia Sinica and the IHÉS, have the special responsability of encouraging their creativity in making available the most appropriate means for its expression in mathematics. It is our duty to convey to them the idea that mathematics is presently a thriving science, and that its future successes largely depend on their engagement.

The development of mathematics requires confrontation of ideas, and for that purpose mathematicians need to meet face-to-face with colleagues from different backgrounds and countries. This is why the IHÉS, that just celebrated its 40th anniversary, has all along its life been conducting an important visitors programme, under the leadership of a very small group of permanent professors. Last year we hosted some 200 visitors from 20 different countries for stays averaging 2.8 months. I understand that the Morningside Centre will follow a comparable route.

I am looking forward to a fruitful collaboration between our two institutes and hope that, in the near future, laureates will consider paying a visit to the IHÉS, which is also a gateway to the Paris mathematical community that counts close to one thousand active members, and more generally to the mathematical Europe.

Let me conclude by presenting to the director of the Morningside Centre, Professor Yang Lo, a volume of the relations between mathematics and theoretical physics, that was produced on the occasion of the IHÉS anniversary. A full collection of the *Publications Mathématiques de l'IHÉS* (87 volumes) is on its way to Beijing as a gift to the library of the Morningside Centre, a happy start of the future exchanges between the Morningside Centre and the IHÉS.

Once again, my sincere congratulations to the laureates and my best wishes for the newly established Centre.

AMS/IP Studies in Advanced Mathematics
Volume 20, 2001

Kun Soo Chang
President of the Korean Mathematical Society

Distinguished Chinese mathematicians.

Congratulations to Morningside Award winners! It is my honor and great pleasure to have this opportunity to address congratulatory speech at the opening ceremony of the first International Congress of Chinese Mathematicians(ICCM) and Morningside Award held here, the Great Hall of the People in Beijing.

On behalf of the Korean Mathematical Society, I would like to extend my sincere thanks to the Organizing Committee for inviting me to participate in this conference.

I am deeply greatful and wish to pay respect to Chinese mathematicians who receives Morningside Award today.

Last August, the general assembly of IMU in Dresden, Germany, decided that 2002 ICM will be held in Beijing. Korean Mathematical Society strongly supported China from the beginning when China applied for hosting ICM. I am sure that China deserves the ability to well organize ICM and to make it very sucessful. As a neighboring country, Korean Mathematical Society will encourage our members to participate in the ICM 2002 in Beijing as many as possible. Also our Society will organize many satelite conferences before or after ICM to make the ICM more attractive to many people. Also last August, in Dreden, 30 deligates from 15 Asian countries attended the first meeting to create AMU (Asian Mathematical Union). We have agreed the foundation of AMU and Korea will host ACM (Asian Congress of Mathematicians) in 2004 in Seoul. This will be another step to promote mathematics in Asia.

Historically and geographically China and Korea have had close relations. We should make effort to establish more friendly relations between two countries, in particular, to develope mathematics.

The Korean Mathematical Society was established in 1946 after World War II with only about 20 members. Half a century has passed since then and the Society has grown remarkably to a big mathematical community in Korea in terms of both quality and quantity. Today, there are more than 2200 members in the Society including 1200 ordinary members. We have 5 publications: 3 research journals on pure and applied mathematics, one journal on mathematical education and a newsletter. Two scientific meetings are organized annually by the Society, one in Spring and one in Fall. Besides, a number of scientific activities are managed by the Society. I strongly hope that every one who participates in this meeting will gain worthwhile experience and the Morningside center will become a corner stone to construct the highest mathematical building in the world as well as in Asia.

It will be my good experience to participate in this conference. I will keep my good memory for the hospitality of Chinese mathematicians.

Thank you very much for your attention.

AMS/IP Studies in Advanced Mathematics
Volume 20, 2001

David Eisenbud
Director,
Mathematical Sciences Research Institute
Berkeley, CA

As Director of the Mathematical Sciences Research Institute in Berkeley, I would like to extend my congratulations to the Morningside Institute and especially to its Founders on this happy event. A mathematical institute can provide enormous benefits to Mathematics and Science for a whole country by encouraging and intensifying research and helping to share resources. I myself have enthusiasm of members here at MSRI suggests that many other people have the same experience.

As a small token of my enthusiasm, I propose that MSRI will contribute copies of its publications, including its book series, reprints, and CDRoms, to the library of the Morningside Institute.

I am sorry that I cannot join you in person today, but I join you in spirit and wish your new institute good health and a long life.

AMS/IP Studies in Advanced Mathematics
Volume 20, 2001

Juergen Jost
Max Planck Society

Ladies and Gentlemen,

It is a great pleasure for me to participate in the first ICCM. I have witnessed the impressive contributions to mathematics made by Chinese scientists over the past two decades, and so it seems very befitting to me that an international congress brings together Chinese mathematicians from different countries and societies, but with a common cultural background, to document some of these achievements and to inspire a new generation of scientists.

In the course of the years, I have closely collaborated with more than 10 Chinese mathematicians, and the dedication to solving difficult problems that I always experienced in these collaborations has been a strong source of inspiration for me. Therefore, there are also personal reasons for me to be very glad to see that this Congress is taking place, and that a new international research center, the Morningside Institute, will be opened in Beijing, thanks to the outstanding scientific achievements and the energy of my friend Shing-Tung Yau and the generosity of Ronnie Chan.

In the name of the German mathematicians and the Max Planck Society, I wish the Congress and the Morningside Institute much success!

AMS/IP Studies in Advanced Mathematics
Volume 20, 2001

Seng Luan Lee
President of Singapore Mathematical Society

First International Congress of Chinese Mathematicians and Official Opening of the Morningside Center for Mathematics

More than four hundred mathematicians from all over the world gathered in Beijing from 12 to 16 December 1998 to attend the First International Congress of Chinese Mathematicians and to witness the Official Opening of the Morningside Center for Mathematics. I am honoured to be invited to deliver a talk at the Congress and to write an essay for its Proceedings. I am not less honoured and happy to meet so many outstanding mathematicians most of whom I have heard of or known through their research and publications but have not had the opportunity to meet before.

Mathematics was an ancient Chinese art. The collection of *Suanjing shi shu* (Ten mathematical classics) dates back to the 4th century AD, although China has lost its lead in modern science and mathematics. Therefore, these double events in Beijing in December 1998 have augured well for mathematics and marked an important milestone in its development. The Congress was a great success. I was very happy to see many excellent young mathematicians who have done great work and who will be leaders of mathematics in the near future, thanks to Professor Yau's leadership. His dedication and devotion to mathematics have inspired many young mathematicians. I was very impressed by the high quality of their work. The Congress has created greater awareness of the high standard of mathematics achieved by the community and inspired others to strive to reach greater heights. It has also provided an opportunity for networking. The establishment of the Morningside Center for Mathematics at the Chinese Academy of Science in Beijing is another big step in the development of mathematics in China. The Morningside Center, which is a generous donation of the Chan Brothers, and similar centers and institutes that have been established in Hong Kong and Taiwan will provide strong impetus to promote mathematical activities, generate more interest in mathematics and attract young talents to mathematics.

Singapore has a small but active group of mathematicians. In the past few years there has also been a tremendous increase in mathematical activities through generous and more flexible funding from the government. We have plugged into the international mathematical networks and have planned to play a bigger role in the development of mathematics.

The award of the Morningside Medals adds glory to the Congress which I hope will become an important triennial international event. May I take this opportunity to congratulate the 1998 award winners: Shouwu Zhang, Chang Shou Lin, Kefeng Liu, Raymond Chan, Chong-Qing Cheng, and Tong Yang.

On behalf of the Singapore Mathematical Society, may I extend my heartiest congratulations and appreciation to Professors S.- T. Yau, Yang Lo, Feng Qi, Huai-Dong Cao and all members of the Organising Committee for doing a wonderful job organising the first ICCM and working towards the establishment of the Morningside Centre of Mathematics. May I wish Morningside Center success

AMS/IP Studies in Advanced Mathematics
Volume 20, 2001

Tohsigazu Sunada
Mathematical Society of Japan

Ladies and Gentlemen,

(More than fifteen years ago, I came to China to participate in a conference held at Shanghai, and had a chance to stay at Beijin for a few days after the conference. I am happy to be here in Beijin again, and to see tremendous and impressive changes of Beijin City after fifteen years. I would like to come back again to Beijin in the occasion of International Congress of Mathematics which you are organizing in 2002, and will see more change.

This is rather my personal comments and let me say a few words for this ceremony on behalf of Japanese mathematicians.)

It is a real pleasure and privilege for me to have this opportunity to address congratulatory greetings to this inauguration ceremony for the Morningside Building of Mathematics and the Morningside Award of Mathematics.

I have learned that the centre is aimed at training young scholars and achieving excellent results in mathematics, and that the award has been established to encourage the outstanding young mathematicians.

It is needless to say that mathematics is a fundemental field in science where a young power is especially exerted for its development. Indeed, history tells us that many outstanding works in mathematics have been done by young people under the age 30. The thing that counts is how to provide a favorable environment to young people with latent talent and to encourage those people in an effective way. For this sake, we need a moral support from the general public and substantial supports from indivisuals and government.

I am delighted to hear that the Chinese Academy of Sciences and Mr. Ronnie Chan and Mr. Gerald Chan of the Morningside Group understand the nature of our field, and have started the new research institution of mathematics and the award. I am very much convinced that this institute will become a world centre of mathematics in coming 21st centuary, and young people from this institute will take active part in mathematical community.

As a representative from Mathematical Society of Japan, I would like to pay my deepest respects to Mr. Ronnie Chan and Mr. Gerald Chan and to all those who have gone to great pains in establishing this institute. I also hope that closer cooperation between China and other Asian countries will be developed through activities in this institute.

Thank you very much for your attention.

AMS/IP Studies in Advanced Mathematics
Volume 20, 2001

M. Taylor
London Mathematical Society

Dear Colleagues!

I bring to you the congratulations of the London Mathematical Society on this *happy and significant occasion.* Just as the Morningside Center is not just for Beijing *but for the whole of China* and indeed the world, so — in spite of its name — the London Mathematical Society *is not just for London*, but is the major learned society for mathematics in the United Kingdom, and it has a world-wide membership. On behalf of all our members, I welcome the foundation of the Morningside Center under the auspices of the renowned Chinese Academy of Sciences, and I congratulate the Chinese Government and the Morningside Group on a collaboration which has already proved so fruitful. The Morningside Mathematical Building will be an inspiration to all who are privileged to work in or to visit it.

Also on behalf of the London Mathematical Society, I should like to offer our congratulations on the founding of the Morningside Awards, and *to congratulate the worthy prizewinners.* They have set a very high standard but, we have no doubt, one which mathematicians will be encouraged to match in the years to come.

Chinese mathematicians are renowned throughout the world and are on the staffs of many of the *world's leading universities.* It is most appropriate to gather this enormous talent to celebrate the opening of the Morningside Mathematical Building with a Congress. The London Mathematical Society is proud to have some of its members amongst the invited speakers. We also recall with pleasure the visit made by Professor L.K. Hua to the United Kingdom in 1979 and his stimulating lecture to the London Mathematical Society.

The London Mathematical Society publishes a number of periodicals. To mark this occasion the Society will be donating some of its periodicals to the Center's library and hopes that this gesture will expand into an exchange of our journals and periodicals for yours.[1] The London Mathematical Society also has a scheme for supporting collaborative work and exchange visits between mathematicians in the UK *and various other countries.* This scheme has now been extended *to include China.* We hope that collaboration between mathematicians from Morningside Center and the *London Mathematical Society will flourish.*

[1] See Council Minute 4E, June 98. I wrote to John Erdos to ask where his negotiations had got to. Here is his reply: My letters to OUP and CUP asking them to send ongoing subscriptions and back copies from the beginning of 1995 went out on 10 July 1998. I have every reason to believe that journals have and are being sent.

The history of China, in both its cultural and mathematical aspects, exceeds that of the United Kingdom by thousands of years.[2] We are certain that mathematics in China has an even more glorious future, and that the Morningside Center will play a major part in forming the glorious future.

[2] *Cambridge History of Archaeology:* Pottery 4,000BC, Civilization 2,000BC, Kung Fu Tsu – Confucius – 551 – 479BC. Needham's *Science and Civilization in China,* Vol. 3, says Chou Pei Suan Ching, which may be as old as 400BC, has the Pythagoras Theorem in it. In the fifth century AD, the Chinese had a value for π which put them ahead of the world for a thousand years, and Pascal's triangle was published by Chu Shih Chieh in 1303, 200 years before it appeared in the West. Wolfgang Ruppert's *Matematik in China,* seems to confirm the Pythagoras date, and I believe says that Greeks did not write the Theorem down until 300BC. (I don't read German.)

COMMITTEES OF ICCM98

HONORARY CHAIRMEN

Shiing–Shen Chern, University of California, Berkeley
Buqin Su, Fudan University

Executive Committee

Ching-Li Chai, University of Pennsylvania
Ben Chow, University of Minnesota
Shui-Nee Chow, Georgia Institute of Technology & National University of Singapore
Fan Chung Graham, University of Pennsylvania
Gu Chaohao, Fudan University
Ka-Sing Lau, Chinese University of Hong Kong
Peter Li, University of California, Irvine
Lin Song Sun, National Chiao Tung University
Gang Tian, Massachusetts Institute of Technology
YANG Lo, Morningside Center of Mathematics
Shing-Tung Yau (Chairman), Harvard University

Scientific Committee

Tony Chan, University of California Los Angeles
Shiu–Yuen Cheng, Hong Kong University of Science and Technology
Charles Chui, Texas A&M University
Ming–Chang Kang, National Taiwan University
Ngaiming Mok, University of Hong Kong
Duong H. Phong, Columbia University
Jian–Shu Li, University of Maryland & Hong Kong University of Science and Technology
Daqian Li, Fudan University
Fanhua Lin, New York University

Tai–Ping Liu, Stanford University
McKenzie Wang, McMaster University
Yuan Wang, Institute of Mathematics, The Academy of Sciences of China
Stephen Yau, University of Illinois at Chicago
Jing Yu, Institute of Mathematics, Acad. Sinica, Taipei
Shing–Tung Yau (Chairman), Harvard University

Communications Committee

Huai-Dong Cao (Chairman), Texas A&M University
Gerard Chang, National Chiao Tung University
Qi Feng, Institute of Mathematics, The Academy of Sciences of China
Song-Ying Li, University of California, Irvine
Yanyan Li, Rutgers University
Michael Ng, Hong Kong University

SPONSORS AND DONORS

Morningside Foundation
(major donor)

The Academy of Sciences of China
Army Research Office of the United States
Epson Foundation
Institute of Mathematical Sciences, Chinese University of Hong Kong

MORNINGSIDE MEDALS

The Morningside Medals were supported by the Morningside group. These medals will be awarded triannually at the International Congress of Chinese Mathematicians. The award recepients will not have advance knowledge of their award.

The nomination committee for the Morningside Medals consists of up to 50 Chinese mathematicians for around the world. They provide names, vita, and qualifications of nominated individuals to the Morningside Medals Committee.

The Morningside Medals Committee will consist of internationally-renowned experts in mathematics of any nationality in order to insure the selection of the top qualified mathematicians. The 1998 Morningside Medals Committee had seven members: L. Carleson, J. Coates, R. Graham, M. Kashiwara, P. Lax, D. Stroock, and Shing-Tung Yau (chairman).

The Morningside Medals, (200,000 yuen for gold medals; 50,000 yuen for silver medals), the banquet, and the Morningside Lectures were funded fully by the Morningside Institute.

Morningside Award Medalists

The works of the Morningside Award gold medalists were presented as follows:

S.-T. Yau: The work of Chang-Shou Lin

J. Coates: The work of Shouwu Zhang

The works of the Morningside Award silver medalists were presented as follows:

T. Chan: The work of Raymond Chan

Z. Xia: The work of Chong-Qing Cheng

S.-T. Yau: The work of Kefeng Liu

T.-P. Lin: The work of Tong Yang

AMS/IP Studies in Advanced Mathematics
Volume 20, 2001

The work of Chang-Shou Lin

S.-T. Yau

C.S. Lin's research work has been concentrating on local isometric embedding problems, scalar curvature equations and semi-linear elliptic equations. The underlying nonlinear partial differential equations are originated from the geometry and closely related to problems in physics, chemistry and engineering. Therefore, the work emphasizes the study of the general property of solutions of these equations which involves lots of delicate analytical techniques.

Some of his important work are listed in the following:

1. Local isometric embedding.

This is an well-known geometry problem: Given a metric ds^2 in the neighborhood of some ds^2 point p, is it possible that the metric ds^2 can be isometrically embedded into $\mathbf{R}^3$ by a C^2 map? When the Gaussian curvature $K(p) \neq 0$, the answer is yes and has been known since last century. As for the general case, it was until the seventies that Pogorelov gave an example to show that the answer is no. In the example, the Gaussian curvature $K \geq 0$ and it is $C^{2,1}$ metric but can not be C^2 embedded into $\mathbf{R}^3$. The following results were proved by C.S. Lin.

(I) Suppose that ds^2 is C^s, $s \geq 10$ and the Gaussian curvature K is nonnegative, then there always exists a C^{s-6} isometrical embedding.

(II) Suppose that $K \in C^s$, $s \geq 6$, $K(p) = 0$,but $\nabla K(p) \neq 0$, then there always exists a C^{s-3} isometrical embedding.

The result in (I) was proved by the method of Nash-Moser. (II) is equivalent to solving a nonlinear Triconi equation.

2. Scalar curvature equation.

It is well known that the scalar curvature has important impact in geometry and topology. In particular, in the end of the seventies, a series of works by Schoen-Yau, especially the theory in the positive mass theorem in relativity and the result of Schoen on the Yamabe problem, proves the importance of scalar curvature in geometry or physics. From analytic point of view, there are still many aspects of the role played by the scalar curvature that deserve further study. After R. Schoen solved the Yamabe problem, there are many important works, for example, Escobar-Schoen, Bahri-Coron, Chang-Gursky-Yang, Schoen-Zhang, Chen-Li, etc., on S^3 and S^4. ¿From the viewpoint of conformal geometry, this is equivalent to solving a nonlinear partial differential equation. In the case that the manifold is

S^n, the equation is when $n \geq 3$,

$$\begin{cases} \Delta u - \frac{n(n-2)}{4} u + K(y) u^{\frac{n+2}{n-2}} = 0 \\ u > 0 \quad \text{on} \quad S^n \end{cases} \tag{1}$$

or when $n = 2$,

$$\Delta u - 1 + K(y) e^{2u} = 0 \quad \text{on} \quad S^2. \tag{2}$$

Equation (1) or (2) also occur in other subjects, for example, (2) often occur in the combustion theory and various vortex theory.

When K is identically a positive number, there exists a sequence of solutions (1) and (2) such that its total energy will be concentrated at some point in S^n. Therefore, for general scalar curvature function K, equations (1) and (2) provide a difficult madel but with lots of interesting phenomenons. The most important questions are how to completely characterize the behavior of a sequence of blowing up solutions near its blow up points and the interaction between different blow up points, to derive apriori estimates of solutions and the existence of solutions. In recent years, Lin has studied these problems thoroughly, particularly in the higher dimensional case. His work includes:

(i) Harnack inequality. He is also able to characterize the blow-up behavior of solutions which is only related to the first term of the Taylor expansion of K. The validity of the Harnack inequality is a major difference comparing to other known nonlinear equations.

(ii) Precisely calculate the interaction between different blow up points, and then derive the apriori estimate and a formula for counting the topological degree of the nonlinear equation (1). The formula is only related to the behavior of K at its critical points. This is an extension of the previous work in S^3 and S^4.

(iii) The role of the dimension. For example, in the case that K is a Morse function, when $n \geq 5$, its topological degree is always zero by using the formula. This verifies the conjecture by **R.** Schoen, that is, in higher dimensions, Morse theory is not as useful as in the lower dimensions. This can also be seen by constructing Morse functions K for $n \geq 5$, such that there does not exist solutions for (1).

(iv) When $n \geq 7$, construct Morse function K such that there exist infinite many solutions for (1) and their energy goes infinity.

(v) Uniqueness of the conformal metric. Construct certain scalar curvature K such that there is only one solution for (1) or (2).

(vi) Proved Liouville theorem. These theorems can be used to derive apriori estimates for solutions in the region where $K \leq 0$.

(vii) Harnack inequality of singular solutions and removable singularity theorem.

(viii) When $n \geq 9$, construct a weak solution of (1) such that its singular set is the equals S^n.

AMS/IP Studies in Advanced Mathematics
Volume 20, 2001

The work of Shouwu Zhang

John Coates

Zhang's work lies in number theory and arithmetical algebraic geometry, a general area of mathematics which is as deep, as beautiful, as difficult, and almost as old as Chinese civilization. It is also a highly evolved part of mathematics, and it is mysterious that Zhang has already made important contributions to three major questions in this field, namely the connexions between L-functions and arithmetic via the Gross-Zagier formula, the Galois distribution of points on abelian varieties, and what one might loosely call pure Arakelov geometry. I shall try and comment briefly on the major new results he has obtained on each of these questions.

Let me begin with Zhang's work on the Gross-Zagier formula. I recall that in the mid 1980's, Gross and Zagier [**1**] established what is probably the most remarkable and fruitful exact formula in the whole of number theory. The first ingredient in the formula is a primitive cusp form $f(z)$ of weight 2 for the congruence subgroup $\Gamma_0(N)$ consisting of all matrices $\left(\begin{smallmatrix} a & b \\ c & d \end{smallmatrix}\right)$ in $SL_2(\mathbb{Z})$ with $c \equiv 0 \mod N$; here N is any integer $\geqslant 1$. The second ingredient is an imaginary quadratic field K, which can vary and is subject only to the condition that every prime which divides N must split in K. We denote the discriminant of K by $-D$, and by χ_K the primitive Dirichlet character modulo D corresponding to K. To these two ingredients, we can attach a complex L-function $L(f, \chi_K, s)$ as follows. If $f(z) = \sum_{n=1}^{\infty} a_n e^{2\pi i n z}$, where $a_1 = 1$, we can define $L(f, \chi_K, s)$ for $R(s) > \frac{3}{2}$ by the product of Dirichlet series

$$L(f, \chi_K, s) = \left(\sum_{n=1}^{\infty} \frac{a_n}{n^s}\right) \cdot \left(\sum_{n=1}^{\infty} \chi_K(n) \frac{a_n}{n^s}\right).$$

This L-function has an entire analytic continuation, satisfies a functional equation relating its values at s and $2 - s$, and has a zero of odd multiplicity at $s = 1$. On the other hand, there is a classical modular curve $X_0(N)$ defined over $\mathbb{Q}$ which classifies elliptic curves endowed with a cyclic subgroup of order N, and, taking a suitable elliptic curve with complex multiplication by the ring of integers of K, we can define canonical points, called Heegner points, on $X_0(N)$ which are rational over the Hilbert class field of K. It should be stressed that these Heegner points are the only examples known at present of the construction of points on a curve over a given number field, other than by naively exhibiting points whose coordinates satisfy the equations of the curve. In a paper of extraordinary technical virtuosity, Gross and Zagier [**1**] proved a simple exact formula for the first derivative at

$s = 1$ of $L(f, \chi_K, s)$ in terms of the Néron-Tate canonical height of the f-isotypical component of the Heegner point attached to K, but now viewed as lying in the H-rational points of the Jacobian of $X_0(N)$; here H denotes the Hilbert class field of K. The remarkable consequences of the Gross-Zagier formula (notably, the solution of the problem of determining all imaginary quadratic fields with given class number via the earlier work of Goldfeld, and its applications to the conjecture of Birch and Swinnerton-Dyer via the work of Kolyvagin) have made its further study of central importance in number theory. I believe that Zhang's work in this direction has been the best contribution since the original work of Gross and Zagier. In a technically brilliant paper echoing the deep calculations of Gross and Zagier, he proved in [**2**] an exact analogue of the Gross-Zagier formula for all primitive cusp forms of even weight $\geqslant 2$, for $\Gamma_0(N)$. In Zhang's formula for forms of weight > 2, the Néron-Tate height of Heegner points on the Jacobian of $X_0(N)$ is replaced by a new global height pairing between Heegner cycles on suitable Kugo-Sato varieties. In another direction, in his joint paper with Goldfeld [**3**], he has made the first substantial simplification in the original proof of Gross and Zagier, by establishing a new analytic continuation of $L(f, \chi_K, s)$ which is better suited to these calculations than the classical Rankin-Selberg convolution. Finally, in [**4**], he has generalized both the original Gross-Zagier formula and its application to the conjecture of Birch and Swinnerton-Dyer given by Kolyvagin, to most primitive cusp forms of weight 2 on GL_2 of the adèle ring of a totally real number field F. I want to stress what an astonishing range of mathematical techniques is used in these three papers of Zhang, including very difficult techniques from classical analytic number theory, Arakelov geometry, and the theory of automorphic forms. I believe that this work alone marks him out as one of the most remarkable young arithmetical algebraic geometers in the world.

In a completely different direction, Zhang [**5**], after earlier collaboration on these questions with Szpiro and Ullmo, discovered a new general phenomena about the distribution of the Galois conjugates of certain infinite sequences of points on any abelian variety A defined over a number field F, which we assume to be embedded in the complex field $\mathbb{C}$. We write $O(x)$ for the set of all Galois conjugates of a point x in $A(\overline{\mathbb{Q}})$. Let $h : A(\overline{\mathbb{Q}}) \to \mathbb{R}$ denote the Néron-Tate height on A with respect to a symmetric and ample line bundle. Let $x_1, x_2, \ldots$ be any infinite sequence of points of $A(\overline{\mathbb{Q}})$ such that $h(x_n) \to 0$ as $n \to \infty$, and no infinite subsequence is contained in the translate of a proper abelian subvariety of A by a torsion point. Then Zhang proved that the sequence of orbits $O(x_1), O(x_2), \ldots$ is equidistributed with respect to the Haar measure dx on $A(\mathbb{C})$, in the sense that the sequence

$$\frac{1}{\sharp(O(x_n))} \sum_{y \in O(x_n)} f(y)$$

converges to the integral $\int_{A(\mathbb{C})} f dx$ for every continuous function f on $A(\mathbb{C})$. This result is non-trivial even when $x_1, x_2, \ldots$ consists of torsion points on A, and it yields a new proof of a theorem of Raynaud, as well as establishing a well known conjecture of Bogomolov. Another interesting consequence of Zhang's theorem is a classical theorem of Imai, which asserts that, for each prime number p, the abelian variety A has only a finite amount of p-power torsion in the field obtained by adjoining all p-power roots of unity to F. The methods used by Zhang rely heavily on arithmetic intersection theory, and his theorem is one of the most striking results to emerge from this theory to date.

Finally, Zhang has a large body of very interesting work in pure Arakelov geometry, extending earlier work of Faltings, Szpiro, Gillet, Soulé, and others. Just one of the striking results he has obtained here is the proof in [**6**] of the analogue of the Nakai-Moishezon theorem for arithmetic surfaces, namely a hermitian line bundle on an arithmetic surface is ample if and only if it is numerically positive.

In a subject as old and as intensely studied as number theory, history has shown that decisive new progress is made by mathematicians with broad vision, who are willing and able to apply new methods to the classical problems. Zhang's work, with its remarkable combination of methods from differential geometry, arithmetic geometry, and analytic number theory, falls exactly into this mould. He richly deserves the honour of being one of the first winners of the Morningside Award.

References

[1] B. Gross, D. Zagier, *Heegner points and derivatives of L-series*, Invent. Math. **84** (1986), 225-320.

[2] S. Zhang, *Heights of Heegner cycles and derivatives of L-series*, Invent. Math. **130** (1997), 99-152.

[3] D. Goldfeld, Z. Zhang, *The holomorphic kernel of the Rankin-Selberg convolution*, to appear.

[4] S. Zhang, *Heights of Heegner points on Shimura curves*, to appear.

[5] S. Zhang, *Equidistribution of small points on abelian varieties*, Annals of Math. **147** (1998), 159-165.

[6] S. Zhang, *Positive line bundles on arithmetic surfaces*, Annals of Math. **136**, 569-587.

AMS/IP Studies in Advanced Mathematics
Volume 20, 2001

The work of Raymond Chan

Tony F. Chan

It is both my pleasure and honor to be asked to introduce my friend and occasional collaborator — Raymond Chan, for the Morningside Silver Award.

In case you are wondering — he and I are not related, even though we have the same family name "Chan", which just happens to be one of the most common Chinese surnames. In particular, I have nothing to do with his being selected for the award!

1. Biography of Raymond Chan

First, I would like to mention a few highlights from Raymond's biography. Raymond was born and raised and educated in Hong Kong, and finished his Bachelor and Master degree at the Chinese University of Hong Kong. He then went abroad to study at the Courant Institute, receiving his PhD under Olof Widlund in 1985. After one year teaching at the University of Massachussetts at Amherst (I remember meeting Raymond during this period), he returned to take up a lecturer position at the University of Hong Kong in 1986 and he has remained in Hong Kong since then. He has moved around quite a bit, having taught at the Hong Kong University of Science and Technology (95-96) before taking up his current position of Professor at the Chinese University of Hong Kong. He is believed to be the only person who has taught at all three of these universities.

Before receiving the Morningside Award at this ceremony, Raymond has already received several honors and awards. Shortly after his PhD degree, he won the Leslie Fox Prize in 1989, offerred by Oxford University for the best singly authored paper by a young numerical analyst. More recently in 1997, he was a co-winner of the inaugural Fang Keng Prize in Scientific Computing, named in honor of one of the founding fathers of scientific computing in China. Together with Prof. S.T. Yau, Raymond now co-manages the Asian Journal of Mathematics.

Raymond has had four (4) PhD students and over 10 M. Phil. students. One of his former PhD students, Prof. Michael Ng of the University of Hong Kong, won honorable mention at the 1996 Householder Prize competition for the best PhD thesis in numerical linear algebra.

2. Raymond Chan's Main Research Area: Iterative Toeplitz Solvers

Next, I'll highlight a few of the major results in Raymond's research. Raymond works in the general area of computational mathematics. He is widely recognized as the leading expert on fast iterative solvers for structured matrix problems (a prime example is the class of Toeplitz matrices), as well as in their applications, including signal and image processing, PDE and integral equations, time series, etc. He and his student Michael Ng published an authoritative SIAM Review article in this area in 1996, which has become a standard reference in this field.

Let me review briefly this research area. Toeplitz matrices are those matrices T with constant entries down the main diagonals; i.e. $T_n = [t_{j-k}]$, for $0 \leq j,k < n$. These matrices arise frequently in applications because they correspond to discretized versions of *translational invariant* operators, such as signal filters, finite difference stencils, etc. Since an $n \times n$ Toeplitz matrix can be specified by only $O(n)$ information, there is a lot of inherent structure in the matrix, which can be exploited when designing computational algorithms involving Toeplitz matrices. For example, it is well known how to perform the matrix-vector multiply Tv, where v is a given vector in R^n, in only $O(n \log n)$ arithmetic operations, instead of the more obvious $O(n^2)$ inner-product based algorithm. On the other hand, it is less obvious how to exploit the Toeplitz structure when solving Toeplitz systems $Tx = b$. There are several reasons for why this is a difficult task. First, T^{-1} is generally not Toeplitz even if T is. Second, most algorithms for solving linear systems, such as Gaussian Elimination and various matrix factorizations, do not preserve the Toeplitz structure. Of course, these classical matrix algorithms can be carefully modified to take advantage of the Toeplitz structure and reduce the operation count. For example, the classical Levinson algorithm (1946) can be considered a special LU factorization of T but which solves Toeplitz systems in $O(n^2)$ time instead of standard Gaussian Elimination's $O(n^3)$ time. Since the 1980's, there have been many attempts to lower this complexity even further, and various *superfast* algorithms with $O(n \log^\alpha n)$ time, where α is a small integer, have been proposed. Unfortunately, many of these algorithms are also numerically unstable.

The lack of stable, superfast, *direct* Toeplitz solvers led to the search for *iterative* Toeplitz solvers, a direction pioneered by Gil Strang, Julia Olkin and Raymond Chan in the mid-1980's. The most popular iterative approach is the Preconditioned Conjugate Gradient method (PCG) and the key step is in designing an effective preconditioner. A preconditioner M for a matrix T should be a good approximation to T and it should also be easy to solve linear systems of the form $Mx = b$. Another way to think about a good preconditioner is that it corresponds to a system that is close to the original system but which can be solved efficiently. The conjugate gradient method provides an optimal way to make use of such an approximating problem in an iterative fashion to solve the original system. Each step of the PCG method requires the matrix-vector product Tv and the solution of $Mx = v$ for a given vector v. As mentioned earlier, for Toeplitz $T's$, Tv can be evaluated very fast, in $O(n \log n)$ time by using the FFT. How well M approximates T controls the convergence rate of the PCG method. It can be shown that the number of iterations required by the PCG method to reduce the initial residual by a constant factor can be bounded proportional to the square root of $\kappa(M^{-1}T)$, where $\kappa(A)$ denotes the condition number of A. In addition to the condition number, the *clustering* of the eigenvalues of the preconditioned system $M^{-1}T$ can also affect the convergence rate of the PCG method. More precisely, if $M^{-1}T$ has k distinct eigenvalues, then the PCG method converges in less than or equal to k iterations. Thus the challenge for iterative Toeplitz solvers is to construct preconditioners $M's$ which are efficient to solve systems with (preferably in $O(n \log n)$ time, commensurate with the time for Tv), but which also cluster the spectrum of $M^{-1}T$ (usually around 1).

Raymond's main work has been in designing such preconditioners and also in analysing their convergence properties. One of his very first piece of work in this area was in proving the convergence rate of Strang's preconditioner M_s. Strang proposed in the mid-80's to use *circulant* matrices as preconditioners for Toeplitz matrices. Circulant matrices have the nice property that they are universally diagonalizable by the Fourier Transform and thus systems with them as coefficient matrices can be solved rapidly in $O(n \log n)$ time using the FFT. The main challenge then is how to construct good circulant approximations to a given Toeplitz matrix T. Strang proposed the following: copy to M half of the diagonals of T, in particular those diagonals closest to the main diagonal, and then extend the remaing diagonals such that M becomes circulant. It can easily be shown that this process gives an unique M. What remains is to identify the classes of $T's$ for which such an M makes a good preconditioner. It should be intuitively clear that if the entries of T decays rapidly away from the main diagonal, then M should be a good approximation of T. This notion can be made precise by introducing the concept of a generating function f for T.

Generating Function (Grenandar and Szegö 84)
Let f be a 2π periodic function and define:

$$a_k = \frac{1}{2\pi} \int_{-\pi}^{\pi} f(0) e^{-ik\theta} d\theta, k = 0, \pm 1, \ldots .$$

Now define the $n \times n$ Toeplitz matrix T_n by:

$$T_n = [a_{j-k}], \quad 0 \leq j, k < n.$$

3. Some Results on Circulant Preconditioners by Raymond Chan

One of Raymond's first results is:

THEOREM 1. **(R. Chan 89)**
If the generating function f corresponding to a class of Toeplitz matrices T is positive and is in the Wiener Class, i.e. $\sum_{k=0}^{\infty} |a_k| < \infty$, then the spectrum $\lambda(M_s^{-1}T)$ clusters around 1 as n tends to ∞.

Thus, systems involving Toeplitz matrices with the underlying generating function belonging to the Wiener Class can be solved efficiently using Strang's preconditioner.

Another popular class of preconditioner is the *optimal* preconditioners M_c proposed by myself. What I did was to define M_c to be:

$$M_c = \arg\min_{C} \|C - T\|_F,$$

where $\|.\|_F$ denotes the Frobenius norm. The idea is to define M_c to be the circulant matrix which is closest to T in the Frobenius norm. Raymond and one of his M. Phil. students (who later finished his PhD under me at UCLA) proved that this way of defining circulant preconditioners has the desirable effects.

THEOREM 2. **(R. Chan and M. Yeung 92)**

1. **If f is positive and 2π periodic, then $\lambda(M_c^{-1}T)$ clusters around 1 as n tends to ∞.**

2. **If f is Lipschitz with order $0 < \nu \leq 1$, then the error after $2q$ iterations of the PCG method using the preconditioner M_c can be bounded as follows:**

$$\frac{\|\mathbf{error}_{2q}\|}{\|\mathbf{error}_0\|} \leq \prod_{k=2}^{q} \frac{c \log^2 k}{k^{2\nu}}.$$

In addition to proving that M_s and M_c have desirable properties, Raymond also extended and unified this whole area by introducing the concept of using different *kernels* in the Fourier space to define circulant preconditioners. For example, he proved that Strang's preconditioner correspons to the Dirichlet kernel and my preconditioner corresponds to the Fejer kernel.

THEOREM 3. **(R. Chan and M. Yeung 92)**

1. **The eigenvalues of M_s can be defined precisely using the generating function f for T and the Dirichlet kernel D as follows:**

$$\lambda_j(M_s) = \left(D_{\lfloor \frac{n}{2} \rfloor} * f\right)\left(\frac{2\pi j}{n}\right), \quad 0 \leq j < n,$$

where

$$D_k(\theta) = \frac{\sin\left(k + \frac{1}{2}\right)\theta}{\sin \frac{1}{2}\theta}.$$

2. **The eigenvalues of M_c can be defined precisely using the generating function f for T and the Fejer kernel F as follows:**

$$\lambda_j(M_c) = (F_n * f)\left(\frac{2\pi j}{n}\right), \quad 0 \leq j < n,$$

where

$$F_k(\theta) = \frac{1}{k}\left[\frac{\sin\left(\frac{k}{2}\theta\right)}{\sin\left(\frac{1}{2}\theta\right)}\right]^2.$$

This link between kernels and preconditioners has led to many other effective circulant preconditioners by making use of classical kernels in function theory.

4. Summary

In summary, I hope I have given you a glimpse of the area that Raymond works in and the mathematical depth of his results. What I don't have time to describe, however is his unique style of contributing significantly and deeply to both the theoretical and the practical aspects of this field. His theoretical results are not just isolated mathematical results but they make us understand the practical algorithms better and make them work better, more efficiently.

Raymond is recognized around the world as the "master" in his field and he is one of the most accomplished "home-grown" mathematicians in HK. I hope you'll agree with me that this MorningSide Award has been given to a most deserving person.

Thank you!!

AMS/IP Studies in Advanced Mathematics
Volume 20, 2001

The works of Chong-Qing Cheng

Zhihong Xia

Chong-Qing Cheng works in the area of dynamical systems and Hamiltonian dynamics. In recent years, he has made several very fundamental and outstanding contributions in various problems related to Kolmogorov-Arnold-Moser (KAM) theory.

Stability has long be a very important problem in dynaimcs. To justify his theory of equilibrium gas dynamics, Boltzmann in the last century conjectured that conservative dynamical systems (volume- preserving diffeomorphisms and symplectic diffeomorphisms) on compact manifolds are typically very unstable and even ergodic (cf. Herman [8], Xia [11]). This conjecture later was made more precise and has bee called the Ergodicity Hypothesis. A weaker conjecture, which has attracted much attention lately, is the so-called Quasi-Ergodicity Conjecture, which states that typical conservative dynamical systems on compact manifolds should admit a dense orbit. The Ergodicity Hypothesis turned out to be false in the case of symplectic diffeomorphisms, as the celebrated KAM (Kolmogorov-Arnold-Moser) theory shows that positive measure sets of invariant tori can survive small perturbations. However, for the volume preserving case, the ergodicity conjecture has been open until recently. In Cheng's joint work with Yi-Sui Sun [5], they proved that near-integrable volume-preserving diffeomorphisms in three dimensions admit a large set of two dimensional invariant tori, hence disproving both Ergodicity Hypothesis and the Quasi-Ergodicity Conjectures for volume-preserving diffeomorphisms. This is a very fundamental and important piece of work and their proof required both a great deal of technical strength and new ways of thinking about invariant tori. In the volume-preserving case, which is different from the symplectic case, no single invariant torus with fixed diophantine rotation number are to survive small perturbations. One needs to show that new invariant tori, probably with different rotation numbers, have to form nearby the disintegrated tori. Cheng and Sun's work has since inspired many other works in the area. A notable example is its generalization to high dimensional spaces, due independently to Herman (cf. Yoccoz [12]) and Xia [10]. Interestingly, the density conjecture for symplectic diffeomorphisms in compact manifold with dimension larger than two is still open and remains to be one of the most important problems in dynamical systems.

In the classical KAM theory, the existence of KAM tori is proved under the nondegeneracy condition on the Hessian of the unperturbed Hamiltonian, or the so-called Kolmogorov condition. An interesting and important problem is to replace the Kolmogorov condition by a weaker condition. Study in this area was lead by Rüssmann, who announced in 1986 [9] that the following condition is sufficient: The image of the frequency map of the unperturbed Hamiltonian is not restricted to any hyperplane. Unaware of Rüssmann's announcement, Cheng and Sun [6] independently proved that a large set of invariant tori still exist under under perturbations on the condition that image of the frequency map contains a curve which twists in all directions. Cheng's condition is equivalent to Russmann's condition and it turned out these are the weakest possible conditions with the required measure estimate. This result has very important applications to various problems such as the Newtonian n-body problem, where the Kolmogorov condition fails.

Cheng's technical strength is further demonstrated in his work on the existence of lower dimensional tori in Hamiltonian systems. KAM theory provides existence of invariant tori of half of the space dimension. Finding lower dimensional tori in Hamiltonian systems has been an interesting and difficult problem. The difficulty comes from the lack of hyperbolicity together with the intrinsic degeneracies of these problems. Recently, great progress has been made in this area by Eliasson [7], Herman, Bourgain [1] and others. So far Cheng's results are among the best. KAM invariant tori all have diophantine rotation numbers that survive under small perturbations. It is well-known that invariant tori with resonant rotation numbers break under typical perturbations. Cheng [2, 3] (see also Cheng & Wang [4]) showed that, in the case where one has only one resonant condition, when the resonant tori break, many lower dimensional invariant tori persist under perturbations. An important feature of Cheng's results is that there is no assumption in the types of perturbation and it gives the existence to both the normally hyperbolic and the normally elliptic lower dimensional tori. Cheng has also been able to obtain the optimal measure estimates on the size of these set of invariant tori.

References

1. Bourgain, J., *Quasi-periodic solutions of Hamiltonian perturbations of 2D linear Schrödinger equations*, Ann. of Math. 148 (1998), no. 2, 363–439
2. Cheng C.Q., *Birkhoff-Kolmogorov-Arnold-Moser tori in convex Hamiltonian systems*, Commun. Math. Phys. 177 (1996) 529-559
3. Cheng, C-Q, *Lower-dimensional invariant tori in the regions of instability for nearly integrable Hamiltonian systems*, Comm. Math. Phys. 203 (1999), no. 2, 385–419
4. Cheng, C-Q & Wang, S., *The surviving of lower-dimensional tori from a resonant torus of Hamiltonian systems*, J. Differential Equations 155 (1999), no. 2, 311–326
5. Cheng, C-Q & Sun, Y.S., *Existence of invariant tori in three-dimensional measure-preserving mappings*, Celest. Mech. Dyn. Astr., **47** (1990), 275-292
6. Cheng, C-Q & Sun, Y.S., *Existence of KAM tori in degenerate Hamiltonian systems. J. Differential Equations*, **114** (1994), no. 1, 288–335.
7. Eliasson, L. H., *Perturbations of stable invariant tori for Hamiltonian systems*, Ann. Scuola Norm. Sup. Pisa Cl. Sci. (4) 15 (1988), no. 1, 115–147 (1989).
8. Herman, M., *Some open problems in dynamical systems*, Proceedings of the International Congress of Mathematicians, Vol. II (Berlin, 1998). Doc. Math. 1998, Extra Vol. II, 797–808

9. Rüssmann, H., *Nondegeneracy in the perturbation theory of integrable dynamical systems*, Stochastics, algebra and analysis in classical and quantum dynamics (Marseille, 1988), 211-223, Math. Appl., 59, Kluwer Acad. Publ., Dordrecht, 1990
10. Xia, Z., *The Existence of the invariant tori in the volume preserving diffeomorphisms*, Ergod. Th. & Dynam. Sys. **12** (1992), 621-631
11. Xia, Z., *Stability Problems in Hamiltonian Dynamical Systems*, these proceedings.
12. Yoccoz, J.C., *Travaux de Herman sur les tores invariants*, Séminaire Bourbaki, No. 754, Astérisque **206** (1992), 311-344

AMS/IP Studies in Advanced Mathematics
Volume 20, 2001

The Work of Kefeng Liu

S.-T. Yau

Dr. Liu's work touches on several central topics in topology, geometry and mathematical physics: His work on elliptic genera rigidity; His work on Mirror conjecture and his work on moduli spaces.

He made important contributions to all these area. What distinguish Liu from others is that he has done much more than just give a proof. His proofs are simple, beautiful and above all are examples of elegance of mathematics.

1. Elliptic Genus

He proved a loop space analogue of the famous Atiyah-Hirzebruch $\hat{A}$-vanishing theorem for group actions and the loop space $\hat{A}$-genus, or the Witten genus. Its proof involves very subtle properties of the Jacobi theta-functions. He proved general vanishing and rigidity theorems for elliptic operators twisted by loop group representations. The proofs use the modular invariance of the characters of Kac-Moody algebras in a substantial way. One consequence of his results is a very simple proof of the famous Witten rigidity conjectures. This is the first time that Kac-Weyl character formula came into geometry and topology.

By using modular forms he generalized a 12-dimensional anomaly cancellation formula of Alveraz-Gaume and Witten to arbitrary dimensions. He studied mod 2 elliptic genera and answered a question raised by Witten about mod 2 partition functions as mod 2 modular forms in quantum field theory.

2. Mirror Principle

In recent joint works with Lian and Yau, they introduced the general notion of Euler data, by using which they obtained a conceptual understanding of the mirror symmetry formula to compute the characteristic classes on the stable map moduli in terms of hypergeometric series. As one of the corollaries they gave a complete proof of the mirror conjecture which relates the counting series of rational curves in a Calabi-Yau quintic manifold to the hypergeometric series of its mirror, as proposed by Candelas and his collaborators. Another consequence is the proof of the local mirror symmetry formula as proposed by Vafa et al. The method extends to computing general characteristic classes on the moduli spaces of stable maps into the so-called balloon manifolds.

3. Heat Kernel and Moduli Spaces

By using explicit formulas of heat kernels on Lie groups, the Reidemeister torsion and symplectic geometry, he derived general formulas for the intersection numbers on the moduli spaces of flat connections over a Riemann surface. As consequences several conjectural formulas of Witten were proved. Such moduli spaces have been studied from many different point of views in the past 50 years. People have used geometric invariant theory, gauge theory and loop groups. His heat kernel method is completely new and generalizes to the cases when the moduli spaces are singular, as well as when the Riemann surface has several boundary components. As another consequence the method gives several very general new vanishing theorems about the characteristic numbers of the moduli spaces.

AMS/IP Studies in Advanced Mathematics
Volume 20, 2001

The work of Tong Yang

Tai-Ping Liu

Dr. Tong Yang of City University of Hong Kong has made seminal contributions to the fundamental well-posedness problem for general systems of hyperbolic conservation laws. He helps, instrumentally, in establishing a satisfactory well-posedness theory. The theory yields, in particular, that the Euler equations for gas dynamics are mathematically complete when supplemented with the second law of thermodynamics in the form of an entropy condition.

Dr. Yang has made significant contributions to other nonlinear evolutionary partial differential equations. These include relaxation phenomena, gas flows with damping, vacuum, spherical waves, boundary behaviors and others. However, his series of ideas for the well-posedness problem by themselves deserve major recognition. Below is a brief description of the resolution of the problem.

The well-posedness problem for partial differential equations is usually the first one to be addressed. Having resolved the problem, one then moves on to the study of the behaviour and the computation of the solutions. For hyperbolic conservation laws, there have been substantial progresses on the solution behaviour, particularly the striking regularity and large-time behaviour of the solutions. On the other hand, the well-posedness problem, notably the smooth dependence of the solutions on the initial data, turned out to be a difficult problem. This is due to the presence of shock waves in the solution and the need to impose the entropy condition on the shock waves.

Starting over five years ago, while Tong Yang and I got stuck with the vacuum problem, we began to discuss other possible topics for our collaboration. At that time, Alberto Bressan and his collaborators were working hard and yielding results on the well-posedness problem using a homotopy approach. For years before that, I had always thought an approach based on robust nonlinear functionals would yield new understanding of the wave coupling. I proposed to Tong Yang this problem. We both understood that success might not come soon, if at all. What was needed was an entirely new understanding of nonlinearity beyond that of Jarne Glimm on wave interactions.

Tong Yang and I were both quite familiar with the fundamental and rich , by then classical, theory of James Glimm. Tong Yang was the product of the vibrant Chinese School of Shock Waves Theory, having learned much on Glimm Scheme from Professor Long-wei Lin of Zhongsan University. After a couple of months of intensive search, we uncovered more hidden difficulties. We also realized gradually that certain strengthening of the well known entropy estimate is required to take advantage of not only the entropy inequality but also the strong nonlinearity of the characteristic fields. This did not come until three years later. In between, we had a clearer idea of wave tracing algorithms which allowed us to concentrate on wave patterns with no nonlinear interactions and to search for ideas orthogonal to those of Glimm's.

It took Tong Yang and I more than five years to come up with a satisfactory well-posedness theory. We have written a series of papers, see my article in this volume. Tong Yang has visited me many times, but mostly worked by himself at City University. It has been my great fortune to collaborate with him. His perseverance and originality are of the first rank. I look forward to benefiting more from our collaboration.

STANFORD MATH DEPARTMENT, STANFORD, CA 94305-2125

Photographs

Opening Ceremony

Lu, Yong-Xiang, President of the Chinese Academy of Science and S.-T. Yau, Professor of Harvard University

Ronnie Chan, President of the Morningside Group

Jean-Pierre Bourguignon, President of European Mathematical Society and Director of Institut des Hautes Études Scientifiques

Ronald Graham, President of the American Mathematical Society and Vice-President of Research, Lucent Technologies

Jürgen Jost, Director of Max Planck Institute

Martin Taylor, President of the London Mathematical Society

Tosigazu Sunada, Treasurer of Japanese Mathematical Society

Morningside Award Recipients, Chairman of the Selection Committee, and Donors, Tong Yang, Kefeng Liu, S.-T. Yau, Chang-Shou Lin, Shou-Wu Zhang, Ronnie Chan, Raymond Chan, Chong-Qing Cheng

Morningside Gold Medalists with Ronnie Chan, Chang-Shou Lin, Ronnie Chan, Shou-Wu Zhang

Mr. Zhu Guang-Ya, Vice Chairman of Chinese People's Political Consultative Committee presenting a gold medal to Chang-Shou Lin

Mr. Chen Siwei, Vice Chairman of Standing Committee of People's Congress of China, presenting a gold medal to Shou-Wu Zhang

Morningside Medal

Morningside Silver Medalists
Tong Yang, Kefeng Liu, Chong-Qing Cheng, Raymond Chan

Prof. S.-T. Yau commenting on the work of Chang-Shou Lin

Prof. John Coates commenting on the work of Shou-Wu Zhang

Prof. Tony Chan commenting on the work of Raymond Chan

Prof. Jeff Xia commenting on the work of Chong-Qing Cheng

Prof. S.-T. Yau commenting on the work of Kefeng Liu

Prof. T.-P. Liu commenting on the work of Tong Yang

Speakers of the Conference

Morningside Lectures

AMS/IP Studies in Advanced Mathematics
Volume 20, 2001

Problems in moonshine

Richard E. Borcherds

The talk at the I.C.C.M. was an introduction to moonshine. As there are already several survey articles on what is known about moonshine ([B94], [B99], [D-M94], [D-M96], [F-H98], [J]), this paper differs from the talk and will concentrate mainly on what we do not know about it.

Moonshine is not a well defined term, but everyone in the area recognizes it when they see it. Roughly speaking, it means weird connections between modular forms and sporadic simple groups. It can also be extended to include related areas such as infinite dimensional Lie algebras or complex hyperbolic reflection groups. Also, it should only be applied to things that are weird and special: if there are an infinite number of examples of something, then it is not moonshine.

We first quickly review the original moonshine conjectures of McKay, Thompson, Conway and Norton [C-N]. The classification of finite simple groups shows that every finite simple group either fits into one of about 20 infinite families, or is one of 26 exceptions, called sporadic simple groups. The monster simple group is the largest of the sporadic finite simple groups, and was discovered by Fischer and Griess [G]. Its order is

$$\begin{aligned}8080, 17424, 79451, 28758, 86459, 90496, 17107, 57005, 75436, 80000, 00000\\ = 2^{46}.3^{20}.5^{9}.7^{6}.11^{2}.13^{3}.17.19.23.29.31.41.47.59.71\end{aligned}$$

(which is roughly the number of elementary particles in the earth). The smallest irreducible representations have dimensions $1, 196883, 21296876, \ldots$. On the other hand the elliptic modular function $j(\tau)$, defined by

$$j(\tau) = \frac{\left(1 + 240\sum_{n>0}\sigma_3(n)q^n\right)^3}{q\prod_{n>0}(1-q^n)^{24}}$$

has the power series expansion

$$j(\tau) = q^{-1} + 744 + 196884q + 21493760q^2 + \ldots$$

where $q = e^{2\pi i\tau}$. John McKay noticed some rather weird relations between coefficients of the elliptic modular function and the representations of the monster as follows:

Supported by NSF grant DMS-9970611.

$$\begin{aligned} 1 &= 1 \\ 196884 &= 196883 + 1 \\ 21493760 &= 21296876 + 196883 + 1 \end{aligned}$$

where the numbers on the left are coefficients of $j(\tau)$ and the numbers on the right are dimensions of irreducible representations of the monster. The term "monstrous moonshine" (coined by Conway) refers to various extensions of McKay's observation, and in particular to relations between sporadic simple groups and modular functions.

McKay and Thompson suggested that there should be a graded representation $V = \oplus_{n\in\mathbf{Z}} V_n$ of the monster, such that $\dim(V_n) = c(n-1)$, where $j(\tau) - 744 = \sum_n c(n)q^n = q^{-1} + 196884q + \cdots$. To characterize V, Thompson suggested looking at the McKay-Thompson series

$$T_g(\tau) = \sum_n Tr(g|V_n)q^{n-1}$$

for each element g of the monster. For example, $T_1(\tau)$ should be the elliptic modular function. Conway and Norton [C-N] calculated the first few terms of each McKay-Thompson series by making a reasonable guess for the decomposition of the first few V_n's into irreducible representations of the monster. They discovered the astonishing fact that all the McKay-Thompson series appeared to be Hauptmoduls for certain genus 0 subgroups of $SL_2(\mathbf{R})$. (A Hauptmodul for a subgroup Γ is an isomorphism from $\Gamma\backslash H$ to $\mathbf{C}$, normalized so that its Fourier series expansion starts off $q^{-1} + O(1)$.)

The module V was constructed explicitly as a representation of the monster group in [F-L-M], and it was shown in [B92] that this module satisfies the moonshine conjectures, using the fact that it has the structure of a vertex algebra. In the rest of this paper we will look at various conjectural ways to generalize this.

Problem 1. Find a "natural" construction for the monster vertex algebra V. All known constructions for it construct it as the sum of several (usually two) different pieces, and it takes a lot of work to show that the vertex algebra structure can be defined on this sum, and to show that the monster acts on it. It would be much nicer to have some sort of construction which gives V as "just one piece".

One idea for doing this might be to "stabilize" the monster vertex algebra by tensoring it with copies of the vertex algebra of the 2-dimensional even Lorentzian lattice $II_{1,1}$. The reason for this is that if we add this lattice to any Niemeier lattice, we get the same answer $II_{25,1}$, so the Niemeier lattices can be recovered from norm 0 vectors in $II_{25,1}$. Moreover the proof of the moonshine conjectures involves taking a tensor product with the vertex algebra of $II_{1,1}$, showing that this is a natural operation. The most naive version of this question would be to ask if the monster vertex algebra tensored with the vertex algebra of $II_{1,1}$ is isomorphic to the vertex algebra of $II_{25,1}$. This seems rather unlikely, though as far as I know it has not been disproved.

Problem 2. Construct a good integral form on the monster vertex algebra V. This should have the property that each of the homogeneous pieces of V has a self dual symmetric bilinear form on it. This integral form might follow from a good answer to problem 1, but might also be possible to prove independently as follows. The usual construction for the monster vertex algebra gives a self dual integral form "up to 2-torsion". If we could carry out a similar construction "up to p-torsion" for some other prime p then it might be possible to splice these together to get a good integral form. There are some constructions [D-M92], [M96] of the monster vertex algebra which look as if they might extend to give a construction up to 3-torsion. One advantage of a good integral form for the monster vertex algebra is that it would make the study of modular moonshine [R], [B-R], [B98] somewhat easier.

Problem 3. Find an affine ind group scheme over the integers for the monster Lie algebra. (Here an affine ind group scheme is roughly a commutative linearly topologized Hopf algebra, in much the same way that an affine group scheme is more or less the same as a commutative Hopf algebra. For infinite dimensional groups, it seems to be necessary to allow linear topologies on the Hopf algebra because there is often no natural map from H to $H \otimes H$, but only to the completion $H \hat{\otimes} H$.) Over the rational numbers it is not hard to find an ind group scheme whose Lie algebra is the monster Lie algebra and whose group of connected components is the monster group. The question is whether one can find a good integral form for this ind group scheme. There are also several intermediate questions: can one find a good integral form for the monster Lie algebra (which would follow from a good integral form for the monster vertex algebra) and can one find a formal group over the integers for the monster Lie algebra? At the moment it is not really clear what one would do with such an ind group scheme. Perhaps it could be used to study infinite dimensional automorphic forms corresponding to the monster Lie algebra, or perhaps its points with values in finite fields might be useful groups. For the related case of the fake monster Lie algebra there is some progress on these questions in [B99a], but the monster Lie algebra seems harder because it is not generated by the root spaces of roots of zero or positive norm.

Problem 4. Prove Norton's generalized moonshine conjectures [N86].

Roughly speaking, these conjectures assign modular functions to pairs of commuting elements of the monster, rather than to elements. The point is that the group $SL_2(\mathbf{Z})$ acts not only on (genus 0) modular functions but also on pairs of commuting elements of any group. Some progress has been made on Norton's conjectures by Dong, Li, and Mason [D-L-M], who proved the generalized moonshine conjectures in the case when g and h generate a cyclic group by reducing to the case when $g = 1$ (the ordinary moonshine conjectures). G. Höhn [H] has made some progress in the harder case when g and h do not generate a cyclic group by constructing the required modules for the baby monster (when g is of type $2A$). It seems likely that his methods would also work for the Fischer group Fi_{24}, but it is not clear how to go further than this. Dong has recently nearly proved the generalized moonshine conjectures. Very roughly speaking, his proof is an extension of Zhu's proof that characters of certain vertex algebras are modular functions to the equivariant case.

Problem 5. Explain the connection (if any) between moonshine and the "Y presentation" of the monster [ATLAS]. The latter is a particularly easy presentation of the monster group conjectured by Conway and Norton and finally proved by Ivanov (see [I]), which looks roughly like the presentation of a Coxeter group with one extra relation. Miyamoto [M95] found a very suggestive relationship between this presentation and the monster vertex algebra, by finding a set of involutions of the monster vertex algebra satisfying the necessary relations. His proof used properties of the E_6^4 Niemeier lattice.

Problem 6. Classify the generalized Kac-Moody algebras that have Weyl vectors and whose denominator functions are automorphic forms (possibly with singularities). The simplest example of such a Lie algebra is the monster Lie algebra itself, with denominator function $j(\sigma)-j(\tau)$. There are plenty of other similar rank 2 Lie algebras, some related to moonshine and sporadic groups, and some which do not seem to be related to sporadic groups. There are also many examples in various ranks up to 26, which is probably the highest possible rank. It is possible to relax the condition on the generalized Kac-Moody algebra that a Weyl vector should exist, and ask just for those whose denominator function is an automorphic form. Gritsenko and Nikulin have classified some of these algebras in low ranks [G-N], [N].

Problem 7. Find all "interesting" hyperbolic reflection groups (this includes the arithmetic ones and some others). This is closely related to the problem of finding interesting generalized Kac-Moody algebras, because the Weyl group of an interesting generalized Kac-Moody algebra is often an interesting hyperbolic reflection group. Nikulin has proved that the number of hyperbolic reflection groups which are close to being arithmetic (in various senses) is finite, and known examples suggest that there are probably a few thousand of them. There seems to be a rather mysterious connection of these hyperbolic reflection groups to genus 0 modular groups; slightly more precisely, most of the interesting hyperbolic reflection groups seem to be associated with certain modular forms with poles of low order at cusps. For example, Conway's reflection group of $II_{1,25}$ is associated with the function $1/\Delta(\tau)$. However this correspondence is not well understood; see [B99b] for some examples.

Problem 8. Find "natural" constructions of the generalized Kac-Moody algebras in the previous problems. All these algebras can be constructed using generators and relations, but this is not a satisfactory way of constructing them; for example, it is very hard to see interesting symmetry groups of these algebras (such as the monster group) in this approach. What we would prefer is a construction in which one can explicitly see the interesting symmetry groups. For example, the monster Lie algebra is constructed from the monster vertex algebra as the Lie algebra of physical states of strings on a 26 dimensional orbifold, so the action of the monster can be seen directly because it acts on the monster vertex algebra. Scheithauer [S] has recently found a similar construction for the "fake monster Lie superalgebra" of rank 10. Harvey and Moore [H-M] have made an exciting suggestion for realizing these algebras using BPS states, or by constructing them using the cohomology groups of moduli spaces of vector bundles of surfaces.

Problem 9. Is there anything like moonshine for the 6 sporadic groups not involved in the monster group? Note that most and probably all the Chevalley groups act naturally on vertex algebras over finite fields, and many of the sporadic groups involved in the monster act naturally on vertex algebras, sometimes over finite fields as in modular moonshine and sometimes over the rational numbers as for the monster. This means that most finite simple groups act naturally on vertex algebras, and an obvious question is whether they all do. Presumably any simple group acting on a natural vertex algebra would have some sort of connection with modular functions. Unfortunately (as far as I know) no one has ever found any serious evidence that the remaining 6 sporadic groups (the Janko groups J_1, J_3, and J_4, the O'Nan group, the Lyons group, and the Rudvalis group) have any connection with vertex algebras or modular functions.

Problem 10. Explain McKay's weird observation (described in [B99] for example) relating the Dynkin diagrams E_8, E_7, and E_6 with the monster, the baby monster, and the group Fi_{24}. The point is that conjugacy classes of pairs of commuting involutions of the monster seem to correspond to vertices of the affine E_8 diagram, and there is a similar connection between the baby monster and the E_7 Dynkin diagram, and between the group $Fi_{24}.2$ and E_6. This may be related to the (well understood) McKay correspondence between Dynkin diagrams and finite rotation groups in 3 dimensions.

Problem 11. Lian and Yau [L-Y] showed that mirror maps for K3 surfaces are sometimes the inverses of Hauptmoduls; for example one of their mirror maps is

$$q - 744q^2 + 356652q^3 + \cdots$$

which is the inverse of the elliptic modular function $j(\tau)$. They found that several other Hauptmoduls of elements turned up. They asked the rather speculative question about whether there is some direct connection between the monster and K3 surfaces. This seems rather wild as there is no obvious way in which the monster could be connected with K3 surfaces, but on the other hand this is what most people said about McKay's original observation connecting the monster and elliptic modular functions!

Problem 12. Is there a monster manifold? More precisely, Hirzebruch asked if there was a 24 dimensional manifold acted on by the monster with Witten genus $j(\tau) - 744$. If so, it might be possible to use this to construct the monster vertex algebra, or at least its underlying space. Hopkins and Mahowald recently constructed a manifold with the correct dimension and Witten genus, but so far it is unclear how to construct an action of the monster on it.

Problem 13. Complex hyperbolic reflection groups. Allcock [A] recently constructed some striking examples of complex hyperbolic reflection groups from the Leech lattice, or more precisely from the complex Leech lattice, a 12 dimensional lattice over the Eisenstein integers. This complex reflection group looks similar in several ways to Conway's real hyperbolic reflection group of the lattice $II_{1,25}$. Allcock also showed that there is an automorphic form on complex hyperbolic space vanishing exactly on the reflection hyperplanes of this reflection group. Several other complex hyperbolic reflection groups found by Allcock are closely related to various moduli spaces, for example the moduli space of cubic surfaces [A-C-T]. Most of these complex hyperbolic reflection groups seem to have something to do with moonshine, though it is hard to be precise about what the relationship is. For example many of them are related to automorphic forms that in turn are related to moonshine. So a general and rather vague question is: what is going on? For some more specific questions we could ask for a classification of arithmetic complex hyperbolic reflection groups, and for each of them we can ask if it is related to some moduli space and some automorphic form. There is some speculation that finite complex reflection groups should be related to some so far unknown algebraic structures provisionally called "spetses" [M99] in the same way that the real reflection groups are related to Lie algebras. As the real hyperbolic reflection groups are closely related to Kac-Moody algebras, an obvious question is to ask if there is an extension of spetses to complex hyperbolic reflection groups.

Problem 14. Is there a "nice" moduli space related to $II_{1,25}$? The denominator function of the rank 10 fake monster Lie superalgebra turns out to be an automorphic form on the period space of Enriques surfaces vanishing exactly along the points of singular Enriques surfaces [B96]. A direct construction of this automorphic form was found by Harvey and Moore [H-M98] and Yoshikawa [Y], following a suggestion of Jorgenson and Todorov [J-T]. (Note that, as pointed out in [Y], the main theorem stated in [J-T] is not correct.) The fake monster Lie superalgebra and the fake monster Lie algebra seem similar in many ways, so we can ask if there is a similar moduli space related to the fake monster Lie algebra and its root lattice $II_{1,25}$. So the period space should be the hermitian symmetric space corresponding to the lattice $II_{2,26}$, with singular points along the divisors of norm -2 vectors, and the moduli space should be the quotient by the automorphism group of the lattice $II_{2,26}$. One can also ask similar questions about other generalized Kac-Moody algebras. Freitag has suggested that maybe most "interesting" moduli spaces should arise in a similar way, related to automorphic forms on the hermitian symmetric space of $\mathbf{R}^{2,n}$ that have an infinite product expansion and all of whose zeros correspond to roots of some lattice. See [F-H] for some examples.

Problem 15. Modular forms with poles at cusps often turn up in the questions in this paper. Is there some useful analogue for these forms of the L-functions or Dirichlet series of classical cusp forms? Note that the obvious Dirichlet series formed by the coefficients does not converge anywhere, so we have to use a different way to define the "L-function". One possibility is to just use the Mellin transform. This does not converge if there are poles at cusps, so it would be necessary to regularize the integral in some way. Another possibility would be to look at functions with poles at some cusps and zeros at others, and just integrate between two zeros. The resulting Mellin transforms would satisfy some functional equation coming from the functional equation of modular forms in the usual way, but it is hard to see what else one can say about them. Perhaps one could look at a set of modular forms with singularities invariant under the Hecke algebra and ask what the corresponding properties of the Mellin transforms are. (In the case of forms without singularities, the Hecke algebra has eigenfunctions and the action of the Hecke algebra on the Eigenfunctions corresponds to an Euler product decomposition on the L-series. However if the modular forms have cusps then the Hecke algebra tends to act freely because Hecke operators make singularities worse, so there are no eigenfunctions.)

References

[ATLAS] J. H. Conway, R. T. Curtis, S. P. Norton, R. A. Parker, R. A. Wilson, Atlas of finite groups. Oxford University Press, Oxford, 1985. ISBN: 0-19-853199-0

[A] D. Allcock, The Leech lattice and complex hyperbolic reflections, 1997 preprint, available from http://www.math.utah.edu/~allcock

[A-C-T] D. Allcock, J. A. Carlson, D. A. Toledo, A complex hyperbolic structure for moduli of cubic surfaces. C. R. Acad. Sci. Paris Sér. I Math. 326 (1998), no. 1, 49–54.

[B92] R. E. Borcherds, Monstrous moonshine and monstrous Lie algebras. Invent. Math. 109, 405-444 (1992).

[B94] R. E. Borcherds, Simple groups and string theory, First European congress of mathematics, Paris July 1992, Ed. A. Joseph and others,vol. 1 p. 411-421, Birkhauser 1994.

[B96] R. E. Borcherds, The moduli space of Enriques surfaces and the fake monster Lie superalgebra. Topology vol. 35 no. 3, 699-710, 1996.

[B98] R. E. Borcherds, Modular moonshine III, Duke Math. J. 93 (1998), no. 1, 129–154.

[B99] R. E. Borcherds, What is moonshine? Proceedings of the International Congress of Mathematicians, Vol. I (Berlin, 1998). Doc. Math. 1998, Extra Vol. I, 607–615.

[B99a] R. E. Borcherds, The fake monster formal group. Preprint math.QA/9805123 Duke Math. J. 100 (1999), no. 1, 139–165.

[B99b] R. E. Borcherds, Reflection groups of Lorentzian lattices. Preprint math.GR/9909123.

[B-R]R. E. Borcherds, A. J. E. Ryba, Modular Moonshine II, Duke Math Journal Vol. 83 No. 2, 435-459, 1996.

[C-N] J. H. Conway, S. Norton, Monstrous moonshine, Bull. London. Math. Soc. 11 (1979) 308-339.

[D-L-M] C. Dong, H. Li, G. Mason, Modular invariance of trace functions in orbifold theory. q-alg/9703016

[D-M92] C. Dong, G. Mason, The construction of the moonshine module as a Z_p-orbifold. Mathematical aspects of conformal and topological field theories and quantum groups (South Hadley, MA, 1992), 37–52, Contemp. Math., 175, Amer. Math. Soc., Providence, RI, 1994.

[D-M94] Moonshine, the Monster, and related topics. Proceedings of the AMS-IMS-SIAM Joint Summer Research Conference held at Mount Holyoke College, South Hadley, Massachusetts, June 18–23, 1994. Edited by Chongying Dong and Geoffrey Mason. Contemporary Mathematics, 193. American Mathematical Society, Providence, RI, 1996.

[D-M96] C. Dong, G. Mason, Vertex operator algebras and Moonshine: a survey. Progress in algebraic combinatorics (Fukuoka, 1993), 101–136, Adv. Stud. Pure Math., 24, Math. Soc. Japan, Tokyo, 1996.

[F-H98] The Monster and Lie algebras. Proceedings of the Special Research Quarter, including the 2nd Monster Conference, held at The Ohio State University, Columbus, OH, May 1996. Edited by J. Ferrar and K. Harada. Ohio State University Mathematical Research Institute Publications, 7. Walter de Gruyter & Co., Berlin, 1998. pp. ISBN: 3-11-016184-2

[F-H] E. Freitag, C. F. Herman, Some modular varieties of low dimension. 1999 preprint, available from http://www.rzuser.uni-heidelberg.de/~t91/skripten/preprints/orth.dvi

[F-L-M] I. B. Frenkel, J. Lepowsky, A. Meurman, Vertex operator algebras and the monster, Academic press 1988. (Also see the announcement A natural representation of the Fischer-Griess monster with the modular function J as character, Proc. Natl. Acad. Sci. USA 81 (1984), 3256-3260.)

[G] R. L. Griess, The friendly giant. Invent. Math. 69 (1982), no. 1, 1–102.

[G-N] V. Gritsenko, Nikulin, Automorphic forms and Lorentzian Kac-Moody algebras, Parts I and II. Preprints alg-geom/9610022 and alg-geom/9611028. Internat. J. Math. 9 (1998), no. 2, 153–199 and 201–275.

[H-M] J. A. Harvey, G. Moore, On the algebras of BPS states. Comm. Math. Phys. 197 (1998), no. 3, 489–519.

[H-M98] J. A. Harvey, G. Moore, Exact gravitational threshold correction in the Ferrara-Harvey-Strominger-Vafa model. Phys. Rev. D (3) 57 (1998), no. 4, 2329–2336.

[H] G. Höhn, Selbstduale Vertexoperatorsuperalgebren und das Babymonster. [Self-dual vertex-operator superalgebras and the Baby Monster] Dissertation, Rheinische Friedrich-Wilhelms-Universität Bonn, Bonn, 1995. Bonner Mathematische Schriften 286. Universität Bonn, Mathematisches Institut, Bonn, 1996.

[I] A. A. Ivanov, Geometry of sporadic groups. I. Petersen and tilde geometries. Encyclopedia of Mathematics and its Applications, 76. Cambridge University Press, Cambridge, 1999. ISBN: 0-521-41362-1

[J] E. Jurisich, Generalized Kac-Moody Lie algebras, free Lie algebras and the structure of the Monster Lie algebra. J. Pure Appl. Algebra 126 (1998), no. 1-3, 233–266.

[J-T] J. Jorgenson, A. Todorov, Enriques surfaces, analytic discriminants, and Borcherds's Φ function. Comm. Math. Phys. 191 (1998), no. 2, 249–264. Erratum in Comm. Math. Phys. 207 (1999), no. 2, 495–497.

[L-Y] B. H. Lian, S.-T. Yau, Mirror symmetry, rational curves on algebraic manifolds and hypergeometric series. (English. English summary) XIth International Congress of Mathematical Physics (Paris, 1994), 163–184, Internat. Press, Cambridge, MA, 1995. Mirror maps, modular relations and hypergeometric series. II. S-duality and mirror symmetry (Trieste, 1995). Nuclear Phys. B Proc. Suppl. 46 (1996), 248–262.

[M99] G. Malle, Spetses. Proceedings of the International Congress of Mathematicians, Vol. II (Berlin, 1998). Doc. Math. 1998, Extra Vol. II, 87–96

[M96] P. S. Montague, Ternary codes and Z_3-orbifold constructions of conformal field theories. Groups, difference sets, and the Monster (Columbus, OH, 1993), 411–432, Ohio State Univ. Math. Res. Inst. Publ., 4, de Gruyter, Berlin, 1996.

[M95] M. Miyamoto, 21 involutions acting on the Moonshine module. J. Algebra 175 (1995), no. 3, 941–965.

[N86] S. P. Norton, Appendix to G. Mason's paper in The Arcata Conference on Representations of Finite Groups (Arcata, Calif., 1986), 181–210, Proc. Sympos. Pure Math., 47, Part 1, Amer. Math. Soc., Providence, RI, 1987.

[N] V. V. Nikulin, On the classification of hyperbolic root systems of rank three, parts I, II and III. Preprints alg-geom/9711032, alg-geom/9712033, math.AG/9905150

[R]A. J. E. Ryba, Modular Moonshine?, In "Moonshine, the Monster, and related topics", edited by Chongying Dong and Geoffrey Mason. Contemporary Mathematics, 193. American Mathematical Society, Providence, RI, 1996. 307-336.

[Y] K. I. Yoshikawa, Enriques surfaces and analytic torsion. Preprint math.AG/9808129

[S] N. R. Scheithauer, Vertex algebras, Lie algebras, and superstrings. J. Algebra 200 (1998), no. 2, 363–403.

Mathematics department, Evans Hall #3840, University of California at Berkeley, CA 94720-3840 U. S. A. www home page www.math.berkeley.edu/~reb

E-mail address: `reb@math.berkeley.edu`

AMS/IP Studies in Advanced Mathematics
Volume 20, 2001

The Ornstein–Uhlenbeck Process on a Riemannian Manifold

Daniel W. Stroock

ABSTRACT. In this article the *Ornstein–Uhlenbeck process* is introduced and discussed on a Riemannian manifold whose Ricci curvature is bounded from below, and various estimates are dervied about the associated heat flow semigroup. For example, it is shown that this semigroup is hypercontractive, a result obtained earlier [**15**] by F–Y Wang using a procedure which is somewhat more involved.

1. A Preliminary Apology

Although this article appears in the proceedings of the first ICCM, its contents bears no relation to the Morningside Lecture which I gave at the ICCM. Indeed, as I explained at the time of its delivery, my lecture in Beijing was on a topic about which I know so little that I am loath to give further, and more lasting, evidence of my ignorance. On the other hand, I am happy and honored to have it known that I participated in the first ICCM. Thus, when I was asked to contribute something to the proceedings, I had no reservations about accepting.

2. Introduction

Perhaps the most famous variant of basic Brownian motion is the one introduced by Ornstein and Uhlenbeck [**14**] in an attempt to provide a physical rationalization for the non-differentiability of Brownian paths. The paths of the Ornstein–Uhlenbeck process are related to Brownian paths via the linear integral equation

$$X(t,x,w) = x + w(t) - \frac{1}{2}\int_0^t X(\tau,x,w)\,d\tau, \quad t \in [0,\infty).$$

That is, given an $x \in \mathbf{R^d}$ and an $\mathbf{R^d}$-valued Wiener path $w \in \mathfrak{W}(\mathbf{R^d}) \equiv \{w \in C([0,\infty);\mathbf{R^d}) : w(0) = 0\}$, the associated Ornstein–Ulenbeck path $X(\,\cdot\,,x,\mathbf{w})$ starting a x is obtained by translating w to x and then introducing a linear restoring force. In terms of generators, the presence of this restoring force results in the

1991 *Mathematics Subject Classification.* Primary 54C40, 14E20; Secondary 46E25, 20C20.
Key words and phrases. Diffusion processes, differential geometry, logarithmic Sobolev ine quality.
Research contained in this article was partially supported by NSF grant DMS #9625782.

generator $\frac{1}{2}\Delta$ of Brownian motion being replaced by the generator $\frac{1}{2}(\Delta - x \cdot \nabla)$. Equivalently, the heat flow semigroup $\{\mathbf{H_t} : \mathbf{t} \geq \mathbf{0}\}$ on $C_b(\mathbf{R^d}; \mathbf{R})$ given by

$$\mathbf{H_t f(x)} = \int_{\mathbf{R^d}} \mathbf{f(y) h_t(y - x)\, dy}, \quad \text{where } \mathbf{h_t(z)} \equiv (\mathbf{2\pi t})^{-\frac{\mathbf{d}}{2}} \exp\left(-\frac{|\mathbf{z}|^2}{\mathbf{2}}\right),$$

which gives the transition mechanism for Brownian motion, gets replaced by the Markov (i.e., both constants as well as non-negativity are preserved) semigroup $\{\mathbf{U_t} : \mathbf{t} > \mathbf{0}\}$ given by

$$\mathbf{U_t f(x)} = \int_{\mathbf{R^d}} \mathbf{f(y) h_{1-e^{-t}}}\left(\mathbf{y} - \mathbf{e}^{-\frac{\mathbf{t}}{2}}\mathbf{x}\right) \mathbf{dy}.$$

Another expression for the same thing is

$$\mathbf{U_t f(x)} = \left(\mathbf{1} - \mathbf{e^{-t}}\right)^{-\frac{\mathbf{d}}{2}} \int_{\mathbf{R^d}} \mathbf{f(y)} \exp\left(-\frac{\mathbf{e^{-t}|x|^2 - e^{-\frac{t}{2}} x \cdot y + e^{-t}|y|^2}}{\mathbf{2(1 - e^{-t})}}\right) \mu(\mathbf{dy}),$$

where $\mu(dy) = h_1(y)\, dy$; and from this latter expression it is clear that each $\mathbf{P_t}$ admits a unique extension as a self-adjoint, trace class contraction on $L^2(\Gamma; \mathbf{R})$. Moreover, because the kernel is positive as well as trace class, one can easily check that the spectrum of $\mathbf{U_t}$ consists of nothing but eigenvalues lying in $(0, 1]$ and that 1 is an isotated, simple eigenvalue. On the other hand, the operators $\mathbf{U_t}$ are less smoothing than one might expect. In fact, they fail to map $L^1(\mu; \mathbf{R})$ into $L^p(\mu; \mathbf{R})$ for any $p > 1$. Equivalently, they map none of the spaces $L^p(\mu; \mathbf{R})$, $p \in [1, \infty)$, into $L^\infty(\mu; \mathbf{R})$. In fact, as E. Nelson (cf. [**8**]) the best that can be shown is that

$$\|\mathbf{U_t f}\|_{\mathbf{L^{q(t)}}(\mu;\mathbf{R})} \leq \|\mathbf{f}\|_{\mathbf{L^2}(\mu;\mathbf{R})} \quad \text{when } \mathbf{q(t)} = \mathbf{1} + \mathbf{e^{2t}}.$$

Moreover, if $q > 1 + e^{2t}$, then, by taking f to be a suitable Gaussian, one can make $\|\mathbf{U_t f}\|_{\mathbf{L^q}(\mu;\mathbf{R})} = \infty$ even though $\|f\|_{L^2(\mu;\mathbf{R})} < \infty$.

Since Nelson's original example, results like the one in (0,1) have come to be known as *hypercontractive estimates*. Although it is seldom true that one can get them to be as sharp as Nelson's, there are, by now, general principles on which their derivation can be based. In particular, first Federbush [**3**] and then, in a more precise form, Gross showed that hypercontractive estimates are equivalent to a minimal sort of Sobolev inequality, which Gross dubbed a *logarithmic Sobolev inequality*. In particular, it has come to be understood that hypercontraction follows from the conjunction of essentially any sort of smoothing with a spectral gap. For example, under very general conditions, one can show (cf. § 6.1 in [**2**]) that if $\{\mathbf{P_t} : \mathbf{t} > \mathbf{0}\}$ is a Markov semigroup which is symmetric with respect to a probability measure μ, then $\{\mathbf{P_t} : \mathbf{t} > \mathbf{0}\}$ will satisfy hypercontractive estimates whenever there exist $T \in (0, \infty)$ and $1 < p < q < \infty$ such that $\mathbf{P_T}$ maps $L^p(\mu; \mathbf{R})$ into $L^q(\mu; \mathbf{R})$ and there exists an $\gamma > 0$ such that

$$\|\mathbf{P_1 f}\|_{\mathbf{L^2}(\mu;\mathbf{R})} \leq \mathbf{e}^{-\gamma} \|\mathbf{f}\|_{\mathbf{L^2}(\mu;\mathbf{R})} \quad \text{for } \mathbf{f} \in \mathbf{L^2}(\mu; \mathbf{R}) \text{ with } \int \mathbf{f}\, \mathbf{d}\mu = \mathbf{0}.$$

Let M denote a complete, connected, d-dimensional Riemannian manifold whose Ricci curvature is bounded below, choose and fix a reference point $x_0 \in M$, and set

$$U(x) = \tfrac{1}{2}\mathrm{dist}(\mathrm{x}, \mathrm{x_0})^2, \quad \mathrm{x} \in \mathrm{M}.$$

A reasonable generalization of the $\mathbf{R^d}$-valued Ornstein–Uhlenbeck process is the diffusion corresponding to the generator whose action is given (somewhat formally) by

$$\mathcal{L}^{(\alpha)} f = \tfrac{1}{2} e^{2\alpha U} \operatorname{div}\big(e^{-2\alpha U} \operatorname{grad} f\big).$$

Indeed, if λ_M is the Riemannian measure on M, then, by standard volume comparison estimates, one can show that there exists a $\alpha \in (0, \infty) \longmapsto \rho^{(\alpha)} \in (0, 1]$, which depends only on d and κ, such that

$$\rho^{(\alpha)} \mathbf{V} \le \mathbf{Z}^{(\alpha)} \equiv \int_{\mathbf{M}} \mathbf{e}^{-2\alpha \mathbf{U}}\, \mathbf{d}\lambda_{\mathbf{M}} \le \frac{\mathbf{V}}{\rho^{(\alpha)}}$$
$$\text{where } \mathbf{V} \equiv \operatorname{vol}(\mathrm{B}_1),$$

and B_1 denotes the Riemannian ball of radius 1 centered at x_0. Moreover, one should suspect that the Markov semigroup $\{\mathbf{P}_{\mathbf{t}}^{(\alpha)} : \mathbf{t} > \mathbf{0}\}$ determined by $\mathcal{L}^{(\alpha)}$ will be symmetric with respect to the probability measure $\mu^{(\alpha)}$ given by

$$\mu^{(\alpha)}(dy) = \frac{1}{Z^{(\alpha)}} e^{-2\alpha U(y)}\, \lambda_M(dy).$$

Besides making sure that all these quantities are well-defined and possess the basic properties alluded to above, the main result of this note is the proof (cf. Theorem 2.9, especially (2.10), below) that the diffusion determined by $\mathcal{L}^{(\alpha)}$ can be dominated on finite intervals in terms of the Brownian motion on M. As a consequence of this domination and standard estimates about heat kernels and derivatives of the distance function on a Riemannian manifold, it is shown (cf. Corollary 3.9) that, for any $\alpha > 0$, $\{\mathbf{P}_{\mathbf{t}}^{(\alpha)} : \mathbf{t} > \mathbf{0}\}$ is hypercontractive. In fact, as my procedure makes explicit, the hypercontractive rate of this semigroup can be made to depend only on dimension, lower bounds on the Ricci curvature, and the size of the largest $\gamma_\alpha > 0$ for which

$$\|\mathbf{P}_{\mathbf{1}}^{(\alpha)} \mathbf{f}\|_{\mathbf{L}^2(\mu^{(\alpha)};\mathbf{R})} \le \mathbf{e}^{-\gamma_\alpha} \|\mathbf{f}\|_{\mathbf{L}^2(\mu^{(\alpha)};\mathbf{R})} \quad \text{for } \mathbf{f} \in \mathbf{L}^2(\mu^{(\alpha)};\mathbf{R}) \text{ with } \int_{\mathbf{M}} \mathbf{f}\, \mathbf{d}\mu^{(\alpha)} = \mathbf{0}.$$

Unfortunately, although we will show (cf. Corollary 3.6) that such a $\gamma_\alpha > 0$ always exists, my method gives no insight into the problem of determining its size.

3. The Ornstein–Uhlenbeck Process

Before going any further, it is necessary to be precise about what we mean by "the Ornstein–Uhlenbeck process associated with the operator $\mathcal{L}^{(\alpha)}$" in (1.3). In fact, we had better begin by saying exactly what operation (1.3) is meant to be describing, and for our purposes it will be best to adopt the following convention. Let $\mathcal{C}$ denote the cut locus of the reference point x_0. Then (cf. (1.2)) U is smooth off of $\mathcal{C}$ and

$$\big\|\operatorname{grad}_{\mathrm{x}} \mathrm{U}\big\|^2 = 2\mathrm{U}(\mathrm{x}) \quad \text{for } \mathrm{x} \in \mathrm{M} \setminus \mathcal{C}.$$

In particular, we can first define the (measurable) vector field Y on M by

$$Y_x = \begin{cases} \operatorname{grad}_{\mathrm{x}} \mathrm{U} & \text{if } x \in M \setminus \mathcal{C} \\ 0 & \text{if } x \in \mathcal{C}, \end{cases}$$

and then interpret the action of $\mathcal{L}^{(\alpha)}$ on $f \in C^2(M; \mathbf{R})$ as

$$\mathcal{L}^{(\alpha)} f(x) = \tfrac{1}{2} \Delta f(x) - \alpha Y_x f.$$

Indeed, it is clear that the right hand side of (2.2) gives the unambiguous meaning of $\mathcal{L}^{(\alpha)} f(x)$ when $x \notin \mathcal{C}$, and, because $\mathcal{C}$ has λ_M-measure 0, it should not matter much what meaning we assign when $x \in \mathcal{C}$.

Next, let $\mathcal{P}(M)$ denote the space of continuous paths $p : [0,\infty) \longrightarrow M$, and turn $\mathcal{P}(M)$ into a complete, separable, metric space with the topology of uniform convergence on compact intervals. Further, use $\mathcal{B}$ to denote the Borel σ-algebra $\mathcal{B}_{\mathcal{P}(M)}$ over $\mathcal{P}(M)$, and, for $t \in [0,\infty)$, let $\mathcal{B}_t$ denote the sub σ-algebra generated by the maps $pp(\tau)$ as τ runs over $[0,t]$. Finally, $\mathbf{M_1}\big(\mathcal{P}(\mathbf{M})\big)$ will be the space of Borel probability measures $\mathbf{P}$ on $\mathcal{P}(M)$, and we think of $\mathbf{M_1}\big(\mathbf{P}(\mathbf{M})\big)$ as a complete, separable space by giving it the *weak topology* (cf. § 3.1 in [**11**]) in which convergence means convergence when tested against elements of $C_{\mathrm{b}}\big(\mathcal{P}(M);\mathbf{R}\big)$.

We will say that $\mathbf{P} \in \mathbf{M_1}\big(\mathcal{P}(\mathbf{M})\big)$ *solves the martingale problem for* $\mathcal{L}^{(\alpha)}$ *starting at* x if $\mathbf{P}\big(\mathbf{p}(\mathbf{0}) = \mathbf{x}\big) = \mathbf{1}$ and, for each $f \in C_{\mathrm{c}}^{\infty}(M;\mathbf{R})$,

$$f\big(p(t)\big) - \int_0^t \mathcal{L}^{(\alpha)} f\big(p(\tau)\big)\, d\tau$$

is a $\mathbf{P}$-martingale relative to $\{\mathcal{B}_t : t \geq 0\}$. Our first goal is to check that, for each $x \in M$, there is precisely one solution starting at x. Actually, uniqueness causes no difficulty. Indeed, as explained in § 6.6 of [**10**], uniqueness for solutions to a martingale problem is an essentially local question. Hence, the fact that we are dealing here with a differentiable manifold makes no difference, and so Theorem 10.1.3 in [**10**] can be used to check that, starting at each $x \in M$, there is at most one solution to the martingale problem for $\mathcal{L}^{(\alpha)}$.[1] Thus, what remains is to check that there is a solution. The easiest way to check this would be to follow the procedure given in Chapter 10 of [**10**] and simply reduce the problem to checking that *explosion* cannot occur. However, because it provides an estimate which will be important in what follows, I have chosen a somewhat more quantitative approach.

In order to describe this approach, it is necessary to review some basic facts about Brownian motion on M. In the first place, it exists. More precisely, for each $x \in M$, there is precisely one solution $\mathbf{P}_{\mathbf{x}}^{(\mathbf{0})}$ to the martingale problem for $\mathcal{L}^{(0)} \equiv \frac{1}{2}\Delta$ starting at x. Again, the only problem is that of possible explosion, and that can (cf. [**16**] or § 8.3.4 in [**12**]) be ruled out by an application of Yau's familiar inequality (cf. § 8.3.3 in [**12**] for a less familiar derivation)

$$\Delta U \leq d + \sqrt{2d\kappa U} \text{ in the sense of distributions,}$$

where

$$\kappa \equiv \sup\Big\{-\langle \mathrm{Ric_x X_x, X_x}\rangle : \mathrm{x \in M \ \&\ X_x \in T_xM \text{ with } \|X_x\| \leq 1}\Big\}$$

is the uniform upper bound on the negative part of the Ricci curvature. In fact, it follows from (8.62) in [**12**] that, for all $(T,x) \in (0,\infty) \times M$ and $r > 0$,

$$\mathbf{P}_{\mathbf{x}}^{(\mathbf{0})}\left(\sup_{\mathbf{t}\in[\mathbf{0},\mathbf{T}]} \mathrm{dist}\big(\mathrm{x}, \mathrm{p(t)}\big) \geq \mathrm{r}\right) \leq (\mathbf{1}+\mathbf{d})! \exp\left(-\frac{\mathbf{r^2}}{\mathbf{2Te^{\kappa T}}}\right).$$

[1]For the reader who is wondering whether a different choice of the vector field Y might have led to a martingale problem with different solutions, the answer is that, because the ambiguity is restricted to a set of λ_M-measure 0, it would not. More precisely, no matter how Y is chosen on $\mathcal{C}$, solutions to the corresponding martingale problem will, almost surely, spend no time in $\mathcal{C}$ and therefore will not see how Y is given on $\mathcal{C}$.

Finally, if $h_t(x,y)$ denotes the heat kernel for the heat equation $\partial_t u = \frac{1}{2}\Delta u$, then[2]

$$\mathbf{E}^{\mathbf{P}_{\mathbf{x}}^{(0)}}\big[\mathbf{f}(\mathbf{p}(\mathbf{t}))\big] = \int_{\mathbf{M}} \mathbf{f}(\mathbf{y})\mathbf{h}_{\mathbf{t}}(\mathbf{x},\mathbf{y})\,\lambda_{\mathbf{M}}(\mathbf{dy}).$$

In particular, from Fubini's theorem and (2.6) it is clear that

$$\lambda_M(\Gamma) = 0 \implies \mathbf{E}^{\mathbf{P}_x^{(0)}}\left[\int_0^\infty 1_\Gamma\big(p(\tau)\big)\,d\tau\right] = 0,$$

and, by combining (2.5) with (2.6), one sees that

$$\lim_{t\searrow 0}\sup_{x\in M} \int_{\mathrm{dist(y,x)\geq r}} h_t(x,y)\,\lambda_M(dy) = 0.$$

Finally, as soon as one knows both existence and uniqueness, one can easily show (cf. Theorems 6.2.1 and 6.3.4 and in [**10**]) that $x \in M \longmapsto \mathbf{P}_x^{(0)} \in \mathbf{M_1}\big(\mathbf{P(M)}\big)$ is continuous and that the family $\{\mathbf{P}_x^{(0)} : x \in M\}$ is strong Markov.

Most of this section will be devoted to the derivation of the following statement.

Theorem 2.9 For each $\alpha > 0$ and $x \in M$, there is precisely one solution $\mathbf{P}_x^{(\alpha)}$ to the martingale problem for $\mathcal{L}^{(\alpha)}$ starting at x. Moreover, $x \in M \longmapsto \mathbf{P}_x^{(\alpha)} \in \mathbf{M_1}\big(\mathbf{P(M)}\big)$ is continuous. Finally, if (cf. (2.4)) $K^{(\alpha)} \equiv d\left(\alpha + \frac{\kappa}{\alpha}\right)$, then

$$\begin{aligned} &e^{-\alpha U(x)}\mathbf{E}^{\mathbf{P}_x^{(\alpha)}}\big[F(p)\big] \\ &\quad\leq e^{K^{(\alpha)}T}\mathbf{E}^{\mathbf{P}_x^{(0)}}\left[\exp\left(-\alpha U\big(p(T)\big) - \frac{3\alpha^2}{2}\int_0^T U\big(p(\tau)\big)\,d\tau\right)F(p)\right]\end{aligned}$$

whenever $T > 0$ and $F : \mathcal{P}(M) \longrightarrow [0,\infty)$ is $\mathcal{B}_T$-measurable.

In what follows, we use B_r and $\bar{B}_r$ to denote, respectively, the open and closed Riemannian balls in M of radius $r > 0$ centered at x_0.

Lemma 2.11 There exists a sequence $\{U_n\}_1^\infty \subseteq C_c\big(M;[0,\infty)\big)$ with the properties that, for each $n \geq 1$,
"(1)" $|U - U_n| \leq \frac{1}{n}$ on $\bar{B}_n$.
"(2)" $\|\mathrm{grad\,U_n}\|^2 \leq \frac{1}{\mathrm{n}} + 2\mathrm{U}$ on $\bar{B}_n$ and $\|\mathrm{gradU_n}\|^2 \longrightarrow 2\mathrm{U}$ λ_M-almost everywhere.
"(3)" $\Delta U_n \leq \frac{1}{n} + d + \sqrt{2d\kappa U}$ on $\bar{B}_n$.

Proof: For each $n \geq 1$, choose $\eta_n \in C_c^\infty\big(M;[0,1]\big)$ so that $\eta_n \equiv 1$ on B_{n+1} and $\eta_n \equiv 0$ off of B_{n+2}. Given $n \geq 1$ and $t > 0$, set

$$U_n(t,x) = \eta_n(x)\int_M (\eta_n U)(y)h_t(x,y)\,\lambda_M(dy).$$

By standard elliptic regularity results, U_n is a smooth function of $(t,x) \in (0,\infty) \times M$, and obviously it is supported in $\bar{B}_{n+2}$. Moreover, after slightly extending a standard application of Bochner's identity (cf. either § 6.2.2 or § 9.3.2 in [**12**]) and using (2.1), one knows that, when $x \in \bar{B}_n$,

$$\big\|\mathrm{grad_x\,U_n(t,\cdot)}\big\| \leq \mathrm{e}^{\kappa \mathrm{t}}\int_{\mathrm{M}} \big(\mathrm{U}\|\mathrm{grad}\,\eta_{\mathrm{n}}\| + \eta_{\mathrm{n}}\sqrt{2\mathrm{U}}\big)(\mathrm{y})\mathrm{h_t(x,y)}\,\lambda_{\mathrm{M}}(\mathrm{dy}).$$

Thus, by an application of (2.8), we can achieve the estimates in (1) and (2) by simply making $t > 0$ sufficiently small.

[2]We use $\mathbf{E}^{\mathbf{P}}[F]$ to denote the $\mathbf{P}$-integral of F.

Turning to the estimate in (3), note that, because $\Delta_x h_t(\,\cdot\,,y) = 2\partial_t h_t(x,y) = \Delta_y h_t(x,\,\cdot\,)$,

Finally, at any $x \notin \mathcal{C}$, every spatial derivative of $U_n(t,\,\cdot\,)$ tends, as $t \searrow 0$, to the corresponding derivative of $\eta_n U$. In particular, $\lim_{t\searrow 0} \mathrm{grad_x\,U_n(t,x)} = \mathrm{grad_x\,U(x)}$ for $x \in \bar{B}_n \setminus \mathcal{C}$. Hence, we now see that we can choose $t_n \searrow 0$ in such a way that the functions $U_n \equiv U_n(t_n,\,\cdot\,)$ will serve. □

Next, given $n \ge 1$ and $\alpha > 0$, refer to Lemma 2.11, and define

$$R_n^{(\alpha)}(t,p) = \exp\left(\alpha U_n\big(p(0)\big) - \alpha U_n\big(p(t)\big) + \int_0^t V_n^{(\alpha)}\big(p(\tau)\big)\,d\tau\right),$$

where

$$V_n^{(\alpha)} \equiv \alpha\Delta U_n - \alpha^2\big\|\mathrm{grad\,U_n}\big\|^2.$$

Because $V_n^{(\alpha)} = -\frac{1}{2}e^{\alpha U_n}\Delta e^{-\alpha U_n}$, it is an easy matter to check (cf. § 7.5 in [**11**]) that $R_n^{(\alpha)}(t,p)$ is a $\mathbf{P_x^{(0)}}$-martingale relative to $\{\mathcal{B}_t : t \ge 0\}$. In fact, more generally, if $f \in C^2_{\mathrm{c}}(M;\mathbf{R})$, then one knows that

$$(Mf)_n^{(\alpha)}(t,p) \equiv R_n^{(\alpha)}(t,p)f\big(p(t)\big) - \int_0^t R_n^{(\alpha)}(\tau,p)\mathcal{L}_n^{(\alpha)}f\big(p(\tau)\big)\,d\tau$$

is a $\mathbf{P_x^{(0)}}$-martingale relative to $\{\mathcal{B}_t : t \ge 0\}$ when

$$\mathcal{L}_n^{(\alpha)}f = \tfrac{1}{2}e^{2\alpha U_n}\mathrm{div}\big(\mathrm{e}^{-2\alpha U_n}\mathrm{grad\,f}\big) = \tfrac{1}{2}\Delta\mathrm{f} - \alpha\langle\mathrm{grad\,U_n}, \mathrm{grad\,f}\rangle.$$

With this information, we can now prove Theorem 2.9.

Proof of Theorem 2.9: Given $r > 0$, define $\zeta_r : \mathcal{P}(M) \longrightarrow [0,\infty]$ by

$$\zeta_r(p) = \inf\big\{t \ge 0 : p(t) \notin \bar{B}_r\big\}.$$

Next, set

$$\hat{R}_n^{(\alpha)}(t,p) = R_n^{(\alpha)}\big(t \wedge \zeta_n(p), p\big),$$

and apply Doob's Stopping Time theorem (cf. Corollary 7.15 in [**11**]) to see that $\hat{R}^{(\alpha)}(t,p)$ is again a $\mathbf{P}_x^{(0)}$-martingale relative to $\{\mathcal{B}_t : t \ge 0\}$. Hence (cf. Theorem 1.3.5 in [**10**]), we can determine a unique $(\mathbf{P}_n^{(\alpha)})_x \in \mathbf{M_1}\big(\mathbf{P(M)}\big)$ by the condition that

$$\mathbf{E}^{(\mathbf{P}_n^{(\alpha)})_x}\big[F(p)\big] = \mathbf{E}^{\mathbf{P}_x^{(0)}}\big[\hat{R}_n^{(\alpha)}(T,p)F(p)\big]$$

for all $T > 0$ and $\mathcal{B}_T$-measurable $F : \mathcal{P}(M) \longrightarrow [0,\infty)$. Moreover, by another application of Doob's theorem, one sees that, for any $0 < r \le n$, $0 \le t_1 < t_2$, $A \in \mathcal{B}_{t_1}$, and any $f \in C^2_{\mathrm{c}}(M;\mathbf{R})$,[3]

We next want to show that $\{(\mathbf{P}_n^{(\alpha)})_x : n \ge 1\}$ is relatively compact in $\mathbf{M_1}\big(\mathbf{P(M)}\big)$, and, by standard criteria (cf. Theorems 3.1.9 and Lemma 3.3.3 in [**11**]), this will be done once we show that, for each $r > 0$ and $T > 0$, there exists a $C(T,r) < \infty$ such that

$$\sup_{n\ge 1}\mathbf{E}^{(\mathbf{P}_n^{(\alpha)})_x}\big[|F(p)|\big] \le C(T,r)\sqrt{\mathbf{E}^{\mathbf{P}_x^{(0)}}\big[F(p)^2\big]}$$

for all $x \in \bar{B}_r$ and $\mathcal{B}_T$-measurable $F \in L^2\big(\mathbf{P}_x^{(0)},\mathbf{R}\big)$.

[3]We use $\mathbf{E}^P[F,\,A]$ to denote the P-integral of F over the set A.

To this end, observe that, by (1) and (2) in Lemma 2.11,

$$\hat{R}_n^{(\alpha)}(T,p) \le \exp\left(\alpha U_n(x) + \alpha^2 \int_0^T \left(\tfrac{1}{n} + d + \sqrt{2d\kappa U\big(p(\tau)\big)}\right) d\tau\right),$$

and therefore that (2.12) is an easy consequence of Schwarz's inequality together with (2.5).

Now suppose that $\mathbf{P}$ is the limit in $\mathbf{M_1}\big(\mathbf{P(M)}\big)$ of some subsequence $\{\mathbf{P}_{n'}\}$ from $\{(\mathbf{P}_n^{(\alpha)})_x : n \ge 1\}$. We want to show that $\mathbf{P}$ solves the martingale problem for $\mathcal{L}^{(\alpha)}$ starting at x. Obviously $\mathbf{P}\big(p(0) = x\big) = 1$. In addition, as a consequence of (2.12), we know that, for each $T > 0$, there is a $C(T) < \infty$ such that

$$\mathbf{E}^{\mathbf{P}}\big[|F(p)|\big] \le C(T)\sqrt{\mathbf{E}^{\mathbf{P}_x^{(0)}}\big[F(p)^2\big]} \quad \text{for all } \mathcal{B}_T\text{-measurable } F \in L^2\big(\mathbf{P}_x^{(0)}, \mathbf{R}\big).$$

Next, let $f \in C_c^2(M;\mathbf{R})$ be given, and set

$$(Nf)^{(\alpha)}(t,p) = f\big(p(t)\big) - \int_0^t \mathcal{L}^{(\alpha)} f\big(p(\tau)\big)\, d\tau.$$

By the preceding together with (2.7) applied to $\Gamma = \mathcal{C}$, $pM(t,p)$ is $\mathbf{P}$-almost surely continuous. Hence (cf. (**vii**) in Theorem 3.1.5 of [**11**]), for any $0 \le t_1 < t_2$ and $F \in C_b\big(\mathcal{P}(M);\mathbf{R}\big)$ which is $\mathcal{B}_{t_1}$-measurable,

To prove that $x\mathbf{P}_x^{(\alpha)}$ is continuous, first note that, by (2.12) and the preceding convergence statement,

Finally, to prove the estimate in (2.10), set

$$\tilde{V}_n^{(\alpha)} = \alpha\Big(\tfrac{1}{n} + d + \sqrt{2d\kappa U}\Big) - \alpha^2\|\mathrm{grad\,U_n}\|^2.$$

Then, by (2) in Lemma 2.11, as long as $r \le n$,

$$\zeta_r(p) > T \implies R_n^{(\alpha)}(T,p) \le \exp\left(\alpha U_n\big(p(0)\big) - \alpha U_n\big(p(T)\big) + \int_0^T \tilde{V}_n^{(\alpha)}\big(p(\tau)\big)\, d\tau\right).$$

Hence, if $F \in C_b\big(\mathcal{P}(M);[0,\infty)\big)$ is $\mathcal{B}_T$-measurable, then, because (cf. (**vi**) in Theorem 1.3.5 in [**11**]) $\{p : \zeta_r(p) > T\}$ is open,

There is one more topic to taken up in this section. Namely, recall the probability measure $\mu^{(\alpha)}$ given in (1.5), and define $\mathbf{P}^{(\alpha)} \in \mathbf{M_1}\big(\mathbf{P(M)}\big)$ by

$$\mathbf{P}^{(\alpha)} = \int_M \mathbf{P}_x^{(\alpha)}\, \mu^{(\alpha)}(dx).$$

We will need to know that $\mathbf{P}^{(\alpha)}$ is *reversible* in the sense made precise by the following statement.

Theorem 2.14 For each $T > 0$, define map $p \in \mathcal{P}(M) \longmapsto \check{p}_T \in \mathcal{P}(M)$ so that $\check{p}_T(t) = p(T - t \wedge T)$. Then, for any $\mathcal{B}_T$-measurable $F : \mathcal{P}(M) \longrightarrow [0,\infty)$,

$$\mathbf{E}^{\mathbf{P}^{(\alpha)}}\big[F(p)\big] = \mathbf{E}^{\mathbf{P}^{(\alpha)}}\big[F(\check{p}_T)\big].$$

Proof: Our proof comes down to showing that $\mathbf{P}^{(\alpha)}$ inherits this property from the analogous property for the σ-finite measure $\mathbf{Q}$ on $\big(\mathcal{P}(M),\mathcal{B}\big)$ given by

$$\mathbf{Q} = \int_{\mathbf{M}} \mathbf{P}^{(\mathbf{0})}_{\mathbf{x}}\, \lambda_{\mathbf{M}}(\mathbf{dx}).$$

Indeed, starting from (2.6) and the fact that the heat kernel $h_t(x,y)$ is a symmetric function of x and y, one can use the Markov property to give an inductive proof that, for all $n\geq 1$, $0=t_0<\cdots<t_n=T$, and measurable $f: M^{n+1}\longrightarrow[0,\infty)$,

$$\mathbf{E}^{\mathbf{Q}}\Big[f\big(p(t_0),\dots,p(t_n)\big)\Big] = \mathbf{E}^{\mathbf{Q}}\Big[f\big(p(T-t_0),\dots,p(T-t_n)\big)\Big].$$

Hence, after an obvious limit procedure, one sees that

$$\mathbf{E}^{\mathbf{Q}}\big[F(p)\big] = \mathbf{E}^{\mathbf{Q}}\big[F(\check{p}_T)\big]$$

for any $\mathcal{B}_T$-measurable $F:\mathcal{P}(M)\longrightarrow[0,\infty)$. In particular, if $\mathbf{Q_{n,T}}$ is given by

$$\mathbf{Q_{n,T}(A)} = \frac{\mathbf{1}}{\mathbf{Z^{(\alpha)}_n}}\int_{\mathbf{M}} \mathbf{e}^{-2\alpha \mathbf{U_n(x)}}\mathbf{E}^{\mathbf{P}^{(0)}_{\mathbf{x}}}\Big[\mathbf{R}^{(\alpha)}_{\mathbf{n}}(\mathbf{T}),\ \mathbf{A}\cap\{\zeta_{\mathbf{n}}(\mathbf{p})>\mathbf{T}\}\Big]\,\lambda_{\mathbf{M}}(\mathbf{dx}),$$

where

$$Z^{(\alpha)}_n \equiv \int_{B_n} e^{-2\alpha U_n(x)}\,\lambda_M(dx),$$

then, because $\zeta_n(p)>T \iff \zeta_n(\check{p}_T)>T$,

$$\mathbf{E}^{\mathbf{Q_{n,T}}}\big[F(p)\big] = \mathbf{E}^{\mathbf{Q_{n,T}}}\big[F(\check{p}_T)\big]$$

for all $\mathcal{B}_T$-measurable $F:\mathcal{P}(M)\longrightarrow[0,\infty)$.

To see how these observations apply, set (cf. the proof of Theorem 2.9)

$$\mathbf{P}^{(\alpha)}_n = \frac{1}{Z^{(\alpha)}_n}\int_{B_n} e^{-2\alpha U_n(x)}\big(\mathbf{P}^{(\alpha)}_n\big)_x\,\lambda_M(dx),$$

and observe that, for any $\mathcal{B}_T$-measurable $F:\mathcal{P}(M)\longrightarrow[0,\infty)$,

$$\mathbf{E}^{\mathbf{P}^{(\alpha)}_n}\big[F(p),\,\zeta_n(p)>T\big] = \mathbf{E}^{\mathbf{Q_{n,T}}}\big[F(p)\big].$$

Thus,

$$\mathbf{E}^{\mathbf{P}^{(\alpha)}_n}\big[F(p),\,\zeta_n(p)>T\big] = \mathbf{E}^{\mathbf{P}^{(\alpha)}_n}\big[F(\check{p}_T),\,\zeta_n(p)>T\big],$$

and so, because $\mathbf{P}^{(\alpha)}_n$ converges to $\mathbf{P}^{(\alpha)}$ in $\mathbf{M_1}\big(\mathbf{P(M)}\big)$ as $n\to\infty$, all that we have to do is note that, because (cf. Theorem 3.1.9 in [**11**]) $\{\mathbf{P}^{(\alpha)}_n : n\geq 1\}$ is relatively compact in $\mathbf{M_1}\big(\mathbf{P(M)}\big)$, $\mathbf{P}^{(\alpha)}_n\big(\zeta_n(p)\leq T\big)\longrightarrow 0$. □

Remark 2.16: The reader may be wondering whether it is not possible to prove that, for all $T>0$ and $A\in\mathcal{B}_T$,

$$\mathbf{P}^{(\alpha)}_x(A) = e^{\alpha U(x)}\mathbf{E}^{\mathbf{P}^{(0)}_x}\left[\exp\left(-\alpha U\big(p(T)\big)+\int_0^T V^{(\alpha)}\big(p(\tau)\big)\,d\tau\right),\ A\right], (*)\text{''}$$

where

$$V^{(\alpha)} = \begin{cases}\alpha\Delta U - 2\alpha^2 U & \text{off } \mathcal{C}\\ 0 & \text{on } \mathcal{C}.\end{cases}$$

Indeed, at first sight, this seems to what the proof of Theorem 2.9 is saying. However, after a little thought, one realizes that without some control on $\Delta U M\setminus\mathcal{C}$ from below, there is no way of justifying the required limits. Thus, for example, (*) does hold when $d\geq 2$ and M has non-positive sectional curvature everywhere, since in this case $\Delta U\geq 0$ and $\mathcal{C}$ has capacity 0. On the other hand, if M is the standard,

1-dimensional unit circle in $\mathbf{R}^2$, then (*) is false. Indeed, elementary considerations show that in this case the exponential must include an additonal term which is basically the local time that Brownian motion spends at the cut locus and which comes from the singular part of ΔU.

4. Analytic Considerations

As a corollary of the existence, uniqueness, and continuity statements in Theorem 2.10, we know (cf. Theorem 6.2.1 in [**10**]) that $\{\mathbf{P}_x^{(\alpha)} : x \in M\}$ is a continuous Markov family. Thus, if $\mathbf{P}_{\mathbf{t}}^{(\alpha)}$ is defined for $t > 0$ on bounded, Borel measurable functions $f : M \longrightarrow \mathbf{R}$ by

$$\mathbf{P}_{\mathbf{t}}^{(\alpha)}\mathbf{f}(\mathbf{x}) = \mathbf{E}^{\mathbf{P}_{\mathbf{x}}^{(\alpha)}}\big[\mathbf{f}(\mathbf{p}(\mathbf{t}))\big],$$

then $\{\mathbf{P}_{\mathbf{t}}^{(\alpha)} : \mathbf{t} > \mathbf{0}\}$ is a Markov semigroup of operators which preserves $C_{\mathrm{b}}(M;\mathbf{R})$. In addition, Theorem 2.14 shows that, for each $t > 0$, $\mathbf{P}_{\mathbf{t}}^{(\alpha)}$ is $\mu^{(\alpha)}$-symmetric in the sense that

$$\int_M f_1 \mathbf{P}_{\mathbf{t}}^{(\alpha)} f_2 \, d\mu^{(\alpha)} = \int_M f_2 \mathbf{P}_{\mathbf{t}}^{(\alpha)} f_1 \, d\mu^{(\alpha)}, \quad f_1, f_2 \in C_{\mathrm{b}}(M;\mathbf{R}).$$

In particular, this means that

$$\|\mathbf{P}_{\mathbf{t}}^{(\alpha)} f\|_{L^2(\mu^{(\alpha)};\mathbf{R})}^2 = \int_M \big(\mathbf{P}_{\mathbf{t}}^{(\alpha)} f\big)^2 \, d\mu^{(\alpha)} \le \int_M \mathbf{P}_{\mathbf{t}}^{(\alpha)} f^2 \, d\mu^{(\alpha)} = \|f\|_{L^2(\mu^{(\alpha)};\mathbf{R})}^2,$$

and therefore each $\mathbf{P}_{\mathbf{t}}^{(\alpha)}$ admits a unique extension $\bar{\mathbf{P}}_t^{(\alpha)}$ as a self-adjoint contraction on $L^2(\mu^{(\alpha)};\mathbf{R})$, and $\{\bar{\mathbf{P}}_t^{(\alpha)} : t > 0\}$ is again a semigroup on $L^2(\mu^{(\alpha)};\mathbf{R})$. Furthermore, because $\mathbf{P}_{\mathbf{t}}^{(\alpha)} f \longrightarrow f$ boundedly and uniformly of compacts for each $f \in C_{\mathrm{b}}(M;\mathbf{R})$, one can easily check that $\bar{\mathbf{P}}_t^{(\alpha)} f \longrightarrow f$ in $L^2(\mu^{(\alpha)};\mathbf{R})$ for each $f \in L^2(\mu^{(\alpha)};\mathbf{R})$. Finally, if $f \in C_{\mathrm{c}}^2(M;\mathbf{R})$, then, from the martingale characterization of the $\mathbf{P}_x^{(\alpha)}$'s, we know that

$$\mathbf{P}_{\mathbf{t}}^{(\alpha)} f - f = \int_0^t \mathbf{P}_\tau^{(\alpha)} \mathcal{L}^{(\alpha)} f \, d\tau$$

and therefore that

In the jargon of the theory of *Dirichlet forms* (cf. [**7**]), the preceding conclusions can be summarized by saying that $\{\bar{\mathbf{P}}_t^{(\alpha)} : t > 0\}$ *is the semigroup of Markov, self-adjoint operators on* $L^2(\mu^{(\alpha)};\mathbf{R})$ *whose Dirichlet form is*

$$\frac{1}{2}\int_M \|\mathrm{grad}\, \mathrm{f}\|^2 \, \mathrm{d}\mu^{(\alpha)}.$$

In particular, because $\mu^{(\alpha)}$ is a locally commensurate with λ_M, one can apply the theory of DiGeorgi, Nash, and Moser (cf. [**13**] or [**1**]) to conclude that there exists a unique, locally Hölder continuous map

$$(t, x, y) \in (0, \infty) \times M \times M \longmapsto h_t^{(\alpha)}(x, y) \in (0, \infty)$$

such that

$$\mathbf{P}_{\mathbf{t}}^{(\alpha)} f(x) = \int_M f(y) h_t^{(\alpha)}(x, y) \, \mu^{(\alpha)}(dy), \quad (t, x) \in (0, \infty) \times M$$

for all $f \in C_b(M;\mathbf{R})$. Moreover, from the symmetry of $\mathbf{P}_{\mathbf{t}}^{(\alpha)}$, it is clear that $h_t^{(\alpha)}(x,y) = h_t^{(\alpha)}(y,x)$, and, from the semigroup property, that the Chapman Kolmogorov equation

$$h_{s+t}^{(\alpha)}(x,y) = \int_M h_s^{(\alpha)}(x,z) h_t^{(\alpha)}(z,y)\,\mu^{(\alpha)}(dz).$$

holds. In particular, by combining these two, one sees that

$$h_{2t}^{(\alpha)}(x,x) = \left\|h_t^{(\alpha)}(x,\,\cdot\,)\right\|_{L^2(\mu^{(\alpha)};\mathbf{R})}^2$$
$$h_{2t}^{(\alpha)}(x,y) \le \sqrt{h_{2t}^{(\alpha)}(x,x) h_{2t}^{(\alpha)}(y,y)}.$$

The main result in this section is the estimate proved in the following.

Theorem 3.5 For each $\alpha > 0$, there exists an $\epsilon_\alpha \in (0,1]$, depending only on d and κ in addition to α, such that

$$h_2^{(\alpha)}(x,x) \le \epsilon_\alpha^{-1} e^{2\alpha(1-\epsilon_\alpha)U(x)}, \quad x \in M.$$

Proof: Let $f \in C_c\big(M;[0,\infty)\big)$ with $\|f\|_{L^1(\lambda_M)} = 1$ be given. Then, by (2.10),

To complete the proof from here, set

$$A = \left\{p:\ \inf_{t\in[0,1]} \operatorname{dist}\big(\mathrm{x}_0,\mathrm{p}(\mathrm{t})\big) \ge \tfrac{1}{2}\operatorname{dist}(\mathrm{x},\mathrm{x}_0)\right\},$$

and write the integral as the sum of the integral I over A and the integral J over A. Obviously,

$$I \le \exp\left(-\frac{3\alpha^2 U(x)}{4}\right) \mathbf{E}^{\mathbf{P}_x^{(0)}}\big[h_1(p(1),x)\big] = \exp\left(-\frac{3\alpha^2 U(x)}{4}\right) h_2(x,x).$$

At the same time, since, by the triangle inequality,

$$p \in A \implies D_x^*(p) \equiv \sup_{t\in[0,1]} \operatorname{dist}\big(\mathrm{x},\mathrm{p}(\mathrm{t})\big) \ge \tfrac{1}{2}\operatorname{dist}(\mathrm{x},\mathrm{x}_0),$$

we know that

$$J \le \mathbf{E}^{\mathbf{P}_x^{(0)}}\Big[h_1(p(1),x),\ D_x^*(p) \ge \sqrt{2^{-1}U(x)}\Big].$$

At this point, we need the Li–Yau estimate, in Corollary 3.1 of [**6**][4], which, together with Bishop's Volume Comparison Theorem (cf. (3.17) in [**6**] and (1.4) here) says that there exist $C < \infty$ and $C' < \infty$, depending only on d and κ, such that

The first part of the following corollary was proved earlier by L. Saloff-Coste in [**9**].

Corollary 3.6 For each $t > 0$, $\bar{\mathbf{P}}_t^{(\alpha)}$ is a compact, self-adjoint operator on $L^2(\mu^{(\alpha)};\mathbf{R})$ for which 1 is a simple eigenvalue with eigenfunction 1. In addition, there exist $T_\alpha > 0$ and $C_\alpha < \infty$, which depend only on α, d, and κ, such that

$$\big\|\bar{\mathbf{P}}_{T_\alpha}^{(\alpha)} f\big\|_{L^4(\mu^{(\alpha)};\mathbf{R})} \le C_\alpha \|f\|_{L^2(\mu^{(\alpha)};\mathbf{R})}, \quad f \in L^2(\mu^{(\alpha)};\mathbf{R}).$$

[4]I am grateful to Laurent Saloff–Coste for pointing out that I had misquoted [**6**] in the first version of this article and abused it in several succeeding versions.

Proof: Because $\{\bar{\mathbf{P}}_t^{(\alpha)} : t > 0\}$ is a continuous semigroup of self-adjoint operators, standard spectral theory tells us that there is a resolution of the identity $\{\mathbf{E}_\lambda : \lambda \geq \mathbf{0}\}$ in $L^2(\mu^{(\alpha)}; \mathbf{R})$ by orthogonal projections such that

$$\bar{\mathbf{P}}_t^{(\alpha)} = \int_{[0,\infty)} e^{-t\lambda}\, d\mathbf{E}_\lambda, \quad \mathbf{t} > \mathbf{0}.$$

In particular, this shows that $\bar{\mathbf{P}}_t^{(\alpha)}$ is compact for all $t > 0$ if it is for some $t > 0$. But, from Theorem 3.5, it is clear that

$$\|\bar{\mathbf{P}}_1^{(\alpha)}\|_{\mathrm{HS}}^2 = \mathrm{Trace}(\bar{\mathbf{P}}_2^{(\alpha)}) = \int_{\mathrm{M}} \mathrm{h}_2^{(\alpha)}(\mathrm{x},\mathrm{x})\, \mu^{(\alpha)}(\mathrm{dx}) < \infty,$$

where $\|\bar{\mathbf{P}}_1^{(\alpha)}\|_{\mathrm{HS}}$ is the Hilbert–Schmidt norm of $\bar{\mathbf{P}}_1^{(\alpha)}$. Hence, $\bar{\mathbf{P}}_1^{(\alpha)}$ is compact.

Next, because $\mathbf{P}_{\mathbf{t}}^{(\alpha)} 1 = 1$, there is no doubt that 1 is an eigenvalue of $\bar{\mathbf{P}}_t^{(\alpha)}$ with eigenfunction 1. In fact, 1 is the top of the spectrum of $\bar{\mathbf{P}}_t^{(\alpha)}$, and so, if $\|f\|_{L^2(\mu^{(\alpha)};\mathbf{R})} = 1$, then $1 \geq (f, \bar{\mathbf{P}}_t^{(\alpha)} f)_{L^2(\mu^{(\alpha)};\mathbf{R})}$ with equality precisely when f is an eigenfunction with eigenvalue 1. In particular, if 1 were not a simple eigenvalue, then we could produce an f satisfying $\|f\|_{L^2(\mu^{(\alpha)};\mathbf{R})} = 1$, $(f,1)_{L^2(\mu^{(\alpha)};\mathbf{R})} = 0$, and $1 = (f, \bar{\mathbf{P}}_t^{(\alpha)} f)_{L^2(\mu^{(\alpha)};\mathbf{R})}$. But, because $h_t^{(\alpha)} > 0$ everywhere, this would lead to the contradiction

Finally, note that, for any $f \in C_{\mathrm{b}}(M; \mathbf{R})$,

$$\left|\mathbf{P}_{\mathbf{1}}^{(\alpha)} f(x)\right| \leq \int_M h_1^{(\alpha)}(x,y) |f(y)|\, \mu^{(\alpha)}(dy) \leq \|h_1^{(\alpha)}(x,\,\cdot\,)\|_{L^2(\mu^{(\alpha)};\mathbf{R})} \|f\|_{L^2(\mu^{(\alpha)};\mathbf{R})}.$$

Hence, by (3.4) and Theorem 3.5,

$$\left|\mathbf{P}_{\mathbf{1}}^{(\alpha)} f(x)\right| \leq \epsilon_\alpha^{-\frac{1}{2}} e^{\alpha(1-\epsilon_\alpha)U(x)} \|f\|_{L^2(\mu^{(\alpha)};\mathbf{R})}.$$

Thus, if $q_\alpha \equiv \frac{4}{2-\epsilon_\alpha}$, then

$$\int_M |\mathbf{P}_{\mathbf{1}}^{(\alpha)} f|^{q_\alpha}\, d\mu^{(\alpha)} \leq \frac{1}{Z^{(\alpha)} \epsilon_\alpha^{\frac{q_\alpha}{2}}} \int_M \exp\left(-\frac{2\alpha\epsilon_\alpha U}{2-\epsilon_\alpha}\right) d\lambda_M = \frac{Z^{(\beta_\alpha)}}{Z^{(\alpha)}} \epsilon_\alpha^{-\frac{q_\alpha}{2}} < \infty,$$

where $\beta_\alpha = \frac{\alpha\epsilon_\alpha}{2-\epsilon_\alpha}$. Hence, we now know that

$$\|\mathbf{P}_{\mathbf{1}}^{(\alpha)} f\|_{L^{q_\alpha}(\mu^{(\alpha)};\mathbf{R})} \leq B_\alpha \|f\|_{L^2(\mu^{(\alpha)};\mathbf{R})} \quad f \in C_{\mathrm{b}}(M; \mathbf{R}),$$

where $q_\alpha \in (2,\infty)$ and $B_\alpha < \infty$ depend only on α, d, and κ. Working by induction, one finds that

$$\|\mathbf{P}_{\mathbf{n}}^{(\alpha)} f\|_{L^{2^{1-n} q_\alpha^n}(\mu^{(\alpha)};\mathbf{R})} \leq B_\alpha^n \|f\|_{L^2(\mu^{(\alpha)};\mathbf{R})} \quad f \in C_{\mathrm{b}}(M; \mathbf{R}).$$

Thus, by choosing n_α to be the smallest integer n such that $2^{1-n} q_\alpha^n \geq 4$ and then taking $T_\alpha = n_\alpha$ and $C_\alpha = B_\alpha^{n_\alpha}$, we achieve (3.7), first for all $f \in C_{\mathrm{b}}(M; \mathbf{R})$ and then for all $f \in L^2(\mu^{(\alpha)}; \mathbf{R})$. □

Corollary 3.9 Define γ_α so that

$$e^{-\gamma_\alpha} = \sup\left\{ \left(f, \bar{\mathbf{P}}_1^{(\alpha)} f\right)_{L^2(\mu^{(\alpha)};\mathbf{R})} : \|f\|_{L^2(\mu^{(\alpha)};\mathbf{R})} = 1 \;\&\; (f,1)_{L^2(\mu^{(\alpha)};\mathbf{R})} = 0 \right\}.$$

Then $\gamma_\alpha > 0$. In addition, there exists a $\sigma_\alpha > 0$, which depends only on α, d, κ, and γ_α, such that, for each $t > 0$, $p \in (1,\infty)$, and $f \in C_{\mathrm{b}}(M; \mathbf{R})$,

$$\|\mathbf{P}_{\mathbf{t}}^{(\alpha)} f\|_{L^q(\mu^{(\alpha)};\mathbf{R})} \leq \|f\|_{L^p(\mu^{(\alpha)};\mathbf{R})} \quad \text{when } q = 1 + (p-1)e^{\sigma_\alpha t}.$$

Proof: The first assertion is an immediate consequence of the spectral properties of $\bar{\mathbf{P}}_1^{(\alpha)}$ proved in Theorem 3.5. Given that $\gamma_\alpha > 0$ and knowing (3.7), (3.10) is an application of general results from § 6.1 in [**2**]. Namely, one first uses Glimm's lemma (Lemma 6.1.5 in [**2**]) to find a $T_p' \geq T_p$, depending only on C_α, T_α and γ_α, such that C_α in (3.7) can be replaced by 1 when T_α is replaced by T_α'. One then applies (**i**) from Theorem 6.1.22 in [**2**] to get a logarithmic Sobolev inequality of the form

$$\int_M f \log f \, d\mu^{(\alpha)} \leq \Lambda_\alpha \int_M \|\mathrm{grad\, f}\|^2 \, \mathrm{d}\mu^{(\alpha)}, \quad \mathrm{f} \in \mathrm{C}_\mathrm{b}^1(\mathrm{M}; \mathbf{R}),$$

where Λ_α depends only on T_α'. Finally, one uses Gross's lemma (Theorem 6.1.14 in [**2**]) to pass from this logarithmic Sobolev inequality to the hypercontraction result in (3.10). □

Remark 3.12: The statements of Corollaries 3.6 and 3.8 are less sharp than they need be. In particular, it would have been preferable to further highlight exactly which estimates do or do not depend on having a lower bound for γ_α. For example, if one looks carefully at the proof of part (**i**) in Theorem 6.1.22 of [**2**] and then combines it with Theorem 6.1.14 there, one realizes that it is possible to pass directly from (cf. the notation in the proof of Corollary 3.6) (3.8) to the conclusion that

$$\|\mathbf{P}_\mathbf{t}^{(\alpha)} f\|_{L^q(\mu^{(\alpha)};\mathbf{R})} \leq (B_\alpha)^{\left(\frac{1}{p}-\frac{1}{q}\right)t} \|f\|_{L^p(\mu^{(\alpha)};\mathbf{R})}, \quad f \in C_\mathrm{b}(M;\mathbf{R})$$

$$\text{for } p \in (1,\infty) \text{ and } q = 1 + (p-1)\exp\left(\frac{2q_\alpha t}{q_\alpha - 2}\right).$$

Notice that, as distinguished from those in (3.10), all the quantities entering (3.11) depend only on α, d, and κ. Thus, the only time the lower bound on γ_α plays a role is when one tries to make the constant factor on the right hand side equal 1.

Remark 3.13: The methodology used is this article is intimately related to that used in [**5**], and I strongly suspect that one can take further advantage of the ideas developed there. In particular, it seems essentially certain that one should be able to show that, for any $\epsilon > 0$ and $\alpha > 0$, the diffusion semigroup determined by the operator

$$\tfrac{1}{2}\Delta - \alpha \mathrm{grad\, U}^{1+\epsilon}$$

is not only hypercontractive relative to the measure $d\mu = \exp\bigl(-2\alpha U^{1+\epsilon}\bigr)\, d\lambda_M$ but is even *ultracontractive* in the sense that it maps $L^1(\mu : \mathbf{R})$ into $C_\mathrm{b}(M;\mathbf{R})$ with a bound which decreases with time.

References

[1] Davies, E.B., **Heat Kernels and Spectral Theory,** Cambridge Univ. Press, 1989.

[2] Deuschel, J.-D. & Stroock, D. **Large Deviations,** Academic Press, Pure and Appl. Math. Series #137, San Diego, 1989.

[3] Federbush, P. *Partially alternative derivation of a result of Nelson,* **J. Math. Physics** vol. 10, 1, 1969; pages 50–52.

[4] Gross, L. *Logarithmic Sobolev inequalities,* **Amer. J. Math.** vol. 97, 1976; pages 1061–1083.

[5] Kusuoka, S. & Stroock, D. *Some boundedness properties of certain stationary diffusion semigroups,* **J. Fnal. Anal.** 1985 vol. 60, #2, pages 243–264.

[6] Li, P. & Yau, S.–T. *On the parabolic kernel of the Schrödinger operator,* **Acta. Math.** vol. 156 1986, 153–201.

[7] Ma, Z.-M. & Röckner, M., **Dirichlet Forms,** Springer–Verlag, NYC, Universitext 1991

[8] Nelson, E. *The free field,* **J. Fnal. Anal.** vol. 12, 1973, pages 211–227.

[9] Saloff-Coste, L., *Uniformly elliptic operators on Riemannian manifolds,* **J. Diff. Geom.** 1992, vol. 36, #2, pages 417–450.

[10] Stroock, D. & Varadhan, S.R.S., **Multidimensional Diffusion Processes,** Springer–Verlag, Grundlehren, #233, 1979, New York.

[11] Stroock, D., **Probability Theory, An Analytic View** Cambridge Univ. Press, 1993, New York.

[12] ———, **An Introduction to the Analysis of Paths on a Riemannian Manifold,** AMS, Math. Surveys & Monographs, # 74, 2000.

[13] ———, *Diffusion semigroups corresponding to uniformly elliptic divergence form operators,* in **Séminaire de Probabililité XXII,** Springer–Verlag LNMS #1321, Azema, J. & Yor, M., 1988, pages 316–348.

[14] Uhlenbeck, G. and Ornstein, L.S., *On the Theory of the Brownian Motion,* **Physical Review,** vol. 15, #1, pages 823–841 (1930).

[15] Wang, F.-Y., *Logarithmic Sobolev Inequalities, Conditions and Counterexamples,* preprint.

[16] Yau, S.–T., *On the heat kernel of a Riemannian manifold,* **J. Math. Pures Appl.** vol. 57, #2, 191–201, 1978.

M.I.T., 2-272, Cambridge, MA 02139-4307, USA
E-mail address: dws@math.mit.edui

Plenary Addresses

AMS/IP Studies in Advanced Mathematics
Volume 20, 2001

Stochastic PDEs in Turbulence Theory

Weinan E

ABSTRACT. This paper reviews the recent progress on stochastic PDEs arising from different aspects of the turbulence theory including the stochastic Navier-Stokes equation, stochastic Burgers equation and stochastic passive scalar and passive vector equations. Issues discussed include the existence of invariant measures, scaling of the structure functions, asymptotic behavior of the probability density functions, dissipative anomaly, etc.

1. Introduction

The problem of hydrodynamic turbulence is a well-known notoriously difficult problem, often viewed as the last unsolved problem in classical physics [**2, 11, 27, 41**]. The present paper does not contain a solution to that problem. Worse than that, in spite of many years of hard work with contributions from some of the best-known names in physics and mathematics of this century, the Mountain Everest [**3**] stands as high as it was almost sixty years ago at the time of Kolmogorov. The repeated experimental and numerical confirmation of Kolmogorov's predictions as well as Landau's objections adds all the more mystery to this old and resilient subject. In the meantime, Kolmogorov's notion of cascade has permeated through several different branches of science.

One thing people realized from this experience is that we should not view turbulence as an isolated problem. Rather it is an example of a wide variety of non-equilibrium processes exhibiting cascades, i.e. a non-trivial flux of energy across scales. Secondly it is important to build intermediate steps on simpler problems so that different aspects of the problem can be understood separately. It is this philosophy that motivated the recent flourish of activity on stochastic Burgers equation and stochastic passive scalar equation. It is also with this philosophy that the present paper is written. The problems to be discussed are:

1. Stochastic Navier-Stokes equation

$$\frac{\partial \boldsymbol{u}}{\partial t} + (\boldsymbol{u} \cdot \nabla)\boldsymbol{u} + \nabla p = \nu \Delta \boldsymbol{u} + \boldsymbol{f}, \qquad (1)$$
$$\nabla \cdot \boldsymbol{u} = 0.$$

2. Stochastic Burgers equation

$$\frac{\partial u}{\partial t} + u\frac{\partial u}{\partial x} = \nu \frac{\partial^2 u}{\partial x^2} + f. \qquad (2)$$

3. Stochastic passive scalar and passive vector equations

(3) $$\frac{\partial T}{\partial t} + (\boldsymbol{u} \cdot \nabla)T = \kappa \Delta T + f,$$

(4) $$\frac{\partial \boldsymbol{B}}{\partial t} + (\boldsymbol{u} \cdot \nabla)\boldsymbol{B} - (\boldsymbol{B} \cdot \nabla)\boldsymbol{u} = \kappa \Delta \boldsymbol{B} + \boldsymbol{f}$$

In (2) (3) $f(x,t)$ is taken to be a zero-mean, Gaussian, statistically homogeneous, and white-in-time random process with covariance

(5) $$\langle f(x,t) f(y,s) \rangle = 2B(x-y)\delta(t-s)$$

where B is smooth function. $\boldsymbol{f}(\boldsymbol{x},t)$ is similarly defined. We will associate a scale L over which B decays. L will be the integral scale in the problem. Another important scale is the dissipation scale denoted by ℓ_d which is defined in terms of ν or κ. $\boldsymbol{u}$ in (3) and (4) is a random velocity field whose statistics will be specified later in Section 4.

It should be emphasized that we are mostly interested in systems involving a dissipation mechanism. Therefore energy has to be supplied (here through the forcing) to maintain a statistical steady state. This excludes equilibrium states that can be constructed for Hamiltonian systems such as the nonlinear Schrödinger [**8**] and the incompressible Euler equations [**13**].

The second emphasis in this paper is on a new look at the issues in stochastic PDEs. A large amount of work has been done in this subject. Much of it takes the point of view that a stochastic PDE is a stochastic ODE in Banach space [**16, 17**]. A lot has been learned from this viewpoint. However, the presence of infinitely many degrees of spatial freedom gives rise to many important new phenomena that are unique to stochastic PDEs. Examples include space-time scaling, universality in local spatial structures, etc. Such characteristics are of special interest in this paper. More precisely, on the physical side, we are interested in the following issues:

(P1). Existence of an inertial range across which energy or some other quantity flows but is not dissipated. If the flux is going toward the small scales, it is called a direct cascade. If the flux is going toward the large scales, it is called an inverse cascade.

(P2). Effective behavior as well as fluctuations at large spatial and temporal scales. For example a often asked question for the passive scalar convection is whether the process is diffusive or super-diffusive at large scales.

(P3). Small scale characteristics. A standard way of measuring the small scale characteristics in a field $\boldsymbol{w}$ is to consider the structure function:

(6) $$S_p(r) = \langle |\boldsymbol{w}(\boldsymbol{x}+\boldsymbol{r},t) - \boldsymbol{w}(\boldsymbol{x},t)|^p \rangle,$$

where $\langle \cdot \rangle$ denotes ensemble (statistical) average, $r = |\boldsymbol{r}|$. The field $\boldsymbol{w}$ is going to be $\boldsymbol{u}$ in (1), u in (2), T in (3) and $\boldsymbol{B}$ in (4). For small r, we typically have

(7) $$S_p(r) \approx C_p r^{\alpha_p}.$$

α_p measures the spatial regularity of the field $\boldsymbol{w}$. For example, if the field $\boldsymbol{w}$ is smooth, we have $\alpha_p = p$. For Wiener process, we have $\alpha_p = \frac{1}{2}p$. These are examples of locally self-similar processes for which α_p is a linear function of p. In the case when α_p is a nonlinear function of p, we say that the field exhibits "anomalous scaling," or "multi-fractal" behavior. In such a case, $\boldsymbol{w}$ has a spectrum of singular behavior, or Hólder exponents.

(P4). A sometimes related but usually more refined question is the asymptotic behavior of PDFs (probability density functions) of certain quantities such as u or $\partial u/\partial x$ in the stochastic Burgers equation, or T and $\delta T(\boldsymbol{r},\boldsymbol{x},t) = T(\boldsymbol{x}+\boldsymbol{r},t)-T(\boldsymbol{x},t)$ in the stochastic passive scalar problem.

On the mathematical side, we are interested in the following questions:

(M1). Existence, uniqueness of statistical steady state or the invariant measure, say μ_ν or μ_κ, and their convergence as $\nu \to 0$ or $\kappa \to 0$. If μ_ν or μ_κ does converge, say to μ_0, then the central object of interest in connection with the questions in (P1-P4) is μ_0. These invariant measures also define the ensemble averages used in (6).

(M2). Characterization of the regularity and singularity behavior of the sample paths supported on the measure μ_0. In particular, the scaling exponents α_p defined in (7) are consequences of the singular structures in the solutions.

In many cases the existence of an invariant measure is guaranteed by the well-known Krylov-Bogoliubov Theorem which states that a dynamical system on a compact state space always admits an invariant measure. This applies also to PDEs whose solution operator is compact. This is the case for many parabolic equations [**17**]. The disadvantage of this approach is that it fails to reveal any insight on the behavior of the solutions supported on the invariant measure and mechanisms for uniqueness or non-uniqueness of the invariant measure.

To put the issue about an invariant measure in a different perspective, let us consider the stochastic ODE

$$dx = b(x)\,dt + \sigma(x)\,dW, \tag{8}$$

where W is the Wiener process, b and σ are smooth functions. Its invariant measure, if exists, is a measure on R^1 with density $\rho(\cdot)$ satisfying

$$\frac{1}{2}\frac{\partial^2}{\partial x^2}\left(\sigma^2(x)\rho\right) - \frac{\partial}{\partial x}(b(x)\rho) = 0, \tag{9}$$

Take the example of the Ornstein-Ulenbeck process for which $b(x) = -x, \sigma(x) = 1$, we have from (9)

$$\rho(x) = \frac{e^{-x^2/2}}{\sqrt{2\pi}}. \tag{10}$$

We see that finding the invariant measure for stochastic ODEs is equivalent to solving an elliptic equation on R^n where n is the dimension of the state space for the stochastic ODE. There is also a formal analogy of this elliptic equation for stochastic PDEs. It is loosely referred to as the Hopf's equation which is an equation satified by the functional PDF of the random field described by the stochastic PDE. It is an (degenerate) elliptic equation in function spaces. For (1)-(3), the Hopf's equations are respectively:

$$\begin{aligned}\frac{\partial}{\partial t}Z[\boldsymbol{u}(\cdot,t)] &= -\int_{R^d}\frac{\delta}{\delta u_\alpha}((-u_\beta\nabla_\beta u_\alpha + \nu\Delta u_\alpha)Z[\boldsymbol{u}(\cdot,t)])d^d\boldsymbol{x}\\ &\quad+\int_{R^d\times R^d} B_{\alpha\beta}(\boldsymbol{x}-\boldsymbol{x}')\frac{\delta^2}{\delta u_\alpha \delta u'_\beta}Z[\boldsymbol{u}(\cdot,t)]d^d\boldsymbol{x}d^d\boldsymbol{x}',\end{aligned} \tag{11}$$

where $u_\alpha = u_\alpha(\boldsymbol{x}, t)$, $u'_\alpha = u_\alpha(\boldsymbol{x}', t)$, $\alpha, \beta = 1, \dots, d$, $Z[\boldsymbol{u}(\cdot, t)]$ is the functional PDF of $\boldsymbol{u}(\cdot, t)$,

$$(12)\quad \begin{aligned} \frac{\partial}{\partial t} Z[u(\cdot,t)] &= -\int_{R^1} \frac{\delta}{\delta u}((-u\frac{\partial u}{\partial x} + \nu\frac{\partial^2 u}{\partial x^2}) Z[u(\cdot,t)])dx \\ &+ \int_{R\times R} B(x-x') \frac{\delta^2}{\delta u \delta u'} Z[u(\cdot,t)]dxdx' \end{aligned}$$

where $u = u(x,t)$, $u' = u(x',t)$, $Z[u(\cdot,t)]$ is the functional PDF of $u(\cdot, t)$,

(13)

$$\begin{aligned} \frac{\partial}{\partial t} Z[T(\cdot,t)] &= -\int_{R^d} \frac{\delta}{\delta T}(\kappa \Delta T + C_{\alpha\beta}(0)\nabla_\alpha\nabla_\beta T) Z[T(\cdot,t)])d^d\boldsymbol{x} \\ &+ \int_{R^d\times R^d} \frac{\delta^2}{\delta T\delta T'}((C_{\alpha\beta}(\boldsymbol{x}-\boldsymbol{x}')\nabla_\alpha T\nabla_\beta T')Z[T(\cdot,t)])d^d\boldsymbol{x}d^d\boldsymbol{x}' \\ &+ \int_{R^d\times R^d} B_{\alpha\beta}(\boldsymbol{x}-\boldsymbol{x}')\frac{\delta^2}{\delta T\delta T'} Z[T(\cdot,t)]d^d\boldsymbol{x}d^d\boldsymbol{x}' \end{aligned}$$

where $T = T(\boldsymbol{x},t)$, $T' = T(\boldsymbol{x}',t)$, $Z[T(\cdot,t)]$ is the functional PDF of $T(\cdot,t)$. Clearly the amount of noise added to the system controls the strength of ellipticity in Hopf's equations.

Before ending this introduction, we should emphasize the fact that (2-4) are of interest to a wide variety of physical problems other than hydrodynamic turbulence. For instance, (3) is often used to model flow in porous media, diffusion of tracer particles and pollutants, etc [**38**]. (4) is also referred to as the kinematic dynamo problem [**51**]. (2) is one of the canonical examples in non-equilibrium statistical physics. As such, it describes the statistical mechanics of strings in a random potential. The string is assumed to be directed, i.e. there exists a time axis such that the realizations of the string can be viewed as (random) graphs over this time axis. Vortex lines in high temperature superconductors [**7**], charge density waves [**25**], directed polymers and stochastic interfaces in $1+1$ dimensional SOS models [**36**] are all examples of such strings. To see this connection, define the partition function for the configurations of the strings over the time interval [0, t] assuming that they are pinned at time t at location $\boldsymbol{x}$:

$$Z(\boldsymbol{x},t) = \left\langle \exp\left(-\beta \int_0^t \left(\tfrac{1}{2}|\dot{\boldsymbol{\xi}}(\tau)|^2 + V(\boldsymbol{\xi}(\tau),\tau)\right) d\tau\right) \middle| \boldsymbol{\xi}(t) = \boldsymbol{x} \right\rangle,$$

where $\beta = 1/kT$, k is the Boltzmann constant, T is the temperature. The first term in the exponent is the elastic energy and the second term is the potential energy with V being the random potential. The associated free energy $\varphi = kT\log Z$ satisfies

$$(14)\qquad \frac{\partial \varphi}{\partial t} + \tfrac{1}{2}|\nabla\varphi|^2 = kT\Delta\varphi + V.$$

In one dimension, let $u = \partial\varphi/\partial x$, then we have

$$\frac{\partial u}{\partial t} + u\frac{\partial u}{\partial x} = \nu\frac{\partial^2 u}{\partial x^2} + \frac{\partial V}{\partial x},$$

with $\nu = kT$. (14) is the well-known KPZ equation [**32**]. In this context, the question of interest is the effective large scale behavior in (14).

2. Stochastic Navier-Stokes Equation

Concerning the problem of hydrodynamic turbulence, the most solid and important theoretical result in turbulence theory is still Kolmogorov's 4/5 law stating that, under the condition of homogeneity and isotropy, the third order structure function for the longitudinal component of the velocity field $u_{\|}$ satisfies

$$\langle (u_{\|}(\boldsymbol{x}+\boldsymbol{r},t) - u_{\|}(\boldsymbol{x},t))^3\rangle \sim -\frac{4}{5}\,\bar{\varepsilon}\,|\boldsymbol{r}| \tag{15}$$

for $\ell_d \ll |\boldsymbol{r}| \ll L$, where $u_{\|}(\boldsymbol{x},t) = \boldsymbol{u}(\boldsymbol{x},t)\cdot\boldsymbol{r}/|\boldsymbol{r}|$, $u_{\|}(\boldsymbol{x}+\boldsymbol{r},t) = \boldsymbol{u}(\boldsymbol{x}+\boldsymbol{r},t)\cdot\boldsymbol{r}/|\boldsymbol{r}|$, and ℓ_d is a dissipation length scale. $\bar{\varepsilon} = \langle \nu|\nabla \boldsymbol{u}|^2\rangle$ is the average rate of energy dissipation. One of the basic assumptions in turbulence theory is that under a fixed external condition such as forcing or boundary condition, $\bar{\varepsilon}$ stays finite in the limit $\nu \to 0$. Hence the limiting invariant measure, if it exists, supports singular Euler flows. This was pointed out by Onsager [**42**] in 1949. In fact Onsager conjectured that the solutions of the 3D incompressible Euler's equation conserve energy if they are spatially Hölder continuous with exponent larger than 1/3, and cease to conserve energy if the exponent is less than 1/3. The first half of this statement was proved in [**14**] in its sharp form formulated in terms of Besov spaces. For simplicity of presentation, we will assume periodic boundary condition on the domain D.

Theorem 2.1 [**14**]. *Let* $\boldsymbol{u} = (u_1, u_2, u_3) \in L^3([0,T], B_3^{\alpha,\infty}(D)) \cap C([0,T], L^2(D))$ *be a weak solution of the 3D incompressible Euler's equation, i.e.*

$$\begin{aligned}
&-\int_0^T\int_D u_\alpha(\boldsymbol{x},t)\frac{\partial}{\partial t}\psi_\alpha(\boldsymbol{x},t)d^3\boldsymbol{x}dt - \int_D u_\alpha(\boldsymbol{x},0)\psi_\alpha(\boldsymbol{x},0)d^3\boldsymbol{x}\\
&-\int_0^T\int_D u_\alpha(\boldsymbol{x},t)u_\beta(\boldsymbol{x},t)\nabla_\alpha\psi_\beta(\boldsymbol{x},t)d^3\boldsymbol{x}dt - \int_0^T\int_D p(\boldsymbol{x},t)\nabla_\alpha\psi_\alpha(\boldsymbol{x},t)d^3\boldsymbol{x}dt = 0,
\end{aligned}$$

for every test function $\boldsymbol{\psi} = (\psi_1,\psi_2,\psi_3) \in C^\infty(D\times[0,T))$ *with compact support. If* $\alpha > \frac{1}{3}$, *then*

$$\int_D |\boldsymbol{u}(\boldsymbol{x},t)|^2 d^3\boldsymbol{x} = \int_D |\boldsymbol{u}(\boldsymbol{x},0)|^2 d^3\boldsymbol{x}, \quad \textit{for}\ \ t\in[0,T).$$

Besov space is the natural setting for formulating this result since its definition closely resembles the definition of structure functions except that the ensemble average is replaced by the spatial average. In fact, in more physical terms, Theorem 2.1 states that if

$$\left(\int_D |u(\boldsymbol{x}+\boldsymbol{r},t) - u(\boldsymbol{x},t)|^3 d^3\boldsymbol{x}\right)^{1/3} \le C|\boldsymbol{r}|^\alpha, \tag{16}$$

for $\alpha > 1/3$, then $\bar{\varepsilon} = 0$. This is the deterministic analog of Kolmogorov's 4/5 law. It is clear from this that Theorem 2.1 is sharp. Furthermore, Theorem 2.1 suggests that in the inviscid limit, turbulent fields live in a space close to $B_3^{1/3,\infty}$.

Next by assuming scale-independence for the skewness factor $S_3(r)/S_2(r)^{3/2}$, Kolmogorov obtained from (16) a prediction for $S_2(r)$:

$$S_2(r) \sim C_2\bar{\varepsilon}^{2/3}r^{2/3}, \qquad \ell_d \ll r \ll L. \tag{17}$$

Translating to Fourier space, we obtain the better known 5/3 law:

$$E(k) = C_K\bar{\varepsilon}^{2/3}k^{-5/3} \tag{18}$$

where $E(k)$ is the energy spectrum, $E(k) = d\mathcal{E}/dk$, $\mathcal{E}(k) = \int_{|\boldsymbol{k}|\le k}\int_{R^3}\langle\hat{\boldsymbol{u}}(\boldsymbol{k})\cdot\hat{\boldsymbol{u}}(\boldsymbol{q})\rangle d^3\boldsymbol{q}d^3\boldsymbol{k}$.

Kolmogorov's theory was immediately challenged by Landau who remarked that due to the intermittent nature of dissipation and the influence by the large scales, there cannot be universal relations such as (18) for all turbulent flows. This ultimately led to the proposal that turbulent fields are multi-fractal in the sense that the function α_p should be a highly nonlinear function. We refer to [**27**] for discussions on phenomenological multi-fractal models.

An alternative picture for accommodating intermittency was put forward by Barenblatt and Chorin [**3, 4, 12**]. Noting that Kolmogorov's picture was based on complete self-similarity, they argue that typical systems in statistical mechanics exhibit complete self-similarity only when they are well-described by the mean field theory. For problems such as turbulent flows for which fluctuations are thought to be important, incomplete self-similarity holds and they proceed to write down an ansatz for the structure functions based on incomplete self-similarity. The Barenblatt-Chorin theory predicts that the classical Kolmogorov theory holds in the vanishing viscosity limit, but is corrected at finite Reynolds numbers by terms depending on $\ln Re$.

Turning to the mathematical issues, the existence and uniqueness of an invariant measure is only proved so far for two dimension under stringent assumptions on the forcing. Assuming periodic boundary condition, if we write

$$\boldsymbol{f}(\boldsymbol{x},t)dt = \sum_{\boldsymbol{k}} \sigma_{\boldsymbol{k}} e^{i\boldsymbol{k}\cdot\boldsymbol{x}} d\boldsymbol{W}_{\boldsymbol{k}}(t), \tag{19}$$

where the $\{\boldsymbol{W}_{\boldsymbol{k}}(\cdot)\}$'s are independent Wiener processes, Flandoli and and Maslowski [**26**] proved the existence and uniqueness of the invariant measure under the assumption that there exist constants c and C such that

$$c|\boldsymbol{k}|^{-1/2} \le |\sigma_{\boldsymbol{k}}| \le C|\boldsymbol{k}|^{-3/8-\varepsilon} \tag{20}$$

for some $\varepsilon > 0$. Conditions of the type (20) are quite unnatural from a physical point of view. Intensive research is now going on to remove this condition. Other interesting aspects of the stochastic Navier-Stokes equation are discussed in [**40**].

3. Stochastic Burgers equation

3.1. Invariant Measures. The existence of an invariant measure for (2) is only understood so far on finite domains. Consider, for example, (2) on $[0, 2\pi]$ with periodic boundary condition. We will sometime identify the domain as S^1, the unit circle. The forcing function can be expressed as:

$$f(x,t)dt = \sum_k f_k(x) dW_k(t), \tag{21}$$

where the $\{W_k(\cdot)\}$'s are independent Wiener processes. We will assume that

$$f_k(\cdot) \in C^3(S^1), \qquad |\frac{\partial f_k}{\partial x}(x)| \le \frac{C}{k^2}, \tag{22}$$

for all k. We will use $(\Omega, \mathcal{F}, P)$ to denote the probability space for the forcing functions f, and $\omega \in \Omega$ to denote a typical realization of the force. $\mathcal{F}_t$ denotes the σ-algebra generated by the forces up to time t.

When $\nu = 0$, (2) is understood in the weak sense with solutions satisfying the entropy condition. The precise definition for the random case is given in [**19**]. In this case, we write (2) as

$$\frac{\partial u}{\partial t} + \frac{1}{2}\frac{\partial}{\partial x}(u^2) = f. \tag{23}$$

A natural phase space for (23) is the Skorohod space D on S^1 which consists of functions admitting only discontinuities of the first kind [**6**]. Let $\mathcal{D}$ be the Borel σ-algebra on D. (23) can then be viewed as a Markov process on D with transition probability

$$P_t(u, A) = \int_\Omega \chi_A(u,\omega) P(d\omega) \tag{24}$$

where $u \in D$, $A \in \mathcal{D}$, and

$$\chi_A(u,\omega) = \begin{cases} 1 & \text{if } S_\omega(t)u \in A \\ 0 & \text{otherwise} \end{cases} \tag{25}$$

Here $S_\omega(t)$ denotes the solution operator of (23) at time t with forcing ω.

Definition. *An invariant measure $\mu_0(du)$ of the Markov process (23) is a measure on $(D, \mathcal{D})$ satisfying*

$$\mu_0(A) = \int_D P_t(u, A)\mu_0(du) \tag{26}$$

for any $A \in \mathcal{D}$ and any $t > 0$.

Theorem 3.1 [**19**]. *(23) admits a unique invariant measure on the space D.*

The proof of Theorem 3.1 was based on the following variational characterization of weak solutions of (23) satisfying the entropy condition. Let $\xi : [t_1, t_2] \to R^1$ be a Lipschitz continuous curve. Define the action functional

$$\begin{aligned} \mathcal{A}_{t_1,t_2}(\xi) &= \int_{t_1}^{t_2} \Big(\tfrac{1}{2}(\dot\xi(s))^2 - \sum_k f_k(\xi(s))\dot\xi(s)(W_k(s) - W_k(t_1)) \Big) ds \\ &\quad + \sum_k f_k(\xi(t_2))(W_k(t_2) - W_k(t_1)) \end{aligned} \tag{27}$$

Then we have for $\tau < t$,

$$u(x,t) = \frac{\partial}{\partial x} \inf_{\xi(t)=x} \left\{ \mathcal{A}_{\tau,t}(\xi) + \int_0^{\xi(\tau)} u(z,\tau)dz \right\} \tag{28}$$

Minimizers of the functional in (27) satisfy the following Euler-Lagrange equation:

$$d\gamma(s) = v(s)ds, \qquad dv(s) = \sum_{k=1}^\infty f_k(\gamma(s))dW_k(s) \tag{29}$$

From (28) one can easily verify the following recipe for constructing solutions of (23). Fix t. For values of x such that the minimizer to the functional (27) is unique, say $\xi(\cdot)$, then $u(\cdot,t)$ is continuous at x, and $u(x,t) = \dot\xi(t)$. On the other hand, for values of x such that the minimizers to the variational problem are not unique, then $u(\cdot,t)$ is discontinuous at x with $u(x+,t) = \inf_\alpha\{\dot\xi_\alpha(t)\}$ and $u(x-,t) = \sup_\alpha\{\dot\xi_\alpha(t)\}$ where $\{\xi_\alpha\}$ denotes the family of minimizers of (27).

This construction is just a reformulation of the method of characteristics for weak solutions. It is a variational formulation of the backward characteristics defined in [**15**]. In particular, the Euler-Lagrange equation (29) is nothing but the equation for the characteristics of (23).

The variational formulation of the method of characteristics allows us to make a connection between weak solutions of the forced Burgers equation and the non-smooth invariant sets constructed in the Aubry-Mather theory [**1, 39**]. This is an extension to weak solutions of the classical Hamilton-Jacobi theory which establishes a connection between smooth solutions of the Hamilton-Jacobi equation and the smooth invariant tori for the corresponding Hamiltonian system [**18, 31, 46**].

Of particular interest to the construction of the invariant measure is a special class of minimizers called the one-sided minimizers (OSM).

Definition. *A piecewise C^1-curve $\xi : (-\infty, 0] \to S^1$ is a one-sided minimizer if for any Lipschitz continuous $\tilde{\xi} : (-\infty, 0] \to S^1$ such that $\tilde{\xi}(0) = \xi(0)$ and $\tilde{\xi} = \xi$ on $(-\infty, \tau]$ for some $\tau < 0$, we have*

$$\mathcal{A}_{s,0}(\xi) \leq \mathcal{A}_{s,0}(\tilde{\xi}), \tag{30}$$

for all $s \leq \tau$. Similarly, we define one-sided minimizers on $(-\infty, t]$, for $t \in R^1$.

Next we ask the following question: Given (x, t), how many OSMs ξ exist such that $\xi(t) = x$? This question is answered by studying the intersection properties of OSMs. As a general fact in the calculus of variations, two different OSMs cannot intersect more than once. In other words, if ξ_1, ξ_2 are two OSMs such that there exist $t_1, t_2 \in R^1, t_1 \neq t_2$, such that $\xi_1(t_1) = \xi_2(t_1), \xi_1(t_2) = \xi_2(t_2)$, then $\xi_1 \equiv \xi_2$ on their common domain of definition [**39**]. However more is true for the random case. If ξ_1 and ξ_2 are two different OSMs, such that $\xi_1(s) = \xi_2(s)$ for some s, then neither ξ_1 nor ξ_2 can be extended as an OSM beyond the interval $(-\infty, s]$. This is because that in the random case, two OSMs always have an effective intersection at $t = -\infty$. The precise formulation of this property is given in [**19**].

These intersection properties have far-reaching consequences. Let us fix $t = 0$. By considering the image of all OSMs at $t = -1$, it is easy to see that the set

$$I_\omega = \{x \in S^1 : \text{there exist more than one OSM } \xi \text{ such that } \xi(0) = x\},$$

can at most be a countable set for almost all $\omega \in \Omega$. Therefore we can define:

$$u_\omega(x, 0) = \dot{\xi}(0), \tag{31}$$

where ξ is the OSM such that $\xi(0) = x$. $u_\omega(\cdot, 0)$ is a well-defined function in $L^\infty(S^1)$ for almost all ω.

Similar construction can be carried out for any $t \in R^1$ which defines $u_\omega(\cdot, t)$. Furthermore it is easy to conclude from the variational principle (28) that

$$u_\omega(\cdot, t) = S_\omega(t) u_\omega(\cdot, 0), \tag{32}$$

for $t > 0$. In other words, u_ω is a solution of (23). From this it is straightforward to check that the distribution of the mapping:

$$\Phi_0 : \omega \to u_\omega(\cdot, 0), \tag{33}$$

is an invariant measure for (23). If we define

$$\Phi_t : \omega \to u_\omega(\cdot, t), \tag{34}$$

then $\{\Phi_t\}$ satisfies the invariance principle:

$$\Phi_t(\omega) = S_\omega(t)\Phi_0(\omega) \tag{35}$$

Therefore we have

Theorem 3.2 [**19**]. *There exists a family of invariant mappings* $\{\Phi_t\} : \Omega \to D$ *satisfying (35). Furthermore the invariant measure* μ_0 *is the distribution of* Φ_0.

Theorem 3.2 states that the solutions supported by the invariant measure are functionals of the forcing. Uniqueness of the invariant measure follows from the fact that the OSMs are largely unique.

It is shown in [**44**] that for $\nu > 0$, (2) also admits a unique invariant mapping Φ_0^ν whose distribution is the unique invariant measure for (2), denoted by μ_ν. We have

Theorem 3.3 [**19**].

$$\Phi_0^\nu(\omega) \to \Phi_0(\omega), \tag{36}$$

in $L^1(S^1)$ *for almost every* $\omega \in \Omega$. *Consequently*

$$\mu_\nu \to \mu_0, \tag{37}$$

weakly.

We remark that (36), (37) are a different kind of statement from standard theorems on inviscid limits of (2) studied in the PDE literature [**45, 47**]. There we are given a sequence of initial data that converge in the inviscid limit, and we ask whether convergence still holds at later times. Here we are not given any initial data, and we proceed to establish convergence with the only information that the solutions are defined for all $t \in R^1$ in a special way using the OSMs. Consequently the techniques used to prove Theorem 3.3 are very different from the ones used in the PDE literature to study inviscid limits. See [**19**].

Our next task is to characterize the solutions supported on the invariant measure. This requires a non-degeneracy condition to the effect that the process (29) starting at any $x \in S^1$ is transitive on S^1. This condition is generically satisfied. However it is violated if the sum in (21) contains only one term. We refer to [**19**] for a detailed formulation and examination of this condition. Under this non-degeneracy condition, we have

Theorem 3.4 [**19**]. *For almost all* ω, u_ω *satisfies the following:*

(1). There exists a unique two-sided minimizer (TSM, defined below) $y_\omega : R^1 \to S^1$ *which is a characteristic of (23) associated with the solution* u_ω.

(2). There exists a unique so-called main shock $\gamma_\omega : R^1 \to S^1$, *which is a continuous shock curve defined for all* $t \in R^1$.

(3). The TSM is a hyperbolic trajectory for the dynamical systems (29).

(4). For any $t \in R^1$, *there exist global stable and unstable manifolds associated with* y_ω *at time* t, *denoted by* $W_\omega^s(t)$ *and* $W_\omega^u(t)$ *respectively, on the phase space* $S^1 \times R^1$.

(5). The graph of $u_\omega(\cdot, t)$ *is a subset of* $W_\omega^u(t)$.

As a corollary, we have that almost surely, u_ω is a piecewise smooth function.

Definition. *A piecewise* C^1*-curve* $\xi : (-\infty, +\infty) \to S^1$ *is a two-sided minimizer if for any Lipschitz continuous* $\tilde{\xi} : (-\infty, +\infty) \to S^1$ *such that* $\tilde{\xi} = \xi$ *away*

from a compact set, we have

$$\mathcal{A}_{-s,s}(\xi) \leq \mathcal{A}_{-s,s}(\tilde{\xi}), \tag{38}$$

for large enough s.

The possibility of establishing hyperbolicity of TSMs in the random case comes from the following:

Basic Collision Lemma. *Assuming the non-degeneracy condition. Then there exists a constant p_0, depending only on the $\{f_k\}$'s, with the following property: Given an arbitrary pair of points $(x_1, x_2) \in [0,1]^2$ at $t = 0$ whose positions are $\mathcal{F}_0$-measurable,*

$$P\{x_1, x_2 \text{ merge before } t = 1\} > p_0. \tag{39}$$

Heuristically two points merge before $t = 1$ if the forward characteristics emanating from them intersect before $t = 1$. This of course depends on the forces as well as the solution at $t = 0$. The lemma states that independent of the solution at $t = 0$, one can always find a set of forces with positive measure under which the two points merge.

The proof of the Basic Collision Lemma relies on PDE techniques and is given in the Appendix D of [**19**].

The Basic Collision Lemma provides the mechanism for the origin of the hyperbolicity. In particular, the uniqueness of the TSM and the main shock is a simple consequenc of the Basic Collision Lemma.

The regularity and structural properties described here are used in [**20, 21, 22**] to study the statistical behavior of the Burgers equation. A summary of these results is given below.

3.2. Statistical Theory. We now address the questions frequently asked in the physics literature regarding (2), building on the regularity results described in Section 3.1. Since we have established the existence of μ_0 which is the statistical steady state at the inviscid limit, we can restrict our attention to this case.

3.2.1. *Structure functions.* The fact that u_ω is piecewise smooth implies easily that

$$S_p(r) = \begin{cases} r^p\langle|\xi|^p\rangle + o(r^p) & \text{if } 0 \leq p < 1, \\ r\rho\langle|s|^p\rangle + o(r) & \text{if } 1 < p, \end{cases} \tag{40}$$

where ρ is the number density of the shocks, which is finite from the results described earlier, s is the jump across the shocks, ξ is the regular part of the velocity gradient:

$$\frac{\partial}{\partial x} u(x,t) = \xi(x,t) + \sum_j s(y_j, t)\delta(x - y_j), \tag{41}$$

with $\xi(\cdot, t) \in L^1(S^1)$.

From (40), α_p is a linear function in p followed by a constant function. This is a reflection of the fact that as far as regularity is concerned, almost everywhere the solution is either Lipschitz continuous or discontinuous. The linear part in the graph of α_p probes the Lipschitz continuous part of the solution. The flat part probes the discontinuous part of the solution. Such a situation is sometimes referred to as a "bifractal".

3.2.2. *Velocity gradient PDF.* More difficult are the questions of PDFs for quantities such as u, $\partial u/\partial x$ and $\delta u(x,t) = u(x+y,t) - u(y,t)$. In particular, the PDF of $\xi = \partial u/\partial x$ (suitably defined), $Q(\xi)$, has attracted a lot of attention in recent years. In the inviscid limit, it is agreed that $Q(\xi)$ has the behavior

$$Q(\xi) \sim \begin{cases} C_-|\xi|^{-\alpha} & \text{as } \xi \to -\infty, \\ C_+\xi^\beta \mathrm{e}^{-\xi^3/(3B_1)} & \text{as } \xi \to +\infty. \end{cases} \tag{42}$$

where C_-, C_+ are constants, $B_1 = -B_{xx}(0)$. But many different values of α and β have been proposed (see [**22, 33**] for a review).

A priori, there is even an issue how to define Q. One can define for the inviscid limit the PDF for the velocity divided difference $(u(x+y,t) - u(y,t))/x$, $Q^\delta(\xi, x)$ and then take $x \to 0$. An alternative is to first define the PDF of $\partial u/\partial x$ for the viscous problem, call it $Q^\nu(\xi)$, and then take the limit as $\nu \to 0$:

$$\underline{Q}(\xi) = \lim_{\nu\to 0} Q^\nu(\xi), \qquad Q(\xi) = \lim_{\delta\to 0} Q^\delta(\xi, x). \tag{43}$$

While a fully rigorous proof was not constructed, [**22**] presented strong evidence that for the case studied here,

$$Q = \underline{Q}. \tag{44}$$

Below we will concentrate on Q.

Using calculus for functions of bounded variation [**49**], we can derive an equation for Q

$$\frac{\partial Q}{\partial t} = \xi Q + \frac{\partial}{\partial \xi}(\xi^2 Q) + B_1 \frac{\partial^2 Q}{\partial \xi^2} + F(\xi, t), \tag{45}$$

where

$$F(\xi, t) = \rho \int_{-\infty}^{0} sV(s, \xi, t) ds \le 0. \tag{46}$$

Here $V(s,\xi,t) = V_+(s,\xi,t) + V_-(s,\xi,t)$, $V_\pm(s,\xi_\pm,t)$ are the PDFs of $(s(y,t), \xi_\pm(y,t) = \partial u(y\pm,t)/\partial x)$ conditional on y being a shock position. This equation should be compared with the equation for Q^ν:

$$\frac{\partial Q^\nu}{\partial t} = \xi Q^\nu + \frac{\partial}{\partial \xi}(\xi^2 Q^\nu) + B_1 \frac{\partial^2 Q^\nu}{\partial \xi^2} - \nu \frac{\partial}{\partial \xi}\Big(\Big\langle \frac{\partial^2 \xi}{\partial x^2}\Big|\xi\Big\rangle Q^\nu\Big), \tag{47}$$

where $\langle \partial^2\xi/\partial x^2|\xi\rangle$ is the ensemble-average of $\partial^2\xi/\partial x^2$ conditional on ξ. We see that

$$F(\xi, t) = -\lim_{\nu\to 0} \nu \frac{\partial}{\partial \xi}\Big(\Big\langle \frac{\partial^2 \xi}{\partial x^2}\Big|\xi\Big\rangle Q^\nu\Big),$$

Even though we are primarily interested in statistical steady states, we have written down the master equations for the more general case including transients.

Integrating (45), we get

$$\frac{d}{dt}\int_{R^1} Q(\xi, t) d\xi = \langle \xi \rangle + \rho \langle s \rangle = 0. \tag{48}$$

Consequently

$$\lim_{\nu\to 0} \int_{R^1} \xi Q^\nu(\xi) d\xi \neq \int_{R^1} \xi Q(\xi) d\xi, \tag{49}$$

since the left hand side vanishes due to homogeneity.

Even though (45) is not a closed equation since the form of F is unknown, we can already obtain from it non-trivial information. As an example, we have

Theorem 3.5 [21, 22].

$$\lim_{|\xi|\to+\infty} |\xi|^3 Q(\xi,t) = 0, \tag{50}$$

i.e. Q *decays faster than* $|\xi|^{-3}$ *as* $\xi \to -\infty$.

This result rules out all the proposed value of α except that of [**20**] which gives $\alpha = 7/2$.

Theorem 3.5 is obtained by combining (45) together with the following:

(1). Take the first moment of (45) gives

$$\frac{d}{dt}\langle\xi\rangle = \left[\xi^3 Q\right]_{-\infty}^{+\infty} + \frac{\rho}{2}\left(\langle s\xi_-\rangle + \langle s\xi_+\rangle\right).$$

(2). Along the shock, we have

$$\frac{d}{dt}(\rho\langle s\rangle) = -\frac{\rho}{2}(\langle s\xi_-\rangle + \langle s\xi_+\rangle),$$

which is a consequence of the equations for the dynamics of the shocks. Since formally

$$\int_{R^1} \xi F(\xi,t)d\xi = \frac{\rho}{2}\left(\langle s\xi_-\rangle + \langle s\xi_+\rangle\right),$$

the process of proving (3.2.2) also proves that $\xi F(\xi,t)$ is absolutely integrable on R^1 for all t.

3.2.3. *Asymptotics for the statistical stationary state.* In statistical steady state, (45) becomes

$$\xi Q + \frac{d}{d\xi}(\xi^2 Q) + B_1\frac{d^2Q}{d\xi^2} + F(\xi) = 0. \tag{51}$$

This is a second order ODE with an inhomogeneous term F. We can write the general solutions of (51) as

$$Q = Q_s + C_1 Q_1 + C_2 Q_2, \tag{52}$$

where Q_1 and Q_2 are two linearly independent solutions of the homogeneous equation, and Q_s is a particular solution. For example, we can take:

$$Q_1(\xi) = \xi \mathrm{e}^{-\Lambda}, \tag{53}$$

$$Q_2(\xi) = 1 - \frac{\xi \mathrm{e}^{-\Lambda}}{B_1}\int_{-\infty}^{\xi} \xi' \mathrm{e}^{\Lambda'} d\xi', \tag{54}$$

$$Q_s(\xi) = \frac{1}{B_1}\int_{-\infty}^{\xi} \xi' F(\xi')d\xi' - \frac{\xi \mathrm{e}^{-\Lambda}}{B_1}\int_{-\infty}^{\xi} \mathrm{e}^{\Lambda'} G(\xi')d\xi', \tag{55}$$

where

$$\Lambda = \frac{\xi^3}{3B_1} \qquad G(\xi) = F(\xi) + \frac{\xi}{B_1}\int_{-\infty}^{\xi} d\xi'\ \xi' F(\xi'). \tag{56}$$

3.2.4. *Bounds from realizability constraints.* Using the realizability constraint $Q \in L^1(R^1)$, $Q \geq 0$, we can show that

$$Q = Q_s, \tag{57}$$

and

$$\lim_{\xi \to +\infty} \xi^{-2} e^{\Lambda} F(\xi) = 0. \tag{58}$$

Indeed, starting from $Q = Q_s + C_1 Q_1 + C_2 Q_2$, we have:

(1). $Q_s \in L^1(R^1)$, $Q_2 \in L^1(R^1)$, but $Q_1 \notin L^1(R^1)$. Therefore C_1 must be zero.

(2). As $|\xi| \to +\infty$, $|Q_2| > |Q_s|$, but $Q_2 > 0$ as $\xi \to -\infty$, and $Q_2 < 0$ as $\xi \to +\infty$. Therefore C_2 must be zero.

(3). As $\xi \to +\infty$, $Q_s \geq 0$, *iff* (58) holds.

These statements can be established by evaluating Q_2, Q_s using Laplace's method. We arrive at:

$$Q(\xi) \sim \begin{cases} |\xi|^{-3} \int_{-\infty}^{\xi} \xi' F(\xi') d\xi' & \text{as } \xi \to -\infty, \\ C_+ \xi e^{-\Lambda} & \text{as } \xi \to +\infty. \end{cases} \tag{59}$$

Once again, we obtain (50). Furthermore we get $\beta = 1$ in (42).

3.2.5. *The exponent 7/2.* Let $W(s, \xi_+, \xi_-, x, t)$ be the PDF of

$$(u(y+x,t) - u(y-x,t), \xi(y+x,t), \xi(y-x,t)),$$

conditional on y being a shock location. Then

$$V_\pm(s, \xi_\pm, t) = \int_{R^1} W(s, \xi_+, \xi_-, 0, t) d\xi_\mp. \tag{60}$$

W satisfies an equation of the form [**22**]

$$\frac{\partial W}{\partial t} = \mathcal{O} W + S, \tag{61}$$

where $\mathcal{O}$ is a differential operator in x, s, $\xi_\pm$, S is a source term accounting for shock creation and collision. Information on the source term S can be obtained using local analysis around shock creation and collision points. Upon using this information in (60), we get:

$$F(\xi) \sim C|\xi|^{-5/2} \qquad \text{as } \xi \to -\infty, \tag{62}$$

Therefore

$$Q(\xi) \sim |\xi|^{-3} \int_{-\infty}^{\xi} \xi' F(\xi') d\xi' = C_- |\xi|^{-7/2} \qquad \text{as } \xi \to -\infty. \tag{63}$$

which confirms the prediction of [**20**].

The analysis in [**22**] that accomplishes this last step is quite involved. The result of that is a confirmation of the geometric picture proposed in [**20**], namely the leading order contribution to the left tail of Q comes from the boundary of the set of the shocks, here the points of shock creation. This geometric picture may have interesting consequences in higher dimension. The analysis in [**22**] is also a success in working with the master equation without making closure assumptions. In a sense closure in [**22**] is achieved through dimension reduction. The PDF of ξ is first expressed in terms of the lower dimensional dissipative structures, here the shocks. The PDF for the shocks and the shock environment is further expressed in terms of the singular structures on the shocks, namely the points of shock creation and collision, which are then amenable to local analysis. Clearly this approach should be applicable to high dimensions and should yield more interesting results.

4. Stochastic Passive Scalar and Passive Vector Equations

4.1. The stochastic passive scalar equation. Turning now to (3). There are two main questions that have been studied in the literature. The first is the effective behavior at large scales, particularly when $\boldsymbol{u}$ contains energy at large scales. We will not discuss this problem here. Instead we refer to the extensive review article [**38**]. For issues with regard to the behavior for the average of temperature, including for the case of white-in-time velocity fields, we refer to [**38**]. Here we will be mainly interested in higher order statistics.

The second question concerns the local characteristics in the temperature field T, such as the scaling of the structure functions. In this context, the most thoroughly studied problem is the so-called Kraichnan's model [**34, 35**] in which $\boldsymbol{u}$ is a Gaussian random process with zero-mean and covariance

$$\langle u_\alpha(\boldsymbol{x},t)u_\beta(\boldsymbol{y},s)\rangle = 2C_{\alpha\beta}(\boldsymbol{x}-\boldsymbol{y})\delta(t-s), \tag{64}$$

The white-in-time nature of the velocity field allows us to significantly simplify the analytical treatment, but is sometimes a poor approximation to the physical situation. For incompressible velocity fields, the tensor C is assumed to satisfy

$$\nabla\cdot C = 0. \tag{65}$$

Following [**2, 41**], we will take

$$C_{\alpha\beta}(\boldsymbol{x}) = F(r)\delta_{\alpha\beta} + \nabla_\alpha\nabla_\beta G(r), \tag{66}$$

where $r = |\boldsymbol{x}|$,

$$F(r) = D_0\int_{R^d} \mathrm{e}^{i\boldsymbol{k}\cdot\boldsymbol{x}}(k_0^2+k^2)^{-(d+\xi)/2}\frac{d^d\boldsymbol{k}}{(2\pi)^d}, \tag{67}$$

and $-\Delta G = F$. Here d is the dimension, $\xi \in (0,2)$ is a parameter, D_0 is a constant, k_0 is an infrared cut-off parameter.

Writing $C_{\alpha\beta}(\boldsymbol{x}) = C_{\alpha\beta}(0) - c_{\alpha\beta}(\boldsymbol{x})$, it can be easily seen that $C_{\alpha\beta}(0) = \bar{D}k_0^{-\xi}\delta_{\alpha\beta}$ for some $\bar{D}\propto D_0$ independent of k_0 (thus $C_{\alpha\beta}(0)$ is divergent as $k_0\to 0$), and

$$c_{\alpha\beta}(\boldsymbol{x}) = D\left((d+\xi-1)\delta_{\alpha\beta} - \xi\frac{x_\alpha x_\beta}{r^2}\right)r^\xi + O(r^2k_0^2), \tag{68}$$

for some $D\propto D_0$ independent of k_0. Thus ξ is a measure the smoothness of the velocity field: Roughly speaking, when $\xi = 0$ the velocity field is spatially independent; when $\xi = 2$ the velocity field is spatially Lipschitz continuous. Note that, since $C_{\alpha\beta}(\boldsymbol{x})\sim 0$ for $r\gg \ell_0 = k_0^{-1}$, we have

$$c_{\alpha\beta}(\boldsymbol{x}) \sim \bar{D}\,\ell_0^\xi\,\delta_{\alpha\beta}, \qquad r\gg\ell_0. \tag{69}$$

However, in the sequel we will be mostly interested in the range $r\ll\ell_0$.

Existence of an invariant measure for $\kappa\geq 0$ and the convergence as $\kappa\to 0$ are established in [**23**]. Below we will study the statistical properties associated with these invariant measures.

4.1.1. *Master equation and the dissipative anomaly.* Assume that the initial temperature profile, $T_0(\boldsymbol{x}) = T(\boldsymbol{x}, 0)$ is statistically isotropic. Define $\theta(\boldsymbol{x}, \boldsymbol{x}', t) = T(\boldsymbol{x}, t) - T(\boldsymbol{x}', t)$ and let $Z^\kappa(\theta, |\boldsymbol{x} - \boldsymbol{x}'|, t)$ be the PDF of $\theta(\boldsymbol{x}, \boldsymbol{x}', t)$ (Z^κ depends on $|\boldsymbol{x} - \boldsymbol{x}'|$ only by isotropy). Then Z^κ satisfies [**9**]

$$\frac{\partial Z^\kappa}{\partial t} = \frac{2}{r^{d-1}} \frac{\partial}{\partial r} \left(r^{d-1} \eta \frac{\partial Z^\kappa}{\partial r} \right) + 2b(r) \frac{\partial^2 Z^\kappa}{\partial \theta^2} - \frac{\partial}{\partial \theta}(H^\kappa Z^\kappa), \tag{70}$$

where $b(r) = B(0) - B(r), \eta(r) = D(d-1)r^\xi$, and

$$H^\kappa(\theta, r, t) = \kappa \langle (\Delta - \Delta')\theta(\boldsymbol{x}, \boldsymbol{x}', t) | \theta \rangle. \tag{71}$$

This is the analog of (47) for the Burgers equation. We are interested in the limit

$$Z(\theta, r, t) = \lim_{\kappa \to 0} Z^\kappa(\theta, r, t). \tag{72}$$

In particular, we are interested in the limit of the dissipation term: $\partial(H^\kappa Z^\kappa)/\partial\theta$. This term does not vanish in the limit as $\kappa \to 0$, giving rise to the dissipative anomaly [**23**]. In contrast, it can be shown that the κ term in the equations for the correlation functions can be neglected in this limit [**24**].

4.1.2. *Equations for the correlation functions.* Instead of working with (70), recent progress hinges on the fact that a closed set of equations for the single time correlation functions can be derived for Kraichnan's passive scalar model.

Consider

$$F_n(\boldsymbol{x}_1, \cdots, \boldsymbol{x}_n, t) = \langle T(\boldsymbol{x}_1, t) \cdots T(\boldsymbol{x}_n, t) \rangle \tag{73}$$

$F_{2n+1} = 0$ by symmetry. F_{2n} satisfies

$$\begin{aligned} \frac{\partial}{\partial t} F_{2n}(\boldsymbol{x}_1, \cdots, \boldsymbol{x}_{2n}, t) &= \sum_{j,k=1}^{2n} C_{\alpha\beta}(\boldsymbol{x}_j - \boldsymbol{x}_k) \nabla_\alpha^j \nabla_\beta^k F_{2n}(\boldsymbol{x}_1, \cdots, \boldsymbol{x}_{2n}, t) \\ &+ 2 \sum_{\substack{j,k=1 \\ j<k}}^{2n} B(\boldsymbol{x}_j - \boldsymbol{x}_k) F_{2n-2}(\boldsymbol{x}_1 \underset{\widehat{j}\ \widehat{k}}{\cdots} \boldsymbol{x}_{2n}, t). \end{aligned} \tag{74}$$

At statistical steady state, these equations reduces to

$$\begin{aligned} &\sum_{j,k=1}^{2n} C_{\alpha\beta}(\boldsymbol{x}_j - \boldsymbol{x}_k) \nabla_\alpha^j \nabla_\beta^k F_{2n}(\boldsymbol{x}_1, \cdots, \boldsymbol{x}_{2n}) = \\ &\qquad -2 \textstyle\sum_{\substack{j,k=1 \\ j<k}}^{2n} B(\boldsymbol{x}_j - \boldsymbol{x}_k) F_{2n-2}(\boldsymbol{x}_1 \underset{\widehat{j}\ \widehat{k}}{\cdots} \boldsymbol{x}_{2n}). \end{aligned} \tag{75}$$

Evaluation of the structures functions $S_{2n}(r) = \langle (T(\boldsymbol{x} + \boldsymbol{r}, t) - T(\boldsymbol{x}, t))^{2n} \rangle$ can in principle be carried out once the solution of (75) is known. However, solving these equations is a very difficult task and so far only perturbative methods have been successful in some regimes.

A simple dimensional argument or balancing $(\boldsymbol{u}\nabla)T$ with f suggests that a Kolmogorov-like theory would predict normal scaling exponents $\alpha_p = \frac{2-\xi}{2}p$. We will see below that this is not true for $p > 2$.

4.1.3. *Zero modes.* Our next task is to analyze the behavior of F_{2n} for small $|\boldsymbol{x}_j - \boldsymbol{x}_k|$. We will restrict ourselves to translation invariant solutions.

From (75), F_{2n} has contributions from the inhomogeneous part F_{2n-2}, and the homogeneous part, which at small distances are solutions of

$$\mathcal{M}_{2n}^{sc}\tilde{F}_{2n} = \sum_{j,k=1}^{2n} d_{\alpha\beta}(\boldsymbol{x}_j - \boldsymbol{x}_k)\nabla_\alpha^j \nabla_\beta^k \tilde{F}_{2n} = 0 \tag{76}$$

where

$$d_{\alpha\beta}(\boldsymbol{x}) = D\left((d+\xi-1)\delta_{\alpha\beta} - \xi\frac{x_\alpha x_\beta}{r^2}\right) r^\xi \tag{77}$$

(76) is obtained as the scaling limit of the homogeneous equation associated with (75). The solutions of (76) are called zero modes [**10, 29**]. This is an important concept that characterizes the origin of the anomalous scaling.

As an example, let us study the behavior of the 2-point function at $d = 3$. F_2 satisfies

$$-\frac{2}{r^2}\frac{\partial}{\partial r}\left(Dr^{2+\xi}\frac{\partial F_2}{\partial r}\right) = B(r). \tag{78}$$

It is easy to obtain from this equation that

$$F_2(r) = C_0 - C_1 r^{2-\xi} + \cdots \tag{79}$$

where C_0, C_1 are constants depending on B and the neglected terms are of higher order. Hence we have for S_2

$$S_2(r) = C_2 r^{2-\xi} + \cdots \tag{80}$$

This implies that S_2 obeys normal scaling. Note that the constant C_2 is in general non-universal.

Let us now look at the 4-point function $F_4(\boldsymbol{x}_1, \boldsymbol{x}_2, \boldsymbol{x}_3, \boldsymbol{x}_4)$ and let $\boldsymbol{x}_{jk} = \boldsymbol{x}_j - \boldsymbol{x}_k, j, k = 1, \cdots 4$. F_4 has contributions from the "Gaussian" part:

$$F_2(\boldsymbol{x}_{12})F_2(\boldsymbol{x}_{34}) + F_2(\boldsymbol{x}_{13})F_2(\boldsymbol{x}_{24}) + F_2(\boldsymbol{x}_{14})F_2(\boldsymbol{x}_{23}), \tag{81}$$

which give rise to normal scaling, as well as the contribution from the zero modes which may be the dominant contribution to S_4. To find the precise form of the zero modes is a very difficult task. Both [**29**] and [**10**] resorted to perturbation technique in either ξ or $1/d$. Here we follow [**29**] and write

$$\tilde{F} = E_0 + \xi G_0 + O(\xi^2), \tag{82}$$

where E_0 is the zero mode for the case when $\xi = 0$. Using the notations $\nabla_{12} = \nabla_{\boldsymbol{x}_{12}}$, etc, we can write down an equation for E_0:

$$-(\Delta_{12} + \Delta_{23} + \Delta_{34} - \nabla_{12}\cdot\nabla_{23} - \nabla_{23}\cdot\nabla_{34})E_0 = 0. \tag{83}$$

Substituting (83) to (76), we obtain an equation for G_0:

$$-\Delta_4 G_0 + \mathcal{L}E_0 = 0$$

where

$$\Delta_4 = \Delta_{\boldsymbol{x}_1} + \cdots + \Delta_{\boldsymbol{x}_4}, \quad \mathcal{L} = \sum_{j\neq k}\left(\delta_{\alpha\beta} - \frac{1}{2}\frac{x_{jk}^\alpha x_{jk}^\beta}{|\boldsymbol{x}_{jk}|^2}\right)\nabla_j^\alpha \nabla_k^\beta - \frac{1}{2}\Delta_4$$

The precise form of G_0 is quite complicated. We refer to [**5, 29**] for details of this calculation from which one obtains that $\tilde{F}$ must be homogeneous of degree

$$\alpha_4 = 4 - \frac{14}{5}\xi + O(\xi^2) = 2\alpha_2 - \rho_2, \tag{84}$$

where $\rho_2 = \frac{4}{5}\xi + O(\xi^2)$ is the anomalous exponent. Numerical results of Frisch et.al. [**27**] has found very good agreement with (84).

4.2. Passive vector equation: The kinematic dynamo problem. Under the same assumptions on the velocity field, the passively advected magnetic fields described by (4) can also be analyzed using similar methods. The general form of the covariance of $\boldsymbol{B}$ is given by [**41**]

$$\langle B_\alpha(\boldsymbol{x}+\boldsymbol{r},t)B_\beta(\boldsymbol{x},t)\rangle = G_1(r,t)\delta_{\alpha\beta} + G_2(r,t)\frac{r_\alpha r_\beta}{r^2}, \tag{85}$$

where $r = |\boldsymbol{r}|$ and the functions G_1 and G_2 are related by

$$\frac{\partial G_1}{\partial r} = -\frac{1}{r^2}\frac{\partial}{\partial r}(G_2 r^2) \tag{86}$$

Following [**48**], it is convenient to work with the trace of the correlation function tensor:

$$H(r,t) = 3G_1(r,t) + G_2(r,t). \tag{87}$$

G_1 and G_2 can then be found through

$$G_1(r,t) = \frac{H(r,t)}{2} - \frac{1}{2r^3}\int_0^r H(s,t)s^2 ds, \tag{88}$$

$$G_2(r,t) = -\frac{H(r,t)}{2} + \frac{3}{2r^3}\int_0^r H(s,t)s^2 ds. \tag{89}$$

In analogy with (78), a closed equation can be derived for H:

$$\begin{aligned} \frac{\partial H}{\partial t} &= \frac{2}{r^2}\frac{\partial}{\partial r}\left((\kappa r^2 + Dr^{2+\xi})\frac{\partial H}{\partial r}\right) + B_L(r) \\ &+ 2D\xi r^{\xi-2}\left((1+2\xi)H + r\frac{\partial H}{\partial r} + \frac{\xi(\xi-2)}{r^3}\int_0^r H(s,t)s^2 ds\right) \end{aligned} \tag{90}$$

where $B_L(r) = \int\langle \boldsymbol{f}(\boldsymbol{x},t)\cdot\boldsymbol{f}(0,0)\rangle dt$. (90) admits a steady state solution in the regime $\xi < 1$ when the dynamo effects are not present. To analyze this steady state solution in the limit of infinite Peclet number, it is convenient to introduce

$$\psi(r) = \frac{\sqrt{Dr^\xi}}{r}\int_0^r H(s)s^2 ds. \tag{91}$$

Then ψ satisfies

$$\frac{d^2\psi}{dr^2} - m(r)U(r)\psi = -\frac{m(r)\sqrt{Dr^\xi}}{r}\int_0^r B_L(s)s^2 ds, \tag{92}$$

for $r \ll \ell_0$, where

$$m(r) = \frac{1}{2Dr^\xi}, \qquad U(r) = \frac{D^2 r^{2\xi}(4 - 3\xi - \frac{3}{2}\xi^2)}{Dr^{2+\xi}}. \tag{93}$$

The zero modes for this problem can be found by solving the homogeneous equation. They are:

$$\tilde{F}_1(r) = r^{s_1}, \qquad \tilde{F}_2(r) = r^{s_2}, \tag{94}$$

where

$$s_{1,2} = \frac{1}{2} \pm \frac{3}{2}\sqrt{1 - \frac{1}{3}\xi(\xi+2)} \tag{95}$$

The general solutions of (92) is given by:

$$\psi = a\tilde{F}_1 + b\tilde{F}_2 - \frac{r^{s_1}}{2\sqrt{D}} \int_0^r \rho^{-2s_1} \int_0^\rho \rho_1^{-s_2-\frac{\xi}{2}} \int_0^{\rho_1} \rho_2^2 B_L(\rho_2) d\rho_2 d\rho_1 d\rho. \tag{96}$$

Clearly b must be zero in order to avoid having a $\tilde{F}_2$-type of singularity at small r. Matching with large scales, we get:

$$a = \frac{r^{s_1}}{2\sqrt{D}} \int_0^\infty \rho^{-2s_1} \int_0^\rho \rho_1^{-s_2-\frac{\xi}{2}} \int_0^{\rho_1} \rho_2^2 B_L(\rho_2) d\rho_2 d\rho_1 d\rho \tag{97}$$

Going back to H we find that the dominant contribution for $r \ll \ell_0$ comes from

$$H(r) \sim \frac{r^\gamma}{2D} \frac{L^{4-s_1-\frac{\xi}{2}}}{s_1 - s_2} \frac{s_1 + 1 - \frac{\xi}{2}}{s_1 - 1 - \frac{\xi}{2}} \int_0^\infty \rho^{3-s_1-\frac{\xi}{2}} B_L(\rho) d\rho \tag{98}$$

where

$$\gamma = -\frac{3+\xi}{2} + \frac{3}{2}\sqrt{1 - \frac{1}{3}\xi(\xi+2)} \sim -\xi - \frac{\xi^2}{3} + O(\xi^3) \tag{99}$$

Note in this case the second order correlation function does not obey the normal scaling.

5. Conclusion

In summary, we conclude that the study of stochastic PDEs in the style described in this paper is a very new, challenging and fruitful area of research. It is at the frontier of several different areas in mathematics as well as physics, ranging from PDE, large deviation theory, stochastic dynamical systems, to quantum field theory. It is at the foundation of many important areas in science and engineering. Fundamental understanding of these questions will likely come from and contribute to our understanding of analysis in infinite dimensional spaces, a recurring theme in this subject.

While a solution to the problem of hydrodynamic turbulence does not seem to be in sight, recent work on related problems of Burgers turbulence, stochastic passive scalar convection and 2D turbulence [**50**] gives renewed hope that definitive progress are being made.

Acknowledgments

I would like to thank my scientific collaborators, P. Constantin, K. Khanin, J. Mattingly, A. Mazel, Ya. Sinai and E. Titi for their contribution which led to some of the results described in this paper. I am particularly grateful to E. Vanden Eijnden, not only for his contribution to our joint work but also for his help in writing up this article. I am also grateful to M. Chertkov and U. Frisch for helpful discussions. This work is supported in part by a Presidential Faculty Fellowship from the National Science Foundation.

References

[1] Aubry, S., and LaDaeron, P. Y., *The discrete Frenkel-Kontorova model and its extensions I: exact results for the ground states*, Physica D, **8** (1983), 381-422.

[2] Batchelor, G. K., *Theory of Homogeneous Turbulence*, Cambridge University Press, England (1953).

[3] Barenblatt, G. I., and Chorin, A. J., *Scaling laws and vanishing viscosity limits for wall-bounded shear flows and for local structures in developed turbulence*, Comm. Pure Appl. Math. **50**, 381-398 (1997).

[4] Barenblatt, G. I., and Chorin, A. J., *Turbulence: an old challenge and new perspective*, Meccanica **33**, 445-468 (1998).

[5] Bernard, D., Gawędzki, K., and Kupiainen, A., *Anomalous scaling in the N-point functions of a passive scalar*, Phys. Rev. E **54**, 2564–2572 (1996).

[6] Billingsley, P., *Convergence of Probability Measures*, John Wiley & Sons, Inc., New York, 1968.

[7] Blatter, G., Feigelman, M. V., Geshkenbein, V. B., Larkin, A. I., and Vinokur, V. M., *Vortices in high-temperature superconductors*, Rev. Modern Phys. **66**, 1125–1388 (1994).

[8] Bourgain, J., *Nonlinear Schrödinger equations*, in *Hyperbolic Equations and Frequency Interactions*, IAS/Park City Mathematics Series, vol.5, Caffarelli and E ed., American Mathematical Society, 1999.

[9] Chen, S., and Kraichnan, R. H., *Simulations of a randomly advected passive scalar*, Phys. Fluids **10**, 2867–2884 (1998).

[10] Chertkov, M., Falkovich, G., Kolokolov, I., Lebedev, V., *Normal and anomalous scaling of the fourth-order correlation function of a randomly advected passive scalar*, Phys. Rev. E, **52**, No. 5, 4924-4941 (1995).

[11] Chorin, A. J., *Vorticity and Turbulence*, Springer, New York, 1994.

[12] Chorin, A. J., *New perspective in turbulence*, Quarterly of Applied Mathematics, **LVI**, No. 4, 767-785 (1998).

[13] Cipriano, F., *The two dimensional Euler equation: A statistical study*, preprint, 1998.

[14] Constantin, P., E, W., and Titi, E., *Onsager's conjecture on the energy conservation for solutions of Euler's equation*, Comm. Math. Phys., **165**, 207-209 (1994).

[15] Dafermos, C., *Generalized characteristics and the structure of solutions of hyperbolic conservation laws*, Indiana Univ. Math. J., **26** (1977), 1097-1119.

[16] Da Prato, G., and Zabczyk, J., *Stochastic Equations in Infinite Dimensions*, Encyclopedia of Mathematics and Its Applications, Cambridge University Press, 1992.

[17] Da Prato, G., and Zabczyk, J., *Ergodicity for Infinite Dimensional Systems*, London Mathematical Society Lecture Notes Series **299**, Cambridge University Press, 1996.

[18] E, W., *Aubry-Mather theory and periodic solutions of forced Burgers equation*, to appear in Comm. Pure Appl. Math..

[19] E, W., Khanin, K., Mazel, A., and Sinai, Ya. G., *Probability distributions functions for the random forced Burgers equation*, Phys. Rev. Lett. **78**, 1904–1907 (1997).

[20] E, W., Khanin, K., Mazel, A., and Sinai, Ya. G., *Invariant measures for the random-forced Burgers equation*, submitted to Ann. Math.

[21] E, W., and Vanden Eijnden, E., *Asymptotic theory for the probability density functions in Burgers turbulence*, chao-dyn/9901006, submitted to Phys. Rev. Lett.

[22] E, W., and Vanden Eijnden, E., *Statistical theory of the stochastic Burgers equation in the inviscid limit*, submitted to Comm. Pure Appl. Math..

[23] E, W., and Vanden Eijnden, E., *Statistical theory of the stochastic passive scalar equation*, in preparation.

[24] Eyink, G., and Xin, J., *Existence and uniqueness of L^2-solutions at zero-diffusivity in the Kraichnan model of a passive scalar*, chao-dyn/9605008 (1996).

[25] Feigelman, M. V., *One-dimensional periodic structures in a weak random potential*, Sov. Phys. JETP **52**, 555–561 (1980).

[26] Flandoli, F., and Maslowski, B., *Ergodicity of the 2-D Navier-Stokes equation under random perturbation*, Comm. Math. Phys., **172**, 119-141 (1995).

[27] Frisch, U., *Turbulence: The Legacy of A. N. Kolmogorov*, Cambridge University Press, 1995.

[28] Frisch, U., Mazzino, A., Noullez, A., and Vergassola, M., *Lagrangian method for multiple correlations in the passive scalar advection*, cond-mat/9810074, submitted to Phys. Fluids (1998).

[29] Gawędzki, K., and Kupiainen, A., *Anomalous scaling of a passive scalar*, Phys. Rev. Lett. **75**, 3834–3837 (1995).

[30] Gurarie, V., and Migdal, A., *Instantons in the Burgers equation*, Phys. Rev. E **54**, 4908–4914 (1996).

[31] Jauslin, H. R., Kreiss, H. O., and Moser, J., *On the forced Burgers equation with periodic boundary conditions*, preprint, 1997.

[32] Kardar, M., Parisi, G., and Zhang, Y. C., *Dynamic scaling of growing interfaces*, Phys. Rev. Lett. **56**, 889–892 (1986).

[33] Kraichnan, R. H., *Note on forced Burgers turbulence*, chao-dyn/9901023.

[34] Kraichnan, R. H., *Dispersion of particle pairs in homogeneous turbulence*, Phys. Fluids **9**, 1937–1943 (1966).

[35] Kraichnan, R. H., *Anomalous scaling of a randomly advected pasive scalar*, Phys. Rev. Lett. **72**, 1016–1019 (1994).

[36] Krug, J., and Spohn, H., *Kinetic roughening of growing surfaces*, 477–582 in: *Solids far from equilibrium*, Godrèche, G. C., ed., Cambridge University Press, England (1992).

[37] Lax, P. D., *Hyperbolic Systems of Conservation Laws and the Mathematical Theory of Shock Waves*, SIAM, Philadelphia (1973).

[38] Majda, A. J., and Kramer, P. R., *Simplified models for turbulent diffusion: Theory, numerical modelling and physical phenomena*, to appear in Phys. Reports.

[39] Mather, J. N., *Existence of quasi-periodic orbits for twist homeomorphisms of the annulus*, Topology, **21** (1982), 457-467.

[40] Mattingly, J., *The stochastic Navier-Stokes equation: Energy estimates and phase space contraction*, thesis, Princeton University, 1998.

[41] Monin, A. S., and Yaglom, A. M., *Statistical Fluid Mechanics. Vol. II*, MIT Press, Cambridge, Massachusetts, and London, England (1975).

[42] Onsager, L., *Statistical hydrodynamics*, Nuovo Cimento (Supplemento), **6**, 279(1949).

[43] Polyakov, A. M., *Turbulence without pressure*, Phys. Rev. E **52**, 6183–6188 (1995).

[44] Sinai, Ya., *Burgers system driven by a periodic stochastic flow*, pp. 347–355 in: *Itô's Stochastic Calculus and Probability Theory*, Springer-Verlag, New York, Berlin (1996).

[45] Smoller, J., *Shock Waves and Reaction Diffusion Equations*, Springer-Verlag, New York, Berlin (1983).

[46] Sobolevskii, A. N., *Aubry-Mather theory and idempotent eigenfunctions of Bellman operator*, to appear in Comm. Contemp. Math..

[47] Tadmor, E., *L^1-stability and error estimates for approximate Hamilton-Jacobi solutions*, UCLA CAM Report 98-21.

[48] Vergassola, M., *Anomalous scaling for passively advected magnetic fields*, Phys. Rev. E, **53**, No. 4, R3021-R3024 (1996).

[49] Vol'pert, A., *The space BV and quasilinear equations*, Math. USSR, Sb. **2** 225–267 (1967).

[50] Yakhot, V., *Two-dimensional turbulence in the inviscid cascade range*, preprint.

[51] Zeldovich, Ya. B., Ruzmaikin, A. A., and Sokoloff, D. D., *Magnetic Fields in Astrophysics*, Gordon and Breach, New York, 1983.

E-mail address: `weinan@cims.nyu.edu`

Courant Institute of Mathematical Sciences, New York University, New York, New York 10012

AMS/IP Studies in Advanced Mathematics
Volume 20, 2001

Recent developments of realization of surfaces in R^3

Hong Jiaxing

Given a smooth n-dimensional Riemannian manifold (M^n,g), can we find a map

$$\phi : M^n \longrightarrow R^p \text{ such that } \phi^* h = g$$

where h is the standard metric in R^p ? Such a map provides a realization of the given Riemannian manifolds in some Euclidean space. This is a long standing problem in Differential Geometry. We also call the map ϕ isometric embedding or isometric immersion if ϕ is embedding or immersion. There is a very nice paper [GR] which contains a detailed survey for known results before 1970 and some developments can be found in [G]. Whereas, what development has been made during the passed decades ? In higher dimensional cases, two directions should be mentioned. The first one is to improve the Nash's theorem so that (M^n, g) has isometric embedding in a Euclidean space of much lower dimension than that given by Nash and meanwhile, to use the contraction mapping principle in place of the theorem of the complicated hard implicit functions, see [H] or [GUN]. Another is concerning local isometric embedding, see [BGY] and this is a remarkable paper although it is deal with local problem. For this direction, also refer to [JA]. In contrast to higher dimensional cases it seems that problems of two dimensional Riemannian manifolds embedded into R^3 attracted much more attention of mathematicians, particularly in the last decade. Besides R^3 is a real space of our life, to a large extent, the lecture note [Y3] by Yau and his several related Problem Sections [$Y1$] [$Y2$] played a crucial role in pushing research forward in this field . The present paper is devoted to a survey on the developments of isometric embedding (or immersion) of surfaces in R^3 in the last decade. For higher dimensional case, refer to the previous papers. Of course the results mentioned in this survey are by no means exhaustive and depend a lot on the author's taste.

Smoothly and isometrically immersing a surface (M, g) into R^3 is equivalent to finding three smooth ($C^s, s \geq 1$) functions $x_\alpha : M \mapsto R^1$, $\alpha = 1, 2, 3$ such that

$$g = dx_1^2 + dx_2^2 + dx_3^2 \tag{1}$$

Although as far as the formulation of (1) be concerned, C^1 regularity is enough and [K] gives a very nice result in this category. In order to see the role of the curvature of surfaces in the problem considered here we prefer to assume $s \geq 2$ throughout the present paper. In the sequel, sometimes we use x, y, z to denote x_1, x_2, x_3. In a local coordinates near a point $p \in M$, the metric g is of the form $g = g_{ij} du^i du^j$.

Then (1) can be written as follows

$$\frac{\partial x_\alpha}{\partial u^i}\frac{\partial x_\alpha}{\partial u^j} = g_{ij}\ i,j = 1,2 \tag{2}$$

(2) is a system composed of three differential equations of first order and hence, this is a determine system. We say that $\vec{r} = (x_1, x_2, x_3)$ is a local smooth (C^s) isometric embedding in R^3 of the given surface if $\vec{r} = (x_1, x_2, x_3)$ is a smooth (C^s) solution to (2) in a neighbourhood of the point p and that $\vec{r}$ is a global smooth (C^s) isometric embedding (immersion) in R^3 if $\vec{r} = (x_1, x_2, x_3)$ is a smooth (C^s) solution to (1) on M and, meanwhile, is an embedding (immersion) into R^3. This survey consists of two parts. The first part includes some recent developments in local isometric embedding, global isometric embedding and boundary value problems for isometric embedding. The second part contains a new result about some counter examples for the existence and rigidity of convex surfaces with some boundary condition

Local Isometric Embedding. The Cartan-Janet theorem tells us that any analytic metric always admits a local analytic isometric embedding in R^{s_n} with $s_n = n(n+1)/2$. This implies that (2) is always locally solvable for any analytic metric g. On the other hand, [P] proves that any smooth surface always has a local smooth isometric embedding in R^4. What obstruction shall we run across while attacking local isometric embedding in R^3? Later we shall see that the difficulty for the local solvability comes from the vanishing of corresponding Gaussian curvature K. In [Y2, No.22] and also refer to [Y1, No. 54] Yau posed to prove that any smooth surface always has a local smooth isometric embedding in R^3. In this direction, it is Lin who first made important breakthrough and his results in [LC1] and [LC2] state

THEOREM 1 (C. Lin). *Any C^s, $s > 10$ nonnegatively curved metric always admits a local C^{s-6} isometric embedding in R^3.*

(2) If g is a C^s, $s > 6$ metric and if its curvature K satisfies

$$K(p) = 0 \text{ and } dK(p) \neq 0$$

then it admits a C^{s-3} isometric embedding in R^3 near p.

By means of Lin's technique the problem for local isometric embedding related to nonpositive curvature metric are also solvable in the following cases.

(3) In [IW], K nonpositive in a neighbourhood of a point p and $d^2K(p) \neq 0$

(4) In [HO1], $K = h^{2q}K_1$ where K_1, h are smooth functions, $K_1(p) < 0$, $dh(p) \neq 0$ and q is an integer.

In what follows let us simply explain what the technique we shall use while attacking the problem of local isometric embedding. Suppose that $\vec{r} = (x, y, z)$ is a smooth solution to (2) in a neighbourhood of the point in question previously. By the Gauss equations we have, in a local coordinate system,

$$\vec{r}_{ij} = \Gamma^k_{ij}\vec{r}_k + \Omega_{ij}\vec{n} \text{ or } \nabla_{ij}\vec{r} = \Omega_{ij}\vec{n}, i, j = 1,2 \tag{3}$$

where subscripts i,j and ∇_{ij} denote Euclidean and covariant derivatives respectively and Ω_{ij} the coefficients of the second fundamental form, Γ^k_{ij} the Christoffel symbols with respect to the metric and $\vec{n}$ the unit normal to $\vec{r}$. For each unit constant vector, for instance, the unit vector $\vec{k}$ of the z axis, taking the scale product of $\vec{k}$ and (3) and using the Gauss equations one can get

$$\det(\nabla_{ij}z) = K\det(g_{ij})(\vec{n}, \vec{k})^2 \tag{4}$$

Notice that

$$(\vec{n},\vec{k})^2 = 1 - \left(\frac{(\vec{r}_1 \times \vec{r}_2) \times \vec{k}}{|\vec{r}_1 \times \vec{r}_2|}\right)^2$$
$$= 1 - g^{ij} z_i z_j = 1 - |\nabla z|^2.$$

Inserting the last expression into (4) we deduce the Darboux equation

$$F(z) = \det(\nabla_{ij} z) - K \det(g_{ij})(1 - |\nabla z|^2) = 0 \tag{5}$$

Obviously each component of $\vec{r}$ does satisfy the Darboux equation. Conversely, we have

LEMMA 1. *For any smooth function ϕ with $|\nabla\phi| < 1$ in the region considered, then the curvature of the metric $g_1 = g - d\phi^2$*

$$K_{g_1} = \frac{1}{(1 - |\nabla\phi|^2)} \left\{ K_g - \frac{\det(\nabla_{ij}\phi)}{(1 - |\nabla\phi|^2)\det(g_{ij})} \right\} \tag{6}$$

Proof: (6) is coordinate free and hence, it is suffices to prove it in any local coordinate system at the point considered. Let us first assume $d\phi \neq 0$ there and take an orthonormal coordinates $u = \phi$ and v in a neighbourhood of this point. And the metric g is of the form

$$g = Edu^2 + Gdv^2 \text{ and } g_1 = (E-1)du^2 + Gdv^2$$

The assumption: $|\nabla\phi| < 1$ implies $|\nabla\phi|^2 = E^{-1} < 1$ and hence, g_1 is a smooth metric too. A direct computation yields

$$\frac{\det(\nabla_{ij}\phi)}{\det(g_{ij})} = \frac{\Gamma^1_{11}\Gamma^1_{22} - (\Gamma^1_{12})^2}{EG} = -\frac{1}{4EG}\left[\frac{E_v^2 + E_u G_u}{E^2}\right] \tag{7}$$

By the Gauss formula we have

$$4EGK_g = \frac{E_v^2 + E_u G_u}{E} + \frac{G_u^2 + E_v G_v}{G} - 2(E_{vv} + G_{uu})$$

$$4(E-1)GK_{g_1} = \frac{E_v^2 + E_u G_u}{E-1} + \frac{G_u^2 + E_v G_v}{G} - 2(E_{vv} + G_{uu})$$

Subtracting the second equation from the first one and combining with (7) one can soon obtain (6). If $d\phi = 0$ at the point considered, by the continuity of all functions involved it is not difficult to prove (6) still valid there. □The proof presented here is due to [SU] and might be one of most simplify proofs. Lemma 1 guarantees that for each smooth solution z to (5) satisfying $|\nabla z| < 1$, $\tilde{g} = g - dz^2$ is a smooth flat metric. Therefore in simply connected domain Ω we can always find a smooth mapping (x, y): $\Omega \mapsto R^2$ such that $dx^2 + dy^2 = g - dz^2$. So the realization of a given metric into R^3 is equivalent to finding a smooth (or C^s) solution to the Darboux equation with a subsidiary condition $|\nabla z| < 1$. If K is positive or negative at the point considered, then (5) is elliptic or hyperbolic Monge-Ampere equation and the local solvability is well known for both of them. But if K vanishes at this point, the situation is very complicated and so far there has been no standard way to deal with such kind of Monge-Ampere equation. Indeed, its linearized operator

$$L_z\xi = \lim_{t \longrightarrow 0} \frac{F(z + t\xi) - F(z)}{t} = F^{ij}\nabla_{ij}\xi + 2K(\nabla z, \nabla\xi) \tag{8}$$

where $F^{ij} = \partial \det(\nabla_{lk} z)/\partial \nabla_{ij} z$. It is easy to see that the type of this linear differential operator completely depends on K. When K vanishes at the point considered, (8) may be degenerate elliptic, hyperbolic or mixed type and its local solvability is not clear. Using a regularized operator instead of (8) and the Nash-Moser procedure Lin succeeded in proving Theorem 1.

The solutions obtained by Lin is only sufficiently smooth. In [CL] and [HZ1] the existence of smooth solutions to (5) for the Lin's case are discussed and particularly, [CL] proves that the solution in the case (2) of Theorem 1 is smooth if s is big enough.

Despite Lin made great progress on local isometric embedding, it is still not clear whether there is obstruction for the local isometric embedding in smooth category even for the nonpositive curvature metric. Some results [EG] on linear degenerate hyperbolic operators of second order which have no local solvability should be noticed.

Global isometric embedding. The first result on global isometric embedding of complete surfaces in R^3 is due to Weyl and Lewy. In 1915 Weyl indicated the following result

(Weyl-Lewy): *Any analytic positive curvature metric defined on* S^2 *always admits an analytic isometric embedding in* R^3.

It was after twenty years Lewy who completed the proof of the above theorem. The generalization of Weyl and Lewy's result to smooth case was independently obtained by Pogorelov and Nirenberg in early 1950s. Their results state

(Pogorelov-Nirenberg): *Any smooth positive curvature metric defined on* S^2 *always admits a smooth isometric embedding.*

For noncompact case, for example, a complete smooth positive curvature metric defined on R^2, this problem was solved by two Russian mathematicians. Olovjanisnikov first found the weak isometric embedding based on the Aleksandrov's theory on convex surfaces and Pogorelov proved the weak solution smooth if the metric smooth.

(Olovjanisnikov-Pogorelov): *Any smooth complete positive curvature metric defined on* R^2 *admits a smooth isometric embedding in* R^3.

The next natural development is to consider the realization in R^3 of nonnegatively curved surfaces. A special case is discussed in [IA]. Recently [GL] and [HZ2] independently obtained the following result

THEOREM 2 (Guan-Li, Hong-Zuily). *Any* C^4 *nonnegative curvature metric defined on* S^2 *always admits a* $C^{1.1}$ *isometric embedding in* R^3.

Olovjanisnikov-Pogorelov's result on complete positively curved plane is also extended to the nonngegatively curved case, see [AM], [HO3]. The result in [HO3] states

THEOREM 3. *Any complete* C^4 *nonnegative curvature metric defined on* R^2 *always admits a* $C^{1.1}$ *isometric embedding in* R^3. *Moreover it is smooth where the metric is smooth and the curvature positive.*

[IA] gives a $C^{2.1}$ convex surface which is not C^3 continuous but realizes analytic metric on S^2 with positive curvature except at one point. On the other hand, Pogorelov gave a $C^{2,1}$ geodesic disk with nonnegative curvature not even admitting a C^2 local isometric embedding in R^3 at the center of this disk. Therefore a natural open question is :

Does there exist a $C^{2.\alpha}$ $(0 < \alpha < 1)$ isometric embedding in R^3 for any sufficiently smooth (even analytic) nonnegatively curved sphere or plane ?

Theorem 2 and Theorem 3 are proved by an approximation of a family of smooth positively curved spheres and positively curved disks to the given nonnegatively curved sphere and plane respectively. The crucial point is the estimates for the mean curvature of convex surfaces which does not depend on the lower bound of their curvature.

The Hilbert theorem [HI] is one of the most important theorems in 3- Euclidean space. This theorem with Efimov's generalization in [EF1] provides a negative answer for the problem of global isometric embedding in R^3 about complete negatively curved surfaces.

(Hilbert-Efimov): *Any complete surface with negative constant curvature (with curvature bounded above by a negative constant) has no C^2 isometric immersion in R^3.*

Evidently, Efimov's result yields a necessary condition for a complete negatively curved surface to embed isometrically in $R^{3,}$,that is, $\inf |K| = 0$ where the infimum is taken over the whole surface. Yau posed the following question [Y1, No.57].

Find a nontrivial sufficient condition for a complete negatively curved surface to embed isometrically in R^3.

He also pointed out that such a nontrivial condition might be the rate of decay of the curvature at infinity. Recently some development in this direction has been made in [HO2]. Let M be a simply connected noncompact complete surface of negative curvature K. By the Hadamard theorem $exp_{:}T_p(M) \longrightarrow M$ is a global diffeomorphism for each point $p \in M$ which induces a global geodesic polar coordinates (ρ, θ) on M centered at p.

THEOREM 4. *Suppose that*

(*a*) *for some $\delta > 0$, $\rho^{2+\delta}|K|$is decreasing in ρ outside a compact set*

and that

(*b*) $\partial_\theta^i \ln |K|$, $i = 1.2$ *and* $\rho \partial_\rho \partial_\theta \ln |K|$ *bounded on* M.

Then M admits a smooth isometric immersion in R^3.i

Remark 1. If $M \in C^{s.1} (s \geq 4)$ and other assumptions in Theorem 4 are fulfilled, then it admits a $C^{s-1.1}$ isometric immersion in R^3.

Remark 2. The assumption (a) implies the rate of decay of the curvature at the infinity

$$|K| \leq \frac{A}{\rho^{2+\delta}}, \text{ for a positive constant } A \tag{9}$$

Can we improve the result on existence in Theorem 4 ? Another result [EF3] by Efimov should be mentioned.

THEOREM 5 (Efimov). *Let M be a smooth complete negatively curved surface with curvature K. Suppose that*

$$\sup_M |K|, \ \sup_M grad(\frac{1}{\sqrt{|K|}}) \leq C \tag{10}$$

for some constant C. Then M has no C^2 isometric immersion in R^3

This theorem tells us that the following negatively curved surface (R^2, g) whose the Gaussian curvature equals

$$-\frac{A}{1+\rho^2} \tag{11}$$

(where ρ is the distance function from some point), always has no C^2 isometric immersion in R^3 for arbitrary positive constant A. The arguments of [EF1] and [EF3] are very genuine and ones expect a more analysis proof for Efimov's results in [EF1]. Anyway, a result [EF2] also by Efimov which is much easier understood, shows that the negatively curved surface mentioned above has no C^2 isometric immersion if in (11) $A > 3$. Comparing (9) with (10) and (11) we can see that the result of Theorem 4 is, nearly, sharp.

Next let us give a brief explain of the idea to attack the realization of complete simply connected negatively curved surfaces in R^3. Let $\vec{r}$ be the position vector of an isometric immersion of the given negatively curved surface (M^2, g). Then in any local coordinate system, the coefficients of second fundamental form, $L = (\vec{r}_{11}, \vec{n})$, $M = (\vec{r}_{12}, \vec{n})$ and $N = (\vec{r}_{22}, \vec{n})$ where $\vec{n}$ denotes the unit normal vector $\vec{r}$, satisfy the Gauss-Codazzi system

$$L_2 - M_1 = L\Gamma^1_{12} + M(\Gamma^2_{12} - \Gamma^1_{11}) - N\Gamma^2_{11} \tag{12}$$

and

$$M_2 - N_1 = L\Gamma^1_{22} + M(\Gamma^2_{22} - \Gamma^1_{21}) - N\Gamma^2_{21} \tag{13}$$

with

$$LN - M^2 = K \det(g_{ij}) \tag{14}$$

If L does not vanish in the region considered. we can solve N from the last equation and then, substitute it to the first two equations. Finally we can get a 2×2 hyperbolic system. Through a computation we can find its Riemann invariants

$$r = \frac{\left[-M - \sqrt{|K| \det(g_{ij})}\right]}{L} \text{ and } s = \frac{\left[-M + \sqrt{|K| \det(g_{ij})}\right]}{L} \tag{15}$$

which are nothing else but the angle coefficients of the asymptotic curve of $\vec{r}$ and some coordinate curve. And the tangent vectors along the asymptotic curves are denoted by $\partial_\beta = \partial_2 + s\partial_1$ and $\partial_\alpha = \partial_2 + r\partial_1$. Moreover, a standard procedure reduces the above 2×2 hyperbolic system to the following diagonal form, with $Q = \ln \sqrt{|K| \det(g_{ij})}$,

$$\partial_\beta r = \frac{1}{2}(r-s)\partial_\alpha Q + A + Br + Cs + Drs + Er^2 + Fr^2 s \tag{16}$$

and

$$\partial_\alpha s = \frac{1}{2}(s-r)\partial_\beta Q + A + Bs + Cr + Drs + Es^2 + Fs^2 r \tag{17}$$

Where A, B, C, D, E and F are completely determined by the Christoffel symbols of the given metric. Conversely, if the system (16) (17) admits a smooth solution (r,s) with $s > r$ in the region considered, then L, M and N which can be uniquely determined by (15) and (14), certainly satisfy the Gauss-Codazzi equations. This is just the start point to attack the realization of complete simply connected negatively curved surfaces in R^3. For the geodesic polar coordinates, $\rho = 0$ is a artificial singularity and it is not convenient for analysis arguments. Instead of the geodesic polar coordinates we use the global semi geodesic coordinate systems on M under which we may assume that $M = R^2$ and $g = dt^2 + B^2 dx^2$, $(x,t) \in R^2$ for some positive $B \in C^\infty(R^2)$. Introduce new unknown functions $r \mapsto Br$ and $s \mapsto Bs$ and still denote them by r and s. and also set $\partial_\alpha = \partial_t + (r/B)\partial_x$, $\partial_\beta = \partial_t + (s/B)\partial_x$, Then (16) (17) are rewritten in the form

$$\partial_\beta r = \frac{r-s}{4}\partial_\alpha(\ln|K|) - s(1+r^2)\partial_t(\ln B) \tag{18}$$

and

$$\partial_\alpha s = \frac{s-r}{4}\partial_\beta(\ln|K|) - r(1+s^2)\partial_t(\ln B) \tag{19}$$

Now the above realization problem in R^3 for the negatively curved surface is reduced to finding a global smooth solution to (18) (19) with the constrain $s > r$ everywhere in $(x,t) \in R^2$. The system (18) (19) is first obtained by [RO]. Surprising, if K is a negative constant, (18) (19) looks much simpler but the Hilbert theorem tells us that it does not admit any global C^1 solution satisfying the subsidiary condition $s > r$ in R^2. [HO2] applies a comparison argument to the above system and completes the estimates of C^0, C^1 norms of (r,s) and the positive lower bound of $s - r$ for the Cauchy problem of (18) and (19) and obtains the expected global smooth solution under the assumptions of Theorem 4.

Boundary value problems for isometric embedding in R^3.

Recently, boundary value problems for isometric embedding of surface in R^3 have attracted much attentions of mathematicians. Various formulations of such questions can be found in [Y2, p286,No.26] and [PO1, p656,No.6]. To author's knowledge this field has not been extensively studied. Let g be a smooth positive curvature metric defined on the closed unit disk $\bar{D}$. Throughout the present paper we always call such surfaces $(\bar{D}, g)$ positive disk. The question ones are now concerning is as follows.

Given a smooth positive disk $(\bar{D}, g)$ and a complete smooth surface $\Sigma \subset R^3$, can we find an isometric embedding $\vec{r}$ of $(\bar{D}, g)$ in R^3 such that $\vec{r}(\partial D) \subset \Sigma$?

As a first step, one can consider the case that

Assume that Σ is a plane. Give a complete description of all isometric embeddings of $(\bar{D}, g)$ satisfying $\vec{r}(\partial D) \subset \Sigma$.

Assume that $\Sigma : \{z = 0\}$. Then we are faced with the following boundary value problem.

(BVP): To find an isometric embedding $\vec{r} = (x, y, z)$ of the given positive disk $(\bar{D}, g)$ such that $z(\partial D) = 0$

It is easy to see that there is some obstruction for the existence of solutions to the above (BVP). Suppose that $\vec{r} = (x, y, z) \in C^2(\bar{D})$ is a solution to the above boundary value problem. Obviously the intersection $\vec{r}(\partial D)$ of $\vec{r}$ and the plane $\{z = 0\}$ is a C^2 planar convex curve. Denoting its curvature by $\tilde{k}$ we have

$$\tilde{k}^2 = k_g^2 + k_n^2 \tag{20}$$

where k_g and k_n are respectively the geodesic curvature and normal curvature of $\vec{r}(\partial D)$. k_n is positive everywhere since $(\bar{D}, g)$ has positive curvature. Notice that the total curvature of $\vec{r}(\partial D)$, as a planar curve, equals 2π. Hence

$$\int_{\vec{r}(\partial D)} |k_g| ds < 2\pi \tag{21}$$

This is a necessary condition in order that the above boundary value problem (BVP) can be solvable. Indeed, this necessary condition is not sufficient for the solvability. There is a counter example which shows that too many changes of the sign of the geodesic curvature of the boundary makes the (BVP) unsolvable.

A counter example for nonexistence. On S^2, take the sphere coordinates $\{\theta \in [0, \pi],\ \phi \in [-\pi, \pi]\}$ and $\theta = \pi$ stands for the North pole, Suppose that D' is the domain on S^2 containing the below hemisphere and enclosed by the equator γ_0, $\gamma_+ : \phi = \beta$ and $\gamma_- : \phi = -\beta$ for some positive constant β to be fixed. Denote the corner points of γ_0 and γ_-, γ_- and γ_+, γ_+ and γ_0 respectively by A, B, C. Use the small smooth arcs A_1A_2, B_1B_2 and C_1C_2 respectively smoothly matching both sides of the corner points A, B and C, Suppose that D stands for the domain on S^2, enclosed by $\Gamma = A_1A_2B_1B_2C_1C_2A_1$. Then $\bar{D}$ equipped with the induced metric ds_0^2 on S^2, is a smooth positive disk. Obviously, the geodesic curvature of A_1A_2 (C_1C_2) is negative and that of B_1B_2 positive with respect to $\bar{D}$. By means of the Gauss-Bonnet formula it is easy to see that

$$\int_{A_1A_2} |k_g| ds = \frac{\pi}{2} + |\Omega_1|, \quad \int_{C_1C_2} |k_g| ds = \frac{\pi}{2} + |\Omega_3| \text{ and } \int_{A_1A_2} |k_g| ds = \pi - \beta + |\Omega_2|$$

where $|\Omega_i|$ $i = 1, 2, 3$ are respectively the areas of the corresponding small curved triangles corresponding to the corner points A, B and C. Hence for any given $\beta > 0$ we can take so short arclength $|A_1A_2|$, $|B_1B_2|$ and $|C_1C_2|$ that $|\Omega_1| + |\Omega_2| + |\Omega_3| < \beta$ and the necessary condition (21) is satisfied for $(\bar{D}, ds_0^2)$ but we have

LEMMA 2. *For $(\bar{D}, ds_0^2)$, (BVP) has no C^2 solution if β is small enough.*

Proof: Suppose that $\vec{r}$ is the required isometric embedding and denote its boundary by $\bar{\Gamma} = \bar{A}_1\bar{A}_2\bar{B}_1\bar{B}_2\bar{C}_1\bar{C}_2\bar{A}_1$. Let $\phi(p)$ stand for the oriented angle between the tangent at $p \in \Gamma$ and the some fixed direction on the plane $z = 0$. Evidently

$$\phi(\bar{A}_2) - \phi(\bar{A}_1) > \int_{A_1 A_2} |k_g| ds = \frac{\pi}{2} + |\Omega_1|$$

also $\phi(\bar{B}_2) - \phi(\bar{B}_1) > \pi - \beta + |\Omega_2|$ and $\phi(\bar{C}_2) - \phi(\bar{C}_1) > (\pi/2) + |\Omega_3|$. Since the total curvature of $\bar{\Gamma}$ equals 2π, it follows that $0 < \phi(\bar{A}_1) - \phi(\bar{C}_2) < \beta$. Letting $\beta \longrightarrow 0$ we have that the arc $\bar{C}_2\bar{A}_1$ approaches to a straight segment whose length equals the limit of the intrinsic arclength of $C_2A_1 = 2\pi$ since the isometric embedding of any curve preserves its arclength. On the other hand the distance between $\bar{C}_2$ and $\bar{A}_1$ is less than the intrinsic distance between C_2 and A_1 on S^2 which approaches to zero as $\beta \longrightarrow 0$ since $\vec{r}$ is a convex surface. This yields a contradiction. □

Lemma 2 tells us that too many changes of the sign of the geodesic curvature of the boundary will make (BVP) unsolvable. So in the sequel we always assume the geodesic curvature

(22) Case a: $k_g > 0$ on ∂D and Case b: $k_g < 0$ on ∂D

It should be mentioned that Pogorelov gave the first solution of (BVP) in [PO1] which states that

Pogorelov: *Let $(\bar{D}, g)$ be a smooth positive disk. Then the boundary value problem admits a solution $\vec{r} \in C^\infty(D) \cap C^{0.1}(\bar{D})$ provided that the geodesic curvature of ∂D with respect to the metric g is nonnegative.*

Pogorelov only obtained a local smooth solution, namely, a solution smooth inside.

One wonder that under what conditions (BVP) always admits a global smooth solution, namely, a solution smooth up to the boundary. Such global smooth solutions are obtained by [DE] for positive curvature case and by [HO4] for some degenerate case if there exists a global C^2 subsolution ψ for the Darboux equation (5) in the unit disk D with ψ =0 on ∂D and $|\nabla\psi|$ strictly less than 1 on $\bar{D}$. It seems that such an assumption on the existence of subsolution is very difficult to check. Recently, [HO5] removes this technique requirement.

THEOREM 6. *For Case a (BVP) always admits a unique solution in $C^\infty(\bar{D})$ if any one of the following assumptions is satisfied (1) $K > 0$ on $\bar{D}$, (2) $K > 0$ in D and $K = 0 \neq |dK|$ on ∂D.*

If the second hypothesis (2) in this theorem is fulfilled and if we smoothly extend the metric g to outside $\bar{D}$, the Gauss curvature will cleanly change its sign. Combining Theorem 5 with the result in [HO1] on existence of smooth solution for degenerate hyperbolic equation of Monge-Ampere type, one can obtain a smooth semiglobal isometric embedding in R^3 for a Riemannian manifold with curvature cleanly changing its sign on a curve. This is a generalization of [LC2]. Under the present case, the region where the curvature is positive, is global and the region where the Gauss curvature is negative, is local. This is the reason why we call "semiglobal".

Obviously, the necessary condition (21) is always satisfied for Case a. Therefore one naturally asks what will happen for Case b if (21) is satisfied. It looks rather complicated. Usually speaking, the construction of isometric embedding of surfaces in R^3 is of two steps. One is concerning some uniform a priori estimates for some extrinsic geometric quantities. Another is the realization of a perturbation of the metric of a given surface in R^3 which is, to a large extent, based on the rigidity of the given surface, namely, the linearized problem of (BVP) has no nontrivial solution. Given a smooth surface $\vec{r}$ in R^3 one can consider its deformation $\vec{r}_t$: $(-\epsilon, \epsilon) \ni t \mapsto R^3$ with $\vec{r}_0 = \vec{r}$. If $t = 0$ is a critical point of the metric $g(t) = d\vec{r}_t^2$, we say that the derivative with respect to t of $\vec{r}_t$ at $t = 0$ gives rise a first order infinitesimal isometric deformation of $\vec{r}$. Denoting this infinitesimal isometric deformation by $\vec{\tau} = (d\vec{r}_t/dt)(0)$ we have

$$\frac{d}{dt}(d\vec{r}_t^2)|_{t=0} = (d\vec{r}, d\vec{\tau}) = 0 \tag{23}$$

Obviously, for arbitrary two constant vectors $\vec{A}$ and $\vec{B}$ in R^3, $\vec{\tau} = \vec{A} \times \vec{r} + \vec{B}$ is a solution of (23) and such solutions are called trivial ones. We say that $\vec{r}$ is of infinitesimal rigidity for first order isometric deformation if (23) has no nontrivial solution. As is well known, for closed surfaces now we only know that closed C^2 convex surfaces are infinitesimally rigid. If $\vec{r}$ has boundary, as usual, we should impose some constrains on the boundary for deformations. For the present boundary value problem we impose that

$$(\vec{\tau}, \vec{k}) = 0 \text{ on } \partial D \tag{24}$$

where $\vec{k}$ is the unit vector of the z axis.

A positive disk of not infinitesimal rigidity. In what follows we shall give a convex surface $\vec{r}$ which is a smooth isometric embedding of a positive disk in Case b but not infinitesimal rigidity. This shows that there exist some difficulties in discussing the existence of (BVP) in Case b. So far the solvability of (BVP) for Case b is still open.

Let us consider a spherical crown $\Sigma_\lambda = \{x^2 + y^2 + z^2 = 1 |\ z \leq \lambda\}$. In the spherical coordinates

$$\Sigma_\lambda = \{(\sin\theta\cos\phi, \sin\theta\sin\phi, -\cos\theta)|0 \leq \phi \leq 2\pi, 0 \leq \theta \leq \theta_* = cos^{-1}(-\lambda)\}$$

where $\theta = 0$ stands for the South pole. Σ_λ is the isometric embedding of the metric $g = d\theta^2 + \sin^2\theta d\phi^2, 0 \leq \theta \leq \theta_*$.If $\lambda > 0$, then Σ_λ contains the below hemisphere.

THEOREM 7. *There is a countable set $\Lambda = \{\theta_1, \theta_2, ...\theta_n, ...\} \subset (\pi/2, \pi)$ with a limit point $\pi/2$ such that Σ_λ is not infinitesimal rigidity if $\theta_* \in \Lambda$.*

Let us study some behavior of solutions to the following ordinary equation, for some positive constant σ,

$$Lu = sht(u_{tt} - \sigma^2 u) - 2u_t = 0 \text{ in } R^1 \tag{25}$$

LEMMA 3. *There is an odd analytic solution U subject to*

$$U^{(i)}(0) = 0, i = 0, 1, 2,\ U^{(3)} = 1 \text{ and}$$

$$|U(t)| \geq \frac{1}{C}e^{\sigma|t|} \text{if } |t| \geq T \text{ for some conatants } C \text{ and } T$$

Proof: Differentiation of (25) two times yields

$$(Lu)_t = sht(u^{(3)} - \sigma^2 u^{(1)}) + (cht - 2)u^{(2)} - \sigma^2 chtu = 0,$$

$$(Lu)_{tt} = sht(u^{(4)} - \sigma^2 u^{(2)}) +$$

$$2(cht - 1)u^{(3)} + shtu^{(2)} - 2\sigma^2 chtu^{(1)} - \sigma^2 shtu = 0, \tag{26}$$

after solving u_t from (25) and inserting it into (26), dividing both sides by sht we can get

$$u^{(4)} + q_1 u^{(3)} + q_2 u^{(2)} + q_4 u = 0 \text{ in } R^1 \tag{27}$$

where

$$q_1 = 2tgh\frac{t}{2},\ q_2 = 1 - \sigma^2(1 + cht) \text{ and } q_4 = \sigma^2(\sigma^2 cht - 1)$$

Obviously, q_i $i = 1, 2, 4$ are all analytic functions in R^1. Conversely, it is easy to see if u is a solution to (27) with $(Lu) = (Lu)_t = 0$ at $t = 0$, then u is a solution to (25). Therefore we consider a solution u to (27) with the Cauchy data: $u^{(3)} = 1$ and $u^{(i)} = 0$ at $t = 0$ for $i = 0, 1, 2$, which guarantees $(Lu) = (Lu)_t = 0$ at $t = 0$. By the uniqueness and existence of solutions to ordinary equation we know that this solution is odd analytic in R^1.

Now we are in a position to prove the second part of the present lemma. It suffices to consider the case $t > 0$. Since near $t = 0$ this solution has the Taylor expansion $u = (t^3/6)$ + higher order term, it implies u, u_t and u_{tt} positive if t is positive and very small. Now we claim

$$u(t),\ u_t(t) \text{ and } u_{tt}(t) \text{ are positive for all } t > 0 \tag{28}$$

Set $t_0 = \sup\{t > 0 | u_{tt} > 0 \text{ on } (0, t]\}$. It is easy to see $t_0 > 0$ and $u(t)$, $u_t(t)$ positive for all $t \in (0, t_0]$. From (25) it follows that $\sigma^2 sht_0 u(t_0) + 2u_t(t_0) = 0$ which gives a contradiction. Thus we have proved $u_{tt} > 0$ for all $t > 0$ and hence, u, $u_t > 0$ for all $t > 0$ too. Combining this conclusion with (25) yields

$$u_{tt} - \sigma^2 u > 0 \text{ for all } t > 0$$

In what follows we shall use the comparison technique to discuss the behavior of u at the infinity. Take $w = \epsilon e^{\sigma t}$ as an auxiliary function for some constant ϵ to be fixed. ϵ is chosen so small that $w(1) \leq u(1)$ and $w_t(1) \leq u_t(1)$. Since

$$(u - w)_{tt} - \sigma^2(u - w) > 0,\ t > 1 \text{ with } (u - w)(1) \geq 0 \text{ and } (u - w)_t(1) \geq 0$$

it infers $u \geq w$ for all $t > 1$. This completes the proof for the present lemma. □

LEMMA 4. *There is a constant σ^* such that for each $\sigma \in [\sigma^*, +\infty)$, (25) admits an even analytic solution $V(t)$ satisfying*

$$V(0) = -1, V^{(1)}(0) = 0,\ V^{(2)}(0) = \sigma^2 \text{ and } V^{(3)}(0) = 0 \tag{29}$$

Moreover, $V(t)$ is monotone-increasing to $+\infty$ as t is sufficiently large, and $V(t)$ has only two zero points $\pm\tilde{t}_\sigma$, $\tilde{t}_\sigma \longrightarrow 0$ as $\sigma \longrightarrow +\infty$.

Proof: Let us first look at a simpler equation

$$t(u_{tt} - u) - 2u_t = 0 \text{ in } R^1 \tag{30}$$

We can directly find two solutions $u = tcht - sht$ and $v = tsht - cht$. And v assumes the initial data (29) for $\sigma = 1$ and is even analytic function, monotone-increasing to $+\infty$. Denote by t_* the unique zero point of v in the positive t axis and

$$v < 0 \text{ as } t \in [0, t_*) \text{ and } v > 0 \text{ as } t > t_* \tag{31}$$

Now based on this solution we proceed to study the behavior of the solution to (25). Set $sht = t(1 + t^2 G(t))$ for some analytic function $G(t)$. A change of variable $s = \sigma t$ gives

$$s(u_{ss} - u) - 2u_s = -\frac{s^3}{\sigma^2} G(\frac{s}{\sigma})(u_{ss} - u)$$

In a similar procedure used in deriving (27) we can get

$$u^{(4)} - 2u^{(2)} + u = \sigma^{-2} f(u^{(4)}, u^{(3)}, u^{(2)}, u^{(1)}, u, s, \frac{1}{\sigma}) \tag{32}$$

where f is linearly with respect to $u^{(i)}$ $i = 0, 1, 2, 3, 4$ and analytic in s and $1/\sigma$. Now let us consider the Cauchy problem for (32) in the interval $[-1 - t_*, t_* + 1]$ with the initial data (29) for $\sigma = 1$.By means of the contraction mapping principle without difficult one can get a solution v_1 to (32) in $[-1-t_*, t_*+1]$ with the above initial data and v_1 is even analytic subject to

$$|v_1 - (sshs - chs)| + |v_1' - schs| \leq \frac{C}{\sigma} \text{ for all } s \in [-1 - t_*, t_* + 1]$$

for some constant C independent of σ. Hence $v_1^{(i)}(t_* + 1) > 0$ for $i = 0, 1$ as $\sigma \geq \sigma_*$ for some sufficiently large positive σ_*. Now we are in a position to prove the second part of the present lemma. By the special choice of the initial data we know $V(t) = v_1(\sigma t)$ is a solution to (25) in $[-(1 + t_*)/\sigma, (t_* + 1)/\sigma]$ and an even function. By solving the Cauchy problem for (25) we can extend this solution to the whole t axis. In view of the fact that $v_1(0) = -1$ and $v_1(t_*+1) > 0$ one can find $s_1 \in (0, t_* + 1)$ such that $v(s_1) = 0$ and $v(s) < 0$ as $s < s_1$. In a similar argument to that in proving (28) one can also conclude that $V(t)$, $V_t(t)$, $V_{tt}(t)$ are positive for all $t > \tilde{t}_\sigma = s_1/\sigma$. Since $V_t(\tilde{t}_\sigma) > 0$ otherwise $V_t(\tilde{t}_\sigma) = V(\tilde{t}_\sigma) = 0$ implies $V(t)$ identically zero by the uniqueness of the Cauchy problem for ordinary equation and $V_t(\tilde{t}_\sigma) < 0$ contradicts the definition of $\tilde{t}_\sigma$. This implies $V(t)$ monotone-increasing to $+\infty$ as $t > \tilde{t}_\sigma$. This ends the proof for the present lemma. □

LEMMA 5. *For each $\sigma \in [\sigma_*, +\infty)$ where σ_* is mentioned in Lemma 4 (25) admits a analytic solution $W(t)$ in R^1 subject to $W(t) < 0$ as $t < 0$, $W(+\infty) = +\infty$ and $|W(t)| \leq Ce^{\sigma t}$ as t approaches $-\infty$. Moreover, $W(t)$ has a unique zero point t_σ which approaches to 0 as $\sigma \longrightarrow +\infty$.*

Proof: Set

$$W = U \int_{-\infty}^{t} U^{-2} \exp[\int_{-\infty}^{t} \frac{2}{sh\tau} d\tau] dt \text{ for } t < 0$$

This is nothing else but a solution of (25) for $t < 0$. Moreover,from Lemma 3 it follows

$$W < 0 \text{ if } t < 0 \text{ and } |W| \leq C_\sigma \int_{-\infty}^{t} e^{\sigma t} dt \leq \frac{C_\sigma}{\sigma} e^{\sigma t} \text{ if } t < -T$$

If $\sigma \in [\sigma_*, +\infty)$, by Lemma 4 and $U(0) = 0$, $V(0) = -1$ there are two positive constants C_1 and C_2 such that

$$W = C_1 U + C_2 V \text{ for all } t \in R^1 \tag{33}$$

Thus we can extend W to the whole t axis by (33) and furthermore, $W \longrightarrow +\infty$ as $t \longrightarrow +\infty$ since both of C_1 and C_2 are positive and both U and V approach to $+\infty$ as t$\longrightarrow +\infty$. Since $W(0) = -C_2 < 0$, by the same argument as that in proving Lemma 4 we can see that $W(t)$ has a unique zero point $t_\sigma > 0$. Notice that t_σ is less than the zero point of $V(t)$, $\tilde{t}_\sigma$. Lemma 4 tells us $t_\sigma \longrightarrow 0$ as $\sigma \longrightarrow +\infty$. This completes the proof of the present lemma. □Let us consider an infinitesimal isometric deformation of Σ_λ, $\vec{r}_\epsilon = \vec{r} + \epsilon\vec{\tau}$ where $\vec{\tau} = (\xi, \eta, \zeta)$ satisfies (23). Notice that $\epsilon = 0$ is the critical point of $g_\epsilon = d\vec{r}_\epsilon^2$ and hence, the differentiation of its Gaussian curvature and Christoffel symbols in ϵ are equal to zero at $\epsilon = 0$. Then differentiation of the Darboux equation (5) for $z + \epsilon\zeta$ with respect to ϵ, letting $\epsilon = 0$, gives

$$\nabla^{ij} z\zeta_{ij} + 2\sin^2\theta(\nabla z, \nabla\zeta) = 0 \tag{34}$$

In the present case

$$\nabla^{22} z = \sin^2\theta\cos\theta, \nabla^{12} z = 0 \text{ and } \nabla^{11} z = \cos\theta$$

Evidently (34) is elliptic as $\theta \neq \pi/2$ and may be written as follows

$$\cos\theta\left[(\sin\theta\zeta_\theta)_\theta + \left(\frac{\zeta_\phi}{\sin\theta}\right)_\phi\right] + 2\sin^2\theta\zeta_\theta = 0, \theta \in (0, \theta_*) \tag{35}$$

with

$$\zeta = 0 \text{ on } \theta = \theta_* \text{ and } \zeta \text{ bounded near } \theta = 0 \tag{36}$$

If (35) admits no nontrivial solution, it is easy to prove that the solution $(\xi, \eta, 0)$ to (23) (24) is the trivial one (for details, see [PO1]) and hence the rigidity of the infinitesimal isometric deformation of Σ_λ is readily proved.

The proof for Theorem 6. A change of variable

$$t = 2\int_{\pi/2}^{\theta} \frac{1}{\sin\theta} d\theta = 2\ln tg\frac{\theta}{2} \text{ from } (0, \theta_*) \text{ onto } (-\infty, 2\ln tg\frac{\theta_*}{2}) \tag{37}$$

reduces (35) to the following one

$$sht(\zeta_{tt} + \frac{1}{4}\zeta_{\phi\phi}) - 2\zeta_t = 0, t \in (-\infty, 2\ln tg\frac{\theta_*}{2}) \tag{38}$$

By the Fourier expansion of ζ in terms of ϕ, we are faced with the following Sturm-Liouville problem, with $\sigma = m/2$ for some nonnegative integer m,

$$sht(\zeta_{tt} - \sigma^2\zeta) - 2\zeta_t = 0, t \in (-\infty, 2\ln tg\frac{\theta_*}{2}) \tag{39}$$

$$\zeta = 0 \text{ as } t = 2\ln tg\frac{\theta_*}{2} \text{ and } \zeta \text{ bounded at the infinity} \tag{40}$$

The boundary conditions of the second part in (40) insures that the final solution to (34) is smooth near $\theta = 0$. It should be emphasized that we are going to discuss the above Sturm-Liouville problem in the smooth category and otherwise, the situation is very complicated. If $\sigma = 0$, the general solutions of (39) are of the form

$$\zeta = C_1\phi(t) + C_2 \text{ where } \phi(t) = \int_0^t \exp[\int_{-\infty}^s \frac{2}{shs_1} ds_1] ds$$

as $t \leq 0$. Therefore when $\sigma = 0$, (39) (40) admits no nontrivial solution. As for the case where $\sigma > 0$, Lemma 5 tells us that if $\sigma = m/2 \in [\sigma_*, +\infty)$ (39) always admits a solution $\zeta(t)$ satisfying the second part of (40) and $\zeta(t_\sigma) = 0$ for some $t_\sigma = t_1(m)$. Denote $\Lambda = \{\theta = 2tg^{-1}(\exp[t_1(m)/2]) | m \geq 2\sigma_*, m \in Z^+\}$ Therefore, if $\theta_* = 2tg^{-1}(\exp[t_1(m)/2])$ for some $m \in Z^+$, then $m^2/4$ is the eigenvalue of (39) (40). Moreover, the eigenfunction

$$|\zeta(t)| \leq Ce^{mt/2} \text{ as } t \longrightarrow -\infty$$

hence,

(41) $$|\zeta(2\ln tg\frac{\theta}{2})| \le C_1\theta^m \text{ as } \theta \longrightarrow 0 \text{ for another constant } C_1$$

Thus with $\tilde{\zeta}(\theta) = \zeta(2\ln tg\frac{\theta}{2})$, $\zeta^1 = \tilde{\zeta}(\theta)\sin m\phi$ and $\zeta^2 = \tilde{\zeta}(\theta)\cos m\phi$ are two solutions to (35) (36), and also to (34) for $\theta > 0$. In view of (41) and the fact that $\zeta^i \in C^\infty(\bar{D}\backslash\{\theta = 0\})$ $i = 1, 2$ by means of the regularity theory of solutions on elliptic equation it is easy to see that both of ζ^i are also smooth solutions to (34) in $\bar{D}$. Still denote ζ^1 (ζ^2) by ζ. Then the remaindering part of the proof is to find other two components (ξ, η) such that with $\vec{\tau} = (\xi, \eta, \zeta)$ $d\vec{\tau} = \vec{A} \times d\vec{r}$ for some smooth vector field $\vec{A}$ on $\bar{D}$. Now the proof for Theorem 6 is completed by the following lemma.

LEMMA 6. *Let $\vec{r} = (x, y, z) : (\bar{D}, g) \longrightarrow R^3$ be a smooth isometric embedding of a positive disk and let $\zeta \in C^\infty(\bar{D})$ be a solution to (34). Then one can find a smooth vector $\vec{A} = (A^1, A^2, A^3) \in C^\infty(\bar{D})$ such that $d\zeta = A^1dy - A^2dx$ and $A^2dz - A^3dy$, $A^3dx - A^1dz$ are two closed $1-$ form.*

Proof: Since $(\bar{D}, g)$ is a smooth positive disk and $\vec{r}$ is its isometric embedding, this implies the set where $dx \wedge dy = 0$ is of zero measure since it is nothing else but the preimage of equator under the Gauss map. Denote by D_+ (D_-) the set of all the preimage of the below (upper) hemisphere under the Gauss map. Evidently D_+ is a simply connect domain. In view of the fact that in D_+, $dx \wedge dy \neq 0$ one can define $A^1 = \zeta_y$, $A^2 = -\zeta_x$ and both of them are smooth in D_+. Write $P = -\zeta_{yx}z_y + \zeta_{yy}z_x$, $Q = -\zeta_{xx}z_y + \zeta_{xy}z_x$. Then a direct computation yields

$$d(Pdy + Qdx) = (P_x - Q_y)dx \wedge dy$$

$$= (\zeta_{yy}z_{xx} + \zeta_{xx}z_{yy} - 2\zeta_{xy}z_{xz})dx \wedge dy = 0$$

since if we regard x and y as the local coordinates, the last identity is equivalent to (34). Thus we have readily proved $Pdy + Qdx$ closed in D_+ and hence, there is $A^3 \in C^\infty(D_+)$ such that $dA^3 = Pdy + Qdx$. On the other hand,

$$d(A^2dz - A^3dy) = dA^2 \wedge dz - dA^3 \wedge dy = Qdx \wedge dy - Qdx \wedge dy = 0$$

$$d(A^1dz - A^3dx) = dA^1 \wedge dz - dA^3 \wedge dx = -Pdx \wedge dy + Pdx \wedge dy = 0$$

Now we proceed to study the behavior of A^i ($i = 1, 2, 3$) near ∂D_+, that is, $\theta = \pi/2$. Indeed, with $\Delta = y_\phi x_\theta - y_\theta x_\phi = \sin\theta\cos\theta$

$$A^1 = \frac{\zeta_\phi x_\theta - \zeta_\theta x_\phi}{\Delta} \text{ and } A^2 = \frac{\zeta_\phi y_\theta - \zeta_\theta y_\phi}{\Delta}$$

Noting (35) and $x_\theta = \cos\theta\cos\phi$, $y_\theta = \cos\theta\sin\phi$ we can see

$$\frac{\zeta_\theta}{\Delta}, \frac{x_\theta}{\Delta}, \frac{y_\theta}{\Delta} \text{ and hence, } A^1 \text{ and } A^2$$

smooth up to ∂D_+ and also in a neighbourhood of ∂D_+. Similarly,

$$\partial_\theta A^3 = -z_\theta \frac{x_\theta \partial_\phi A^2 - y_\theta \partial_\phi A^1}{\Delta}$$

is smooth up to ∂D_+ and in a neighbourhood of ∂D_+ too. It remains to discuss

$$\partial_\phi A^3 = -z_\theta \frac{x_\phi \partial_\phi A^2 - y_\phi \partial_\phi A^1}{\Delta}$$

$$= \frac{z_\theta}{\Delta}\left\{\zeta_{\phi\phi} + \frac{\zeta_\theta}{\cos\theta}\sin\theta\right\}$$

In view of (35), $\cos\theta\zeta_{\phi\phi} + \zeta_\theta \sin\theta = 0$ at $\theta = \pi/2$ and also

$$\partial_\theta(\cos\theta\zeta_{\phi\phi} + \zeta_\theta\sin\theta) = \zeta_{\theta\theta} - \zeta_{\phi\phi} = 0 \text{ at } \theta = \frac{\pi}{2}$$

So A^i, $i = 1, 2, 3$ are smooth functions defined on $\bar{D}_+$ and in a neighbourhood of D_+ too. By the same argument we can smoothly extend A^i, $i = 1, 2, 3$ to $\bar{D}$ and the requirements in the present lemma are fulfilled. This completes the proof of the present lemma. □

Acknowledgements.

The present paper is partially supported by the NSF of China and the STDF of Shanghai. Part of this work was done at the Institute of Mathematical Sciences, the Chinese University of Hong Kong and the author would like to express his appreciation for its hospitality. and also to thank Q. Zhow for his suggestion in constructing the counter example of Lemma 2.

References

[AL] Aleksandrov,A.D., Intrinsic geometry of convex surfaces, Moscow, 1948, German transl., Academie Verlag Berlin, 1955.

[AM] Amano,K., Global isometric embedding of a Riemannian 2-manifold with nonnegative curvature into a Euclidean 3-space, J.Diff.Geometry, **35**,49-83(1991).

[BG] Bryant, R. and Griffiths,P., The Gauss equation and rigidity of isometric embeddings, Duke Math. Journal, 50(1983), 803-892.

[BGY] Bryant, R., Griffiths,P. and Yang, D., Characteristics and existence of isometric embeddings, Duke Math. Journal, 50(1983), 893-994.

[CL] Chen,S.X. and Liu, L.Q., Fourier integral operators with finite smoothness and their applications, Chin. Ann. of Math., 6(A)(1985),539-559.

[DE] Delanoe,Ph., Relations globalement regulieres de disques strictement convexes dans les espaces d'Euclide et de Minkowski par la methode de Weingarten, Ann.Sci.Ec Norm.Super.,IV. **21**(1988), 637-652.

[EF1] Efimov,N.V., Generalization of singularities on surfaces of negative curvature, Mat. Sb., 64(1964), 286-320.

[EF2] Efimov,N.V., Surfaces with slowly varying negative curvature, Russ. Math. Survey, 21(1966), 1-55.

[EF3] Efimov,N,V,. A Criterion of homemorphism for some mappings and its applications to the theory of surfaces, Mat. Sb., 76(1968), 499-512.

[EG] Egorov,Y., Sur un exemple d'equation lineaire hyperbolique du second ordre n'ayant pas de solution, Journees Equations aux Derivees Partielles, Saint-Jean-de-Monts, 1992.

[G] Gromov.M., Partial differential relations, Springer- Verlage, Berlin Heidelberg, 1986

[GR] Gromov,M. and Rohklin,V.A., Embeddings and Immersions in Riemannian Geometry, Russian Mathematical Surveys, 25(1970), 1-57.

[GUN] Gunther, M., Isometric embeddings of Riemannian manifolds, Proceeding of the International Congress of Mathematicians, Kyoto, Japan, 1990.

[GL] Guan, F. and Li, Y., The weyl problem with nonnegative Gauss curvature, J. Diff. Geometry, 39(1994), 331-342.

[H] Hormander,L., On the Nash-Moser implicit function theorem, Ann. Acad.Sci of Fenn., 10. 255-259.

[HI] Hilbert,D., Uber Plachen von konstanter Gausscher krummung, Trans. Amer. Math. Soc., 2(1901),87-99.

[HO1] Hong,J.X.,Cauchy problem for degenerate hyperbolic Monge-Ampere equations, J.Partial Diff. Equations, **4**(1991),1- 18.

[HO2] Hong,J.X., Realization in R^3 of complete Riemannian manifolds with negative curvature, Communications Anal. Geom., 1(1993), 487-514.

[HO3] Hong,J.X., Isometric embedding in R^3 of complete noncompact nonnegatively curved surfaces, Manuscripta Math., 94(1997), 271-286.

[HO4] Hong,J.X., Dirichlet problems for general Monge-Ampere equations, Math. Z., 209(1992), 289-306.

[HO5] Hong,J.X., Darboux equations and isometric embedding of Riemannian manifolds with nonnegative curvature in R^3, Chin. Ann. of math., 2(1999),123-136.

[HZ1] Hong, J.X. and Zuily,C., Existence of C^∞ local solutions for the Monge-Ampere equations, Invention Math., 89(1987), 645-661.

[HZ2] Hong, J.X. and Zuily, C., Isometric embedding of the 2-sphere with non negative curvature in R^3, Math.Z., 219(1995), 323-334.

[IA] Iaia, J.A., Isometric embeddings of surfaces with nonnegative curvature in R^3, Duke Math. J., 67(1992) 423-459.

[IW] Iwasaki, N., The strongerly hyperbolic equation and its applications, Proc. Sym. Pure Math., 44(1985).

[JA] Jacobowitz, H., Local isometric embeddings, Seminar on Differential Geometry, Princeton University Press, 1982.

[JX] De M.Jorge, L. and Xavier,F., A complete minimal surface in R^3 between two parallel planes, Ann. of Math.,

[K] Kuiper,N.H., On C^1-isometric embeddings, I, II, Indag. Math., (1955), 545-556, 683-689.

[LE2] Lewy,H., A priori limitation for solutions of Monge- Ampere equations, Trans. Amer.Math.Soc., 37(1935), 417-434.

[LE1] Lewy, H., On the existence of a closed convex surface realizing a Riemannian metric, Proc. Nat. Acad. Sci., 24(1936),104- 106.

[LC1] Lin, C.S., The local isometric embedding in R^3 of 2-dimensional Riemannian manifolds with nonnegative curvature, J. Diff. Geometry, 21(1985), 213-230.

[LC2] Lin,C.S., The local isometric embedding in R^3 of two dimensional Riemannian manifolds with Gaussian curvature changing sign clearly, Comm. Pure Appl.Math., 39(1986). 307-326.

[LF] Lin, F.H., On Darboux system, Nonlinear Analysis and Microlocal Analysis, Proceeding Internat. Conf., Nankai Institute, World Scientific, Singapore, 1992.

[RO] Rozhdestvenski, B., Quasilinear hyperbolic systems for theory of surfaces, Dokl. Akad. Nauk. USSR., 143(1962), 50-52.

[NI1] Nirenberg,L., The Weyl and Minkowski problems in Differential Geometry in the large, Comm. Pure Appl.Math., 6(1953), 337-394.

[NI2] Nirenberg,L., In Nonlinear problems, editor by R.E.Langer, University of Wisconsin Press, Madison, 1963, 177-193.

[P] Poznjak, E., Isometric immersions of 2-dimensional metrics into Euclidean spaces, Uspechy 28(1970), 47-76.

[PO1] Pogorelov,A.V., Extrinsic geometry of convex surfaces (transl. Math. Monogr. Vol 35), Providence, RI, Am.Math.Soc., 1973.

[PO2] Pogorelov,A.V., An example of a two dimensional Riemannian metric not admitting a local realization in R^3, Dokl. Akad. Nauk. USSR., 198(1971), 42-43.

[Y1] Yau, S.T., Problem Section, Seminar on Differential Geometry, Princeton University Press, 1982.

[Y2] Yau, S.T., Open problems in Geometry, Chern-A great Geometer on the Twentieth Century, International Press, 1992.

[Y3] Yau, S.T., Lecture on Differential Geometry, in Berkeley, 1977.

INSTITUTE OF MATHEMATICS, FUDAN UNIVERSITY, SHANGHAI CHINA

AMS/IP Studies in Advanced Mathematics
Volume 20, 2001

Numerical Study of Interfacial Problems with Small Surface Tension

Hector D. Ceniceros and Thomas Y. Hou

ABSTRACT. This paper reviews some recent advances in developing stable and efficient boundary integral methods for solving interfacial flows with surface tension. Both spatial and temporal stability issues are discussed in detail. We then use these highly accurate and stable methods to investigate the peculiar regularization effect of small surface tension in overturning water waves, unstably stratified two-fluid interfaces, and Hele-Shaw flows. Several interesting and surprising phenomena are revealed in the limit of small surface tension. This paper is not intended to be a general survey of the subject and the discussion is limited by both the taste and expertise of the authors.

1. Introduction

Surface tension plays a fundamental role in the development of several important physical phenomena. Examples vary from materials science and engineering applications such as pattern formation in crystal growth and dendritic solidification to nonlinear phenomena in fluids such as capillary waves, droplet formation, fingering instabilities in porous media, and the appearance of finite-time singularities to name only a few.

Surface tension at an interface between two immiscible fluids arises from the imbalance of their intermolecular cohesive forces. It is thus, an ever-present force although it could be very small. But even very small surface tension can have dramatic effects on the nonlinear evolution of a fluid or material (as in the case of crystal growth) interface. The numerical study of small surface tension has both physical and mathematical implications. On the physical stand, it contributes to the understanding of important observable phenomena. On the mathematical stand, it provides valuable information about the structure and behavior of the solutions, and about the dependence of the solutions on the small parameter. Last but not least, the knowledge of the limiting behavior of the interface solutions as surface tension tends to zero, constitutes a critical test to the validity of the simplified zero-surface-tension modeling equations. This has a profound implication. Neglecting seemingly small terms can lead to an unphysical model.

1991 *Mathematics Subject Classification.* Primary 65M12, 76B15.

Research supported in part by a NSF Grant DMS-9704976 and by an ARO Grant DAAD19-99-1-0141.

Numerically, one can approach the investigation of interfacial evolution with small surface tension with different techniques. For example, phase field models, volume-of-fluid methods, level set methods, front tracking, and boundary integral/element methods. Boundary integral methods outperform other numerical methods when it comes to the accurate resolution of the small scale and often very singular structures characteristic of surface tension effects. Our discussion here focuses exclusively on boundary integral methods.

One of the advantages of using boundary integral methods is that it reduces the two dimensional problem into a one dimensional problem involving quantities along the interface only. Consequently, their use avoids the difficulty of differentiating discontinuous fluid quantities across the fluid interface. However, numerical simulations using boundary integral methods also suffer from the sensitivity to numerical instabilities because the underlying problems are very singular [**LHC76, Rob83, Dol92**]. Straight-forward discretizations may lead to numerical instabilities. It turns out that a certain *compatibility* is required between the choice of the quadrature rule for the singular velocity integral and the choice of the spatial derivative. This compatibility ensures that a delicate balance of terms that holds on the continuous level is preserved on the discrete level. This balance is crucial for maintaining numerical stability. A stability analysis by Beale, Hou, and Lowengrub [**BHL96**] for a boundary integral method for two-dimensional water waves reveals that certain Fourier filtering is needed to maintain the compatibility condition at the discrete level. With this filtering, determined by the choices of the quadrature rule and of the approximation for the spatial derivative, one can prove stability of the boundary integral method for water waves [**BHL96**] and for multi-fluid interfaces if surface tension is included [**CH98**].

While the spatial discretizations are proved to be stable and convergent, stability of the time discretization is very difficult to obtain in the presence of surface tension. Surface tension introduces a large number of spatial derivatives through the local curvature. If an explicit time integration method is used, these high order derivative terms induce strong stability constraints on the time step. For example, the time step stability constraint for the Hele-Shaw flows is given by $\Delta t \leq Ch^3$, where Δt is the time step, and h is the minimum particle spacing. These stability constraints are time dependent and become more severe by the differential clustering of points along the interface.

Hou, Lowengrub, and Shelley [**HLS94**] have successfully removed the severe stability constraint (stiffness) by designing an efficient semi-implicit scheme based on a new reformulation of the problem. This reformulation employs a dynamic change of variables from the Cartesian coordinates (x, y) to the interface tangent angle and arclength metric. Using a *small scale decomposition* technique, one can separate the leading order operators that contribute to the stiffness. Moreover, the leading order singular terms are shown to be linear and have constant coefficients (in space). Thus, a Crank-Nicholson type discretization can be used to eliminate the stiffness of the time discretization. This reformulation greatly relaxes the stability constraint. Many interfacial problems that were previously unobtainable are now solvable using this method and new phenomena have been discovered.

In this paper, we apply this highly accurate and stable method to investigate the effect of small surface tension for several important interfacial flows. The first example is about the surface tension regularization effect for unstably stratified two-fluid interfaces. Without any physical regularization, the interface is subject to the Rayleigh-Taylor instability and a curvature singularity may develop in finite time (see, e.g. [**DR81, BMO80**]). Our numerical results indicate that surface tension indeed regularizes the curvature singularity induced by the Rayleigh-Taylor instability and the interface rolls up into a spiral. However, the roll-up is terminated by the collision of the interface with itself, forming trapped bubbles of fluid at the core of the spiral. This process of bubble formation through self-intersection of a fluid interface has been observed by Hou, Lowengrub, and Shelley [**HLS97**] for a vortex sheet. Such pinching singularity is commonly observed in 3-D liquid jets where the azimuthal component of surface tension is believed to play an essential role in the collapsing process. However, this component is completely absent in a 2-D interface, making the appearance of such pinching singularities very surprising.

Our second example deals with the effect of small surface tension on the generation and structure of capillary waves on the surface of overturning water waves. The understanding of these short waves is important in the remote sensing of sea surface because the fine structure associated with short wavelengths scatters electromagnetic radiation. The phenomenon of capillary waves has been studied extensively in the literature (see, e.g. [**LH95, LH96, Cra70, Tul96**]). By using the stable spectrally accurate method discussed here, we are able to capture with high accuracy not only up to the appearance of the capillary jump [**LH96**], but also to follow the subsequent development of the small scale structure produced by the capillaries. Our numerical study also reveals certain scaling behavior of the capillary wavelength which agrees well with experimental observation [**DQPW99**]. The small-surface-tension solution is found to converge strongly to the zero-surface-tension solution.

Our third example is about the small surface tension limit of Hele-Shaw flows with suction. It is well known from laboratory experiments [**Pat81**] that Hele-Shaw flows with suction can develop long 'fingers' that encroach upon the viscous fluid being sucked. In the absence of surface tension, such fingering is not observed. Instead, solutions of the Hele-Shaw equations with suction are known to form finite-time cusp singularities before the fluid interface reaches the sink [**HLO88, Ric72**].

Here, we would like to investigate the limiting behavior of the small surface tension solution past the cusp singularity time t_c of the zero-surface-tension solution. Our computations show two surprising results. First, we find that an asymptotic corner angle is selected at the finger tip as it reaches the sink in the limit as surface tension tends to zero. This seems to be counter intuitive at first glance since one may think the tip angle would decrease with decreasing surface tension. Secondly, we find that the finger bulges and develops a neck singularity before it forms a wedge at the sink. This neck appears at a height close to that of the zero-surface-tension cusp. It is possible that the formation of this neck singularity is due to the influence of the zero-surface-tension singularity, and this pair of neck singularities may be responsible for the selection of the limiting wedge angle at the sink.

Our last example is on the singular perturbation of surface tension for Hele-Shaw problems. A natural conjecture is that the small-surface-tension solution would converge to the zero-surface-tension solution as surface tension tends to zero, provided that the zero-surface-tension solution is smooth (see e.g. [**HLO88**]). However, the asymptotic theory of Tanveer [**Tan93**] and Siegel, Tanveer, and Dai [**STD96**] suggests that the small surface tension limit could be singular for certain initial data even if the limiting zero-surface-tension solution is perfectly smooth. This possibility is quite surprising.

The numerical investigation of this question is very challenging. Since the zero-surface-tension problem is ill-posed, the numerical solution is extremely sensitive to noise for small surface tension [**DS93, STD96**]. We overcome this difficulty by using (i) a compact parametrization of the interface, (ii) a scaled equation for the perturbed quantity to factor out the small surface tension coefficient, and (iii) very high precision arithmetic. Using these techniques, we are able to compute accurately for extremely small values of surface tension. Our computations show convincingly that the small surface tension limit is indeed singular for certain initial data. The reason is that surface tension can introduce a new complex singularity at $t = 0^+$ which could subsequently impinge the physical domain at a time when the zero-surface-tension solution is still analytic. The impact of this so-called daughter singularity is to cause an indentation in the interface which subsequently evolves into a bulged finger. We also find that surface tension defines a length scale of the finger width.

All these calculations demand stable and highly accurate numerical methods. We also need to be extremely careful in controlling the noise effect to compute accurately the small surface tension limit and to capture the singular behavior of the solution. Although boundary integral methods are not general-purpose methods, they are especially suited for these types of challenging problems where extremely high resolution is needed. Grid-based methods, on the other hand, could have a lot of difficulties in achieving the same resolution. In addition, the inherent numerical diffusion or other artificial numerical regularizations could overshadow the small surface tension effect. In this sense, boundary integral methods provide an edge over other methods for accurate solution of this type of problems.

The rest of this paper is divided as follows. In Section 2, we describe the evolution equations of a fluid interface corresponding to two-fluid flows, water waves, and Hele-Shaw flows. The equations are presented in a boundary integral representation. Then, in Section 3, we discuss the stability of boundary integral methods for spatial discretizations. The technique to remove the surface tension induced stiffness is reviewed in Section 4. The four examples of numerical computations of interfacial flows with small surface tension are presented in Section 5.

2. General two-fluid interfaces

In this section, we provide the evolution equations in a boundary integral representation of three important examples of interfacial flows: water waves, stratified two-fluid interfacial flows, and Hele-Shaw flows. In the presence of small surface tension, intriguing and fascinating phenomena develop dynamically in this type of flows. The understanding of these phenomena has a fundamental impact in many physical and engineering applications. For example, capillary waves produced by surface tension are considered one of the key mechanisms for wave breaking and for the onset of turbulence. Moreover, the understanding of the fine scale structure generated by capillary waves is important in the remote sensing of the sea surface. On the other hand, stratified interfacial flows have been used as models to understand mixing of fluids, separation of boundary layers, generation of sounds (in bubbly flows), and coherent structures in turbulence models. Finally, theoretical and numerical studies of Hele-Shaw flows have received renewed interests and increasing attention in recent years because of the rich phenomena in the physical solutions and the potential applications in pattern formation and materials science.

2.1. Stratified two-fluid interfaces. We first consider the motion of general two-fluid interfacial flows in two space dimensions. The fluid on each side of the interface is assumed to be inviscid, incompressible, and irrotational. We denote the fluid quantities above the interface with the subscript 2 and those below the interface with subscript 1. In each fluid, the flow satisfies the Euler equations

$$\rho_i[\partial_t \mathbf{u}_i + (\mathbf{u}_i \cdot \nabla)\mathbf{u}_i] = -\nabla p_i - \rho_i g \hat{\jmath}, \quad i = 1, 2,$$

where ρ_i, $\mathbf{u}_i$, and p are the density, velocity, and pressure respectively. The constant g is the gravity acceleration. The incompressibility and irrotationality constraints imply that $\nabla \cdot \mathbf{u}_i = 0$ and $\nabla \times \mathbf{u}_i = 0$. At the interface, which we denote by Γ, we impose the Laplace-Young boundary condition which relates a pressure jump to the curvature of the interface κ by

$$[p]|_\Gamma = \tau\kappa,$$

where $[p]|_\Gamma$ denotes the jump of pressure across the interface Γ and τ is the surface tension coefficient. The normal velocity is assumed to be continuous across the interface. The tangential velocity at the interface in general is not uniquely defined since it may have a jump discontinuity across the interface. At any time t, we represent the interface parametrically by $(x(\alpha,t), y(\alpha,t))$, where α is a Lagrangian variable. The interface governing equation can be put in a convenient form by introducing the complex position variable $z(\alpha,t) = x(\alpha,t) + iy(\alpha,t)$. Of special interest is the case when the interface is a periodic perturbation of the flat interface, i.e. $z(\alpha,t) = \alpha + s(\alpha,t)$, where $s(\alpha,t)$ is a 2π-periodic function of α. Then, taking the tangential velocity to be the average of the limiting velocities above and below the interface, the interface evolves according to the so-called Birkhoff-Rott equation

$$\frac{d\bar{z}}{dt} = \frac{1}{2\pi i}\int_{-\pi}^{\pi} \gamma(\alpha') \cot\frac{1}{2}(z(\alpha) - z(\alpha'))d\alpha', \tag{2.1}$$

where we have used the periodic structure in s. Here, $\bar{z}$ denotes the complex conjugate of z. The above integral should be understood as the Cauchy principal-value integral. The variable γ is the unnormalized vortex sheet strength, whose the evolution equation can be obtained from the Euler equations in both sides of the interface [**BMO82**]:

$$\frac{d\gamma}{dt} = -2A\left(Re\left\{\frac{d^2\bar{z}}{dt^2}z_\alpha\right\} + \frac{1}{8}\partial_\alpha\left(\frac{\gamma^2}{|z_\alpha|^2}\right) + gy_\alpha\right) + S\kappa_\alpha, \tag{2.2}$$

where $A = (\rho_1 - \rho_2)/(\rho_1 + \rho_2)$ is the Atwood number and $S = \tau/(\rho_1 + \rho_2)$ is a scaled surface tension parameter. The curvature κ is given by

$$\kappa = \frac{x_\alpha y_{\alpha\alpha} - x_{\alpha\alpha}y_\alpha}{(x_\alpha^2 + y_\alpha^2)^{3/2}}.$$

Equation (2.2) is a Fredholm integral equation of the second kind for $d\gamma/dt$. An important feature of this formulation is that it has a globally convergent Neumann series [**BMO82**]. As a result, the integral equation for $d\gamma/dt$ can be solved effectively by iteration. Equations (2.1) and (2.2) completely determine the motion of the interface.

2.2. Water wave equations. In the special case of water waves, we have $\rho_1 = 0$. The above derivation needs to be modified. Only the Bernoulli equation in the fluid region is needed to obtain an evolution equation for the potential. The first boundary integral method for water waves was proposed by Longuet-Higgins and Cokelet [**LHC76**] who used a single layer representation. This gives rise to a Fredholm integral equation of the first kind for $\frac{\partial\phi}{\partial n}$, which determines the normal velocity of the interface. The disadvantage of this approach is that the matrix associated with a Fredholm integral of the first kind is usually not well-conditioned and the number of iterations required increases rapidly with the number of the mesh points. The double layer representation introduced by Baker, Meiron, and Orszag [**BMO82**], on the other hand, does not have this disadvantage. Following [**BMO82**], we obtain a system of evolution equations given by:

$$\bar{z}_t = \frac{1}{4\pi i}\int_{-\pi}^{\pi}\gamma(\alpha')\cot\frac{1}{2}(z(\alpha)-z(\alpha'))d\alpha' + \frac{\gamma(\alpha)}{2z_\alpha(\alpha)} \equiv u - iv, \tag{2.3}$$

$$\phi_t = \frac{1}{2}(u^2 + v^2) - gy\,, \tag{2.4}$$

$$\phi_\alpha = \frac{\gamma}{2} + \mathrm{Re}\left(\frac{z_\alpha}{4\pi i}\int_{-\pi}^{\pi}\gamma(\alpha')\cot\frac{1}{2}(z(\alpha)-z(\alpha'))d\alpha'\right). \tag{2.5}$$

Here ϕ is the velocity potential and again γ denotes the vortex sheet strength. Equations (2.3)-(2.5) completely determine the motion of the system. Like the two-fluid interface problem, the integral equation for γ is a Fredholm integral equation of second kind which has a globally convergent Neumann series. Thus, γ can be solved efficiently by fixed point iteration.

2.3. Hele-Shaw flows. A closely related problem is the Hele-Shaw flow which describes a viscosity dominant creeping flow confined between two closely-spaced plates. The theoretical study in the West begins with Saffman and Taylor [**ST58**]. They found a family of exact self-similar finger-like solutions for the interface in a channel geometry without surface tension. Subsequent works have mostly focused on the role of surface tension in the selection of the finger width. Significant progress has been achieved in understanding steady states and their linear stability (see [**Pel88**] for a review). The dynamical behavior of Hele-Shaw flows has received a lot of interests inspired by the complex patterns formed by an expanding bubble. However, there are still many mysteries regarding the selection mechanism of finger widths for small surface tension.

Consider an interface Γ which separates two Hele-Shaw fluids of different viscosities and densities in a radial geometry. The velocity in each fluid is given by Darcy's law, together with the incompressibility constraint:

$$\mathbf{u}_j = -\frac{b^2}{12\mu_j}\nabla p_j, \quad \nabla\cdot\mathbf{u}_j = 0, \qquad \text{for } j = 1,2$$

Here b is the gap of the Hele-Shaw cell, μ_j, p_j, and ρ_j are the viscosity, pressure, and density on each side of the fluid interface. Nonlinearity comes through the boundary conditions. We impose again the Laplace-Young boundary condition which states that surface tension causes a jump in the pressure across the interface proportional to the local mean curvature. In addition, the normal velocity should be continuous at the interface.

We assume that there is a point sink (or a source) at the origin, inside a fluid blob. For large distances away from the sink, the velocity field tends to the simple radial flow:

$$\mathbf{u}(\mathbf{x}) \to Q\frac{\mathbf{x}}{|\mathbf{x}|^2}, \quad \text{as} \quad |\mathbf{x}| \to \infty. \tag{2.6}$$

Here, Q is the suction rate (or pumping rate respectively) which is assumed constant and negative (positive).

Again, we represent the interface by the complex position variable $z(\alpha,t) = x(\alpha,t) + iy(\alpha,t)$. Unlike the fluid interface case, here we assume $z(\alpha,t)$ to be 2π-periodic in α. This implies that the interface Γ is a close curve. Denote the complex conjugate velocity by $W(\alpha,t) = u(\alpha,t) - iv(\alpha,t)$. This interface velocity can be represented by a boundary integral plus the sink contribution. Following [**TA83, DS93**], we can derive a boundary integral formulation given by:

$$(2.7)\ \bar{z}_t = -\frac{1}{z(\alpha,t)} + \frac{1}{2\pi i}\int_0^{2\pi} \frac{\gamma(\alpha',t)}{z(\alpha,t)-z(\alpha',t)}d\alpha',$$

$$(2.8)\ \gamma = 2A_\mu \mathrm{Re}\left(-\frac{z_\alpha(\alpha,t)}{z(\alpha,t)} + \frac{z_\alpha(\alpha,t)}{2\pi i}\int_0^{2\pi} \frac{\gamma(\alpha',t)}{z(\alpha,t)-z(\alpha',t)}d\alpha'\right) + S\kappa_\alpha,$$

where $A_\mu = (\mu_1-\mu_2)/(\mu_1+\mu_2)$ is the viscosity Atwood ratio and $S = \frac{b^2}{12\bar{\mu}}\tau$ is a scaled surface tension parameter. We nondimensionalize the equations of motion by taking the initial blob radius to be 1 and by setting $Q = -1$.

3. Stability of boundary integral methods

In this section, we review some stability issues of boundary integral methods for interfacial flows. We use the water wave problem as an example to illustrate the main ideas. The complete stability analysis of the boundary integral method for water waves was performed by Beale, Hou, and Lowengrub [**BHL96**]. Ceniceros and Hou analyzed the more general two-fluid interfacial flow with surface tension [**CH98**].

Numerical methods can be readily derived from the boundary integral formulation of water waves by choosing a discretization for the principal-value integral and for the spatial derivative. A classical discretization for the integral is the point vortex method introduced by Rosenhead [**Ros32**], while a popular choice for the derivative is the cubic spline approximation. However, numerical simulations using boundary integral methods are known to be sensitive to numerical instabilities [**LHC76, Rob83, Dol92**]. This includes some of the existing boundary integral methods. In order to avoid numerical instability, a certain compatibility between the choice of quadrature rule for the singular integral *and* the discrete derivatives must be satisfied. This compatibility ensures that a delicate balance of terms that holds on the continuous level is preserved on the discrete level. This point will be further illustrated below.

Let $z_j(t)$ be the numerical approximation of $z(\alpha_j,t)$, where $\alpha_j = jh$, $h = 2\pi/N$ and N is the number of discrete points on the interface. $\phi_j(t)$ and $\gamma_j(t)$ are defined similarly. To approximate the velocity integral, we use the alternating point trapezoidal rule:

$$\int_{-\pi}^{\pi} \gamma(\alpha') \cot \frac{1}{2}(z(\alpha_j) - z(\alpha'))d\alpha' \simeq \sum_{\substack{k=-N/2+1 \\ (j-k)\text{ odd}}}^{N/2} \gamma_k \cot \frac{1}{2}(z_j - z_k)2h \,.$$

The advantage of using this quadrature is that the approximation is spectrally accurate(see, e.g. [**She92**]). Let us denote by D_h the discrete derivative operator. It is convenient for the analysis to define D_h through its spectrum property: $\widehat{(D_h)}_k = ik\rho(kh)$ where the over-caret stands for the Fourier transform and ρ is some nonnegative even function. The specific form of $\rho(kh)$ depends on the approximation. For example, we have $\rho(kh) = 3\sin(kh)/(kh(2+\cos(kh)))$ for the cubic spline approximation and $\rho(kh) = 1$ for the pseudo-spectral derivative.

Now we can present our numerical algorithm [**BHL96**] for the water wave equations (2.3)-(2.5) as follows:

$$\frac{d\bar{z}_j}{dt} = \frac{1}{4\pi i} \sum_{\substack{k=-N/2+1 \\ (j-k)\text{ odd}}}^{N/2} \gamma_k \cot \frac{1}{2}(z_j^{(\rho)} - z_k^{(\rho)})2h + \frac{\gamma_j}{2D_h z_j} \equiv u_j - iv_j, \tag{3.1}$$

$$\frac{d\phi_j}{dt} = \frac{1}{2}(u_j^2 + v_j^2) - gy_j \,, \tag{3.2}$$

$$D_h\phi_j = \frac{\gamma_j}{2} + Re\left(\frac{D_h z_j}{4\pi i} \sum_{\substack{k=-N/2+1 \\ (j-k)\text{ odd}}}^{N/2} \gamma_k \cot \frac{1}{2}(z_j^{(\rho)} - z_k^{(\rho)})2h\right), \tag{3.3}$$

where $z^{(\rho)}$ denotes a Fourier filtering applied on z and defined as $z^{(\rho)} = \alpha + s^{(\rho)}$ with $\widehat{(s^{(\rho)})}_k = \hat{s}_k\rho(kh)$. The filtered quantity $z^{(\rho)}$ in (3.1) and (3.3) is to balance the high frequency errors introduced by D_h. This will become apparent in the discussion of stability below.

Theorem 1[**BHL96**]. *Assume that the water wave problem is well-posed and has a smooth solution in C^{m+2} ($m \geq 3$) up to time T. Then if D_h corresponds to a r-th order derivative approximation, we have for $0 < h \leq h_0(T)$*

$$\|z(t) - z(\cdot,t)\|_{l^2} \leq C(T)h^r \,. \tag{3.4}$$

Similar convergent results hold for ϕ_j and γ_j. Here $\|z\|_{l^2}^2 = \sum_{j=1}^{N} |z_j|^2 h$.

3.1. Discussion of stability analysis. Here, we discuss some of the main ingredients in the stability analysis of the scheme given by (3.1)-(3.3). We will mainly focus on linear stability. Once linear stability is established, nonlinear stability can be obtained relatively easily by using the smallness of the error and an induction argument. The reader is referred to [**BHL96**] for details.

The goal is to derive evolution equations for the errors $\dot{z}_j(t) \equiv z_j(t) - z(\alpha_j, t)$, etc., and to estimate their growth in time. To simplify the stability analysis, we expand the periodic sum into an infinite sum over the whole line, i.e.

$$\frac{1}{2} \sum_{\substack{k=-N/2+1 \\ (j-k)\text{ odd}}}^{N/2} \gamma_k \cot \frac{1}{2}(z_j^{(\rho)} - z_k^{(\rho)})2h = \sum_{(j-k)\text{ odd}} \frac{\gamma_k}{z_j^{(\rho)} - z_k^{(\rho)}} 2h \equiv W_j.$$

Let us now define

$$\dot{W}_j = W_j - \sum_{(j-k)\ \text{odd}} \frac{\gamma(\alpha_k)}{z(\alpha_j)^{(\rho)} - z(\alpha_k)^{(\rho)}} 2h. \tag{3.5}$$

$\dot{W}_j$ can be written as the sum of linear terms in $\dot{z}_j$ and $\dot{\gamma}_j$ plus nonlinear terms. We find that

$$\begin{aligned} \dot{W}_j &= \frac{h}{\pi i} \sum_{(j-k)\ \text{odd}} \frac{\dot{\gamma}_k}{z(\alpha_j)^{(\rho)} - z(\alpha_k)^{(\rho)}} \\ &- \frac{h}{\pi i} \sum_{(j-k)\ \text{odd}} \frac{\gamma(\alpha_k)(\dot{z}_j^{(\rho)} - \dot{z}_k^{(\rho)})}{(z(\alpha_j)^{(\rho)} - z(\alpha_k)^{(\rho)})^2} + \text{nonlinear terms.} \end{aligned} \tag{3.6}$$

Next, we use the Taylor expansion for the singular kernel to obtain the most singular contributions and write

$$\frac{1}{z(\alpha_j) - z(\alpha_k)} = \frac{1}{z_\alpha(\alpha_j)(\alpha_j - \alpha_k)} + f(\alpha_j, \alpha_k) , \tag{3.7}$$

where f is a smooth function. Thus, the leading order contribution to the first term in (3.6) is $(2iz_\alpha)^{-1} H_h \dot{\gamma}_j$, where H_h is the discrete Hilbert transform defined as

$$H_h(\dot{\gamma}_j) \equiv \frac{1}{\pi} \sum_{(j-k)\ \text{odd}} \frac{\dot{\gamma}_k}{\alpha_j - \alpha_k} 2h . \tag{3.8}$$

The remaining term involving the smooth kernel f becomes a smoothing operator A_{-1} of order one, i.e., $D_h A_{-1} = A_0$ and $A_{-1} D_h = A_0$, where A_0 denotes a bounded operator form l^p to l^p. Similarly, the most important contribution to the second term in (3.6) is $-\gamma(2iz_\alpha^2)^{-1}\Lambda_h(z_j^{(\rho)})$, where Λ_h is defined as follows:

$$\Lambda_h(\dot{f}_j) \equiv \frac{1}{\pi} \sum_{(j-k)\ \text{odd}} \frac{\dot{f}_j - \dot{f}_k}{(\alpha_j - \alpha_k)^2} 2h . \tag{3.9}$$

Denote by H and Λ the corresponding continuous operators for H_h and Λ_h respectively, i.e.

$$H(f) = \frac{1}{\pi} \int \frac{f(\alpha')}{\alpha - \alpha'} d\alpha', \quad \Lambda(f) = \frac{1}{\pi} \int \frac{f(\alpha) - f(\alpha')}{(\alpha - \alpha')^2} d\alpha'.$$

It is easy to verify using integration by parts that

$$\Lambda(f) = H(D_\alpha f) , \tag{3.10}$$

where D_α is the continuous derivative operator. Our analysis indicates that a similar compatibility condition must be satisfied among the discrete operators H_h, Λ_h, and D_h to maintain numerical stability, i.e.

$$\Lambda_h(\dot{z}_i) = H_h D_h(\dot{z}_i) , \tag{3.11}$$

for $\dot{z}$ satisfying $\widehat{\dot{z}}_0 = \widehat{\dot{z}}_{N/2} = 0$. Unfortunately, many existing boundary integral methods violate this compatibility condition, and consequently are subject to numerical instability.

An important question is how to modify our discretization to obtain a stable method. One way to do this is to introduce appropriate Fourier filtering in the approximations of the velocity integral as we have done in (3.1) and (3.3). By doing this, we can ensure that a variant of the compatibility condition (3.11) is satisfied:

$$\Lambda_h(\dot{z}_j^{(\rho)}) = H_h D_h(\dot{z}_j) \; . \tag{3.12}$$

We can verify this from the spectrum properties of H_h and Λ_h and the definitions of D_h and the ρ filtering. With this modified compatibility condition, we can obtain stability of our boundary integral method.

To complete our stability analysis, we need to make use of the following properties of the discrete Hilbert transform: (i) $H_h^2 = -I$, (ii) $\Lambda_h(z^{(\rho)}) = H_h D_h(z)$, (iii) the commutator, $[H_h, f] \equiv H_h f - f H_h$, is a smoothing operator, i.e. $[H_h, f](\dot{z}^{(\rho)}) = A_{-1}(\dot{z})$ for smooth f. Following the linear well-posedness analysis, we can derive an error equation for $\dot{z}_j$ as follows (see [**BHL96**])

$$\frac{d\dot{z}_j}{dt} = \frac{1}{z_\alpha(\alpha_j)}(I - iH_h)D_h\dot{F}_j + A_0(\dot{z}_j) + A_{-1}(\dot{\phi}_j) + O(h^r),$$

where $\dot{F}_j = \dot{\phi}_j - u(\alpha_j)\dot{x}_j - v(\alpha_j)\dot{y}_j$. At this point, it is essential to project the error equation into the local tangent and normal vectors of the underlying interface, $z(\alpha,t)$. In this local coordinate, the error equations are greatly simplified and the stability analysis becomes apparent. Alternatively, the local tangent angle to the interface has also been shown [**CH98**] to be a natural variable for the stability analysis, specially in the presence of surface tension.

Denote by $\dot{z}_j^N$ and $\dot{z}_j^T$ the normal and tangential components of $\dot{z}_j$, with respect to the underlying interface $z(\alpha_j,t)$ and let $\hat{\mathbf{n}}$ be the outward unit normal. We further make the following change of variables $\dot{\delta}_j = \dot{z}_j^T + H_h\dot{z}_j^N$. We obtain after some simplification that

$$\frac{d\dot{\delta}_j}{dt} = A_{-1}(\dot{F}_j) + A_0(\dot{z}_j), \tag{3.13}$$

$$\frac{d\dot{z}_j^N}{dt} = \frac{1}{|z_\alpha(\alpha_j)|}H_h D_h\dot{F}_j + A_{-1}(\dot{F}_j) + A_0(\dot{z}_j), \tag{3.14}$$

$$\frac{d\dot{F}_j}{dt} = -c(\alpha_j,t)\dot{z}_j^N + A_{-1}(\dot{z}_j), \qquad c(\alpha,t) = (u_t, v_t + g)\cdot\hat{\mathbf{n}}, \tag{3.15}$$

where equation (3.15) is obtained by performing error analysis on Bernoulli equation and by using the Euler equations. In this form it is clear that only the normal component of $\dot{z}_j$ is important. This is consistent with the physical property of interfacial dynamics. The leading error equations now become so simple that an energy estimate can be easily carried out. Note that $H_h D_h$ is a positive operator with a Fourier symbol $\rho(kh)|k|$. The discretization is stable if the water wave problem is well-posed, i.e., the sign condition $c(\alpha,t) > 0$ is satisfied. We refer to [**BHL96**] for details.

Generalization of the above stability analysis to 3-D interfacial flows proves to be quite challenging due to the non-removable branch point singularity in the 3-D boundary integral kernel. Nevertheless, substantial progress has been made recently by Beale [**Bea98**] and by Hou and Zhang [**HZa, HZb**] in designing stable 3-D boundary integral methods for water waves. These analyses also shed new light into the 2-D interface problems. Since these stabilized methods are quite sophisticated, we will omit their discussion here.

4. Removing the stiffness of surface tension for interfacial flows

Surface tension at an interface between two immiscible fluids arises from the imbalance of their intermolecular cohesive forces. It is one of the most commonly used physical regularizations in the study of interfacial flows. Surface tension plays an important role in understanding fluid phenomena such as pattern formation in Hele-Shaw cells, the motion of capillary waves on free surfaces, and the formation of fluid droplets. Although surface tension provides a natural physical regularization for fluid interfaces, it also induces a severe time step stability constraint for explicit numerical methods. This severe stability constraint, which we refer to as stiffness, is introduced by surface tension through the high order spatial derivatives in the local curvature. These high order derivatives couple to the interface evolution equation in a nonlinear and nonlocal manner. If an explicit time integration method is used, these terms produce strong stability constraints on the time step (for the Hele-Shaw flows, the time stability constraint is $S\Delta t \leq \min_\alpha |z_\alpha| h^3$). Moreover, these constraints are generally time dependent, and become more severe by the differential clustering of points along the interface. This has been one of the major difficulties in simulating interfacial flows with surface tension.

Hou, Lowengrub, and Shelley (HLS) [**HLS94**] have developed an efficient method to remove the stiffness induced by surface tension for 2-D fluid interfaces. Their method is based on a *small scale decomposition* technique and on a reformulation of the problem in terms of the local tangent angle θ at the interface and arclength metric $s_\alpha = \sqrt{x_\alpha^2 + y_\alpha^2}$. The main motivation of using these new dependent variables is that curvature has a very simple expression in these variables: $\kappa = \theta_\alpha / s_\alpha$. Another important observation is that the stiffness only arises at small scales. Moreover, the leading order contribution of the singular integral operator at small scales can be expressed in terms of the Hilbert transform, which is diagonalizable in Fourier space. By treating only the leading order terms implicitly, but treating the lower order terms explicitly, an efficient semi-implicit discretization which effectively removes the stiffness of surface tension can be obtained.

We will now give a brief description of the method. Given an equation of motion of a free interface,

$$(x_t(\alpha, t), y_t(\alpha, t)) = U\hat{\mathbf{n}} + T\hat{\mathbf{s}},$$

it can be recast in terms of the variables θ and s_α, and the following evolution equations can be easily obtained:

$$(s_\alpha)_t = T_\alpha - \theta_\alpha U \,, \tag{4.1}$$

$$\theta_t = \left(\frac{1}{s_\alpha}\right)(U_\alpha + \theta_\alpha T) \,, \tag{4.2}$$

where $\hat{\mathbf{n}}$ and $\hat{\mathbf{s}}$ are the unit local normal and tangential vectors respectively. U and T are the local normal and tangential components of the interface velocity. Note that only the normal velocity is physically prescribed. The tangential velocity would determine the parametrization of the interface, but it does not affect the shape of the interface. This degree of freedom in choosing T is exploited to derive a simpler evolution equation for s_α and θ. One possible choice of T is to enforce α to be an arclength variable, which implies that s_α is independent of α if it is so initially. To achieve this, the tangential velocity T is selected as follows:

$$T(\alpha,t) = T(0,t) + \int_0^{\alpha} \theta_{\alpha'} U d\alpha' - \frac{\alpha}{2\pi}\int_0^{2\pi} \theta_{\alpha'} U d\alpha'. \tag{4.3}$$

This equation expresses T entirely in terms of θ and U. With this tangential velocity, the evolution equations for θ and s_α reduce to

$$(s_\alpha)_t = -\frac{1}{2\pi}\int_0^{2\pi} \theta_\alpha U d\alpha. \tag{4.4}$$

$$\theta_t = \left(\frac{1}{s_\alpha}\right)(U_\alpha + \theta_\alpha T) . \tag{4.5}$$

This system should be solved together with the evolution equations governing other dynamical variables, such as the vortex sheet strength or the velocity potential. This is a complete reformulation of the interface evolution problem.

The next step in the method of HLS is to extract the leading order singular operators that contribute to the numerical stiffness. Again, a Taylor expansion of the singular kernel is performed as in (3.7). For smooth interfaces, only the singular term on the right hand side of (3.7) contributes to the stiffness. The smooth kernel f gives rise to a smoothing operator which decays rapidly in high frequency, consequently it does not contribute to the stiffness. Thus, the leading order of the normal velocity U at small scales becomes

$$U \sim \frac{1}{2}\left(\frac{2\pi}{s_\alpha}\right)\mathcal{H}[\gamma] .$$

In the case of two-fluid interface motion, we need to get the leading order term for the vortex sheet strength γ. This term clearly comes form the curvature which introduces the highest order spatial derivatives. Therefore

$$\gamma_t \sim S\kappa_\alpha .$$

After singling out the leading order terms, the system for θ and γ is decomposed as follows:

$$\theta_t = \left(\frac{1}{2s_\alpha^2}\right)\mathcal{H}[\gamma_\alpha] + P , \tag{4.6}$$

$$\gamma_t = \left(\frac{S}{2s_\alpha}\right)\theta_{\alpha\alpha} + Q , \tag{4.7}$$

where P and Q represent the smoother and lower order terms. They are obtained by subtracting off the leading order terms from the right hand sides of the θ and γ equations respectively. We refer to this technique as the *Small Scale Decomposition* (SSD). A similar but simpler, SSD can be found for Hele-Shaw flows. Hou, Klapper, and Si [**HKS98**] have extended this technique to remove the stiffness of curvature in 3-D filaments.

It is important to note that the leading order linear operators can be diagonalized under Fourier transform. An efficient semi-implicit scheme can be easily designed by discretizing the leading order stiff terms implicitly using a Crank-Nicholson discretization, and leap-frogging on the lower order terms. Let $L(t) = 2\pi s_\alpha$ which represents the total arclength of the interface. In Fourier space, (4.6) and (4.7) become

$$\frac{\hat{\theta}^{n+1}-\hat{\theta}^{n-1}}{2\Delta t} = \frac{|k|}{4}\left((\frac{2\pi}{L^{n+1}})^2\hat{\gamma}^{n+1}+(\frac{2\pi}{L^{n-1}})^2\hat{\gamma}^{n-1}\right)+\hat{P}^n(k) \tag{4.8}$$

$$\frac{\hat{\gamma}^{n+1}-\hat{\gamma}^{n-1}}{2\Delta t} = -\frac{S}{2}k^2\left(\frac{2\pi}{L^{n+1}}\hat{\theta}^{n+1}+\frac{2\pi}{L^{n-1}}\hat{\theta}^{n-1}\right)+\hat{Q}^n(k)\ . \tag{4.9}$$

Note that there is no stiffness in the ODE for $L(t)$ so we can discretize L explicitly. Once L^{n+1} is updated, $\hat{\theta}^{n+1}(k)$ and $\hat{\gamma}^{n+1}(k)$ can be found by inverting a 2×2 matrix. This gives rise to a semi-implicit scheme which is at the same cost of an explicit scheme but with no high order stability constraint. It is also possible to design a fourth order semi-implicit discretization based on the same idea. This has been used effectively in many applications, including those to be described in the next section.

5. Numerical Study of Interfacial Flows

We now apply the methods reviewed in the previous sections to investigate numerically interfacial flows with small surface tension. We consider three examples. In the first one, we investigate the surface tension stabilizing effect for unstably stratified two-fluid interfaces. In the second example we study the dynamic generation of capillary waves in overturning water waves with small surface tension. The third example is devoted to the study of a Hele-Shaw flow with suction in the limit of small surface tension. In addition, we review an innovative method to study the singular perturbation of surface tension and the effect of noise in Hele-Shaw flows and present surprising numerical results. All these studies reveal some fascinating singular limiting behaviors of the interfacial flows as surface tension tends to zero.

5.1. Unstably stratified two-fluid interfacial flows. One of the classical examples of hydrodynamic instability occurs when a layer of heavy fluid sits on top of a lighter fluid, or equivalently when the interface is accelerated toward the heavy fluid. This is called Raleigh-Taylor Instability [**DR81**]. Rayleigh-Taylor instability is a fundamental instability of incompressible fluid flow at high Reynolds number. The idealization of an unstably stratified shear layer as a two-fluid interface separating two regions of potential flow has often been used as a model to study mixing properties, boundary layers and coherent structures of fluids.

Pullin [**Pul82**] was one of the first to study numerically the stabilizing effect of surface tension for unstably stratified two-fluid interfaces. But due to the inherent numerical instability, Pullin's computations were not conclusive. Rangel & Sirignano [**RS88**] also studied the effect of surface tension and density ratio on the nonlinear growth of the Kelvin-Helmholtz instability. To overcome the time-step stiffness constraint, they redistributed their grid points every time step. While this technique can effectively relax the time step stability constraint, it also introduces excessive numerical diffusion which could overshadow the effect of surface tension.

In [**CH98**], we computed several examples of two-fluid interfacial flows in the presence of surface, including the particular case we present now. We consider here an unstably stratified two-fluid interfacial flow (corresponding to $Ag < 0$) with the following initial data

$$\begin{aligned} z(\alpha,0) &= \alpha+0.1i\cos(2\pi\alpha),\\ \gamma(\alpha,0) &= 0. \end{aligned}$$

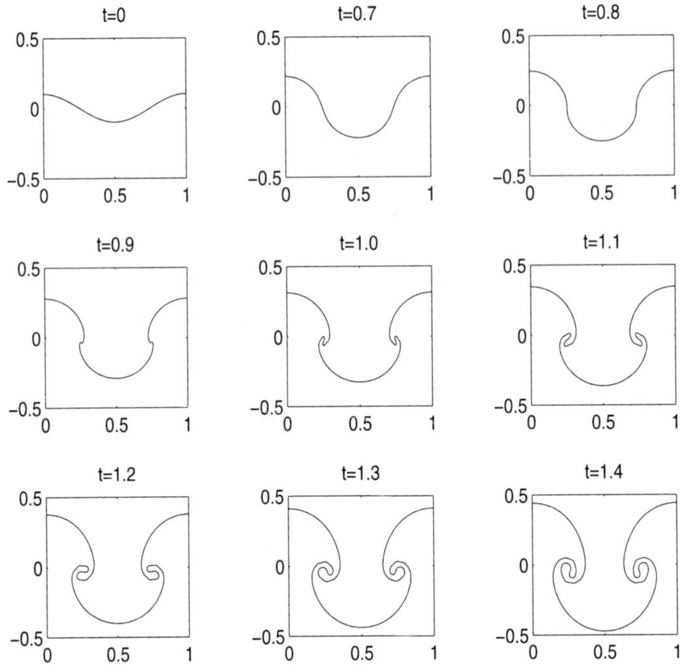

FIGURE 1. Unstably stratified flow: Time evolution of the interface with $A = -0.1$ and $S = 0.005$. $N = 1024$ and $\Delta t = 2.5 \times 10^{-4}$.

In Figure 1 we plot the time evolution of an interface with $A = -0.1$, $g = 10$, and $S = 0.005$. For this calculation we use $N = 1024$ and $\Delta t = 2.5 \times 10^{-4}$. We observe two small fingers that appear around $t = 0.9$, and the interface begins to roll up. One can also see some capillary waves that are generated around $t = 1.2$ and move outwards from the centers of roll-up. To ensure that our calculations are accurate, we perform a careful resolution study. We find that up to $t = 1.4$, the accuracy stays at about 4.66 digits for $N = 1024$ and about 7.2 digits for $N = 2048$. With $N = 2048$, we can continue our computations with reasonably good accuracy. In Figure 2, we plot a sequence of the interface position in time from $t = 1.5$ to $t = 1.75$. Note that the finger tips broaden as they continue to roll, and that the interface bends towards the finger tips. At $t = 1.75$, the minimum distance between the finger tip and the opposite side of the interface is about 0.016. The minimum distance continues to approach zero and by $t = 1.785$ (Figure 3(a)), is approximately 5×10^{-4}.

This process of bubble formation through self-intersection of a fluid interface has been observed by HLS [**HLS97**] for a vortex sheet. The vortex sheet problem is relatively easier to compute than the two-fluid interface here because the vortex sheet strength, γ, can be determined dynamically without having to invert an integral equation iteratively. HLS also used adaptive mesh refinement which improved the local resolution near the singularity. Based on their numerical results and a local model, HLS have suggested that the interface minimum separation decreases to zero according to $(t_c - t)^{2/3}$. In Figure 3(b), we compare the computed minimum distance d_i, represented as circles in the plot, and a fitted curve of the form $d(t) = C(t_c - t)^{2/3}$ as a solid line. We can clearly see a good agreement between

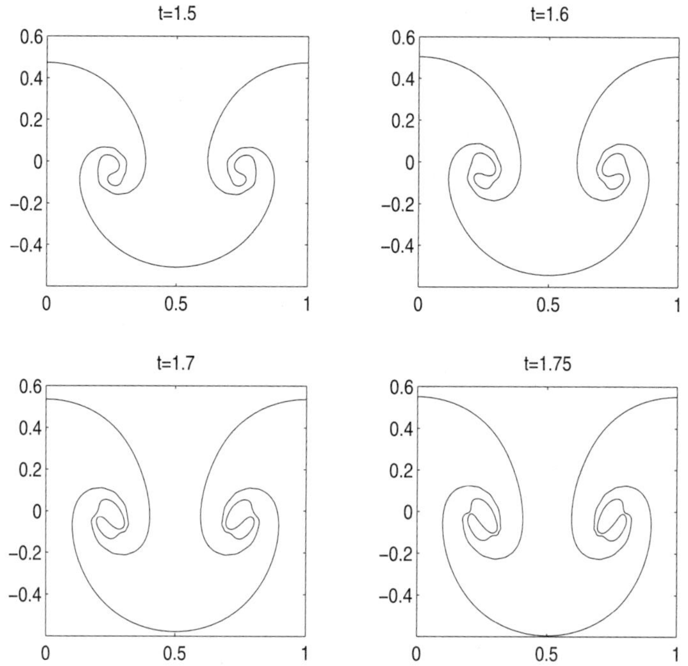

FIGURE 2. Unstably stratified flow: $A = -0.1$ and $S = 0.005$. Sequence of interface positions. $N = 2048$ and $\Delta t = 1.25 \times 10^{-4}$.

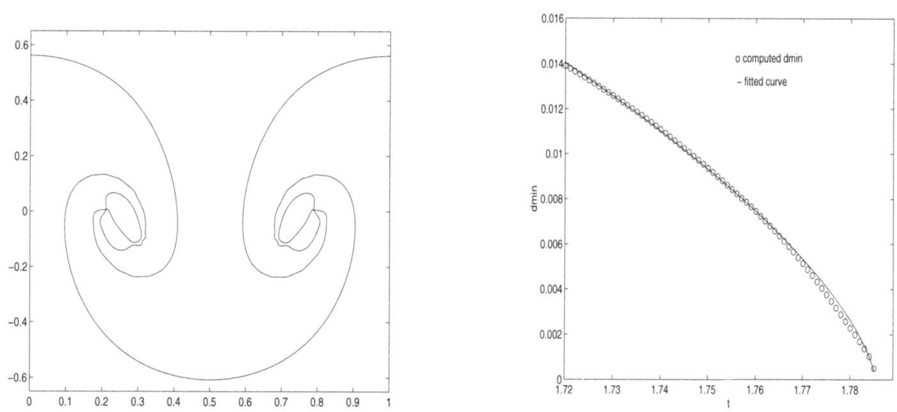

FIGURE 3. Unstably stratified flow: $A = -0.1$ and $S = 0.005$. (a) Interface position at $t = 1.785$ and (b) minimum separation against time. The circles represent the computed values and the solid line a fitted curve of the form $d(t) = C(t_c - t)^{2/3}$.

the values d_i and $d(t_i)$ up to $t = 1.76$. Beyond which, our calculations are not well resolved.

If the interface indeed goes through a self-intersection, this constitutes a singularity in the evolution. We believe that the collision of the interface is caused by the creation of intense localized jets produced by surface tension. Our numerical results indicate that the curvature blows up at the time of the collision and the interface forms corners. This collision may signal an imminent change in the topology of the flow. Such topological singularity is commonly observed in interfacial flows, e.g. in three-dimensional liquid jets. For three-dimensional liquid jets, the azimuthal component of surface tension is believed to play an essential role in driving the jet to form a pinching singularity. However, this component is completely absent in the two-dimensional unstably stratified interfacial flow we study here, making the appearance of such pinching singularities very surprising.

5.2. Dynamic generation of capillary waves. Unsteady motion of water waves is one of the most familiar examples of free surfaces in our everyday experience. It is also the source of many interesting nonlinear phenomena. An important event is the generation of capillary waves on the forward wave front [**DQPW99, EKT87, PLT93**]. Surface tension is believed to play a fundamental role in generating these capillary waves.

The phenomenon of capillary waves generated by steady steep gravity waves has been studied systematically by a number of researchers (see e.g. [**LH95, LH96, Cra70**]). The effects of surface tension on breaking waves have recently been addressed by Tulin [**Tul96**]. The careful simulations by Tulin show the appearance of a capillary jump, as defined by Longuet-Higgins [**LH96**], near the wave crest. Recent computations by the authors [**CH99**] using the stable spectrally accurate method discussed here, reveal a more detailed structure of the capillaries. As we show next, we are able to compute with high accuracy not only up to the appearance of the capillary jump as observed by Tulin, but also to follow the subsequent development of the small scale structure produced by the capillaries. We also examine the limiting behavior of the capillary waves as surface tension decreases to zero.

Our initial data is the following:

$$z(\alpha,0) = \alpha + 0.1i\cos(2\pi\alpha) \,, \tag{5.1}$$

$$\gamma(\alpha,0) = -1 + 0.1\sin(2\pi\alpha) \,. \tag{5.2}$$

The nonzero mean in γ corresponds to a nonzero velocity (equal to -1/2) at $y = -\infty$. This gives the wave an impulse to cause it to overturn. We take $g = 10$ and vary the surface tension coefficient S throughout the simulations. Our first simulation is for $S = 0.001$. In Figure 4, we present the interface profile at times $t = 0$, 0.30, and 0.45, computed using $N = 2048$ and $\Delta t = 5 \times 10^{-5}$. We can see that the interface becomes vertical at $t = 0.30$ and a capillary wave appears soon after the wave begins to overturn. Figure 5(a) gives a close-up of a neighborhood of the crest at time $t = 0.45$. The curvature in this region is plotted in Figure 5(b). We can clearly see a capillary wave train. We also examine closely the onset of this capillary wave and find that at $t = 0.26$, the curvature develops a spike right after the wave tip. Such spike is not present in the case $S = 0$ for which the curvature varies smoothly in the entire region. This spike subsequently develops into the capillary wave train.

A closer look at the interface profiles near $t = 0.45$ (Figure 5(a)) shows that the capillary trough gets narrower in time, suggesting a possible formation of a trapped bubble through self-intersection of the interface. We also find that both the curvature and its derivative have maxima (in absolute value) in a small neighborhood about the edge.

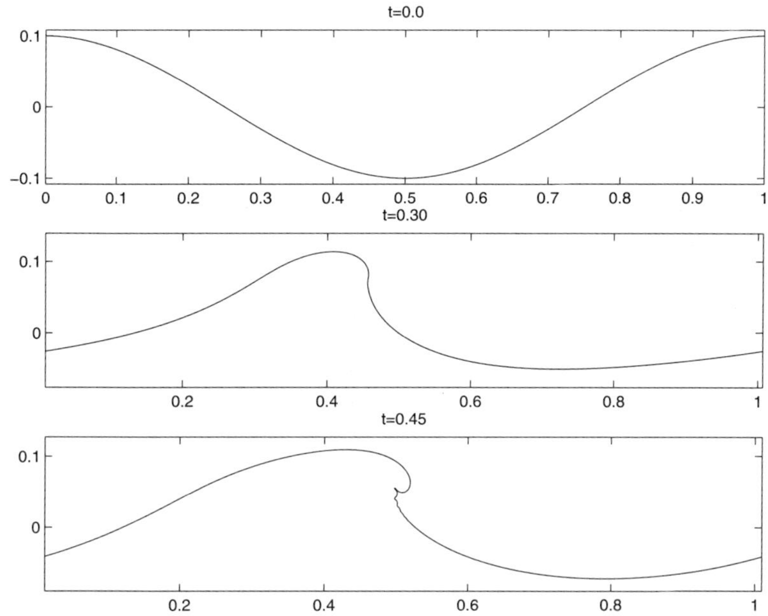

FIGURE 4. Breaking wave profiles for $S = 0.001$ at times $t = 0$, 0.30, and 0.45. $N = 2048$ and $\Delta t = 5 \times 10^{-5}$.

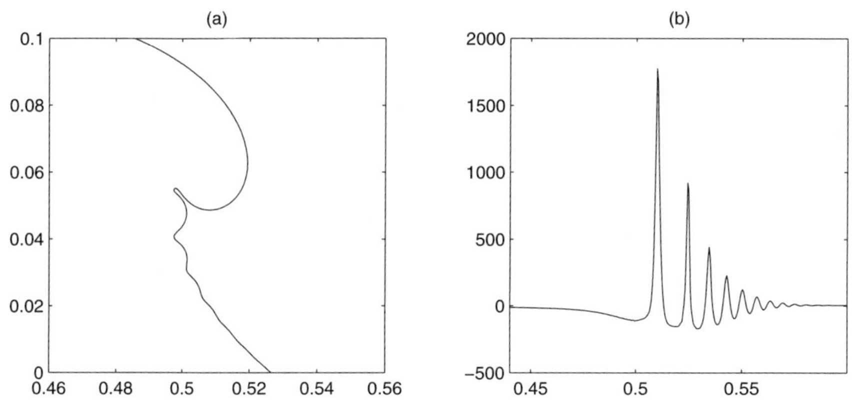

FIGURE 5. (a) Close-up of the interface at $t = 0.45$ and (b) the curvature plotted against the Lagrangian parameter α for $S = 0.001$, $N = 2048$, and $\Delta t = 5 \times 10^{-5}$.

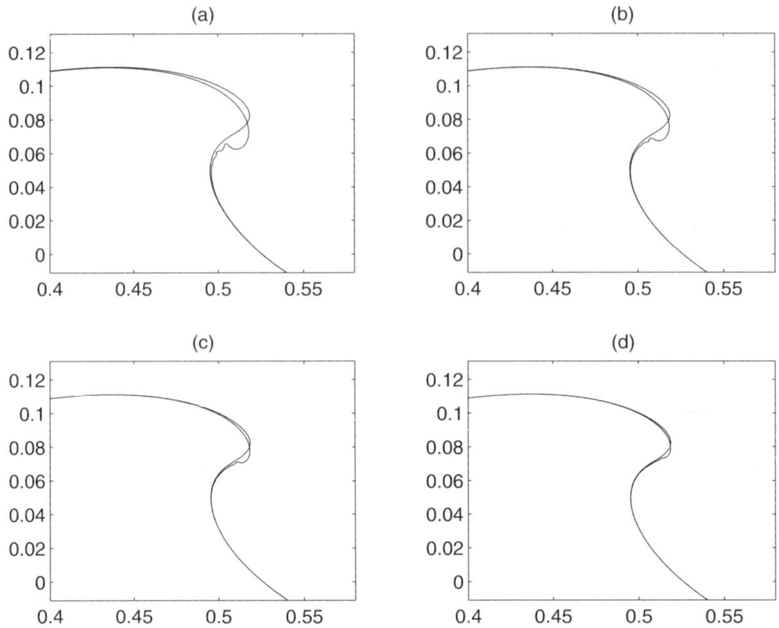

FIGURE 6. Comparison of the zero surface tension interface profile with the $S > 0$ profiles as S is decreased at $t = 0.45$. (a) $S = 2.5 \times 10^{-4}$. (b) $S = 1.25 \times 10^{-4}$. (c) $S = 6.25 \times 10^{-5}$. (d) $S = 3.125 \times 10^{-5}$.

We investigate how the capillary wavelength, denoted as λ_S, and its amplitude, denoted as a_S, scale with S as S is decreased. The capillary wavelength is defined as the the distance between the two largest values of $|\kappa|$. Although it is difficult to obtain an accurate scaling for λ_S and a_S, we find that for a fixed time, both quantities decrease nonlinearly as S decreases to zero. The scaling for λ_S is roughly $O(\sqrt{S})$. This seems to be in agreement with the experimental results of Duncan *et al.* [**DQPW99**]. The monotonely decreasing behavior of λ_S and a_S suggests that the limiting solutions converge strongly to the $S = 0$ profile for a fixed time. This is indeed the case. We see from Figure 6 that the interface profiles with decreasing surface tensions converge to the zero-surface-tension profile at the fixed time $t = 0.45$. This should be expected since the limiting zero-surface-tension water wave problem is well-posed, even after the wave overturns [**Wu97**]. Surface tension acts as a regular perturbation in this case. As we will see in the last subsection, this conclusion does not apply to the ill-posed Hele-Shaw problem.

5.3. Hele-Shaw flow with suction. Here, we would like to investigate the small surface tension limit of the Hele-Shaw flow with suction. This is a challenging problem because the zero-surface-tension problem is ill-posed. There are limited existence results of Hele-Shaw solutions in the presence of surface tension. Short-time existence has been obtained by Duchon-Robert [**DR84**]. There is also a long-time existence result but near equilibrium and without suction by Constantin and Pugh [**CP93**]. Tian [**Tia95**] shows that singularity formation is inevitable if the center of the viscous blob is not at the sink. However, the type of singularity is still unknown. On the other hand, there has been some attempt to perform a perturbation analysis of the small surface tension solution using the knowledge of the zero-surface-tension solutions. In [**HLO88**], Howison, Lacey, and Ockendon propose an asymptotic model by assuming the existence of a self-similar steady-state solution. Their analysis predicts that small surface tension could cause the interface in the neighborhood of the cusp to propagate rapidly as a narrow jet, analogous to a thin crack. They termed this the 'crack' model. However, since existence of this self-similar steady-state solution is unknown, the validity of this asymptotic result is still in doubt.

To investigate the asymptotic behavior as surface tension tends to zero, we consider an initially circular blob of viscous fluid surrounded by inviscid fluid, i.e., $A_\mu = 1$. The center of the initial blob is located at $(0, -0.1)$ and the sink is placed at the origin. All the computations presented here are performed in double precision. Due to the ill-posedness of the zero-surface-tension Hele-Shaw problem, it is necessary to use Krasny filtering [**Kra86**] to suppress the spurious growth of the round-off error. Krasny filtering is a simple but effective numerical filter which sets to zero all the Fourier modes of the solution whose magnitude are below certain filter level. In our computations we set this filter level to be 10^{-12}.

Due to the singular nature of the small surface tension solution, extremely high space and time resolutions are required to obtain accurate numerical solutions. Initially, we use $N = 2048$ for most of the computations. We double N as soon as the magnitude of the highest frequency mode of the tangent angle $\theta(\alpha, t)$ is greater than the filter level. Since the interface propagates rapidly to the sink, very small values of time step Δt are required to compute accurately the interface motion as it approaches the sink. At the latest stage of the motion, the number of grid points typically increases to $N = 8192$ or $N = 16384$ and $\Delta t = 2 \times 10^{-7}$. We remark that Kelly and Hinch [**KH97**] have studied numerically the effects of surface tension on the Hele-Shaw flow with suction. By using a different boundary integral method with a maximum of 200 grid points which are redistributed dynamically, their computations lack the high accuracy necessary to capture the interface behavior for sufficiently small surface tension. Nie and Tian [**NT98**] have performed a careful numerical study of this problem and found that that the flow develops a curvature singularity (in the form of a corner) when the interface reaches the sink. However, their study is restricted to large surface tension case, and does not address the limiting behavior of the small-surface-tension solution.

In the absence of surface tension, our computation shows that the interface develops a cusp singularity at $t_c = 0.2842$ and is located at $(0, 0.2305)$ for this particular initial data. With surface tension $S = 5 \times 10^{-5}$, we can compute beyond the cusp singularity of the zero-surface-tension solution. Figure 7 shows the interface shape for $S = 5 \times 10^{-5}$. We clearly see that the finger bulges and develops a well-defined neck before it forms a wedge. It is interesting to note that this neck appears at a height close to that of the zero-surface-tension cusp. It is reasonable to conjecture that the formation of the neck and the bulging of the finger are due to the influence of the zero-surface-tension singularity.

We consider now the limiting behavior of the interface past t_c as surface tension is decreased. We find with surprise that an asymptotic corner angle is selected in the limit as surface tension tends to zero when the finger tip is about to reach the sink. To validate this finding, we compare the interfaces for a set of decreasing values of surface tension. Since the velocity of the interface depends on surface tension, it

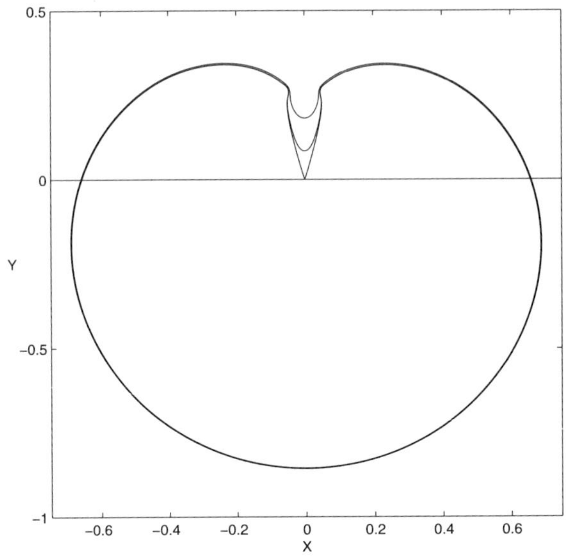

FIGURE 7. Evolution of the initially circular fluid blob past t_c, for $S = 5 \times 10^{-5}$. The interface is plotted at $t = 0.2880$, 0.2900, and 0.29181. $N = 16384$ and $\Delta t = 2 \times 10^{-7}$ for the last stage of the motion

S	Wedge angle	Variation
8×10^{-4}	0.67459	–
4×10^{-4}	0.65719	0.0174
2×10^{-4}	0.64399	0.0132
1×10^{-4}	0.63660	0.0074
5×10^{-5}	0.63359	0.0030

TABLE 1. The angle of the wedge (in radiants) for a decreasing set of surface tensions. The variation (third column) is the difference between consecutive angles, corresponding to surface tensions S and $2S$.

is more reasonable to compare the interfaces when their finger tips reach the same level above the sink. As surface tension is reduced, the finger tip reaches the given level faster. Figure 8 provides some indication of the asymptotic trend of the fingers as surface tension is successively halved from $S = 8 \times 10^{-4}$ to $S = 5 \times 10^{-5}$. We choose the fixed level to be $y = 0.01$ which is very close to the sink. Table 1 shows that the difference between consecutive angles (corresponding to surface tensions S and $2S$) decreases as surface tension is reduced. This strongly suggests that an asymptotic angle is selected for the wedge as it touches the sink.

But, what is the limiting behavior of the interface in the vicinity of the neck? Clearly, a potential neck singularity can only develop past t_c, the singularity time of the zero-surface-tension solution. Before t_c, we find that the small surface tension solution converges strongly to the zero-surface-tension solution. The interesting

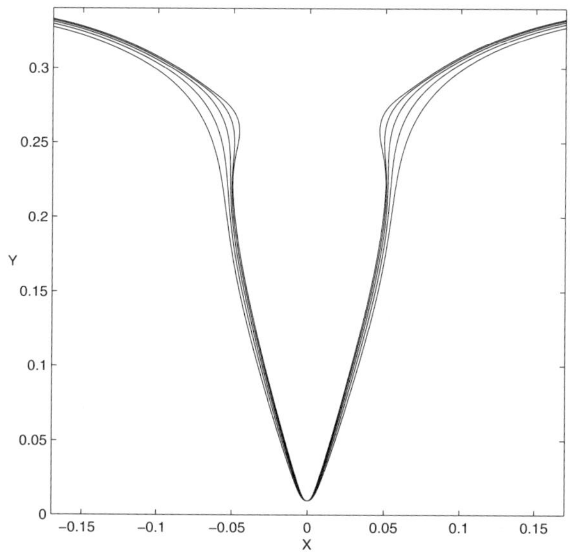

FIGURE 8. Comparison of the interface finger for a sequence of surface tensions. From the outer curve inwards, the fingers correspond to the surface tension values $S = 8\times10^{-4}$, 4×10^{-4}, 2×10^{-4}, 1×10^{-4}, and 5×10^{-5}. Each interface is plotted when the tip of the finger reaches the fixed level $y = 0.01$ at $x = 0$. $N = 16384$ and $\Delta t = 2 \times 10^{-7}$.

question is whether the small surface tension solution forms a neck singularity past t_c but before the time at which the limiting wedge reaches the sink. We find strong evidence that the small surface tension solution indeed forms a neck (corner) singularity in the limit as surface tension tends to zero. A pronounced neck with two corners can be clearly observed for $S = 5\times10^{-6}$ in Figure 9. A close inspection of its tangent angle and curvature also suggests formation of two corner singularities at the neck. Unfortunately, well-resolved computations for surface tensions smaller than this value become extremely difficult due to interface singular behavior and to growth of the round-off error noise.

The above numerical results provide us with a new understanding of the limiting behavior of small surface tension Hele-Shaw flow with suction. The two necking singularities past t_c are surprising in a sense. They seem to be responsible for the selection of limiting wedge angle. The fact that there is a well-defined limiting wedge angle defies our intuition (one may think that the wedge angle should decrease with surface tension). This also contradicts the assumption of the crack model in which a self-similar profile of the finger is assumed. In [**CHS99**], we also investigate the limiting behavior for two-phase Hele-Shaw flows, i.e. $A_\mu < 1$. We find that larger viscosity in the exterior fluid prevents the formation of the neck and leads to the development of thinner fingers. It is observed that the asymptotic wedge angle of the fingers decreases as the viscosity ratio is reduced, apparently towards the zero angle (cusp) of the zero-viscosity-ratio solution. For more discussions, we refer the reader to [**CHS99**]. It is natural to ask how the behavior of the interface would

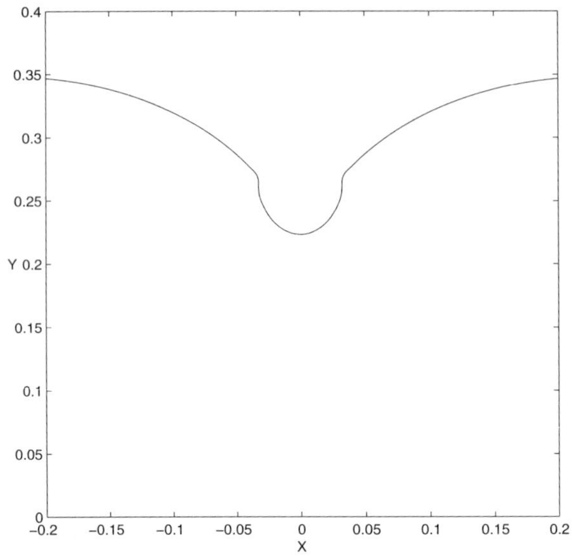

FIGURE 9. Close-up of the interface around $x = 0$ at $t = 0.2857$ (just past t_c) for $S = 5 \times 10^{-6}$. $N = 8192$ and $\Delta t = 5 \times 10^{-6}$.

be in the corresponding 3-D Darcy flow. Ceniceros and Si [**CS99**] address this problem numerically and find with surprise that the azimuthal component of the surface tension does not play an important role in this case. The limiting behavior is very similar to the 2-D flow studied here and no pinching singularity is found. This is in contrast with 3-D liquid jets. It is conceivable that the suction force plays a more dominant role than the azimuthal component of surface tension in this case.

5.4. Finger selection and noise effects in Hele-Shaw cells. Morphological instabilities are common to pattern formation problems. The non-equilibrium growth of crystals and directional solidification are some of the best known examples. Due to the underlying Mullins-Sekerka instability, very small perturbations caused by noise originate convoluted interfacial patterns when surface tension is small. The generic mechanisms in the formation of these complex patterns are present in the simpler problem of a Hele-Shaw interface. The purpose of this study is to investigate the impact of small surface tension on the pattern formation and selection of fingers.

We first formulate the Hele-Shaw problem in a way which is convenient for analysis. We rely on the conformal mapping technique on which most analyses are based. To this end, we represent the interface motion by a time-dependent conformal map $z(\zeta, t)$ that takes the interior of the unit circle in the ζ-plane onto the physical domain of the viscous fluid in the z-plane. The unit circle itself is mapped to the Hele-Shaw interface. Because of the presence of the source at the origin, z has the following form

$$z(\zeta, t) = \frac{a(t)}{\zeta} + f(\zeta, t), \tag{5.3}$$

where f is an analytic function inside the disk, and $a(t)$ is real and positive. We nondimensionalize all variables by taking the pumping rate $Q = 2\pi$ and $a(0) = 1$. Letting $Z(\alpha, t) = z(e^{i\alpha}, t)$, we obtain the following evolution equation (for a derivation see for example [**CK91**])

$$Z_t = F(Z_\alpha) + SG(Z_\alpha), \tag{5.4}$$

where

$$F(Z_\alpha) = -iZ_\alpha(I - iH)|Z_\alpha|^{-2}, \tag{5.5}$$

and

$$G(Z_\alpha) = iZ_\alpha(I - iH)[|Z_\alpha|^{-2}H\mathcal{K}_\alpha]. \tag{5.6}$$

Here $\mathcal{K}$ is the mean curvature. The subscript α means differentiation with respect to that variable. H is the periodic Hilbert transform defined as

$$Hf(\alpha) = \frac{1}{2\pi}\int_0^{2\pi} \cot\frac{1}{2}(\alpha - \alpha')f(\alpha')d\alpha'. \tag{5.7}$$

Many exact solutions have been found for the zero surface tension problem (see, e.g. [**Saf59, How86a, How86b, How86c, SB84**]). Here we are interested in comparing non-zero S solutions with those zero-surface-tension solutions for which $z_\zeta(\zeta, 0)$ has at least one zero in the extended complex domain $|\zeta| > 1$. For concreteness, we consider the following expanding bubble with three-fold symmetry:

$$z(\zeta, t) = \frac{A(t)}{\zeta}\left[1 + \frac{\zeta^3}{2\zeta_0^3(t)}\right], \tag{5.8}$$

where $A(t)$ and $\zeta_0(t)$ have analytical expressions (see [**SB84**]). The parameters $A(0)$ and $\zeta_0(0)$ are real numbers satisfying $A(0) > 0$ and $|\zeta_0(0)| > 1$. The derivative z_ζ has three zeros in $|\zeta| > 1$ which approach the unit disk and impinge on it at a time t_c. This particular set of initial data has been used by Siegel, Tanveer, and Dai [**STD96**]. In our study, we choose $A(0) = 1$ and $\zeta_0(0) = 1.2$. With this choice of initial data, $t_c = 0.3301$. At t_c, the zero-surface-tension solution forms three cusp singularities.

5.4.1. *The daughter singularity.* Before we proceed to our numerical study, we need to describe briefly the perturbation analysis of Tanveer [**Tan93**] and Siegel, Tanveer, and Dai [**STD96**]. The perturbation analysis requires to extend the equations of motion to $|\zeta| > 1$. First, equation (5.4) can be analytically continued into the domain $|\zeta| < 1$ by the use of Poisson integral formula. The analytic continuation to $|\zeta| > 1$ is achieved by contour deformation.

In the domain $|\zeta| > 1$, the extended evolution equation has the following form

$$z_t = q_1 z_\zeta + q_2 + S(q_3(z_\zeta)_{\zeta\zeta}^{-1/2} + r), \tag{5.9}$$

where q_1, q_2, and q_3 are analytic functions of ζ in $|\zeta| > 1$. The term $(z_\zeta)_{\zeta\zeta}^{-1/2}$ contains the leading order contribution from surface tension. The function r contains those surface tension terms which are less singular than $(z_\zeta)_{\zeta\zeta}^{-1/2}$ in the neighborhood of a zero of $z_\zeta(\zeta, 0)$. We do not write the explicit forms of these functions since they play no role in the leading order asymptotic analysis.

When performing a perturbation analysis, it is important to choose a function space and an associated norm so that the limiting problem is well-posed in this function space. Although the zero-surface-tension problem is ill-posed in the physical space, it becomes well-posed in the extended complex domain. Thus, it makes sense to perform a perturbation expansion in the extended $|\zeta| > 1$ domain for small S:

$$z(\zeta, t) = z_0(\zeta, t) + S z_1(\zeta, t) + \cdots, \tag{5.10}$$

where z_0 is the zero-surface-tension solution [**Tan93**]. It is easy to see that z_0 and z_1 satisfy the following equations:

$$z_{0t} = q_{10} z_{0\zeta} + q_{20} \tag{5.11}$$

$$z_{1t} = q_{10} z_{1\zeta} + q_{30} (z_{0\zeta})_{\zeta\zeta}^{-1/2} + \cdots, \tag{5.12}$$

where the subscript 0 on any q_i term denotes its evaluation using the corresponding $S = 0$ solution.

It is worth noting that the leading order surface tension term enters the z_1 equation as a known forcing function. This singular forcing function contains a branch point singularity when $z_{0\zeta} = 0$ in $|\zeta| > 1$. Thus, if the initially given z has a zero in z_ζ in the extended $|\zeta| > 1$ domain, i.e.

$$z_\zeta(\zeta, 0) \sim D(0)(\zeta - \zeta_0(0)) \quad \text{for } |\zeta_0(0)| > 1, \tag{5.13}$$

then the singular forcing term $(z_{0\zeta})_{\zeta\zeta}^{-1/2}$ will generate a new complex singularity at $t = 0^+$ at $\zeta = \zeta_0(0)$. The solution for z_1 will have the form

$$z_1 \sim A_0(t)(\zeta - \zeta_0(t))^{-3/2} + A_d(t)(\zeta - \zeta_d(t))^{-3/2} \tag{5.14}$$

with $A_0(0) + A_d(0) = 0$. The spawn daughter singularity $\zeta_d(t)$ moves according to the following equation

$$\dot{\zeta}_d(t) = -q_{10}(\zeta_d(t), t) \quad \text{with } \zeta_d(0) = \zeta_0(0), \tag{5.15}$$

whereas the motion of the zero $\zeta_0(t)$ is governed by

$$\dot{\zeta}_0(t) = -q_{10}(\zeta_0(t), t) - q_{20\zeta}(\zeta_0(t), t)[z_{\zeta\zeta}(\zeta_0(t), t)]^{-1}. \tag{5.16}$$

Thus, $\zeta_0(t)$ and $\zeta_d(t)$ travel at very different speeds. Depending on the initial data, it is possible that $\zeta_d(t)$ could reach the unit disk first when the zero is still far from it. For future reference, we define t_d to be the time at which the solution to (5.15) reaches the unit disk. For the data we consider here, we can solve the above ODE numerically to obtain $t_d = 0.0463$, which is much smaller than $t_c = 0.3301$. At t_d, ζ_d reaches the unit disk and the daughter singularity becomes a physical singularity for z_1. According to the prediction of Tanveer and Siegel, Tanveer, and Dai, the nonzero surface tension solution can deviate significantly from the corresponding zero-surface-tension solution no matter how small the surface tension is, even though the zero-surface-tension solution is still analytic at t_d. Further inner asymptotic analysis of [**Tan93**] suggests that the initial singularity $\zeta_d(0)$ is transformed by the presence of surface tension into cluster of an infinite number of $-4/3$ singularities. The cluster is localized around $\zeta_d(t)$ before it breaks up and disperse near the unit disk.

5.4.2. *An effective numerical method to study the limit.* The daughter singularity theory is quite surprising and is a bit counter intuitive at first glance. Numerical results in [**STD96**] give partial confirmation of this singular perturbation of small surface tension to Hele-Shaw flows. But due to the extreme noise sensitivity, Siegel, Tanveer, and Dai find it very difficult to compute for $S < 10^{-5}$, even if quadruple (128-bit) precision was used.

In [**CH00**], we develop a novel numerical method to overcome the difficulties mentioned above. Our method relies on three key observations. The first one is that noise can be significantly reduced by using a parametrization that yields a compact representation of the solution in Fourier space. Out of many choices (Lagrangian, equal-arclength, etc.), we find that the conformal map representation gives the most compact parametrization of the interface. This compact parametrization has also been used by Dai and Shelley [**DS93**] and by Siegel, Tanveer, and Dai [**STD96**]. With this parametrization, we can use effectively Krasny filtering [**Kra86**] to separate noise from the physical solution for the majority of high to intermediate modes. Moreover, we can also perform selective filtering based on the symmetry of the solution.

The second observation is to derive a scaled equation for the leading order perturbation. The reason for doing this is because the daughter singularity is only born at $t = 0^+$. For times of $O(\Delta t)$, the amplitude of the daughter singularity is $O(S\Delta t)$. Moreover, since the daughter singularity is still far from the unit disk, the interface is analytic and its Fourier modes decay exponentially, in a rate of $\frac{S\Delta t}{|\zeta_d|^k}$, where k is the wave number. Therefore, many digits are required to capture the presence of the singularity in the complex plane.

In order to effectively capture the complex singularity, we need to factor out the small surface tension coefficient in front of the complex singularity. One way to do this is to derive a scaled equation for the perturbed quantity. Specifically, we obtain an evolution equation for the scaled deviation, $\tilde{Z} = (Z - Z_0)/S$, where Z_0 is the known zero-surface-tension solution. Note that Z_0 satisfies the equation $Z_{0t} = F(Z_{0\alpha})$. Subtracting this equation from (5.4) we get

$$\tilde{Z}_t = \frac{1}{S}[F(Z_{0\alpha} + S\tilde{Z}_\alpha) - F(Z_{0\alpha})] + G(Z_{0\alpha} + S\tilde{Z}_\alpha), \tag{5.17}$$

which can be further simplified to [1]

$$\tilde{Z}_t = iZ_\alpha(I - iH)h(\alpha) + i\tilde{Z}_\alpha(I - iH)|Z_{0\alpha}|^{-2} + G(Z_{0\alpha} + S\tilde{Z}_\alpha), \tag{5.18}$$

where

$$h(\alpha) = \frac{2\mathrm{Re}(\tilde{Z}_\alpha/Z_{0\alpha}) + S|\tilde{Z}_\alpha/Z_{0\alpha}|^2}{|Z_\alpha|^2}. \tag{5.19}$$

Thus, the small surface tension coefficient is factored out from the scaled equation completely.

Our third observation is that we need to perform a resolution study in the precision digits and the filter level. To achieve this, we use very *high precision arithmetic* to solve numerically the scaled equation (5.17). High precision has the additional benefit of reducing the amplitude of the round-off noise. Variable high precision (up to 80 digits) was accomplished in this work with the use of the multi-precision package developed by Bailey [**Bai90**]. All these three ingredients are essential in capturing the singular behavior of the small surface tension solution.

5.4.3. *Numerical results.* We first present our computations performed for $S = 10^{-7}$, 10^{-8}, and 10^{-9} (the computations presented here actually used the original $\tilde{Z}$ equation presented in [**CH00**]). In our numerical computations, we start with $N = 4096$. N is doubled to 8196 when the magnitude of the highest Fourier mode exceeded the filter level. The time-step is then reduced by a factor of 8.

[1]This improved formula was pointed out to us by an anonymous referee.

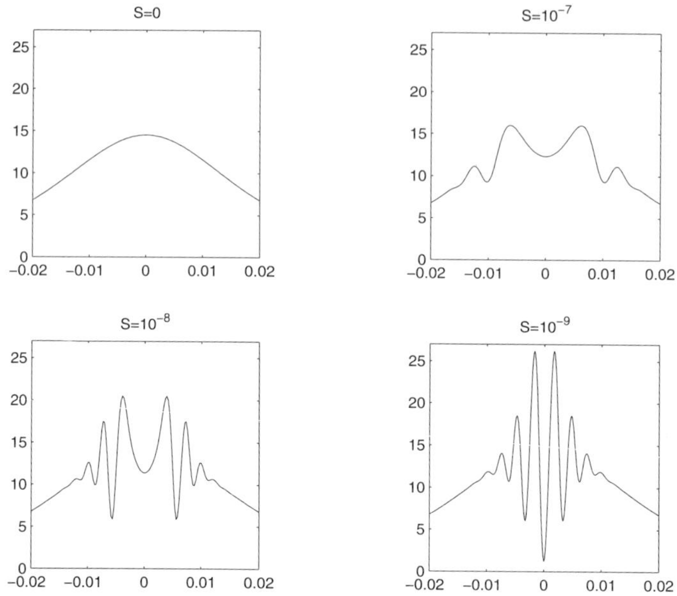

FIGURE 10. Curvature vs $\alpha/2\pi$ around one tip of the interface at $t = 0.048$ for $S = 0$, 10^{-7}, 10^{-8}, and 10^{-9}. $N = 8192$. Precision level 60. Filter level 10^{-56}.

We use the 60-digit computations for all the values of S which are found to be in good agreement with the 80-digit computations. The singular nature of the asymptotic solution can be seen clearly in the deviations of the $S > 0$ curvatures from the corresponding zero-surface-tension curvature. As observed in Figure 10, the deviations are significant and rather localized at the fixed time $t = 0.048$. Indeed, the smaller the surface tension the larger the deviation.

Next, we present first a sequence of pictures of the interface curvature near one tip at early times near t_d (Figure 11), now for a fixed surface tension. The computations correspond to $S = 10^{-8}$. One can already feel the effect of the daughter singularity in the physical domain at the time $t = 0.043$ around the interface tips. At $t = 0.043$, the tip curvature flattens in a very localized finite region. This is in agreement with the theoretical estimate of the daughter singularity impact time. After $t = 0.046$ on, the curvature develops large deviations from the zero-surface-tension curvature. The affected physical regions near the three tips also spread in time. This behavior is consistent with the asymptotic theory which implies that daughter singularity cluster will disperse once it gets sufficiently close to the unit disk. It is worth pointing out that the length scale decreases in time while the amplitude grows in time. During this process, the impact of the daughter singularity on the physical domain becomes more pronounced.

We would like to perform a long time calculation to identify the impact of the daughter singularity on the interface. To do this, we first compute up to t_d using 60-digit arithmetic. Shortly after that, we switch to double precision when the highest mode of $\tilde{Z}$ is about 10^{-15}. The number of points is doubled whenever the spectrum becomes under-resolved. Finally we stop the computation with $N =$

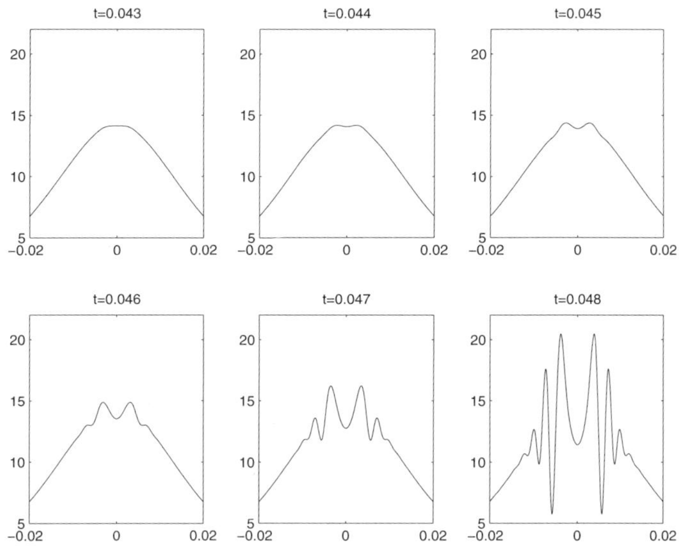

FIGURE 11. Tip curvature for $S = 10^{-8}$ at early times near $t_d = 0.0463$.

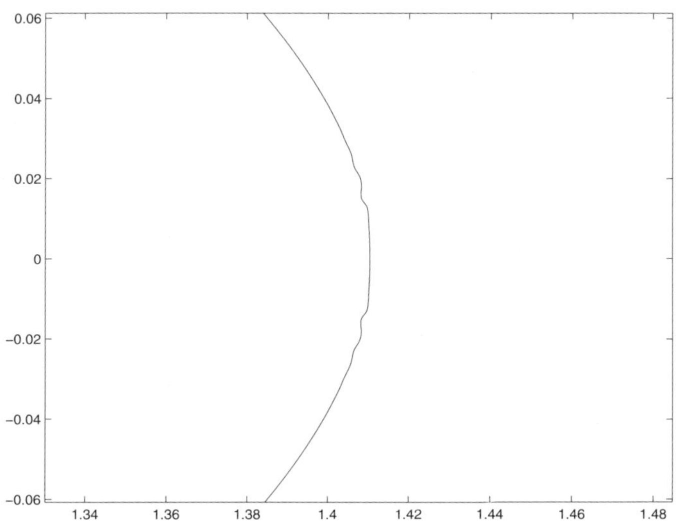

FIGURE 12. Close-up of the interface at $t = 0.0502$ for $S = 10^{-8}$.

32768 and $\Delta t = 10^{-7}$. In this calculation, we find that the curvature grows rapidly at subsequent times. In a very short time interval, the maximum of curvature has grown more than 10 times its value at $t = 0.049$, and the singular regions continue to spread in time. In Figure 12, we can see that the singular effects are already visible in the interface. The small indentations near the interface tip correspond to the largest curvature transition. These side indentations are a signature of surface tension.

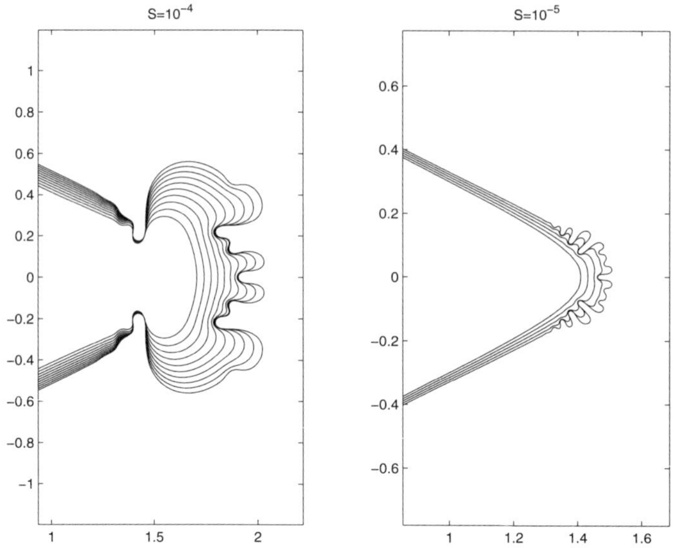

FIGURE 13. Close-ups of the Hele-Shaw interface at different times for $S = 10^{-4}$ (t = 0.23-0.50, with a 0.03 time difference between profiles) and $S = 10^{-5}$ (t = 0.05-0.09, with a 0.01 time difference between profiles). Computation performed with the equal-arclength method.

To better illustrate the impact of daughter singularity on the physical domain, we present two more computations with larger surface tensions using the the equal-arclength method of HLS [**HLS94**]. The method presented in this subsection cannot compute accurately for larger surface tension due to severe time-stepping limitations. Figure 13 shows the close-up of the evolution of the interface near one tip for $S = 10^{-4}$ and $S = 10^{-5}$ respectively. The symmetric indentations are clearly formed before the noise-induced tip splitting occurs. For $S = 10^{-5}$, the side indentations produced by surface tension begin to be visible in the second curve from left to right. Shortly after that the tip-splitting process and the finger formation begin. The above calculations also suggest that surface tension defines a length scale of the finger width. According to the asymptotic theory, this length scale is related to the minimum distance of the complex singularity to the unit disk.

References

[Bai90] D. H. Bailey, *MPFUN: A portable high perfomance multiprecision package*, Tech. Report RNR-90-022, NASA Ames Research Center, 1990.

[Bea98] J. T. Beale, *A convergent boundary integral method for 3-D water waves*, preprint, submitted to Math. Comput., 1998.

[BHL96] J. T. Beale, T. Y. Hou, and J. S. Lowengrub, *Convergence of a boundary integral method for water waves*, SIAM J. Numer. Anal. **33** (1996), no. 5, 1797–1843.

[BMO80] G. R. Baker, D. I. Meiron, and S. A. Orszag, *Vortex simulations of the Rayleigh-Taylor instability*, Phys. Fluids **23** (1980), 1485–1490.

[BMO82] G. R. Baker, D. I. Meiron, and S. A. Orszag, *Generalized vortex methods for free-surface flow problems*, J. Fluid Mech. **123** (1982), 477–501.

[CH98] H. D. Ceniceros and T. Y. Hou, *Convergence of a non-stiff boundary integral method for interfacial flows with surface tension*, Math. Comput. **67** (1998), 137–182.

[CH99] H. D. Ceniceros and T. Y. Hou, *Dynamic generation of capillary waves*, Phys. Fluids **11** (1999), no. 5, 1042–1050.

[CH00] H. D. Ceniceros and T. Y. Hou, *The singular perturbation of surface tension in Hele-Shaw flows*, J. Fluid Mech. **409** (2000), 251–272.

[CHS99] H. D. Ceniceros, T. Y. Hou, and H. Si, *Numerical study of Hele-Shaw flow with suction*, Phys. Fluids **11** (1999), no. 9, 2471–2486.

[CK91] P. Constantin and L. Kadanoff, *Dynamics of a complex interface*, Physica D **47** (1991), 450–460.

[CP93] P. Constantin and M. Pugh, *Global solutions for small data to the Hele-Shaw problem*, Nonlinearity **6** (1993), 393–415.

[Cra70] G. D. Crapper, *Non-linear capillary waves generated by steep gravity waves*, J. Fluid Mech. **40** (1970), 149–159.

[CS99] H. D. Ceniceros and H. Si, *Computation of axi-symmetric suction flow through porous media in the presence of surface tension*, preprint, submitted to J. Comput. Phys., 1999.

[Dol92] J. W. Dold, *An efficient surface-integral algorithm applied to unsteady gravity waves*, J. Comput. Phys. **103** (1992), 90–115.

[DQPW99] J. H. Duncan, H. Qiao, V. Philomin, and A. Wenz, *Gentle spilling breakers:crest profile evolution*, J. Fluid Mech. **379** (1999), 191–222.

[DR81] P. G. Drazin and W. H. Reid, *Hydrodynamic stability*, Cambridge monographs on mechanics and applied mathematics, Cambridge University Press, New York, 1981.

[DR84] J. Duchon and R. Robert, *Évolution d'une interface par capillarité et diffusion de volume I. Existence locale en temps*, Ann. Inst. Henri Poincaré **1** (1984), no. 5, 361–378.

[DS93] W.-S. Dai and M. J. Shelley, *A numerical study of the effect of surface tension and noise on an expanding Hele-Shaw bubble*, Phys. Fluids A **5** (1993), no. 9, 2131–2146.

[EKT87] N. Ebuchi, H. Kawamure, and Y. Toba, *Fine structure of laboratory wind-waves surfaces studied using an optical method*, Boundary Layer Met. **39** (1987), 133–151.

[HKS98] T. Y. Hou, I. Klapper, and H. Si, *Removing the stiffness of curvature in computing 3-D filaments*, J. Comput. Phys. **143** (1998), 628–664.

[HLO88] S.D. Howison, A.A. Lacey, and J.R. Ockendon, *Hele-Shaw free-boundary problems with suction*, Q. JI Mech. Appl. Math. **41** (1988), 183–193.

[HLS94] T. Y. Hou, J. S. Lowengrub, and M. J. Shelley, *Removing the stiffness from interfacial flows with surface tension*, J. Comput. Phys. **114** (1994), 312–338.

[HLS97] T. Y. Hou, J. S. Lowengrub, and M. J. Shelley, *The long-time motion of vortex sheets with surface tension*, Phys. Fluids **9** (1997), no. 7, 1933–1954.

[How86a] S. D. Howison, *Bubble growth in porous media and Hele-Shaw cells*, Proc. R. Soc. Edinb. A **102** (1986), 141–148.

[How86b] S. D. Howison, *Cusp development in Hele-Shaw flow with a free surface*, SIAM J. Appl. Math. **46** (1986), 20–26.

[How86c] S. D. Howison, *Fingering in Hele-Shaw cells*, J. Fluid Mech. **167** (1986), 439–435.

[HZa] T. Y. Hou and P. Zhang, *A new stabilizing technique for boundary integral methods*, Math. Comput., to appear.

[HZb] T. Y. Hou and P. Zhang, *Stability of a boundary integral method for 3-D water waves*, preprint, submitted to SIAM J. Numer. Anal.

[KH97] E.D. Kelly and E.J. Hinch, *Numerical simulations of sink flow in the Hele-Shaw cell with small surface tension*, Euro. Jnl. of Applied Maths **8(6)** (1997), 533–550.

[Kra86] R. Krasny, *A study of singularity formation in a vortex sheet by the point vortex approximation*, J. Fluid Mech. **167** (1986), 65–93.

[LH95] M. S. Longuet-Higgins, *Parasitic capillary waves: a direct calculation*, J. Fluid Mech. **301** (1995), 79–107.

[LH96] M. S. Longuet-Higgins, *Capillary jumps on deep water*, Journal of Physical Oceanography **26** (1996), 1957–1965.

[LHC76] M. S. Longuet-Higgins and E. D. Cokelet, *The deformation of steep surface waves on water I. A numerical method of computation*, Proc. R. Soc. Lond. A. **350** (1976), 1–26.

[NT98] Q. Nie and F. R. Tian, *Singularities in Hele-Shaw flows*, SIAM J. Appl. Math **58(1)** (1998), 34–54.

[Pat81] L. Paterson, *Radial fingering in a Hele-Shaw cell*, J. Fluid Mech. **113** (1981), 513–529.

[Pel88] P. Pelcé, *Dynamics of curved fronts*, Academic Press, Inc., San Diego, 1988.

[PLT93] M. Perlin, H. Lin, and C.-L. Ting, *On parasitic capillary waves generated by steep gravity waves: an experimental investigation with spatial and temporal measurements*, J. Fluid Mech. **255** (1993), 597–620.

[Pul82] D. I. Pullin, *Numerical studies of surface-tension effects in nonlinear Kelvin-Helmholtz and Rayleigh-Taylor instability*, J. Fluid Mech. **119** (1982), 507–532.

[Ric72] S. Richardson, *Hele-Shaw flows with a free boundary produced by the injection of fluid into a narrow channel.*, J. Fluid Mech. **56** (1972), 609–618.

[Rob83] A. J. Roberts, *A stable and accurate numerical method to calculate the motion of a sharp interface between fluids*, IMA J. Appl. Math. **31** (1983), 13–35.

[Ros32] L. Rosenhead, *The point vortex approximation of a vortex sheet*, Proc. Roy. Soc. London Ser. A **134** (1932), 170–192.

[RS88] R.H. Rangel and W.A. Sirignano, *Nonlinear growth of Kelvin-Helmholtz instability: Effect of surface tension and density ratio*, Phys. Fluids **31** (1988), no. 7, 1845–1855.

[Saf59] P. G. Saffman, *Exact solutions for the growth of fingers from a flat interface between two fluids in a porous medium*, Q. J. Mech. Appl. Maths **12** (1959), 146–150.

[SB84] B. Shraiman and D. Bensimon, *Singularities in nonlocal interface dynamics*, Phys. Rev. A. **30** (1984), no. 5, 2840–2842.

[She92] M. J. Shelley, *A study of singularity formation in vortex sheet motion by a spectrally accurate vortex method*, J. Fluid Mech. **244** (1992), 493–526.

[ST58] P. G. Saffman and G. I. Taylor, *The penetration of fluid into a porous medium or Hele-Shaw cell*, Proc. Roy. Soc. A. **245** (1958), 312–329.

[STD96] M. Siegel, S. Tanveer, and W.-S. Dai, *Singular effects of surface tension in evolving Hele-Shaw flows*, J. Fluid Mech. **323** (1996), 201–236.

[TA83] G. Tryggvason and H. Aref, *Numerical experiments in Hele-Shaw flows with a sharp interface*, J. Fluid Mech. **136** (1983), 1–30.

[Tan93] S. Tanveer, *Evolution of Hele-Shaw interface for small surface tension*, Phil. Trans. R. Soc. Lond. A **343** (1993), 155–204.

[Tia95] F.R. Tian, *On the breakdown of Hele-Shaw solutions with non-zero surface tension*, J. Nonlinear Sci. **5** (1995), 479–484.

[Tul96] M. P. Tulin, *Breaking of ocean waves and downshifting*, Waves and Nonlinear Processes in Hydrodynamics, Kluwer Academic Publishers, 1996, pp. 177–190.

[Wu97] S. J. Wu, *Well-posedness in Sobolev spaces of the full water wave problem in 2-D*, Invent. Math. **130** (1997), 39–72.

Applied Mathematics, California Institute of Technology, Pasadena, CA 91125
E-mail address: `hdc@ama.caltech.edu`

Applied Mathematics, California Institute of Technology, Pasadena, CA 91125
E-mail address: `hou@ama.caltech.edu`

AMS/IP Studies in Advanced Mathematics
Volume 20, 2001

Selected Topics in Moduli spaces

Jun Li

The development of low dimensional topology and mathematical physics have initiated a major research effort in studying moduli spaces in algebraic geometry in the last decade. The moduli spaces are those of auxiliary algebraic objects associated to an algebraic variety. The two mostly studied examples are the moduli spaces of stable vector bundles (or stable sheaves) over algebraic varieties and the moduli spaces of stable maps from curves to algebraic varieties.

It is safe to say that the first intensively studied moduli space is the moduli of stable curves. Another example is the moduli space of stable vector bundles over a smooth curve of genus $g \geq 2$. They share the common feature that the deformations of stable curves and of stable vector bundles over a smooth curves are unobstructed. As the result, the moduli spaces of stale curves and of the stable vector bundles have at most quotient singularities.

In 70's, the moduli of stable sheaves over a smooth algebraic variety was constructed. The study of this moduli space took off after the discovery of Donaldson's polynomial invariants. The moduli of stable sheaves over (a general) surface differs from that over curves in that the former is usually singular due to the presence of non-trivial obstructions to their deformations. Thus studying its obstruction theory becomes an integral part in studying such moduli space.

The recent class of moduli spaces under much study is the moduli of stable morphisms from curves to algebraic varieties. This grew out of the interest in enumerating curves in Calabi-Yau manifolds, which plays an essential role in the recent development of Super-String theories. Similar to the moduli of vector bundles over surfaces, this class of moduli spaces is also obstructed in general.

Looking forward, studying moduli spaces will continue to be a main research direction in algebraic geometry, partly because of the major research effort in Super-Sting theories. As expected, many of the interesting moduli spaces are obstructed. Thus developing tools to study such moduli spaces becomes a challenging task. In this note, we will review the development of many advanced techniques in studying moduli spaces with obstructed deformations. As is usually the case, the obstruction will complicate the study of the moduli spaces. We will show that with newly developed techniques, we can still attack such moduli spaces with success. We will do this by studying the moduli spaces of vector bundles over surfaces and the

Supported partially by NSF grant and Terman fellowship.

moduli spaces of stable morphisms. Some interesting future research directions are discussed in the end of this note.

The organization of this note is as follows: In the first two sections, we will briefly discuss the construction of the moduli spaces and the deformation theory that is the main tool in studying the singularities of moduli spaces. In the next three sections, some aspects of the moduli of stable sheaves over surfaces are reviewed. Most of the topics included are related to the asymptotic properties of the moduli spaces. They are chosen because obstruction theory played essential role in their studies. In the last three sections, we will discuss the construction of virtual cycles of moduli spaces. This is central to the study of Donaldson type invariants of moduli spaces. Such cycles were constructed by G. Tian and the author [**LT1**] (c.f. [**Beh, BFa**] for an alternative construction).

Many recent developments in the area of studying moduli spaces are not mentioned here. Understandably it will take many more pages to discuss a few more of them. The selection of topics covered in this note reflects author's the research interest. A list of references is given at the end. They represent the progress in this field in different stages. This is a partial list of the papers that are directly related to the topics discussed in this note.

In this note, we will work with the complex number field $k = \mathbb{C}$. Though most of the result works over any algebraically closed field of characteristic 0.

1. Construction of moduli spaces

We begin with our discussion with the moduli spaces of stable sheaves over smooth algebraic surfaces.

Let X be a smooth algebraic surface with a polarization H. We fix a choice of $(r, I, d) \in \mathbb{Z}^+ \times \operatorname{Pic}(X) \times \mathbb{Z}$. We consider the moduli functor

$$\mathfrak{F} : (\mathbb{C}-\text{Schemes}) \longrightarrow (\text{Set})^0$$

associating to any scheme S to the set of equivalence classes of flat families of (Gieseker)-stable[1] sheaves $\mathcal{E}$ over $X \times S$ such that $\operatorname{rank}\mathcal{E} = r$, $\det \mathcal{E} \cong p_X^*\mathcal{O}(I) \otimes p_S^*\mathcal{L}$ and $p_{S*}c_2(\mathcal{E}) = d[S] \in A_*S$. Here $\mathcal{L}$ is an invertible sheaf of $\mathcal{O}_S$-modules. Two such families are equivalent if they differ by tensoring a pullback of an invertible sheaf of $\mathcal{O}_S$-modules. The first step to study such moduli problem is to investigate if the moduli functor $\mathfrak{F}$ is represented by a scheme. We say $\mathfrak{F}$ is represented by a scheme M if the functor $\mathfrak{F}$ is isomorphic to $\operatorname{Mor}(-, M)$. In case $\mathfrak{F}$ is represented by M, then there is a universal family $\mathcal{E}$ over $X \times M$ so that to each scheme S and family $\mathcal{F} \in \mathfrak{F}(S)$ there is a morphism $\rho : S \to M$ such that $(i_X \times \rho)^*\mathcal{E}$ is equivalent to $\mathcal{F}$. Such moduli space will be called *fine moduli spaces*.

In many instances, the construction of moduli spaces relies on Grothendieck's Quot-scheme. By a result of Kleiman, the set $\mathfrak{F}(\operatorname{Spec} k)$ (of stable sheaves of prescribed topological data) is bounded. Here by a set is bounded we mean it is parameterized by finitely many schemes of finite type. Because of this, for sufficiently large n any $\mathcal{E} \in \mathfrak{F}(\operatorname{Spec} k)$ has $H^i(\mathcal{E}(n)) = 0$ for $i > 0$ and $\mathcal{E}(n)$ is generated by global sections. We fix such an n once and for all and let $m = \chi_{\mathcal{E}}(n) = h^0(\mathcal{E}(n))$. We then form the Grothendieck's Quot-scheme $\mathbf{Q}$ parameterizing all quotient sheaves

[1] For any sheaf $\mathcal{E}$ of $\mathcal{O}_X$-modules, we define its Poincare polynomial to be the polynomial $\chi_{\mathcal{E}}$ in n so that $\chi_{\mathcal{E}}(n) = h^0(\mathcal{E}(n))$ for $n \gg 0$. Here we abbreviate $\mathcal{E} \otimes \mathcal{O}(nH)$ to $\mathcal{E}(n)$. We let $p_{\mathcal{E}} = \chi_{\mathcal{E}}/r$. We say $\mathcal{E}$ is stable (resp. semistable) if to each proper subsheaf $\mathcal{F} \subset \mathcal{E}$ with $\operatorname{rank}\mathcal{F} > 0$ we have $p_{\mathcal{F}}(n) < p_{\mathcal{E}}(n)$ (resp. $\leq$) for $n \gg 0$.

$\mathcal{O}(-nH)^{\oplus m} \xrightarrow{\tau} \mathcal{E}$ such that $\operatorname{rank} \mathcal{E} = r$, $\det \mathcal{E} = I$ and $c_2(\mathcal{E}) = d$. Since being stable (or semistable) is an open condition, the subset of stable (or semistable) quotient sheaves[2] is an open subset of $\mathbf{Q}$. We denote them by $\mathbf{Q}^s$ and $\mathbf{Q}^{ss}$ respectively. Obviously, the reductive group $SL(m)$, which acts on $\mathcal{O}(-nH)^{\oplus m}$ in the obvious way, acts on $\mathbf{Q}$. Clearly, all $g = c\mathbf{1}_m$ in $SL(m)$ acts trivially on $\mathbf{Q}$ and hence the $SL(m)$ action descends to a $PGL(m)$ action. For simplicity, we let $G = PGL(m)$. Both $\mathbf{Q}^s$ and $\mathbf{Q}^{ss}$ are G-invariant open subschemes of $\mathbf{Q}$.

THEOREM 1.1. *The G-scheme $\mathbf{Q}^s$ admits a geometric quotient $\mathbf{Q}^s/G$. The G-scheme $\mathbf{Q}^{ss}$ admits a GIT quotient $\mathbf{Q}^{ss}//G$.*

The Theorem was proved by Gieseker [**Gi1**] for surfaces and by Maruyama [**Mar**] for higher dimensional varieties. Simpson has generalized the notion of stable sheaves and constructed their moduli spaces using GIT theory [**Sim**].

Following the convention, we will denote by $\mathfrak{M}_d(r, I)$ the moduli space of semistable sheaves of topological type

$$\Delta = (r, I, d),$$

which is $\mathbf{Q}^{ss}//G$, and denote by $\mathfrak{M}_d(r, I)^s$ the moduli space of stable sheaves, which is $\mathbf{Q}^s/G$. When the choice of (r, I) is understood, we will abbreviate $\mathfrak{M}_d(r, I)$ to $\mathfrak{M}_d$. In general, $\mathfrak{M}_d(r, I)$ is a coarse moduli scheme (see [**FM**]). Part of the definition of coarse moduli scheme is that whenever there is a family of semistable sheaves $\mathcal{E}$ on $X \times S$ over S with topological data Δ, then there is a natural morphism $\phi: S \to \mathfrak{M}_d(r, I)$ so that for any point $\xi \in S$ the image $\phi(\xi)$ is the point parameterizing the S-equivalence class[3] of the sheaf $\mathcal{E}|_{X \times \{\xi\}}$.

For the topological data Δ, there is an associated Poincare polynomial

$$\chi(n) = \frac{r}{2} n^2 H^2 + (I \cdot H - \frac{r}{2} K_X \cdot H) n + (\frac{1}{2} I^2 - d - \frac{1}{2} I \cdot K_X) + r\chi(\mathcal{O}_X) \tag{1.1}$$

which can be written as $\sum_{i \geq 0} a_i \binom{n+i}{i}$ for some integers a_i. Maruyama showed that if $g.c.d(a_0 : i \geq 0) = 1$, then $\mathfrak{M}_d(r, I)$ admits a universal family. In this case $\mathfrak{M}_d(r, I)$ is a fine moduli space [**Mar**]. Note that in this case $\mathfrak{M}_d(r, I) = \mathfrak{M}_d(r, I)^s$. In general, $\mathfrak{M}_d(r, I)$ may not admit universal families. However, since the automorphism groups of all stable sheaves (of fixed determinant) are identical, which is $SL(r)$, locally (in analytic topology or étale topology) $\mathfrak{M}_d(r, I)^s$ does admit universal family. Hence in studying local properties of $\mathfrak{M}_d(r, I)^s$ we can treat it as if it admits a universal family.

We now come to the second class of moduli spaces: The moduli spaces of stable morphisms to a projective scheme X. This is a natural generalization of Deligne-Mumford stable curves to the setting of morphisms. Recall that an n-pointed connected curve is stable if it is nodal and if there are no non-trivial vector fields of C that vanish along the nodal points and the marked points of C. The notion of stable morphism was first introduced by Kontsevich [**Kon**]. Let C be a pointed connected nodal curve and $f : C \to X$ a morphism. We call f with C a stable morphism if there are no non-trivial vector fields of C that vanishes along the nodal points and the marked points of C and are annihilated by $f_*: T_C \to T_X$. Given integers g and n and a class $A \in H_2(X, \mathbb{Z})$, we intend to form a moduli

[2] We call $\mathcal{O}(-nH)^{\oplus m} \to \mathcal{E}$ a stable quotient sheaf if $\mathcal{E}$ is stable.

[3] Any torsion free sheaf $\mathcal{E}$ admits a filtration $0 = \mathcal{E}_1 \subset \cdots \subset \mathcal{E}_k = \mathcal{E}$ so that $\mathcal{E}_{i+1}/\mathcal{E}_i$ are stable. Then we denote by $\mathrm{pr}\mathcal{E} = \oplus \mathcal{E}_{i+1}/\mathcal{E}_i$. Two sheaves $\mathcal{E}$ and $\mathcal{F}$ are S-equivalent if $\mathrm{pr}(\mathcal{E}) \cong \mathrm{pr}(\mathcal{F})$.

space of stable morphisms from connected n-pointed genus g curves C to X so that $f_*([C]) = A$. Here we say two stable morphisms $f: C \to X$ and $f': C' \to X$ are equivalent if there is an isomorphism $\phi: C \to C'$ as pointed curves so that $f = f' \circ \phi$. When $f = f'$, then such ϕ forms a group, called the automorphism group of f and denoted by $\mathrm{Aut}(f)$. Note that unlike the case of stable sheaves, the automorphism groups $\mathrm{Aut}(f)$ may vary with f. With this said, we see that the natural framework to study this moduli space is the notion of Delign-Mumford stacks. For the definition of DM-stacks please see [**DMu, Vis**]. In this note, we will treat the moduli space as a stack. For those who are more familiar with orbifolds, he can simply view it as a $\mathbb{Q}$-scheme (as singular algebraic orbifolds), which is a scheme with a collection of charts keeping the orbifold information, without losing any generality.

THEOREM 1.2. *Let X be a projective scheme and let (n, g, A) be a set of topological data as before. Then the moduli functor of stable morphisms from n-pointed genus g curves to X of fundamental class A is represented by an algebraic stack* $\mathfrak{M}_{g,k}(X, A)$.

I will comment on the construction of moduli spaces before I close this section. The moduli space of stable curves $\mathfrak{M}_g$ and of stable vector bundles over curves $\mathfrak{M}_d(r, I)$ can be constructed using the GIT theory developed by Mumford [**Mum, Gi3**]. If GIT theory can be worked out, it almost always provides the best possible setting to studying the moduli space. However, technically checking the stability condition can be very involved, if not impossible. Two examples are Gieseker's and Vieweg's work on moduli of varieties [**Gi4, Vie**]. An alternative approach is based on J. Kollár's work on positive line bundles. In this approach, one first constructs the moduli space as an algebraic space. One then construct certain universally constructed line bundles on this moduli space and proves that it is ample, thus showing that it is indeed projective. For instance, one can construct a coarse moduli space of stable morphisms following this strategy. This has been worked out by [**Ale**]. This moduli space also exists as a Deligne-Mumford stack. The notion of stack is more powerful in certain instance in that it admits a universal family [**FP**].

2. Local structure of moduli spaces

We shall investigate the local structure of the moduli spaces of stable sheaves and of stable morphisms. Following Grothendieck, this can be studied by the formal deformation theory. That is, the functor $\mathfrak{D}ef$ that assigns to an Artin local k-algebra R the set of isomorphism classes of flat families of studied objects over $\mathrm{Spec}\, R$. We say a complete local k-algebra $\hat{O}$ *pro-represents* the functor $\mathfrak{D}ef$ of formal deformation if the functor $\mathfrak{D}ef$ is isomorphic to the functor $R \mapsto \mathrm{Hom}_k(\hat{O}, R)$. If $\hat{O}$ is isomorphic to $k[[z_1, \cdots, z_n]]$, then the functor $\mathfrak{D}ef$ is said to be unobstructed.

Since the moduli spaces we are interested are obstructed, we need to work with their obstruction theories. Let $\mathfrak{D}ef$ be a formal deformation functor. In the case of deformation of a sheaf $\mathcal{E}_0$ of $\mathcal{O}_X$-modules of fixed determinant I, $\mathfrak{D}ef(R)$ is the set of all isomorphism classes of pairs $(\mathcal{E}, \rho)$, where $\mathcal{E}$ are sheaves of $\mathcal{O}_{X \times \mathrm{Spec}\, R}$ modules, flat over R with determinant pull back of I, and ρ is an isomorphism $\mathcal{E}|_{X \times \{0\}} \cong \mathcal{E}_0$. The formal deformation functor for stable morphisms is similar.

DEFINITION 2.1. *The functor* $\mathfrak{Def}$ *is said to have an obstruction theory with coefficients in* V, *where* V *is a finite dimensional* k*-linear space, if the following holds: There is a function* Ob *that associates to any triple* (A, I, ξ), *where* A *is an Artin* k*-algebra,* $I \subset A$ *is an ideal annihilated by the maximal ideal* $\mathfrak{m} \subset A$ *and* $\xi \in \mathfrak{Def}(A/I)$, *the value* $\mathrm{Ob}(A, I, \xi) \in V \otimes_k I$ *of which the following hold:*
(1) $\mathrm{Ob}(A, I, \xi) = 0$ *if and only if there is a* $\bar{\xi} \in \mathfrak{Def}(A)$ *that restricts to* ξ, *i.e. the image of* $\bar{\xi}$ *under the natural transformation* $\mathfrak{Def}(A) \to \mathfrak{Def}(A/I)$ *is* ξ.
(2) This function $\mathrm{Ob}(A, I, \xi)$ *satisfies the natural base change property.*

The exact statement of the base change property can be found in most of the reference concerning obstruction theory (cf. [**Ill, Lau, Li2, LT1**]).

An easy example of this is when the deformation functor is the deformation functor of a point in a scheme.

EXAMPLE 2.2. *Let* W *be a smooth scheme and* $X \subset W$ *be a subscheme defined by the vanishing of a section* s *of a vector bundle* E. *Let* $p \in X$. *We define the deformation functor of* p *in* X *to be*

$$\mathfrak{Def}_{\hat{p}}(R) = \mathrm{Mor}((\mathrm{Spec}\, R, 0), (X, p)).$$

Here the right hand side is the set of all morphisms $\mathrm{Spec}\, R \to X$ *that map* $0 \in \mathrm{Spec}\, R$ *to* p. *We let*

$$V = \mathrm{Coker}\{ds(p) : T_p W \longrightarrow E|_p\}. \tag{2.1}$$

Then s *induces an obstruction theory of* $\mathfrak{Def}_{\hat{p}}$ *with coefficients in* V.

It is easy to construct this obstruction theory. Let (A, I, ξ) be a triple given in Definition 2.1. Since W is smooth, $\xi : \mathrm{Spec}\, A/I \to X$ extends to $\eta : \mathrm{Spec}\, A \to W$. Since $s \circ \eta(\mathrm{Spec}\, A/I) = 0$ and $\mathfrak{m} \cdot I = 0$, $s \circ \eta$ induces an element in $E|_p \otimes_k I$. We denote by

$$\mathrm{Ob}(A, I, \xi) \in V \otimes_k I$$

its image in $V \otimes_k I$ under the quotient map $E|_p \otimes_k I \to V \otimes_k I$. It is straightforward to show that this assignment is an obstruction theory to the functor $\mathfrak{Def}_{\hat{p}}$.

Now let $\mathfrak{Def}$ be any formal deformation functor. Let $R_2 = \mathrm{Spec}\, k[t]/(t^2)$. The set $\mathfrak{Def}(R_2)$ admits a natural k-vector space structure. We call it the tangent space of $\mathfrak{Def}$ and denote it by T_1. Assume $\mathfrak{Def}$ admits an obstruction theory taking values in T_2. For simplicity, we denote by $\mathrm{Sym}^\bullet(V)$ the space $\oplus_{n=0}^\infty \mathrm{Sym}^n(V)$.

LEMMA 2.3. *Let* $\mathfrak{Def}$ *be a formal deformation functor as before with tangent and obstruction spaces* T_1 *and* T_2 *respectively. Suppose both* T_1 *and* T_2 *have finite* k*-dimensions. Then there is a homomorphism*

$$\mathcal{K} : \mathrm{Sym}^\bullet(T_2^\vee) \to \mathrm{Sym}^\bullet(T_1^\vee)$$

so that $\mathfrak{Def}$ *is represented by* $\mathrm{Sym}^\bullet(T_1^\vee)/(\mathcal{K})$, *where* $(\mathcal{K})$ *is the ideal generated by the image of* $\mathcal{K}$. *Further, the obstruction theory of* $\mathfrak{Def}$ *is isomorphic to the obstruction theory to the deformation of* 0 *in* $\mathrm{Spec}\,\mathrm{Sym}^\bullet(T_1^\vee)/(\mathcal{K})$ *induced by the defining equation* $\mathcal{K}$. $\mathcal{K}$ *is called a Kuranishi map of the obstruction theory.*

Kuranishi map was first constructed by Kuranishi in analytic category. After this the algebraic construction is straight forward, can can be found in [**Lau, Li2, LT1, Kol**].

It is instructive to work out an explicit example. Let $s(z) = 2z^3 + z^4$. It defines the subscheme $X = \{z^3 = 0\}$ in $\mathbf{C}$. For triple $A = \mathbb{C}[t]/(t^3)$, $I = (t^2)$ and

$\varphi_2(t) = z : \operatorname{Spec} A/I \to X$, the obstruction class $\mathrm{Ob}(A, I, \varphi) = 0$. Obviously φ_2 extends to $\varphi_3(t) = z : \operatorname{Spec} A \to X$. Now let $A' = \mathbb{C}[t]/(t^4)$ and $I' = (t^3)$. We compute $\mathrm{Ob}(A', I', \varphi_3) = 2$, the coefficient of the leading non-zero term in s. This reveals the fact that the leading non-trivial terms in a Kuranishi map is determined by the obstruction theory, and the *non-uniqueness* of Kuranishi maps comes from the freedom of higher order terms. In this case, any function $\Phi(z) = 2z^3 + O(z^4)$ is a Kuranishi map. We will investigate the uniqueness of the Kuranishi maps in the later section.

We now work out the obstruction theories of the two mentioned classes of moduli spaces.

Let $\mathcal{E}$ be a stable sheaf of $\mathcal{O}_X$-modules and $\mathfrak{Def}$ the formal deformation functor of $\mathcal{E}$ with fixed determinants. Let $\operatorname{Ext}^i(\mathcal{E}, \mathcal{E})^0$ be the traceless part of the extension group $\operatorname{Ext}^i(\mathcal{E}, \mathcal{E})$.

PROPOSITION 2.4. *The functor $\mathfrak{Def}$ has tangent space $\operatorname{Ext}^1(\mathcal{E}, \mathcal{E})^0$ and admits an obstruction theory taking values in the vector space $\operatorname{Ext}^2(\mathcal{E}, \mathcal{E})^0$.*

We sketch the proof for the case of locally free sheaf $\mathcal{E}$. The proof uses standard ideas in deformation theory. For simplicity, we look at the case $R = k[t]/(t^{k+1})$ and $I = (t^k)$. We let $A_k = \operatorname{Spec} R/I$ and $A_{k+1} = \operatorname{Spec} R$. Let $\mathcal{E}_k$ be a sheaf of $\mathcal{O}_{X \times A_k}$-modules extending $\mathcal{E}$ and flat over R/I. We cover $X \times A_{k+1}$ by affine open subsets U_α so that over each U_α we have locally free sheaf $\mathcal{E}_\alpha$ which restricts to $\mathcal{E}_k$ on $U_k \cap X \times A_k$. Let $\phi_{\alpha\beta}$ be isomorphisms of $\mathcal{E}_\beta$ with $\mathcal{E}_\alpha$ over $U_\alpha \cap U_\beta$ which extend the identity map on $\mathcal{E}_k$ when restricted to $U_\alpha \cap U_\beta \cap X \times A_k$. Let

$$\Psi_{\alpha\beta\gamma} = \mathrm{Id} - \phi_{\alpha\gamma} \circ \phi_{\gamma\beta} \circ \phi_{\beta\alpha}.$$

Now $\psi_{\alpha\beta\gamma}$ is an endomorphism of $\mathcal{E}_\alpha$ over $U_\alpha \cap U_\beta \cap U_\gamma = U_{\alpha\beta\gamma}$. Because $\phi_{\alpha\beta}$ restricts to the identity homomorphism on $U_\alpha \cap U_\beta \cap X \times A_k$, $\phi_{\alpha\beta\gamma}$ is a homomorphism of sheaves $\mathcal{E}_\alpha \to \mathcal{E}_\alpha \otimes \mathcal{O}_{X \times A_{k+1}}(-X \times A_k)$ on $U_{\alpha\beta\gamma}$. So we can regard it as a section of $\operatorname{End}(\mathcal{E}) \otimes I$. It is direct to check that $\bar{\psi} = \{\psi_{\alpha\beta\gamma}\}$ is a cocycle and so defines an element

$$\bar{\psi} \in \operatorname{Ext}^2(\mathcal{E}, \mathcal{E}) \otimes I.$$

Let $\operatorname{Tr} : \operatorname{Ext}^2(\mathcal{E}, \mathcal{E}) \to H^2(\mathcal{O}_X)$ be the trace homomorphism. It can be shown that $\operatorname{Tr} \bar{\psi} = 0$. This follows from the fact that $\operatorname{Tr} \bar{\psi}$ is the obstruction to deforming the invertible sheaf $\det \mathcal{E}$ to $X \times A_{k+1}$. This is zero, either by the fact that $\det \mathcal{E}$ is the pull back of the sheaf $\mathcal{O}_X(I)$ or the Picard scheme of X is smooth. Hence $\operatorname{Tr} \bar{\psi} = 0$ and then $\psi \in \operatorname{Ext}^2(\mathcal{E}, \mathcal{E})^0 \otimes I$, the traceless part of the extension group.

In [**Mar**], Maruyama showed that the obstruction to the deformation of $\mathcal{E}$ lies in the group $\operatorname{Ext}^2(\mathcal{E}, \mathcal{E})$. In [**Muk**], Mukai discovered that such obstruction indeed lies in the traceless part of the extension group for sheaves over a $K3$ surface, which was generalized to all smooth varieties by Atamkin [**Art**]. Our presentation here follows that of Gieseker [**Gi2**].

The obstruction to deformation of stable morphisms also lies in an extension group, as follows:

PROPOSITION 2.5. *Let $f : C \to X$ be a stable morphism from a pointed curve C to X with $D \subset C$ the divisor of marked points. Let $\mathfrak{Def}$ be the formal deformation functor of f. Then its tangent space and obstruction space are the extension groups*

$$\operatorname{Ext}^1([f^*\Omega_X \to \Omega_C], \mathcal{O}_C) \quad \text{and} \quad \operatorname{Ext}^2([f^*\Omega_X \to \Omega_C], \mathcal{O}_C).$$

Here $[f^*\Omega_X \to \Omega_C]$ is the obvious complex with degree in $[-1, 0]$.

There are several different ways to describe the obstruction theory of stable morphisms. Z. Ran in [**Ran**] worked out a non-commutative extension group that works for a more general class of deformation problems. A general framework of obstruction theory was given in [**Ill**]. The obstruction space described here appeared in [**LT1**].

The obstruction theory of the Grothendieck's Quot-scheme $\mathbf{Q}^{ss}$ is identical to that in Proposition 2.4.

COROLLARY 2.6. *The obstruction to deformations of semistable quotient sheaf in* $\mathbf{Q}^{ss}$ *is the traceless extension group* $\operatorname{Ext}^2(\mathcal{E},\mathcal{E})^0$.

3. Moduli spaces of stable sheaves

We will study the moduli spaces of stable sheaves in this section.

Let X be a smooth surface with polarization H and let $\mathfrak{M}_d(r,I)$ be the moduli spaces of stable sheaves of $\mathcal{O}_X$-modules of topological type $\Delta = (r,I,d)$. Let $\mathcal{E}$ be a stable sheaf representing a point ξ in $\mathfrak{M}_d(r,I)$ and let $\mathfrak{Def}$ be the formal deformation functor of $\mathcal{E}$ of fixed determinant. Then by Lemma 2.3 and Proposition 2.4, $\mathfrak{Def}$ is represented by a formal scheme

$$\operatorname{Spec} k[[z_1,\cdots,z_{n_1}]]/(f_1,\cdots,f_{n_2}), \tag{3.1}$$

where $n_i = \dim \operatorname{Ext}^i(\mathcal{E},\mathcal{E})^0$. Since locally $\mathfrak{M}_d(r,I)^s$ admits local universal family (in étale topology), we have the following.

COROLLARY 3.1. *Let the notation be as before. Then the formal completion of* $\mathfrak{M}_d(r,I)^s$ *along* ξ, *denoted* $\hat{\xi}$, *is isomorphic to (3.1). It follows that the dimension of* $\mathfrak{M}_d(r,I)^s$ *at* ξ *is at least* $n_1 - n_2$. *When the dimension of* $\mathfrak{M}_d(r,I)^s$ *is exactly* $n_1 - n_2$, *then* $\mathfrak{M}_d(r,I)^s$ *has local complete intersection singularity.*

In some special cases, the Serre duality

$$\operatorname{Ext}^2(\mathcal{E},\mathcal{E})^0 \cong \operatorname{Ext}^0(\mathcal{E},\mathcal{E}\otimes K_X)^0$$

ensures that the obstruction to deforming any stable sheaf $\mathcal{E}$ vanishes automatically. Let X be a smooth surface so that $H\cdot K_X < 0$, then $\mathcal{E}$ stable implies that $\operatorname{Ext}^0(\mathcal{E},\mathcal{E}\otimes K_X)^0 = 0$. This can be seen as follows: Let $\sigma \in \operatorname{Ext}^0(\mathcal{E},\mathcal{E}\otimes K_X)^0$ viewed as a traceless homomorphism. Then $H\cdot K_X < 0$ implies that $\det\sigma \in H^0(K_X^{\otimes r}) = 0$. Hence $\mathcal{E}$ has a quotient sheaf $\operatorname{Im}(\sigma)$ which is also a proper subsheaf of $\mathcal{E}\otimes K_X$. This is impossible since $\mathcal{E}$ is stable and $H\cdot K_X \le 0$. For a K3 surface, $\operatorname{Ext}^2(\mathcal{E},\mathcal{E})^0 \cong \operatorname{Ext}^0(\mathcal{E},\mathcal{E})^0$, which is zero by the stability of $\mathcal{E}$.

For general surfaces, many sheaves $\mathcal{E}$ have obstructed deformation theories. Examples of such kind can be found in [**Gi2**]. To study these moduli spaces, our first order of task is to investigate the quantity $n_1 - n_2$ and n_2, where $n_i = \dim \operatorname{Ext}^i(\mathcal{E},\mathcal{E})^0$. By the Hirzebruch-Riemann-Roch theorem,

$$\eta(r,I,d) := n_1 - n_2 = 2rd + -(r-1)I^2 - (r^2-1)\chi(\mathcal{O}_X).$$

Note that it increases linearly in d. On the other hand, an easy argument shows that n_2 is bounded from above by a constant independent of I and d [**GL2**, Lemma 1.8].

LEMMA 3.2. *There is a constant* N *depending only on* X *and* r *such that for any rank* r *stable sheaf* $\mathcal{E}$ *we have* $\dim \operatorname{Ext}^2(\mathcal{E},\mathcal{E})^0 \le N$.

Another simple observation that serves to explain the asymptotic behavior of the moduli space $\mathfrak{M}_d$. Let $\mathcal{E}$ be any stable sheaf and $x \in X$ be any point. We look at a quotient homomorphism $\tau : \mathcal{E} \to k_x$ and its kernel $\mathcal{E}'$. When x and τ are general, then $\mathcal{E}'$ will be stable and $\operatorname{Ext}^2(\mathcal{E}', \mathcal{E}')^0$ is a proper subspace of $\operatorname{Ext}^2(\mathcal{E}, \mathcal{E})^0$. If we pick sufficiently many points on X to form a quotient homomorphism $\mathcal{E} \to \oplus k_{x_i}$, then its kernel will be a stable sheaf with unobstructed deformation.

The above discussion set the stage for the introducing the notion of asymptotic property of $\mathfrak{M}_d$. We begin with the non-emptiness of $\mathfrak{M}_d$.

PROPOSITION 3.3. *For any surface (X, H) and any choice (r, I) there is a constant N so that whenever $d \geq N$ the the moduli space $\mathfrak{M}_d$ is non-empty.*

This was proved by Maruyama in [**Mar**]. A better bound for N was derived by Gieseker [**Gi2**] and Qin [**Qi2**].

We now state the asymptotic smoothness of $\mathfrak{M}_d$.

THEOREM 3.4. *Let (X, H) be any smooth polarized surface. Then there is a function $\rho(d)$ depending on (X, H, r, I) satisfying $\lim_{d \to \infty} \eta(r, I, d) - \rho(d) = \infty$ such that*

$$\dim \operatorname{Sing}(\mathfrak{M}_d^s) \leq \rho(d).$$

COROLLARY 3.5. *Let (X, H) be as in Theorem 3.4. For any (r, I) there is an N so that whenever $d \geq N$ then the moduli space $\mathfrak{M}_d^s$ is normal and has pure dimension $\eta(r, I, d)$. Further, by increasing N if necessary, the moduli space $\mathfrak{M}_d^s$ has $\mathbb{Q}$-factorial singularity.*

We sketch the proof of the Corollary. According to the Theorem we can find N so that when $d \geq N$

$$\dim \operatorname{Sing}(\mathfrak{M}_d^s) \leq \eta(r, I, d) - 2.$$

By Lemma 2.3, $\dim \mathfrak{M}_d^s \geq \eta(r, I, d)$ and is non-empty, which follows from Proposition 3.3. Therefore $\mathfrak{M}_d^s$ has pure dimension $\eta(r, I, d)$ and is smooth at its general points. Then by Corollary 3.1 again, we know $\mathfrak{M}_d^s$ has local complete intersection singularity. Because $\operatorname{Codim}(\operatorname{Sing}(\mathfrak{M}_d^s), \mathfrak{M}_d^s) \geq 2$, $\mathfrak{M}_d^s$ must be normal. If we have $\operatorname{Codim}(\operatorname{Sing}(\mathfrak{M}_d^s), \mathfrak{M}_d^s) \geq 4$, which will be true if we increase N, then $\mathfrak{M}_d^s$ will have $\mathbb{Q}$-factorial singularity. This completes the proof of the Corollary.

The Theorem was first proved by Donaldson for rank two vector bundles and $\rho(d) = 3d + C_1\sqrt{d} + C_2$ [**Don**]. Later Friedman and K. Zuo [**Zu1**] provided different proofs. For moduli of arbitrary rank, the theorem was proved by the author and Gieseker [**GL2**] for the existence of such function and by O'Grady [**OG3**] for an explicit function $\rho(d) = c_1 d + c_2$ with $c_1 < 2r$.

All current proofs of Donaldson's general smoothness result hinge on the fact that the moduli space is smooth at a stable sheaf $\mathcal{E}$ unless $\operatorname{Ext}^2(\mathcal{E}, \mathcal{E}) \neq \{0\}$. Using Serre duality, the non-vanishing of this group provides us homomorphisms $\mathcal{E} \to \mathcal{E}(K_X)$. By exploiting such homomorphisms, one can bound the size of the set of such stable sheaves. Then combined with the lower bound of the dimension of the moduli space, one obtains the bound in the Theorem. Similar to other success stories in studying surfaces, the group we intend to control is isomorphic to a group we can work with, using Serre duality.

Much of the statement holds true for the full moduli scheme $\mathfrak{M}_d$.

COROLLARY 3.6. *Let the notation be as in Corollary 3.5. Then there is an N so that when $d \geq N$ the moduli scheme $\mathfrak{M}_d$ is normal. Further, $\mathfrak{M}_d$ is $\mathbb{Q}$-factorial.*

The proof follows from the study of the singularity of the Quot-scheme $\mathbf{Q}^{ss}$ [**Li4**]. First, following the proof of Corollary 3.5, $\mathbf{Q}^{ss}$ is normal, has local complete intersection singularity and is $\mathbb{Q}$-factorial. Then the GIT quotient $\mathfrak{M}_d = \mathbf{Q}^{ss}//G$ is necessarily normal. We now show that $\mathfrak{M}_d$ has $\mathbb{Q}$-factorial. Let L be any line bundle on the smooth part of $\mathfrak{M}_d$. Let $\pi : \mathbf{Q}^{ss} \to \mathfrak{M}_d$ be the projection and π^*L the pull back line bundle. Since $\mathbb{Q}^{ss}$ has $\mathbb{Q}$-factorial singularity, a tensor power $\pi^*L^{\otimes k}$ extends to a line bundle $\bar{L}$ on $\mathbf{Q}^{ss}$. Then using the technique of Drezet and Narasimhan in their study of vector bundles over curves [**DN**], we can show that $\bar{L}$ descends to $\mathfrak{M}_d$. This completes the proof of the Corollary.

Another topic we like to mention is the Uhlenbeck's compactification of the moduli of stable vector bundles over surfaces. This compactification is to compactify the moduli of stable vector bundles by analytic mean. The resulting space is then called Uhlenbeck's compactification, where the moduli space of Gieseker stable sheaves are often called Gieseker compactification of the moduli of stable vector bundles. Since the Donaldson's polynomial invariants were first defined using Uhlenbeck's compactification, it was important to understand the relation of these two classes of compactifications. In [**Li1**] the author showed that the Uhlenbeck's compactification is homeomorphic to a contraction of the moduli schemes of semi-stable sheaves. (Around the same time Morgan [**Mog**] showed that there is a continuous map from Gieseker's compactification to Uhlenbeck's compactification.) One key technique is the study of linear series on the moduli space $\mathfrak{M}_d$ [**LP1, Li1, Qi2**]. So far it is unclear how important the Uhlenbeck's compactification is to the study of the moduli of vector bundles. There are some indications that this might be so [**LQ1, Bar**].

4. Asymptotic property of moduli of bundles

In this section, we will study the asymptotic property of the moduli schemes $\mathfrak{M}_d(r, I)$. As shown in the previous section, for fixed (r, I) the moduli scheme

$$\mathfrak{M}_d = \mathfrak{M}_d(r, I)$$

has pure dimension $\eta(r, I, d)$ when d is sufficiently large and $\mathrm{Codim}(\mathrm{Sing}(\mathfrak{M}_d), \mathfrak{M}_d)$ become arbitrarily large then d tends to infinity. This suggests that the asymptotic property of $\mathfrak{M}_d$ should not be hindered by the singularity of $\mathfrak{M}_d$.

To have a glimpse of the kind of asymptotic property $\mathfrak{M}_d$, we shall look no further than the moduli space of rank one stable sheaves. In this case, all moduli scheme $\mathfrak{M}(1, I, d)$ are isomorphic to the Hilbert scheme of zero-subschemes of X of length d, which we denote by $\mathrm{S}^{[d]}X$.

PROPOSITION 4.1. *$\mathrm{S}^{[d]}X$ is a smooth projective variety of dimension $2d$. The Poincare polynomial of $\mathrm{S}^{[d]}X$ is*

$$\sum_{d=0}^{\infty} p\left(\mathrm{S}^{[d]}X, z\right) t^d = \prod_{n=1}^{\infty} \frac{(1+z^{2n-1}t^n)^{b_1}(1+z^{2n+1}t^n)^{b_3}}{(1-z^{2n-2}t^n)^{b_0}(1-z^{2n}t^n)^{b_2}(1-z^{2n+2}t^n)^{b_4}}. \tag{4.1}$$

The Picard group of $\mathrm{S}^{[d]}X$ is isomorphic to $X^{[2]}$.

Here the Poincare polynomial $p(W, z)$ of W is $\sum b_n(W)z^n$, where $b_n(W)$ are the Betti numbers of W. The Poincare polynomials of $\mathrm{S}^{[d]}X$ was found by Göttsche [**Go1**]. Note that the polynomials $p(\mathrm{S}^{[d]}X, z)$ stabilize when $d \to \infty$. Indeed, it is easy to see that for any n

$$p(\mathrm{S}^{[d]}X, z) - p(X^{[d+1]}, z) \equiv 0 \mod z^{n+1}$$

for $d \geq n$. Here the bound is far from optimal. One can get a better bound for large n.

The folklore conjecture among experts working in moduli spaces of stable sheaves over surfaces is that asymptotically the Poincare polynomials of $\mathfrak{M}_d$ coincide with that of the Hilbert scheme $\mathrm{S}^{[d]}X$.

CONJECTURE 4.2. *Let (r, I) be fixed and let $\mathfrak{M}_d$ be the moduli scheme of semistable sheaves as before. Let $p(\mathfrak{M}_d, z)$ be the Poincare polynomial of $\mathfrak{M}_d$. Then there is a non-decreasing function $\eta(d)$ with $\lim_{d\to\infty}\eta(d) = \infty$ such that*

$$p(\mathfrak{M}_d, z) - p(\mathrm{S}^{[d]}X, z) \equiv 0 \mod z^{\eta(d)}.$$

In rank two case, this conjecture was partially verified for some surfaces, including $\mathbf{P}^2$ and ruled surfaces [**Yo1, Yo2**], the K3 surface [**GH**] and some elliptic surfaces [**Yo3**]. For general surfaces, this conjecture is largely open.

We now review the progress toward this conjecture for general surfaces. Along this line, the first task is to show that the moduli scheme $\mathfrak{M}_d$ is irreducible for large d. This was proved for the case of rank two by Gieseker and the author in [**GL1**]. The case for higher rank was proved in [**GL2, OG3**].

THEOREM 4.3. *Let (X, H) be a polarized smooth surface and (r, I) be as before. Then there is a constant N depending only on (X, H, r) so that the moduli space $\mathfrak{M}_d$ is irreducible provided $d \geq N$.*

The proof is based on a stabilization argument, which we now outline. The crucial step to prove this theorem is to show that for large d, all irreducible components of $\mathfrak{M}_d$ contains non-locally free sheaves. This was proved in [**GL1, GL2, OG2, OG3**]. To prove this result, one needs first to find sheaves from each irreducible component of $\mathfrak{M}_d$ that are sufficiently unstable when restricted to an ample divisor. This is achieved by studying the degree of certain determinant line bundle of the moduli spaces. After that, one needs to deform these sheaves globally to non-locally free stable sheaves. With this done, one then apply the stabilization technique: Let Σ_d be the set of irreducible components of $\mathfrak{M}_d$. For each stable sheaf $\mathcal{E}$ in $V \in \Sigma_d$, we can form a new sheaf $\mathcal{E}'$ that is the kernel of $\mathcal{E} \to k_x$, for some general $x \in X$. Note that for large d the sheaf $\mathcal{E}'$ belongs to a unique irreducible component $V' \in \Sigma_{d+1}$. This way, we define a map

$$\rho : \Sigma_d \to \Sigma_{d+1}.$$

The fact that for large d each irreducible component of $\mathfrak{M}_d$ contains non-locally free stable sheaves implies that then $\rho(\Sigma_d) = \Sigma_{d+1}$. It remains to show that $\rho^r(\Sigma_d)$ is a single point set for r sufficiently large. Let V_i, $i = 1, 2$ be two components in Σ_d. We pick $\mathcal{E}_i \in V_i$. Then we can find sufficiently ample invertible sheaf $\mathcal{L}$ so that $\mathcal{E}_i$ fits into the exact sequence

$$0 \longrightarrow \oplus^{r-1}\mathcal{L}^{-1} \longrightarrow \mathcal{E}_i \longrightarrow \mathcal{L}^{\otimes(r-1)} \otimes \mathcal{O}(I) \otimes \mathcal{I}_{Z_i} \longrightarrow 0$$

where Z_i is a zero-scheme of X. Without loss of generality we can assume Z_1 and Z_2 are disjoint. Then we can find new subsheaves $\mathcal{E}_i' \subset \mathcal{E}_i$ so that they fit into the exact sequence

$$0 \longrightarrow \oplus^{r-1}\mathcal{L}^{-1} \longrightarrow \mathcal{E}_i' \longrightarrow \mathcal{L}^{\otimes(r-1)} \otimes \mathcal{O}(I) \otimes \mathcal{I}_{Z_1\cup Z_2} \longrightarrow 0$$

Thus $\mathcal{E}_1'$ and $\mathcal{E}_2'$ belong to a family of sheaves parameterized by the total space of

$$\mathrm{Ext}^1(\mathcal{L}^{\otimes(r-1)} \otimes \mathcal{O}(I) \otimes \mathcal{I}_{Z_1 \cup Z_2}, \oplus^{r-1}\mathcal{L}^{-1})$$

and hence belong to the same component of the moduli space. This proves that eventually, $\rho^r(\Sigma_d)$ contains a single point, thus the asymptotic irreducibility of $\mathfrak{M}_d$.

The question of the higher Betti numbers of the moduli scheme $\mathfrak{M}_d$ is largely open. The only known result along this line is for b_1 and b_2 of the moduli scheme $\mathfrak{M}(2, I, d)$ proved by the author [**Li3**].

THEOREM 4.4. *Let the notation be as before. Then there is a constant N depending only on (X, H) such that for $d \geq N$ the first and the second Betti numbers of $\mathfrak{M}_d$ coincide with the first and second Betti numbers of $\mathrm{S}^{[d]}X$. They are $b_1(X)$ and $b_2(X) + b_1(X)(b_1(X) - 1) + 1$.*

Combining this result and a careful study of the singularity of the moduli scheme $\mathfrak{M}_d$, the author also determined the Picard group of the moduli scheme $\mathfrak{M}_d$ up to torsions.

THEOREM 4.5. *Let the notation be as before. Then there is a constant N depending on (X, H) so that for $d \geq N$ the Picard groups $\mathrm{Pic}(\mathfrak{M}_d) \otimes_{\mathbb{Z}} \mathbb{Q}$ is isomorphic to $\mathrm{Pic}(X^{[2]} \otimes_{\mathbb{Z}} \mathbb{Q})$.*

Recently, Yoshioka [**Yo4**] has determined the Picard group for ruled surfaces.

To understand the cohomology group of the moduli space $\mathfrak{M}_d$, the diagonal technique of Ellingsrud and Stromme might be useful. The diagonal technique states that if W is a smooth variety then the rational cohomology group of W is generated by the Kunneth component of the diagonal $\Delta \subset W \times W$. Beauville used this technique to study the topology of the moduli spaces of stable sheaves. In [**Bea**] he observed that for surfaces X with $K_X \cdot H < 0$, the diagonal class $\Delta \subset \mathfrak{M}_d \times \mathfrak{M}_d$ of the moduli space $\mathfrak{M}_d$ of stable sheaves on X can be expressed in terms of the Kunneth components of the Chern class of the universal sheaf over $X \times \mathfrak{M}_d$. Here the condition is quite strict that it only applies to rational ruled surfaces. The idea is to look at the relative extension sheaf, a perfect complex,

$$\mathcal{E}xt^{\bullet}_{\pi}(p_1^*\mathcal{E}, p_2^*\mathcal{E}) \tag{4.2}$$

where $\pi : X \times \mathfrak{M}_d \times \mathfrak{M}_d \to \mathfrak{M}_d \times \mathfrak{M}_d$, $\mathcal{E}$ is a universal family on $X \times \mathfrak{M}_d$ and p_i are the projections $X \times \mathfrak{M}_d \times \mathfrak{M}_d \to X \times \mathfrak{M}_d$. Here we assume that there are no strictly semistable sheaves in $\mathfrak{M}_d$. The condition $K_X \cdot H < 0$ implies that for any pair E, $F \in \mathfrak{M}_d$, $\mathrm{Ext}^0(E, F)$ is 0 unless $E = F$, in which case it is $\mathbb{C}$. Thus the complex (4.2) has the diagonal built into it. Applying the Geothendieck-Riemann-Roch formula, we can express the diagonal class of Δ in terms of the Kunneth components of the Chern classes of the universal family on $\mathfrak{M}_d \times X$.

A simple argument shows that this diagonal argument implies the Conjecture 4.2 (cf. [**Yo3**]). Since the moduli space $\mathfrak{M}_d$ has local complete intersection singularity, and since the local embedding of $\mathfrak{M}_d$ is at most the dimension of $\mathfrak{M}_d$ plus a universal constant, the singularity of $\mathfrak{M}_d$ should not complicate the asymptotic behavior of the Betti number of $\mathfrak{M}_d$. Naturally it is interesting to find a way to make this argument work for moduli of sheaves for general surfaces X. We can still consider the same complex but then the other terms, especially $\mathrm{Ext}^2(E, F)$ can be non-zero for a large set of sheaves in $\mathfrak{M}_d$. Hence a better understanding of the singularity of $\mathfrak{M}_d$ as well as the terms $\mathrm{Ext}^2(E, F)$ is required to attack this problem.

5. The S-duality conjecture

It is impossible to avoid mentioning the S-duality conjecture of Vafa-Witten [**VW**] on the Betti numbers of the moduli space $\mathfrak{M}_d(r, I)$ while discussing the topology of the moduli spaces of stable sheaves.

We will state the S-duality conjecture for K3 surfaces since this is the case that is well understood. We consider the moduli spaces of gauge equivalent classes of Anti-Self-Dual connections on a principle-G bundle over X with $G = SU(2)$ or $SO(3)$. In algebraic geometry, they corresponds to moduli spaces of rank 2 stable sheaves of Chern class c_1 and c_2. By tensoring sheaves by an invertible sheaf $\mathcal{I}$ we obtain a birational map

$$\mathfrak{M}(2, c_1, c_2) - - \rightarrow \mathfrak{M}(2, \tilde{c}_1, \tilde{c}_2),$$

where

$$\tilde{c}_1 = c_1 + 2c_1(\mathcal{I}) \quad \text{and} \quad \tilde{c}_2 = c_2 + c_1(\mathcal{I}) \cdot (c_1(\mathcal{I}) + c_1) \tag{5.1}$$

By a result of Göttsche and Heybrecht [**GH**], these two moduli spaces are deformation equivalent, and hence they have identical Poincare polynomial. (A similar statement can be said on their Hodge numbers.) Following the convention of gauge theory, we will view the moduli space $\mathfrak{M}(2, c_1, c_2)$ as a compactification of the moduli space of equivalent classes of Anti-Self-Dual SO(3)-connections of the Stiefel-Whitney class w_1 and the instanton number k, with

$$w_1 \equiv c_1 \mod 2 \quad \text{and} \quad k = c_2 - c_1^2/4. \tag{5.2}$$

Note that (w,k) is invariant under the transformation given in (5.1). For any choice of (w_1, k), we denote by $\mathfrak{M}(w_1, k)$ one of the birational model $\mathfrak{M}(2, c_1, c_2)$ with relation (5.2). As mentioned, the Poincare polynomial of $\mathfrak{M}(w_1, k)$ is well-defined.

Following Vafa-Witten, we form the generating function

$$Z_X(\tau, \mathrm{SO}(3)) = \sum_{w,k} \chi(\mathfrak{M}(w_1, k))q^k, \quad q = e^{2\pi i \tau} \tag{5.3}$$

where the sum is over all $w_1 \in H^2(X, \mathbb{Z}_2)$ and $k \in \mathbb{Z}[\frac{1}{2}]$. For the group SU(2), we define

$$Z_X(\tau, \mathrm{SU}(2)) = \frac{1}{2} \sum_k \chi(\mathfrak{M}(0, k))q^k, \quad q = e^{2\pi i \tau}. \tag{5.4}$$

Here we insert $\frac{1}{2}$ because 2 is the order of the center of SU(2). Note that the moduli space $\mathfrak{M}(0, 2k)$ is singular and its Euler number is not well-understood yet. The S-duality conjecture predicts the transformation law under $\tau \to -1/\tau$ to be

$$Z_X(-1/\tau, \mathrm{SU}(2)) = 2^{-11} \left(\frac{\tau}{\sqrt{-1}} \right)^{w/2} Z_X(\tau, \mathrm{SO}(3))$$

for some integer w. A similar pattern for moduli spaces of sheaves over general surfaces was also suggested by Vafa-Witten.

This is an amazing conjecture on the moduli spaces of vector bundles over surfaces. For K3 surfaces, this conjecture was largely confirmed, by the work of [**GH**] It also yields a conjectured formula for $\chi(\mathfrak{M}(0, 2k))$. Note that this moduli space is singular with quotient singularity by groups $\mathbb{C}^*$ or $PGL(2)$. The conjectured formula of $\chi(\mathfrak{M}(0, 2k))$ contains rational numbers, indicating that the Euler number will not be the conventional one. It is interesting to investigate exactly what kind of Euler number is proposed in in this conjecture. For general surface, it is still not clear what is the exact form of the S-duality conjecture. There are a few works along this line, notably the blowing-up formula of W.P. Li and Qin [**LQ1, LQ2, LQ3**], the mysterious relation with affine Lie algebra [**Nak**] and the recent work [**Bar**].

6. Moduli spaces and Donaldson type invariants

Enumerative geometry is a classical research area in algebraic geometry. In a broader sense, all such problems can be viewed as an intersection theory on some moduli spaces. The discovery of Donaldson type invariants injects new life to this century old research area.

Let us first recall the general setting of the Donaldson type invariants. Given a variety X, one can form moduli spaces of certain objects associated to X. Let us denote by $\mathfrak{M}$ one such moduli space. Usually $\mathfrak{M}$ carries with it a collection of tautological topological classes. We assume they are cohomological classes, say $\alpha_1, \cdots, \alpha_k \in H^*(\mathfrak{M}, \mathbb{Q})$. Assume $\mathfrak{M}$ is smooth, complete and has the expected dimension, we can integrate the tautological class over $\mathfrak{M}$ to obtain a functional[4]

$$\varphi(t_1, \cdots, t_k) = \int_{\mathfrak{M}} \exp(t_1\alpha_1 + \cdots + t_k\alpha_k). \tag{6.1}$$

When $\mathfrak{M}$ is only complete and has the expected dimension, we can replace the integration by an operational cohomological theory of schemes:

$$\varphi(t_1, \cdots t_k) = \deg\big(\exp(t_1\alpha_1 + \cdots t_k\alpha_k)[\mathfrak{M}]\big)_0. \tag{6.2}$$

(Here we implicitly assume that the $\alpha_i \in A^*X$ are algebraic cocycles.) This still defines the desired functional. For instance, when X is an algebraic surface and $\mathfrak{M}$ the moduli spaces of rank two semistable sheaves, this defines the Donaldson polynomial invariant [**Don**]: Let (X, H) be a polarized smooth surface and $\mathfrak{M}(2, I, d)$ be the moduli space of semi-stable sheaves of rank 2, determinant I and second Chern class d. As we know, $\mathfrak{M}(2, I, d)$ is projective. For simplicity, we assume $\mathfrak{M}(2, I, d)$ admits a universal family, say $\mathcal{E}$ over $\mathfrak{M}(2, I, d) \times X$. Then the combination $c_2(\mathcal{E}) - I^2/4$ defines a map

$$\mu : H_*(X, \mathbb{Q}) \longrightarrow H^*(\mathfrak{M}(2, I, d), \mathbb{Q}).$$

Let $\alpha_1, \cdots, \alpha_n$ be a basis of $H_*(X)$. Assume $\mathfrak{M}(2, I, d)$ has the expected dimension, then

$$\varphi(t_1, \cdots, t_n) =< \exp(\mu(\alpha_1)t_1 + \cdots + \mu(\alpha_n)t_n), [\mathfrak{M}(2, I, d)] >$$

is the generating function of the Donaldson polynomial invariants.

The Gromov-Witten invariants were defined similarly [**KMa, LT1, RT, Rua**]. Let X be a smooth projective variety and $\mathfrak{M}_{g,k}(X, A)$ the moduli space of stable morphisms $f : D \subset C \to X$ from n-pointed (arithmetic) genus g nodal curve $D \subset C$ to X of degree $f_*([C]) = A$. The moduli space $\mathfrak{M}_{g,k}(X, A)$ admits a natural evaluation map

$$\text{ev} : \mathfrak{M}_{g,k}(X, A) \longrightarrow X^n$$

associating to each stable map f the image of the n-marked points under f. When the moduli space $\mathfrak{M}_{g,k}(X, A)$ has expected dimension, is smooth at general points

[4] $t_1, \cdots, t_k$ are formal variables and $\exp(\cdot)$ is the formal power series of the exponential function.

and the subset of maps with smooth domains is dense, then the GW invariant is

$$\gamma^X_{A,g,k}(\alpha) = \int_{[\mathfrak{M}_{g,k}(X,A)]} \mathrm{ev}^*(\alpha), \qquad \alpha \in H^*(X^k, \mathbb{Q}). \tag{6.3}$$

The equivalence of this with the one based on counting is apparent from Poincare duality.

However, when $\mathfrak{M}$ does not have the expected dimension, we run into serious trouble if we keep the formula (6.2). For one, when X is a member of a smooth family of varieties, the associated moduli spaces may not necessarily form a flat family, and hence the resulting functional likely will change as X varies. Also, many of these functionals can be defined using (6.1) with $\mathfrak{M}$ replaced by moduli of corresponding analytic objects, yielding interesting topological invariants. These invariants might will differ from (6.2) when $\mathfrak{M}$ does not have the expected dimension. The solution to this difficulty is to find a *virtual moduli cycle* and to define the functional using (6.2) by replacing $[\mathfrak{M}]$ with this cycle. Morally speaking, one should view the moduli space $\mathfrak{M}$ as defined by certain relations Ψ. The situation when $\mathfrak{M}$ has bigger than expected dimension occurs when these relations are not in general position. If we perturb these relations so they are in general position. Then their solutions space, which can be viewed as a perturbation of the original moduli space $\mathfrak{M}$, defines a moduli space whose fundamental class is the virtual moduli cycle of $\mathfrak{M}$. The final formula for the Donaldson type invariant then takes the form

$$F(t_1, \cdots, t_k) = \deg\big(\exp(t_1\alpha_1 + \cdots + t_k\alpha_k)[\mathfrak{M}]^{\mathrm{vir}}\big)_0 .$$

7. Algebraic construction of virtual cycles

We now discuss the algebraic construction of virtual cycles. Such cycles were first developed Tian and the author [**LT1**]. An alternative construction was provided by Behrend-Fantechi [**Beh, BFa**]. We will follow the construction in [**LT1**].

We begin with an example that will illustrate this construction. Let

$$\begin{matrix} E \\ s \uparrow\downarrow \\ W \end{matrix} \tag{7.1}$$

be a section of a vector bundle over a smooth projective variety W. Let $X = s^{-1}(0)$ be the subscheme of W. We think of W as the ambient space and s the defining equation of X. Let $n = \dim W$ and $m = \operatorname{rank} E$. Since $X \subset W$ was cut out by m equations, the expected dimension

$$\exp.\dim X = n - m.$$

In case X has pure dimension $n - m$, we will take its fundamental class $[X]$ its virtual cycle $[X]^{\mathrm{vir}}$. Note that this is the Euler class of E, viewed as a class in $A_{n-m}W$.[5] In general, we demand $[X]^{\mathrm{vir}} \in A_*X$ to be the class so that it will remain constant if we vary X by varying the section s. This leaves us the only option that $[X]^{\mathrm{vir}}$ is the Euler class $e(E)$. To make it a class in A_*X, we will take the localized Euler class

$$e_{\mathrm{loc}}(E, s) \in A_{n-m}s^{-1}(0) = A_{n-m}X$$

[5] Z_*W is the group of algebraic cycles and A_*X is Z_*X modulo the rational equivalence relation.

thanks to the section s. The localized Euler class is constructed as follows [**Fu**]: Let $N_{X/W}$ be the normal cone to X in W. It is canonically a subcone of $E|_X$. Let $\eta_E : X \to E|_X$ be the zero section and $\eta_E^* : A_*E|_X \to A_*X$ the Gysin map. Then

$$e_{\text{loc}}(E,s) = \eta_E^*[N_{X/W}] \in A_*X.$$

Analytically, $[N_{X/W}]$ can be viewed as the limit current (cycle) of Γ_{ts} when $t \to \infty$, where Γ_s is the graph of s in E, and $\eta_E^*[N_{X/W}]$ can be viewed as a class in $H_*(X)$ resulting from intersecting $N_{X/W}$ with a generic section of $E|_X$. This interpretation shows that $\eta_E^*[N_{X/W}]$ is the usual Euler class when viewed as an element in $H_*(W)$.

It is this localized Euler class that we define to be the virtual cycle $[X]^{\text{vir}}$.

To push this construction to a general moduli space, say M, we face the improbable task of embedding it *naturally* into some ambient space W as the vanishing locus of a section s of a vector bundle E over W. Here we stress the naturality of such choice since otherwise the virtual moduli cycle will not be a naturally defined object. For instance, if $M \subset W$ is the zero locus of a section S of E, a section of $E \oplus \mathbb{C}_W$, then M will be the zero locus of $S \oplus 0$, where 0 is the zero section of the trivial line bundle $\mathbb{C}_W$. However, $e(E) \neq e(E \oplus \mathbb{C}_W)$. To satisfies this "naturality" condition, we ask ourselves *what is the canonical defining equation(s) of a moduli space.* The answer is that the canonical defining equations are the *Kuranishi maps of the obstruction theory of the moduli spaces.*

Here is our strategy: *The moduli space should come with an obstruction theory; Its associated Kuranishi maps will serve as its defining equation; After that we will use the Kuranishi maps to construct a virtual cone in a vector bundle E; The virtual cycle is then the intersection of this cone with the zero section of the vector bundle E.*

We now demonstrate how to carry out this program. We continue our investigation in the set up of (7.1). According to the plan, we intend to reconstruct the normal cone $N_{X/W}$ near $p \in X$ by the Kuranishi maps of the obstruction theory. Let (ι, Φ) be a Kuranishi map of the obstruction theory of $(\hat{X}, p)$, as in Lemma 2.3. Here $\hat{X}$ is the formal completion of X along p, Φ is an element in $E^\vee|_p \otimes k[[z]]$ with (z) a multi-variable dual to T_pW and ι is an isomorphism $\operatorname{Spec} k[[z]]/(\Phi) \to \hat{X}$ ((Φ) is the ideal generated by components of ϕ). We consider the normal cone

$$N_\Phi := N_{\hat{X}/\operatorname{Spec}\mathbb{C}[[z]]} \subset V \times \hat{X} \tag{7.2}$$

which is canonically a subcone of $V \times \hat{X}$. The cone N_Φ is closely related to the normal cone $N_{X/W}$ at p.

LEMMA 7.1. *Let the notation be as before and let $\Phi_1, \Phi_2 \in V \otimes \mathbb{C}[[z]]$ be any two Kuranishi maps of $(\hat{X}, p)$. Then there is a vector bundle automorphism ϕ of $V \times \hat{X}$ (as bundle over $\hat{X}$) that restricts to the identity homomorphism at the closed fiber $V \times \{p\}$ such that $\phi^{-1}(N_{\Phi_1}) = N_{\Phi_2}$.*

The upshot of this Lemma is that though Kuranishi maps are not unique, the normal cones they produce are almost unique. The only ambiguity comes from those automorphisms whose restrictions to the closed fiber are the identity automorphisms.

We now investigate the choice of the vector bundle E. We first note that the sheaf

$$\mathcal{T} = \operatorname{Coker}\{\, \Omega_W^\vee|_X \to \mathcal{O}_X(E)\,\}, \tag{7.3}$$

has the property that at each point $p \in X$ the space $\mathcal{T} \otimes_{\mathcal{O}_X} \mathbb{C}_p$ is the obstruction space to deforming p in X. Also $\mathcal{T}$ is a quotient sheaf of $\mathcal{O}_X(E)$. We call $\mathcal{T}$ the obstruction sheaf. We remark that not all obstruction theories have this property. We phrase this property in the the following definition.

DEFINITION 7.2. *A scheme Y is said to have a perfect obstruction theory if there is a perfect complex $\mathcal{T}^\bullet = [\mathcal{T}_1 \to \mathcal{T}_2]$ in the derived category $\mathcal{D}^b(X)$ of which the following holds: Associating to each triple (A, I, φ), where A is a $\mathbb{C}$-algebra, $I \subset A$ an ideal so that $I^2 = 0$ and $\varphi : \operatorname{Spec} A/I \to X$ is a morphism, there is a function* $\operatorname{Ob}(A, I, \varphi) \in \mathfrak{h}^2(\mathcal{T}^\bullet \otimes_{\mathcal{O}_X} I)$ *such that*
(1) $\operatorname{Ob}(A, I, \varphi) = 0$ *if and only if there is an extension* $\tilde{\varphi}: \operatorname{Spec} A \to X$ *of* φ.
(2) If $\operatorname{Ob}(A, I, \varphi) = 0$ *then the space of all extensions $\tilde{\varphi}$ are parameterized by* $\mathfrak{h}^1(\mathcal{T}^\bullet \otimes_{\mathcal{O}_X} I)$.
(3) The function $\operatorname{Ob}(\cdot, \cdot, \cdot)$ *satisfies the base change property.*

This definition is almost the same as Definition (2.1) except that now A can be any $\mathbb{C}$-algebra. Thus this covers deformation of *all* points in the sense of Grothendieck. For those who are unfamiliar with the derived category, they can replace $[\mathcal{T}_1 \to \mathcal{T}_2]$ with a complex of locally free sheaves of $\mathcal{O}_X$ modules $\mathcal{E}^\bullet = [\mathcal{E}_1 \to \mathcal{E}_2]$, and $\mathfrak{h}^j(\mathcal{T}^\bullet \otimes_{\mathcal{O}_X} I)$ by the kernel and the cokernel of $\mathcal{E}_1 \otimes_{\mathcal{O}_X} I \to \mathcal{E}_2 \otimes_{\mathcal{O}_X} I$ for $i = 1$ and 2 respectively.

Let Y be any projective scheme endowed with a perfect obstruction theory with associated complex $\mathcal{T}^\bullet$. Let E be a vector bundle on Y so that the $\mathfrak{h}^2(\mathcal{T}^\bullet)$ is a quotient sheaf of $\mathcal{O}_X(E)$. For any $p \in Y$, we denote by $\hat{Y}$ the formal completion of Y along p and V_p the vector space $\mathfrak{h}^2(\mathcal{T}^\bullet \otimes \mathbb{C}_p)$. The later is the obstruction space to deformation of p in $\hat{Y}$.

BASIC CONSTRUCTION 7.3. *Let the notation be as before. Then there is a cone scheme N so that for each $p \in Y$ there is a vector bundle homomorphism*

$$\phi : E|_{\hat{Y}} \longrightarrow V_p \times \hat{Y}$$

that extends the given homomorphism $E|_p \to V_p$ so that

$$N|_{\hat{Y}} = \phi^{-1}(N_\Phi). \tag{7.4}$$

In the case of (7.1), the cone so constructed is exactly the normal cone $N_{X/W}$ when using the obstruction theory induced by s with the associated complex $\Omega_W^\vee|X \to \mathcal{O}_X(E)$ and the vector bundle E.

DEFINITION-THEOREM 7.4. *Let X be any projective scheme endowed with a perfect obstruction theory with associated complex $\mathcal{T}^\bullet = [\mathcal{T}_1 \to \mathcal{T}_2]$. Then X has a canonical virtual cycle $[X]^{\mathrm{vir}}$ constructed as follows:*
(1) We pick a vector bundle E over X so that the sheaf cohomology $\mathfrak{h}^2(\mathcal{T}^\bullet)$ is a quotient sheaf of $\mathcal{O}_X(E)$.
(2) We form a cone $N \subset E$ according to the Basic Construction 7.3.
(3) We define the virtual cycle $[X]^{\mathrm{vir}} = \eta_E^*[N]$.

Here is a few remarks. First, since $\mathcal{T}^\bullet$ is a perfect complex with two terms, its cokernel is a well-defined sheaf of $\mathcal{O}_X$-modules. Because X is projective, the vector bundle E always exists; Second, the existence of the cone N satisfying rule (7.4) does not follow immediately from the existence of the Kuranishi maps. The existence is proved by constructing a relative Kuranishi family. Once such cone exists, then it will by unique, following (7.4); Thirdly, the cycle $[X]^{\mathrm{vir}}$ so defined is independent of the choice of E. One caution, the virtual cycle only exists as a class in A_*X. It does not have a canonical cycle representative, just like the Euler class has no canonical submanifold representative.

8. Gromov-Witten Invariants

We now show that this machinery can be applied to construct the GW invariants of all smooth projective varieties. Following the blueprint of the construction, we need to describe the canonical obstruction theory of the moduli space $\mathfrak{M}_{g,k}(X, A)$ of stable morphisms, to workout a global vector bundle that makes the obstruction sheaf its quotient sheaf and then use the Gysin map to define the virtual cycle.

Before getting into the details, a few comments on the moduli space $\mathfrak{M}_{g,k}(X, A)$ is in order. Because in general there are stable morphisms f that have non-trivial automorphisms, $\mathfrak{M}_{g,k}(X, A)$ is not a fine moduli space. It is a coarse moduli space and is projective, as shown by [**Ale**]. However, by working with the $\mathfrak{M}_{g,k}(X, A)$ as coarse moduli space we lose information, some essential to our program. Recall that constructing virtual cycles requires a global obstruction theory, which relies on the universal family of the moduli spaces. There are several ways to remedy this. One way is to view $\mathfrak{M}_{g,k}(X, A)$ as a categorical quotient of a G schemes, where G is reductive, so that over this scheme there is a G equivariant universal family. This approach is most convenient if the moduli space is a GIT quotient of this scheme. So far we only know that $\mathfrak{M}_{g,k}(X, A)$ is a categorical quotient of G. This is good enough to carry out the construction of virtual cycles. The other is to view $\mathfrak{M}_{g,k}(X, A)$ as a $\mathbb{Q}$-schemes, which is a generalization of the notion of orbifolds. The last option is to work with $\mathfrak{M}_{g,k}(X, A)$ as a stack.

The notion of stack is widely accepted nowadays among algebraic geometers working with moduli problems. It was developed by Deligne-Mumford in attacking the moduli of stable curves [**DMu**]. The underlining idea behind the notion of *stack* is as follows: It consists of a collection of schemes S_α with families of stable morphisms f_α over S_α; For any pair α and β, there is a subscheme $S_{\beta\alpha} \subset S_\alpha$ and a subscheme$S_{\alpha\beta} \subset S_\beta$ so that the families f_α and f_β over them are isomorphic. If we identify$S_{\alpha\beta}$ with $S_{\beta\alpha}$, we end up with a scheme, the coarse moduli scheme. However, for the moduli $\mathfrak{M}_{g,k}(X, A)$ we can not patch the families f_α and f_β to obtain a universal family over the coarse moduli space. With stack, we view it as the collection of open charts S_α with the information how to identify $S_{\alpha\beta}$ with $S_{\beta\alpha}$. By stoping short of actually identifying them, we can construct a universal family over the stack. The collection S_α is called an *atlas* and each S_α is called a chart of the stack. In the following, we will view $\mathfrak{M}_{g,k}(X, A)$ as a stack.

One essential precondition for constructing virtual cycle is the existence of a perfect obstruction theory. Let S be a chart of $\mathfrak{M}_{g,k}(X, A)$ with a tautological family $f : D \subset C \to X$ over S. T Then S has a perfect obstruction theory with the associated complex

$$\mathcal{E}xt^1_{C/S}(\mathcal{A}_f, \mathcal{O}_C) \xrightarrow{\times 0} \mathcal{E}xt^2_{C/S}(\mathcal{A}_f, \mathcal{O}_C), \tag{8.1}$$

where $\mathcal{A}_f$ is the complex $[f^*\Omega_X \to \Omega_{X/S}(D)]$. It is direct to check that this is a perfect complex in the derived category.

We are now ready to construct the virtual moduli cycle of $\mathfrak{M}_{g,k}(X, A)$. We pick a collection of charts of $\mathfrak{M}_{g,k}(X, A)$, say $\{(S_\alpha, f_\alpha)\}$, with $f_\alpha : D_\alpha \subset C_\alpha \to X$ the tautological family over S_α. The associated complex of the obstruction theory of f_α is $\mathcal{E}xt^{\bullet}_{C_\alpha/S_\alpha}(\mathcal{A}_\alpha, \mathcal{O}_{S_\alpha})$, where $\mathcal{A}_\alpha = [f_\alpha^* X \to \Omega_{C_\alpha/S_\alpha}(D_\alpha)]$. Following the recipe, we need to find a global vector bundle over $\mathfrak{M}_{g,k}(X,A)$ that makes the obstruction sheaves its quotient sheaves. Working with stack, it amounts to find a collection of vector bundles E_α over S_α, and quotient homomorphisms

$$\mathcal{O}_{S_\alpha}(E_\alpha) \longrightarrow \mathcal{E}xt^2_{C_\alpha/S_\alpha}(\mathcal{A}_\alpha, \mathcal{O}_{C_\alpha}),$$

that are compatible over $S_{\alpha\beta} \cong S_{\beta\alpha}$. In our case, such a vector bundle can be constructed explicitly [**LT1**]. We still denote it by $\{E_\alpha\}$. Then from the Basic Construction 7.3, we can construct (virtual normal) cone cycle $N_\alpha \subset E_\alpha$. Since such construction is unique, $N_\alpha \subset E_\alpha$ are compatible.

In the end, we need to intersect this collection of cone-cycles $[N_\alpha] \in Z_* E_\alpha$ with the collection of zero sections of E_α. We denote by E the collection $\{E_\alpha\}$ viewed as a vector bundle over the stack. We denote by $[N] \subset E$ the corresponding virtual normal cone cycle. We let η_E be the zero section of E and define the virtual moduli cycle to be

$$[\mathfrak{M}_{g,k}(X,A)]^{\mathrm{vir}} = \eta_E^*([N]) \in A_* \mathfrak{M}_{g,k}(X,A),$$

where η_E^* is the Gysin homomorphism of the zero section η_E, defined in [**Vis**]. The image of this Gysin homomorphism takes values in the group of cycles with rational coefficient.

Once we have the virtual moduli cycle, we define the GW invariant to be

$$\gamma^X_{A,g,k}(\alpha) =< [\mathfrak{M}_{g,k}(X,A)]^{\mathrm{vir}}, \mathrm{ev}^*(\alpha) > \in \mathbb{Q}, \qquad \alpha \in A^* X^k.$$

This completes the task of constructing the GW invariants of smooth projective varieties.

9. Some Remarks

The author will close this article by bring attention to some challenging research directions.

One challenging problem is the *GW-invariants of singular varieties*. This is important in regards of Super-String theories. Recently, the question to how GW-invariants vary under deformations (degenerations) and/or resolutions of varieties become central to understanding various super-String theories. A closely related question is the following: Let $i : X \to Y$ be a subvariety of a smooth variety. Then $\mathfrak{M}_{g,k}(X,A)$ is a substack of $\mathfrak{M}_{g,n}(Y, i_* A)$, and hence $[\mathfrak{M}_{g,k}(X,A)]^{\mathrm{vir}}$ is a cycle in $A_* \mathfrak{M}_{g,n}(Y, i_* A)$. The question is how the cycle $[\mathfrak{M}_{g,k}(X,A)]^{\mathrm{vir}}$ related to the virtual moduli cycle $[\mathfrak{M}_{g,n}(Y, i_* A)]^{\mathrm{vir}}$, preferably relying only on the data associated to the embedding $X \to Y$. This is known for the special case when $X \subset Y$ is the vanishing locus of a section of an ample vector bundle and $g = 0$. The main obstacle to this problem is how to construct the virtual cycle, if it ever exists, when the obstruction theory is not perfect.

The other research direction is Donaldson polynomial invariant of surfaces of general type. Kronheimer and Mrowka in [**KM**] found a surprising close formula of the generating function of this invariant under the condition that the four manifolds are of simple type. As to surfaces of general type, it is not clear at the moment if they belong to this class of manifolds. Along the same line, Witten in [**Wi3**] argued that surfaces of general type should have simple type. As mentioned in this note, due to Donaldson's general smoothness result, for large second Chern class the Donaldson polynomial invariants are the integration of the tautological topological classes over the moduli spaces. The conjectured closed form of Donaldson polynomial invariants implies that all (except some special case) invariants of the moduli spaces with non-negative expected dimensions are non-trivial. It is expected that these invariants coincide with the one defined using virtual cycle technique. If all these can be confirmed, then at least we have an existence theorem of stable vector bundles of low second Chern class which is better than those currently known.

References

[Ale] V. Alexeev, *Moduli spaces $M_{g,n}(W)$ for surfaces.* Higher-dimensional complex varieties (Trento, 1994), 1-22, de Gruyter, Berlin, 1996.

[Art] I.V. Artamkin, *On deformation of sheaves.* Math. USSR. Izvestiya **32**(1989) No. 3, 663-668.

[Bar] V. Baranovsky, *Moduli of sheaves on surfaces and action of the oscillator algebra.* math.AG/9811092.

[Bea] A. Beauville, *Sur la cohomologie de certains espaces de modules de fibr es vectoriels.* Geometry and analysis (Bombay, 1992), 37–40, Tata Inst. Fund. Res., Bombay, 1995.

[Beh] K. Behrend, *Gromov-Witten invariants in alebraic geometry.* Invent. Math. **127**(1997), no. 3, 601-617.

[BFa] K. Behrend and B. Fantechi, *The intrinsic normal cone.* Invent. Math. **128**(1997), no. 1, 45-88.

[DMu] P. Deligne and D. Mumford, *The irreducibility of the space of curves of given genus.* Publ. I.H.E.S. **45**(1969), 101-145.

[Don] S.K. Donaldson, *Polynomial invariants for smooth four-manifolds* Topology **29**(1986) No. 3, 257-315.

[DN] J.M. Drezet and M.S. Narasimhan, *Group de Picard des varietes de modules de fibres semi-stables sur les coubes algebriques.* Invent. Math. **97**(1989), 53-94.

[FM] J. Fogarty, and D. Mumford, *Geometric Invariant Theory.* Springer-Verlag, 1982.

[FP] W. Fulton and R. Pandaripandas, *Notes on stable maps and quantum cohomology.* Algebraic geometry—Santa Cruz 1995, 45–96, Proc. Sympos. Pure Math., 62, Part 2, Amer. Math. Soc., Providence, RI, 1997.

[Fr1] R. Friedman, *Rank two vector bundles over regular elliptic surfaces* Invent. Math. **96**(1989), 283-332.

[Fr2] R. Friedman, *Algebraic surfaces and holomorphic vector bundles.* Springer, 1991.

[FQ1] R. Friedman and Z-B. Qin, *Flips of moduli spaces and transition formulas for Donaldson polynomial invariants of rational surfaces.* Comm. Anal. Geom. **3**(1995), no. 1-2, 11–83.

[FQ2] R. Friedman and Z-B. Qin, *On complex surfaces diffeomorphic to rational surfaces.* Invent. Math. **120**(1995), no. 1, 81–117.

[Fu] W. Fulton, *Intersection theory.* Ergebnisse der Math. und ihrer Grenzgebiete 3. Folge Band 2, Springer 1984.

[GH] L. Gottsche and D. Huybrechts, *Hodge numbers of moduli spaces of stable bundles on $K3$ surfaces.* Internat. J. Math. **7**(1996), no. 3, 359–372.

[Gi1] D. Gieseker, *On the moduli of vector bundles on an algebraic surface.* Ann. Math. **106**(1977), 45-60.

[Gi2] D. Gieseker, *A construction of stable bundles on an algebraic surface.* J. Diff. Geom. **27**(1988) 137-154.

[Gi3] D. Gieseker, *Geometric invariant theory and the moduli of bundles.* Gauge theory and the topology of four-manifolds (Park City, UT, 1994), 5–21, IAS/Park City Math. Ser., 4, Amer. Math. Soc., Providence, RI, 1998.

[Gi4] D. Gieseker, *Global moduli for surfaces of general type.* Invent. Math. **43**(1977), no. 3, 233–282.

[GL1] D. Gieseker, and J. Li, *Irreducibility of moduli of rank two vector bundles.* J. Diff. Geom. **40**(1994), 23-104.

[GL2] D. Gieseker, and J. Li, *Moduli of high rank vector bundles over surfaces.* J. Amer. math. soc., **9** no. 1 (1996) 107-151.

[Go1] L. Göttsche, *The Betti numbers of the Hilbert scheme of points on a smooth projective surface.* Math. Ann. **286**(1990), no. 1-3, 193–207.

[Go2] L. Göttsche, *Modular forms and Donaldson invariants for 4-manifolds with $b_+ = 1$.* J. Amer. Math. Soc. **9**(1996), no. 3, 827–843.

[Gro] A. Grothendieck, *Techniques de construction et théorèmes d'existence en géométrie algébrique IV, les schémas de Hilbert.* Séminaire Bourbaki **221**(1960-61).

[HL] D. Huybrechts and M. Lehn, *The geometry of moduli spaces of sheaves.* Aspects of Mathematics, E31. Friedr. Vieweg and Sohn, Braunschweig, 1997.

[HLi] Y. Hu and W-P. Li *Variation of the Gieseker and Uhlenbeck compactifications.* Internat. J. Math. **6**(1995), no. 3, 397–418.

[Ill] L. Illusie, *Complexe cotangent et deformations, i.* Lecture Notes in Math. **239**, Springer, Berlin, 1971.

[Kol] J. Kollár, *Rational curves on algebraic varieties.* Ergebnisse der Mathematik und ihrer Grenzgebiete. 3. Folge. A Series of Modern Surveys in Mathematics, 32. Springer-Verlag, Berlin, 1996.

[KM] P.B. Kronheimer and T.S. Mrowka, *Embedded surfaces and the structure of Donaldson's polynomial invariants.* J. Differential Geom. **41**(1995), no. 3, 573–734.

[KMa] M. Kontsevich and Y. Manin, *Gromov-Witten classes, quantum cohomology, and enumerative geometry.* Comm. Math. Phys. **164**(1994), no. 3, 525-562.

[KMu] F. Knudsen and D. Mumford, *The projectivity of the moduli space of stable curves I, preliminaries on 'det' and 'div'.* Math. Scand. **39**(1976), 19-55.

[Kon] M. Kontsevich, *Enumeration of rational curves via torus actions.* The Moduli Space of Curves, edited by R.Dijkgraaf, C.Faber, G. van der Geer, Progress in Mathematics vol. 129, Birkhauser, 1995.

[Kre] A. Kresch, *Cycle groups for Artin stacks.* Invent. Math. **138**(1999), no. 3, 495–536.

[Lau] O. Laudal, *Formal moduli of algebraic structures.* Lecture Notes in Mathematics, 754. Springer, Berlin, 1979.

[LP1] J. Le Potier, *Fibré déterminant et courbes de saut sur les surfaces algb́riques.* Complex projective geometry (Trieste, 1989/Bergen, 1989), 213–240, London Math. Soc. Lecture Note Ser., 179, Cambridge Univ. Press, Cambridge, 1992.

[LP2] J. Le Potier, *Lectures on vector bundles.* Translated by A. Maciocia. Cambridge Studies in Advanced Mathematics, 54. Cambridge University Press, Cambridge, 1997.

[Li1] J. Li, *Algebraic geometric interpretation of Donaldson's polynomial invariants.* J. of Diff. Geom. **37**(1993), 417-466.

[Li2] J. Li, *Kodaira dimension of moduli space of vector bundles on surfaces.* Invent. Math. **115**(1994), 1-40.(with D. Gieseker) Irreducibility of moduli of rank two vector bundles on surfaces, *J. Diff. Geom.*, **40** (1994) 23-104.

[Li3] J. Li, *The first two Betti numbers of moduli of vector bundles over surfaces. Comm. in Analysis and Geometry*, **5** no. 4 (1997) 625-684.

[Li4] J. Li, *Picard groups of the moduli spaces of vector bundles over algebraic surfaces.* Moduli of vector bundles (Sanda, 1994; Kyoto, 1994), 129–146, Lecture Notes in Pure and Appl. Math., 179, Dekker, New York, 1996.

[LT1] J. Li and G. Tian, *Virtual moduli cycles and Gromov-Witten invariants of algebraic varieties.* J. Amer. Math. **11**(1998), no. 1, 119-174.

[LT2] J. Li and G. Tian, *Virtual moduli cycles and Gromov-Witten invariants of general symplectic manifolds.* Topics in symplectic 4-manifolds (Irvine, CA, 1996), 47–83, First Int. Press Lect. Ser., I, Internat. Press, Cambridge, MA, 199.

[LT3] J. Li and G. Tian, *Algebraic and symplectic geometry of virtual moduli cycles.* Algebraic geometry—Santa Cruz 1995, 143–170, Proc. Sympos. Pure Math., 62, Part 2, Amer. Math. Soc., Providence, RI, 1997.

[LT4] J. Li and G. Tian, *Comparison of algebraic and symplectic Gromov-Witten invariants.* Asian J. Math. **3**(1999) no. 3, 689-728.

[LT5] J. Li and G. Tian, *A brief tour of Gromov-Witten invariants.* To appear.

[Lw] W-P. Li, *Relations between moduli spaces of stable bundles over P_2 and rationality.* J. Reine Angew. Math. **484**(1997), 201–217.

[LQ1] W-P.. Li and Z. Qin, *On blowup formulae for the S-Duality conjecture of Vafa and Witten I.* Invent. Math. **136**(1999), no. 2, 451–482.

[LQ2] W-P. Li and Z-B. Qin, *Vertex operator algebras and the blowup formula for the S-duality conjecture of Vafa and Witten.* Math. Res. Lett. **5**(1998), no. 6, 791–798.

[LQ3] W-P. Li and Z-B. Qin, *On blowup formulae for the S-duality conjecture of Vafa and Witten. II. The universal functions.* Math. Res. Lett. **5**(1998), no. 4, 439–453.

[Mar] M. Maruyama, *Moduli of stable sheaves I.* J. Math. Kyoto Univ. **17**(1977), 91-126. *II.* J. Math. Kyoto Univ. **18**(1978), 557-614.

[Mog] J. Morgan, *Comparison of the Donaldson polynomial invariants with their algebro-geometric analogues.* Topology **32**(1993), no. 3, 449–488.

[Muk] S. Mukai, *Symplectic structure on the moduli space of sheaves on an Abelian or K3 surfaces.* Inve. Math. **77**(1984), 101-116.

[Mum] D. Mumford, *Geometric Invariant Theory.* Ergebnisse der Mathematik und ihrer Grenzgebiete, Neue Folge, Band 34 Academic Press, Inc., New York; Springer-Verlag, Berlin-New York 1965.

[Nak] H. Nakajima, *Heisenberg algebra and Hilbert schemes of points on projective surfaces.* Ann. of Math. (2) **145**(1997), no. 2, 379–388.

[OG1] F.G. O'Grady, *Algebro-geometric analogues of Donaldson's polynomials.* Invent. Math. **107**(1992), 351-395.

[OG2] K.G. O'Grady, *The irreducible components of moduli spaces of vector bundles on surfaces.* Invent. Math. **112**(1993) 585-613.

[OG3] K.G. O'Grady, *Moduli of vector bundles on projective surfaces: some basic results.* Invent. Math. **123**(1996), no. 1, 141–207.

[OSS] C. Okonek, M. Schneider and H. Spindler, *Vector Bundles on Complex Projective Spaces.* Progress in Mathematics, 3 Birkhauser Boston-Basel-Stuttgart, 1980.

[Qi1] Z. Qin, *Birational properties of moduli spaces of stable locally free rank-2 sheaves on algebraic surfaces.* Manuscripta Math., **72**(1991), 163-180.

[Qi2] Z. Qin, *On the existence of stable rank-*2 *sheaves on algebraic surfaces.* J. Reine Angew. Math. **439**(1993), 213–219.

[Qi3] Z-B. Qin, *Complex structures on certain differentiable* 4*-manifolds.* Topology **32**(1993), no. 3, 551–566.

[Qi4] Z-B. Qin, *On smooth structures of potential surfaces of general type homeomorphic to rational surfaces.* Invent. Math. **113**(1993), no. 1, 163–175.

[PW] T. Parker and J. Wolfson, *Pseudo-holomorphic maps and bubnle trees.* J. Geom. Anal. **3**(1993), no. 1, 63-98.

[Ran] Z. Ran, *Deformations of maps.* Algebraic curves and projective geometry, E. Ballico, C. Ciliberto, eds, Lecture notes in Math. 1389.

[Rua] Y. Ruan, *Topological Sigma model and Donaldson type invariants in Gromov theory.* Duke Math. J. **83**(1996), no. 2, 461–500.

[RT] Y.B. Ruan and G. Tian, *A mathematical theory of quantum cohomology.* J. Diff. Geom. **42**(1995), no. 2, 259-367.

[Sim] C. Simpson, *Some families of local systems over smooth projective varieties.* Ann. of Math. **138**(1993), 337-425.

[Vaf] C. Vafa, *Extending mirror conjecture to Calabi-Yau with bundles.* Commun. Contemp. Math. **1**(1999), no. 1, 65–70.

[Vie] E. Viehweg, *Weak positivity and the stability of certain Hilbert points I, II, III.* Invent. Math. **96**(1989), no. 3, 639–667; **101**(1990), no. 1, 191–223; **101**(1990), no. 3, 521–543.

[Vis] A. Vistoli, *Intersection theory on algebraic stacks and on their moduli spaces.* Invent. Math. **97**(1989), no. 3, 613-670.

[VW] C. Vafa and E. Witten, *A strong coupling test of S-duality.* Nuclear Phys. B **431**(1994), no. 1-2, 3–77.

[Wi1] E. Witten, *Topological sigma models,* Comm. Math. Phys. **118**(1988), no. 3, 411-449.

[Wi2] E. Witten, *Two dimensional gravity and intersection theory on moduli space,* Surveys in differential geometry (Cambridge, MA, 1990), 243–310, Lehigh Univ., Bethlehem, PA, 1991.

[Wi3] E. Witten, *Monopoles and four-manifolds.* Math. Res. Lett. **1**(1994), no. 6, 769–796.

[Yo1] K. Yoshioka, *The Betti numbers of the moduli space of stable sheaves of rank* 2 *on* $\mathbf{P}^2$. J. Reine Angew. Math. **453**(1994), 193–220.

[Yo2] K. Yoshioka, *The Betti numbers of the moduli space of stable sheaves of rank* 2 *on a ruled surface.* Math. Ann. **302**(1995), no. 3, 519–540.

[Yo3] K. Yoshioka, *Betti numbers of moduli of stable sheaves on some surfaces.* S-duality and mirror symmetry (Trieste, 1995). Nuclear Phys. B Proc. Suppl. **46**(1996), 263–268.

[Yo4] K. Yoshioka, *The Picard group of the moduli space of stable sheaves on a ruled surface.* J. Math. Kyoto Univ. **36**(1996), no. 2, 279–309.

[Zu1] K. Zuo, *Generic smoothness of the moduli of rank two stable bundles over an algebraic surface.* Math. Z. **207**(1991), no. 4, 629-643.

[Zu2] K. Zuo, *Regular 2-forms on the moduli space of rank-2 stable bundles on an algebraic surface.* Duke Math. J. **65**(1992), no. 1, 45–84.

DEPARTMENT OF MATHEMATICS, STANFORD UNIVERSITY, STANFORD, CA 94305

AMS/IP Studies in Advanced Mathematics
Volume 20, 2001

A Tour of Mirror Symmetry

Bong H. Lian and Shing-Tung Yau

ABSTRACT. This is a brief survey of some recent developments toward understanding Mirror Symmetry. Emphasis is put on the construction of the large radius limit, the mirror formula and its many generalizations, and a general theory of characteristic classes of vector bundles on stable map moduli spaces.

1. Introduction

A major current problem in string physics is to understand how and why a plethora of apparently different string theories are related, often in an unexpected and remarkable way. All of this goes under the rubric of "String Duality". Mirror symmetry is in fact only a special case of String Duality. Here, string theories associated to a certain pair of apparently unrelated Calabi-Yau manifolds turn out to be equivalent. From the mathematical point of view, the ramifications of this equivalence is no less than remarkable. This short note is meant to explain some of the recent progress toward understanding the mathematical ramifications. The survey is by no means complete, and is skewed toward some of the authors' own mathematical interests. We will begin with an outline of our topics described in laymen terms without details. We will point out the topics we shall discussed later in more details.

Mirror symmetry was discovered by physicists [**Dixon**][**LVW**]in the late 80s as a means to relate string theories built on two distinct Calabi-Yau threefolds, known as mirror manifolds. Here a CY manifold X is a compact Kähler manifold of complex dimension n, whose holonomy group is $SU(n)$. The many important reasons why these manifolds are essential will not concern us here. Suffices it to say that a theorem of [**Yau**] ensures that there is a Ricci flat Kähler metric on X, and this metric turns out to be an essential ingredient for building a string theory on X. While two mirror threefolds, say X, Y, may be distinct, they are related topologically in a straightforward way. Namely, their Betti numbers satisfy

$$b_3(X) = 2b_2(Y) + 2, \ b_3(Y) = 2b_2(X) + 2.$$

The first example with this topogical property was constructed in [**GreenePlesser**], followed by many generalizations. [**Morrison**][**KlemmSchimm**][**LT**][**Batyrev**] [**BatyrevBorisov**]One reason for those identities of Betti numbers is that complex structure deformation of X must correspond to the Kähler structure deformation of Y. The number of deformation parameters for X turns out to be $1\frac{2}{b}{}_3(X) - 1$, while the number of deformation parameters of Y is $b_2(Y)$. Thus the two numbers being equal yield the identity $b_3(X) = 2b_2(Y) + 2$ above.

Actually, mirror symmetry goes far beyond a correspondence between deformations of the two manifolds. One must also compactify the deformation spaces appropriately, and there must also be a mapping between the compactified deformation spaces, at least near an infinity point. This mapping is known as the mirror map, and the infinity point is known as the large radius limit, indicating that the volume of one manifold Y becomes large as its Kähler structure approaches this limit. On the deformation space of X, there is a unique vanishing 3-cycle in X near this limit. From the viewpoint of Hodge theory of X, the limit is characterized by the property that all but one period blow up near this point. Constructing a suitable compactification and this limit point for a given Calabi-Yau manifold therefore poses an important problem. We'll discuss a general solution of [**HLY**] to this problem in section 2, when X is a hypersurface in a toric manifold. We'll also describe the mirror map in this case. Mirror symmetry, in fact, goes even further. It turns out that the variation of the periods of X under deformations encode important geometric information about Y as well. Roughly, these periods of X count the (virtual) number of rational curves of any given degree in the manifold Y. The fundamental problem of enumerating rational curves in a manifold can be thought of as an analogue of counting zeros of polynomial equations in number theory. The generating function for our virtual numbers is known as the prepotential of Y. In section 2, we shall discuss a beautiful formula, known as the mirror formula of [**CDGP**], relating the prepotential of Y to the periods of X, and its generalization in [**HKTY, HLYII**]. There is also a far reaching generalization of such counting formulas due to [**BCOV**], for enumerating curves of all genus in a Calabi-Yau threefold.

The remarkable mirror formula and its generalizations allow one to study the prepotential as a function. For example, one can use this formula to construct explicit nonlinear differential equations for this prepotential. These differential equations have been conjectured to exists by physicists. The construction has been carried out in many cases in [**LianYauODE**]. These differential equations have rather remarkable properties. For example, they can be reduced modulo a prime number. Using them, we have in turn deduced many divisibility properties of the prepotential and their Fourier coefficients. One interesting example is a general quintic Y in $\mathbf{P}^4$. The degree d virtual number n_d of rational curves in Y turns out to be divisible by 5^3, as proven in [**LianYauI**] (for $5 \not| d$). This divisibility has been conjectured in [**Clemens**].

Remarkably, mirror symmetry gives an answer to the counting problem in enumerative geometry before the problem has been formulated mathematically. The problem of *defining*, let alone computing, the virtual numbers turns out to be a deep problem, in itself. This problem has recently been solved by [**RuanTian**] in the context of pseudo-holomorphic maps [**SacksUhlenbeck**][**Gromov**][**ParkerWolfson**] [**Ye**] following ideas of [**Vafa**][**WittenI**].

In the algebraic category, the problem has been solved in [**LiTianII**] by constructing a virtual fundamental cycle in the context of stable maps [**KontsevichI**] [**BehrendManin**]. A similar construction has been done in [**BehrendFantachi**] [**FukayaOno**]. In the symplectic case, the virtual cycle has also been studied by [**LiTianIII**][**SiebertI**][**Ruan**]. A remarkable result in [**LiTianIV**] (see also [**SiebertII**]) says that the symplectic and the algebraic approaches are essentially equivalent on algebraic manifolds. We shall follow the second approach in section 3. We refer the readers for a discussion of both approaches to the lectures of J. Li and G. Tian in this Congress. Once the proper definitions of the virtual numbers are established, the prepotential, the mirror formula, and its generalizations can now be studied from a mathematical viewpoint [**RuanTian**][**KontsevichI**]. In section 3, we shall describe a general theory in [**LLYI**][**LLYII**][**LLYIII**], for characteristic classes of vector bundles on a stable map moduli space. The the mirror formula and its many generalizations are proved as a consequence. For a more detailed survey see [**LLYSurvey**][**CoxKatz**]. Another approach considers a mirror formula from the point of view of "quantum differential equations". For more details on this approach, see [**GiventalI**][**GiventalII**][**BDPP**][**Pandharipande**].

There is another interesting geometrical approach to mirror symmetry, which we will mention without going into details here. In [**SYZ**], the authors propose to construct a mirror manifold geometrically using the theory of special Lagrangian fibrations. A further generalization in [**Fukaya**][**KontsevichII**][**VafaII**] considers mirror correspondence as a kind of equivalence between two geometrical categories. This approach is currently in its exciting developmental stages.

Acknowledgement. We thank the organizers of the First International Congress of Chinese Mathematicians for their hospitality and for providing a stimulating atmosphere at the Congress. We also thank our collaborators S. Hosono and K. Liu for their cooperation. B.H.L.'s research is supported by NSF grant DMS-9619884. S.T.Y. is supported by DOE grant DE-FG02-88ER25065 and NSF grant DMS-9803347.

2. The Large Radius Limit and The Mirror Formula

Let X be a Calabi-Yau manifold. Thus it is a compact Kähler manifold X whose canonical bundle K_X is trivial. The simplest example of a Calabi-Yau manifold is an elliptic curve:

$$y^2 = x^3 + g_2 x + g_3.$$

More generally, if M be a compact Kähler manifold and the bundle K_M^{-1} has a nonzero section f such that its zero locus X is nonempty and nonsingular, then by the adjunction formula, we have

$$K_X = (K_M \otimes [X])|_X = 0.$$

In this case, the hypersurface X in M is a Calabi-Yau manifold. For example for $M = \mathbf{P}^n$, we have $K_M^{-1} = \mathcal{O}(n+1)$, and a generic section of this bundle cuts out a Calabi-Yau manifold.

Let X be a Calabi-Yau n-fold. Since K_X is trivial, there is a nowhere vanishing holomorphic n-form ω which is unique up to a scalar. From this, we get a map $H_n(X) \to \mathbf{C}$, $\gamma \mapsto \int_\gamma \omega$. We will be interested in the family version of this map.

Let $\mathcal{X} \to \mathcal{M}$ be a flat family whose generic fiber X_z is a Calabi-Yau manifold. There is a flat bundle on $\mathcal{M}$ (away from the singular X_z) whose fiber at z is the vector space $H^n(X_z, \mathbf{C})$. In it, there is a line bundle $\mathcal{L}$ whose fiber is $H^0(X_z, \wedge^n T^* X_z)$. Thus a section of $\mathcal{L}$ gives a family of holomorphic n-forms ω_z. Integrating this against a class $\gamma \in H_n(X_z)$, we get a (multi-valued) function

$$f_\gamma(z) = \int_\gamma \omega_z.$$

This is called the period associated to γ.

Example. Consider the family of cubic curves

$$\sigma_z(x) := x_0^3 + x_1^3 + x_2^3 + zx_0x_1x_2 = 0, \ z \in \mathbf{P}^1$$

defined in $\mathbf{P}^2$. One way to construct the periods is by means of Poincaré residue as follows. Take a meromorphic 2-form Ω_z on $\mathbf{P}^2$ with a first order pole along the curve $\sigma_z(x) = 0$, and take a primitive cycle γ (which is a circle) in the curve. Take a circle bundle Γ over γ inside the normal bundle of $\sigma_z(x) = 0$ in $\mathbf{P}^2$, so that each fiber in Γ goes around the curve once. Now integrate the 2-form Ω_z against Γ, and get a function $f_\gamma(z)$. Explicitly, we can choose

$$\Omega_z = \frac{\sum_i (-1)^i x_i dx_0 \wedge \cdots \wedge \hat{dx_i} \wedge \cdots \wedge dx_2}{\sigma_z(x)}.$$

Thus we get the periods

$$f_\gamma(z) = \int_\Gamma \Omega_z.$$

These functions turn out to be a Gauss hypergeometric function.

For a general Calabi-Yau manifold X, a theorem of Bogomolov-Tian-Todorov says that the deformation space of X is unobstructed. In fact, the periods determine the local variation of the complex structure of X completely.

Definition: *A point $z_0 \in \mathcal{M}$ is called a large radius limit if there is a choice of n-forms ω_z such that the periods have the following property: the subspace $\{\gamma| \ |f_\gamma(z)|$ is bounded near $z_0\} \subset H_n(X)$ is one dimensional.*

Note that near a generic point $z \in \mathcal{M}$, each function $f_\gamma(z)$ is bounded, and so the subspace above will be all of $H_n(X)$. Near a large radius limit, on the other hand, there are as many periods blowing up there as possible. Intuitively, a large radius limit is a point of deepest degeneration in the family. We note that there is an alternative to the above notion, in which one uses the monodromy around z_0 to characterize the deepest degeneration (cf. [**MorrisonII**]).

Example. In the case of the 1-parameter family $\sigma_z(x) = 0$ of cubic curves above, there are exactly three points on $\mathbf{P}^1$ in the descriminant locus, where the fibers are apparently singular. By computing the periods $f_\gamma(z)$ near those points, it is easy to show that $z = \infty$ is the only large radius limit.

As alluded in the introduction, the construction of a LRL is important for realizing the mirror correspondence between two families of Calabi-Yau manifolds. We now discuss a construction of [**HLY**] for a family of Calabi-Yau hypersurfaces in a toric manifold. For background on toric manifolds, see [**Oda**][**FultonToric**]. For a detailed discussion of mirror manifolds in toric varieties, see [**Batyrev**].

Theorem 1: [**HLY**] Let M be a toric manifold with $c_1(M) \geq 0$. Then The moduli space $\mathcal{M}$ of Calabi-Yau hypersurfaces in M admits a large radius limit.

Note that the sections of K_M^{-1} parameterize Calabi-Yau hypersurfaces in M. Let T be the algebraic torus acting on M. Then we have an induced linear action on $H^0(M, K_M^{-1})$. By "the moduli space" $\mathcal{M}$ above, we mean the quotient

$$\mathcal{M} := \mathbf{P}H^0(M, K_M^{-1})/T.$$

(For more details, see [**HLY**].) Note that $\mathcal{M}$ is itself a toric variety. Strictly speaking, the LRL lies in some resolution of singularity of the moduli space.

The proof of the theorem first reduces the problem to a question about a system of holonomic system of PDEs on $\mathcal{M}$. It turns out that these systems, known as a GKZ system, have been studied in a long series of papers by Gel'fand-Graev-Kapranov- Retakh-Zelevinsky. The problem we have to solve is the existence and uniqueness of a certain holomorphic solution to the GKZ system.

It turns out the space

$$\mathcal{M} := \mathbf{P}H^0(M, K_M^{-1})/T$$

has a set of canonical coordinates $(a_1, ..., a_p)$ upstairs which correspond to toric divisors $D_1, ..., D_p$ on another toric manifold W. For convenience, we write $e^{t_i} := -a_i$. Let $[D_i] \in H^2(W, \mathbf{Z})$ be the cohomology classes Poincare dual to the divisors D_i. In [**HLY**], the following formula for solutions to the GKZ system was given:

$$\omega(t) := \sum_{l \in \Lambda} \frac{\prod_{m=1}^{\langle l,[\mathcal{L}]\rangle}([\mathcal{L}]+m) \times \prod_{\langle l,D_i\rangle<0} \prod_{m=0}^{-\langle l,[D_i]\rangle-1}([D_i]-m)}{\prod_{\langle l,D_i\rangle\geq 0}\prod_{m=1}^{\langle l,[D_i]\rangle}([D_i]+m)} e^{\langle l,t\rangle+t}.$$

Here $t := \sum t_i[D_i]$, $[\mathcal{L}] := c_1(K_W^{-1})$, and $\Lambda := \mathcal{K}' \cap H_2(W, \mathbf{Z})$ where $\mathcal{K}'$ is the cone in $H_2(W, \mathbf{R})$ dual to the Kähler cone $\mathcal{K} \subset H^2(W, \mathbf{R})$.

Note that $\omega(t)$ is vector-valued series, ie. it takes values in $H^*(W, \mathbf{C})$. To ensure convergence of $\omega(t)$, we must restrict t so that $Re\langle l, t\rangle < 0$ all nonzero $l \in \Lambda$. This means that

$$t \in -\mathcal{K} + \sqrt{-1}H^2(W, \mathbf{R}).$$

The set on the right hand side is called complexified Kähler cone W. The components of the vector-valued function $\omega(t)$ give the periods of the Calabi-Yau hypersurfaces X in M.

Consider now the case when both toric manifolds M, W are 4 dimensional. When $W = \mathbf{P}^4$, the mirror formula of [**CDGP**] is an explicit formula giving the prepotential Ψ of the quintic in W in terms of the function $\omega(t)$. After [**CDGP**], there is also long series of other examples [**Morrison**][**LT**][**HKTY**][**BatyrevStraten**] [**CDFKM**][**BKK**]. Finally, the general form of the mirror formula for general toric manifold was given in [**HLY**]. In the notation above, this formula is

$$\frac{1}{\omega_0(t)} \int_W [\mathcal{L}] \cdot \omega(t) = 2\Psi(T) - \sum_k T_k \frac{\partial \Psi(T)}{\partial T_k}$$

where the $T_k = T_k(t)$ are functions defined by the asymptotic form

$$\omega(t) = \omega_0(t) + \sum_k T_k(t) H_k + (deg \geq 2 \ terms).$$

Here $H_1, .., H_m$ is a fixed basis of Kähler classes in $H^2(W)$. Note that

$$(*)\ T_k(t) = \langle H_k, t\rangle + exponential\ term.$$

Exponentiating this, we have

$$q_k = z_k(1 + O(z))$$

where $q_k := e^{T_k}$ and $z_k := e^{\langle H_k, t\rangle}$. The z_k can be thought of as coordinates on $\mathcal{M}$ near infinity, and the q_k as the coordinates on the domain $-\mathcal{K} + \sqrt{-1}H^2(W, \mathbf{R}/\mathbf{Z})$. Then the relations (*) can be thought of as a mapping from $\mathcal{M}$ to $-\mathcal{K} + \sqrt{-1}$ $H^2(W, \mathbf{R}/\mathbf{Z})$ near infinity. This mapping is called the mirror map.

Even though the prepotential $\Psi(T)$ was not known exactly before the advent of mirror symmetry, on general ground physicists have known for a long time that it must have the form

$$\Psi(T) = \frac{1}{3!}\sum \int_W H_i H_j H_k \cdot [\mathcal{L}]\ T_i T_j T_k + (deg\ 2\ polynomial) + O(q).$$

The term of order $O(q)$ is known as worldsheet instanton correction, and it should be a sum including contributions from rational curves of all degrees in the Calabi-Yau manifold $Y \subset W$. The form of these contributions has been greatly clarified by [**CDGP**]: they are in fact counting the "virtual" number of rational curves in Y. We will discuss their precise mathematical meaning in section 3.

3. Mirror Principle

Let X be a projective n-fold, and $d \in H^2(X, \mathbf{Z})$. Let $M_{0,k}(d,X)$ denote the moduli stack of k-pointed, genus 0, degree d, stable maps $(C, f, x_1, .., x_k)$ on X[**KontsevichI**]. (Note that our notation is without the bar.) By [**LiTianII**], each nonempty $M_{0,k}(d,X)$ admits a homology cycle $LT_{0,k}(d,X)$ of degree $dim\ X + \langle c_1(X), d\rangle + k - 3$. This cycle plays the role of the fundamental class in topology, hence $LT_{0,k}(d,X)$ is called the virtual fundamental class.

Let V be a convex vector bundle on X. (ie. $H^1(\mathbf{P}^1, f^*V) = 0$ for every holomorphic map $f : \mathbf{P}^1 \to X$.) Then V induces on each $M_{0,k}(d,X)$ a vector bundle V_d, with fiber at $(C, f, x_1, .., x_k)$ given by the section space $H^0(C, f^*V)$. Let b be any multiplicative characteristic class [**Hirzebruch**]. (ie. if $0 \to E' \to E \to E'' \to 0$ is an exact sequence of vector bundles, then $b(E) = b(E')b(E'')$.) The problem we study here is to compute the characteristic numbers

$$K_d := \int_{LT_{0,0}(d,X)} b(V_d)$$

and their generating function:

$$\Phi(t) := \sum K_d\ e^{d\cdot t}.$$

When X is a toric manifold with $c_1(X) \geq 0$, and V is the anticanonical bundle, then Φ amounts to what physicists call (the exponential part of Ψ at the end of section 2) the prepotential of the Calabi-Yau hypersurface cut out by a section of V. There is a similar and equally important problem if one starts from a concave vector bundle V [**LLYI**]. (ie. $H^0(\mathbf{P}^1, f^*V) = 0$ for every holomorphic map $f : \mathbf{P}^1 \to X$.) More generally, V can be a direct sum of a convex and a concave bundle. Important progress made on these problems has come from mirror symmetry. All of it seems to point toward the following general phenomenon that there is a mirror formula for $\Phi(t)$. This is what we call *the Mirror Principle.*

We now sketch our main ideas for computing the classes $b(V_d)$.

Step 1. Localization on the linear sigma model. Consider the moduli spaces $M_d(X) := M_{0,0}((1,d), \mathbf{P}^1 \times X)$. The projection $\mathbf{P}^1 \times X \to X$ induces a map $\pi : M_d(X) \to M_{0,0}(d,X)$. Moreover, the standard action of S^1 on $\mathbf{P}^1$ induces an S^1 action on $M_d(X)$. We first study a slightly different problem. Namely consider the classes $\pi^* b(V_d)$ on $M_d(X)$, instead of $b(V_d)$ on $M_{0,0}(d,X)$. First, there is a canonical way to embed fiber products (see below)

$$F_r = M_{0,1}(r,X) \times_X M_{0,1}(d-r,X)$$

each as an S^1 fixed point component into $M_d(X)$. Let $i_r : F_r \to M_d(X)$ be the inclusion map. Second, there is an evaluation map $e : F_r \to X$ for each r. Third, suppose that there is a projective manifold W_d with S^1 action, that there is an equivariant map $\varphi : M_d(X) \to W_d$, and embeddings $j_r : X \to W_d$, such that the diagram

$$\begin{array}{ccc} F_r & \xrightarrow{i_r} & M_d(X) \\ e \downarrow & & \downarrow \varphi \\ X & \xrightarrow{j_r} & W_d \end{array}$$

commutes. Let α denote the weight of the standard S^1 action on $\mathbf{P}^1$. Then applying the localization formula [**AB**], this diagram allows us to recast our problem to one of studying the S^1-equivariant classes

$$Q_d := \varphi_! \pi^* b(V_d)$$

defined on W_d. Moreover we can expand the class

$$A_d := \frac{j_0^* Q_d}{e_{S^1}(X_0/W_d)}$$

on X in powers of α^{-1}, and find that it is of order α^{-2}.

The spaces W_d in the commutative diagram above are called the linear sigma model of X. They have been introduced in [**MP**] following [**WittenIII**] when X is a toric manifold,

Step 2. Gluing identity. Consider the vector bundle $\mathcal{U}_d := \pi^* V_d \to M_d(X)$, restricted to the fixed point components F_r. A point in (C, f) in F_r is a pair $(C_1, f_1, x_1) \times (C_2, f_2, x_2)$ of 1-pointed stable maps glued together at the marked points, ie. $f_1(x_1) = f_2(x_2)$. From this, we get an exact sequence of bundles on F_r:

$$0 \to i_r^* \mathcal{U}_d \to U_r' \oplus U_{d-r}' \to e^* V \to 0.$$

Here $i_r^* \mathcal{U}_d$ is the restriction to F_r of the bundle $\mathcal{U}_d \to M_d(X)$. And U_r' is the pullback of the bundle $U_r \to M_{0,1}(d, X)$ induced by V, and similarly for U_{d-r}'. Taking the multiplicative characteristic class b, we get the identity on F_r:

$$e^* b(V) b(i_r^* \mathcal{U}_d) = b(U_r') b(U_{d-r}').$$

This is what we call the *gluing identity.* This may be translated to a similar quadratic identity, via Step 1, for Q_d in the equivariant cohomology groups $H_{S^1}^*(W_d)$. The new identity is called the Euler data identity.

Step 3. Linking theorem. The construction above is functorial, so that if X comes equipped with a torus T action, then the entire construction becomes $G = S^1 \times T$ equivariant and not just S^1 equivariant. In particular, the Euler data identity is an identity of G-equivariant classes on W_d. Our problem is to first compute the G-equivariant classes Q_d on W_d satisfying the Euler data identity, and with the property that $A_d \sim \alpha^{-2}$. Note that the restrictions $Q_d|_p$ to the T fixed points p in $X_0 \subset W_d$ are polynomials functions on the Lie algebra of G. Suppose that X is a balloon manifold. This is a complex projective T-manifold satisfying the following conditions [**GKM**]:

- (i) The T fixed points are isolated.
- (ii) Let p be a T fixed point. Then the T weights $\lambda_1, .., \lambda_n$ of the isotropic representation on the tangent space T_pX are pairwise linearly independent.

Then it can be shown that the classes Q_d are uniquely determined by the values of the $Q_d|_p$, when α is some scalar multiple of a weight λ_i These values of $Q_d|_p$ can be computed explicitly by exploiting the moment map [**Atiyah**][**GS**] as well as certain structure of a balloon manifold [**GZ**].

Once these values are known, it is often easy to manufacture explicit G-equivariant classes $\tilde{Q}_d$ with the restrictions $\tilde{Q}_d|_p$ having the above same values, and satisfying the Euler data identity. In this case, we say that the data $\tilde{Q}_d$ are linked to the data Q_d. By a suitable change of variables, one can also arrange that $\frac{j_0^*\tilde{Q}_d}{e_{S^1}(X_0/W_d)} \sim \alpha^{-2}$. By the preceding discussion, we get $Q_d = \tilde{Q}_d$.

Step 4. Computing $\Phi(t)$. Once the classes $Q_d = \varphi_! \pi^* b(V_d)$ are determined, we can unwind the many maps use in Step 1. The preceding computations can be done simply in the form of power series. This finally computes the generating function $\Phi(t)$.

For a large class of balloon manifolds (defined later), it has been proven in [**LLYI**][**LLYII**] that the 4-step procedure above yield a mirror formula for $\Phi(t)$. For a general projective manifold without T action, we expect that this formula continues to hold. In this note, for clarity, we restrict ourselves to the case when the tangent bundle of X is convex. In the nonconvex case, we must replace $M_{0,k}(d, X)$ by Li-Tian's virtual fundamental cycle [**LiTianII**] for the purpose of localization and integration. A soon-to-appear paper, Mirror Principle III, joint with K. Liu will be devoted entirely to dealing with the added technicality arising from this replacement. All the results in this paper can be generalized with only slight modifications resulted from this replacement.

We now formulate our main theorem precisely. Let X be balloon manifold We write

$$M_d(X) := M_{0,0}((1, d), \mathbf{P}^1 \times X).$$

Since X is assumed to be convex, $M_d(X)$ is an orbifold [**BehrendManin**]. The standard S^1 action on $\mathbf{P}^1$ together with the T action on X induce a $G = S^1 \times T$ action on $M_d(X)$.

Here is a description of the S^1 fixed point components F_r, labelled by $0 \preceq r \preceq d$, inside of $M_d(X)$. Let F_r be the fiber product

$$F_r := M_{0,1}(r, X) \times_X M_{0,1}(d - r, X)$$

More precisely, consider the map

$$ev_r \times ev_{d-r} : M_{0,1}(r, X) \times M_{0,1}(d - r, X) \to X \times X$$

given by evaluations at the marked points; and

$$\Delta : \ X \to X \times X$$

the diagonal map. Then

$$F_r = (ev_r \times ev_{d-r})^{-1} \Delta(X).$$

Note that $F_d = M_{0,1}(d, X) = F_0$ by convention. The set F_r can be identified with an S^1 fixed point component of $M_d(X)$ as follows. Consider the case $r \neq 0, d$ first. Given a point $(C_1, f_1, x_1) \times (C_2, f_2, x_2)$ in F_r, we get a new curve C by gluing C_1, C_2 to $\mathbf{P}^1$ with x_1, x_2 glued to $0, \infty \in \mathbf{P}^1$ respectively. The new curve C is mapped into $\mathbf{P}^1 \times X$ as follows. Map $\mathbf{P}^1 \subset C$ identically onto $\mathbf{P}^1$, and collapse C_1, C_2 to $0, \infty$ respectively; then map C_1, C_2 into X with f_1, f_2 respectively, and collapse the

$\mathbf{P}^1$ to $f(x_1) = f(x_2)$. This defines a point (C, f) in $M_d(X)$. For $r = 0$, we glue (C_1, f_1, x_1) to $\mathbf{P}^1$ at x_1 and 0. For $r = d$, we glue (C_2, f_2, x_2) to $\mathbf{P}^1$ at x_2 and ∞. We will identify F_r as a subset of $M_d(X)$ as above, and let

$$i_r : F_r \to M_d(X)$$

denotes the inclusion map. Clearly, we also have an evaluation map

$$e_r : F_r \to X$$

which sends a pair in F_r to the value at the marked point. In the following, we will simply write e_r as e without causing any confusion.

We call a compact manifold or orbifold W_d with $G = S^1 \times T$ action a *linear sigma model* of degree d for X, if the following conditions are satisfied:

- 1. The S^1 action on W_d has fixed point components given by X_r, labelled by $0 \preceq r \preceq d$, and each X_r is T-equivariantly isomorphic to X.
- 2. There is a G-equivariant birational map φ from $M_d(X)$ to W_d, such that $\varphi|_{F_r} = e$, and $\varphi^{-1}(X_r) = F_r$.
- 3. All equivariant cohomology classes in $H^2_G(W_d)$ are lifted from $H^2_T(X)$, and the lift $\hat{D} \in H^2_G(W_d)$ of $D \in H^2_T(X)$ restricts to $D + \langle D, r\rangle\alpha$ on X_r.
- 4. The G-equivariant Euler class of the normal bundle of X_0 in W_d has the form

$$e_G(X_0/W_d) = \prod_a \prod_{m_a} (D_a - m_a\alpha)$$

where the m_a's are positive integers and the D_a's are classes in $H^2_T(X)$, such that at a given T fixed point p in X, the nonzero $D_a(p)$'s are multiples of distinct weights of T_pX.

Note W_d need not be unique. We identify X_r with X by assumption 1, and denote by

$$j_r : X_r \to W_d$$

the inclusion map.

We call a balloon manifold X a *admissible* if it has a linear sigma model W_d for each d, and and that for any equivariant class $c \in H^2_T(X)$ we have $c|_p \neq c|_p$ for any two distinct fixed points p, q in X. For simplicity, we state the main result in the case when $V^{\pm}$ are direct sum of line bundles. on an n dimensional admissible balloon manifold X, and the multiplicative class b is the Chern polynomial. For more general case see [**LLYII**].

We denote by $e_{S^1}(X_0/W_d)$ the equivariant Euler class of the normal bundle of X_0 in W_d. Let

$$V = V^+ \oplus V^-,\ V^+ := \oplus L_i^+,\ V^- := \oplus L_j^-$$

satisfying $c_1(V^+) - c_1(V^-) = c_1(X)$, and $rk\ V^+ - rk\ V^-(n-3) \geq 0$. where the $L_i^\pm$ are respectively convex/concave line bundles on X. Let

$$\begin{aligned}
\Omega &= B_0 := c(V^+)/c(V^-) = \prod_i (x + c_1(L_i^+))/\prod_j (x + c_1(L_j^-)) \\
B_d &:= \frac{1}{e_{S^1}(X_0/W_d)} \times \prod_i \prod_{k=0}^{\langle c_1(L_i^+),d\rangle} (x + c_1(L_i^+) - k\alpha) \\
&\quad \times \prod_j \prod_{k=1}^{-\langle c_1(L_j^-),d\rangle - 1} (x + c_1(L_j^-) + k\alpha). \\
HG[B](t) &:= \sum B_d e^{d\cdot t} \\
\Phi(t) &:= \sum K_d e^{d\cdot t}.
\end{aligned}$$

Theorem 2: *There exist unique power series $f(t), g(t)$ such that the following formula holds:*

$$\frac{1}{s!}\left(\frac{d}{dx}\right)^s |_{x=0} \int_X \left(e^{f/\alpha} HG[B](t) - e^{-H\cdot\tilde{t}/\alpha}\Omega\right) = \alpha^{-3} x^{-s} (2\Phi(\tilde{t}) - \sum_i \tilde{t}_i \frac{\partial\Phi(\tilde{t})}{\partial\tilde{t}_i}).$$

where $s := rk\ V^+ - rk\ V^- - (n-3)$, $\tilde{t} := t + g$. Moreover, f, g are determined by the condition that the integrand on the left hand side is of order $O(\alpha^{-2})$.

Note that when $x \to 0$, the formula above reduces to the case when b is the Euler class.

Example. A complete intersection in $\mathbf{P}^1 \times \mathbf{P}^2 \times \mathbf{P}^2$.

The complete intersection of degrees $(1,3,0)$, $(1,0,3)$ in this 5-dimensional toric manifold X has been studied in [**HKTY**] using mirror symmetry, and in [**HSS**] computing some of the intersection numbers K_d for the Euler class $b = e$ in terms of modular forms.

From our point of view, that complete intersection correspond to the following choice of convex bundle:

$$V = \mathcal{O}_1(1) \otimes \mathcal{O}_2(3) \oplus \mathcal{O}_1(1) \otimes \mathcal{O}_3(3)$$

where $\mathcal{O}_i(l)$ denotes the pullback of $\mathcal{O}(l)$ from the ith factor. The Kähler cone of X is abviously generated by the hyperplanes H_1, H_2, H_3 from the three factors of X, and hence $K^\vee$ can be identified with the set of $d = (d_1, d_2, d_3) \in \mathbf{Z}^3_{\geq 0}$. We consider intersection numbers K_d for the Euler class $b = e$ as before. In this case, we get $\Omega = e(V) = (H_1 + 3H_2)(H_1 + 3H_3)$, and

$$\begin{aligned}
B_d &= \frac{1}{e_{S^1}(X_0/W_d)} \times \prod_{k=0}^{d_1+3d_2} (H_1 + 3H_2 - k\alpha) \times \prod_{k=0}^{d_1+3d_3} (H_1 + 3H_3 - k\alpha) \\
e_G(X_0/W_d) &= \prod_{m=1}^{d_1} (H_1 - m\alpha)^2 \prod_{m=1}^{d_2} (H_2 - m\alpha)^3 \prod_{m=1}^{d_3} (H_3 - m\alpha)^3.
\end{aligned}$$

Now we can easily write down the hypregeometric series and all the K_d can be computed at once using the obvious intersection form on X, given by the relations:

$$\int_X H_1 H_2^2 H_3^2 = 1,\ H_1^2 = 1,\ H_2^3 = 1,\ H_3^3 = 1.$$

Once we have the hypergeometric series, the corresponding Picard-Fuchs equation can be easily written down as given in [**HKTY**].

References

[Atiyah] M. Atiyah, *Convexity and commuting Hamiltonians*, Bull. London Math. Soc. 14 (1982) 1-15.

[AB] M. Atiyah and R. Bott, *The moment map and equivariant cohomology*, Topology 23 (1984) 1-28.

[Batyrev] V. Batyrev, *Dual polyhedra and mirror symmetry for Calabi-Yau hypersurfaces in toric varieties*, J. Alg. Geom. 3 (1994) 493-535.

[BatyrevBorisov] V. Batyrev and L. Borisov, *On Calabi-Yau complete intersections in toric varieties*, alg-geom/9412017.

[BatyrevStraten] V. Batyrev and D. van Straten, *Generalized hypergeometric functions and rational curves on Calabi-Yau complete intersections in toric varieties*, Comm. Math. Phys. 168 (1995) 495-533.

[BehrendFantachi] K. Behrend and B. Fantachi, *The intrinsic normal cone*, Invent. Math. 128 (1997) 45-88.

[BehrendManin] K. Behrend and Y. Manin, *Stacks of stable maps and Gromov-Witten invariants*, alg-geom/9506023.

[BKK] P. Berglund, S. Katz and A. Klemm, *Mirror Symmetry and the Moduli Space for Generic Hypersurfaces in Toric Varieties*, hep-th/9506091, Nucl. Phys. B456 (1995) 153.

[BH] P. Berglund and T. Hübsch, Nucl. Phys. B393(1993) 377.

[BCOV] M. Bershadsky, S. Cecotti, H. Ooguri and C. Vafa (with an appendix by S.Katz), Nucl. Phys. B405 (1993) 279.

[BDPP] G. Bini, C. De Concini, M. Polito, and C. Procesi, *Givental's work relative to mirror symmetry*, math.AG/9805097.

[Bott] R. Bott, *A residue formula for holomorphic vector fields*, J. Diff. Geom. 1 (1967) 311-330.

[CDGP] P. Candelas, X. de la Ossa, P. Green, and L. Parkes, *A pair of Calabi-Yau manifolds as an exactly soluble superconformal theory*, Nucl. Phys. B359 (1991) 21-74.

[CDFKM] P. Candelas, X. de la Ossa, A. Font, S. Katz and D. Morrison, *Mirror symmetry for two parameter models I*, hep-th/9308083, Nucl. Phys. B416 (1994) 481-538.

[Clemens] H. Clemens, in *Proceedings of ICM 1986*, at Berkeley.

[Cox] D. Cox, *The Functor of a Smooth Toric Variety*, alg-geom/9312001.

[CoxKatz] D. Cox, and S. Katz, *Mirror Symmetry and Algebraic Geometry*, Mathematical Surveys and Monograph, Volume 68.

[Dixon] L. Dixon, in *Superstrings, Unified Theories, and Cosmology 1987*, G. Furlan et al eds., World Scientific (1988) 67.

[FultonToric] W. Fulton, *Introduction to Toric Varieties*, Princeton Univ. Press 1993.

[FukayaOno] K. Fukaya and K. Ono, *Arnold conjecture and Gromov-Witten invariants*, preprint, 1996.

[Fukaya] K. Fukaya, *Morse homotopy, A^∞-category, and Floer homologies*, in proceedings of 1993 GARC Workshop on Geometry and Topology, H.J. Kim ed., Seoul National Univ.

[GKZ] I. Gel'fand, M. Kapranov and A. Zelevinsky, *Hypergeometric functions and toral manifolds*, Funct. Anal. Appl. 23 (1989) 94-106.

[GiventalI] A. Givental, *Equivariant Gromov-Witten invariants*, alg-geom/9603021.

[GiventalII] A. Givental, *A mirror theorem for toric complete intersections*, alg-geom/9701016.

[GreenePlesser] B. Greene and R. Plesser, *Duality in Calabi-Yau moduli space*, Nucl. Phys. B338 (1990) 15-37.

[Gromov] M. Gromov, *Pseudo holomorphic curves in symplectic manifolds*, Invent. Math. 82 (1985) 307-347.

[GKM] M. Goresky, R. Kottwitz and R. MacPherson, *Equivariant cohomology, Koszul duality and the localization theorem*, Invent. Math. 131 (1998) 25-83.

[GS] V. Guillemin and S. Sternberg, *Convexity properties of the moment mapping*, Invent. Math. 67 (1982) 491-513.

[GZ] V. Guillemin and C. Zara, *Equivariant de Rham Theory and Graphs*, math.DG/9808135.

[Hirzebruch] F. Hirzebruch, *Topological methods in algebraic geometry*, Springer-Verlag, Berlin 1995, 3rd Ed.

[HKTY] S. Hosono, A. Klemm, S. Theisen and S.T. Yau, *Mirror symmetry, mirror map and applications to complete intersection Calabi-Yau spaces*, hep-th/9406055, Nucl. Phys. B433 (1995) 501-554.

[HLYII] S. Hosono, B.H. Lian and S.T. Yau, *GKZ-generalized hypergeometric systems and mirror symmetry of Calabi-Yau hypersurfaces*, alg-geom/9511001, Commun. Math. Phys. 182 (1996) 535-578.

[HLY] S. Hosono, B.H. Lian and S.T. Yau, *Maximal Degeneracy Points of GKZ Systems*, alg-geom/9603014, Journ. Am. Math. Soc., Vol. 10, No. 2 (1997) 427-443.

[HSS] S. Hosono, M. Saito and J. Stienstra, *On mirror conjecture for Schoen's Calabi-Yau 3-folds*, preprint 1997.

[KlemmSchimm] A. Klemm and R. Schimmrigk, Nucl. Phys. B411 (1994) 559.

[KontsevichI] M. Kontsevich, *Enumeration of rational curves via torus actions.* In: The Moduli Space of Curves, ed. by RḊijkgraaf, CḞaber, Gv̇an der Geer, Progress in Mathv̇ol1̇29, Birkhäuser, 1995, 335–368.

[KontsevichII] M. Kontsevich, in Proceedings of 1994 ICM I, Birkäuser, Zurich 1995, p120; alg-geom/9411018.

[LVW] W. Lerche, C. Vafa, and N. Warner, Nucl. Phys. B324 (1989) 427.

[LiTianII] J. Li and G. Tian, *Virtual moduli cycle and Gromov-Witten invariants of algebraic varieties*, JAMS 11, No. 1 (1998) 119-174.

[LiTianIII] J. Li and G. Tian, *Virtual moduli cycle and Gromov-Witten invariants of general symplectic manifolds. Topics in symplectic* 4*-manifolds (Irvine, CA, 1996), 47-83, First Int. Press Lect. Ser. I*, Intl. Press, Cambridge, MA, 1998.

[LiTianIV] J. Li and G. Tian, *Comparison of algebraic and symplectic GW-invariants*, to appear in Asian J. Math. 1998.

[LLYI] B. Lian, K. Liu and S.T. Yau, *Mirror Principle I*, Asian J. Math. Vol. 1, No. 4 (1997) 729-763.

[LLYSurvey] B. Lian, K. Liu and S.T. Yau, *Mirror Principle, A Survey*, to appear in Proceedings of Current Development in Mathematics 1998, Intl. Press.

[LLYII] B. Lian, K. Liu and S.T. Yau, *Mirror Principle II*, to appear in Asian J. Math.

[LLYIII] B. Lian, K. Liu and S.T. Yau, *Mirror Principle III*, in preparation.

[LianYauI] B. Lian, and S.T. Yau, *Arithmetic Properties of Mirror Map and Quantum Coupling*, Commun. Math. Phys. 176, no. 1 (1996).

[LianYauODE] B. Lian, and S.T. Yau, *Differential Equations from Mirror Symmetry*, preprint, to appear in Survey in Diff. Geom (1999).

[LT] A. Libgober, and J. Teitelboim, *Lines on Calabi-Yau complete intersections, mirror symmetry, and Picard-Fuchs equations*, Int. Math. Res. Notices 1 (1993) 29-39.

[MorrisonII] D. Morrison, *Mirror symmetry and rational curves on quintic threefolds: a guide for mathematicians*, AMS Journ. 6 (1993) 223-247.

[Morrison] D. Morrison, *Picard-Fuchs Equations and Mirror Maps for Hypersurfaces*, in *Essays on Mirror Manifolds I*, S.T. Yau ed., Intl. Press (1992) 241.

[MP] D. Morrison and R. Plesser, *Summing the instantons: quantum cohomology and mirror symmetry in toric varieties*, alg-geom/9412236.

[Oda] T. Oda, *Convex Bodies and Algebraic Geometry*, Series of Modern Surveys in Mathematics 15, Springer Verlag (1985).

[Pandharipande] R. Pandharipande, *Rational curves on hypersurfaces (after givental)*, math.AG/9806133.

[ParkerWolfson] T. Parker and J. Wolfson, *A compactness theorem for Gromov's moduli spaces*, J. Geom. Anal. 3 (1993) 63-98.

[Ruan] Y.B. Ruan, *Virtual neighborhoods and pseudo-holomorphic curves*, math.AG/9611021.

[RuanTian] Y.B. Ruan and G. Tian, *A mathematical theory of quantum cohomology*, Journ. Diff. Geom. Vol. 42, No. 2 (1995) 259-367.

[SacksUhlenbeck] J. Sacks and K. Uhlenbeck, *The existence of minimal immersion of 2-sphere*, Annals. Math. 113 (1981) 1-24.

[SiebertI] B. Siebert, *Gromov-Witten invariants for general symplectic manifolds*, alg-geom/9608005.

[Sturmfels] B. Sturmfels, *Grobner bases and convex polytopes*, Univ. Lecture Series 8, AMS, Providence RI 1996.

[SiebertII] B. Siebert, *Algebraic and symplectic Gromov-Witten invariants coincide*, math.AG/9804108.

[SYZ] A. Strominger, S.T. Yau, and E. Zaslow, Nucl. Phys. B479 (1996) 243.

[Vafa] C. Vafa, *Topological mirrors and quantum rings*, in *Essays on Mirror Manifolds I*, S.T. Yau ed., Intl. Press (1992).

[VafaII] C. Vafa, *Extending Mirror Conjecture to Calabi-Yau with Bundles*, hep-th/9804131.

[WittenI] E. Witten, *Two dimensional gravity and intersection theory on moduli space*, Surveys Diff. Geom. 1 (1991) 243-310.

[WittenII] E. Witten, *Phases of N=2 theories in two dimension*, hep-th/9301042.

[WittenIII] E. Witten, *Mirror Manifolds and Topological Field Theory*, in *Essays on Mirror Manifolds I*, S.T. Yau ed., I ntl. Press 1992.

[Yau] S.T. Yau, *Calabi conjecture and some new results in algebraic geometry*, PNAS Vol. 74, No. 5 (1977) 1798-1799.

[Ye] R. Ye, *Gromov's compactness theorem for pseudo-holomorphic curves*, Trans. Amer. Math. Soc. 1994.

DEPT OF MATHEMATICS, BRANDEIS UNIV, WALTHAM MA 02154, USA.
E-mail address: `lian@brandeis.edu`

DEPT OF MATHEMATICS, HARVARD UNIV, CAMBRIDGE MA 02138, USA.
E-mail address: `yau@math.harvard.edu`

AMS/IP Studies in Advanced Mathematics
Volume 20, 2001

The Scalar Curvature Equation on S^n

Chang-Shou Lin

1. Introduction

For the past twenty years, one of major difficult problems in nonlinear elliptic equation is to study so-called "the phenomenon of concentrations", where the total energy of solutions could be concentrated at a finite set of points. Two well-known examples are the scalar curvature equation on a compact Riemannian manifold of dimension $n \geq 2$ and the mean field equation on a compact Riemannian surface of genus g. In this lecture, I would like to describe my recent joint work with Professor Chiun-Chuan Chen on the scalar curvature equation on S^n. For the mean field equation, we refer the readers to [CL8], [DJLW], [Li3], [Ln3], [NT] and the references therein.

Let (M, g_0) be a Riemannian manifold of dimension $n \geq 3$ and $R(y)$ a smooth function on M. One would like to ask whether there is a metric g conformal to g_0 such that $R(y)$ is the scalar curvature of the new metric g. Setting $g = v^{\frac{4}{n-2}} g_0$ for some positive smooth function v on M, then v satisfies

$$(1) \qquad \frac{4(n-1)}{n-2} \Delta_{g_0} v - k(y)v + R(y)v^{\frac{n+2}{n-2}} = 0 \quad \text{on } M,$$

where Δ_{g_0} is the Laplace-Beltrami operator associated with the metric g_0 and $k(y)$ is the scalar curvature of g_0. When (M, g_0) is not conformally equivalent to the unit sphere S^n and $R(y) \equiv$ a constant, the problem is called the Yamabe problem.

In the sixties, Yamabe asked the question that for a Riemannian manifold, does there exist a conformal metric g such that the scalar curvature of the new metric is a constant? For $n = 2$, this is the theorem of uniformization. For $n \geq 3$, the answer to the Yamabe problem is proved to be affirmative through important contributions of Yamabe [Y], Trudinger [T], Aubin [A] and Schoen [S1]. In [S1], R. Schoen applied the positive mass theorem to solve the difficultiest case of the problem and his proof shows how the Green function of the conformal Laplacian operator is related to the concentration of solutions. After R. Schoen's work, many mathematicians have turned their attentions to equation (1) for a nonconstant function $R(y)$, in particular when (M, g_0) is the n-dimensional unit sphere S^n.

For the Yamabe problem, R. Schoen solved equation (1) by proving the inequality:

$$\inf_{v\in H^1} \frac{\int_M |\nabla v|^2 + \frac{n-2}{4(n-1)} kv^2}{(\int_M v^{\frac{2n}{n-2}})^{\frac{n-2}{2}}} < S_n,$$

when (M, g_0) is not conformally equivalent to S^n. Here S_n is the best constant of Sobolev's embedding in R^n. This inequality holds due to the positive mass theorem. By the inequality, R. Schoen can use the direct method of the calculus of variation to find a solution with the least energy. Also, due to the positive mass theorem, this inequality holds again for the situation when all of the derivatives of R up to order $n-2$ vanish for (M, g_0) not conformal to S^n. See [ES]. But, for a nonconstant smooth function R on S^n, this inequality does not hold and for equation (1) on S^n, therefore there are no solutions in the level of the least energy. To overcome the difficulty, Bahri-Coron [BC] developed an "algebraic-topology" method to find critical points at infinity. They have succeeded on S^3. Schoen-Zhang [SZ] studied equation (1) by the approach of subcritical exponents, and they applied a blowing-up argument to develop a Morse-type theory for generic R on S^3 and obtained a similar result as Bahri-Coron. Later, Chang-Gursky-Yang [CGY] proved the uniform bound for equation (1) on S^3 and derived a formulas for counting topological degree of equation (1). Later, Y.Y. Li [Li 1,2]combined these techniques to extend previous works on S^3 to S^4, where the degree counting formulas adds some effects due to the interaction of multiple blow-up points. In this talk, we only consider the case of S^n and will extend these previous works to the dimension $n \geq 5$.

When $R(y) \equiv$ a positive constant on S^n, it is well-known that equation (1) possesses a family of solutions such that the total energy of solutions can be concentrated in a small neighborhood of some point $y_0 \in S^n$. Thus, from the analytic viewpoint, it is rather interesting to study the solution structure for equation (1) when R is a nonconstant smooth function. New phenomenons are expected, in particular when (M, g_0) is the standard unit sphere S^n, where there is a large group of conformal transformations. By a straightforward computation, equation (1) is invariant under conformal transformations. We want to understand the role played by conformal transformations for equation (1). Indeed, the conformal transformation plays a fundamental role in our work.

When (M, g_0) is the unit sphere S^n, the scalar curvature $k(y) \equiv n(n-1)$ and then (1) becomes

$$(2) \qquad \Delta_{g_0} v - \frac{n(n-2)}{4} v + R(y) v^{\frac{n+2}{n-2}} = 0 \quad \text{on} \quad S^n.$$

Let π denote the stereographic projection from S^n onto R^n and set

$$u(x) = \left(\frac{2}{1+|x|^2}\right)^{\frac{n-2}{2}} v(y) \quad \text{with} \quad x = \pi(y).$$

Then $u(x)$ satisfies

$$(3) \qquad \begin{cases} \Delta u + K(x) u^{\frac{n+2}{n-2}} = 0 & \text{on} \quad R^n, \\ u(x) = O(|x|^{2-n}) & \text{at} \quad \infty, \end{cases}$$

where $K(x) \equiv R(y)$. When $K(x) \equiv n(n-2)$, any solution of (3) has the form of $U_\lambda(x-x_0) = \left(\frac{\lambda}{1+\lambda^2|x-x_0|^2}\right)^{\frac{n-2}{2}}$. See [GNN] and [CGS]. As $\lambda \to +\infty$, the total energy of $U_\lambda(x)$, is concentrated in a small neighborhood of the origin. It is natural to ask whether the blowup behavior of solutions of (2) is similar to the case when $R \equiv a$ positive constant. We discuss it in the next section.

2. The Spherical Harnack Inequality

In this section, we study the behavior of local solutions near a blowup point. A point $x_0 \in \Omega$ is called a *blow-up* point for a sequence of solution u_j of

$$\Delta u_j + K_j(x) u_j^{p_j} = 0 \quad \text{for} \quad x \in B_1, \tag{4}$$

where B_1 is the unit ball and $1 < p_j \le \frac{n+2}{n-2}$ with $\lim_{j\to\infty} p_j = \frac{n+2}{n-2}$. $\lim_{j\to+\infty} u_j(x_j)$ $= +\infty$with $\lim_{j\to+\infty} x_j = x_0$. First, we can locate the position of blowup points.

Theorem 2.1. *Suppose $K_j(x)$ converges to $K(x)$ in $C^1(\Omega)$. Then $\bigtriangledown K(x_0) = 0$ for any blowup point x_0 of u_j.*

Theorem 2.1 was proved by Schoen-Zhang [SZ] and Y.Y. Li [Li1]when K is positive and u_j is a global solution defined on S^n. For the proof of the general case, see [**CL2**] and [**Ln1**]. For the simplicity, we assume $K_j(x) \equiv K(x)$ is a positive C^1 function such that the following condition holds.

[K1] In a neighborhood of P where $\bigtriangledown K(P) = 0$, K can be written as

$$K(x) = K(P) + Q(x-P) + \text{remainder terms}, \tag{5}$$

where Q is a homogeneous C^1 function of degree $\beta > 1$ such that

$$c_1|y|^{\beta-1} \le |\bigtriangledown Q(y)| \le c_2|y|^{\beta-1}$$

for some $\beta > 1$ and for two positive constants c_1 and c_2.

The first question concerning the type of concentration is to ask whether it is possible to have simultaneously more then one bubbles with different centers near a blowup point. The answer is no by the following *spherical* Harnack inequality.

Theorem 2.2. ([CL3]) *Assume that the origin is the only critical point of K in the unit ball B_1 and [K1] holds for K at the origin. Let u_j be a solution of (4). Then there is a constant $c > 0$ such that*

$$\max_{|x-x_j|=r} u_j(x) \le c \min_{|x-x_j|=r} u_j(x) \tag{6}$$

for $0 < r \le \frac{1}{2}$ where x_j is the local maximum point of u_j,

$$M_j = u_j(x_j) = \max_{|x|\le\frac{1}{2}} u_j(x). \tag{7}$$

The spherical Harnack inequality plays an important role for us to understand better the blowup behavior of u_j near a blowup point. Furthermore, the following result shows that the behavior of u_j depends only on the flatness β of K at the blowup point.

Theorem 2.3. **([CL3])** *Let K be a positive C^1 function and satisfy the assumption of Theorem 2.2 and u_j is a solution of (4) with $p_j = \frac{n+2}{n-2}$ such that the maximum $M_j \to +\infty$. Then*
(i) The origin is simple if and only if $\beta \geq n-2$.
(ii) Let

$$m_j = \inf_{\bar{B}_1} u_j.$$

Then $m_j \to 0$ if and only if $\beta > \frac{n-2}{2}$. Moreover,

$$m_j \sim M_j^{-\gamma}, \tag{8}$$

where $\gamma = \min\{1, \frac{2\beta}{n-2} - 1\}$.
(iii) The total energy of u_j tends to infinity if and only if $\beta < \frac{n-2}{2}$.

Here a blowup point is called *simple* if u_j satisfies

$$u_j(x) \leq c\left(\frac{M_j}{1+M_j^2|x-x_j|^2}\right)^{\frac{n-2}{2}} \tag{9}$$

where the right-hand side is the solution of (3) with $K(x) \equiv a$ positive constant. Note that the notion of simple blowup points is different from the one given by R. Schoen. But these two definitions are equivalent. When the flatness of K at $0 \geq n-2$, the result that 0 is a simple blowup point was proved by Y.Y. Li ([Li1]) by using the blowup argument due to Schoen. Later, it was also proved by the authors ([CL2]) by using the method of moving planes. We remark that part (ii) is important when we come to study the interaction between different blowup points, because the local information of u_j can be passed through it. Another way to reveal the local information is through the Pohozaev identity. See the discussion in section 4. When K is a positive Morse function, part (iii) of Theorem 2.3 says that for $n \geq 7$, the total energy also blows up whenever the super-norm of solutions blows-up.

3. Apriori estimates

In this section, we assume u_j is a global solution of (3) with a fixed C^1 function K. Here K is assumed to be *positive* and the north pole of S^n is *not* a critical point of R. At each critical point P_k, K satisfies [K1] with the homogeneous function Q_k of degree β_k. For each P_k, we assume one of the following conditions according to β_k

[K2] (i) If $\beta_k < n$, then Q_k sataisfies

$$\begin{pmatrix} \int_{\mathbf{R}^n} \nabla Q_k(y+\xi) U_1^{\frac{2n}{n-2}}(x)dx \\ \int_{\mathbf{R}^n} Q_k(y+\xi) U_1^{\frac{2n}{n-2}}(x)dx \end{pmatrix} \neq \begin{pmatrix} 0 \\ 0 \end{pmatrix}, \tag{10}$$

for any $\xi \in \mathbf{R}^n$, where $U_1(x) = (1+|x|^2)^{-\frac{n-2}{2}}$.

(ii) If $\beta_k = n$, the Q_k satisfies either

$$\int_{\mathbf{R}^n} \nabla Q_k(x+\xi) U_1^{\frac{2n}{n-2}}(x)dx \neq 0, \forall \xi \in \mathbf{R}^n,$$

or

$$\begin{cases} \int_{\mathbf{R}^n} \nabla Q_k(x+\xi) U_1^{\frac{2n}{n-2}}(x)dx = 0 \quad \text{for} \quad \xi \neq 0, \\ \int_{S^{n-1}} Q_k(x) d\sigma \neq 0. \end{cases}$$

(iii) If $\beta_k > n$, then

$$\int_{\mathbf{R}^n} < x - P_k, \nabla K > |x-P_k|^{-2n} dx \neq 0.$$

Definition. *We say $P_k \in \Gamma^-$ if K satisfies one of the following conditions according to β_k.*

: *(i) If $\beta_j < n$, there exists $\xi \in \mathbf{R}^n$ such that*

$$\begin{cases} \int_{\mathbf{R}^n} \nabla Q_k(x+\xi) U_1^{\frac{2n}{n-2}}(x)dx = 0 \\ \int_{\mathbf{R}^n} Q_k(x+\xi) U_1^{\frac{2n}{n-2}}(x)dx < 0 \end{cases}$$

: *(ii) If $\beta_k = n$,*

$$\int \nabla Q_k(x) U_1^{\frac{2n}{n-2}}(x)dx = 0 \quad \textit{and}$$

$$\int_{S^{n-1}} Q_k(x) U_1^{\frac{2n}{n-2}}(x) d\sigma < 0.$$

: *(iii) If $\beta_k > n$,*

$$\int_{\mathbf{R}^n} < x - P_k, \nabla K > |x-P_k|^{-2m} dx < 0.$$

Theorem 3.1. **([CL6])** *Suppose K is a positive C^1 function and satisfies the nondegenerate condition [K1] and [K2] at each critical points P_k of K. Assume*

$$\beta_k > \frac{n-2}{2}, \quad \textit{and} \tag{11}$$

$$\frac{1}{\beta_j^*} + \frac{1}{\beta_k^*} \neq \frac{2}{n-2} \tag{12}$$

for all $P_j \neq P_k \in \Gamma^-$ where $\beta_k^* = \min\{\beta_k, n-2\}$. Then there exists a constant C such that

$$C^{-1} \leq v(y) \leq C \quad \textit{for} \quad y \in S^n$$

holds for any solution v of (2).

We prove it by contradiction. Assume that there is a sequence of blowing up solutions. For simplicity, we assume that there are only two blow-up points P_1 and P_2 only. For the general case, the proof requires more refined estimates than in section 2. See [CL6]. Since one of them is a simple blow-up point, we may assume P_1 is simple. Applying the conformal transformation, we assume that P_1 is the south pole and P_2 is the north pole. Recall $u_j(x) = \left(\frac{2}{1+|x|^2}\right)^{\frac{n-2}{2}} v_j(y)$, $x = \pi(y)$ where π is the stereographic projection. Let $M_{j,1}$ and $M_{j,2}$ be the local maximum of u_j near P_i and P_2 respectively. For the simplicity, we assume both β_1 and $\beta_2 < n$. Since $\beta_2 > \frac{n-2}{2}$, we have by (2.5),

$$M_{j,1}^{-1} \sim M_{j,2}^{1-\frac{2\beta_2}{n-2}}. \tag{13}$$

By the Pohozaev identity

$$\int_{|x|\le 1} (x\cdot \nabla K) u_j^{\frac{2n}{n-2}}(x)dx = -\int_{|x|\ge 1} (x\cdot \nabla K) u_j^{\frac{2n}{n-2}}(x)dx,$$

and the scaling, both sides can be computed straightforwardly

$$M_{j,1}^{-\frac{2\beta_1}{n-2}} \sim M_{j,2}^{-\frac{2\beta_2}{n-2}}. \tag{14}$$

Together with (13) and (14), $\frac{\beta_2}{\beta_{j,1}} = -1 + \frac{2\beta_2}{n-2}$ i.e., a contradiction to (12). Thus, the apriori bound is proved.

We remark that Theorem 3.1 still holds without the assumption (10). See Theorem 10.3 in [CL6] for precise statement. For a positive Morse function R, the apriori bound was proved in [CGY] for $n = 3$ and in [Li2] for $n = 4$.

4. Asymptotic behavior of maximums

As shown in the last section, in order to prove the uniform bound for solutions, we have to calculate very precisely each term in the Pohozaev identity. This requires more refined estimates than in section 2, in particularly, for the case when the sequence of solutions u_j has only one simple blowup point only (but, have more than one blowup points) . Also, after the apriori bound is established, we want to count the Leray-Schauder degree for the scalar curvature equation on S^n. To do it, we deform the given scalar curvature R by $R_t = tR + (1-t)$ for $0 < t \le 1$. By Theorem 3.1, the Leray-Schauder degree for equation (2) with R_t is independent of t. So, we have to consider (3) for the situation $K_j = n(n-2) + t_j\hat{K}$ with $t_j \to 0$. This is the degenerate case. For local solutions, estimates in section 2 obviously do not hold for the degenerate case. But, due to the interaction between different blowup points, the spherical Harnack inequality near each blowup point still holds for *global* solutions, and then with the help of the Pohozaev identity, we can prove that for sequence of blowup solutions, there are at least two blowup points and at least one of them should be simple. To prove the Harnack inequality, we require the same refined estimates also. An important consequence of these refined estimates is about the asymptotic behavior of local maximyms of u_i near each blow-up point P_k. To state the result, we set

$$M_{j,k} = u_j(x_{j,k}) \max_{|x-P_k|\le\delta} u_j(x).$$

Theorem 4.1. **([CL6])** *Suppose $\hat{K}$ satisfies [K1] and [K2] at any critical point $P \in \Gamma^-$ with the degree β of $Q \ge \frac{n-2}{2}$, and u_j be a sequence of solutions of (3)*

with $K_j = n(n-2) + t_j\hat{K}$. Let $\{P_k\}_{k=1}^N$ be the set of blowup points for u_j such that $P_1, \dots, P_l$ are those simple blowup points. Then we have $\beta_1 = \beta_2 = \dots \beta_l > \beta_k$ for $k > l$. Furthermore, the followings hold.
(i) If $l \geq 2$, then $M_{j,k}$ satisfies

$$t_j M_{j,k}^{-\frac{2\beta_1}{n-2}} = (1+o(1)) \sum_{\substack{i=1\\ i\neq k}}^{l} \eta_{ki} M_{j,k}^{-1} M_{j,i}^{-1} \quad \textit{for} \;\; 1 \leq j \leq l,$$

and

$$t_j M_{j,k}^{-\frac{2\beta_k}{n-2}} = (1+o(1)) \sum_{i=1}^{l} \eta_{ki} M_{j,k_i}^{-1} M_{i,k}^{-1} \quad \textit{for} \;\; l+1 \leq j \leq N.$$

(ii) Assume $l = 1$. We index q_k according to the order of $\beta_j : \beta_1 > \beta_2 = \dots = \beta_{l_1} > \beta_{l_1+1} \geq \dots \geq \beta_N$. Then

$$\left.\begin{array}{ll} t_j M_{j,1}^{\frac{-2\beta_1^*}{n-2}} & \textit{if} \;\; \beta_1 \neq n \\ t_j M_{j,1}^{-\frac{2n}{n-2}} \log M_{i,1} & \textit{if} \;\; \beta_1 = n \end{array}\right\} = (1+o(1)) \sum_{i=2}^{l_1} \eta_{1i} M_{j,i}^{-1} M_{j,1}^{-1},$$

and

$$t_j M_{j,k}^{-\frac{2\beta_j}{n-2}} = (1+o(1)) \eta_{k1} M_{j,k}^{-1} M_{j,1}^{-1} \quad \textit{for} \; 2 \leq k \leq N.$$

Here η_{ki} are constants depending on depends on the dimension n, the distance $|P_j - P_k|$ and Q_k only. For their precise expressions, see [**CL6**]. In [**CL6**], we can even locate the position of $x_{j,k}$ in the micro-scale. Set $\xi_{j,k} = (x_{j,k} - P_k) M_{j,k}^{\frac{2}{n-2}}$, and $\xi_k = \lim_{j\to+\infty} \xi_{j,k}$. Then for these P_k such that $\beta_k < n$, we have

$$\int_{\mathbf{R}^n} \nabla Q_k(x+\xi_k) U_1^{\frac{2n}{n-2}}(x)dx = 0, \quad \text{and}$$
$$\int_{\mathbf{R}^n} Q_k(x+\xi_k) U_1^{\frac{2n}{n-2}}(x)dx < 0.$$

Thus, $P_k \in \Gamma^-$. These information of $M_{j,k}$ and $x_{j,k}$ are essential when we want to construct blowup solutions with some given datas.

5. Counting the Leray-Schauder degree

In this section, $K(x)$ satisfies [K1] and [K2]. We say $A \in \Lambda^-$ if $A \subseteq \Gamma^-$ and satisfies the following conditions.

1. The number of the elements in $A \geq 2$.
2. For any $P_j \neq P_k$ in A, the exponent β_j and β_k satisfies

$$\frac{1}{\beta_j^*} + \frac{1}{\beta_k^*} > \frac{2}{n-2}.$$

For each $P_k \in \Gamma^-$, define $F_k = \nabla Q_k$. By [K1], F_k is a continuous map from S^{n-1} to $\mathbf{R}^n \backslash \{0\}$. Denote the topological degree of F_k by $\deg F_k$. Then our main result can be stated as follows.

Theorem 5.1. ([**CL7**]) *Suppose K satisfies the assumption of Theorem 3.1. Let d denote the Leray-Schauder degree for the nonlinear map $w + L^{-1}\left(Rw^{\frac{n+2}{n-2}}\right)$ on $C^{2,\alpha}(S^n)$ with $0 < \alpha < 1$. Then d satisfies*

$$(15) \qquad d = -\left[1 + \sum_{k\in\Gamma^-} (-1)^{n+1} \deg F_k + \sum_{A\in\Lambda^-} \prod_{k\in A} ((-1)^{n+1} \deg F_k)\right].$$

In the proof of Theorem 5.1, the formulas d can be written as

$$d = -\underbrace{\left[1 + \sum_{P_j\in\Gamma^-} (-1)^{n+1} \deg(\nabla Q_j)\right.}_{\substack{\text{solutions which}\\ \text{remainds bounded}\\ \text{as} t_j \longrightarrow 0}} + \underbrace{\left.\sum_{A\in\Lambda^-} \Pi_{P_j\in A} \deg(\nabla Q_j)\right]}_{\substack{\text{solutions with}\\ \text{more than one}\\ \text{blow-up points}}}$$

The first part of the right hand side is exactly the degree computed in [**CY**], although [CY] did not give such a explicit expression as ours.

For S^3 and a positive Morse function R, the degree counting formulas was first obtained in [**CGY**]. When the right-hand side of (15) does not vanish, the existence of solutions was first proved by Bahri-Coron [**BC**] and also by Schoen-Zhang [**SZ**]. A formulas counting the Leray-Schauder degree also can be derived without the assumption (3.2). See [CL7]. This generalizes the results of Y.Y. Li [Li2] on S^4.

Theorem 5.2. *Suppose that R is a positive More function on S^5 such that $\Delta R(y) \neq 0$ for any critical point y of R. Then the Leray-Schauder degree d vanishes always.*

Proof. Clearly, $\Gamma^- = \{P \mid \Delta R(P) < 0 \text{ and } \nabla R(P) = 0\}$. Since any $\beta_k = 2$, we have

$$\frac{1}{\beta_k} + \frac{1}{\beta_j} = 1 > \frac{2}{3} = \frac{2}{n-2}$$

for any $P_k \neq P_j \in \Gamma^-$. Thus $A \in \Lambda^-$ if and only if A contains at least two elements of Γ^-. Thus, by (5.1)

$$d = -\prod_{k\in\Gamma^-} (1 + (-1)^{n+1} \deg F_k) = 0,$$

because $\deg F_{k_0} = (-1)^n$ when P_{k_0} is a maximum point of R. Q.E.D.

When every β_k are positive integers, the degree counting formulas has a simpler form.

Theorem 5.3. *Suppose that $n \geq 6$ and K satisfies the assumption of Theorem 3.1 and β_k are positive integers for $1 \leq k \leq N$. Then the Leray-Schauder degree d satisfies*

$$d = -\left[1 + \sum_{\substack{K\in\Gamma^-\\ \beta_k \geq n-2}} (-1)^{n+1} \deg F_k\right] \prod_{\substack{k\in\Gamma^-\\ \beta_k < n-2}} (1 + (-1)^{n+1} \deg F_k). \tag{16}$$

Proof. It is elementary to prove that for $j, k \in \Gamma^-$ with $\beta_K < n-2$ and $\beta_j \geq n-2$, we have for $n \geq 6$

$$\frac{1}{\beta_k} + \frac{1}{\beta_j^*} > \frac{2}{n-2}.$$

Here we use (12) and the fact that β_j and β_k are positive integer. Thus, $A \in \Lambda^-$ if and only if A contains two elements of Γ^- at least such that there are at most $P_k \in A$ with $\beta_k \geq n-2$. Then (16) follows immediately form (15). Q.E.D.

6. Liouville Theorem

In this section, we discuss the situation when R is allowed to change signs. First, we have the following Liouville Theorem.

Theorem 6.1. ([Ln1]) *Let $Q(x)$ be a homogeneous C^1 function of order $\beta > 0$ satisfying the following nondegenerate condition*

$$c_1|x|^{\beta-1} \leq |\bigtriangledown Q(x)| \leq c_2|x|^{\beta-1} \tag{17}$$

for $x \in \mathbf{R}^n\backslash\{0\}$ and for two positive constants c_1 and c_2. Then the equation,

$$\Delta u(x) + Q(x)u^{\frac{n+2}{n-2}}(x) = 0 \quad in \ \mathbf{R}^n, \tag{18}$$

possesses no positive solutions.

Proof. The proof in [Ln1] holds only for $\beta \leq 2$. we take this opportunity to correct the original proof for $\beta > 2$. Let u be a positive solution of (17). By Lemma 3.1 in [Ln1], we have

$$\begin{aligned} u(x) &\leq c|x|^{-\frac{(n-2)(2+\beta)}{4}} \quad \text{and} \\ |\bigtriangledown u(x)| &\leq c|x|^{-(\frac{(n-2)(2+\beta)}{4}+1)} \end{aligned}$$

for $|x|$ large. Thus, $\bigtriangledown u(x) \in L^2(\mathbf{R}^n)$ for any $\beta > 0$. By applying the Pohozaev identity on a large ball with radius R, we have

$$\int_{|x|\leq R} (x \cdot \bigtriangledown Q(x))u^{\frac{2n}{n-2}}(x)dx = \text{the boundary terms}$$

$\to 0$ as $R \to +\infty$. Thus,

$$\begin{aligned} \int_{\mathbf{R}^n} |\bigtriangledown u|^2 &= \int_{\mathbf{R}^n} Q(x)u^{\frac{2n}{n-2}}(x)dx \\ &= \frac{1}{\beta}\int (x \cdot \bigtriangledown Q(x))u^{\frac{2n}{n-2}}(x)dx \\ &= 0, \end{aligned}$$

which yields a contradiction. Q.E.D.

With the help of Theorem 6.1, results in section 3 and section 5 can be extended to include those K which in a neighborhood of a critical point p with $K(p) = 0$, $K(x)$ can be written as $K(x) = Q(x-p)+$Remainder term, where (17) holds for Q, besides the conditions [K1] and [K2] for the region $K(x) > 0$. The degree counting formulas should be stated as

$$d = -\left[1 + \sum_{p\in\tilde{\Gamma}^-} (-1)^n \deg F_k + \sum_{A\subseteq\tilde{\Gamma}^-} \left(\prod_{p_k\in A} (-1)^n \deg F_k\right)\right]. \tag{19}$$

where $\tilde{\Gamma}^- = \{p \in \Gamma^- \mid K(p) > 0\}$. For example,

Theorem 6.2. ([Ln1]) *Suppose R is a Morse function on S^3, satisfying $\Delta R(y) \neq 0$ whenever $\nabla R(y) = 0$ and $R(y) > 0$. Then there is a constant c such that for any solution v of (2),*

$$c^{-1} \le v(y) \le c \quad for \quad y \in S^3$$

holds. Furthermore, the Leray-Schauder degree d is given by

$$d = -\left(1 - \sum_{P_k\in\Gamma^-} (-1)^{m_k}\right),$$

where $\Gamma^- = \{P \mid P$ is a critical point of R with $R(P) > 0$ and $\Delta R(P) < 0\}$ and m_k is the Morse index of the Hessian of K at P_k.

7. Singular solution

Let u be a C^2 positive soslution of

$$\Delta u(x) + K(x)u^{\frac{n+2}{n-2}} = 0 \quad \text{in} \quad B_1\backslash\{0\}. \tag{20}$$

Here, a singular solution $u(x)$ of (20) means that the origin is a nonremovable singularity for u. For $K(x) \equiv 1$, the famous result by Gidas-Caffarelli-Spruch [GCS] says that for any singular solution u of (20) with $K(x) \equiv 1$, there exists a entire singular solution $u_0(x) = u_0(|x|)$ of

$$\Delta u_0(x) + u_0^{\frac{n+2}{n-2}} = 0 \quad \text{in} \quad \mathbf{R}^n\backslash\{0\} \tag{21}$$

such that

$$u(x) = u_0(|x|)(1 + o(1)). \tag{22}$$

We want to extend the result of asymptotic symmetry to nonconstant $K(x)$. Our assumption is that in a neighborhood of the origin, $K(x)$ can be written as

$$K(x) = K(0) + Q(x) + \text{the remainder term}, \tag{23}$$

where $Q(x) \in C^1(\mathbf{R}^n\backslash\{0\})$ is a homogeneous function of degree $\beta > 0$ and satisfies

$$0 < c_1|x|^{\beta-1} \le |\nabla Q(x)| \le c_2|x|^{\beta-1}. \tag{24}$$

Theorem 7.1. **([Ln2])** *Let u be a singular positive solution of (20) where K is positive and satisfies (23). Then there exists a constant $c > 0$ such that*

$$u(x) \leq c|x|^{-\frac{n-2}{2}} \quad for \quad |x| \leq \frac{1}{2}, \tag{25}$$

where c depends on K and the lower bound of $u(x)$. Furthermore, if $Q(x)$ additionally satisfies

$$\int \nabla Q(x+\xi) U_1^{\frac{2n}{n-2}}(x)dx \neq 0 \quad for \quad \xi \neq 0,$$

then for any singular solution $u(x)$, there is a entire positive singular solution $u_0(x) = u_0(|x|)$ of (21) such that (22) holds.

By applying the Kelvin transformation, an immediate consequence of Theorem 7.1 is the following result

Theorem 7.2. **([Ln2])** *Suppose $K(x)$ satisfies for $|x|$ large,*

$$0 < c_1|x|^{-\beta} \leq |\nabla K(x)| \leq c_2|x|^{-\beta}$$

for some $\beta > 1$ and $K(\infty) = \lim_{|x|\to+\infty} K(x) > 0$. Then for any positive solution $u(x)$ of

$$\Delta u(x) + K(x)u^{\frac{n+2}{n-2}} = 0 \quad in \quad \mathbf{R}^n, \tag{26}$$

the inequality

$$u(x) \leq c|x|^{-\frac{n-2}{2}} \quad for \quad |x| \geq 1,$$

holds for some constant c. In particular, any positive entire solution of (7.5) is bounded.

To conclude the survey, we have the following remarks.

(i) For the apriori bound (Theorem 3.1) and the degree counting formulas (Theorem 5.1), condition [K2] sometimes can be removed. For example, consider $n \geq 5$ and a Morse function R which $\Delta R(y_0) = 0$ for some critical point y_0 of R. By Taylor expansion of R, Q is the next nonvanishing homogeneous term of order $\beta > 2$ after the quadratic term and satisfies

$$\begin{pmatrix} \int_{\mathbf{R}^n} \nabla Q(x) U_1^{\frac{2n}{n-2}}(x)dx \\ \int Q(x) U_1^{\frac{2n}{n-2}}(x)dx \end{pmatrix} \neq \begin{pmatrix} 0 \\ 0 \end{pmatrix},$$

where $\beta < n$ is assumed. By a further analysis, we could prove that β is the *effective* order at y_0 instead of 2, that is, β should be used in (11), (12) and (15) in stead of 2 at the critical point y_0. By this replacement, for any $n \geq 5$ we could construct a C^∞ function $R(y)$ such that the Leray-Schauder degree $\neq 0$. This R is a Morse function except at the maximum point R must have the flatness order $\geq n-2$. Of course, R satisfies $\Delta R(y) = 0$ for critical points of R.

(ii) The nondegenerate condition [K1] is the essential assumption for us to prove the spherical Harnack inequality (Theorem 2.2) and the inequality (25). it is a very interesting problem to ask whether they still hold without (10).

(iii) As the author knows, there is only one paper [DN] to discuss the existence of entire positive solutions with the decay $|x|^{-\frac{n-2}{2}}$ at infinity. This kind of solutions is not in $H^1(\mathbf{R}^n)$ and usually can not be obtained by variational method. Related problems are the existence of singular solutions. For $K \equiv$ constant, see [KMPS] and the references therein.

References

[A] T. Aubin, *Équations différentielles non linéaries et probléme de Yamabe concernant la courbure scalaire*, J. Math. Pures Appl. 55 (1976), 269-296.

[B1] A. Bahri, *An invariant for Yamabe-type flows with applicatiaons to scalar-curvature problems in high diemnsion*, Duke Math. J. 81 (1996), 323-466.

[BC] A. Bahri and J.M. Coron, *The scalar-curvature problem on standard three-dimensional sphere*, J. of Func. Anal. 95 (1991), 106-172.

[BCH] M. Benayed, Y. Chen, H. Chtioui, and M. Hammami, *On the prescribed scalar curvature problem on 4-manifolds*, Duke Math. J. 85(1996), 633-677.

[Bi] G. Bianchi, *Non-existence and symmetry of solutions to the scalar curvature equation*, Comm. Partial Differential Equatiaons 21 (1996), 633-677.

[CGS] L. Caffarelli, B. Gidas and J. Spruck, *Asymptotic symmetry and local behavior of semilinear elliptic equations with critical Sobolev growth*, Comm. Pure Appl. Math. 42 (1989), 271-291.

[CY] S.Y. Chang and P. Yang, *A perturbation result in prescribing scalar curvature on S^n*, Duke Math. J. 64 (1991), 27-69.

[CGY] S.Y. Chang, M.J. Gursky and P. Yang, *The scalar curvature equation on 2- and 3-sphere*, Calc, Var. Partial Differential Equations 1 (1993), 205-229.

[CL1] C.C. Chen and C.S. Lin, *Local behavior of singular positive solutions of semilinear elliptic equations with Sobolev exponent*, Duke Math. J. 78 (1995), 315-334.

[CL2] C.C. Chen and C.S. Lin, *Estimates of the conformal scalar curvature equation via the method of moving planes*, Comm. Pure Appl. Math. 50 (1997), 971-1017.

[CL3] C.C. Chen and C.S. Lin, *Estimates of the conformal scalar curvature equation via the method of moving planes, II*, J. Differential Geometry 49 (1998), 115-178.

[CL4] C.C. Chen and C.S. Lin, *Blowing up with infinite energy of conformal metrics on S^n*, Comm. in PDEs 24(1999),, 785-799.

[CL5] C.C. Chen and C.S. Lin, *On axisymmetric solutions of the conformal scalar curvature equatiaons on S^n*, Adv. in Diff. Eq., 5 (2000), 121-146.

[CL6] C.C. Chen and C.S. Lin, *Prescribing scalar curvature on S^n, Part 1: apriori estimates.* Preprint.

[CL7] C.C. Chen and C.S. Lin, *On prescribing scalar curvature on S^n, Part 2: counting the Leray-Schauder degree*, in preparation.

[CL8] C.C. Chen and C.S. Lin, *Concentration of the mean field equation on S^2: the case of three blowup points.* Preprint.

[CLi] W. Chen and C. Li, *A priori estimates for prescribing scalar curvature equations*, Ann. of Math. 145 (1997), 547-564.

[DJLW] W. Ding, J. Jost, J. Li and G. Wang, *The differential equation $\Delta u = 8\pi - 8\pi h e^u$ on a compact Riemann surface*, Asian J. Math. 1 (1997), 230-248.

[DN] W.Y. Ding and W.M. Ni, *On the elliptic equation $\Delta u + K u^{\frac{n+2}{n-2}} = 0$ and relatied topics*, Duke Math. J., 52 (1985), 485-506.

[ES] J.F. Escobar and R. Schoen, *Conformal metrics with prescribed scalar curvature*, Invent. math. 86 (1986), 243-254.

[GNN] B. Gidas, W.M. Ni and L. Nirenberg, *Symmetry and related properties via the maximym principle*, Comm. Math. Phys. 68 (1979), 209-243.

[H] Z.C. Han, *Asymptotic approach to singular solutions for nonlinear elliptic equations involving critical Sobolev exponent*, Ann. Inst. H. Poincaré, Analyse Non-Lineáire 8 (1991), 159-174.

[KMPS] N. Korevaar, R. Mazzeo, F. Pacard and R. Schoen, *Refined asymptotics for constant curvature metrics with isolated singularities*, Inventiones Math., 135 (1999), 233-272.

[LP] J. Lee and T. Yarker, *The Yamabe problem, Bull.* Amer. Math. Soc. 17 (1987), 37-91.

[Li1] Y.Y. Li, *Prescribing scalar curvature on S^n and related problems*, Part I, J. Diff. Equ. 120 (1995), 319-410.

[Li2] Y.Y. Li, *Prescribing scalar curvature on S^n and related problems*, Part II, Comm. pure Appl. math. 49 (1996), 541-597.

[Li3] Y.Y. Li, *Harnack type inequality: the method of moving planes*, Comm. Math. Phys. 200 (1999), 421-444.

[Ln1] C.S. Lin, *On Liouville theorem and apriori estimates for the scalar curvature equations*, Ann. Scuola Norm. Sup. Pisa Cl. Sci. 27 (1999), 107-130.

[Ln2] C.S. Lin, *Estimates of the conformal scalar curvature equation via the method of moving planes, III.* Comm. Pure and Appl. Math., LIII (2000), 611-646.

[Ln3] C.S. Lin, *Topological degree for mean field equations on S^2*.Duke Math. J., to appear.

[NT] M. Nolasco and G. Tarantello, On a sharp Sobolev-type inequality on two-dimensional compact manifolds, Arch. Rational Mech. Anal. 145 (1998), 161-195.

[O] M. Obata, *The conjecture on conformal transformations of Riemannian manifolds*, J. Diff. Geom. 6 (1971), 247-258.

[S1] R. Schoen, *Conformal deformation of a Riemannian metric to constant scalar curvature*, J. Diff. Geom. 20 (1984), 479-495.

[S2] R. Schoen, *Variational theory for the total scalar curvature functional for Riemannian metrics and related topics, Topics in Calculus of Variations*, Lecture notes in mathematics, No. 1365, edited by M. Giaquinta, Springer-Verlag (1989), 120-154.

[S3] R. Schoen, *On the number of constant scalar curvature metrics in a conformal class*, Differential Geometry: A symposium in honor of Manfredo Do Carmo (H.B. Lawson and K. Tenenblat, eds), Wiley (1991), 311-320.

[S4] R. Schoen, *The existence of weak solutions with prescribed singular behavior for a conformally invariant scalar equation.* Comm. Pure and Appl. Math., XLI (1988), 317-392.

[SY1] R. Schoen and S.T. Yau, *Conformally flat manifolds, Kleinian groups, and scalar curvature*, Invent. Math. 92 (1988), 47-71.

[SZ] R. Schoen and D. Zhang, *Prescribed scalar curvature on the n-sphere*, Calc. Var. Partial Differential Equations 4 (1996), 1-25.

[T] N. Trudinger, *Remarks concerning the conformal deformation of Riemannian structures on compact manifolds*, Ann. Scuola Norm. Sup. Cl. Sci 22 (1968), 265-274.

[Y] H. Yamabe, *On a deformation of Riemannian structures on compact manifolds*, Osaka Math. J. 12 (1960), 21-37.

DEPARTMENT OF MATHEMATICS, CHUNG-CHENG UNIVERSITY, MINGHSIUNG, CHIA-YI 621, TAIWAN
E-MAIL:cslin@math.ccu.edu.tw

AMS/IP Studies in Advanced Mathematics
Volume 20, 2001

Symplectic Isotopy in Four Dimension

Gang Tian

In this note, I will discuss the isotopic problem for symplectic surfaces in rational 4-dimensional symplectic manifolds. This problem has inspired many activities in recent research of symplectic 4-manifolds and is related to many other basic questions in symplectic topology, such as symplectic Lefschetz fibrations of 4-manifolds (cf. [Do], [Au], [ST], [Sm]).

Our discussion here is an expansion of the last part of my talk at 1998 ICCM in Beijing. Almost all results here were taken from my joint works with B. Siebert and J.Y. Chen. I would like to thank them both for making this note possible because of our enjoyable collaboratopns.

In the following, unless specified, we shall let M be a compact symplectic 4-manifold and ω be a symplectic form on it. We will always denote by J an almost complex structure on M compatible with ω, that is for any tangent vectors u, v in TM, we have

$$\omega(Ju, Jv) = \omega(u, v) \text{ and } \omega(u, Ju) \geq 0, \tag{0.1}$$

moreover, the equality in the second equation holds if and only if $u = 0$. We will use g to denote the riemannian metric defined $g(u, v) = \omega(u, Jv)$.

1. Symplectic isotopy Problem

Two smoothly embedded surfaces Σ_1 and Σ_2 in M are isotopic if there is a smooth family of embedded surfaces joining Σ_1 to Σ_2. If both Σ_1 and Σ_2 are symplectic, we say that they are symplectically isotopic if they can be connected by a smooth family of embedded symplectic surfaces.

The **symplectic isotopy problem** in 4-dimension can be stated as follows: *For a fixed 2-dimensional homology class α in a simply connected symplectic 4-manifold M, up to smooth isotopy, how many connected smoothly embedded symplectic surfaces representing α.*

If M is a Kähler surface and there is a holomorphic line bundle L with $c_1(L) = \alpha$, then holomorphic surfaces in M representing α are mutually isotopic to each other through holomorphic surfaces. This is because holomorphic surfaces with homology class α are parametrized by a projective space $PH^0(M, L)$ and the subset of $PH^0(M, L)$ corrsponding to singular surfaces is a subvariety of real codimension at least two. Here $H^0(M, L)$ denotes the space of all holomorphic sections of L.

Supported partially by NSF grants and a Simons fund.

However, it is no longer true in the symplectic case. Fintushel and Stern have constructed a simply-connected symplectic 4-manifold which contains infinitely many nonisotopic smoothly embedded symplectic tori. Therefore, we have to impose further condition on M so that the symplectic isotopy problem still holds. The correct condition seems to be the positivity of the first Chern class $c_1(M)$ of M. In fact, we believe (cf. [ST])

CONJECTURE 1.1. *If M is an algebraic surface with positive first Chern class, then the isotopy class of embedded symplectic surfaces with a fixed homology class is unique.*

Note that such an M is called a Del-Pezzo surface in algebraic geometry. In the following, we will give some evidences on this conjecture. but we will first discuss some motivations.

A differential Lefschetz fibration (DLF) of X is a proper C^∞-map $q : X \mapsto S^2$ with finitely many critical points $Q_1, \cdots, Q_\mu$ in disjoint fibers and such that locally near each Q_r, there are complex coordinates z, w of X and t of S^2 with $t = q(z, w) = zw$. If z, w and t can be chosen compatible with orientations of X and S^2, then $q : X \mapsto S^2$ is called a symplectic Lefschetz fibration (SLF).

We will assume that all fibers of $q : X \mapsto S^2$ are irreducible and generic ones are of genus $g \geq 2$. Put $s_i = q(Q_i)$, then it is well known that $q : X \mapsto S^2$ induces a representation of $\pi_1(S^2 \backslash \{s_1, \cdots, s_\mu\})$ in the mapping class group MC_g of genus g. This representation is called the monodromy representation of $q : X \mapsto S^2$. It was shown by Kas that the monodromy representation determines the SLF structure of X.

Recall that $q : X \mapsto S^2$ is called hyperelliptic if up to inner automorphisms of MC_g, the monodromy representation takes values in the hyperelliptic mapping class group $HMC_g \subset MC_g$.

THEOREM 1.2. *Hyperelliptic SLF's are in one-to-one correspondence with double coverings of rational ruled surfaces branched along embedded surfaces which are symplectic with respect to some symplectic structures. Furthermore, the equivalence classes of hyperelliptic SLF's are in one-to-one correspondence with isotopy classes of branched surfaces.*

This was shown in [ST] and the case of genus 2 was also proved by I. Smith [Sm]. An alternative proof was given by Fuller in [Fu]. This theorem indicates the importance of the isotopy problem for symplectic surfaces in rational ruled surfaces. If the conjecture on symplectic isotopy is true, then we have that every hyperelliptic SLF's are equivalent to holomorphic fiberations.

2. Pseudo-holomorphic curves and their dformation

In this section, we will describe an approach to the symplectic isotopy conjecture by using pseudo-holomorphic curves.

Let Σ be a symplectic surface in M with positive first Chern class $c_1(M)$. Then one can introduce an almost complex structure J of M such that Σ is J-holomorphic. It was observed by Gromov that the space of almost complex structures on M compatible with ω is path-connected, therefore, we can choose a generic path J_t $(0 \leq t \leq 1)$ with $J_0 = J$ and J_1 being the standard complex structure on the rational ruled surface M. We denote by $\mathcal{M}_a$ the set of all (Σ, t), where $t \in [0, 1]$, such that Σ is an embedded J_t-holomorphic curve in M with homology class a.

Since the path $\{J_t\}$ is generic, $\mathcal{M}_a$ is a smooth manifold (maybe empty). There is a natural projection map $\pi : \mathcal{M}_a \mapsto [0,1]$ sending (Σ, t) to t. In general, this map may have critical points. However, we have

LEMMA 2.1. *Let (M, ω) be a compact symplectic manifold with positive first Chern class. Then the projection map $\pi : \mathcal{M}_a \mapsto [0,1]$ is a submersion.*

Note that by $c_1(M)$ being positive, we mean that there is φ representing $c_1(M)$ such that $\varphi \geq \lambda\omega$ for some positive constant λ.

Since $\pi^{-1}(1)$ is the set of all embedded holomorphic curves in the rational ruled surface M, it is path connected. It follows that if $\mathcal{M}_a$ is compact, then the above lemma implies that $\mathcal{M}_a$ is path-connected and consequently, the isotopy class of symplectic surfaces with homology class a is unique. If a is the homology class of complex lines in $\mathbb{C}P^2$, then one can show that $\mathcal{M}_a$ is compact, so every symplectic surface homologous to a complex line is isotopic to a complex line.

However, the moduli space $\mathcal{M}_a$ is not compact in general. We can compactify it by adding all cusp J_t-holomorphic curves of homology class a ([Gr], [Pa], [PW], [RT]). Denote by $\overline{\mathcal{M}}$ the resulting space. This $\overline{\mathcal{M}}$ can be thought as the moduli space of all unparametrized J_t-holomorphic curves (possibly singular). We then need to worry about structure of $\overline{\mathcal{M}}$ near any $(\Sigma, t) \in \overline{\mathcal{M}}$ with Σ being singular.

There are two basic problems: 1. For t' sufficiently close to t, is there an embedded surface $(\Sigma', t') \in \mathcal{M}$ close to (Σ, t)? 2. If (Σ', t') and (Σ'', t'') of $\overline{\mathcal{M}}$ are sufficiently close to (Σ, t), are they isotopic to each other?

To solve the first problem, one has to extend the above lemma to singular curves. It is possible, at least for a very large class of J-holomorphic cusp curves. It seems that the second problem is harder. Of course, if one can show that for a generic path J_t, $\overline{\mathcal{M}}$ is smooth near any singular curves, then the second problem is solved. But the usual transversality arguments do not work near singular surfaces.

Nevertheless, Siebert and I can prove

THEOREM 2.2. *If M is a rational ruled surface, then there is only one isotopy class of embedded symplectic surfaces of genus no more than three.*

If M is holomorphic to $S^2 \times S^2$, we can prove the uniqueness for larger classes of embedded symplectic manifolds. Also the cases for genus no more than one were already shown by Mcduff, Sikarov, T.J. Li, etc.. In those cases, one does not need to worry about genuine singularities by imposing constraints. However, when the genus is higher, one has to study smoothing complicated singularities. We believe that our method can be pushed further to cover symplectic surfaces of much higher degree.

An interesting question which arose from our instigation is the local version of the isotopy problem: *Let C_0 be a J-holomorphic curve in $\mathbb{R}^4$ (J may not be integrable) with isolated singularity at the origin. Assume that C and C' are two smooth J-holomorphic curves and sufficiently close to C_0. Are C and C' isotopic to each other.*

Again, if J is integrable, then this is trivial. However, it seems to be difficult to solve this problem in case of non-integrable almost complex structure J. The reason is that the Cauchy-Riemann equation is nonlinear when J is not integrable.

If the multiplicity of C_0 is two, such as for ordinary cusps, then the answer to the above problem is affirmative.

3. Related problems

Though we investigate more time in solving the isotopy problem by using pseudo-holomorphic curves, there may be other methods which are worth being explored.

Let M be a Kähler surface with a Kähler-Einstein metric ω, say $\mathrm{Ric}(\omega) = R\omega$. By a symplectic minimal surface, we mean a minimal surface $\Sigma \subset M$ such that $\omega|_\Sigma$ is a positive multiple of the induced volume form. By a Lagrangian minimal surface, we mean a minimal surface $\Sigma \subset M$ such that $\omega|_\Sigma = 0$.

It follows from [CT] and [Wo]

THEOREM 3.1. *Let (M, ω) be as above. We have (1) if R is positive, then any symplectic minimal surface is holomorphic; (2) If $R = 0$, then any minimal surface is holomorphic with respect to some complex structure compatible with the given symplectic form ω; (3) If $R < 0$, then any minimal surface is either symplectic or Lagrangian if and only if the adjunction formula holds, that is the Euler number and the self-intersection number of the surface add up to the integral of $c_1(M)$ on the surface, where $c_1(M)$ denotes the first Chern class.*

This can be proved by applying the Maximum principle to the following equation: Write $\omega|_\Sigma = \cos\alpha dv$, where dv denotes the induced volume form. The α measures the deviation of $T\Sigma$ from complex subspaces. Then

$$\Delta \cos\alpha = R\sin^2\alpha\cos\alpha + |\nabla J_\Sigma|^2\cos\alpha, \tag{3.1}$$

where Δ and ∇ are the Laplacian and the gradient of the induced metric, J_Σ is the induced almost complex structure of TM along Σ (see [CT] for its definition) such that both $T\Sigma$ and $N\Sigma$ are complex subbundles with compatible orientation.

We would like to point out that if $R < 0$, a symplectic minimal surface may not be holomorphic. C. Arezzo has constructed a number of examples of symplectic minimal surfaces which are non-holomorphic ([Ar]).

A natural problem is *how to deform a surface Σ in M satisfying the adjunction formula to a minimal surface Σ' with the same property.* If such a Σ' can be constructed, is it isotopic to Σ? Since all symplectic surfaces satisfy the adjunction formula, we may ask if *any symplectic surface can be deformed to a minimal symplectic surface through symplectic surfaces.* One possible approach to this problem is to study the mean curvature flow for surfaces in M:

$$\frac{\partial F}{\partial t} = H_{F(\Sigma,t)}, \tag{3.2}$$

where $F : \Sigma \times [0, T) \mapsto M$ is a smooth family of immersions and H denotes the mean curvature. Write $F_t(x) = F(x, t)$. J.Y. Chen and I observed that along the mean curvature flow, the $\cos\alpha$ of $F_t(\Sigma)$ satisfies

$$(\frac{\partial}{\partial t} - \Delta)\cos\alpha = R\sin^2\alpha\cos\alpha + |\nabla J_\Sigma|^2\cos\alpha. \tag{3.3}$$

Clearly, this is simply the dynamic version of (3.1). By the Maximum principle, one can easily show that any $F_t(\Sigma)$ is symplectic so long as $F_0(\Sigma)$ is symplectic. If one is extremally lucky and can show that the flow (3.3) exists for all t, then one can deform $F_0(\Sigma)$ to a minimal symplectic surface. If one can further assure that all surfaces $F_t(\Sigma)$ are embedded, then one can solve the isotopy problem for rational ruled surfaces we discussed in last section. However, the life seems to be hard. It is likely that the flow (3.3) will develop singularities at a finite time. For instance, if $F_0(\Sigma)$ is an immersed surface with negative nodes (there are ample examples of such surfaces), then one can not expect to deform it into a symplectic minimal surface. So the flow may develop singularities at finite time. Nevertheless, it might be a very interesting analytic problem to study singularities of the mean curvature flow in this symplectic case. It may be worth mentioning that codimension two mean curvature flows arise naturally from blow-ups of the heat flow for harmonic maps (cf. [LT]).

References

[Ar] C. Arezzo, *On the Torelli theorem for minimal immersions.*, Internat. J. Math. **10** (1999), no. 6, 697–706.

[Au] D. Auroux, *Symplectic 4-manifolds as branched coverings of* $\mathbf{CP}^2$., Invent. Math. **139** (2000), no. 3, 551–602.

[CT] J.Y. Chen and G. Tian, *Minimal surfaces in Riemannian 4-manifolds.*, Geom. Funct. Anal.**7** (1997), no. 5, 873–916.

[Do] S. Donaldson, *Symplectic submanifolds and almost-complex geometry.*, J. Differential Geom. **44** (1996), no. 4, 666–705J.

[Fu] T. Fuller: *Lefschetz fibrations and 3-fold branched covering spaces.*, preprint *math.GT/9806010.*

[Gr] M. Gromov, *Pseudoholomorphic curves in symplectic manifolds*, Invent. Math. **82** (1985), no. 2, 307-347.

[LT] J.Y. Li and G. Tian, *Blow-up locus for heat flow of harmonic maps,* Acta Math. Sinica, vol. 1 (2000).

[PW] T. Parker and J. Wolfson, *Pseudo-holomorphic maps and bubble trees*, J. Geom. Anal. **3** (1993), no. 1, 63-98.

[RT] Y.B.Ruan and G. Tian, *A mathematical theory of quantum cohomology.*, J. Diff. Geom. **42**(1995), no. 2, 259-367.

[Sm] I. Smith, *Lefschetz fibrations and the Hodge bundle.* , Geom. Topol. **3** (1999), 211–233.

[ST] B. Siebert and G. Tian, *On hyperelliptic symplectic Lefschetz fibrations of four manifolds,* Communications Cont. Math. **1** (1999) 255–280.

[SU] J. Sacks and K. Uhlenbeck, *The existence of minimal immersions of 2 spheres*, Ann. of Math. **113**(1981), 1-24.

[Wo] J. Wolfson, *Minimal Lagrangian diffeomorphisms and the Monge-Ampre equation.*, J. Differential Geom. **46** (1997), no. 2, 335–373.

DEPARTMENT OF MATHEMATICS, MASSACHUSETTS INSTITUTE OF TECHNOLOGY, CAMBRIDGE, MA 02139

AMS/IP Studies in Advanced Mathematics
Volume 20, 2001

Stability in Hamiltonian Dynamical Systems

Zhihong Xia

ABSTRACT. One of the most important problems in Hamiltonian dynamics concerns the stability for near integrable Hamiltonian systems and symplectic diffeomorphisms. On one hand, the celebrated KAM theory guarantees measure stability, or stability for a large measure set of orbits. While on the other hand, it is widely believed and conjectured that typical Hamiltonian systems of higher dimensions are topologically unstable. In this talk, we give a brief survey and discuss some of the recent developments in this problem.

1. Introduction

Perhaps one of the most important problems in dynamical systems is the stability of near integrable Hamiltonian systems and symplectic diffeomorphisms. This problem ahs a very long history, which probably goes back to the times of Newton, Lagrange, Laplace and Poincaré. It is interesting to note that the results or widely believed conjectures to this problem has two seemingly opposite sides, depending on one's point of view. On one hand, KAM (Kolmogorov-Arnold-Moser) theory asserts stability for typical near integrable systems from measure theoretic point of view and on the other hand, strong evidence suggests that typical high dimensional near integrable systems are topologically instable. In this lecture, I will provide a brief survey on some selected topics in the subject and discuss some of the recent progresses in the problem.

A large part of modern dynamical systems originated from Poincaré's study of the Newtonian n-body problem. The solar system, with masses of planets being small comparing to that of the Sun, provides a natural example of a near integrable Hamiltonian system. The question of stability of our solar system, or more generally the n-body problem, has fascinated generations of astronomers and mathematicians and it is still an unsolved problem. Herman [12] called this problem the oldest open question in dynamical systems. Newton certainly believed that the n-body problem, $n \geq 3$, is topologically instable, while Lagrange and Laplace believed in the stability of the solar system. Birkhoff [6] (cf. Herman [12]) went much further to conjecture that in the 3-body problem the set of all bounded orbits is nowhere dense, which implies that an open and dense set of initial conditions lead eventually to escape. It is a remarkable fact that, when two of the three masses are relatively small, the set of bounded orbits, though conjectured to be nowhere dense by Birkhoff, has

Research supported in part by National Science Foundation.

positive and large Lebesgue measure. This result is due to Arnold [1] (cf. Herman [12]) and shows that indeed one has stability, but only in the sense of measure. It is a known fact that the Newtonian n-body problem is a very complicated system and some other very chaotic solutions may also occur (cf. Xia [24]).

We return to general Hamiltonian systems. Even though we will be mostly restricted to near integrable systems on the cotangent bundle of $\mathbb{T}^n$, it is sometimes very convenient to consider Hamiltonian systems on general symplectic manifolds M^{2n}, equipped with a non-degenerate closed two-form ω. A Hamiltonian function on M^{2n} is a real valued function H on M^{2n} and the Hamiltonian flow for H is generated by the vector field $J(dH)$ where J is the canonical isomorphism from T^*M to TM given by the symplectic form ω.

Closely related to Hamiltonian flow is the symplectic diffeomorphism. Tt is sometime very convenient to state results and problems in terms of symplectic diffeomorphisms. Let M^{2n} be a symplectic manifold with a symplectic form ω. A symplectic diffeomorphism of M^{2n} is a diffeomorhism that preserves the symplectic form ω. Poincaré maps and time-one maps of a Hamiltonian flow are natural examples of symplectic diffeomorphisms.

A symplectic diffeomorphism $f : M^{2n} \to M^{2n}$ is said to be *exact symplectic* if there is a one-form α such that $d\alpha = \omega$ and $f^*\alpha - \alpha$ is an exact one-form. There is a more generalized definition of exact symplectic diffeomorphism that makes, for example, the standard map defined on $\mathbb{T}^2$ an exact symplectic diffeomorphism, even though $\mathbb{T}^2$ itself is not exact symplectic. We omit the details here.

In the following section, we will present KAM theory and results on stability and in the rest of this lecture, we will discuss results and conjectures related to instability, particularly Arnold diffusion and Aubry-Mather theory.

Unless otherwise specified, all the functions considered in this paper are C^∞ smooth.

2. Stability and KAM theory

In this section, we will restrict ourselves to near integrable systems of the following type:

$$\dot{I} = -\frac{\partial H_\epsilon}{\partial \theta}, \quad \dot{\theta} = \frac{\partial H_\epsilon}{\partial I} \tag{1}$$

where the Hamiltonian function is of the form:

$$H_\epsilon(I, \theta) = G(I) + \epsilon H^1(I, \theta, \epsilon),$$

where $I \in \mathbb{R}^n$ and $\theta \in \mathbb{T}^n$ ($\mathbb{T} = \mathbb{R}/\mathbb{Z}$), ϵ is a small real parameter, $G(I)$ and $H^1(I, \theta, \epsilon)$ are smooth function on all their variables. (I, θ) are called action-angle variables.

If we set $\epsilon = 0$, the Hamiltonian equation reduces to

$$\begin{aligned} \dot{I} &= 0 \\ \dot{\theta} &= \frac{\partial G(I)}{\partial I} \end{aligned}$$

This is a so-called integrable system and the equations can be solved explicitly. We have: $I = I_0$ and $\theta = G'(I_0)t + \theta_0 (\mod 1)$, where I_0 and θ_0 are initial conditions. We see that every solution move in an n-dimensional invariant torus given by $I = I_0$ and if all the components of $G'(I_0)$ are rationally independent, then any orbit on the torus fills densely on the whole torus.

To state the basic theorem in KAM theory, we need to make the following assumption: $|\partial^2 G(I)/\partial I_i \partial I_j| \neq 0$. This is the so-called Kolmogorov condition. Under this condition, the image of the frequency map $\partial G(I)/\partial I : \mathbb{R}^n \to \mathbb{R}^n$ covers an open set in $\mathbb{R}^n$. We let $\omega_i = G_{I_i}(I_0)$, $i = 1, \cdots, n$ be the frequencies of the rotation on the torus. We say that $(\omega_1, \cdots, \omega_n)$ is Diophantine with Diophantine constant (c, μ), if

$$\left|\sum_{k=1}^{n} j_k \omega_k\right| \geq c|j|^{-\mu}$$

for all integers j_k with $|j| = \sum_{k=1}^{n} |j_k| > 0$. We let $D(c, \mu)$ be the set of all Diophantine numbers with Diophantine constant (c, μ). c and μ are both real positive constants.

We can state the following theorem:

THEOREM 2.1. *Fix c and μ. There exists $\epsilon_0 > 0$ such that if $|\epsilon| \leq \epsilon_0$ and $(\omega_1, \cdots, \omega_n) \in D(c, \mu)$, there exist solutions to the corresponding Hamiltonian system of the form*

$$\begin{aligned} I_k &= \phi_k + F_k(\phi, \epsilon),\ \theta_k = G_k(\phi_k, \epsilon) \\ \phi_k &= \omega_k t + \phi_k(0),\ k = 1, \ldots, n \end{aligned}$$

where F_k and G_k are small functions, periodic in $\theta_1, \ldots, \theta_n$. In other words, there exists solutions for I_k and θ_k that are quasi-periodic in t.

Moreover, these quasi-periodic solutions form a set of positive measure. More precisely, let $V \subset \mathbb{R}^n$ be a bounded open set and let $U = V \times \mathbb{T}^n$. Then for any $\delta > 0$, there exists $\epsilon_0 > 0$ such that for $|\epsilon| \leq \epsilon_0$, the above type of quasi-periodic solutions in U fill out a closed set S with the measure $m(U - S) < \delta m(U)$, where m is a given smooth measure on $\mathbb{R}^n \times \mathbb{T}^n$.

The above theorem is due to Kolmogorov and Arnold [14, 1] for the cases where all the functions involved are analytic. For finite smoothness case, it is due to Moser [20].

The non-degeneracy or Kolmogorov condition for the above theorem can also be replaced by another independent condition. Instead of requiring the frequencies $\partial G(I)/\partial I$ to vary over an n-dimensional open set, we may ask that on a fixed energy surface $G(I) = h$ the ratio G_{I_i}/G_{I_1}, $(i = 2, \ldots, n)$ with $G_{I_1} \neq 0$ vary over an $(n-1)$ dimensional set. This leads to the assumption

$$\begin{vmatrix} G_{I_k I_1} & G_{I_k} \\ G_{I_1} & 0 \end{vmatrix} \neq 0$$

In various applications, such as in the Newtonian n-body problem, neither of the above two non-degeneracy conditions are satisfied. It is interesting to see how much these conditions can be relaxed. Recently, Rüssmann [22] and Cheng & Sun [8] independently showed that if the image of the frequency map $\partial G(I)/\partial I$ is not contained in any linear hyper-plane (even if the image is just a curve), then the majority of the perturbed orbits for small perturbations are quasi-periodic. We remark that under this much weaker condition, we no longer have any persistence result as stated in the theorem. In fact, any and all invariant tori with Diophantine rotation numbers can be perturbed away with this weak assumption. However, new invariant tori with different rotation number have to form and majority of orbits are still on some invariant tori.

KAM theory asserts that for near integrable Hamiltonian system, the majority orbits are quasi-periodic and for these solutions, their action variables never deviate much from their original values, i.e.,

$$\sup_{t\in\mathbb{R}} ||I(t) - I(0)|| \to 0 \quad \text{as } \epsilon \to 0,$$

for a large measure set of initial condition (in any pre-fixed bounded subset of $\mathbb{R}^n$). Moser calls this type of stability as *stability in measure*. If the above is true for all initial conditions in any bounded subset, then the system is said to be *topologically stable*. For $n = 2$, invariant tori are two dimensional, they divide the three-dimensional energy surface. Therefore, stability in measure in this case leads to topological stability. However, for $n \geq 3$, stability in measure does not imply topological stability. In fact, one expects topological instability in generic situations. This is the subject of the next two sections.

3. Instability and Arnold diffusion

As opposite to stability, it has long been believed, particularly among physicists, that typical conservative systems are unstable. This types of believes at least go back to times of Boltzmann and Maxwell, who actually conjectured a stronger notion of instability, i.e., ergodicity. To justify his equilibrium theory of gas dynamics, Boltzmann conjectured, in terms of modern language, typical symplectic diffeomorphisms and volume preserving diffeomorphisms on compact manifolds are ergodic. This was later called Ergodicity Hypothesis. This of course turned out to be false. For the symplectic case, KAM theory, as we have discussed in the last section, shows that the invariant tori form a positive measure set for near integrable systems, therefore these systems can not be ergodic. Moreover, by the same reason, any system with a non-degenerate elliptic periodic point can not be ergodic either, because every non-degenrate elliptic periodic points is surrounded by a set of invariant tori with positive measure. For the volume-preserving diffeomorphisms on compact manifolds, the falsity of the ergodicity hypothesis was only established recently: Cheng & Sun [7] proved that, for three dimensional near integrable volume preserving maps, there also exists large positive measure set of invariant tori of, this time, co-dimension one. The same result for all dimensions was proved independently by Herman [cf. 27] and Xia [23].

Since the disproval of ergodicity hypothesis by KAM theory, a weaker conjecture has become widely accepted. This weaker version is called Quasi-Ergodicity Hypothesis by Herman. To make the conjecture plausible, one has to take a few precautions on the neccessary assumptions. It is convenient to state the conjecture for symplectic diffeomorphisms. First, we assume that the symplectic manifold is compact and has a dimension larger or equal to four. If the dimension is lower than that, the KAM theory guarantees topological stability for near integrable systems. Second, we assume that map is exactly symplectic. Herman [11] has a very interesting example of non-exact Hamiltonian system that has topological stability. Supprisingly, the systems is even structurely stable! Under these two assumptions, the quasi-ergodicity hypothesis states that typical such systems have dense orbits. This conjecture remains one of the most important open problem in Hamiltonian dynamics. We remark that the similar conjecture for volume preserving diffeomorphisms are false. This is again because of the existence of codimension one invariant tori for near integrable systems, which implies topological stability (Cheng & Sun [7], Herman [Cf. 27], Xia [23]).

Arnold [2] made a fundamental progress in the problem of instability by giving a remarkable example where instability can be proved. Arnold considered the following Hamiltonian function:

$$\begin{aligned} H &= H_0 + \epsilon H_1 \\ H_0 &= \frac{1}{2}(I_1^2 + I_2^2) \\ H_1 &= (\cos\theta_1 - 1) + \mu P(I, \theta, t) \\ P &= (\cos\theta_1 - 1)(\sin\theta_2 + \cos t) \end{aligned}$$

where both ϵ and μ are small parameters. For $\epsilon = 0$, the system is integrable and satisfies the non-degeneracy condition. Therefore, KAM theory shows that for $\epsilon \neq 0$, the system is stable in measure. However, Arnold proved the following theorem:

THEOREM 3.1 (Arnold). *Assume $0 < A < B$. For every $\epsilon > 0$, there exists a $\mu_0 > 0$ such that for $0 < \mu < \mu_0$, the above system is unstable: there exist a trajectory which connects the region with $I_2 < A$ with the region $I_2 > B$.*

The mechanism that causes the instability in the above theorem was later termed *Arnold diffusion.*

Motivated by the above example, Arnold [3] conjectured that above situation occurs in typical near integrable Hamiltonian systems. The conjecture can be regarded as a weaker version of the quasi-ergodicity conjecture. More precisely,

CONJECTURE 1 (Arnold). *Consider the Hamiltonian function:*

$$H_\epsilon(I, \theta, t) = G(I) + \epsilon H^1(I, \theta, t, \epsilon),$$

where $I \in \mathbb{R}^n$ and $\theta \in \mathbb{T}^n$ and H_ϵ is periodic in t. Then, the typical case for $n \geq 2$ is topological instability*: through an arbitrarily small neighborhood of any point there pass phase trajectories along which the action variables drift away from the initial values by a quantity of order 1, independent of ϵ.*

Since Arnold's example, little progress has been made until very recently in Arnold diffusion. Arnold's mechanism uses invariant lower dimensional tori, together with their stable and unstable manifolds to build something called transition chain. Then, by the concept of *obstructing set*, one shows that orbits can move freely, or diffuses, in the transition chain. The size of diffusion, or the size the instability region, obtained this way is limited by the size of the transition chain. It is a well-known fact that typical perturbations can break invariant tori, leaving gaps among invariant tori which are larger than the splitting sizes of the stable manifolds and unstable manifolds of these tori. This implies that the transition chain can be broken by typical perturbations, thus the required large size instability has to be obtained by alternative methods. this problem is known as the *gap problem*. In his example, Arnold chose a very special function which does not perturb any invariant torus on the transition chain, hence avoiding the gap problem which normally appears in general perturbations. Previous works on Arnold diffusion (e.g. [13], [19], [25], [9]) are all limited by the gap problem.

In recent years, great progress has been made in Aubry-Mather theory and variational methods in Hamiltonian dynamics. It turns out that gap problem can be solved with this method (Xia [26]). Instead of using lower dimensional invariant tori as basis of transition as in Arnold's example, one can use invariant cantor sets, called cantori, and their connecting orbits for diffusion. We call this mechanism Mather's mechanism. The advantage of using invariant cantori is that they never break under small perturbations. In higher dimensions, a generalized version of cantori, the minimal measure and action-minimizing orbits, has to be used.

To state our results in Arnold diffusion, we consider the so-called *a priori* unstable systems:

$$H_\epsilon(x, I, \theta, t) = F(x) + G(I) + H^1(x, I, \theta, t),$$

where $x \in M^2$, a surface, $I \in \mathbb{R}^n$ and $\theta \in \mathbb{T}^n$ and

$$H_\epsilon(x, I, \theta, t) \equiv H_\epsilon(x, I, \theta, t+1)$$

Where H^1 is a small perturbation term. We assume that the Hamiltonian flow on M^2 generated by $F(x)$ has a hyperbolic rest-point whose stable and unstable manifolds coincide and we also assume that $G(I)$ is positive definite in I with super-linear growth. This condition will be made precise in the next section.

A priori unstable systems typically are results of first averaging of near integrable systems. Arnold's example is an *a priori* unstable system. For the *a priori* unstable systems, we have the following theorem.

THEOREM 3.2. *Fix two open sets U and V in $\mathbb{R}^n$. There exists an open set $\mathcal{A}$, containing the zero function, in the set of all C^r, $r \geq 2$, functions $H^1 : M^2 \times \mathbb{R}^n \times \mathbb{T}^n \times \mathbb{T} \to \mathbb{R}$ and an open and dense subset $\mathcal{B}$ of $\mathcal{A}$, such that if $H^1(x, I, \theta, t) \in \mathcal{B}$, then there is an orbit that connects the region $I \in U$ with the region $I \in V$.*

Usually the perturbation function H^1 takes the form of $H^1 = \epsilon P(x, I, \theta, t, \epsilon)$, where ϵ is a small parameter. For $n = 1$, there is an easy criterion for the results of the theorem to hold, for ϵ small. The criterion uses the so-called *Melnikov function*, which is defined by an improper integral:

$$M = \int_{-\infty}^{\infty} P(x(t), I(t), \theta(t), t, \epsilon)dt,$$

where the integration is along the homoclinic orbits of the unperturbed invariant tori, with proper care taken for the convergence of the improper integral. It turns out that Melnikov function $M = M(\theta_0, I_0)$ is a real valued function defined on $\mathbb{T}^n \times \mathbb{R}^n$. For the theorem to hold, it suffices to assume that there is an arc γ from U to V such that for each $I_0 \in \gamma$, the minimum set of $M(\theta_0, I_0)$ is contractible.

For $n \geq 2$, the author believes that the Melnikov function should also work. However, due to complications with minimal measures, it remains to be shown.

As an application of our theorem, we can show that if one replaces $P = (\cos\theta_1 - 1)(\sin\theta_2 + \cos t)$ in Arnold's example with $P = \cos\theta_1(\sin\theta_2 + \cos t)$, then the same results hold. In fact, Arnolds result can be generalized to an open and dense set of such functions.

The proof of the above theorem is an application and extension of Mather's results ([16], [17], [18]) on minimal measures in Lagrangian and Hamiltonian systems. In the next section, we will give a review on Aubry-Mather theory and Mather's remarkable work on construction of connecting orbits in Hamiltonian systems.

4. Aubry-Mather theory and variational method

The variational method works in the framework of typical Lagrangian systems coming from classical mechanics. In this section, we use Lagrangian formulation instead of the Hamiltonian formulation. These two formulation are interchangeable. It is straight forward to convert a Lagrangian systems to a Hamiltonian systems and vice versa, by using Legendre transformations.

We consider a smooth compact manifold M and a C^2 real valued function L defined on $TM \times \mathbb{T}$, where TM is the tangent bundle of M and $\mathbb{T} = \mathbb{R}/\mathbb{Z}$. L is defines a Lagrangian system on M and it is call the *Lagrangian.*

The basic principle in Lagrangian formulation is that a true trajectory of a mechanical system must be an extremal of action, i.e., a trajectory $\gamma(t)$ must satisfy the following fixed endpoint variational equation:

$$\delta \int_a^b L(\gamma(t), \gamma'(t), t)dt = 0$$

In local coordinates, trajectories are solutions of the following Euler-Lagrange equations:

$$\frac{d}{dt}\frac{\partial L}{\partial v}(x, \dot{x}, t) = \frac{\partial L}{\partial x}(x, \dot{x}, t)$$

We need to make the following three hypothesis for L:

- *Positive definiteness*: For each $x \in M$, $v \in T_xM$ and $t \in \mathbb{T}$, the restriction of L to $TM_x \times t$ is strictly convex in the sense that its Hessian second derivative $(\partial^2 L(x, v, t)/\partial v_i \partial v_j)$ is everywhere positive definite.
- *Super-linear Growth*:
$$\frac{L(x, v, t}{||v||} \to +\infty, \quad \text{as } ||v|| \to \infty$$
where $|| \ ||$ is a Riemannian metric on M and x ranges over M and $t \in \mathbb{T}$.
- *Completeness*: All the solutions of the Euler-Lagrange equation can be extended for all $t \in \mathbb{R}$.

Let μ be a probability measure on $TM \times \mathbb{T}$, invariant under the Euler-Lagrange flow. We can define its *average action*

$$A(\mu) = \int L d\mu.$$

This integral is bounded below and it may be $+\infty$. When it is finite, we may associate to μ its *rotation vector*, $\rho(\mu) \in H_1(M, \mathbb{R})$. Let λ_c be a closed one form on M whose cohomology class is $c \in H^1(M, \mathbb{R})$. The rotation vector $\rho(\mu)$ is uniquely characterized by the following equation:

$$< c, \rho(\mu) > = \int \lambda d\mu$$

Where the bracket $< >$ is the canonical pairing of $H^1(M, \mathbb{R})$ and $H_1(M, \mathbb{R})$.

Let $U_L = \{(\rho(\mu), A(\mu)) \mid \mu \in \mathcal{M}(L)\}$, where $\mathcal{M}(L)$ is the set of all invariant probability measures on $TM \times \mathbb{T}$ satisfying $A(\mu) < \infty$. Since the set of all invariant probability measures is convex and A and ρ are linear functions on the set, U_L is convex. It turns out (Mather [16]) that the projection of U_L on $H_1(M, \mathbb{R})$ is surjective. This together with the fact that L is bounded below implies that U_L is the epigraph of a convex function $\beta = \beta_L : H_1(M, \mathbb{R}) \to \mathbb{R}$. For each $h \in H_1(M, \mathbb{R})$, we will call $\beta(h)$ the *minimal average action* of the rotation vector h. The invariant measure that achieves the minimal average action for its rotation vector are called *minimal measure* or *action-minimizing measure*.

We emphasize that for every rotation vector in $H_1(M, \mathbb{R})$, there exists at least one minimal measure associated to that rotation vector. However, for manifolds other than $\mathbb{T}$, the structure of the minimal measures are very complicated and so far, we have little understanding of these minimal measures. For the case $M = \mathbb{T}$, because of the special order property, much more can be obtained. These results on $\mathbb{T}$ are collectively referred to as the Aubry-Mather theory.

It is easy to see that β has super-linear growth. Let $\alpha : H^1(M, \mathbb{R}) \to \mathbb{R}$ be the conjugate of β in the sense of convex analysis, or,

$$\alpha(c) = \max\{< c, h > -\beta(h) : h \in H_1(M, \mathbb{R})\},$$

for $c \in H^1(M, \mathbb{R})$. Since β has super-linear growth, the maximum is achieved and α is everywhere defined. α also has super-linear growth.

Let λ_c be a closed 1-form on M whose de Rham cohomology class is c. λ_c can be regarded as a real valued function on $TM \times \mathbb{T}$. The function $L - \lambda_c$ can be regarded as another Lagrangian that satisfies the conditions of positive definiteness, super-linear growth, and completeness. In fact, one can easily verify that its Euler-Lagrange flow is exactly the same as that of L. However, by adding or subtracting a closed form to the Lagrangian, we obtain different global minimal measures and global minimal actions. We remark that adding a constant to the Lagrangian makes no effect to the Euler-Lagrangian flow.

A curve γ in M is said to be a c-minimizer (where $c \in H^1(M, \mathbb{R})$) if it satisfies the following condition. For any interval $[a, b]$ and any curve $\gamma_1 : [c, d] \to M$ such that $c - a \in \mathbb{Z}$ and $d - b \in \mathbb{Z}$, we have:

$$\int_a^b (L - \lambda - \alpha(c))(\gamma, \dot{\gamma}, t)dt \leq \int_c^d (L - \lambda - \alpha(c))(\gamma_1, \dot{\gamma}_1, t)dt$$

where α is the conjugate function of β.

We need a few more definitions before we can state Mather's main results [18]. Let $\Sigma_c \subset M \times \mathbb{T}$ be the set of all point $\{(\gamma(t), t) \in M \times \mathbb{T} \mid \gamma(t) \text{ is a } c\text{-minimizer}\}$. Let $\Sigma_c^0 = \Sigma_c \cap (M \times 0)$. Finally we let W_L be the set of all $c \in H^1(M, \mathbb{R})$ that satisfies the following property: there exists an open neighborhood U of Σ_c^0 in M such that the inclusion map $H_1(U, \mathbb{R}) \to H_1(M, \mathbb{R})$ is the zero map.

We are ready to state two remarkable theorems due to Mather [18], which are in some sense a generalization Mather's work on twist maps [16].

THEOREM 4.1 (Mather). *Suppose c_0 and c_1 are in the same connected component of W_L. Then there is trajectory of the Euler-Lagrange flow whose α-limit set lies in Σ_{c_0} and whose ω-limit set lies in Σ_{c_1}.*

THEOREM 4.2 (Mather). *Consider a bi-infinite sequence $(\cdots, c_i, \cdots)$ of cohomology classes, each of which lies in the same connected component of W_L. Let $(\cdots, \omega_i, \cdots)$ be a sequence of positive numbers. Then there is a trajectory of the Euler-Lagrange flow which passes within a distance of ω_i of each of the set Σ_{c_i} in turn.*

The proof of the above theorems relies on good choices of the so-called barrier functions. For details, see Mather [18].

The proof of Theorem 3.2 of the last section is based on the techniques that Mather developed to prove the above two theorems. One key point is to extend the theorem to include a larger set than the original W_L. We need to cover all the cohomology classes c where there is an open set $U \in M$ containing Σ_c^0 such that the inclusion map $H_1(U, \mathbb{R}) \to H_1(M, \mathbb{R})$ is not an isomorphism. We remark that for the case $n = 1$, the proof is somewhat easier (cf. Xia [26]).

We end this lecture by remarking that in order to prove Conjecture 1 of Arnold, or to prove a similar theorem as Theorem 3.2 in the *a priori* stable situation, a great deal of understanding on minimal measures are required. Mañé's work [15] provided good examples of generic properties for minimal measures.

References

1. Arnold, V.I., *Small denominators and problems of stability of motions in classical celestial mechanics*, Russ. Math. Survey **18** 6 (1963), 85-191
2. Arnold, V.I., *Instabilities of dynamical systems with several degrees of freedom*. Sov. Math. Dokl., **5**, 581-585(1964).
3. Arnold, V.I. (ed), *Dynamical Systems III*, Encyclopaedia of Mathematical Sciences, Vol. 3, Springer-Verlag, New York/Berlin, 1998.
4. Bangert, V., *Mather sets for twist maps and gedesics on tori*, Dyn. Rep., **1**, 1-56(1988).
5. Bessi, U., *An approach to Arnold diffusion through calculus of variations*, Nonlinear Analysis, **26**(6), 1115-1135(1996).
6. Birkhoff, G.D., *Dynamical Systems*, Amer. Math. Soc. Colloquium Pub. Vol. IX (1966)
7. Cheng, C-Q & Sun, Y.S., *Existence of invariant tori in three-dimensional measure-preserving mappings*, Celest. Mech. Dyn. Astr., **47** (1990), 275-292
8. Cheng, C-Q & Sun, Y.S., *Existence of KAM tori in degenerate Hamiltonian systems. J. Differential Equations*, **114** (1994), no. 1, 288–335.
9. Chierchia, L. & Gallavotti, G., *Drift and diffusion in the phase space*. Ann. de la Inst. H. Poincaré Phys. Th., bf 60(1), 1-144 (1994).
10. Douady, R., *Stabilité ou instabilité des points fixes elliptiques*, Ann. scient. Éc. norm. sup., **21**(4), 1-46(1988).
11. Herman, M., *Exemples de flots Hamiltoniens dont aucune perturbation en topologie C^∞ n'a d'orbites périodiques sur un ouvert de surfaces d'énergie*, C.R. Acad. Sci. Paris, t. **312**(1991), 989-994
12. Herman, M., *Some open problems in dynamical systems*, Proceedings of the International Congress of Mathematicians, Vol. II (Berlin, 1998). Doc. Math. 1998, Extra Vol. II, 797–808
13. Holmes, P., & Marsden, J., *Melnikov method and Arnold diffusion for perturbation of integrable Hamiltonian systems*, J. Math. Phys., **234**, 669-675(1982)
14. Kolmogorov, A.N., *Théorie générale des systémes dynamique et mécanique classique*, Proc. Int. Congress of Math. **1** (1954), 315-333, North Holland.
15. Mañé, R., *Generic properties and problems of minimizing measures of Lagrangian systems*, Nonlinearity, **9**, 273-310(1996).
16. Mather, J., *Action minimizing invariant measures for positive definite Lagrangian systems*, Math. Z., **207**, 169-207(1991).
17. Mather, J., *Variational construction of orbits of twist diffeomorphisms*, J. Amer. Math. Soc., **4**, 207-263(1991).

18. Mather, J., *Variational construction of connecting orbits*, Ann. Inst. Fourier, Grenoble, **43**5, 1349-1386(1993).
19. Moeckel, R., *Transition tori in five-body problem*, J. Diff. Eqns., **129**, 290-314(1996).
20. Moser, J., *On invariant curves of area-preserving mappings of an annulus*, Nachr. Akad. Wiss. G ̈ttingen, math-phys. Kl. 1962, 1-10
21. Robinson, C., *Horseshoes for autonomous Hamiltonian systems using Melnikov integrals,* Ergodic Theory Dynamical Systems, **8**, 395-409(1988).
22. Rüsmann, H., *Nondegeneracy in the perturbation theory of integrable dynamical systems*, to appear.
23. Xia, Z., *The Existence of the invariant tori in the volume preserving diffeomorphisms*, Ergod. Th. & Dynam. Sys. **12** (1992), 621-631
24. Xia, Z., *The Existence of the Non-Collision Singularities in Newtonian Systems*, Annals of Mathematics, 135 (1992) 411-468
25. Xia, Z., *Arnold diffusion and oscillatory solutions in plannar three-body problem*, J. Diff. Eqns., **110**, 289-321 (1994).
26. Xia, Z., *Arnold Diffusion: A Variational Construction*, in Proceedings of International Congress of Mathematicians, Berlin 1998. Documenta Mathematica, Extra Volume ICM'98, 1998
27. Yoccoz, J.C., *Travaux de Herman sur les tores invariants*, Séminaire Bourbaki, No. 754, Astérisque **206** (1992), 311-344

DEPARTMENT OF MATHEMATIC, NORTHWESTERN UNIVERSITY, EVANSTON, ILLINOIS 60208, AND, DEPARTMENT OF MATHEMATICS, PEKING UNIVERSITY, BEIJING, PEOPLE'S REPUBLIC OF CHINA

E-mail address: xia@math.nwu.edu

AMS/IP Studies in Advanced Mathematics
Volume 20, 2001

On The Behavior of Solutions to The Compressible Navier-Stokes Equations

Zhouping Xin*

CONTENTS

1. Introduction

The motions of a d-dimensional compressible viscous, heat-conductive, isotropic Newtonian fluid can be governed by the following system of Navier-Stokes Equations [4,17]

$$(1)\qquad \begin{cases} \partial_t \rho + div(\rho u) = 0 \\ \partial_t(\rho u) + div(\rho u \otimes u) + \nabla p = div(T) \\ \partial_t(\rho E) + div(\rho u E + up) = div(uT) + k \bigtriangleup \theta \end{cases}$$

where $(x,t) \in \mathbb{R}^d \times \mathbb{R}_+$, and for $d \geq 1$, $\rho = \rho(x,t)$,$u = (u_1, ..., u_d) \in \mathbb{R}^d$, p, e and θ denote the density, velocity, pressure, internal energy and absolute temperature respectively, $E = e + \frac{1}{2}|u|^2$ is the total energy, and T is the stress tensor given by $T = \mu(\nabla u + (\nabla u)^t) + \mu'(div\ u)I$ with I being the identify matrix, μ and μ' being the coefficient of viscosity and the second coefficient of viscosity respectively, k denotes the coefficient of the heat conduction. It is usually assumed that $\mu \geq 0$, $\mu' + \frac{2}{d}\mu \geq 0$, and $k \geq 0$. The relation between p, θ, ρ, and e is given by the equation of state and the second law of thermo dynamics. In the special case of polytropic fluids, the equations of state in given by

*This work is supported partially by Zheng Ge Ru Foundation, UGC Earmarked Grant CUHK–4129/99P, NSF Grants DMS-96-00137, and DMS-99-71978, and DOE Grant DE-FG02-88ER-25053.

$$p = R\rho\theta, \quad e = c\theta, \quad p = Ae^{\frac{S}{c}}\rho^{\gamma} \tag{2}$$

where $A > 0$ is a constant, $\gamma(> 1)$ is the ratio of specific heat, S is the entropy, and $C = R/(\gamma - 1)$.

For ideal inviscid fluids, $\mu' = \mu = k = 0$, then system (1) becomes the well-known compressible Euler equations:

$$\begin{cases} \partial_t \rho + div(\rho u) = 0 \\ \partial_t(\rho u) + div(\rho u \otimes u) + \nabla p = 0 \\ \partial_t(\rho E) + div(\rho u E + up) = 0 \end{cases} \tag{3}$$

which is one of the most important examples of systems of nonlinear hyperbolic conservation laws, and has been the main focus and the driving force for the mathematical theory of shock waves [18,34].

Tremendous progress has been made in solving systems (1) and (2) both theoretically and numerically, and in understanding and interpreting the behavior of solutions to these systems. This is particularly so in the case of one-space-dimension and for "small" solutions [7,5,8,12,16,21,28]. Two major issues have been investigated extensively. One is the global (in time) well-posedness of smooth solutions and their large time asymptotic behavior for (1) with fixed viscosity and heat conduction. This has been successfully treated in the case for small smooth non-vacuum data whose long time behavior is shown to be governed by the linear diffusion waves [16,28]. This theory has been generalized even for small discontinuous initial data in multi-dimensions [12]. The other issue is to understand the effects of the small scale dissipations on the behavior of large scale physical flows, i.e. to discribe the asymptotic relationship between the inviscid and viscous flows in the presence of discontinuities and physical boundaries for either small (but non-zero) dissipation or large time (with fixed dissipation). This is a very important issue not only for the understanding of the asymptotic behavior of the solutions to the Navier-Stokes system, but also crucial for the theory of the inviscid Euler equations, which is due to the fact that it is the dissipative mechanism which selects the entropy weak solutions to the Euler system from other nonphysical solutions. An almost complete theory on the nonlinear large time asymptotic stability toward basic viscous waves (viscous shock profile, rarefaction waves, contact discontinuities, and diffusive waves) has been developed for one-space-dimension [9,15,22-24,26,29,35,30]. And the asymptotic equivalence between the viscous system (1) and inviscid Euler equations (2) has also been understood in one-space-dimension for special flows [5,10,13,21,39]. More recently, the global (in time) existence of large amplitude weak solutions to either Cauchy problem or initial-boundary value problems of the isentropic compressible Navier-Stokes equations with some special equations of state has been established [19], and some important properties of these weak solutions, such as boundness of the total energy, the zero-mach number limit to the incompressible Navier-Stokes equations, and large time asymptotic behavior toward steady states etc, have been investigated [19,20]. For the corresponding inviscid Euler system (2), substantial progress has been made in the one-dimensional theory of shock wave for solutions of small total variation [1-3,8,18,25,34].

Despite these important achievements, there remain many significant challenging outstanding issues to be settled. One of these is the well-posedness and large time asymptotic behavior of smooth solutions of finite amplitudes to (1) in arbitrary space dimensions. Can a smooth solutions to the Navier-Stokes system (1) develop singularities in finite time? A related problem is the regularities of the weak solutions constructed by P. L. Lions [19]. It should be noted that in the case where the initial data is not periodic, the existence of weak solution [19] requires that the initial far fields must be vacuum, and then the weak solutions cannot be regular in general even for smooth initial due to the fact that the total pressure will decay sufficiently fast as will be seen in the next section (see also [36]). Another issue with considerable interest is the boundary layer theory for the compressible Navier-Stokes system (1). There have been extensive studies on the zero dissipation limit for the Navier Stokes system in the presence of the physical boundaries due to its importance in the applications to high speed flows [33,4,17]. In the case where the physical boundaries are uniformly non-characteristic, the multi-scale structure of the viscous solutions can be revealed by matched asymptotic analysis and the strong convergence of the Navier-Stokes solutions to the one of the Euler system away from boundaries can be obtained by nonlinear stability analysis of the boundary layer before shock formations provided the boundary layer is suitably weak [37,40,30,11,6]. For the well-known no-slip boundary conditions for (1), which is uniformly characteristic, the formal Prandtl's boundary layer theory has existed for a long time, yet its validity has been proved rigorously only in the analytical setting [32]. The rigorous justification of the Prandtl's boundary layer theory for the linearized flow will be given in section 3 (see also [38]).

2. An A-Priori Estimate on The Total Pressure And Some Applications

The main result of this section is the following a priori estimate on the total pressure for the viscous polytropic fluid. Consider the Cauchy problem for the Navier-Stokes system (1) with initial data:

$$(\rho, u, S)(x, t = 0) = (\rho_0, u_0, S_0)(x) \tag{4}$$

Then one has that

Theorem 1 Assume that

$$m_2 = \int_{\mathbb{R}^d} |x|^2 \rho_0(x) dx < +\infty \tag{5}$$

Let (ρ, u, S) be a smooth solution to Navier-Stokes (1) with initial data (4), such that

$$(\rho, u, S - \bar{s}) \in C^1([0, T), H^m(\mathbb{R}^2)), \quad m \geq [\frac{d}{2}] + 2, \tag{6}$$

where $\bar{s}$ is a constant, and $0 < T \leq +\infty$. Then there exists a positive constant C_0 (independent of T) such that

(7) $$\int_{\mathbb{R}^d} p(x,t)dx \leq \begin{cases} \frac{\gamma-1}{2} m_2 t^{-2}, & \gamma \geq 1 + \frac{2}{d} \\ \frac{\gamma-1}{2} C_0 (1+t)^{-(\gamma-1)d}, & 1 < \gamma < 1 + \frac{2}{d} \end{cases}$$

for all $t \in [0,T)$, where

$$C_0 = m_2 - \int_{\mathbb{R}^d} (x \cdot u_0(x))\rho_0(x)dx + 2\int_{\mathbb{R}^d} \rho_0 E_0 dx < +\infty \quad .$$

A few remarks are in order:

Remark 1 The key assumption in Theorem 1 is (5), so that the initial far fields are in the vacuum. In this case, the fast decay in time of the total pressure given in (7) indicates that the pressure behaves dispersively instead of diffusively. Indeed, for the monotonic fluid in 3-dimension, $d=3, \gamma = \frac{5}{3}$, hence (7) becomes

(8) $$\int_{\mathbb{R}^3} p(x,t)dx \leq \frac{m_2}{3} t^{-2}$$

This shows that the pressure decays faster than any diffusive scales in 3-dimension.

Remark 2 The rigorous derivation of the a-priori estimate (7) is given by studying a differential inequality for the functional

(9) $$I_\gamma(t) \equiv \begin{cases} \int_{\mathbb{R}^d} |x - u(x,t)(t+1)|^2 \rho(x,t)dx + \\ \frac{2}{\gamma-1}(t+1)^2 \int_{\mathbb{R}^d} p(x,t)dx, \quad 1 < \gamma < 1 + \frac{2}{d}, \\ \\ \int_{\mathbb{R}^d} |x - u(x,t)t|^2 \rho(x,t)dx + \\ \frac{2}{\gamma-1} t^2 \int_{\mathbb{R}^d} p(x,t)dx, \quad 1 + \frac{2}{d} \leq \gamma < +\infty. \end{cases}$$

We refer the readers to [36] for details. However, the special case, (8), can be seen formally by a microscopic-kinetic description of the fluids as follows. Let $f(x,t,\xi)$ be the density function of particles at position x at time t with velocity ξ. Assume that $f(x,t,\xi)$ satisfies the Boltzmann type equation of the form.

$$\partial_t f + \xi \nabla_x f = \frac{1}{\varepsilon} Q(f,f)$$

where Q denotes the collision operator and ε is the mean free path. It is expected that as the mean free path, ε, tends to zero, f approaches a local Maxwellian, whose moments

$$\rho(x,t) = \int_{\mathbb{R}^3} f(x,t,\xi)d\xi \quad , \quad (\rho u)(x,t,\xi) = \int_{\mathbb{R}^3} \xi f(x,t,\xi)d\xi \quad ,$$
$$(\rho E)(x,t) = \frac{1}{2}\int_{\mathbb{R}^3} |\xi|^2 f(x,t,\xi)d\xi$$

are governed by the macroscopic Navier-Stokes system (1). On the other hand, for any reasonable description such that the principles of conservation of mass, momentum, and energy hold at the microscopic level, it is always true that $\int_{\mathbb{R}^3}\int_{\mathbb{R}^3} |x -$

$\xi t|^2 f(x,+,\xi)d\xi dx$ is a time invariant. Thus

$$\int_{\mathbb{R}^3} |x-u(x,+)t|^2\rho(x,t)dx + 2t^2\int_{\mathbb{R}^3}(\rho e)(x,t)dx \\ = \int_{\mathbb{R}^3}|x|^2\rho_0(x)dx = m_2 \tag{10}$$

(10) implies the desired estimate (8) immediately since $p = \frac{2}{3}\rho e$ for the polytropic fluids.

We now give some simple applications of the a-priori estimate. We first recall the classical theory on the well-posedness of "small" as "non-vacuum" solutions to the Navier-Stokes equations [28,16,12].

Theorem (Classical Theory) Let $\bar\rho$ be a positive constant, $m = [\frac{d}{2}]+2$. Then there exists a suitably small constant δ such that if

$$\begin{cases} (p_0-\bar\rho, u_0, S_0) \in H^m(\mathbb{R}^d)\cap L^1(\mathbb{R}^d) \\ \|(\rho_0-\bar\rho, u_0, S_0)\|_{H^m} \le \delta \end{cases} \tag{11}$$

Then, these exists a unique global solution to the Navier-Stokes equations (1) with initial data (4) with the properties that

$$\begin{cases} (p-\bar\rho, u, S) \in C^1([0,\infty), \quad H^m(\mathbb{R}^d)) \\ \|u(\cdot,t)\|_{L^\infty(\mathbb{R}^d)} \le C(1+t)^{-\beta}, \quad \beta=\beta(d) \end{cases} \tag{12}$$

where the rate of decay is exactly same as that of heat kernel.

It should be noted that for the classical theory, the key assumption is that the initial data is a "small" perturbation of a non-vacuum state $(\bar\rho,0,0)$. On the other hand, it is shown by P. L. Lions recently [19] that for some special isentropic fluids, a global weak solutions exists without smallness assumption provided that

$$\rho_0 \in L^q(\mathbb{R}^d), \quad d=1,2,3, \quad q=q(d,\gamma). \tag{13}$$

Note that (13) implies that the initial far fields must be in vacuum. Two natural questions arise:

- Can the classical theory hold true in the case that the initial far fields are in vacuum?
- What is the regularity of the weak solution obtained in [19]?

Applying the a priori estimate (7), we can easily derive some negative answers to both of these two questions in the case that

$$supp \quad \rho_0(x) \subseteq B_{R_0}, \quad 0<R_0<+\infty \tag{14}$$

Where B_{R_0} is the ball centered at origin with radius R_0, i.e., we assume that initial density has compact support. Indee, we have the following results [36]:

Corollary 1 Assume (14). Then there exists no non-trivial solution $(\rho, u, S) \in C^1([0,\infty) : H^m(\mathbb{R}^d))$ to the Cauchy problem for the Navier-Stokes system (1) with initial data (4) such that (12) holds.

Corollary 2 Assume (14), and

$$\mu > 0, \quad \mu' + \frac{2}{d}\mu > 0, \quad \mu \geq 0.$$

Then there exists no non-trivial solution $(\rho, u, S) \in C^1([0,\infty); H^m(\mathbb{R}^d))$ to (1) with initial data (4), which is bounded in time.

Furthermore, one can estimate the life span of the smooth solution as follows:

Corollary 3 Assume (11),

$$\mu > 0, \quad \mu' + \frac{2}{d}\mu > 0, \quad \text{and}$$

either $k = 0$, or the entropy will not increase in time.

Let $(u, \rho, S) \in C^1([0,T), H^m(\mathbb{R}^d))$ be a solution to the Navier-Stokes system (1). Then there exists a $T^* > 0$, $T^* < +\infty$, and T^* depends only on the initial data, d, γ, and R_0 such that

$$T \leq T^*,$$

in other words, every smooth solution blows up in finite time!

Remark 3 The theory and method also apply to the Euler system (2) to show that there exists no non-trivial solution in $C^1([0,\infty); H^m(\mathbb{R}^d))(m > [\frac{d}{2}] + max(1, \frac{4}{\gamma-1}))$ to (2) with compact initial density and velocity. This was shown before in [27] by a different method.

Remark 4 In contrast to the classical theory, there is no requirement on the amplitude of the initial data in corollaries 1-3.

Remark 5 These corollaries follow easily from the faster decay of the total pressure, (7), the conservation of mass, and the slow growth of the support of density, for details, see [36].

3. Boundary Layer Theory

We now consider the boundary layer problem for the compressible Navier-Stokes system. For simplicity in presentation, we will only discuss a 2-Dimensional viscous isentropic fluid. In this case, the Navier-Stokes system, (1), becomes

$$(15) \quad \begin{cases} \partial_t \rho + div(\rho \vec{u}) = 0 \\ \partial_t(\rho \vec{u}) + div(\rho \vec{u} \otimes \vec{u}) + \nabla p = div \quad (T) \end{cases} \quad \text{on} \quad \Omega \times (0,t)$$

with

$$p = p(\rho), \quad p'(\rho) > 0, \quad \forall \rho > 0, \tag{16}$$

where Ω is a regionl is $\mathbb{R}^2$ with smooth boundary. We assume that

$$\mu = \varepsilon^2 \quad (\varepsilon > 0), \quad \mu' = 0(1)\varepsilon^2, \tag{17}$$

and ε is small.

We supplement (15) with initial data

$$(\rho, \vec{u})(x, t = 0) \equiv (\rho_0, \vec{u}_0)(x) \tag{18}$$

and boundary data

$$\vec{u}\,(x,t)|_{\partial\Omega\times|0,T)} = \vec{f}\,(x,t). \tag{19}$$

The associated initial-boundary value problem for the corresponding inviscid Euler equations is

$$\begin{cases} \partial_t \rho + div(\rho\,\vec{u}) = 0 \\ \partial_t(\rho\,\vec{u}) + div(\rho\,\vec{u}\otimes\vec{u}) + \nabla p = 0 \end{cases} \qquad \Omega \times (0,t) \tag{20}$$

$$(\rho, \vec{u})(x, t = 0) \equiv (\rho_0, \vec{u_0})(x) \quad , \tag{21}$$

$$\vec{u}\cdot\vec{n} = \vec{f}\cdot\vec{n} \quad \text{on} \quad \partial\Omega \times (0,t) \quad , \tag{22}$$

where $\vec{n}$ is the unit out normal of $\partial\Omega$.

We will discuss two distinct types of boundary conditions:

(A) Non-characteristic boundary conditions, i.e.

$$\vec{f}\,(x,t)\cdot\vec{n} \neq 0 \quad \text{on} \quad \partial\Omega \times (0,T) \quad . \tag{23}$$

In this case, the boundary condition (22) is uniformly non-characteristic for the Euler system (20).

(B) No-slip boundary condition, i.e.

$$\vec{f}\,(x,t) \equiv 0 \quad \text{on} \quad \partial\Omega \times (0,T) \quad . \tag{24}$$

In this case, the boundary condition (22) is uniformly-characteristic for the Euler system (20).

Due to the disparity of boundary conditions (19) and (22), the important pheonema, boundary layers, will appear for high Reynolds number flows. One would like to understand the asymptotic structures of solutions to the initial-boundary values problems, (15), (18)-(19), as $\varepsilon \to 0^+$, in particular, its relation to the solutions to the problem (20)-(22). Formally, one would expect (at least for short time) that

(A) for non-characteristic boundary, there appears a sharp changing gradient in region near the boundary of width of order $0(\varepsilon^2)$ (boundary layer), and away from the boundary layer, the flow is governed by the inviscid Euler equation. Furthermore, the boundary layer is governed by a system of ordinary differential equations.

(B) in the case of no-slip boundary condition, the width of the boundary layer is of order $0(\varepsilon)$, the flow near the boundary is governed by the Prandtl's boundary layer equations (which is a degenerate elliptic-parabolic system of partial differential equations) [33], and away from the boundary layer, the solution converges to inviscid one uniformly.

One of the main problems is to justify such formal theory rigorously. We will present some preliminary results here.

3.1. Non-Characteristic Boundary Condition. Under the assumption that the boundary is uniformly non-characteristic i.e., (23) holds. Then any "weak" boundary layers are nonlinearly stable and the visious solution converge uniformly to the corresponding inviscid solutions away from the boundary as $\varepsilon \to 0^+$ before shock formation.

To be precise, we consider the special case,

(25) $$\Omega = \mathbb{R}^1_+ \times \mathbb{R} = \{(x_1, x_2) \in \mathbb{R}^2, \quad x_1 \geq 0\} \quad ,$$

and the boundary condition (22) becomes

(26) $$u_1(x_1 = 0, x_2, t) = f_1(x_2, t)$$

It can be shown that under suitable constraints on f_1, the initial-boundary value problem, (20),(21), and (26), has a unique smooth solution$(\rho^0, \vec{u}^{\,0})(x_1, x_3, t)$ with

(27) $$(\rho^0 - \underline{\rho}, \vec{u}^{\,0}) \in C^l([0, T_*); H^{l+3}(\mathbb{R}^1_+ \times \mathbb{R}^1))$$

where $\underline{\rho}(>0)$ is a constant, $l \geq 1$, and $[0, T_\alpha)$ is the maximal interval of such a solution. Set

(28) $$\delta(T) = max_{\{0 \leq t \leq T\}} \quad ||u_2^0(0, \cdot, t) - f_2(\cdot, t)||_{L^\infty(\mathbb{R}^1)}, \forall T < T_* \quad .$$

Then, one has that (see [40])

Theorem 2 Let $(\rho^0, \vec{u}^{\,0})$ be the solution to the initial-boundary value problem (20), (21) and (26) such that (27) holds true with $l \geq 2$. Let $T \in (0, T_*)$ be fixed. Then there exist positive constants δ_0 and ε_0 independent of ε such that for

(29) $$\delta(T) \leq \delta_0 \quad ,$$

there exists a unique solution $(\rho^\varepsilon, \vec{u}^{\,\varepsilon})$ to the viscous initial-boundary value problem, (15), (18), and (19), such that

(30) $$(\rho^\varepsilon - \underline{\rho}, \vec{u}^{\,\varepsilon}) \in C^1([0, T]; H^4(\mathbb{R}^1_+ \times \mathbb{R}^1))$$

and

(31) $$sup_{\{0 \leq t \leq T, x_1 \geq h > 0\}} ||(\rho^\varepsilon - \rho^O, \vec{u}^{\,\varepsilon} - \vec{u}^{\,0})(x_1, \cdot, t)||_{L^\infty(\mathbb{R}^1)} \leq C_h \mu$$

provided that $\varepsilon \in (0, \varepsilon_0]$. Furthermore, for any given $n > 0$, there exists a smooth bounded function $(\tilde{\rho}, \tilde{u})(x, t, n, \varepsilon)$ such that

(32) $$sup_{\{0 \leq t \leq T\}} \quad ||(\rho^\varepsilon, \vec{u}^{\,\varepsilon})(\cdot, t) - (\tilde{\rho}, \tilde{u})(\cdot, t)||_{L^\infty(\Omega)} \leq C_n \varepsilon^{2n} \quad ,$$

where

$$(33)\qquad \begin{aligned}&(\tilde\rho,\tilde u)(x,t)\\ &\equiv (\rho^0,\vec{u}^{\,0})\quad (x,t)+\textstyle\sum_{i=1}^n \mu^{i-1}\,\vec{B_i}\,(\frac{x_1}{\mu},x_2,t)+\sum_{i=1}^n \mu^i\,\vec{I_i}\,(x,t)\end{aligned}$$

with B_i and I_i are smooth functions.

Remark 6 Due to the compatibility of the initial and boundary data, (29) holds for T small. So the asymptotic convergence for short time is a simple consequence of Theorem 2.

Remark 7 (32) and (33) give detailed asymptotic behavior of the solution to the Navier-Stokes system up to any given order, since the asymptotic ansatz $(\tilde\rho,\tilde u)$ in (33) can be constructed explicitly by the method of matched asymptotic expansions [40]. Similar idea has been used for uniform parabolic systems in [37, 30, 11, 14].

Remark 8 The stability of strong boundary layer is still open.

We now turn to even more difficult case.

3.2. No-Slip boundary condition. The asymptotic behavior of the solutions to Navier-Stokes system for small viscosity in this case turns out to be one of most challenging problems in fluid dynamics. Very few rigorous results exist for the dynamical boundary layer behavior of the compressible Navier-Stokes system. To indicate the difficulties in treating this problem, we consider the special case that Ω is the half plane given in (25). Then the no-slip boundary condition becomes

$$(34)\qquad \vec{u}\,(x_1=0,x_2,t)\equiv 0\quad .$$

and the corresponding boundary condition for the Euler equation, (22) is

$$(35)\qquad u_1(x_1=0,x_2,t)\equiv 0\quad .$$

Applying Prandtl's ansatz

$$(36)\qquad \begin{pmatrix}\vec{u}\\ P\end{pmatrix}(x_1,x_2,t)=\begin{pmatrix}\varepsilon u_1\\ u_2\\ p\end{pmatrix}(\frac{x_1}{\varepsilon},x_2,\quad t)\quad ,$$

one can derive the leading order governing system is the following Prandtl's type boundary layer equations [33]

$$(37)\qquad \begin{cases}\partial_t u_2+u_1\partial_{y_1}u_2+\partial_{x_2}(\frac12 u_2^2)=\partial_{y_1}^2 u\\[2ex] \partial_{y_1}u_1+\partial_{x_2}u_2=0(1)\end{cases}$$

where $y_1=\frac{x_1}{\varepsilon}$, and $o(1)$ does not involve derivative of $\vec{u}$. (37) is a degenerate parabolic-elliptic system. Even short time well-posedness in a sobolev space is not known (except some special cases [31]). Even in the case that (37) can be solved in a class of smooth functions, the justification of the formal boundary layer theory is still very difficult due to the stiffness in the convection speed. Here we will present a justification of the Prandtl's boundary layer theory for linearized Navier-Stokes equations [38]. For the simplicity in presentation, we still consider the half plane problem so $\Omega = \mathbb{R}^1_+ \times \mathbb{R}^1$. Let (ρ', u_1', u_2') be a given smooth flow with the properties:

$$\left\{ \begin{array}{l} (\rho', u_1', u_2')(x_1, x_2, t) \in C^l([0,T]; H^{l+2}(\Omega)) \quad , \\ \\ \rho' \geq C_0 > 0 \quad \text{on} \quad \Omega \times (0,T) \quad , \\ \\ u_1'(x_1 = 0, x_2, t) \equiv 0 \quad . \end{array} \right. \tag{38}$$

Set

$$V = \begin{pmatrix} p(\rho) \\ u_1 \\ u_2 \end{pmatrix} \tag{39}$$

Then the linearized initial-boundary value problem is

$$A_0(V')\partial_t V^\varepsilon + \sum_{j=1}^{2} A_j(V')\partial_j V^\varepsilon \quad = B(\varepsilon^2, C\varepsilon^2)V^\varepsilon \tag{40}$$

$$\text{on} \quad \Omega \times (0,T)$$

$$M^+ V^\varepsilon = 0 \quad , \qquad \text{on} \quad \partial\Omega \times [0,T] \tag{41}$$

$$V^\varepsilon(x,0) = (P_0, u_{10}, u_{20})^t \equiv V_0(x) \quad , \qquad x \in \Omega \tag{42}$$

where

$$A_0 = \begin{pmatrix} \alpha & 0 & 0 \\ 0 & \beta & 0 \\ 0 & 0 & \beta \end{pmatrix} \quad , \quad A_j = \begin{pmatrix} \alpha u_j' & e_j^t \\ e_j & \beta u_j' I \end{pmatrix} \quad , \quad j = 1.2 \quad ,$$

$$B(\mu, \mu')V = \mu \begin{pmatrix} 0 & 0 & 0 \\ 0 & 1 & 0 \\ 0 & 0 & 1 \end{pmatrix} \triangle V + \mu' \begin{pmatrix} 0 \\ \nabla \end{pmatrix} (0, \nabla^t)V \quad , \tag{43}$$

$$\alpha = \frac{\rho_p'}{\rho'}, \quad e_1 = (1,0)^t, \quad e_2 = (0,1)^t, \quad \text{and} \quad \beta = \rho' \quad .$$

The corresponding linearized inviscid problem is

$$\left\{ \begin{array}{l} A_\partial(V')\partial_t V^0 + \sum_{j=1}^2 A_j(V')\partial_j V^0 = 0 \quad \Omega \times (0,T) \\ \\ M^0 V^0 = 0 \qquad \text{on} \quad \partial\Omega \times [0,T] \\ \\ \bar{V}^0(x, t=0) = V_0(x) \end{array} \right. \tag{44}$$

where

$$M^0 = (0,1,0) \quad . \tag{45}$$

Assume that the initial data V_0 is compatible with the boundary conditions up to high order [38]. Then the well-posedness of both problems (39)-(41) and (42) in the Sobolev space $C^l([0,T]; H^m(\Omega))$ can be established easily. Our goal is to give a pointwise estimate between V^ε and V^0. One has the following result [38].

Theorem 3 Assume V^0 is compatible with the boundary condition $M^+V = 0$. Let $V^0 \in C^l([0,T]; H^m(\Omega))$ be the unique solution to the initial-boundary value problem (42) for any T, $0 < T < \infty$. Then the unique solution V^ε to the viscous problem (39)-(41) admits the following estimate

$$\{|\rho^\varepsilon - \rho^0|(x_1, x_2, t) + \sum_{j=1}^{2} |u_j^\varepsilon - u_j^0|(x_1, x_2, t)\} \le C_h^\delta \varepsilon \tag{46}$$

holds uniformly for $t \in [0, T], x_1 \ge h\varepsilon^{1-\delta}$, and $x_2 \in \mathbb{R}^1$,
where $h > 0$ is any finite number, $0 < \delta < 1$, $0 < \varepsilon < 1$, and C_h^δ is a positive constant independent of ε. Furthermore, the following asymptotic expansion

$$V^\varepsilon(x, t) \sim V^0 + \sum_{i \ge 1} \varepsilon^i I^i(x_1, x_2, t) + \sum_{j \ge 0} \varepsilon^j B^j(\frac{x_1}{\varepsilon}, x_2, t) \tag{47}$$

can be constructed and justified up to any given order.

Remark 9 The main tools used in the proof of Theorem 2 are matched asymptotic analysis and weighted energy estimates. One of key elements is the analysis is to show that the linearized Prandtl's system is well-posed in a weighted Sobolev space [38]. However, these estimates are not enough to handle the nonlinear problem. Thus the boundary layer theory for the nonlinear Navier-Stokes system with no-slip boundary condition is far from being established.

References

[1] Bressan, A., Crasta, G., and Piccoli, B., Well-posedness of the Cauchy problem for nxn systems of conservation laws, preprint, 1997.

[2] Bressan, A., Liu, T. P. and Yang, T. L^1 - stability estimates for nxn systems of conservation laws, preprint, 1998.

[3] Chen, G. and Glimm, J., Global solution to the compressible Euler equations with geometrical structures, Comm. Math. Phys. 179 (1996), 153-193.

[4] Courant, R. and Friedrichs, K. O., Supersonic Flow And Shock Waves, Wiley-Interscience, 1948.

[5] Diperna, R., Convergence of the viscosity method for isentropic gas dynamics, Comm. Math. Phys., 91 (1983), 1-30.

[6] Gisclon, M. and Serre, D., Study of boundary conditions for a strictly hyperbolic system via parabolic approximation, C. R. Acad. Sci. Paris Ser. I Math. 319 (1994). No. 4, 377-382.

[7] Glimm, J., Solution in the large for nonlinear hyperbolic systems of equation, Comm. Pure Appl. Math. 18 (1965), 95-105.

[8] Glimm, J. and Lax, P. D., Decay of solutions of systems of nonlinear hyperbolic conservation laws, AMS Memoir 701 (1920).

[9] Goodman, J., Nonlinear asymptotic stability of viscous shock profiles for conservation laws, Arch. Rat. Mech. Anal. 95 (1986), 325-344.

[10] Goodman, J. and Xin, Z. P., Viscous limit to precewise smooth solutions of systems of conservation laws, Arch. Rat. Mech. Anal. 121 (1992), 235-265.

[11] Grenier, E., and Gues, O., On the inviscid limit of noncharacteristic nonlinear parabolic systems, preprint, 1998.

[12] Holf, D., Global solution of the Navier-Stokes equations for multidimensional compressible flow with discontinuous initial data, J. Diff. Equations, 120 (1995), 215-254.

[13] Holf, D., and Liu, T., The inviscid Limit for the Navier-Stokes equations of compressible isentropic flow with shock data, Indiana Univ. J. 38 (4) (1989), 861-915.

[14] Joseph, K. T., Boundary Layers in approximating solutions, Transac. AMS 314 (1989), 709-726.

[15] Kawashima, S. and Matsumura A., Asymptotic stability of traveling wave solutions of systems of one-dimensional gas motion, Comm. Math. Phys. 10 (1985), 97-127.

[16] Kawashima, S. and Nishida, T., The initial-value problems for the equations of viscous compressible and perfect compressible fluids, RIMS, Kokyuroku 428, Kyoto University, Nonlinear Functional Analysis, June 1981, 34-59.

[17] Landau, L. and Lifschitz, E., Fluid Mechanics, Addison-Wesley, New York , 1953.
[18] Lax, P. D., Hyperbolic Systems of Conservation Laws And The Mathematical Theory of Shock Waves, Conf. Board Math. Sci. 11, SIAM, 1973.
[19] Lions, P. L., Mathematical Topics In Fluid Mechanics, Vol. 2, Compressible Models, Lecture Series in Mathematics And Its Application, V. 6, Clarendon Press, Oxford, 1998.
[20] Lions, P. L. and Masmudi, N., Incompressible limit for a viscous compressible fluid, J. Math. Pure Appl. 77 (1998), 585-627.
[21] Lions, P. L. Perthame, B. and Souganidis, P. E., Existence and stability of entropy solutions for the hyperbolic systems of isentropic gas dynamic in Eulerian and Lagrangian coordinates, Comm. Pure Appl. Math., 49 (1996), 599-638.
[22] Liu, T. P., Pointwise convergence to shock waves for systems of visious conservation laws, preprint, to appear at Comm. Pure Applied Math.
[23] Liu, T. P. and Xin, Z. P., Pointwise decay to contact chscontinuities for systems of conservation laws, Asia J. Math., 1 (1997), 701-730.
[24] Liu, T. P. and Xin, Z. P. Nonlinear stability of rarefaction waves for compressible Navier-Stokes equations, Comm. Math. Phys. 118 (1988), 451-465.
[25] Liu, T. P. and Yang, T., L_1-well-posedness of Glrmm solution for systems of conservation laws, preprint, 1998.
[26] Liu, T. P. and Zeng, Y., Large time behavior of general systems of hyperbolic-parabolic conservation laws, AMS Memoir, 1997.
[27] Makino, T., Ukai, S. and Kawashima, S., Sur solutions a support de L'equation d'Euler Compressible, Japan J. Appl. Math. 3 (1986), 249-257.
[28] Matsumura, A. and Nishicla, T., The initial-value problem for the equations of motion of viscous and heat-conductive gases, J. Math. Kyoto Univ. 20 (1980), 67-104.
[29] Matsumura, A. and Nishihara, K., On the stability of traveling waves of a one-dimensional model system for compressible viscous gas, Japan J. Appl. Math. 2 (1985), 17-25.
[30] Michaelson, D., Initial-boundary value problems for incomplete singular perturbations of hyperbolic systems, J. Analyse Math., 53 (1989), 1-138.
[31] Oleinik, O. A., The Prandtl system of equations in boundary layer theory, Dokl. Akad. Nank S.S.S.R., 150 4 (3) (1963), 583-586.
[32] Sammartino, M. and Caflish, R. E., Zero viscosity limit for analytic solutions of the Navier-Stokes equations on a half space, (I and II), preprint, 1997.
[33] Schlichting, H., Boundary-Layer Theory, 7th Edition, McGraw-Hill, 1987.
[34] Smoller, J. A., Shock Waves And Reaction-Diffusion Equations, Berlin-Heidallarg-New York: Springer-Verlay, New York, 1984.
[35] Szepessy, A., and Xin, Z. P., Nonlinear stability of viscous rarefaction craves, Arch. Rat. Mech. Anal. 122 (1993), 53-103.
[36] Xin, Z. P., On the blow-up of smooth solutions to the compressible Navier-Stokes equations with compact density, Comm. Pure Appl. Math., 51 (1998), 229-240.
[37] Xin, Z. P., Viscous boundary layers and their stability (I), T. Partial Diff. Eqs., 11 (1998), 497-541.
[38] Xin, Z. P. and Yanagisawa, T., Viscosity limit of the linearized Navier-Stokes equations for compressible viscous fluid in the half plan, Comm. Pure Appl. Math., Vol VII (1999), 497-541.
[39] Xin, Z. P., Zero dissipation limit to rarefaction waves for the one-dimensional Navier-Stokes equations for compressible isentropic gas, Comm. Pure Appl. Math. 86 (1993), 1499-1533.
[40] Xin Z. P., Zero dissipation limit for the compressible Navier-Stokes equations with non-characteristic boundary conditions, preprint, 2000.

Courant Institute, NYU, USA, and, IMS and Department of Mathematics, The Chinese University of Hong Kong, HK

AMS/IP Studies in Advanced Mathematics
Volume 20, 2001

Scaling Limit of Particle Systems, Incompressible Navier-Stokes Equation and Boltzmann Equation

Horng-Tzer Yau

ABSTRACT. We review some basic ideas concerning classical, stochastic and quantum dynamics of large systems under various scalings and related analysis in infinite dimensions. The limiting classical equations obtained in this lecture are the Boltzmann, Euler and incompressible Navier-Stokes equations.

1. Introduction

Although the fundamental equations for a fluid such as the Euler equations or the Navier-Stokes equations have been known and widely used for centuries, basic issues concerning these equations remain unanswered. The equations are on the macroscopic scale and are derived from a continuum formulation of conservation of mass and momentum. But one can ask the deeper question of how to obtain these equations directly from microscopic models, in particular from many body Hamiltonian systems, which in principle governs all classical phenomenon. Therefore, the fundamental question in dynamics is to understand the general many-body microscopic Newtonian (or Schrodinger) equations. Closed solutions to these equations for large interacting systems are unfortunately impossible. A more moderate goal is thus to solve these equations well enough so that conclusions about some basic quantities can be made. The fundamental observation of Boltzmann states that the typical behavior for classical Newtonian systems in equilibrium is governed by ensemble (statistical) averages, the *Gibbs states* in today's language. This avoids the difficulties of solving directly the Newton equations by postulating statistical averages, though the hypothesis remains to be proved. Boltzmann's formulation (or the *Boltzmann-Gibbs formulation*) concerns systems in equilibrium, or in other words, behavior of systems as the time approaches infinity. At the other end is the kinetic theory for short time behavior when the systems are near free dynamics (i.e., typical particles undergo at most a few collisions). The fundamental observation of kinetic theory, the idealization of the collision processes, is again due to Boltzmann in his celebrated work on the *Boltzmann equation.* For systems neither in equilibrium nor near free dynamics, that is, for time scale short for using equilibrium theory but longer than the scale for kinetic theory, the most useful descriptions

Research partially supported by NSF grants DMS-9703752.

are still the classical macroscopic equations such as the Euler equations or the Navier-Stokes equations.

In this lecture, we review recent progress on the foundation of these assertions: the Boltzmann-Gibbs formulation, the Boltzmann equation, and the continuum formulation. There are intrinsic microscopic questions such as fluctuations, time-dependent correlations and behavior of tagged particles which are naturally formulated only on the microscopic level. Due to the length of this lecture, we shall not go into these directions. We shall use the term classical (Hamiltonian) dynamics for systems evolving according to the Newton equations; the quantum dynamics for systems according to the Schrodinger equations.

The Boltzmann-Gibbs principle states that all stationary measures of the classical dynamics (i.e., measures invariant under the classical dynamics) are Gibbs. It has lead to the development of modern ergodic theory but its basic assertion on the stationary measures of the classical dynamics remains a fundamental question to be addressed. Due to the existence of singular stationary measures, the Boltzmann-Gibbs principle is incorrect in its full generality. These singular stationary measures are generally believed to have no contribution to the statistical behavior of the systems. We believe that a regularity condition, the finite specific entropy condition, is sufficient to eliminate these singular stationary measures [**17**]. Our claim is backed by the theorem [**17**] stating that all stationary measures of classical dynamics with finite specific entropy are Gibbs provided that there are no correlations in their velocity distributions (correlations in the positions are allowed in this theorem). We call a stationary measure regular if it has finite specific entropy. Therefore, to prove the Boltzmann-Gibbs principle it remains to prove that all regular stationary measures have no correlations in the velocity distributions. We call the last property the *weak Boltzmann-Gibbs principle.* Notice that there is *no reference to Gibbs measures* in this formulation.

The Boltzmann equation is rather easy to derive if we assume the *molecular chaos* assumption, i.e., the 'incoming particles' of a collision process have no correlation. This assumption is very subtle since the 'outgoing particles' of a collision remember the collision and thus possess correlations. Therefore, the molecular chaos assumption in principle selects a direction of the time. This was a novel concept and there were many debates on the validity of the Boltzmann equation and its conclusion, the Boltzmann H-theorem. It was not until H. Grad in the fifties that a mathematical meaning to the Boltzmann equation from classical dynamics was clearly given. Grad observed that in a suitable space-time rescaling, the Boltzmann-Grad limit (or simply the Grad limit), the probability density of finding a particle at a fixed position with a given velocity satisfies the Boltzmann equation. In other words, while the Newtonian mechanics govern the dynamics of the particles and are universal, *the Boltzmann equation is valid only on the macroscopic scales.* The Grad limit prescribes the precise relation between the microscopic and the macroscopic scales and is characterized by the typical number of collisions per particle being finite. It is thus restricted to *relatively short time* and the *low density gas.* Grad's claim on the classical dynamics was proved rigorously in the seventies by O. Lanford [**14**] for the hard core billiards in short time.

A mathematical formulation of the classical continuum equations from the classical dynamics poses many conceptual questions which should be properly addressed. The Euler equations or the Navier-Stokes equations are equations for the density, velocity and energy. Though these quantities have phenomenological meanings, they have to be defined rigorously from microscopic quantities. The obvious definition for, say, velocity at a macroscopic location, is simply the average of the velocities of the particles in a 'small neighborhood' of the given location. Similarly we can define density and energy. In order that these definitions are useful notions so that they assume their macroscopic deterministic values, we need these definitions to be at least independent of the sizes (and shapes) of these small neighborhoods. In a more mathematical formulation, we need these quantities to obey *laws of of large numbers at each location in space.* We are, on the other hand, given a *deterministic* dynamical system of many particles and there is no indication that these laws of large numbers hold in any way. It was first observed by C. Morrey [**16**] in the sixties that the Euler equations become 'exact' in the Euler limit, provided that the solution to the classical dynamics are 'locally' in equilibrium. Notice that mathematical formulations of Gibbs states were only developed in the sixties and thus Gibbs states were not precisely defined at that time. Morrey's work was thus vague and there was also no mention of a law of large numbers. In today's language, the meaning of the Euler equations is as follows:

(1) *The microscopic Newtonian dynamics are 'locally' in equilibrium.*
(2) *The local equilibria are parameterized by density, velocity (momentum) and energy, the Boltzmann-Gibbs principle.*
(3) *The parameters of the local equilibria evolve according to the macroscopic hydrodynamical equations, the Euler equations.*

Notice that the Boltzmann equation involved only kinematics quantities and no statistical concept. It has been surprisingly difficult to carry out Morrey's program even for highly simplified systems. Both the local equilibrium property and the Boltzmann-Gibbs principle are difficult to establish for many-body dynamics. As most tools for relaxation to equilibrium are valid only for systems with dissipation, the systems under discussion since C. Morrey have almost always been dissipative, namely, stochastic dynamics. The stochastic dynamics was pioneered by M. Kac in his work of the Boltzmann equation in the Grad limit. It was developed later on by, among others, H. Mckean, R. Dobrushin, J. Lebowitz, F. Spitzer C. Kipnis, E. Presutti, H. Spohn, and S. Varadhan, a subject generally known as scaling limits of interacting particle systems.

To understand Morrey's original problem for classical dynamics, C. Boldrighini, L. Bunimovich, R. Dobrushin, Y. Suhov and Y. Sinai [**2**], [**1**] constructed certain one dimensional systems and considered their scaling limits. The Euler equations were proved in the special case of one dimensional hard rods systems with elastic collisions [**2**]. This case is unusual because there are infinitely many conserved quantities and the systems are integrable. The basic question to estimate the relaxation time to equilibrium for the classical dynamics remained to be answered. It turns out that if we assume the weak Boltzmann-Gibbs principle, a rigorous estimate [**17**] for the relaxation time to equilibrium can be obtained. We remark that *the (weak) Boltzmann-Gibbs principle imposes no condition on the convergence rate to equilibrium*; it is strictly a statement on the classification of the stationary measures to the classical dynamics. The estimate obtained [**17**] is strong enough to prove the convergence from the classical dynamics to the Euler equations. We also construct from the (strong) solution to the Euler equations an approximate solution to the Newton equations. It is a local Gibbs state (i.e., a local equilibrium measure) with local density, velocity and energy given by the solution to the Euler equations. The Euler equations can thus be interpreted as giving *the leading asymptotics to the Newton equations.*

The derivation of the Navier-Stokes equations from classical dynamics presents a deeper conceptual question relating to the origin of diffusivity in classical physics. The diffusion equation, though well-accepted, is *not* a basic equation of classical dynamics, which are time reversible and have no dissipation. Even for a classical particle in an environment of random obstacles, the Lorentz gas, the diffusivity is a basic open problem. The viscosity in the Navier-Stokes equations is a measure of collective diffusivity of many particles and is thus far more difficult to establish than the diffusivity of a single classical particle. This was particularly emphasized by, e.g., R. Dobrushin, J. Lebowitz and H. Spohn. On the physics side, there was a lattice formulation of classical dynamics [**10**]. The viscosity was given in this work by a Green-Kubo formula as a time integral of current-current correlation functions. Contrary to the common belief, the Green-Kubo formula does not explain the origin of the viscosity; it is simply an expression for the viscosity *under the assumption that the system is diffusive and has Gaussian fluctuations.*

Thus we do not aim to derive the Navier-Stokes equations from the classical dynamics rigorously. Our goal is rather to understand the viscosity from simpler dynamics which capture the basics of the classical dynamics. These simpler models will be stochastic lattice gases which are systems of random walks with collisions. It was proved that these models give rise to the correct Navier-Stokes equations in the incompressible limits [**9**], [**18**] for space dimension $d \geq 3$. Although our models contain random walks and are thus dissipative apriori, we are able to show that the viscosity is strictly larger than the viscosity of the symmetrized dynamics. The latter is interpreted as the artificial viscosity (dissipation) we put into the dynamics as a regularization. The difference of these two viscosities, which was proved to be positive, corresponds to the true viscosity in the Newtonian dynamics.

Our next topic is the quantum dynamics. The most basic objects here are the electrons and the radiation or phonon fields. The full electron-field dynamics are governed by suitable Schrodinger equations which are systems of infinitely coupled equations. Our main objective is to solve these systems reasonably precisely so that meaningful conclusions can be made on the dynamics of the electrons. The precise form of the external fields, though crucial for many purposes, are not very important for a few basic questions. So we shall take for example the phonon field. A natural model is thus the dynamics of electrons in disordered media. The disordered media are described by a random potential modeling obstacles (doping) and a phonon field modeling lattice vibrations. Thus the full dynamics contain electron-electron, electron-obstacle and electron-phonon interactions. Even for this much simplified model, we do not know very much about its dynamics. We shall first restrict ourselves to the kinetic region where the dynamics are near the free dynamics. Our initial goal is to prove the Boltzmann equation from the quantum dynamics [**6**], [**7**]. Due to the particle-wave duality of the Schrodinger equation and internal structures such as statistics or spins of quantum particles, the Boltzmann equation will take different forms in quantum dynamics. A quantum system is described by a complex-valued wave function, or more general, a quantum state. The fundamental mechanism for quantum dynamics is through 'interferences' of waves determined by the complex phases of the wave function, a mechanism completely different from the classical or the stochastic ones.

We now define the scalings used in this lecture. Denote coordinates by lower case letters (x,t) in the microscopic scale; by capital letters (X,T) in the macroscopic scale. We put the system in a cube of size L in d-dimensional space with periodic boundary condition and we will usually assume $d=3$. Denote the particles by $(x_1,\cdots,x_N,v_1,\cdots,v_N)$ with the density (in the microscopic unit, i.e., number of particles per microscopic unit volume) $\rho = N/L^d$. Let ε be the ratio between the microscopic unit and the macroscopic unit (say, $\varepsilon \sim 10^{-8}$). There are typically three choices of scalings:

$$\left\{\begin{array}{ll} \text{Grad} & \rho=\varepsilon,(X,T):=(x\varepsilon,t\varepsilon) \\ \text{Euler} & \rho=1,(X,T):=(x\varepsilon,t\varepsilon) \\ \text{Diffusive} & \rho=1,(X,T):=(x\varepsilon,t\varepsilon^2) \end{array}\right\} \Longrightarrow \left\{\begin{array}{lcc} \text{collisions} & & \text{finite} \\ & : & \varepsilon^{-1} \\ \text{per particle} & & \varepsilon^{-2} \end{array}\right\}$$

The Euler and diffusive limits will be referred to as hydrodynamic limits. The typical number of collisions is finite for the Grad limit; infinite in the hydrodynamic limits. Hence the Grad limit is the closest to free motion without collisions. Essentially due to this feature, O. Lanford [**14**] proved the convergence of the hard core billiards to the Boltzmann equation in the Grad limit in short time based on the BBGKY hierarchy.

2. Euler Equations

At present there is no rigorous derivation of Euler equations from Hamiltonian mechanics. Unlike the Grad limit, the Euler limit involves an infinite number of collisions and the typical behavior is governed by the stationary (equilibrium, invariant) states, which are assumed to be Gibbs in the famous Boltzmann Hypothesis. More precisely:

Boltzmann Hypothesis : The invariant (stationary) measures of many body classical dynamics are *Gibbs* $\sim e^{-\beta H}$. In particular, the typical velocity distributions of different particles are uncorrelated (Weak Boltzmann Hypothesis).

The Boltzmann Hypothesis is strictly speaking *incorrect* because there are singular invariant measures. We believe that these singular invariant measures can be removed by regularity assumption such as *finite specific entropy*, i.e., entropy per microscopic unit volume is finite. The following theorem is a joint work with S. Olla and S. Varadhan [**17**].

THEOREM 2.1. *Assume the weak Boltzmann Hypothesis holds for invariant measures with finite specific entropy. Suppose the Euler equation has a smooth solution in* $[0,T]$. *Then the empirical density, velocity. and energy converge to the solution of the Euler equations in* $[0,T]$ *with probability one.*

Recall that classical dynamics are characterized by a Hamiltonian

$$H(x,v) = \frac{1}{2}\sum_{\alpha=1}^{N} \|v_\alpha\|^2 + \sum_{\alpha<\beta\leq N} V(x_\alpha - x_\beta)H)$$

with V a two-body potential and the Liouville equation

$$\partial_t f_{N,t}(x_1,\cdots,x_N,v_1,\cdots,v_N) = \mathcal{L}^* f_{N,t}li)$$

where $f_{N,t}$ is the density (w.r.t. the standard Lebesgue measure) of the system at time t and the Liouville operator is given by

$$-\mathcal{L}^* = \mathcal{L} = \sum_{\alpha=1}^{N} \left[\frac{\partial H}{\partial v_\alpha} \frac{\partial}{\partial x_\alpha} - \frac{\partial H}{\partial x_\alpha} \frac{\partial}{\partial v_\alpha} \right]$$

with the adjoint taken w.r.t. the standard Lebesgue measure.

For a given configuration $\omega = (x_1, \cdots, x_N, v_1, \cdots, v_N)$ the empirical density and velocity (which rigorously speaking are measures) are defined by

$$\widehat{\rho}_{\epsilon,\omega}(X) = N^{-1} \sum_{\alpha=1}^{N} \delta(X - \varepsilon x_\alpha) ,$$

$$\widehat{v}_{\varepsilon,\omega}(X) = N^{-1} \sum_{\alpha=1}^{N} \delta(X - \varepsilon x_\alpha) v_\alpha ,$$

where δ is the standard delta function on Euclidean space. We shall say $\widehat{\rho}_{\varepsilon,\omega(T/\varepsilon)}(X)$ has a density $\rho(X,T)$ if for any test function J on the unit torus,

$$\int J(\varepsilon x) \widehat{\rho}_{\epsilon,\omega(T/\varepsilon)}(X) dX = N^{-1} \sum_{\alpha=1}^{N} J(\,\varepsilon\, x_\alpha(T/\varepsilon)\,) \to \int J(X) \rho(X,T) dX$$

as $\varepsilon \to 0$. Similarly for the velocity,

$$N^{-1} \sum_{\alpha=1}^{N} J(\,\varepsilon\, x_\alpha(T/\varepsilon)\,)\, v_\alpha(T/\varepsilon) \to \int J(X)\,(\rho v)(X,T) dX.$$

To obtain the Euler equation, we differentiate the velocity

$$\frac{d}{dT} \int J(X) \rho(X,T) v(X,T) dX \sim \varepsilon^{-1} \frac{d}{dt} N^{-1} \sum_{\alpha=1}^{N} J(\varepsilon x_\alpha) v_\alpha$$

$$= -(2N)^{-1} \sum_{\alpha=1}^{N} \varepsilon^{-1} J(\varepsilon x_\alpha) \frac{\partial H}{\partial x_\alpha} + \cdots$$

$$= -(2N)^{-1} \sum_{\alpha=1}^{N} \nabla J(\varepsilon x_\alpha) \underbrace{\sum_{\beta \neq \alpha} \frac{x_\alpha - x_\beta}{\varepsilon} \cdot (\nabla V) \left(\frac{x_\alpha - x_\beta}{\varepsilon} \right)}_{\textbf{micro current}} + \cdots mi)$$

(the micro current appearing here is only a main term for illustration of the idea). Recall the Euler equations:

$$\begin{array}{lcl} \frac{d\rho}{dt} + \nabla(\rho v) & = & 0 \\ \frac{d(\rho v)}{dt} + \nabla\{\rho v \otimes v + P\} & = & 0 \\ \frac{d(\rho e)}{dt} + \nabla\{\rho e\, v - v\, P\} & = & 0 \end{array}$$

Here the pressure P is a function of density, velocity and energy and is determined by the equation of state from the equilibrium Gibbs measure. So in order to obtain the Euler equations we need to show that

$$\textbf{micro current} \to \textbf{macro current}\{= P(\widehat{\rho}_{\varepsilon,\omega}, \widehat{v}_{\varepsilon,\omega}, \widehat{e}_{\varepsilon,\omega})\}$$

in the limit $\varepsilon \to 0$. This equality is understood in the sense of law of large numbers w.r.t. the density of the systems $f_{N,t}$ at time t, i.e.,

$$N^{-1} \int f_{N,t}(\omega) \Big| \sum_{\alpha=1}^{N} \nabla J(\varepsilon x_\alpha) \times \Big[\sum_{\beta \neq \alpha} \tfrac{x_\alpha - x_\beta}{\varepsilon} \cdot (\nabla V) \left(\tfrac{x_\alpha - x_\beta}{\varepsilon} \right) - P(\widehat{\rho}_{\varepsilon,\omega}, \widehat{v}_{\varepsilon,\omega}, \widehat{e}_{\varepsilon,\omega}) \Big] \Big| d\omega \to 0$$

where $d\omega = dx_1 dv_1 \cdots dx_N dv_N$.

The density $f_{N,t}$ satisfies the Liouville equation. At the present time we have essentially no estimate on this equation and the required identity has not been proved. To appreciate the difficulties, we list a few comments on the Liouville equation:

- It conserves L^p norm and positivity (thus f_N can be considered as a probability density) but L^p norm is not useful since $\| f_N \|_p \sim e^{CN}$ which is a huge number.
- There is no elliptic theory for classical dynamics.
- The BBGKY method works only for perturbation of free dynamics and thus is only useful for the Grad limit for which $\rho \sim \varepsilon$.

Instead of approaching it via elliptic estimates or L^p theory, a useful way to establish 4key) is to consider the ergodic property of the Hamiltonian systems. The key observation, due to Morrey [**16**], is that 4key) holds if we replace $f_{N,t}$ by any Gibbs measure (with Hamiltonian H 4H)), or more generally, if "locally" $f_{N,t}$ is a Gibbs measure of the Hamiltonian H. If we can prove that "locally" $f_{N,t}$ is a equilibrium measure with finite specific entropy, we have proved 4key) provided that we assume the Boltzmann Hypothesis. So the proof of Theorem 2.1 consists of two main ingredients: 1. Prove that the weak Boltzmann hypothesis implies the Boltzmann hypothesis. 2. Clarify the precise meaning of the word "locally" and eliminate the possibility of meso-scale fluctuation. The method we used for 2 is the relative entropy method.

Recall that for any two probability densities the relative entropy is defined by

$$S(f|g) = \int f \log(f/g) d\omega$$

Suppose f_t is a solution of the Liouville equation and ψ_t is any density. Then

$$\partial_t S(f_t|\psi_t) = \int f_t \left\{ \psi_t^{-1} \mathcal{L}^* - \partial_t \right.$$

$\psi_t d\omega$

This identity can be checked easily from the Liouville equation. It also has a version for Markov processes:

$$\partial_t S(f_t|\psi_t) = -D(f_t|\psi_t) + \int f_t \left\{ \psi_t^{-1} \mathcal{L}^* - \partial_t \right.$$

$\psi_t d\omega$

where $D(f|\psi)$ is the Dirichlet form of f w.r.t. ψ and is nonnegative [**23**], [**17**]. Now recall the entropy inequality (or the Jensen inequality) which states that for any function W,

$$\int fW d\omega \leq S(f|\psi) + \log \int \psi \exp{(W)}\ d\omega$$

Thus from 4r1),

$$\partial_t S(f_t|\psi_t) \leq S(f_t|\psi_t) + \log \int \psi_t \exp\{\ \psi_t^{-1}\{\mathcal{L}^* - \partial_t\}\psi_t\ \} d\omega$$

If we have

$$N^{-1} \log \int \psi_t \exp\{\ \psi_t^{-1}\{\mathcal{L}^* - \partial_t\}\psi_t\ \} d\omega \to 0 key2)$$

then the relative entropy can be controlled on the relevant time scale and this will imply the estimate 4key) and thus the Euler equations. Note that the left hand side of 4key2) is independent of f_t so the remaining argument in [**17**] can be summarized as showing that 4key2) holds iff ψ_t is a local Gibbs state with density, velocity and energy chosen according to the Euler equations (Note: As it is, 4key2) is incorrect; some arguments using ergodicity of the Hamiltonian dynamics are also needed). This is essentially a dynamical variational approach because we solve the problem by guessing a good trial function which in this case is the local Gibbs state.

3. The incompressible Navier-Stokes equations

The Navier-Stokes equations are the next order corrections to the Euler equations. In order to derive them one needs to show that

$$\textbf{micro current} \to \textbf{macro current} + \varepsilon\nu\nabla\widehat{v}_{\varepsilon,\omega} + o(\varepsilon) mmc)$$

where the currents are given by 4mi) and 4mm) and ν is the viscosity. Since there is an ε appearing in the viscosity term, 4mmc) is in a sense *the next order correction to the Boltzmann hypothesis*! From the expression for the micro current in 4mi), it is hard to even imagine how the viscosity correction arises. This difficulty was recognized decades ago by Dobrushin, Lebowitz, and Spohn, among others. Recent work [**22**], [**8**], [**9**], [**15**] has given us a good understanding of the nature of 4mmc), though a rigorous proof from the Hamiltonian dynamics is still very far off.

The equation for the leading order terms of 4mmc) is 4mm) and it holds w.r.t. Gibbs measures in the sense of law of large numbers. The difficulty to justify 4mm) rigorously for Hamiltonian dynamics (i.e. 4key)) is to prove that the solutions to the Liouville equation are locally stationary and all stationary measures are Gibbs. On the other hand, one can check easily that 4mmc) is *incorrect* w.r.t any Gibbs measures with Hamiltonian H. Indeed, 4mmc) is a "dynamical identity". It can be interpreted physically via the linear response theory or the Green-Kubo formula (see [**19**] for an account). A more mathematical interpretation is through the fluctuation-dissipation equation which we now explain.

Roughly speaking, the fluctuation-dissipation equation states that

$$\textbf{micro current} \to \textbf{macro current} + \varepsilon\nu\nabla\widehat{v}_{\varepsilon,\omega} + \varepsilon\mathcal{L}g + o(\varepsilon) mmc1)$$

for some function $g(\omega)$, where $\mathcal{L}$ is the Liouville operator. In other words, 4mmc) is correct only up to a quotient of the image of the Liouville operator. The image of the Liouville operator is understood as fluctuation, negligible in the relevant scale after *time average*: for any bounded function g

$$\varepsilon \int_0^t ds \int f_{s,N}(\omega)\,(\varepsilon\mathcal{L}g)(\omega)\ d\omega = \varepsilon^2 \int [f_{t,N} - f_{0,N}](\omega)\,g(\omega)\ d\omega \sim \varepsilon^2$$

and is thus negligible to the first order in ε, the relevant scale.

It is difficult to work on "next order correction" and thus we turn to the incompressible Navier-Stokes (INS) equations

$$\frac{\partial u}{\partial t} + u \cdot \nabla u = -\nabla p + \nabla \nu \nabla u, \qquad \nabla \cdot u = 0. \tag{1}$$

The INS equations are invariant under the *incompressible scaling*,

$$x \to \varepsilon x, \quad t \to \varepsilon^2 t, \quad u \to \varepsilon^{-1} u, \quad p \to \varepsilon^2 p, \tag{2}$$

under which mmc1 becomes

$$\textbf{micro current} \to \textbf{macro current} + \nu \nabla \widehat{v}_{\varepsilon,\omega} + \mathcal{L} g \tag{3}$$

Notice that both the viscosity and the function g are unknown and 4mmcc) determines both. We interpret mmcc as a decomposition of the space of microscopic currents into a direct sum of the space of macroscopic currents, the gradient of the velocity representing the dissipation and the image of the Liouville operator representing the fluctuation.

Equation mmcc is extremely difficult to solve as it requires inversion of the Liouville operator. A class of more manageable *stochastic lattice gas* models were introduced in a joint work with R. Esposito and R. Marra [**9**]. Even for these, 4mmcc) requires the inversion of a nonsymmetric operator in infinite dimensions with a complex interaction. If the generator $\mathcal{L}$ is symmetric, i.e., the dynamics is reversible, 4mmcc) can be solved by formulating the problem in an appropriate space so that it reduces to inverting a self-adjoint operator. This formulation, due to S. Varadhan [**22**], is already quite sophisticated since the terms appearing in 3 do not live in a natural space. On the other hand, in order to obtain the INS equations, the dynamics has to retain essential features of the Hamiltonian dynamics; this forces us into nonzero drifts and therefore nonreversibility. The invertibility in the nonsymmetric case is very subtle [**15**]. Dimension comes into play, and we believe that 3 has no solution at all for dimension $d \leq 2$.

In the models of [**9**] particles have velocities in a chosen finite set and at each site of the lattice at most one particle of each velocity is allowed. The dynamics consists of two parts: Random walks and binary collisions between particles. The random walk part of the dynamics requires only that particles with velocity v should have the mean drift v. The binary collisions conserve momentum. Note that conservation of energy is not important here because the INS equations are equations of velocity alone. The combined dynamics should have good ergodic properties and also restore rotational symmetry in the limit. The restoration of the rotational symmetry is not trivial because the lattice structure breaks the symmetry. Sets of velocities and dynamics satisfying all the requirements can be found in [**9**].

The main result in [**9**] states that 3 has a solution (in a suitable sense) for $d \geq 3$ and if the INS equations have a strong solution up to a fixed time T then the rescaled empirical velocity densities (measures)

$$v_{\varepsilon,\omega}(X) := \varepsilon^{d-1} \sum_x \delta(X - \varepsilon x) \sum_v v\eta(x, v) \tag{4}$$

converges to that solution. Here $\eta(x, v) \in \{0, 1\}$ is the number of particles of velocity v at site x. Notice the blowup of the velocity by ε^{-1} in accordance with the scaling 4scaling).

The assumption that the INS equations have a strong solutions has a long history in their derivation from more basic models. Derivations of the INS equations from the Boltzmann equation go back already a century to Chapman, Enskog and Hilbert, and were made rigorous in the seventies [**4**],[**5**]. However the removal of the smoothness assumption has not been so easy. A program [**3**] of deriving the weak (Leray) solutions from the DiPerna-Lions solutions of the Boltzmann equation remains far from complete, due to a lack of good estimates. Though it was believed that the analysis of particle systems would be even more difficult because they are essentially infinite dimensional, in a joint work with J. Quastel [**18**] we have been able to remove this obstacle.

THEOREM 3.1. *Let P_ε be the distributions of the empirical momentum densities 41.13). Then P_ε are precompact (as a set of probability measures with respect to a suitable topology) and any weak limit is supported entirely on weak solutions of the INS equations satisfying the energy inequality.*

Theorem 3.1 is proven only for $d = 3$. The restriction $d \leq 3$ is for technical reasons; *the restriction $d \geq 3$, however, is intrinsic.* Since the macroscopic velocity is defined through the law of large numbers in statistical physics, it inherits a small fluctuation from the central limit theorem, which is of order $\varepsilon^{d/2}$. When we blow up the velocity by ε^{-1} in the incompressible limit 41.13), this term becomes of order one or larger for dimensions $d \leq 2$ and thus the macroscopic velocity is not well defined in this limit. Note that this argument applies to any dynamics including the Hamiltonian dynamics.

Though 4mmcc) determines the viscosity, it is important to have an independent characterization, traditionally expressed as a time integral of current-current correlation functions, which up to constants is given by:

$$\nu = \int_0^\infty \big\langle \textbf{ micro current } (t=0); \textbf{ micro current } (t=s) \big\rangle \, ds \tag{5}$$

where $\langle f; g\rangle = \langle fg\rangle - \langle f\rangle\langle g\rangle$ is the correlation function and the expectation is w.r.t. lattice gas dynamics starting from *equilibrium*. This is called the Green-Kubo formula and is proved rigorously in [**15**], [**9**] for $d \geq 3$. For dimension $d \leq 2$, the Green-Kubo formula 4gk) diverges, 4mmcc) has no solution, and the time scaling is faster than diffusive. We are thus forced to conclude that the two dimensional INS equations cannot be obtained as the incompressible limit of *any* microscopic dynamics.

A large deviation principle was also given in [**18**]. One main step in [**18**] is a proof of the energy estimate for the INS equations directly from the lattice gas dynamics by implimenting a renormalizatin group analysis. The techincally most demanding points, the large field problems in the standard field theory and the large fluctuation here, are controlled by the entropy method [**12**] and and the logarithmic Sobolev inequality [**24**]. The entropy method is an infinite dimensional version of the energy method in PDE; the logarithmic Sobolev inequality plays the role of the usual Sobolev inequalities.

4. Quantum Dynamics

Most problems concerning classical or stochastic dynamics have corresponding quantum versions. They are however often too difficult to study. The classical or stochastic dynamics are governed by the evolution of a probability density; the quantum dynamics by a complex wave function. Although both dynamics are linear, the physics in the quantum case is given by the square of the wave function, breaking the superposition law. Furthermore, the evolution of a wave function is determined by its phase which is very hard to control. We mention here a result on the quantum Lorentz gases **[6]**, **[7]** to give some flavor of quantum dynamics.

Classical Lorentz gases model a classical particle in an environment of fixed scatterers distributed randomly (or periodically). The question is the time evolution of this particle for a typical configuration of the scatterers. Denote by $\omega = (x_\alpha), \alpha = 1, \cdots, N$, the configuration of scatterers in a cube of width L. We are interested in the Grad limit 4scale) with $\varrho = N/L^d$ denoting the density of the scatterers. The typical number of collisions is now of the order $t\rho \sim 1$. It was proved in **[11]**,**[21]**, **[1]** that its time evolution converges to a linear Boltzmann equation

$$\partial_T F_T(X,V) + V \cdot \nabla_X F_T(X,V) = \int \sigma(U,V) \left[F_T(X,U) - F_T(X,V)\right] dU, 1.8)$$

where F is the phase space density and $\sigma(U,V)$ is the scattering cross section.

The quantum Lorentz gases can be obtained by simply replacing the classical dynamics by the quantum dynamics. More precisely, let $V_0(x)$ be a fixed "nice" function. The Schrödinger equation governing the quantum particle is given by

$$i\partial_t \psi_t = H_{N,L}\psi_t, \quad \psi_{t=0} = \psi_0, scheq)$$

where the Hamiltonian is given by

$$H_{N,L} = H := -\Delta/2 + V_\omega, \qquad V_\omega(x) = \sum_{\alpha=1}^{N} V_0(x - x_\alpha).schop)$$

The classical phase space density of a wave function ψ is defined through the Wigner transform:

$$W_\psi(x,v) := \int \overline{\psi(x + z/2)}\psi(x - z/2)e^{ivz}dz.$$

The scaling is the same as in the classical case,

$$W_\psi^\varepsilon(X,V) := W_\psi(X/\varepsilon, V).1.7)$$

Notice that the velocity is not rescaled. The Wigner transform typically has no definite sign, and the associated Husimi function or coherent states are needed to define a positive density, but we will not go into these details here.

Let $\psi_{\omega,t}^\varepsilon$ be the solution to the Schrödinger equation 4scheq), 4schop) with initial data ψ_0^ε. Suppose that the initial data is of the following form

$$\psi_0^\varepsilon(x) = \varepsilon^{3/2} h(\varepsilon x) e^{iu_0 x},$$

for some smooth functions h so that as $\varepsilon \to 0$ the rescaled Wigner transform $W_{\psi_0^\varepsilon}^\varepsilon(X,V)dXdV$ converges weakly to $|h(X)|^2\delta(V - u_0)dXdV =: F_0(X,V)dXdV$ as probability measures on $\boldsymbol{R}^{2d}$. Then in dimension $d = 3$ and for V_0 small enough (so that there is no bound state) our main result with L. Erdős **[6]**, **[7]** is that *for any $T > 0$,*

$$\mathbf{E}W_{\psi_{\omega,T/\varepsilon}^\varepsilon}^\varepsilon W(X,V)dXdV \to F_T(X,V)dXdV$$

weakly as $\varepsilon \to 0$ and $F_T(X,V)$ satisfies the linear Boltzmann equation 41.13) with initial data $F_0(X,V)$ and effective collision kernel σ given by the quantum scattering operator of the potential V_0.

A simple example illustrates the difference between the classical and the quantum dynamics. Suppose that the particle in a Lorentz gas has one collision. Classically we simply choose a scatterer and the particle collides with it. In quantum mechanics, we have from the Duhamel formula

$$\psi_t = e^{-itH}\psi_0 = e^{-itH_0}\psi_0 - i\int_0^t e^{-i(t-s)H_0}Ve^{-isH_0}\psi_0\, ds + \cdots$$

where V_ω is the potential given in 4schop) and $H_0 = -\Delta/2$. The one collision term is the second term on the right hand side which, for simplicity, we write as $\sum_{\alpha=1}^N \psi_{t,\alpha}$. Notice that instead of collision with a scatterer in classical dynamics, it is now a sum of collisions with all scatterers! Since we have to square the wave function to get physical quantities, we need to show that the overlaps (or interference) of off-diagonal terms

$$\langle\, \psi_{t,\alpha},\ \psi_{t,\beta}\,\rangle$$

are very small. Stationary phase methods show they are small, but the number of the off-diagonal terms is much larger than that of diagonal terms. The analysis of this competition is very simple in this first term but very complicated in higher order terms. It nevertheless can be carried out rigorously to all orders [**20**], [**13**]. However such results are restricted to the weak coupling limit (a semiclassical limit) and short time. Instead we renormalize the perturbation theory so that we can consider the Grad limit to obtain the quantum scattering kernel. Furthermore, we truncate the Duhammel formula and estimate the error terms to remove the short-time restriction and thus we obtain results global in time [**6**], [**7**].

5. Acknowledgement:

It is a great pleasure to thank J. Quastel for critical readings and commments of the manuscript.

References

[1] C, Boldrighini, L. Bunimovich and Y. Sinai: J. Stat. Phys. **32**, 477-501, (1983).
[2] C, Boldrighini, R. L. Dobrushin and Y. M. Suhov: J. Stat. Phys. 31 (1983) 577–616.
[3] C. Bardos,F. Golse and D. Levermore: Comptes Rendus de l'Acad. Sci., Ser. 1, **11**, 727–732, (1989).
[4] R. E. Caflisch and G. C. Papanicolaou: Comm. Pure Appl. Math., **32**, 589–616, (1979).
[5] A. De Masi, R. Esposito and J. L. Lebowitz: Commun. Pure and Appl. Math., **42**,1189–1214, (1989).
[6] L. Erdős and H. T. Yau: Linear Boltzmann equation as Scaling limit of quantum Lorentz gas, to appear in GT-UAB Conf. Proceedings.
[7] L. Erdős and H. T. Yau: Linear Boltzmann Equation as weak coupling limit of a quantum particle in random field, preprint.
[8] R. Esposito and R. Marra: Jour. Stat. Phys., **74**, 981, (1993).
[9] R. Esposito, R. Marra and H.-T. Yau: Rev. Math. Phys., **6**, 1233-1267 and Commun. Math. Phys, 182, 396-456, 1996.
[10] U. Frisch, D. d'Humieres, B. Hasslacher, P. Lallemand, Y. Pomeau and J. P. Rivet: *Lattice Gas Hydrodynamics in Two and Three Dimensions* Complex Systems **1**, 649–707, (1987).
[11] G. Gallavotti: Nota inteerna n. 358, Univ. di Roma (1970).
[12] Guo,M. Papanicolau, G.C., and Varadhan, S. R. S.: Comm. Math. Phys. **118**, 31–59, (1988).
[13] T. G. Ho, L. J. Landau and A. J. Wilkins: Rev. Math. Phys. **5**, 209-298 (1992).
[14] O. E. Lanford: Astérisque **40**, 117-137 (1976).
[15] C. Landim and H.-T. Yau: Prob. Related Field, **108**, 321-356 (1997)
[16] C. B. Morrey: Comm. Pure Appl. Math., **8**, 279-290, (1955).
[17] S. Olla, R. Varadhan and H.-T. Yau: Commun. Math. Phys., **155**, 523–560, (1993).

[18] J. Quastel and H.-T.Yau: Lattice gases, large deviations, and the incompressible Navier-Stokes equations, to appear in Ann. Math.
[19] H. Spohn: *Large Scale Dynamics of Interactin Particles*, Springer-Verlag New York (1991).
[20] H. Spohn: J. Stat. Phys., **17**, p385 (1977).
[21] H. Spohn: Commun. Math. Phys. **60**, 277-290 (1978).
[22] S. R. S. Varadhan, Taniguchi Symp.,p 75-130, ed. K. D. Elworthy and N. Ikeda, Longman Scientific & Technical, (1990).
[23] H.-T. Yau: Lett. Math. Phys. **22**, 63-80 (1991).
[24] H.-T. Yau: Commun. Math. Phys, **181**, 367-408, 1997.

E-mail address: yau math.nyu.edu

COURANT INSTITUTE OF MATHEMATICAL SCIENCES NEW YORK UNIVERSITY 251 MERCER STREET NEW YORK, NY 10012

AMS/IP Studies in Advanced Mathematics
Volume 20, 2001

Geometry of algebraic points

Shouwu Zhang

Introduction

In this note, I will try to give an elementary introduction to the subject of Diophantine geometry for people who are not familiar with either number theory or algebraic geometry. My basic goal is to explain first of all, the important question and their current solutions, and secondly how geometry (both in language and technique) is used to address and answer these questions. This note consists of three parts: rational points, algebraic points, and the ABC and discriminant conjectures. We now informally discuss the contends of each of these sections.

The language used in the first part is completely elementary. Thus we don't assume any advanced knowledge of algebraic geometry. The basic question is how to solve Diophantine equations in two variables in integers? Equivalently, in the language of algebraic geometry, how to find rational points on a plane algebraic curve? It turns out that the nature of the solution sets really depends on the degree of the equation. In the degree 1 and 2 case, once one solution is found, then all other solutions can be found by using a *ruler*. This ruler method has an obstruction for curves with higher degree which are precisely given by a new object, *the Jacobian variety*, and the abelian group structure on the rational solution on the Jacobian variety. This Jacobian variety is of fundamental importance in the study of Diophantine questions related to curves. The *Mordell-Weil Theorem* and *Faltings Theorem* give the finiteness of rational points on these Jacobian varieties and the original curves respectively. Besides algebraic geometry, the proof of these theorems uses the *theory of heights*. The heights are quantities that measure the complexity (= logarithmic size) of points and varieties. A crucial question which remains today is the effectivity of solutions: *what is the maximal size of the solutions?* We will present so called effective Mordell conjecture at the end of this section.

In the second part, we will study certain sets of *algebraic points* in the *abelian varieties* (a generalized notion of Jacobian varieties). First of all, we restrict ourselves to either torsion points, or division points which is a mixed notion of torsion points and rational points. (Notice that the structure of the rational solutions is the subject in the first part.) *Raynaud's Theorem* shows that Faltings' theorem can be extended to these division points. Secondly, we consider to points which

Research supported by a grant from the National Science Foundation, and a Fellowship from the Sloan Foundation.

are close to division points, which we call *almost division points*. The new phenomenon is that the Galois orbits of these points in abelian varieties are *uniformly distributed, or equidistributed* with respect to the invariant probability measure on the associated complex torus. Of course the property to be equidistributed is much stronger than the property to be *Zariski dense*. (In the curves case, this means to be *infinite*.) Thus, the equidistribution property implies Raynaud's result. (Both Raynaud's result and equidistribution result extend Faltings' theorem but do not imply it as their proof use Faltings' theorem.) Like the study of rational points, the study of almost division points relies on a renovated theory of heights: *Arakelov geometry*. In this theory, the *Grothendieck's theory of schemes* is naturally combined with the *differential geometry of Kähler manifolds* to provide a very fine tool to study algebraic points with small heights and integral sections with small norms.

In the last part, we will discuss the *ABC – conjecture* which provides effective bound on the heights of rational points for a variety of equations, including diagonal equations like the Fermat equation. Moreover, the work of Szpiro, Moret-Bailly, and Elkies suggests that some strong form of the ABC conjecture is actually equivalent to the effective Mordell conjecture discussed in the first part of this note. The ABC conjecture originated in *Szpiro's discriminant conjecture* via Frey's construction. The analog of the discriminant conjecture in function field is a theorem of Szpiro. The original proof of this theorem makes essential use of the differentials of the function field. Thus it is impossible to move to the number field case. In this last part of the note, we will provide a different proof for the case where the function field has characteristic 0. (The original ideal of this proof is due to Bogomolov etc., at least for case where the base curves are rational.) Our proof does not use the differential of the function field but make essential use of the fact that the fundamental group of the base acts on the cohomology of the generic fiber with coefficients in *integers*. Such an action does exist in number fields (or function fields of positive characteristic) but with coefficients in *ℓ-adic integers*. Thus our proof extends to neither the number field nor function field case of positive characteristic.

1. Rational Points

One of the major goals of number theory is to study the solution set of a system of Diophantine equations. The simplest case is when this system is given by a single equation in two variables:

$$f(x, y) = 0 \tag{1.1}$$

where $f(x,y) = \sum_{i,j} a_{i,j} x^i y^j$ is an polynomial with integer coefficients $a_{i,j} \in \mathbb{Z}$ in the variables x and y. Thus the main problem is to describe the set

$$\left\{(x, y) \in \mathbb{Z}^2 \mid f(x, y) = 0\right\}.$$

To describe what we know about this question, we consider the homogenized form

$$F(x_0, x_1, x_2) := \sum_{i,j} a_{i,j} x^i y^j z^{n-i-j} = 0, \tag{1.2}$$

where $n = \deg f$, and assume for simplicity that

$$\frac{\partial F}{\partial x} = \frac{\partial F}{\partial y} = \frac{\partial F}{\partial z} = 0 \tag{1.3}$$

has only the trivial solution $(x, y, z) = (0, 0, 0)$ in $\mathbb{C}^3$. Now the equation (1.2) defines a curve X in the projective plane $\mathbb{P}^2$. Here, for any natural number n, the n-dimensional projective space is defined by

$$\mathbb{P}^n(\mathbb{C}) := \mathbb{C}^{n+1} - \{0\}/\sim, \quad \text{where} \quad v_1 \sim v_2 \Longleftrightarrow \mathbb{C}v_2 = \mathbb{C}v_1.$$

Condition (1.3) says that X is a smooth curve. Now the question becomes *how to describe the set $X(\mathbb{Q})$ of rational points on X?*

Here $X(\mathbb{Q})$ stands for points with rational coordinates. The answer to this question depends on the degree n of F.

Case where $n = 1$. In this case, X is defined by a linear equation

$$ax + by + cz = 0$$

with one nonzero coefficient, say a. Thus we can define a map

$$X(\mathbb{Q}) \xrightarrow{\sim} \mathbb{P}^1(\mathbb{Q}), \quad (x, y, z) \mapsto (y, z)$$

Case where $n = 2$. In this case, $X(\mathbb{Q})$ could be empty. For example, if X is defined by the equation $x^2 + y^2 + z^2 = 0$. Assume that $X(\mathbb{Q})$ is not empty and choose an element $O = (\alpha, \beta, \gamma)$. Without loss of generality, we assume that $\alpha \neq 0$. Then we have a map

$$X(\mathbb{Q}) \xrightarrow{\sim} \{\text{rational lines} \ni O\}, \qquad P \mapsto \text{line } OP,$$

where OP stands for the line passing through both P and O if $P \neq O$, and the tangent line passing through O if $P = O$. Notice that each line passing through O is defined by an equation

$$L_{a,b,c} : ax + by + cz = 0 \quad \text{with} \quad a\alpha + b\beta + c\gamma = 0.$$

Since $\alpha \neq 0$, we have a map

$$\{\text{rational lines} \ni O\} \xrightarrow{\sim} \mathbb{P}^1(\mathbb{Q}), \qquad L_{a,b,c} \mapsto (b, c).$$

Thus in case $n = 2$ we have the following conclusion:

$$\text{either } X(\mathbb{Q}) = \emptyset \text{ or } X(\mathbb{Q}) \xrightarrow{\sim} \mathbb{P}^1(\mathbb{Q}).$$

Hasse's principle. When does $X(\mathbb{Q})$ contain a rational point? By Hasse's principle, to see if $X(\mathbb{Q}) \neq \emptyset$ one needs only check if $X(\mathbb{R}) \neq \emptyset$, and if $X(\mathbb{Q}_p) \neq \emptyset$ for every *odd prime* p dividing the discriminant of the quadratic form $F(x_0, x_1, x_2)$. More precisely, after some linear transformation with coefficients in $\mathbb{Q}$, we may assume that X is defined by an equation:

$$ax^2 + by^2 = z^2. \tag{1.4}$$

where a and b are two square free integers. For every odd prime p, we define the Hilbert symbol $(a, b)_p \in \{\pm 1\}$ which equals 1 if and only if (1.4) has solution for every power of p. Then $X(\mathbb{Q}) \neq \emptyset$ if and only if

1. one of a and b is positive,
2. $(a, b)_p = 1$ for all odd prime number p dividing ab.

For $a = \alpha p^m$ and $b = \beta p^n$, one has the formula

$$(a, b)_p == \left(\frac{\alpha}{p}\right)^m \left(\frac{\beta}{p}\right)^n \left(\frac{-1}{p}\right)^{mn}.$$

Where $\left(\frac{x}{p}\right) = \pm 1$ is the residue quadratic symbol, which equals 1 if and only if x is a square modulo p. Using the quadratic reciprocity law, one can compute $(a, b)_p$ quite effectively. (The reason I don't care about the prime 2 is because of the product formula for Hilbert symbols. The above two conditions together will imply that $(a, b)_2 = 1$.)

For more details about quadratic forms, we refer to Serre's book [**4**].

Case where $n = 3$. Again $X(\mathbb{Q})$ could be empty, for example in the case where X is defined by equation

$$3x^3 + 4y^3 + 5z^3 = 0.$$

Assume $X(\mathbb{Q}) \neq 0$ and pick a point $O \in X(\mathbb{Q})$ inside, then $X(\mathbb{C})$ has a unique algebraic group structure such that O is the unit element. (For example, after a certain transformation, we may assume that O is a *reflection point* on X- the tangent line at O does not meet other points on X, then we may define the group rule so that $P + Q + R = 0$ whenever P, Q, R are three collinear points in $X(\mathbb{C})$.) The curve X together with the point O is called an elliptic curve. We let E denote X with this distinguished point O.

As a Lie group, $E(\mathbb{C})$ has dimension 2 and is isomorphic to $(\mathbb{R}/\mathbb{Z})^2$. But as an abstract group, $E(\mathbb{C})$ is very large. Indeed,

- the subgroup $E(\mathbb{C})_{\text{tor}}$ of elements of finite order is isomorphic to $(\mathbb{Q}/\mathbb{Z})^2$,
- the non torsion part $E(\mathbb{C})/E(\mathbb{C})_{\text{tor}}$ is an $\mathbb{Q}$-vector space of infinite dimension.

It is not difficult to show that $E(\mathbb{Q})$ is closed under the group operation, and thus defines a subgroup of $E(\mathbb{C})$. The following theorem tells us the structure of this group:

THEOREM 1.1 (Mordell). *The group $E(\mathbb{Q})$ is finitely generated, or equivalently, $E(\mathbb{Q})_{\text{tor}}$ is finite and*

$$E(\mathbb{Q})/E(\mathbb{Q})_{\text{tor}} \simeq \mathbb{Z}^r$$

where r is a nonnegative integer.

Infinite descent. The proof this theorem uses *infinite descent*, a technique used in Fermat's own proof of his last theorem for the exponent 4. In our case, this technique is a combination of Kummer's theory and Néron-Tate height theory. More precisely, the proof has two steps:

1. $E(\mathbb{Q})/2E(\mathbb{Q})$ is finite,
2. $E(\mathbb{Q})_{\text{tor}}$ is finite and $E(\mathbb{Q})$ is *discrete* in $E(\mathbb{Q}) \otimes \mathbb{R}$.

The second step implies that the rank of $E(\mathbb{Q})/2E(\mathbb{Q})$ over $\mathbb{Z}/2\mathbb{Z}$ is equal to or great than the rank of $E(\mathbb{Q})$ over $\mathbb{Z}$.

To prove the first step, one embeds $E(\mathbb{Q})/2E(\mathbb{Q})$ into the Galois cohomology group $H^1(\mathbb{Q}, E[2])$ using Kummer's sequence,

$$0 \longrightarrow E[2] \longrightarrow E \xrightarrow{2\times} E \longrightarrow 0$$

where $E[2]$ is the subgroup of 2-torsion points in $E(\mathbb{C})$ which is actually algebraic and thus admits an action from the Galois group over $\mathbb{Q}$. Then one finds that the image is unramified except for a finite set of primes. Thus the image is finite, so is $E(\mathbb{Q})/2E(\mathbb{Q})$.

For the second step, let us assume that O is a reflection point on X. Then we define the height of a point $p \in E(\mathbb{Q})$ by

$$h(P) = \log\max\{|a_i| : \quad i = 0, 1, 2\}$$

where P has homogeneous coordinates a_0, a_1, a_2 which are relatively prime integers. Then one can show that this function on $E(\mathbb{Q})$ is almost quadratic: the function $h(2p) - 4h(p)$ on $E(\mathbb{Q})$ is bounded. Thus one can normalize this height by taking the limit:

$$\widehat{h}(p) := \lim_{n\to\infty} 4^{-n}h(2^n p).$$

Now it is easy to show that $\widehat{h}$ on $E(\mathbb{Q})$ satisfies the following properties:

1. for any positive integer H, the set of points $p \in E(\mathbb{Q})$ with height $h(p) \leq H$ is finite;
2. $\widehat{h}(x)$ is nonnegative on $E(\mathbb{Q})$, and $h(x) = 0 \Longleftrightarrow x \in E(\mathbb{Q})_{\mathrm{tor}}$;
3. $\widehat{h}(p)$ is quadratic on $E(\mathbb{Q})$.

By these properties, one can show that $E(\mathbb{Q})_{\mathrm{tor}}$ is finite and $\widehat{h}$ induces a positive and quadratic norm on $E(\mathbb{Q}) \otimes \mathbb{R}$. Again, the first property implies that $E(\mathbb{Q})/E(\mathbb{Q})_{\mathrm{tor}}$ can't have a limit point in $E(\mathbb{Q}) \otimes \mathbb{R}$. Thus $E(\mathbb{Q})$ is discrete in $E(\mathbb{Q}) \otimes \mathbb{R}$.

Tate-Shafarevich group. When is $X(\mathbb{Q})$ nonempty? Unlike the case where $n = 2$, Hasse's principle is false in the case $n = 3$. Indeed, by a theorem of Hasse, every X has a solution in $\mathbb{R}$ and $\mathbb{Q}_p$ for every prime number p. There is even an analogue of Galois theory which gives a group that allows one to measure how difficult it is for X to have a solution. Indeed, one may assign to X an elliptic curve E, the Jacobian of X, which acts on X. This makes X a principal homogeneous space. Thus E has a rational point over $\mathbb{Q}$ and E is isomorphic to X over $\mathbb{C}$. Let Ш denote the group of isomorphism classes of principal homogeneous spaces of E over $\mathbb{Q}$ which have rational points over $\mathbb{R}$ and $\mathbb{Q}_p$ for every prime p. Then X defines an element in Ш. The group Ш is torsion and is called the *Tate-Shafarevich group*. It may be nontrivial but is conjectured to be finite.

For more details about elliptic curves, we refer to Silverman's book [**5**].

Case where $n \geq 4$. In this case, $X(\mathbb{C})$ is not a Lie group any more. But it can be embedded into a complex Lie group J of dimension $g = (n-1)(n-2)/2$ which is called the Jacobian of X. Thus, as a real Lie group,

$$J(\mathbb{C}) \simeq (\mathbb{R}/\mathbb{Z})^{2g}, \qquad J(\mathbb{C})_{\mathrm{tor}} \simeq (\mathbb{Q}/\mathbb{Z})^{2g}.$$

The group J is an *abelian variety*, this means that

1. J is defined by some equations with rational coefficients in a higher dimensional projective space,
2. the origin of J has coordinates in $\mathbb{Q}$,
3. the addition operator is defined by functions with coefficients in $\mathbb{Q}$.

Thus, we can talk about the set $J(\mathbb{Q})$ of rational points, which is again a subgroup of $J(\mathbb{C})$. The natural generalization of Mordell's theorem is the following:

THEOREM 1.2 (Weil). *The group $J(\mathbb{Q})$ is finitely generated. More generally, for any number field F, $J(F)$ is finitely generated.*

The set $X(\mathbb{Q})$ could be empty or nonempty. If it is nonempty and contains a point O then we can define an algebraic morphism $X \to J$ which takes O to be the unit element. This morphism takes rational points to rational points. In this way we have a picture:

$$X(\mathbb{Q}) = X(\mathbb{C}) \cap J(\mathbb{Q}) \tag{1.5}$$

THEOREM 1.3 (Faltings). *The set $X(\mathbb{Q})$ is finite. More generally, for any subvariety Y of an abelian variety A defined over a number field F which is not a translate of an abelian subvariety, $Y(F)$ is not Zariski dense.*

The Jacobian variety J as a generalization of an elliptic curve, is a very important companion of the curve X. There are two proofs of Faltings' theorem for the curves case. In this case, it was called *Mordell's conjecture.*

The first proof which works only for the case of curves uses the fact that J is faithful to X in the sense that if two curves have the same Jacobian then these two curves should be same. Thus, using some construction of Parshin, Mordell's conjecture is reduced to *the Shafarevich conjecture* for abelian varieties. This can be further reduced to *Tate's conjecture* and Weil's theorem for the *Riemann-Hypothesis for Abelian varieties over finite fields.* Many steps of the proof has been reduced to some boundedness of *Faltings' heights* for Abelian varieties with semiabelian reductions. For more details, we refer to Cornell and Silverman's book [**2**] for the first proof and background material.

The second proof, originally due to Vojta for the case of curves and extended to the general case by Faltings, uses the fact that J includes X as a subvariety, (i.e., the formula (1.5)), and performs the Diophantine approximation method on J. The basic technique includes finding small sections of a hermitian line bundle which is not very vanishing at a given point. (This will implies that the bundle has small degree at the given point.) Of course, as Roth did in the classic case, the lower vanishing index can be reached only by considering the product of varieties. Again the heights for subvarieties of an abelian variety have been defined and used in the whole proof. For the second proof, we refer to Faltings' original paper [**3**]

Effectivity. How to effectively find $X(\mathbb{Q})$? To make this question more precise for $n \geq 4$, we define the heights of X and points P in $\mathbb{P}^2(\mathbb{Q})$ as follows:

$$\begin{aligned} h(X) =& \text{complexity of } X \\ =& 1 + \log \max\{|a_{i,j}| : \quad i,j \geq 0\}, \end{aligned}$$

It is the general belief that the following should be true:

CONJECTURE 1.4 (Effective Mordell). *For each natural number $n \geq 4$, there exists a positive number $c(n)$ such that all points in $X(\mathbb{Q})$ have heights bounded by $c(n)h(X)$.*

We refer to Vojta's book [**7**] for a detailed discussion for conjectured bounds of rational points.

2. Algebraic points

Let $\bar{\mathbb{Q}}$ be the field of algebraic numbers in $\mathbb{C}$. For a projective space $\mathbb{P}^N$, let $\mathbb{P}^N(\bar{\mathbb{Q}})$ denote the points P with some homogeneous coordinates $(a_0, \cdots, a_N)$ in $\bar{\mathbb{Q}}^{N+1}$. Similarly we may define $Y(\bar{\mathbb{Q}})$ for subvarieties Y of $\mathbb{P}^N$. Let Y be a subvariety of an abelian variety A defined over a number field F in $\bar{\mathbb{Q}}$. For example,

Y is a curve of genus $g \geq 2$ and A is the Jacobian variety of Y. *What can we say about $Y(\bar{F})$ and $A(\bar{F})$?*

The first answer to this question, which is almost trivial, is that both sets are very large. First of all, $Y(\bar{F})$ is infinite. For example, when Y is defined by $f(x, y) = 0$, one may solve x in $f(x, y) = 0$ for any given $y \in \bar{F}$. Secondly, since the operations on A are defined by algebraic equations, and since the origin is an algebraic point we see that all torsion points of $A(\mathbb{C})$ are in $A(\bar{F})$. Thus

$$A(\bar{F})_{\text{tor}} = A(\mathbb{C})_{\text{tor}} \simeq (\mathbb{Q}/\mathbb{Z})^{2g}.$$

Finally, one may show that $A(\bar{F})/A(\bar{F})_{\text{tor}} = A(\bar{F}) \otimes \mathbb{Q}$ is infinite dimensional. To give some less trivial answer to this question, we have to restrict ourselves to some special class of points in $A(\bar{F})$.

Division points. A point P in $A(\bar{F})$ is called a division point if some positive multiple of it nP is in $A(F)$. Let $A(\bar{F})_{\text{div}}$ denote the subgroup of $A(\bar{F})$ of division points, then we have an exact sequence

$$0 \to A(\bar{F})_{\text{tor}} \to A(\bar{F})_{\text{div}} \to A(F)_F \to 0$$

where $A(F)_{\mathbb{Q}}$ stands for $A(F) \otimes \mathbb{Q}$. A natural generalization of Faltings theorem is the following

THEOREM 2.1 (Raynaud). *If Y is a subvariety of A defined over a number field F which is not a translate of an abelian variety by a torsion point, then $Y(\bar{F}) \cap A(\bar{F})_{\text{div}}$ is not Zariski dense in Y.*

COROLLARY 2.2. *In particular, the set $Y(\bar{F}) \cap A(\bar{F})_{\text{tor}}$ is not Zariski dense.*

It is obvious that Raynaud's theorem is more general than Faltings' theorem and the above Corollary. But Raynaud actually first proved the above corollary and then used a Galois theory argument to obtain his theorem from this corollary and Faltings' theorem.

Almost division points. Now we want to treat points which are very close to division points with distance defined by the Neron-Tate height pairing on $A(\bar{F})$. Recall that the Neron-Tate height pairing on $A(\bar{F})$ is a bilinear, symmetric pairing:

$$\langle \cdot, \cdot \rangle : \quad A(\bar{F}) \times A(\bar{F}) \to \mathbb{R} \tag{2.1}$$

such that

1. $\langle x, x \rangle \geq 0$,
2. $\langle x, x \rangle = 0$ if and only if $x \in A(\bar{F})_{\text{tor}} = (\mathbb{Q}/\mathbb{Z})^{2g}$.

Thus the pairing in (2.2) defines a pre-Hilbert space structure on the $\mathbb{Q}$-vector space $A(\bar{F})/A(\bar{F})_{\text{tor}} = A(\bar{F}) \otimes \mathbb{Q}$. Now for any two points x and y in $A(\bar{F}) \otimes \mathbb{Q}$ we can define the distance

$$\|x - y\| = \langle x - y, x - y \rangle^{1/2}.$$

In particular for $x \in A(\bar{F})$, we can define its distance to the group of division points:

$$d(x) = \inf \left\{ \|x - y\| : \quad y \in A(\bar{F})_{\text{div}} \right\}.$$

So $d(x) = 0$ if and only if $x \in A(\bar{F})_{\text{div}}$.

Raynaud's theorem can be generalized to the following

THEOREM 2.3 (Poonen, Zhang). *There is positive number $\epsilon > 0$ such that the subset*

$$\{x \in Y(\bar{F}) : \quad d(x) < \epsilon\}$$

is not Zariski dense in Y.

COROLLARY 2.4 (Bogomolov conjecture). *There is positive number $\epsilon > 0$ such that the subset*

$$\{x \in Y(\bar{F}) : \quad \|x\| < \epsilon\}$$

is not Zariski dense in Y.

Again, Theorem 2.3 clearly implies Faltings' theorem and the above Corollary 2.4. But all proofs of Theorem 2.3 we know actually make use of Faltings theorem and Bogomolov's conjecture. To prove Bogomolov's conjecture, one first develops a theory of positive line bundles in Arakelov geometry, then deduces an equidistribution theorem about Galois orbits of small points. We should mention that in some cases of curves, Szpiro made the first link between Bogomolov conjecture and Arakelov geometry, and then Ullmo found a clever use of some equidistribution theorem. We refer to our paper [**8**] for some historical remarks about the proof of Bogomolov's conjecture.

Distribution of almost division points. If X contains one point $x \in A(\bar{F})$ then it contains every conjugate x^σ $(\sigma \in \mathrm{Gal}(\bar{F}/F))$. One idea to prove Corollary 2.2 and 2.4 is that the Galois orbits of torsion points or small points tend to be uniformally distributed with respect to the Haar measure on $A(\mathbb{C})$. Thus X can't contain infinitely many such points. In the following we want to explain this idea more precisely.

Let x_n $(n = 1, 2, \cdots)$ be a sequence of points in $A(\bar{F})$. We say x_n is a sequence of *almost division points* if $d(x_n)$ converges to 0 as $n \to \infty$. We say that the Galois orbits

$$O(x_n) := \{x_n^\sigma : \quad \sigma \in \mathrm{Gal}(\bar{F}/F)\}$$

are *equidistributed* with respect to a measure $d\mu$ on $A(\mathbb{C})$, if the uniform probability measure

$$\frac{1}{\#O(x_n)} \sum_{y \in O(x_n)} \delta_y$$

converges to $d\mu$ as $n \to \infty$. Here, δ_y is the Dirac measure at y. Equivalently, this means that for any continuous function f on $A(\mathbb{C})$,

$$\lim_{n\to\infty} \frac{1}{\#O(x_n)} \sum_{y \in O(x_n)} f(y) = \int_{A(\mathbb{C})} f(y) d\mu(y).$$

THEOREM 2.5. *Let $(x_n, n \in \mathbb{N})$ be a sequence of almost division points. There is a subsequence $(y_n, n \in \mathbb{N})$ of $(x_n, n \in \mathbb{N})$, an abelian subvariety B of A defined over F, and a finite subset T of $A(\mathbb{C})$ such that the Galois orbits of y_n are equidistributed to the $B(\mathbb{C})$-invariant uniform measure on $T + B(\mathbb{C})$. Moreover, $B = 0$ if and only if there is a finite extension F' of F such that all x_n are rational over F'.*

Now it is easy to see the following

- Theorem 2.5 + Faltings' theorem $\Longrightarrow$ Theorem 2.3.
- Theorem 2.5 $\Longrightarrow$ Corollary 2.4.

But to prove Theorem 2.5, one must prove Bogomolov's conjecture first. See our paper [**9**] for more details.

Scatteredness of big points. What can we say about the points which are far away from division points? For a point $x \in A(\bar{F})$, instead of the distance $d(x)$ to $A(\bar{F})_{\text{div}}$ we consider the angle $\theta(x)$ to $A(\bar{F})_{\text{div}}$:

$$\theta(x) = \sin^{-1}(d(x)/\|x\|).$$

Then from Faltings' proof of Theorem 1.3 for higher dimensional case, we have

THEOREM 2.6. *Assume that Y is a subvariety of A defined over F which is not a translate of an abelian subvariety. Then there is a positive number ϵ such that the subset*

$$\{x \in Y(\bar{\mathbb{Q}}) : \quad \theta(x) < \epsilon\}$$

is not Zariski dense.

3. The ABC conjecture and the discriminant Conjecture

When discussing Diophantine equations, it is unavoidable to mention the *ABC-Conjecture:*

CONJECTURE 3.1 (Masser, Osterlé). *Let A, B, C be relative prime positive integer satisfying $A + B = C$, then for any $\epsilon > 0$,*

$$C \leq \kappa(\epsilon) \left(\prod_{p|ABC} p \right)^{1+\epsilon}$$

where $\kappa(\epsilon)$ is a number depending only on ϵ.

For application purposes, it suffices to have a *weaker form where $1+\epsilon$ is replaced by any fixed constant.*

There are two ways to link this conjecture to Diophantine equations. First of all this conjecture applies directly to the diagonal (or generalized Fermat's) equation:

$$aX^n + bY^n = cZ^n \tag{3.1}$$

where a, b, c are positive and relative prime integers, and n is a positive integer. The conjecture implies that there is a constant $C(a, b, c)$ such that the equation has no nontrivial solution if

$$n \geq C(a, b, c).$$

The second link is that this conjecture can be generalized to a form which holds for an arbitrary number field. Then the generalized conjecture is equivalent to a certain effective Mordell conjecture like Conjecture 1.4. This follows from some work of Szpiro, Moret-Bailly, and Elkies. We won't discuss this matter here.

We want to discuss a third link that relates the ABC-conjecture to an early conjecture of Szpiro on discriminants of elliptic curves.

Szpiro's conjecture. Recall that an elliptic curve E over $\mathbb{Q}$ is a projective curve of genus 1 with a rational point O. Any elliptic curve E can be defined by a Weierstrass equation

$$E_{\mathbb{Z}} : \qquad y^2 + a_1xy + a_3y = x^3 + a_2x^2 + a_4x + a_6 \tag{3.2}$$

where $a_i \in \mathbb{Z}$. The origin O is at infinity. One may use various affine transformations to obtain different form of Weierstrass equation. For such an equation one can define a discriminant Δ which a positive integer with the property that $p \mid \Delta$ if and only if equation (3.2) modulo p defines a singular curve. We say a Weierstrass equation is minimal if Δ is minimal. In this case, we call $E_{\mathbb{Z}}$ the minimal integral model of E and write Δ by Δ_E.

Recall also that an elliptic curve $E_{\mathbb{Z}}$ given by a minimal Weierstrass equation $f(x, y) = 0$ over $\mathbb{Z}$ has stable reduction at a prime p, if this equation modulo p defines a smooth curve (which we called a good reduction), or a curve with node (which we called a bad reduction). In the stable but bad reduction case, in a strict henselian neighborhood node, the definition equation of $E_{\mathbb{Z}}$ has the form $xy = p^{n_p}$, where n_p is a positive integer. If $E_{\mathbb{Z}}$ has stable reduction everywhere, the discriminant of $E_{\mathbb{Z}}$ has the form $\prod_p p^{n_p}$ where p runs through the set of bad reductions. Write $N = \prod_{p|\Delta} p$.

CONJECTURE 3.2 (Szpiro). *Let E be an elliptic curve defined over $\mathbb{Q}$ whose minimal model $E_{\mathbb{Z}}$ has stable reduction everywhere. Then for any $\epsilon > 0$,*

$$\Delta_E \leq \kappa(\epsilon)(N_E)^{6+\epsilon}$$

where $\kappa(\epsilon)$ is a constant depending only on ϵ.

Again for application, it suffices to have a *weaker form where $6+\epsilon$ is replaced by any fixed constant ϵ.* The link between the ABC-conjecture and Szpiro's conjecture is given by the following construction of Frey.

Frey curve. Let A, B, C be three relative prime, positive integers such that $A+B = C$, and such that $16 \mid A$ and $B \equiv -1 \pmod 4$. Consider the curve $E_{A,B,C}$ defined by the equation

$$y^2 = x(x + A)(x - B).$$

Frey proves the following:

1. The minimal Weierstrass equation of $E_{A,B,C}$ is given by
$$y^2 + xy = x^3 + \frac{A - B - 1}{4}x^2 - \frac{AB}{4}x.$$
2. The above minimal model of $E_{A,B,C}$ has stable reduction modulo every prime p.
3. Under the hypothesis of (1), the discriminant of $E_{A,B,C}$ is equal to $(ABC/16)^2$.

If you apply Szpiro's conjecture then you immediately obtain the ABC-conjecture, at least in their weak forms.

Szpiro's Theorem. Szpiro proposed this based on his theorem over a function field. Let $f : E \to C$ be a proper and flat morphism from a surface E, smooth over algebraically closed field k, to a curve C, smooth and projective of genus g and geometrically connected over k. Suppose the generic fiber of f is a smooth and geometrically connected elliptic curve over the function field of C. Suppose in addition that f is not isotrivial and the degenerate fibers are stable. Then we have a nonconstant morphism

$$j : C \to \mathbb{P}^1_k$$

by taking j-invariants.

Let Δ_E denote the discriminant divisor of f, and let S denote the set of points on C over which f is not smooth. Let p be the characteristic of k and p^e be the degree of inseparability of the morphism j. (If k has characteristic 0, we write $p^e = 1$ for convenience).

THEOREM 3.3 (Szpiro).

$$\deg \Delta_E \leq 6p^e(2g - 2 + \#S)$$

Szpiro's proof makes essential use of the derivative map induced by j on $C - S$. We refer to Szpiro's paper [**6**] for his conjecture and his theorem.

In the following we want to present a proof using the monodromy action on homology. We will assume that $k = \mathbb{C}$, because we will make essential use of integral coefficients in homology. When the base C is a Riemann sphere, this proof is due to J. Amorós, F. Bogomolov, L. Katzarkov, and T. Pantev [**1**].

Monodromy action on homology. We let $f : E \to C$ be as above with $k = \mathbb{C}$. Fix a point p on C then we have a monodromy action

$$\rho : \qquad \pi_1(C - S, p) \longrightarrow \mathrm{SL}(H_1(E_p, \mathbb{Z}))$$

where the fundamental group means topological fundamental group and SL means automorphisms with determinant 1. Let U be a simply connected neighborhood in $C(\mathbb{C})$ of p in C containing S. For each $s \in S$, let $c_s \in \pi_1(C - S, p)$ be an element which can be represented by a simple loop around only s in U with positive orientation. There are a_i, b_j $(i, j = 1, \cdots, g)$ in $\pi_1(C - S, p)$ such that $\pi_1(C - S, p)$ is generated by a_i, b_j, c_s with a single relation

$$\prod_{i=1}^{g} [a_i, b_j] \cdot \prod_{s \in S} c_s = 1.$$

Recall that the intersection pairing gives an alternating pairing

$$\langle \cdot, \cdot \rangle : \qquad H_1(E_p, \mathbb{Z}) \otimes H_1(E_p, \mathbb{Z}) \longrightarrow \mathbb{Z}.$$

Fix a basis e_1 and e_2 such that $\langle e_1, e_2 \rangle = 1$. This defines an isomorphism $\mathrm{SL}(H_1(E_p, \mathbb{Z})) \simeq \mathrm{SL}_2(\mathbb{Z})$ which is unique up to conjugation by elements in $\mathrm{SL}_2(\mathbb{Z})$ (not only $\mathrm{GL}_2(\mathbb{Z})$!) Now we have the following equality in $\mathrm{SL}_2(\mathbb{Z})$:

$$\prod_{i=1}^{g} [\rho(a_i), \rho(b_j)] \cdot \prod_{s \in S} \rho(c_s) = 1. \tag{3.3}$$

We will actually show that Szpiro's inequality follows from (3.3) and the following monodromy theorem:

THEOREM 3.4. *For each $s \in S$, $\rho(c_s)$ is conjugate in $\mathrm{SL}_2(\mathbb{Z})$ to*

$$\begin{pmatrix} 1 & n_s \\ 0 & 1 \end{pmatrix}$$

where $n_s = \mathrm{ord}_s(\Delta_E)$.

Proof: This is a well known theorem but we provide here a proof for the convenience of the reader. Let q be the local parameter of C at s. We may enlarge q such that $D := \{q \in \mathbb{C}, |q| \leq 1/e\}$ is embedded in C. Let D^* denote $D - \{0\}$. Then the restriction of E on D^* is a Tate curve with the form

$$E^* \simeq D^* \times \mathbb{C}^\times / q^{n_s \mathbb{Z}}. \tag{3.4}$$

Obviously, the theorem does not depend on the choice of p in C. Thus we may assume that $p = 1/e$ and c_s is a simple loop in D^* with counter clockwise orientation:

$$c_s(t) = e^{2\pi i t - 1} \qquad (0 \leq t < 1).$$

By (3.4), the fiber E_t over $c_s(t)$ has the form

$$E_t = \mathbb{C}^*/e^{2\pi i t-1} = \mathbb{C}/(\mathbb{Z} + (i+t)n_s\mathbb{Z})$$

Thus if we use $e_1 = 1$ and $e_2 = in_s$ as a base of $H_1(E_p, \mathbb{Z})$, then they are moved to 1 and $(i+t)n_s$, and they have limits 1 and $(i+1)n_s$ as $t \to 1$. Thus, we have shown that

$$c_s \cdot (e_1, e_2) = (e_1, n_s e_1 + e_2) = (e_1, e_2) \cdot \begin{pmatrix} 1 & n_s \\ 0 & 1 \end{pmatrix}.$$

Groups $\widetilde{\mathrm{SL}}_2(\mathbb{R})$ and $\widetilde{\mathrm{SL}}_2(\mathbb{R})$. The action of $\mathrm{SL}_2(\mathbb{R})$ on $\mathbb{R}^2$ induces a topological action on

$$\mathbb{R}/2\pi\mathbb{Z} \simeq \mathbb{R}^2/\mathbb{R}^+$$

where $t \in \mathbb{R}/2\pi\mathbb{Z}$ corresponds to $(\cos t, -\sin t)\cdot\mathbb{R}^+$ and $\mathbb{R}^+$ is the group of positive numbers. Let $\widetilde{\mathrm{SL}}_2(\mathbb{R})$ be the group of topological homeomorphisms $\widetilde{\gamma}$ of $\mathbb{R}$ which induce elements γ in $\mathrm{SL}_2(\mathbb{R})$. Thus for all $t \in \mathbb{R}$,

$$\widetilde{\gamma}(t) \mod 2\pi = \gamma(t \mod 2\pi).$$

LEMMA 3.5. *The homomorphism from $\widetilde{\mathrm{SL}}_2(\mathbb{R})$ to $\mathrm{SL}_2(\mathbb{R})$ induced by $\widetilde{\gamma} \to \gamma$ is surjective with kernel a free group generated by z^2, where $z(t) = t + \pi$. Thus we have an exact sequence:*

$$1 \to < z^2 > \to \widetilde{\mathrm{SL}}_2(\mathbb{R}) \to \mathrm{SL}_2(\mathbb{R}) \to 1.$$

Moreover $\widetilde{\mathrm{SL}}_2(\mathbb{R})$ has center generated by z.

Proof: We need only prove that the homomorphism $\widetilde{\mathrm{SL}}_2(\mathbb{R}) \to \mathrm{SL}(\mathbb{R})$ is surjective. Let $\gamma \in \mathrm{SL}_2(\mathbb{R})$. We want to explicitly construct a *canonical lifting* $\widetilde{\gamma}$ which commutes with z.

Case where $\gamma = -I$. We take an obvious lifting $\widetilde{\gamma}(t) = z$.

Case where γ is parabolic. This means that γ has a fixed point θ on $\mathbb{R}/2\pi\mathbb{Z}$ or equivalently, γ has a positive eigenvalue. In this case, γ has a unique lifting $\widetilde{\gamma}$ fixing pointwise all points in the preimage of θ.

General case. We claim that γ is a product of two elements in the above cases. This is clear if γ is either upper triangular or lower triangular. If this is not the case, then there is a unique x such that

$$\det(\gamma - u_x) = 0, \quad \text{where} \quad u_x = \begin{pmatrix} 1 & x \\ 0 & 1 \end{pmatrix}.$$

Thus $\gamma = \gamma_1 \cdot u_x$ where γ_1 has an eigenvalue equal to 1. So $\widetilde{\gamma}_1\widetilde{u}_x$ gives a lifting for γ.

It is well known that $\mathrm{SL}_2(\mathbb{Z})$ is generated by the following elements and relations:

$$u = \begin{pmatrix} 1 & 1 \\ 0 & 1 \end{pmatrix}, \qquad v = \begin{pmatrix} 1 & 0 \\ -1 & 1 \end{pmatrix}, \qquad (uv)^6 = I, \qquad uvu = vuv$$

Let $\widetilde{A}$ and $\widetilde{B}$ be canonical liftings of A and B constructed in the proof of Lemma 3.5.

LEMMA 3.6. *Let $\widetilde{\mathrm{SL}}_2(\mathbb{Z})$ be the inverse image of $\mathrm{SL}_2(\mathbb{Z})$ in $\widetilde{\mathrm{SL}}_2(\mathbb{R})$. Then*

1. *$\widetilde{\mathrm{SL}}_2(\mathbb{Z})$ is generated by $\widetilde{u}$ and $\widetilde{v}$ with the relation $\widetilde{u}\widetilde{v}\widetilde{u} = \widetilde{v}\widetilde{u}\widetilde{v}$.*

2. $z = (\widetilde{u}\widetilde{v})^3$.

Proof: It can be checked directly that

$$0 \leq \widetilde{u}x - x < \frac{\pi}{3}, \qquad 0 \leq \widetilde{v}x - x < \frac{\pi}{3} \tag{3.5}$$

Thus

$$0 \leq (\widetilde{u}\widetilde{v})^3 x - x < \frac{\pi}{3} \times 6 = 2\pi. \tag{3.6}$$

$$0 \leq \widetilde{u}\widetilde{v}\widetilde{u}x - x < \pi, \qquad 0 \leq \widetilde{v}\widetilde{u}\widetilde{v}x - x < \pi \tag{3.7}$$

As $(\widetilde{u}\widetilde{v})^3$ is a lifting of $-I$, then $(\widetilde{u}\widetilde{v})^3$ must be equal to kz with k an odd integer. By (3.6), $k = 1$, or equivalently $z = (\widetilde{u}\widetilde{v})^3$. The same reasoning gives the equality $\widetilde{u}\widetilde{v}\widetilde{u} = \widetilde{v}\widetilde{u}\widetilde{v}$.

As the subgroup $< \widetilde{u}, \widetilde{v} >$ contains the center $< z >$ and maps surjectively to $\mathrm{SL}_2(\mathbb{Z})$, this subgroup must be $\widetilde{\mathrm{SL}}_2(\mathbb{Z})$. Since $\widetilde{u}\widetilde{v}$ has infinite order in $\widetilde{\mathrm{SL}}_2(\mathbb{Z})$, $\widetilde{\mathrm{SL}}_2(\mathbb{Z})$ is generated by $\widetilde{u}$ and $\widetilde{v}$ with only one relation $\widetilde{u}\widetilde{v}\widetilde{u} = \widetilde{v}\widetilde{u}\widetilde{v}$.

Proof of Szpiro's inequality. We start with a lifting of equation (3.3) to $\widetilde{\mathrm{SL}}_2(\mathbb{Z})$. Let α_i, β_j, and γ_s be canonical liftings of some liftings of $\rho(a_i)$ and $\rho(b_j)$ and $\rho(c_s)$ respectively in $\widetilde{\mathrm{SL}}_2(\mathbb{Z})$ as constructed in the proof of Lemma 3.5. Then we have a new equation in $\widetilde{\mathrm{SL}}_2(\mathbb{Z})$:

$$\prod_{i=1}^{g} [\alpha_i, \beta_j] \cdot \prod_{s \in S} \gamma_s = z^{2m} \tag{3.8}$$

where m is some integer.

Lets determine m. By Lemma 3.6, there is a degree homomorphism

$$\deg : \quad \widetilde{\mathrm{SL}}_2(\mathbb{Z}) \longrightarrow \mathbb{Z} \quad \text{such that} \quad \deg \widetilde{u} = \deg \widetilde{v} = 1.$$

Applying this degree map to equation (3.8), we obtain $12m = \sum_{s \in S} \deg \gamma_i$. By Theorem 3.4, $\rho(c_s)$ is conjugate to u^{n_s}. As γ_i is the canonical lifting of $\rho(c_i)$, γ_s is conjugate to $\widetilde{u}^{n_s}$ in $\widetilde{\mathrm{SL}}_2(\mathbb{Z})$. Thus $\deg \gamma_s = n_s$. So we obtain

$$12m = \sum_{s \in S} n_s = \deg \Delta_E. \tag{3.9}$$

For any $\alpha \in \widetilde{\mathrm{SL}}_2(\mathbb{R})$, define its *length* by

$$\ell(\alpha) = \sup_{x \in \mathbb{R}} |\alpha(x) - x| \in \mathbb{R}^+.$$

(The right hand side is finite, because $\alpha z = z\alpha$, or equivalently $\alpha(x+\pi) = \alpha(x)+\pi$.) This length function has properties:

$$\ell(\alpha\beta) \leq \ell(\alpha) + \ell(\beta), \qquad \ell(z^n) = \pi n.$$

Thus equation (3.8) gives us

$$2m\pi \leq \sum_{i=1}^{g} \ell\left([\alpha_i, \beta_i]\right) + \sum_{s \in S} \ell(\gamma_s). \tag{3.10}$$

LEMMA 3.7.

$$\ell(\gamma_s) < \pi, \qquad \ell\left([\alpha_i, \beta_i]\right) \leq 2\pi.$$

Proof: The first one is easy, as γ_s has fixed points on $\mathbb{R}$ with period π. For the second one, from the construction in the proof of Lemma 3.5, upto a factor z which does not change the bracket, α_i is a product of two elements conjugate in $\widetilde{\mathrm{SL}}_2(\mathbb{R})$ to elements $\widetilde{u}_x$ and $\widetilde{u}_y$. Thus $[\alpha_i, \beta_i] = \alpha_i \cdot (\beta_i \alpha_i \beta_i^{-1})$ is a product of four elements conjugate to $\widetilde{u}_x$, $\widetilde{u}_{-x}$, $\widetilde{u}_y$, $\widetilde{u}_{-y}$. Notice that two of these four elements will move points in positive direction, and two of them will move points in negative direction. All of them have length bounded by π. Thus $[\alpha_i, \beta_i]$ has length bounded by 2π.

By (3.9) and (3.10) we have $2m \leq 2g - 1 + s$. Combining with (3.8), we have

$$\deg \Delta_E \leq 6(s + 2g - 1) \tag{3.11}$$

This inequality is not as sharp as Szpiro's. But for $g > 0$, then one may apply this equality to an unramified base change

$$E' = E \times_C C' \to C'$$

where $\pi : C' \to C$ is an unramified extension. Then we have equalities

$$\begin{aligned} \deg \Delta_{E'} &= \deg \pi \cdot \deg \Delta_E, \\ 2g(C') - 2 &= \deg \pi \cdot (2g - 2), \\ s' &= \deg \cdot \pi s \end{aligned}$$

where s' is the number of bad fibers of $E' \to C'$. Now (3.11) for $E' \to C'$ gives

$$\deg \Delta_E \leq 6(s + 2g - 2 + 1/\deg \pi).$$

As π can be chosen such that $\deg \pi$ arbitrary large, one obtains Szpiro's inequality. However this argument doesn't apply to the case where $g = 0$.

References

[1] J. Amorós, F. Bogomolov, L. Katzarkov, and T. Pantev, *Symplectic Lefschetz fibration with arbitrary fundamental groups,* Preprint, Math.GT/9810042 v2 25 Nov 1998.

[2] G. Cornell and J. H. Silverman (eds), *Arithmetic geometry,* Springer-Verlag, 1986

[3] G. Faltings, *Diophantine approximation on abelian varieties,* Ann. Math. **133** (1991), pp. 549-576.

[4] J. -P. Serre, *A course in arithmetic,* GTM **7**, Springer-Verlag, New York, 1973

[5] J. H. Silverman, *The arithmetic of elliptic curves,* GTM **106**, Springer-Verlag, New York, 1986

[6] L. Szpiro, *Discriminant et conducteur des courbes elliptiques,* In: *Séminaire sur les pinceaux de courbes elliptiques: à la recherche de Mordell effectif.* Astérisque **183**, 1990.

[7] P. Vojta, *Diophantine approximations and value distribution theory,* LNM **1239**, Springer-Verlag, Berlin, 1987

[8] S. Zhang, *Small points and Arakelov theory,* Proceedings of ICM, Berlin 1998, Vol II, pp. 217-225.

[9] ———, *Distribution of almost division points,* to appear in Duke Math. J.

DEPARTMENT OF MATHEMATICS, COLUMBIA UNIVERSITY, NEW YORK, NEW YORK 10027
E-mail address: `szhang@@math.columbia.edu`

Invited Addresses

AMS/IP Studies in Advanced Mathematics
Volume 20, 2001

Jordan structures in Banach manifolds

Cho-Ho Chu

ABSTRACT. We give an exposition of Jordan structures in infinite dimensional Banach manifolds and introduce some recent applications of Jordan theory to symmetric Banach manifolds.

1. Introduction

Since Jordan algebras were first introduced by Jordan, von Neumann and Wigner in the 1930's [**9**] for an axiomatic study of quantum mechanics, Jordan structure has found many applications in other areas of mathematics. Jordan theory has been used, for instance, to classify the bounded symmetric domains and Banach manifolds. In this article, we give an exposition to show how the theory of Jordan algebras and more generally, Jordan triple systems, may be applied to the study of the geometry of symmetric Banach manifolds.

Symmetric Banach manifolds are infinite dimensional generalization of the Hermitian symmetric spaces classified by E. Cartan [**4**] in the 1930's using Lie theory. A connection of this classification to Jordan algebras was later pioneered by Koecher [**15**]. Jordan triple systems became prominent when Loos [**16**] showed the correspondence, in finite dimensions, between bounded circular domains and Jordan triples. The full generalization of this correspondence to that between infinite dimensional symmetric Banach manifolds and Jordan triples was developed by Kaup in [**10, 11, 12, 13**].

We recall that a *Hermitian symmetric space* is a finite dimensional complex manifold M with a Hermitian metric Φ such that each point in M is an isolated fixed point of a symmetry of M, where a symmetry is a biholomorphic map $\sigma\ M \to M$ which is an isometry for Φ and σ^2 is the identity map.

A Hermitian symmetric space M is *non-compact* if it has constant negative sectional curvature, and M is *compact* if it has constant positive sectional curvature. The *Euclidean* Hermitian symmetric spaces are the ones with zero sectional curvature.

The non-compact Hermitian symmetric spaces are exactly the bounded symmetric domains in $\mathbb{C}^n$ which have been classified by Cartan [**4**]. Up to biholomorphic equivalence, the *irreducible* ones are the open unit balls of one of the following six types of finite dimensional complex normed vector spaces of matrices:

type I	$m \times n$ complex matrices,
type II	$m \times m$ skew symmetric complex matrices,
type III	$m \times m$ symmetric complex matrices,
type IV	spin factor,
type V	$M_{1,2}(\mathbb{O}) = 1 \times 2$ matrices over the Cayley algebra $\mathbb{O}$,
type VI	$M_3^8(\mathbb{O}) = 3 \times 3$ Hermitian matrices over $\mathbb{O}$.

Spin factor will be defined later. We refer to [**6, 8, 23**] for further details on symmetric spaces and classical domains.

One can associate a Jordan triple system, called a *JB^*-triple*, to each symmetric Banach manifold, and using this device, Cartan's classification can be extended to the infinite dimensional case. We will explain this association and introduce some recent results on infinite dimensional manifolds associated with JB*-triples.

2. Jordan triple systems

We will only consider algebras over the complex field. A *Jordan algebra* is a nonassociative and commutative algebra whose elements satisfy the Jordan identity

$$a(ba^2) = (ab)a^2.$$

A *Jordan triple system* is a complex vector space V with a *Jordan triple product* $\{\cdot,\cdot,\cdot\} : V \times V \times V \longrightarrow V$ which is symmetric and linear in the outer variables, conjugate linear in the middle variable and satisfies the Jordan triple identity

$$\{a, b, \{x, y, z\}\} = \{\{a, b, x\}, y, z\} - \{x, \{b, a, y\}, z\} + \{x, y, \{a, b, z\}\}.$$

A Jordan algebra is a Jordan triple system with the usual Jordan triple product

$$\{a, b, c\} = (ab)c + (bc)a - (ac)b.$$

The spaces of matrices that appeared above in Cartan's classification are all Jordan triple systems with natural triple products which will be described later in their infinite dimensional generalization.

In the study of Jordan triple systems, the following operators and the Peierce decomposition are useful tools :

Quadratic operator	$Q(x) : v \in V \mapsto \{x, v, x\} \in V$;
Box operator	$x y : v \in V \mapsto \{x, y, v\} \in V$;
Bergman operator	$B(x, y) : V \to V,\ B(x, y) = \mathsf{Id} - 2x y + Q(x)Q(y)$.

The Peierce decomposition is induced by a *tripotent* which is an element e in V satisfying $\{e, e, e\} = e$. Nonzero tripotents may or may not exist. To each tripotent e, there are associated Peierce projections

$$\begin{aligned} P_0(e) &= B(e, e); \\ P_1(e) &= 2((e e - Q(e)^2); \\ P_2(e) &= Q(e)^2. \end{aligned}$$

We have $\mathsf{Id} = P_0(e) + P_1(e) + P_2(e)$ and the *Peierce decomposition* $V = V_0(e) \oplus V_1(e) \oplus V_2(e)$ where $V_j(e) = P_j(e)V$ $(j = 0, 1, 2)$ are the *Peierce spaces*, subject to the following computation rules:

$$\{V_i(e), V_j(e), V_k(e)\} \subset V_{i-j+k}(e)$$

where $V_\ell(e) = \{0\}$ for $\ell \neq 0, 1, 2$.

A complex Banach space V is called a *J^*-triple* if it is a Jordan triple system such that for every $\alpha \in V$, the box operator $\alpha\alpha : V \longrightarrow V$ is Hermitian, that is, $\|e^{i\lambda\alpha\alpha}\| \leq 1$ for all $\lambda \in \mathbb{R}$. Further, a J*-triple is called a *JB^*-triple* if every box operator $\alpha\alpha$ has non-negative spectrum with $\|\alpha\alpha\| = \|\alpha\|^2$. A JB*-triple which is a dual Banach space is called a *JBW^*-triple.* There is a useful recent survey of JB^*-triples in [**19**]. See also [**3**].

3. Symmetric Banach manifolds

A *complex Banach manifold* is a manifold modelled locally on open subsets of complex Banach spaces such that the coordinate transformations are biholomorphic mappings.

We use *manifold* to mean a *connected* complex Banach manifold in this section. Given a manifold M, one usually studies the group G of all biholomorphic maps from M to itself. Unfortunately, G may fail to be a Lie group. On the other hand, it is known that if M has a 'norm' on its tangent bundle TM then the group of all biholomorphic maps that preserve this norm does have a Lie group structure. For this reason, we restrict ourselves to manifolds having such a norm. To be precise, we say that a mapping $\gamma\, TM \to \mathbb{R}^+$ is a *tangent norm* on M if

(i) $\gamma|_{T_pM}$ is a norm, for every tangent space T_pM, $p \in M$ and
(ii) for each p in M there is a neighbourhood U of p in M, $U \subset Z$ a complex Banach space, and constants $0 < c < C$ such that

$$c\|v\| \leq \gamma(z, v) \leq C\|v\| \text{ for all } (z, v) \in TU \cong U \times Z.$$

A manifold M with a tangent norm γ is called a *normed manifold* and a holomorphic map $f\, M \to M$ is said to be an *isometry* for γ if

$$\gamma(f(z), f'(z)v) = \gamma(z, v) \text{ for all } (z, v) \in TM.$$

We define the *automorphism group* Aut(M) to be the group of all biholomorphic maps from M to M that are isometries for γ. Then Aut(M) is a real Lie group and its Lie algebra aut(M) consists of complete holomorphic vector fields on M.

Definition. A (normed) manifold M is said to be *symmetric* if to each a in M there exists an involution σ_a in Aut(M) having a as an isolated fixed point. The mapping σ_a is called the *symmetry* of M at a.

By a *domain,* we mean an open connected subset of a complex Banach space. A domain is a Banach manifold and a *symmetric domain* is one that is symmetric in the above definition. We refer to [**22**] for a useful reference for symmetric manifolds and Jordan algebras.

Let M be a symmetric manifold. Fix a in M and let σ_a be the symmetry of M at a. The natural action of the involution σ_a on the group Aut(M), $g \mapsto \sigma_a g \sigma_a^{-1}$, induces the adjoint action $\mathrm{Ad}(\sigma_a)$ on $\mathrm{aut}(M)$. Since $\sigma_a^2 = I$, $\mathrm{Ad}(\sigma_a)$ can be seen to have eigenvalues ± 1 and $\mathrm{aut}(M)$ has an eigenspace decomposition

$$\mathrm{aut}(M) = \mathfrak{k} \oplus \mathfrak{p}$$

where $\mathfrak{k}$ is the $+1$ eigenspace and $\mathfrak{p}$ is the -1 eigenspace of $\mathrm{Ad}(\sigma_a)$. Let $Z = T_a M$ be the tangent space at a. In a suitable local chart of M at a, for each $\alpha \in Z$, there is a unique vector field X_α in $\mathfrak{p}$ with $X_\alpha(a) = \alpha$ and $X_\alpha = \left(\alpha - q_\alpha(z)\right)\frac{\partial}{\partial z}$, where $q_\alpha : Z \longrightarrow Z$ is a homogeneous polynomial of degree 2, that is, $q_\alpha(z) = f(z,z)$ for some continuous (symmetric) bilinear map $f : Z \times Z \longrightarrow Z$. We define

$$\{z, \alpha, w\} = \frac{1}{2}(q_\alpha(z+w) - q_\alpha(z) - q_\alpha(w)) \quad \text{for } z, w \in Z.$$

Then $(Z, \{\cdot,\cdot,\cdot\})$ is a *J^*-triple*. In this manner, one associates a J^*-triple to each symmetric manifold. Conversely, given a J^*-triple Z one can associate to it a symmetric manifold M. Further, M is a bounded symmetric domain if, and only if, for each $\alpha \in Z$, the operator $\alpha\alpha$ has non-negative spectrum and the norm $\|\alpha\|_\infty := \sqrt{\|\alpha\alpha\|}$ is an equivalent norm on Z, in which case, M is biholomorphically equivalent to $\{\alpha \in Z : \|\alpha\|_\infty < 1\}$ which is the open unit ball of the JB*-triple $(Z, \|\cdot\|_\infty)$. We summarise this in the following theorem of Kaup [**13**].

THEOREM 3.1. *Every bounded symmetric domain is biholomorphically equivalent to the open unit ball of a JB^*-triple and the open unit ball of every JB^*-triple is a bounded symmetric domain.*

We will see that this theorem generalizes Cartan's classification of finite dimensional bounded symmetric domains. The JB^*-triples form a large class of Banach spaces. They include for instance, C^*-algebras, Hilbert spaces and spaces of rectangular matrices. Indeed, given any norm closed subspace Z of the algebra $\mathcal{L}(H)$ of bounded linear operators on a Hilbert space H, Z is a JB^*-triple if for any $z \in Z$, one has $zz^*z \in Z$, where z^* denotes the adjoint of z. In this case, the triple product is given by $\{x,y,z\} = \frac{1}{2}(xy^*z + zy^*x)$ and for each b in its open unit ball B, the mapping $g_b\, B \to B$ given by

$$g_b(z) = (1 - bb^*)^{-\frac{1}{2}}(z+b)(1+b^*z)^{-1}(1-b^*b)^{\frac{1}{2}}$$

is an automorphism of B satisfying $g_b(0) = b$ and $g_b^{-1} = g_{-b}$ [**5**]. The symmetry of B at b is then $g_b \circ s \circ g_{-b}$, where $s(z) = -z$ is the symmetry at 0. We recall that C*-algebras are closed *-subalgebras of $\mathcal{L}(H)$.

A JBW*-triple Z is called a *factor* if it contains no proper weak*-closed ideal where a subspace J of Z is called an *ideal* if $\{Z,J,Z\} + \{Z,Z,J\} \subset J$. It has been shown recentlty in [**1**] that a JBW*-triple Z is a factor if, and only if, $\|Q(x,y)\| \geq \frac{1}{6}\|x\|\|y\|$ for all $x, y \in Z$, where $Q(x,y)$ is the operator $Q(x,y)z = \{x,z,y\}$. The factors are the prime JBW*-triples. Prime Jordan triple systems have been classified by Zelmanov [**24**]. We call a symmetric domain *irreducible* if it is not biholomorphically equivalent to a direct product of two symmetric domains.

LEMMA 3.2. *Let B be the open unit ball of a JBW^*-triple Z. The following conditions are equivalent:*

i: *B is an irreducible symmetric domain;*
ii: *Z is a factor.*

Proof. $(i) \Rightarrow (ii)$. If Z contains a proper weak*-closed ideal J, then $Z = J \oplus J^{\perp}$ (ℓ_∞-sum) where $J^{\perp} = \{z \in Z : \{z, J, Z\} = \{0\}\}$ is a weak*-clsoed ideal. Hence $B \simeq B_J \times B_{J^{\perp}}$ is reducible.
$(ii) \Rightarrow (i)$. This follows from the fact that the tangent space at $(a, b) \in B_1 \times B_2$ is $T_a B_1 \oplus T_b B_2$ for any symmetric domains B_1 and B_2. □

Since a Banach space is a dual space if its closed unit ball is compact in some locally convex topology, the following result is immediate from Theorem 3.1 and Lemma 3.2.

PROPOSITION 3.3. *Let M be an irreducible bounded symmetric domain whose norm-closure is compact in some locally convex topology. Then M is biholomorphically equivalent to the open unit ball of a factor.*

There are the following six types of atomic factors, called the *Cartan factors*:

type I	$\mathcal{L}(H, K)$ *with triple product* $\{x, y, z\} = \frac{1}{2}(xy^*z + zy^*x)$
type II	$\{z \in \mathcal{L}(H) : z^t = -z\}$
type III	$\{z \in \mathcal{L}(H) : z^t = z\}$
type IV	spin factor
type V	$M_{1,2}(\mathbb{O})$ *with triple product* $\{x, y, z\} = \frac{1}{2}(x(y^*z) + z(y^*x))$
type VI	$M_3^8(\mathbb{O})$

where $\mathcal{L}(H, K)$ is the Banach space of bounded linear operators between complex Hilbert spaces H and K, and z^t is the transpose of z induced by a conjugation on H. Cartan factors of type III and VI are Jordan algebras with the usual Jordan product $x \circ y = \frac{1}{2}(xy + yx)$. A *spin factor* is a closed subspace V of $\mathcal{L}(H)$ of dimension at least 3 such that $z \in V$ implies that $z^* \in V$ and z^2 is a scalar multiple of the identity.

The finite dimensional factors are atomic and are exactly the six types of matrix spaces arising from Cartan's classification, and in finite dimensions Proposition 3.3 is exactly Cartan's classification.

4. Grassmann manifolds

The classical Grassmann manifolds can be regarded as manifolds of projections in spaces of matrices. The infinte dimensional analogues are the manifolds of projections in C*-algebras. Since projections are tripotents, the concept of Grassmann manifolds can be extended to include the manifolds of tripotents in JB*-triples (cf. [**2, 7, 14, 17, 18, 20**]). In this section, we discuss some recent results on these infinite dimensional manifolds, in particular, the manifolds of finite-rank tripotents. We will describe the affine connections, geodesics and symmetries in these manifolds as well as the sectional curvature in the rank-one case.

Let Z be a JB*-triple with a tripotent e and let

$$Z = Z_2(e) \oplus Z_1(e) \oplus Z_0(e)$$

be the Peirce decomposition. We note that $Z_2(e)$ is a Jordan algebra with Jordan product $a \circ b = \{a, e, b\}$ and involution $a^\# = \{e, a, e\}$. The *rank* of $Z_2(e)$ or e is the cardinality of any maximal orthogonal family of nonzero tripotents in $Z_2(e)$, where two elements x and y are *orthogonal* if $x\,y = 0$. Let

$$A(e) = \{z \in Z_2(e) : z^{\#} = z\}.$$

Then we have $Z_2(e) = A(e) \oplus iA(e)$. Write

$$Z = [iA(e) \oplus Z_1(e)] \oplus [A(e) \oplus Z_0(e)] = V \oplus W.$$

For $v = iv_2 + v_1 \in iA(e) \oplus Z_1(e)$, we define $\tilde{v} = \frac{i}{2}v_2 + 2v_1$ and the map

$$k_v = \tilde{v}\,e - e\,\tilde{v} : Z \longrightarrow Z.$$

We note that $\exp k_v$ is in the connected component $\mathrm{Aut}\,(Z)^\circ$ of the identity in $\mathrm{Aut}\,(Z)$. Let $\mathcal{T}$ be the subspace of Z consisting of tripotents. It has been shown in [**21**, 4.4] that $\mathcal{T}$ is a real analytic direct submanifold of Z with local chart $z \mapsto \exp k_z(e)$ at $e \in \mathcal{T}$.

Let $\mathcal{M}$ be a connected component of $\mathcal{T}$. Then every element in $\mathcal{M}$ has the same rank and the tangent space $T_e\mathcal{M}$ at $e \in \mathcal{M}$ is identified with $iA(e) \oplus Z_1(e)$ [**2**]. Thus a vector field Y on $\mathcal{M}$ can be regarded as a Z-valued map. Let $\mathfrak{X}(\mathcal{M})$ be the Lie algebra of analytic vector fields on $\mathcal{M}$.

DEFINITION 4.1. Let $E(e) : Z \longrightarrow iA(e)$ be the projection

$$E(e) = \frac{1}{2}(Q^2(e) - Q(e)).$$

Note that for $z \in Z$, one has $Q^2(e)z \in Z_2(e)$ and $Q(e)z = (Q^2(e)z)^{\#}$, so $E(e)z \in iA(e)$.

DEFINITION 4.2. Let $\mathcal{M}$ be a connected component of $\mathcal{T}$. We define an affine, torsionfree, $\mathrm{Aut}\,(Z)^\circ$-invariant connection ∇ on $\mathcal{M}$ by

$$(\nabla_X Y)_e = (E(e) + P_1(e))(Y_e'(X_e)), \qquad e \in \mathcal{M}, \qquad X, Y \in \mathfrak{X}(\mathcal{M})$$

where $Y_e' : Z \longrightarrow Z$ is the Fréchet derivative of Y at e.

Let $e \in \mathcal{M}$ and let $v \in T_e\mathcal{M} = iA(e) \oplus Z_1(e)$. Define $\gamma_{e,v} : \mathbb{R} \longrightarrow \mathcal{M}$ by

$$\gamma_{e,v}(t) = (\exp tk_v)(e) \qquad (t \in \mathbb{R}).$$

We have $\gamma_{e,v}(0) = e$ and $\dot{\gamma}_{e,v}(0) = v$. The following description of the ∇-geodesics has been proved in [**2**], using Jordan arithmetic.

THEOREM 4.1. *Let $\mathcal{M}$ be the connected component of a finite-rank tripotent in $\mathcal{T}$. Let $e \in \mathcal{M}$ and $v \in T_e\mathcal{M}$ with $v = iv_2 + v_1 \in iA(e) \oplus Z_1(e)$. Then $\gamma_{e,v}$ is a geodesic for ∇ if, and only if, either $v_2 = 0$ or $v_1 = 0$.*

EXAMPLE 4.3. Let Z be a Hilbert space. It is a JB*-triple with triple product $\{x, y, z\} = \frac{1}{2}(\langle x, y\rangle\, z + \langle z, y\rangle\, x)$. The nonzero tripotents of Z are exactly the set $\mathcal{M}$ of extreme points of its closed unit ball and are all rank 1. The Peirce spaces for a tripotent e are $Z_2(e) = \mathbb{C}e$, $Z_1(e) = \{e\}^\perp$ and $Z_0(e) = \{0\}$. The tangent space $T_e\mathcal{M}$ is $i\mathbb{R}e \oplus \{e\}^\perp$ and $\gamma_{e,v}$ is a geodesic if, and only if, $v \in i\mathbb{R}e$ or $v \in \{e\}^\perp$.

EXAMPLE 4.4. Let $Z = C(\Omega)$ be the C*-algebra of complex continuous functions on a compact Hausdorff space Ω. It is a JB*-triple with triple product $\{fgh\} = f\bar{g}h$. The tripotents are the set $\mathcal{T} = \{e \in C(\Omega) : e(1 - |e|^2) = 0\}$ and the Peirce spaces are $Z_2(e) = \{f \in C(\Omega) : e^{-1}(0) \subset f^{-1}(0)\}$ and $Z_1(e) = \{0\}$. Let $\mathcal{M}$ be a connected component of $\mathcal{T}$. The tangent space $T_e\mathcal{M}$ is $iA(e) = i\{f \in Z_2(e) : fe = ef\}$. For $v = if \in iA(e)$, we have $k_v = i(fe)$ and $\gamma_{e,v}$ is a geodesic.

Every tripotent $e \in \mathcal{M}$ induces an involutory map $\sigma_e : Z \longrightarrow Z$, defined by $\sigma_e = \mathsf{Id} - 2P_1(e)$, which preserves the manifold $\mathcal{T}$ and, as $\sigma_e(e) = e$, it also preserves the connected component $\mathcal{M}$. Thus σ_e is an involutory real analytic diffeomorphism of the manifold $\mathcal{M}$. But σ_e need not be a symmetry of $\mathcal{M}$ as e need not be an *isolated* fixed-point in general. However, if $\mathcal{M}$ is a connected component of a projection (that is, a self-adjoint idempotent) in the set $\mathcal{P}$ of projections in a C*-algebra $\mathcal{A}$, then it is a symmetric manifold with symmetry σ_e at $e \in \mathcal{M}$. Indeed, let $e \in \mathcal{P}$ with Peierce decomposition

$$\mathcal{A} = \mathcal{A}_2(e) \oplus \mathcal{A}_1(e) \oplus \mathcal{A}_0(e).$$

Then the Peirce subspaces are self-adjoint: $\mathcal{A}_k(e)^* = \mathcal{A}_k(e)$ where $\mathcal{A}_2(e) = iA(e) \oplus A(e)$ as before, and if $\mathcal{M}$ is the connected component of e in $\mathcal{P}$, then $\mathcal{M}$ is a real analytic manifold and the tangent space $T_e\mathcal{M}$ at the point e is the self-adjoint part $\mathcal{A}_1(e)_s$ of $\mathcal{A}_1(e)$ with $z \mapsto \exp k_z(e)$ as a local chart of $\mathcal{M}$ at e [**2**]. We have $P_1(e)(\mathcal{A}_1(e)_s) \subset \mathcal{A}_1(e)_s$ and we can define an affine connection ∇ on $\mathcal{M}$ in the same way as we did for manifolds of tripotents.

DEFINITION 4.5. For real analytic vector fields $X, Y : \mathcal{M} \to T\mathcal{M}$, we define

$$(\nabla_X Y)_e = P_1(e)(Y'_e(X_e)), \qquad e \in \mathcal{M}.$$

Then ∇ is a torsionfree $\mathrm{Aut}(\mathcal{A})^\circ$-invariant connection on $\mathcal{M}$.

PROPOSITION 4.2. *Let $\mathcal{M}$ be a connected component of a projection in a C*-algebra. Then the geodesics of ∇ in $\mathcal{M}$ are the curves $t \mapsto \gamma_{e,u}(t) = \exp tk_u(e)$ where $t \in \mathbb{R}$, $e \in \mathcal{M}$ and $u \in \mathcal{A}_1(e)_s$.*

PROPOSITION 4.3. *A connected component $\mathcal{M}$ of a projection in a C*-algebra is a symmetric manifold.*

Proof. For $e \in \mathcal{M}$, $\sigma_e : \mathcal{M} \to \mathcal{M}$ is a symmetry at e. Indeed, σ_e clearly preserves $\mathcal{A}_1(e)_s$ which is the tangent space $T_e\mathcal{M}$. Moreover $\sigma_e|\mathcal{A}_1(e)_s = -\mathsf{id}$ and the following diagram is commutative

$$\begin{array}{ccc} \mathcal{M} & \xrightarrow{\sigma_e} & \mathcal{M} \\ {\scriptstyle \exp k_{(\cdot)}(e)}\uparrow & & \uparrow{\scriptstyle \exp k_{s(\cdot)}(e)} \\ \mathcal{A}_1(e)_s & \xrightarrow{-\mathsf{id}} & \mathcal{A}_1(e)_s. \end{array}$$

Locally exp is a homeomorphism and so (locally) the set of fixed points for σ_e is the image by exp of the set of fixed points of $-\mathsf{id}$. Hence σ_e is a symmetry. □

We conclude with an illustrative example by applying our results to the real analytic manifold $\mathcal{N}$ of rank-one projections in the C*-algebra $\mathcal{A} = \mathcal{L}(H)$ of bounded linear operators on a Hilbert space H.

Given $a \in \mathcal{N}$, we have the Peierce decomposition $\mathcal{A} = \mathcal{A}_2(a) \oplus \mathcal{A}_1(a) \oplus \mathcal{A}_0(a)$ and there exists a unit vector $\alpha \in H$ such that $a(\cdot) = \langle \cdot, \alpha \rangle \alpha$. There is a real linear homeomorphism $x \in T_a\mathcal{N} = \mathcal{A}_1(a)_s \mapsto x(\alpha) \in \{\alpha\}^\perp$ between the tangent space and the orthogonal complement [**2**]. Hence $T_a\mathcal{N}$ is a real Hilbert space with inner product and equivalent norm

$$\langle\langle x, y\rangle\rangle = 2\,\mathrm{Re}\,\langle x(\alpha), y(\alpha)\rangle, \qquad |x|_a = \sqrt{2}\|x(\alpha)\| \qquad (x, y \in T_a\mathcal{N}).$$

The above does not depend on the choice of α representing a. We can therefore define a Riemannian metric g on $\mathcal{N}$ by

$$g(X, Y)_a = \langle\langle X(a), Y(a)\rangle\rangle, \qquad X, Y \in \mathfrak{X}(\mathcal{N}), \qquad a \in \mathcal{N}.$$

The affine connection ∇ given in Definition 4.5 is compatible with g and is therefore the unique Riemann connection of g on $\mathcal{N}$. We note that $\mathcal{N}$ can be identified, as a real manifold, with the complex projective space $\mathbb{P}(H) = \{L \subset H : \dim L = 1\}$ via the real bianalytic map $a \in \mathcal{N} \mapsto \mathbb{C}\alpha \in \mathbb{P}(H)$ where $a(\cdot) = \langle\cdot, \alpha\rangle\alpha$. If $\dim H < \infty$, then the metric g is equivalent to the Fubini-Study metric on $\mathbb{P}(H)$. Nevertheless, one can also compute directly the Riemann distance and the sectional curvature of $\mathcal{N}$, using Jordan techniques as in [**17, 18**]. The details of the following computation can be found in [**2**].

For $a = \langle\cdot, \alpha\rangle\alpha$, the state $\varphi_a\colon \mathcal{A} \to \mathbb{C}$ defined by

$$\varphi_a(z) = \langle z(\alpha), \alpha\rangle, \qquad z \in \mathcal{A}$$

is real analytic. Given $b \in \mathcal{N}$, we have $0 \le \varphi_a(b) \le 1$ and it equals 1 if, and only if, $a = b$. Also $\varphi_a(b) = 0$ if, and only if, $a \circ b = \frac{1}{2}(ab + ba) = 0$.

One can show that any two distinct points $a, b \in \mathcal{N}$, with $a \circ b \neq 0$, can be joined by a geodesic $\gamma_{a,\,x}$, given in Proposition 4.2, which is minimizing with

$$x = -2\left(\sqrt{\frac{\varphi_a(b)}{1 - \varphi_a(b)}}\right) a + \frac{2a \circ b}{\sqrt{\varphi_a(b)(1 - \varphi_a(b))}} \in \mathcal{A}_1(a)_s,$$

$\gamma_{a,x}(0) = a$ and $\gamma_{a,x}(\theta) = b$, where $\cos^2\theta = \varphi_a(b)$. As $|x|_a = \sqrt{2}$ and $k_x = 2(x \,\square\, a - a \,\square\, x)$, we have $\gamma_{a,x}(t) = (\exp tk_x)(a) = (\cos 2t)a + \frac{1}{2}(\sin 2t)x + \frac{1}{2}(1 - \cos 2t)x^2$. Therefore the Riemann distance between a and b is given by

$$d(a, b) = \int_0^\theta |\dot\gamma_{a,x}(t)|_a dt = \sqrt{2}\theta$$

where $|\dot\gamma_{a,x}(t)|_a = \sqrt{2}\|\dot\gamma_{a,x}(t)(\alpha)\| = \sqrt{2}$.

For $a \circ b = 0$, we have the same result by taking x with $x^2 = a + b$ and observing that $b = \gamma_{a,x}(\frac{\pi}{2})$ and $a \circ \gamma_{a,x}(t) \neq 0$ for $0 < t < \pi/2$ which, by previous argument, gives $d(a, b) = \lim_{t\to\pi/2} d(a, \gamma_{a,x}(t)) = (\sqrt{2}/2)\pi$.

As in [**2**], one can also compute the sectional curvature κ_a of $\mathcal{N}$ at $a = \langle\cdot, \alpha\rangle\alpha$. For $x, y \in T_a\mathcal{N}$ with $|x|_a = |y|_a = 1$ and $x \perp y$, we have

$$\kappa_a(x, y) = \langle\langle R_a(x, y)y, x\rangle\rangle$$

where $R_a(x, y)y = 4(x \circ y^2 - y \circ (x \circ y))$. Since $x = \langle\cdot, \alpha\rangle\xi + \langle\cdot, \xi\rangle\alpha$ and $y = \langle\cdot, \alpha\rangle\eta + \langle\cdot, \eta\rangle\alpha$ with $\xi = x(\alpha), \eta = y(\alpha) \in \{\alpha\}^\perp$, we have

$$\kappa_a(x,\, y) = \frac{1}{2} - 2\,\mathrm{Re}\,(\langle\eta, \xi\rangle^2) = \frac{1}{2} + 2\,(\mathrm{Im}\,\langle\eta,\, \xi\rangle)^2 \le 1$$

where $\mathrm{Re}\,\langle\eta,\,\xi\rangle = \langle\langle x,y\rangle\rangle = 0$ and $|\langle\eta,\,\xi\rangle|^2 \leq \|\eta\|^2\|\xi\|^2 = \frac{1}{4}$.

References

[1] L.J. Bunce, C-H. Chu, L. Stacho and B. Zalar, *On prime JB*-triples*, Quart. J. Math. Oxford **49** (1998) 279-290.

[2] C-H. Chu and J.M. Isidro, *Manifolds of tripotents in JB*-triples*, Math. Z. (to appear)

[3] C-H. Chu and P. Mellon, *Jordan structures in Banach spaces and symmetric manifolds*, Expos. Math. **16** (1998) 157-180.

[4] E. Cartan, *Sur les domaines bornés homogènes de l'espace de n variables complexes*, Abh. Math. Semin. Univ. Hamburg **11** (1935) 116–162.

[5] L.A. Harris, *Bounded symmetric domains in infinite dimensional spaces*, Lecture Notes in Math. **364**, Springer-Verlag (1974) 13–40.

[6] S. Helgason, Differential geometry, Lie groups and symmetric spaces, Academic Press 1980.

[7] U. Hirzebruch, *Über Jordan-Algebren und kompakte Riemannsche symmetrische Räume von Rang 1*, Math. Z. **90** (1965) 339-354.

[8] L.K. Hua, Harmonic analysis of functions of several complex variables in the classical domains, Translations of Math. Monographs 6, Amer. Math. Soc. Providence (1963).

[9] P. Jordan, J. von Neumann and E. Wigner, *On an algebraic generalisation of the quantum mechanical formalism*, Ann. of Math. **36** (1934) 29–64.

[10] W. Kaup, *Algebraic characterization of symmetric complex Banach manifolds*, Math. Ann. **228** (1977) 39-64.

[11] W. Kaup, *Über die Klassifikation der symmetrischen hermiteschen Mannigfaltigkeiten unendlicher Dimension I*, Math. Ann. **257** (1981) 463–483.

[12] W. Kaup, *Über die Klassifikation der symmetrischen hermiteschen Mannigfaltigkeiten unendlicher Dimension II*, Math. Ann. **262** (1983) 57–75.

[13] W. Kaup, *A Riemann mapping theorem for bounded symmetric domains in complex Banach spaces*, Math. Z. **138** (1983) 503–529.

[14] W. Kaup, *Über die Automorphismen Grassmannscher Mannigfaltigkeiten unendlicher Dimension*, Math. Z. **144** (1975) 75–96.

[15] M. Koecher, *Jordan algebras in differential geometry*, Proc. Internat. Congress I (Nice 1970) 279-283.

[16] O. Loos, *Bounded symmetric domains and Jordan pairs*, Lecture Notes, University of California at Irvine, 1977.

[17] T. Nomura, *Manifold of primitive idempotents in a Jordan-Hilbert algebra*, J. Math. Soc. Japan **45** (1993) 37-58.

[18] T. Nomura, *Grasmann manifold of a JH-algebra*, Ann. Global Analysis and Geometry, **12** (1994) 237-260.

[19] B. Russo, *Structure of JB^*-triples*, Proceedings of the 1992 Oberwolfach conference on Jordan algebras, Walter de Gruyter, Berlin (1994) 208-280.

[20] N. Salinas, *The Grassmann manifold of a C*-algebra and hermitian holomorphic bundles*, Oper. Theory, Adv. Appl. **28** (1988) 267-289.

[21] J. Sauter, *Randstrukturen beschränkter symmetrischer Gebiete*, Ph.D Dissertation, Universität Tübingen 1995.

[22] H. Upmeier, Symmetric Banach manifolds and Jordan C^*-algebras, North-Holland, Amsterdam, 1985.

[23] J. Wolf, *Fine structure of Hermitian symmetric spaces*, in Symmetric Spaces, Marcel Dekker (1972) 271–357.

[24] E.I. Zelmanov, *On prime Jordan triple systems*, II, III, Siberian Math. J. **24** (1983) 23-27; **25** (1984) 50-61; **26** (1985) 71-82.

GOLDSMITHS COLLEGE, UNIVERSITY OF LONDON, LONDON SE14 6NW, ENGLAND
E-mail address: maa01chc@gold.ac.uk

AMS/IP Studies in Advanced Mathematics
Volume 20, 2001

THE RATIONALITY PROBLEM OF FINITE GROUP ACTIONS

Ming-chang Kang

ABSTRACT. Let K be any field, $K(x_1, \cdots, x_n)$ the rational function field of n variables over K, and G a finite group acting on $K(x_1, \cdots, x_n)$. The main question of this article is to study whether the fixed field $K(x_1, \cdots, x_n)^G$ is rational (=purely transcendental) over K. First of all, we shall consider the case when G is the symmetric group. We shall introduce Noether's problem and related results also. Multiplicative actions will be emphasized.

1. Symmetric group actions

Let K be any field, $K[x_1, \cdots, x_n]$ be the polynomial ring in n variables over K, $K(x_1, \cdots, x_n)$ the quotient field of $K[x_1, \cdots, x_n]$. Suppose that S_n is the symmetric group of degree n acting naturally on $K[x_1, \cdots, x_n]$, i.e.

$$\sigma \cdot x_i = x_{\sigma(i)} \qquad \forall\, \sigma \in S_n \quad \forall_1 \leq i \leq n\ .$$

Thus S_n acts on $K(x_1, \cdots, x_n)$ also. It is well-known that the ring of invariants $K[x_1, \cdots, x_n]^{S_n} := \{g \in K[x_1, \cdots, x_n] : \sigma(g) = g\ \forall \sigma \in S_n\}$ is a polynomial ring and the fixed field $K(x_1, \cdots, x_n)^{S_n} := \{h \in K(x_1, \cdots, x_n) : \sigma(h) = h\ \forall \sigma \in S_n\}$ is rational (= purely transcendental) over K. In fact, if $f_1, \cdots, f_n$ denote the elementary symmetric functions in $x_1, \cdots, x_n$, i.e.

$$\begin{aligned} f_1 &:= x_1 + x_2 + \cdots + x_n\ , \\ f_2 &:= \textstyle\sum_{1 \leq i < j \leq n} x_i x_j\ , \\ &\cdots \\ f_n &:= x_1 x_2 \cdots x_n\ , \end{aligned}$$

then

$$\begin{aligned} K[x_1, \cdots, x_n]^{S_n} &= K[f_1, \cdots, f_n]\ , \text{ and} \\ K(x_1, \cdots, x_n)^{S_n} &= K(f_1, \cdots, f_n)\ . \end{aligned}$$

We may wonder whether the answer to the similar question for the alternating group is also affirmative. Namely,

Question 1.1. Let A_n be the alternating group acting on $K(x_1, \cdots, x_n)$ by

$$\sigma \cdot x_i = x_{\sigma(i)} \qquad \forall\, \sigma \in A_n \quad \forall\ 1 \leq i \leq n\ .$$

Is the fixed field $K(x_1, \cdots, x_n)^{A_n}$ rational over K ?

It is not difficult to show that $K(x_1, \cdots, x_n)^{A_n}$ is rational over K for any field K and for $n = 3, 4$. In fact, if char $K \neq 3$ and K contains a primitive cubic root of unity ω, then Burnside (1907) showed that

$$K(x_1, x_2, x_3, x_4)^{A_4} = K(s, \frac{a}{b}, \frac{c}{b}, \frac{d}{b})$$

where s, a, b, c, d were defined by

$$\begin{aligned} s &:= x_1 + x_2 + x_3 + x_4 , & z_1 &:= x_1 + x_2 - x_3 - x_4 , \\ z_2 &:= x_1 - x_2 + x_3 - x_4 , & z_3 &:= x_1 - x_2 - x_3 + x_4 , \\ a &:= z_1^2 + z_2^2 + z_3^2 , & b &:= z_1 z_2 z_3 , \\ c &:= z_1^4 + z_2^4 + z_3^4 , & d &:= (z_1^2 + \omega z_2^2 + \omega^2 z_3^2)^3 . \end{aligned}$$

On the other hand, it is late until 1989 that Maeda and Kervaire and Vust [M; KV; HK3, Example 4] independently established the rationality of the fixed field $K(x_1, \cdots, x_5)^{A_5}$. It is still unknown whether $K(x_1, \cdots, x_n)^{A_n}$ is rational or not, if $n \geq 6$. For a discussion of this problem, see [J, Exercises 3 and 4, pages 524-525].

More generally, if $\rho : S_n \to GL(V)$ is a linear representation of S_n over K and $K(V)$ is the function field of V(if $x_1, \cdots, x_n$ is a basis of the dual space V^* of V, then $K(V)$ is just the rational function field $K(x_1, \cdots, x_n)$), we may want to know under what situation the fixed field $K(V)^{S_n}$ will be rational. It is easy to show that $K(V)^{S_n}$ is rational when $n \leq 4$. By using a variant of Hilbert's Theorem 90 [HK1, Theorem 1], we can reduce the problem, for $n \geq 5$, to the following

Question 1.2. Let K be any field, $\rho : S_n \to GL(V)$ be an irreducible linear representation of S_n. Is $K(V)^{S_n}$ rational over K ?

When $K = \mathbb{C}$, Shepherd-Barron showed that $\mathbb{C}(V)^{S_5}$ was rational for any irreducible representation $\rho : S_5 \to GL(V)$[SB1]. As far as we know, the answer to Question 1.2. is still open when $n \geq 6$.

For the rationality of $K(V)^{S_n}$ for some other (reducible) representations, see [HK3; HK4]. The reader may consult [Ke] for the actions of some groups of Lie type.

Both Questions 1.1. and 1.2. are just special cases of the following rationality problem:

Rationality Problem: Let K be any field, $K(x_1, \cdots, x_n)$ the rational function field of n variables over K. If L is a field with $K \subset L \subset K(x_1, \cdots, x_n)$, is it necessary that L is rational over K ?

We would like to remark that the answer to the rationality problem is affirmative when (i) trans $\deg_K L = 1$ (Lüroth's Theorem), or (ii)$n = 2$, K is algebraically closed, and $K(x_1, x_2)$ is separable algebraic over L (Castelnuovo-Zariski's Theorem). Counter-examples exist when $K = \mathbb{C}$ and trans $\deg_{\mathbb{C}} L = 3$. For the details, see [MT; D; SB2; P] and the references therein.

2. Noether's problem

Now we turn to the case of finite group actions.

Noether's Problem. Let K be any field, G a finite abelian group, $\rho : G \to GL(V)$ the regular representation of G over K. Is $K(V)^G$ rational over K ? In particular, when $K = \mathbb{Q}$ and $G =< \sigma >$ is the cyclic group of order n, we will like to know whether $\mathbb{Q}(x_1, \cdots, x_n)^{<\sigma>}$ is rational over $\mathbb{Q}$ where

$$\sigma : x_1 \mapsto x_2 \mapsto x_3 \mapsto \cdots \mapsto x_n \mapsto x_1 .$$

Note that if K contains enough roots of unity, the answer to Noether's problem is affirmative (Fischer's Theorem).The first counter-example to Noether's problem was constructed by Swan [Sw1]. In particular, he showed that $\mathbb{Q}(x_1, \cdots, x_{47})^{<\sigma>}$ was not rational over $\mathbb{Q}$. Later, Lenstra showed that $\mathbb{Q}(x_1, \cdots, x_8)^{<\sigma>}$ was not rational either. For an account of further developments of Noether's problem due to Voskresenskii, Lenstra, Endo, Miyata, Colliot-Thélène, Sansuc, Saltman etc, see the survey papers by Swan and Colliot-Thélène [Sw2; Sw3; CT]. To conclude this section, we mention three results.

Theorem 2.1. (Saltman [Sa1; Sh]) For any prime number p, define a finite p-group G as follows: G is generated by four elements $\alpha_1, \alpha_2, \alpha_3, \alpha_4$ with relations

$$\alpha_i^p = [\alpha_i, \alpha_j]^p = [[\alpha_i, \alpha_j], \alpha_k] = [\alpha_1, \alpha_2][\alpha_3, \alpha_4] = 1$$

where $1 \leq i, j, k \leq 4$. Then the unramified Brauer group of $\mathbb{C}(V)^G$ over $\mathbb{C}$ is not zero and $\mathbb{C}(V)^G$ is not rational over $\mathbb{C}$ where $\rho : G \to GL(V)$ is any faithful representation of G over $\mathbb{C}$.

Remarks. (i) The unramified Brauer group of a field L over $\mathbb{C}$, denoted by $Br_{nr}(L/\mathbb{C})$, is defined as

$$Br_{nr}(L/\mathbb{C}) = \bigcap Br(R)$$

where R runs over all the DVR's R with $\mathbb{C} \subset R \subset L$ and the quotient field of R being L. In particular, if L is the function field of a smooth projective variety X over $\mathbb{C}$, then $Br_{nr}(L/\mathbb{C}) \simeq H^2_{\text{ét}}(X, \mathbb{G}_m)_{\text{torsion}}$; moreover, if X is unirational, then $Br_{nr}(L/\mathbb{C}) \simeq H^3(X, \mathbf{Z})_{\text{torsion}}$.

(ii) E. Beneish proves that if G is a non-abelian group of order p^3, $\rho : G \to GL(V)$ the regular representation, then $\mathbb{C}(V)^G$ is stably rational over $\mathbb{C}$. In fact, we can prove that $\mathbb{C}(W)^G$ is rational over $\mathbb{C}$ for any linear representation $G \to GL(W)$, where G is any group of order p^3 or p^4.

Theorem 2.2. (Hajja [H1]) Let G be a finite group containing a normal abelian subgroup N such that G/N is a cyclic group of order < 23. If $\rho : G \to GL(V)$ is a complex linear representation of G, then $\mathbb{C}(V)^G$ is rational.

Theorem 2.3. [AHK2] Let K be any field, V a K-vector space of dimension 2 or 3, and $\sigma \in GL(V)$. Then both $K(V)^{<\sigma>}$ and $K(\mathbb{P}(V))^{<\sigma>}$ are rational over K.

Note that Theorem 2.3. would become trivial if K contains all the eigenvalues of σ. Thus the ingredients of its proof consist of (i) Galois descent, and (ii) reducing the problem to the rationality of 2-dimensional algebraic tori (Voskresenskii's Theorem [V]). Note that Theorem 2.3. will not be valid if $\dim_K V \geq 4$, because $\mathbb{Q}(x_1, x_2, x_3, x_4)^{<\sigma>}$ is not rational over $\mathbb{Q}$ where

$$\sigma : x_1 \mapsto x_2 \mapsto x_3 \mapsto x_4 \mapsto -x_1 .$$

3. Multiplicative actions

In the proof of Theorem 2.2., multiplicative actions arise naturally.

Definition 3.1. Let K be any field, G a finite subgroup of $GL_n(\mathbf{Z})$. Then G gives rise to a multiplicative action on $K(x_1, \cdots, x_n)$ by K-automorphisms defined by

$$\sigma(x_i) := a_i(\sigma) \prod_{j=1}^{n} x_j^{\alpha_{ij}}$$

where $\sigma = (\alpha_{ij})_{1 \le i,j \le n} \in G \subset GL_n(\mathbf{Z})$ and $a_i(\sigma) \in K \backslash \{0\}$. Such an action is called a monomial action of G. If $a_i(\sigma) = 1$ for any $\sigma \in G$, any $1 \le i \le n$, then it is called a purely monomial action of G.

Example 3.2. Let a, $b \in K\backslash\{0\}$ and σ a K-automorphism acting on $K(x, y)$ by

$$\sigma(x) := \frac{a}{x}, \quad \sigma(y) := \frac{b}{y} .$$

It can be shown that the fixed field $K(x, y)^{<\sigma>}$ is rational over K [HK2, (2.7) Lemma]. In fact,

$$K(x,y)^{<\sigma>} = K\left(\frac{x - \frac{a}{x}}{xy - \frac{ab}{xy}}, \frac{y - \frac{b}{y}}{xy - \frac{ab}{xy}} \right) .$$

Note that, when a, $b \in K^2$, the rationality of $K(x, y)^{<\sigma>}$ can be established easily as follows : By replacing x and y by $\frac{x}{\sqrt{a}}$ and $\frac{y}{\sqrt{b}}$, we may assume that $\sigma(x) = \frac{1}{x}$, $\sigma(y) = \frac{1}{y}$. If char $K \neq 2$, let $u := \frac{1-x}{1+x}$, $v := \frac{1-y}{1+y}$; then $\sigma(u) = -u$, $\sigma(v) = -v$. If char $K = 2$, let $u := \frac{1}{1+x}$, $v := \frac{1}{1+y}$; then $\sigma(u) = u + 1$, $\sigma(v) = v + 1$.

We do not know the answer of the following question:

Question 3.3. Let K be any field, $a, b, c \in K\backslash 0$ and σ a K-automorphism acting on $K(x, y, z)$ by

$$\sigma(x) := \frac{a}{x}, \sigma(y) := \frac{b}{y}, \sigma(z) := \frac{c}{z} .$$

Is $K(x, y, z)^{<\sigma>}$ rational over K ?

Theorem 3.4. (i) (Hajja [H2]) For any monomial action of a finite group G on $K(x_1, x_2)$, the fixed field $K(x_1, x_2)^G$ is rational over K.

(ii) [HK1; HK2] For any purely monomial action of a finite group G on $K(x_1, x_2, x_3)$, the fixed field $K(x_1, x_2, x_3)^G$ is rational over K except for those subgroups which are conjugate in $GL_3(\mathbf{Z})$ to the subgroup generated by

$$\begin{pmatrix} 1 & 1 & 0 \\ -2 & -1 & -1 \\ 0 & 0 & 1 \end{pmatrix} \text{ and } \begin{pmatrix} -1 & -1 & -1 \\ 0 & 0 & 1 \\ 0 & 1 & 0 \end{pmatrix} .$$

We still do not know whether the fixed field under the action of this "exceptional" group is rational or not.

Note that $\mathbb{Q}(x_1, x_2, x_3)^{<\sigma>}$ is not rational over $\mathbb{Q}$ where

$$\sigma : x_1 \mapsto x_2 \mapsto x_3 \mapsto -\frac{1}{x_1 x_2 x_3} .$$

Similarly, $\mathbb{Q}(x_1, x_2, x_3, x_4)^{<\sigma>}$ is not rational either where

$$\sigma : x_1 \mapsto x_2 \mapsto x_3 \mapsto x_4 \mapsto \frac{1}{x_1} .$$

Hence the result in Theorem 3.4. (ii) cannot be strengthened.

Using previous results in Noether's problem, it is easy to see, for example, that $\mathbb{Q}(x_1, \cdots, x_{46})^{<\sigma>}$ is not rational over $\mathbb{Q}$ where

$$\sigma : x_1 \mapsto x_2 \mapsto \cdots \mapsto x_{46} \mapsto \frac{1}{x_1 x_2 \cdots x_{46}} .$$

Theorem 3.5. (Farkas [Fa1]) Let G be a finite subgroup of $GL_n(\mathbf{Z})$. Consider the purely monomial action of G on $\mathbb{C}(x_1, \cdots, x_n)$. Then $\mathbb{C}(x_1, \cdots, x_n)^G$ is rational provided that G is a finite reflection group when we regard it as a subgroup of $GL_n(\mathbb{Q})$.

Question 3.6. Let K be any field, S_n the symmetric group of degree n acting on $K(x_1, \cdots, x_n)$ with $x_1 x_2 \cdots x_n = 1$ by

$$\sigma(x_i) = \frac{1}{x_{\sigma(i)}} \qquad \forall \sigma \in S_n \backslash A_n \quad \forall 1 \leq i \leq n .$$

Is $K(x_1, \cdots, x_{n-1})^{S_n}$ rational over K ?

When $n = 4$, the above question was solved by Formanek [Fo1, Section 4] and [HK2, pages 850-855]. Formanek solved this question in order to show that the center of the ring of 4×4 generic matrices is rational. We do not know the answer of Question 3.6. when $n \geq 5$. For a survey of the rationality problem of the center of the ring of $n \times n$ generic matrices, see [Fo2; L]. For the convenience of the reader we record one of its main problems as the following Question 3.7.

Question 3.7. Let K be any field, M and N be free abelian groups defined as

$$M := \bigoplus_{1 \leq i \leq n} \mathbf{Z} \cdot x_i , \qquad N := \bigoplus_{1 \leq i,j \leq n, i \neq j} \mathbf{Z} \cdot (y_i \otimes y_j)$$

Let the symmetric group S_n acting on M and N by

$$\sigma \cdot x_i = x_{\sigma(i)} , \sigma \cdot (y_i \otimes y_j) = y_{\sigma(i)} \otimes y_{\sigma(j)} \quad \forall \sigma \in S_n .$$

Define a morphism

$$\begin{aligned} \varphi : \quad & N \longrightarrow M \\ & y_i \otimes y_j \mapsto x_i - x_j \end{aligned} \tag{1}$$

and denote by $L := \mathrm{Ker}\,(\varphi)$.

Consider the purely monomial action of S_n on $K(M \oplus L)$. Is $K(M \oplus L)^{S_n}$ rational over K ? Is it stably rational over K ?

Question 3.8. Let K be any field, $a \in K \backslash \{0\}$, and the symmetric group S_n acting on $K(x_1, \cdots, x_n)$ by

$$\sigma(x_i) := \frac{a}{x_{\sigma(i)}} \qquad \forall \sigma \in S_n \backslash A_n \quad \forall 1 \leq i \leq n .$$

Is $K(x_1, \cdots, x_n)^{S_n}$ rational over K ?

We do not know the answer of the above question even in the case $n = 3$. However, if $a \in K^2$, the answer follows easily from [HK4].

Remark 3.9. The purely monomial actions considered in this section is quite similar to, and is related to the function field of an algebraic torus. In fact, if T is a n-dimensional algebraic torus defined over K and split over L where L is a finite Galois extension of K with Galois group G, then the function field of T is the fixed field $L(x_1, \cdots, x_n)^G$ where $\rho : G \to GL_n(\mathbf{Z})$ is an integral representation and

$$\sigma(x_j) = \prod x_{ij}^{\alpha_{ij}}$$

if $\rho(\sigma) = (\alpha_{ij})_{1 \le i,j \le n} \in GL_n(\mathbf{Z})$.

All the 2-dimensional algebraic tori are rational [V], while the birational classification and Zariski cancellation of 3-dimensional algebraic tori were established by Kunyavskii [Ku]. Although all fixed fields of the 3-dimensional purely monomial actions are rational except possibly for only one case (Theorem 3.4. (ii)), there are several non-rational 3-dimensional algebraic tori.

For other directions of multiplicative group actions, the reader is recommended to consult Saltman's paper [Sa2]. For some other problems not discussed in this article, the reader may see, for examples, [AHK1; Fa2; H3; HKO].

Finally we shall give a linearization criterion for multplicative actions.

Theorem 3.10. Let G be a finite abelian group with exponent e, K be any field with char $K = 0$ or char $K = p \not| e$, M a G-lattice and $\mathbf{Z}$ the trivial G-lattice. Suppose that K contains a primitive e-th root of unity.

(i) If the action of G on $K(M \bigoplus \mathbf{Z}^r)$ can be linearized, then there exist permutation G-lattices P_1 and P_2 so that the following is a short exact sequence

$$0 \to M \to P_1 \to P_2 \to 0 \ .$$

(ii) Conversely, if there exist permutation G-lattices P_1 and P_2 so that the following is a short exact sequence

$$0 \to M \to P_1 \to P_2 \to 0 \, ,$$

then the action of G on $K(M \bigoplus \mathbf{Z}^r)$ can be linearized where $r =$ rank $(P_1)-$ rank P_2. In particular, $K(M)^G$ is stably rational over K.

Remarks. (i) The action of G on $K(N)$ can be linearized, if there is a K-vector space $\bigoplus_{i=1}^n Ku_i$ contained in $K(N)$ such that $K(N) = K(u_1, \cdots, u_n)$ with $n =$ rank (N) and G sends $\bigoplus_{i=1}^n Ku_i$ into itself. A G-lattice N is called a permutation lattice if $N \cong \bigoplus \mathbf{Z}[G/H]$ for some subgroups H. (Note that it is unnecessary to assume that H is a normal subgroup.)

(ii) The proof of the above theorem is essentially the same as those in Lemma 7 and Lemma 8 of [Sw1].

(iii) Let $G = <\sigma, \tau>$ act on $K(x, y, z)$ by

$$\sigma(x) = \tfrac{1}{x} \, , \ \sigma(y) = \tfrac{1}{z} \, , \ \sigma(z) = \tfrac{1}{y} \, ,$$
$$\tau(x) = \tfrac{1}{x} \, , \ \tau(y) = \tfrac{z}{x} \, , \ \tau(z) = \tfrac{y}{x} \, .$$

It can be shown that $K(x, y, z)^G$ is rational over K, but the action of G on $K(x, y, z)$ cannot be linearized.

References

[AHK1] H. Ahmad, M. Hajja and M. Kang, Negligibility of projective linear automorphisms, J. Alg. 199 (1998) 344-366.

[AHK2] H. Ahmad, M. Hajja and M. Kang, Rationality of some projective linear actions, to appear in "J. Alg.".

[CT] J. L. Colliot-Thélène, Birational invariants, purity and the Gersten conjecture, in "K-theory and algebraic geometry, Santa Barbara, 1992", Proc. Symp. Pure Math. vol. 58, Amer. Math. Soc., 1995, Providence.

[D] I. V. Dolgachev, Rationality of fields of invariants, in "Algebraic geometry, Bowdoin, 1985", Proc. Symp. Pure Math. vol. 46, Amer. Math. Soc., 1987, Providence.

[Fa1] D. R. Farkas, Reflection groups and multiplicative invariants, Rocky Mount. J. Math. 16 (1986) 215-222.

[Fa2] D. R. Farkas, Birational invariants of crystals and fields with a finite group of operators, Math. Proc. Cambridge Phil. Soc. 107 (1990) 417-424.

[Fo1] E. Formanek, The center of the ring of 4×4 generic matrices, J. Alg. 62 (1980) 304-319.

[Fo2] E. Formanek, Rational function fields, Noether's problem and related questions, J. Pure Appl. Alg. 31 (1984) 28-36.

[H1] M. Hajja, Rational invariants of meta-abelian groups of linear automorphisms, J. Alg. 80 (1983) 295-305.

[H2] M. Hajja, Rationality of finite groups of monomial automorphisms of $K(x, y)$, J. Alg. 109 (1987) 46-51.

[H3] M. Hajja, Linear and monomial automorphisms of fields of rational functions, in "Algebra and number theory, Fès, 1997", Marcel Dekker, 2000, New York.

[HK1] M. Hajja and M. Kang, Finite group actions of rational function fields, J. Alg. 149 (1992) 139-154.

[HK2] M. Hajja and M. Kang, Three-dimensional purely monomial group actions, J. Alg. 170 (1994) 805-860.

[HK3] M. Hajja and M. Kang, Some actions of symmetric groups, J. Alg. 177 (1995) 511-535.

[HK4] M. Hajja and M. Kang, Twisted actions of symmetric groups, J. Alg. 188 (1997) 626-647.

[HKO] M. Hajja, M. Kang and J. Ohm, Function fields of conics as invariants subfields, J. Alg. 163 (1994) 383-403.

[J] N. Jacobson, Basic algebra II, 2nd ed, Freeman, 1989, San Francisco.

[Ke] G. Kemper, A constructive approach to Noether's problem, Manu. Math. 90 (1996) 343-363.

[Ku] H. Kunyavskii, Three-dimensional algebraic tori, Selecta Math. Soviet. 9 (1990) 1-21.

[KV] M. Kervaire and T. Vust, Fractions rationnelles invariantes par un groupe fini: quelques exemples, in "DMV Sem. 13: Algebraische Transformationsgruppen und Invariantentheorie", Birkhäuser, 1989, Basel.

[L] L. Lebruyn, Centers of generic division algebras, the rationality problem 1965-1990, Israel J. Math. 76 (1991) 97-111.

[M] T. Maeda, Noether's problem for A_5, J. Alg. 125 (1989) 418-430.

[MT] Y. I. Manin and M. A. Tsfasman, Rational varieties: algebra, geometry and arithmetic, Russian Math. Survey 41 (1986) 51-116.

[P] A. V. Pukhlikov, Birational automorphisms of higher-dimensional algebraic varieties, in "Proc. ICM Berlin 1998", DMV, 1998, Berlin.

[Sa1] D. J. Saltman, Noether's problem over an algebraically closed field, Invent. math. 77 (1984) 71-84.

[Sa2] D. J. Saltman, Twisted multiplicative field invariants, Noether's problem and Galois extensions, J. Alg. 131 (1990) 535-558.

[SB1] N. I. Shepherd-Barron, Invariant theory for S_5 and the rationality of M_6, Comp. Math. 70 (1989) 13-25.

[SB2] N. I. Shepherd-Barron, Rationality of moduli spaces via invariant theory, in "Topological methods in algebraic transformation groups", Birkhäuser, 1989, Basel.

[Sh] I. R. Shafarevich, On Lürith's problem, Proc. Steklov Inst. Math. 183 (1991) 241-246.

[Sw1] R. G. Swan, Invariant rational function and a problem of Steenrod, Invent. math. 7 (1969) 148-158.

[Sw2] R. G. Swan, Galois theory, in "Emmy Noether, a tribute to her life and work", Marcel Dekker, 1981, New York.

[Sw3] R. G. Swan, Noether's problem in Galois theory, in "Emmy Noether in Bryn Mawr", Springer-Verlay, 1983, New York.

[V] V. E. Voskresenskii, Two-dimensional algebraic tori, Amer. Math. Soc. Transl. Ser. (2) 73 (1968) 190-195.

Added in proof. Saltman has solved Question 3.3.

Theorem. *Let k be a field with char $K \neq 2$ and $\sigma : K(\chi_1, \cdots, \chi_n) \longrightarrow K(\chi_1 \cdots, \chi_n)$ be a K-automorphism defined by $\sigma(\chi_i) = \alpha_i/\chi_i$ where $\alpha_i \in K - \{0\}$ for $1 \leq i \leq n$. Then $k(\chi_1, \cdots, \chi_n)^{<\sigma>}$ is rational over K if and only if* $\left[K(\alpha_1^{\frac{1}{2}}, \alpha_2^{\frac{1}{2}}, \cdots, \alpha_n^{\frac{1}{2}}) : K\right] \leq 4$. For the details, see: D. J. Saltman, *A nonrational field, answering a question of Hajja, in "Algebra and Number Theory."* edited by M. Boulagouz and J.-P. Tignol, *Lecture Notes in Pure and Appl. Math.* vol. 28, Marcel Dekker, Inc., 2000, New York.

Department of Mathematics, National Taiwan University, Taipei, Taiwan
E-mail address: kang@math.ntu.edu.tw

AMS/IP Studies in Advanced Mathematics
Volume 20, 2001

Hecke Algebras and Representation ring of Hopf Algebras

Yongchang Zhu

1. Introduction

For a group G and a subgroup B satisfying the condition that every double coset of B contains only finitely many left cosets of B, one may construct a Hecke algebra $H(G//B)$. Among various important examples; we like to mention the following three cases. $G = G(F_q)$ is a reductive algebraic group over a finite field, B is a Borel subgroup; $G = G(\mathbf{Q}_p)$ is a reductive group over a p-adic field, B is a Iwahori subgroup of G; $G = G(\mathbf{Q})$ is the group of $\mathbf{Q}$-points of a reductive algebraic group, B is a congruence subgroup. A distinguished feature of the Hecke algebra $H(G//B)$ is that it has a basis given by the double cosets of B under which the multiplication table has non-negative integer coefficients. In this work, we construct for a pair (G, B) a tensor category $\mathcal{C}(G, B)$ and prove that $H(G//B)$ is a quotient of the Grothendick ring of $\mathcal{C}(G, B)$ and the double coset basis of $H(G//B)$ are images of simple objects in $\mathcal{C}(G//B)$. In the case G is a finite group, we construct a quasi-Hopf algebra $A(G, B)$ for the pair (G, B) and prove that the module category of $A(G, B)$ is equivalent to $\mathcal{C}(G, B)$. Moreover we prove that a certain tensor sub-category of the module category of $A(G, B)$ has its Grothendick ring isomorphic to the Hecke algebra $H(G//B)$. In other words, we obtain a categorification of $H(G//B)$ in the sense of [**CF**]. In an independent work [**Sc**], a construction of quasi-bialgebra $A(G, B)$ was given under a more general setting including the situation that G is a quasibialgebra and B is a finite Hopf subalgebra (the terminology used in [**Sc**] is dual to that used here). We also discuss the case $G = GL_n(F_q)$ and B is a Borel subgroup in detail and obtain an interesting extension of $H(G//B)$. We hope to study the third case in a separated work.

Throughout this paper, we assume that (G, B) satisfies the condition at the beginning. The main results of this paper was given in author's talk at ICCM Beijing 1998.

2. A tensor category related to Hecke algebras

We first recall the definition of the Hecke algebra $H(G//B)$ (c.f.[**Sh**]). Let $G = Bw_1B \cup Bw_2B \cup \dots$ be a double coset decomposition of G. $H(G//B)$ is a $\mathbf{Z}$-algebra with a basis $[Bw_iB]$. To define the multiplication, we decompose Bw_iB

and Bw_jB as the disjoint union of left cosets:

$$\begin{aligned} Bw_iB &= u_1B \cup u_2B \cup \cdots \cup u_sB \\ Bw_jB &= v_1B \cup v_2B \cup \cdots \cup v_tB \end{aligned}$$

let c_{ij}^k be the number of pairs (u_α, v_β) such that $u_\alpha v_\beta \in w_kB$. One can prove that c_{ij}^k depends only on the double cosets Bw_iB, Bw_jB and Bw_kB but not on the choices of various representatives and for given i, j, c_{ij}^k are all 0 except finitely many k's. The multiplication is given by

$$[Bw_iB]\,[Bw_jB] = \sum_k c_{ij}^k [Bw_kB]. \tag{1}$$

It is clear that $[B] = [BeB]$ is the identity element.

We sketch a proof of that the above multiplication rule is associative. Our proof is more complicated than the standard proof such as in [**Sh**], but it will lead to a construction of a tensor category $\mathcal{C}(G, B)$ related to $H(G//B)$.

We choose a set of the double coset representatives $W = \{w_1, w_2, \dots\}$ with $e \in W$. So we have the decomposition

$$G = Bw_1B \cup Bw_2B \cup \dots .$$

We also choose a set of left coset representatives $R = \{u_1, u_2, \dots, \}$ such that $W \subset R$. Every element $x \in G$ can written as as

$$x = [x]\{x\}^{-1}$$

for unique $[x] \in R$ and $\{x\} \in B$. The following identities are used often later.

$$[x[y]] = [xy] \tag{2}$$

$$\{xb\} = b^{-1}\{x\} \tag{3}$$

$$\{b^{-1}[bu]\} = \{bu\}^{-1} \tag{4}$$

(2) holds because G acts on R from the left (this is the same action as the G-action on G/B). (3) is straightforward to prove. (4) can be proved using $[bu] = bu\{bu\}$.

For $w \in W$, we put $B_w = wBw^{-1} \cap B$, and for simplicity we write B_i for B_{w_i}. B_w is the isotropy subgroup of w for the B-action on R. We define for each $w_i, w_j, w_k \in W$, a set C_{ij}^k

$$C_{ij}^k = \{(g_1, g_2) \in B^2 \mid [g_1w_i][g_2w_j] \in w_kB\}$$

It clear that $B_i \times B_j$ acts on C_{ij}^k from the right by

$$(g_1, g_2) \cdot (b_i, b_j) = (g_1b_i, g_2b_j)$$

and B_k acts on C_{ij}^k from the left by

$$b \cdot (g_1, g_2) = \left(bg_1,\ \{b[g_1w_j]\}^{-1}g_2\right). \tag{5}$$

The formula (5) is derived by the following consideration, for $(g_1, g_2) \in C_{ij}^k$ and $b \in B_k$, we have $[g_1w_i][g_2w_j] \in w_kB$, therefore $b[g_1w_i][g_2w_j] \in w_kB$, using

$$b[g_1w_i][g_2w_j] = [bg_1w_i]\{b[g_1w_j]\}^{-1}[g_2w_j] \in [bg_1w_i]\left[\{b[g_1w_j]\}^{-1}g_2w_j\right]B$$

we see that the right hand side of (5) is in C_{ij}^k. It is easy to prove that the right action of $B_i \times B_j$ and the left action of B_k commute. So C_{ij}^k is a B_k-set-$B_i \times B_j$. Notice that C_{ij}^k is isomorphic to a disjoint union of copies $B_i \times B_j$ as a right

$B_i \times B_j$-set and is isomorphic to a disjoint union of copies B_k as a left B_k-set. We have

$$c_{ij}^k = |C_{ij}^k / G_i \times G_j|$$

The associativity of (1) follows the fact that, for given i, j, k, t, there exists an isomorphism of sets

$$\bigcup_s C_{sk}^t \times_{B_s} C_{ij}^s \simeq \bigcup_s C_{is}^t \times_{B_s} C_{jk}^s. \tag{6}$$

where symbol $\times_{B_s}$ means that for a right B_s-set U and a left B_s-set V, $U \times_{B_s} V$ is the set of the equivalence classes in $U \times V$ with the equivalence relation generated by $(xg, y) \sim (x, gy)$ for $(x, y) \in U \times V$ and $g \in B_s$.

Now lets give an explicit isomorphism of the sets in (6). For given i, j, k, t, we introduce two sets,

$$\begin{aligned} C_{(i,j),k}^t &= \{(g_1, g_2, g_3) \in B^3 \mid [[g_1 w_i][g_2 w_j]]\,[g_3 w_k] \in w_t B\} \\ C_{i,(j,k)}^t &= \{(g_1, g_2, g_3) \in B^3 \mid [g_1 w_i]\,[[g_2 w_j][g_3 w_k]] \in w_t B\} \end{aligned}$$

$C_{(i,j),k}^t$ and $C_{i,(j,k)}^t$ are B_t-sets-$B_i \times B_j \times B_k$, where the right $B_i \times B_j \times B_k$-actions are the multiplication from the right, the left action of B_t on $C_{(i,j),k}^t$ is given by

$$b \cdot (g_1, g_2, g_3) = (bg_1, \{b[g_1 w_i]\}^{-1} g_2,\ \{b[[g_1 w_i][g_2 w_j]]\}^{-1} g_3\})$$

and the left action of B_t on $C_{i,(j,k)}^t$ is given by

$$b \cdot (g_1, g_2, g_3) = (bg_1, \{b[g_1 w_i]\}^{-1} g_2,\ \left\{\{b[g_1 w_i]\}^{-1}[g_2 w_j]\right\}^{-1} g_3\})$$

We derive these two formulas in a similar way as driving (5).

We define a map $C_{sk}^t \times C_{ij}^s \to C_{(i,j),k}^t$, for $(g, g_3) \in C_{sk}^t$ and $(g_1, g_2) \in C_{ij}^s$, since

$$[g_1 w_i][g_2 w_j] \in w_s B$$

we have

$$[g[g_1 w_i][g_2 w_j]] = [g w_s]$$

therefore

$$[g[g_1 w_i][g_2 w_j]]\,[g_3 w_k] \in w_t B \tag{7}$$

(7) implies that $(gg_1, \{g[g_1 w_i]\}^{-1} g_2, g_3) \in C_{(i,j),k}^t$. So we have a map $C_{sk}^t \times C_{ij}^s \to C_{(i,j),k}^t$ given by

$$((g, g_3), (g_1, g_2)) \mapsto (gg_1, \{g[g_1 w_i]\}^{-1} g_2, g_3)$$

It is easy to prove this map induces an isomorphism of B_t-sets-$B_i \times B_j \times B_k$

$$f_1 : \bigcup_s C_{sk}^t \times_{B_s} C_{ij}^s \to C_{(i,j),k}^t.$$

Similarly, we can prove that the map $C_{is}^t \times C_{jk}^s \to C_{i,(j,k)}^t$ given by

$$((g_1, g), (g_2, g_3)) \mapsto (g_1, gg_2, \{g[g_2 w_j]\}^{-1} g_3)$$

induces an isomorphism of B_t-sets-$B_i \times B_j \times B_k$

$$f_2 : \bigcup C_{is}^t \times_{B_s} C_{jk}^s \to C_{i,(j,k)}^t$$

Finally the map $f : C^t_{(i,j),k} \to C^t_{i,(j,k)}$ given by

$$(g_1, g_2, g_3) \mapsto (g_1, g_2, \{[g_1 w_i][g_2 w_j]\} g_3)$$

is also an isomorphism of B_t-sets-$B_i \times B_j \times B_k$. Therefore

$$\Phi = f_2^{-1} \circ f \circ f_1 : \ \cup_s C^t_{sk} \times_{B_s} C^s_{ij} \to \cup_s C^t_{is} \times_{B_s} C^s_{jk}. \tag{8}$$

is an isomorphism of B_t-sets-$B_i \times B_j \times B_k$. This implies the associativity $\sum_s c^s_{ij} c^t_{sk} = \sum_s c^t_{is} c^s_{jk}$.

In fact, we get more from the above proof. Let $\mathcal{C}(G,B)$ be the direct sum of the categories of finite dimensional representations of B_i. So an object in $\mathcal{C}(G,B)$ is of the form $V_1 \oplus V_2 \oplus \ldots$, where V_i is a finite dimensional representation of B_i and all V_i except finitely many are 0. A morphism from $V_1 \oplus V_2 \oplus \ldots$ to $W_1 \oplus W_2 \oplus \ldots$ is $f_1 \oplus f_2 \oplus \ldots$ where f_i is a morphism from V_i to W_i of representations of B_i.

For a representation V_i of B_i, we use the same symbol V_i to denote $\cdots \oplus 0 \oplus V_i \oplus 0 \oplus \ldots$. Sets C^k_{ij} are used to define a tensor product $\bar{\otimes}$ in $\mathcal{C}(G,B)$. For a representation V_i of B_i and a representation V_j of B_j, let

$$V_i \bar{\otimes}_k V_j = \mathbf{C}C^k_{ij} \otimes_{\mathbf{C}B_i \times B_j} (V_i \otimes V_j) \tag{9}$$

where $\mathbf{C}C^k_{ij}$ is a $\mathbf{C}B_k$-module-$\mathbf{C}B_i \times B_k$ by the linearization. $V_i \bar{\otimes}_k V_j$ is a representation of B_k. We define a tensor product $\bar{\otimes}$ in $\mathcal{C}(G,B)$ by

$$V_i \bar{\otimes} V_j = \oplus_k V_i \bar{\otimes}_k V_j \tag{10}$$

THEOREM 2.1. *$\mathcal{C}(G,B)$ is a tensor category with the above tensor product. The associativity constraint is induced by Φ in (8).*

It is easy to prove that the trivial representation of B plays the role of a unit in this tensor category. And the isomorphism Φ in (8) gives an isomorphism from

$$\Phi : \ (V_i \bar{\otimes} V_j) \bar{\otimes} V_k \to V_i \bar{\otimes} (V_j \bar{\otimes} V_k). \tag{11}$$

To prove Φ satisfy the pentagon identity, we may follow the remarks after the proof of Theorem 3.2 below. And if we make a different choice for W and R, the resulting tensor category is equivalent to the present one, see also the remarks in next section.

PROPOSITION 2.2. *The following map gives a surjective homomorphism from the Grothendick ring of $\mathcal{C}(G,B)$ to the Hecke algebra $H(G//B)$,*

$$V_i \mapsto dim(V_i)\,[Bw_iB]$$

where V_i is a finite dimensional representation of B_i.

We remark that if we replace $\mathcal{C}(G,B)$ by the direct sum of the categories of B_i-sets (or finite B_i-sets) and define the tensor product in a similar way as (9) and (10), we again get a tensor category.

3. Quasi-Hopf algebras $A(G,B)$

Let G be a finite group and $B \subset G$ be a subgroup. We construct a quasi-Hopf algebra $A(G,B)$ for the pair (G,B) and show that the module category of $A(G,B)$ is equivalent to $\mathcal{C}(G,B)$ constructed in the last section. Our construction for $A(G,B)$ also makes sense for infinite groups, if one interprets certain infinite sums properly.

We refer to [**Sc**] for a more general construction and conceptual reasons behind it. And it is expected that under some favorable conditions, the construction in [**Sc**] also has an antipode.

We first recall the definition of a quasi-Hopf algebra, which is appears first in [**Dr1**] and [**Dr2**].

DEFINITION 3.1. A quasi-bialgebra over $\mathbf{C}$ is an algebra H over $\mathbf{C}$ together with homomorphisms $\epsilon : H \to \mathbf{C}$ (the counit) and $\Delta : H \to H \otimes H$ (comultiplication), and an invertible element $\Phi \in H \otimes H \otimes H$ (the coassociator), satisfying the following conditions

$$(1 \otimes \Delta)(\Delta(a)) = \Phi \cdot (\Delta \otimes 1)(\Delta(a)) \cdot \Phi^{-1} \tag{12}$$

for all $a \in H$.

$$(1 \otimes 1 \otimes \Delta)(\Phi) \cdot (\Delta \otimes 1 \otimes 1)(\Phi) = (1 \otimes \Phi) \cdot (1 \otimes \Delta \otimes 1)(\Phi) \cdot (\Phi \otimes 1); \tag{13}$$

$$(\epsilon \otimes 1) \circ \Delta = (1 \otimes \epsilon) \circ \Delta = 1; \tag{14}$$

$$(1 \otimes \epsilon \otimes 1)(\Phi) = 1. \tag{15}$$

A quasi-Hopf algebra is a quasi-bialgebra H equipped with a map $S : H \to H$ (antipode), and elements $\alpha, \beta \in H$, such that S is an algebra anti-morphism and for all $x \in$ with $\Delta x = x_{(1)} \otimes x_{(2)}$, we have

$$\sum S(x_{(1)})\alpha x_{(2)} = \epsilon(x)\alpha, \ \sum x_{(1)}\beta S(x_{(2)}) = \epsilon(x)\beta; \tag{16}$$

and

$$\sum x_i \beta S(y_i)\alpha z_i = 1, \text{ where } \Phi = \sum x_i \otimes y_i \otimes z_i = 1, \tag{17}$$

$$\sum S(\bar{x}_i)\alpha \bar{y}_i \beta S(\bar{z}_i) = 1, \text{ where } \Phi^{-1} = \sum \bar{x}_i \otimes \bar{y}_i \otimes \bar{z}_i = 1. \tag{18}$$

There exists a common set of representatives $R = \{u_1, u_2, \ldots, u_N\}$ of both left coset G/B and right coset $B\backslash G$ and we may put $u_1 = e$. We fix such a choice of R once for all. Every element $x \in G$ can written as as

$$x = [x]\{x\}^{-1}$$

for unique $[x] \in R$ and $\{x\} \in B$. Since R is a set of representatives for both left cosets and right cosets, the map $u_i \mapsto [u_i^{-1}]$ is a permutation on R. The action of G on G/B induces an action of G on R, this action is given by $x \cdot [y] = [xy]$. We identify $\mathbf{C}R$ with the space of the functions on R, then $\mathbf{C}R$ is an algebra with multiplication $u_i u_j = \delta_{i,j} u_i$. The G-action on R induces a G-action on $\mathbf{C}R$, and this action preserves the algebra structure. The cross product algebra $\mathbf{C}R \otimes \mathbf{C}B$ has a basis $R \times B = \{(u, a) = u \times a \,|\, u \in R, a \in B\}$ and multiplication

$$(u, a)(v, b) = \delta_{u,[av]}(u, ab) = \delta_{uB,avB}(u, ab). \tag{19}$$

The identity element is $1 = \sum(u_i, e)$.

We define on $\mathbf{C}R \otimes \mathbf{C}B$ the comultiplication Δ, counit ϵ and antipode S as follows. For $(u, a) \in R \times B$,

$$\begin{aligned}
\Delta(u,a) &= \sum_{i=1}^{N} (u_i, a) \otimes ([u_i^{-1}u], \{a^{-1}u_i\}), \\
\epsilon(u,a) &= \delta_{u,e}, \\
S(u,a) &= ([[a^{-1}u]^{-1}], \{a^{-1}u\}^{-1}).
\end{aligned}$$

THEOREM 3.2. *$\mathbf{C}R \otimes \mathbf{C}B$ is a quasi-Hopf algebra with the above multiplication, comultiplication, counit, antipode and the following structural elements:*

$$\Phi = \sum_{i,j,k} (u_i, e) \otimes (u_j, e) \otimes (u_k, \{u_i u_j\}),$$

$$\alpha = 1, \quad \beta = \sum_i (u_i, \{u_i^{-1}\})$$

We denote the quasi-Hopf algebra in the theorem by $A(G, B, R)$. And we remark that if R is a set of left coset representatives but not a set of right coset representatives, $A(G, B, R)$ is still a quasi-Hopf algebra with the same formulas for multiplication, comultilplication, counit and Φ, but the formulas for S, α and β take more complicated form. For two choices R and R', $A(G, B, R)$ and $A(G, B, R')$ are quasi-isomorphic.

Proof of Theorem 3.2. We first prove Δ is an algebra morphism. It is clear that $\Delta 1 = 1 \otimes 1$. For $(u, a), (b, v) \in R \times B$, we have

$$\begin{aligned}
&\Delta(u,a)\Delta(v,b) \\
= &\left(\sum_i (u_i, a) \otimes ([u_i^{-1}u], \{a^{-1}u_i\})\right)\left(\sum_j (u_j, b) \otimes ([u_j^{-1}v], \{b^{-1}u_j\})\right) \\
= &\sum \delta_{[[au_j]^{-1}u]B, \{a^{-1}[au_j]\}[u_j^{-1}v]B} \, ([au_j], ab) \otimes ([[au_j]^{-1}u], \{a^{-1}[au_j]\}\{b^{-1}u_j\})
\end{aligned}$$

Using (4), it is easy to prove

$$\delta_{[[au_j]^{-1}u]B, \{a^{-1}[au_j]\}[u_j^{-1}v]B} = \delta_{uB, avB},$$

this together with the identity

$$\{au_i\}^{-1}\{b^{-1}u_j\} = \{b^{-1}a^{-1}[au_i]\}$$

imply that

$$\Delta(u,a)\Delta(v,b) = \sum \delta_{uB,avB} \, (u_i, ab) \otimes ([u_i^{-1}u], \{b^{-1}a^{-1}u_i\}).$$

The right hand side is precisely $\Delta((u, a)(v, b))$. So Δ is an algebra morphism.

To verify (12), we note that Φ has inverse

$$\Phi^{-1} = \sum (u_i, e) \otimes (u_j, e) \otimes (u_k, \{u_i u_j\}^{-1}).$$

We first get

$$\Phi \cdot (\Delta \otimes 1)\Delta(u,a) = \sum (u_i, a) \otimes (u_j, \{a^{-1}u_i\}) \otimes ([u_j^{-1}u_i^{-1}u], \{a^{-1}u_i u_j\}),$$

then we compute

$$\Phi \cdot (\Delta \otimes 1)\Delta(u,a) \cdot \Phi^{-1} = \sum (u_i, a) \otimes (u_j, \{a^{-1}u_i\}) \otimes$$
$$([u_j^{-1}u_i^{-1}u], \{a^{-1}u_iu_j\}\{[a^{-1}u_i][\{a^{-1}u_i\}^{-1}u_j]\}^{-1}). \tag{20}$$

It is easy to prove that

$$\{a^{-1}u_iu_j\}\{[a^{-1}u_i][\{a^{-1}u_i\}^{-1}u_j]\}^{-1} = \{\{a^{-1}u_i\}^{-1}u_j\}.$$

The right hand side of (20) is $(1 \otimes \Delta)\Delta(u,a)$. This proves (12).

A direct calculation shows that the both sides of (13) is

$$\sum (u_i, e) \otimes (u_j, e) \otimes (u_k, \{u_iu_j\}) \otimes (u_l, \{u_iu_ju_k\}).$$

And it is easy to prove (14) and (15).

The first identity of (16) is easy to verify. To prove the second of (16), for $x = (u,a) \in R \times B$, we have

$$x_{(1)}\beta S(x_{(2)}) =$$
$$\sum \delta_{u_iB, a\{[a^{-1}u_i]^{-1}\}[\{a^{-1}u_i\}^{-1}[u_i^{-1}u]]^{-1}B}\, (u_i, a\{[a^{-1}u_i]^{-1}\}\{\{b^{-1}u_i\}^{-1}[u_i^{-1}u]\}^{-1}).$$

When $u = e$,it can be proved that

$$a\{[a^{-1}u_i]^{-1}\}[\{a^{-1}u_i\}^{-1}[u_i^{-1}u]]^{-1}B = u_iB. \tag{21}$$

For any given $u_i \in R$ and $a \in B$, the map

$$uB \mapsto a\{[a^{-1}u_i]^{-1}\}[\{a^{-1}u_i\}^{-1}[u_i^{-1}u]]^{-1}B.$$

is a permutation on G/B. So (21) implies that

$$\delta_{u_iB, a\{[a^{-1}u_i]^{-1}\}[\{a^{-1}u_i\}^{-1}[u_i^{-1}u]]^{-1}B} = \delta_{u,e}.$$

From this, it is easy to prove

$$x_{(1)}\beta S(x_{(2)}) = \sum (u_i, \{u_i^{-1}\}) = \delta_{u,e} = \beta\epsilon(x).$$

And it is easy to verify (17) and (18). This completes the proof.

If G is an infinite group, assume there is a choice of R that is a set of representatives for both left coset and right coset. Let $\mathbf{C}R$ be the space of finite linear combinations of elements in R, and $\mathbf{C}R$ is an algebra with multiplication $u_iu_j = \delta_{i,j}u_i$. We again let $A(G,B) = \mathbf{C}R \otimes \mathbf{C}B$ be the cross product algebra. And we define all the formulas as before. Though now Δ, Φ and β invovle infinite sums, the identities in the definition of a quasi-Hopf algebra make sense and hold for them. If M is a finite dimensional $A(G,B)$-module, then all (u_i, e) but finitely many act on M as 0. So the actions of Δ, Φ and β on finite dimensional modules (or tensor products of them) are finite sums. So $A(G,B)$ is an infinite dimensional quasi-Hopf algebra in certain more general sense, in particular, the category of finite dimensional $A(G,B)$-modules is a tensor category.

THEOREM 3.3. *The category of finite dimensional $A(G,B)$-modules is equivalent to $\mathcal{C}(G,B)$ as a tensor category.*

The equivalence is given as follows. For a finite dimensional B_i-module V_i, let $\bar{V}_i = \mathbf{C}B \otimes_{\mathbf{C}B_i} V_i$. It clear that $\bar{V}_i$ is finite dimensional and $A(G,B)$ acts as

$$(u,b) \cdot (b' \otimes v) = \delta_{uB, bb'w_iB}\, bb' \otimes v.$$

Conversely, for a finite dimensional $A(G,B)$-module V, let $V_i = (w_i, e)V$, then V_i is a B_i-module, and all V_i except finitely many are 0, so $\oplus_i V_i$ is an object in $\mathcal{C}(G,B)$.

Lets return to the case G is finite. Let L_i denote a $\mathbf{C}B_i$-module isomorphic to $\mathbf{C}B_i$ (with the $\mathbf{C}B_i$-action given by the left multiplication), let $\bar{L}_i$ denote the corresponding $A(G,B)$-module.

THEOREM 3.4. *Let $\mathcal{H}(G,B)$ denote the sub-category of the module category of $A(G,B)$ which has objects isomorphic the direct sums of $\bar{L}_i$'s, then $\mathcal{H}(G,B)$ is a tensor sub-category, and its Grothendick ring is isomorphic to $H(G//B)$ by the map*

$$\bar{L}_i \mapsto |B_i|\,[Bw_iB].$$

In particular, the multiplication table for the basis $\{|B_i|[Bw_iB]\}$ has coefficients in $\mathbf{Z}_{\geq 0}$.

The category of $\mathcal{H}(G,B)$ is a categorification of thee Hecke algebra $H(G//B)$. This result can generalized to the Hecke algebras in [**BLM**].

4. Examples

In this section, We study the case $G = GL_n(F_q)$ and B is the subgroup consisting of the invertible upper triangular matrices. Let $T \subset B$ be the subgroup consisting of invertible diagonal elements. All the results in this section can be generalized to other reductive groups split on F_q. We shall give a description of certain quotient ring of the representation ring of $A(G,B)$.

Let $W \subset GL_n(F_q)$ be the set of permutation matrices. Recall we have the Bruhat decomposition: $G = \cup_{w\in W} BwB$. The Hecke algebra $H(G//B)$ can be described by the generators and relations (c.f.[**Iw**]). $H(G//B)$ is generated by $\sigma_1, \sigma_2, \ldots, \sigma_{n-1}$ (σ_i is the double coset of the permutation $(i, i+1)$), with the relations

$$\sigma_i^2 = q + (q-1)\sigma_i,$$
$$\sigma_i\sigma_j = \sigma_j\sigma_i, \quad \text{if} \quad |i-j| > 1$$
$$\sigma_i\sigma_{i+1}\sigma_i = \sigma_{i+1}\sigma_i\sigma_{i+1}.$$

For each $w \in W$, It is easy to show that $B_w = wBw^{-1} \cap B$ is given by

$$B_w = \{A \in B \,|\, a_{ij} = 0 \text{ if } i < j \text{ and } w^{-1}(i) > w^{-1}(j)\}$$

Put

$$N_w = \{A \in B \,|\, a_{ii} = 1; \text{ and}\, a_{ij} = 0 \text{ if } i < j \text{ and } w^{-1}(i) < w^{-1}(j)\}$$

N_w is a subgroup of B, and every $x \in B$ can be uniquely written as $x = n_w b_w$ for $n_w \in N_w$ and $b_w \in B_w$. Therefore $R = \cup_{w\in W} N_w w$ is a set of representatives of G/B. Under this choice of R, we have a quasi-Hopf algebra $A(G,B) = A(G,G,R)$ as in Section 3.

$A(G,B)$ has a subalgebra $A(G,B)_T$ given by

$$A(G,B)_T = \mathbf{C}\{(u,t) \mid u \in R,\, t \in T\}.$$

It is easy to verify that

$$\Delta A(G,B)_T \subset A(G,B)_T \otimes A(G,B)_T. \tag{22}$$

Though $A(G,B)_T$ is not a quasi-Hopf subalgebra (since Φ is not in $A(G,B)_T^{\otimes 3}$), we can define an equivalence relation on $A(G,B)$-modules, for $A(G,B)$-modules W,V, $W \sim V$ iff W and V are isomorphic as $A(G,B)_T$-module. Because of (22), we have if $W_1 \sim W_2$ and $V_1 \sim V_2$ then $W_1 \otimes V_1 \sim W_2 \otimes V_2$. So the Representation ring of $A(G,B)$ has a quotient ring which we denote by $H_T(G//B)$.

This is easy to describe the set of the equivalence classes. First we identify the set of classes of irreducible representations of T with $L/(q-1)L$, where $L = \mathbf{Z}^n$ is the weight lattice. For each $v \in L/(q-1)L$, let M_v be the corresponding 1-dimensional T-module, and we view M_v as a B_w-module via the obvious morphism $B_w \to T$. As in Section 3, B_w-module M_v gives arise to a $A(G,B)$-module which we denote by $M(w,v)$. It is easy to prove that every $A(G,B)$-module M is isomorphic to the direct sums of $M(w,v)$ as a $A(G,B)_T$-module. Therefore $H_T(G//B)$ has a basis given by the set of pairs (w,v) ($w \in W$, $v \in L/(q-1)L$), under this basis the multiplication table has coefficients in $\mathbf{Z}_{\geq 0}$. Using the similar method as in [**Iw**], we can prove

THEOREM 4.1. *$H_T(G//B)$ is generated by $T(v)$ ($v \in L/(q-1)L$), $\sigma_i (i = 1,2,\ldots,n-1)$ with relations*

$$\begin{aligned}
&T(v_1)T(v_2) = T(v_1+v_2),\\
&\sigma_i T(v) = T(\sigma_i(v))\sigma_i,\\
&\sigma_i^2 = \sum_{k=0}^{q-1} T(k\lambda_i) + \sum_{k=1}^{q-1} \sigma_i T(k\lambda_i),\\
&\sigma_i\sigma_j = \sigma_j\sigma_i \quad \text{if } |i-j|>1,\\
&\sigma_i\sigma_{i+1}\sigma_i = \sigma_{i+1}\sigma_i\sigma_{i+1}.
\end{aligned}$$

Where λ_i is the i-th simple root, and σ_i acts on $L/(q-1)L$ by interchanging the i-th and $(i+1)$-th components.

Note that there is a surjective morphism $H_T(G//B) \to H(G//B)$ given by $T(v) \mapsto 1$, $\sigma_i \mapsto \sigma_i$. This morphism is compatible with the morphism in Proposition 2.2.

References

[BLM] Beilinson,A., Lusztig, G., MacPherson,R., A geometric setting for the quantum eformations of GL_n. *Duke Math. J.* **61** (1990), 655-77.

[CF] Crane, L., Frenkel, I., Four dimensional topological quantum field theory, Hopf categories, and the canonical bases, *J. Math. Phys.* **35**(10) (1994), 5136-45.

[DPR] Dijkgraaf,R., Pasquier, V., Roche,R., Quasi-quantum groups related to orbifold models, in *Integral systems and quantum groups, edited by M.Carfora, M.Martinelli, A.Marzuoli,* World Scientific. (1992).

[Dr1] Drinfel'd, V., Quasi-Hopf Algebras and the Knizhnik-Zamolodchiknov equations, in *Problems in modern quantum field theory, edited by Belavin, Klimyk, Zamolodchikov* **1** (1989), Springer.

[Dr2] Drinfel'd, V., Quasi-Hopf Algebras, *Leningrad Math. J.* **1** (1990), 1419-57.

[Iw] Iwahori,N., Generalized Tits systems (Bruhat decomposition) on p-adic semisimple groups, *Algebraic groups and discontinuous subgroups, Proc. Sym. Pure Math.* **9** (1966), AMS.

[Sc] Scha··enburg, P. *Hopf Bimodules, Coquasibialgebras, and an Exact Sequence of Kac* , Ph.D. thesis, Universität München, 1999.

[Sh] Shimura, G. *Introduction to the Arithmetic Theory of Automorphic Functions* (1971), Princeton U. Press.

DEPT. OF MATH., HKUST, CLEAR WATER BAY, HONG KONG

AMS/IP Studies in Advanced Mathematics
Volume 20, 2001

On the Euler numbers of threefolds

Mei-Chu Chang

Let X be a smooth, connected, compact, complex manifold of dimension n, and let T_X be the tangent bundle of X. The ith *Chern class* of X, by definition, is the ith Chern class of T_X, i.e.

$$c_i(X) := c_i(T_X) \in H^{2i}(X, \mathbb{Z}).$$

For Chern classes with complementary codimensions, the cup product gives a *Chern number* in $H^{2n}(X, \mathbb{Z}) \cong \mathbb{Z}$. The *Euler number* is the top Chern class of X, i.e.

$$e(X) := c_n(T_X) \in H^{2n}(X, \mathbb{Z}) \cong \mathbb{Z}.$$

The *canonical bundle* of X is $K_X := \Lambda^n(T_X^*)$. If $\Gamma(mK_X)$ is large, then it defines a map $\varphi_m : X \to \varphi_m(X) \subset \mathbb{P}^N$. The *Kodaira dimension* of X is

$$\kappa(X) := \begin{cases} \max_m \varphi_m(X) & \text{if } \varphi_m(\dot{X}) \neq \emptyset \text{ for some } m \\ -\infty & \text{otherwise} \end{cases}$$

We say X is of *general type* if $\kappa(X) = \dim X$. From now on, we assume that X is a threefold (i.e. $n = 3$).

DEFINITION 1. *X is a Calabi-Yau if X is simply connected and $K_X = 0$.*

The bounding of the Euler number of a threefold X is the finiteness problem if X is Calabi-Yau, and is the threefold geography if X is of general type. In either case, very little has been done.

1. The finiteness problem of Calabi-Yau threefolds.

The interest of bounding the Euler numbers of Calabi-Yau threefolds was started by the physicists, because in superstring theory, the number of generations of quarks is $\frac{|e(X)|}{2}$. The finiteness problem is the following

Problem. Is $e(X)$ bounded by a constant?

S. Katz conjectured that this constant is 960, because Philip Candelas computed the Hodge numbers of 7555 Calabi-Yau threefolds which are desingularizations of hypersurfaces in the weighted projective four space and the largest Euler number on the list is 960. (Other similar computations were done by Klemm-Schimmvigk and Kreuzer-Skarke). From the algebraic geometric viewpoint, the following related problem is also interesting [R].

Problem. Are there only finitely many components of the moduli space of Calabi-Yau threefolds?

M. Gross [G] proved that it is true for Calabi-Yau threefolds with elliptic fibrations (i.e. those mapped to a surface with elliptic curves as general fibers).

In [CK] and [CL] we study Calabi-Yau threefolds in $\mathbb{P}^N$ for $N \geq 6$, (Ballico and Chiantini proved that Calabi-Yau threefolds in $\mathbb{P}^5$ are complete intersections), and bound the Euler number by a constant multiple of N. More precisely, we have the following

THEOREM 1. [CL] *Let X be a smooth irreducible Calabi-Yau threefold and L a globally generated line bundle on X such that the map $\varphi : X \to \mathbb{P}H^0(L)$ is birational and its image has a singular locus of dimension at most one. Then*

$$e(X) \geq -\min\{2d + 58N + 80 + 4\varepsilon, 8d + 28N + 92 + 6\varepsilon\},$$

where $d = L^3$ *and* $\varepsilon = \begin{cases} 0 & \text{if } N \geq 7 \\ 1 & \text{if } N = 6. \end{cases}$

If L does not map any divisor to a point, then

$$e(X) \leq 2\rho(X) \leq 4d + 20N + 24 \leq \min\{44N + 48, \frac{4}{3}(8d + 53)\},$$

where $\rho(X)$ is the Picard number of X.

The idea is to look at the Euler characteristic $\chi(T_X) = h^0(T_X) - h^1(T_X) + h^2(T_X) - h^3(T_X) = \frac{1}{2}c_3$, then obtain an upper bound for $h^i(T_X), 0 \leq i \leq 3$. To bound $h^i(T_X)$, we study the tangent-normal bundle sequences for general surface sections and curve sections via their geometry. Doing the same for $T_X \otimes L^j$ for various j, we obtain different bounds. A better understanding of points imposing conditions on the sections of bundles can certainly sharpen the bounds.

2. Threefold geography

The geography problem for algebraic varieties of general type is the problem of the possible Chern numbers. Explicitly, it is to find inequalities among the Chern numbers, and then for each set of integers satisfying all the inequalities, find a variety with those integers as Chern numbers. The surface geography was well-studied ([Hi], [P1], [P2], [S], [X]). Particularly, Sommese showed that the slopes of lines in the plane for (c_2, c_1^2) are all the rational numbers in $[\frac{1}{5}, 3]$. (A line through (c_2, c_1^2) and the origin has slope $\frac{c_1^2}{c_2}$.) In order to have better-behaved threefolds, we impose the condition that the canonical bundle K_X is ample (i.e. $KC > 0$ for any irreducible C). We have the following fundamental facts:

$c_1^3 < 0, c_1c_2 < 0$ (K_X being ample),

$c_1c_2 \equiv 0(24)$ (Schwarzenberger),

$\frac{c_1^3}{c_1c_2} \leq \frac{8}{3}$ (Bogomolov, Miyaoka-Yau [M1]).

An easier problem is the study the Chern ratios $(\frac{c_1^3}{c_1c_2}, \frac{c_3}{c_1c_2})$ of threefolds. Hunt [H] was the first to find a 2-dimensional region in the affine plane $(\frac{c_1^3}{c_1c_2} \frac{c_3}{c_1c_2})$ by desingularizing the blow up of an arrangement of planes in $\mathbb{P}^3$ (see the two triangles in the attached chart). Liu [L] generalized the idea and used double cover of $\mathbb{P}^1 \times F_n, F_n$ being the Hirzebruch surface, to get a much larger region (the shaded area in the chart attached).

In [C1], we study the Chern ratios of complete intersections of threefolds in the projective spaces. Since all invariants of complete intersections of hypersurfaces are symmetric functions of the degrees of the hypersurfaces, this is an application of the multiple Waring's problem by Hua [Hu]. The study really shows that it is not possible to describe the Chern ratios explicitly. (In fact, there is no "region" for them.) What is feasible is to describe the limit points (SCI in the attached chart). This inspires the study of the *limit points* of the Chern ratios of threefolds in $\mathbb{P}^5$ [C2].

The study of surfaces in $\mathbb{P}^4$ and threefolds in $\mathbb{P}^5$ has received a lot of attention in the past 30 years since Hartshorne conjectured that for $N \geq 6$, the only codimension two subvarieties in $\mathbb{P}^N$ are complete intersections. Our study of the limit points confirms that threefolds in $\mathbb{P}^5$ are "scarce". In fact, the limit points of their Chern ratios all lie on the line segment $x + y = 2, 1 \leq x \leq 2$. On the other hand, applying the multiple Waring's problem, we show that the limit points of Chern ratios of "determinantal threefolds" in $\mathbb{P}^5$ (i.e., drop rank loci of m by $m+1$ matrices whose entries are homogeneous polynomials in 6 variables) are all rational numbers on the same line with $1 \leq x \leq \frac{17}{12}$.

However the same arguments (generalized Hodge index theorem, generalized double point formula ...) don't shed any new light for the slopes of surfaces in $\mathbb{P}^4$, not even if we only consider the limit points. On the other hand, from the Chern numbers of the zero loci of the pull-backs of the Horrocks-Mumford bundle on $\mathbb{P}^4$ (combined with some more integer theory), we can extend the line segment of slopes of determinantal surfaces [C3]. Perhaps the discovery of a "new" bundle on $\mathbb{P}^5$ can complete the study of Chern ratios of threefolds in $\mathbb{P}^5$. In [CKN] and [CL] we study an abstract threefold X. We map X birationally to some projective space by rK_X (Chen, Ein and Lazarsfeld [C], [EL] showed that r can be taken 5). Similar to arguments in the Calabi-Yau case, we give upper and lower bounds of c_3 by linear functions of c_1c_2, c_1^3 and r. Clearly, besides a more careful study of the geometry of the surface and curve sections, a better r can also improve the bounds.

Another interesting aspect of bounding the Euler numbers is to give an upper bound on the numbers of nodes. The problem of finding the maximal number of nodes of hypersurfaces was studied by Severi, Segre, Basset etc. and more recently, Bruce, Miyaoka, Varchenko, Van Straten and Givental ([Br], [M2], [V], [VS], and [Gi]). Our bound on the numbers of nodes of nodal complete intersection threefolds improves Kleiman-Gaffney's bound in some cases.([Ga], [K]).

3. Acknowledgment

We would like to thank Sheldon Katz and Albrecht Klemm for helpful communications while preparing this note. We also thank the organizers of ICCM for their hospitality.

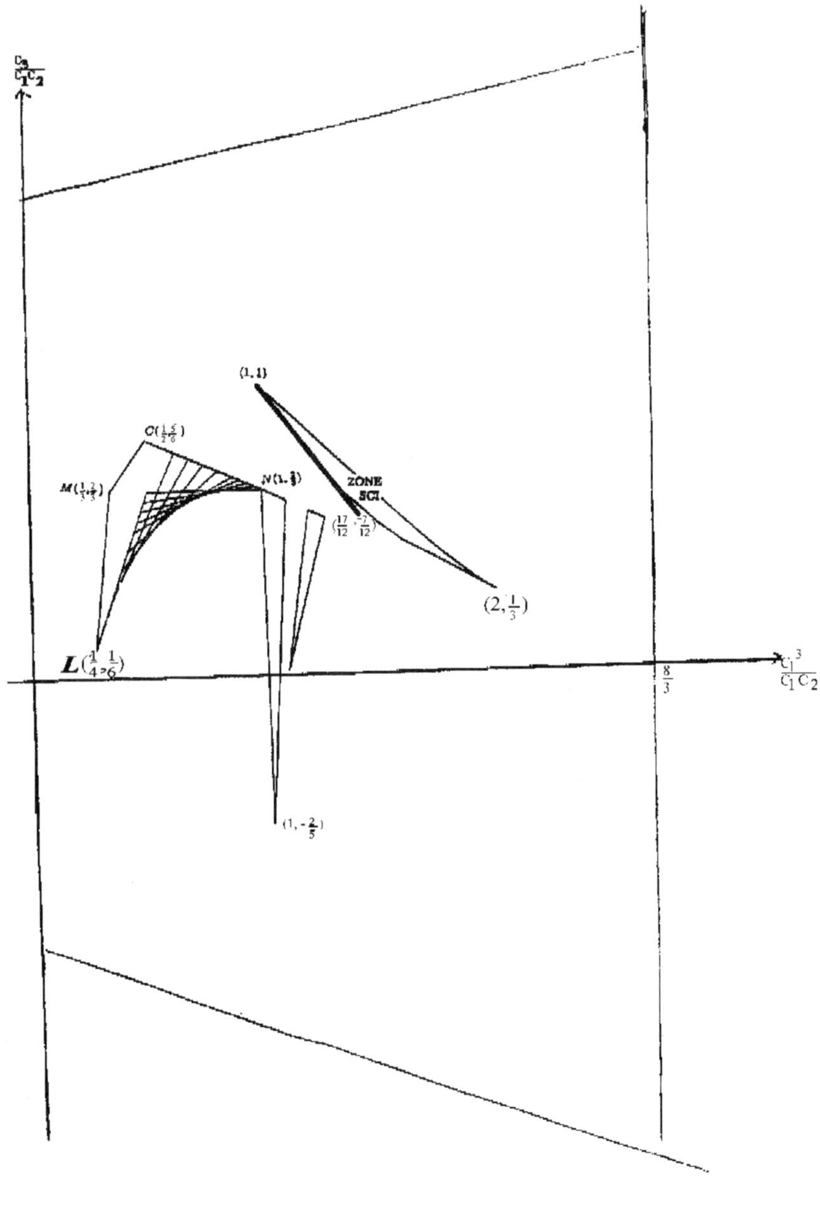

FIGURE 1

References

[1] [BC]" E. Ballico and L. Chiantini, On smooth subcanonical varieties of codimension 2 in $\mathbb{P}^n$, Ann. Mat. Pura Appl. (4) **135** (1983) 99-117.

[2] [Br]" J. Bruce, An upper bound for the number of singularities on a pr ojective hypersurface, Bull. London Math. Soc. **13**, (1981) 47-50.

[3] [C1]" M. Chang, Distributions of Chern numbers of complete intersection threefolds, GAFA, **7**(1997) 861-872.

[4] [C2]" ________, On the density of ratios of Chern numbers of embedded threefolds, Communications in Algebra, **27** No. 8, (1999) 3771-3776.

[5] [C3]" ________, On the Chern numbers of surfaces and 3-folds of codimension 2, Tokyo J. Math., **19** No. 2, (1996), 369-386.

[6] [CK]" M. Chang and H. Kim, The Euler number of primitive Calabi-Yau Threefolds, Math. Proc. of Cambridge Phil. Soc. (to appear).

[7] [CKN]" M. Chang, H. Kim and S. Nollet, Bounds on c_3 for threefolds, Manuscripta Math. **97**, (1998) 135-141.

[8] [CL]" M. Chang and A. Lopez, (in preparation).

[9] [EL]" L. Ein and R. Lazarsfeld, Global generation of pluricanonical and adjoint linear series on a smooth projective threefold, Jour. A.M.S. **6**(1993) 875-903.

[10] [Ga]" T. Gaffney, Multiplicity and equisingularity of ICIS germs. Invent. Math. J. **123**, (1996) 209-220.

[11] [Gi]" A. Givental, The maximum number of singular points, Functional Anal. Appl. **17**, No. 3,(1983), 223-225.

[12] [G]" M. Gross, A finiteness theorem for elliptic Calabi-Yau threefolds.

[13] [H]" B. Hunt, Complex Manifold Geography in dimension 2 and 3, J. Diff. Geom. **30**(1989) 51-153.

[14] [Hi]" F. Hirzebruch, Arrangements of lines and algebraic surfaces, in "Arithmetic and geometry", Progr. in Math. **36**-II (1983), 113-140.

[15] [Hu]" L. Hua, ENZ. MATH. WISS. B and 12 Heftr 13, Teil. (1959).

[16] [Le]" S. Lee, Remarks on pluricanonical and adjoint linear series, (preprint).

[17] [Li]" X. Liu, On the geography of threefolds, Tohoku Math. J. **49** (1997).

[18] [M1]" Y. Miyaoka, The maximal number of quotient singularities on surfaces with given numerical invariants, Math. Ann. **68** (1984) 159-171.

[19] [M2]" ________, The Chern classes and Kodaira dimension of a minimal v ariety, In: Algebraic Geometry, Sendai 1985. Adv. Stud. Pure Math. **10**. North-Holland: Amsterdam-New York (1987) 449-476.

[20] [P1]" U. Persson, Double coverings and surfaces of general type (Proc. algebraic geometry, Norway, 1977), LNM **687**, Springer, Berlin, 1978.

[21] [P2]" ________, Chern invariant of surfaces of general type, Compositio Math. **43** (1981) 3-58.

[22] [R]" M. Reid, The moduli space of 3-folds with $K = 0$ may nevertheless be irreducible, Math. Ann. **278** (1987), 329-334.

[23] [S]" A. Sommese, On the density of ratios of Chern numbers of algebraic surfaces, Math. Ann. **268** (1984) 207-221.

[24] [VS]" D. Van Straten, A quintic hypersurface in P^4 with 130 nodes, Topology. **32** (1993), No. 4, 857-864.

[25] [V]" A. Varchenko, On semi-continuity of the spectrum and an upperbound for the number of singular points of projective hyperfaces, Dokl. Akad. Nauk. USSR **270** (1983).

[26] [X]" G. Xiao, Surfaces fibrees en courbes de genre deux, LNM, vol.1137, Springer-Verlag, 1985.

[27] [Y]" S.-T. Yau, On the Ricci curvature of a compact Kähler manifold and the complex Monge-Ampère equation. I, Comm. Pur Appl. Math. **31** (1978) 339-411.

MATH DEPARTMENT, UCR

AMS/IP Studies in Advanced Mathematics
Volume 20, 2001

Vector bundles, the S-duality conjecture of Vafa and Witten, and vertex algebras

Wei-ping Li and Zhenbo Qin

ABSTRACT. This article surveys recent progress toward the S-duality conjecture formulated by Vafa and Witten [**24**] and the role played by vertex algebras.

1. Introduction

We start with the n-dimensional complex projective space $\mathcal{P}^n$. It is well-known that the topological Euler number $\chi(\mathcal{P}^n)$ of $\mathcal{P}^n$ is $(n+1)$. So the generating function of these Euler numbers is:

$$\sum_{n=0}^{+\infty} \chi(\mathcal{P}^n) q^n = \sum_{n=0}^{+\infty} (n+1) q^n = \frac{1}{(1-q)^2}. \tag{1.1}$$

In a more general setting, given a sequence of topological spaces $\{X_n\}_{n\geq 0}$, we can consider the generating function $\sum\limits_{n=0}^{+\infty} \chi(X_n) q^n$. Naturally one knows little about the generating function if the spaces X_n's are quite arbitrary. However, if these spaces share similar properties, the generating function may have explicit and simple expression. An example of such kind is Hilbert schemes of points on smooth varieties. Let X be a smooth compact complex algebraic variety. The Hilbert scheme $\mathrm{Hilb}^n(X)$ parameterizes all zero-dimensional subschemes of length n, and admits a natural morphism to the n-th symmetric product of X [**14**]: $\mathrm{Hilb}^n(X) \to \mathrm{Sym}^n(X)$. When $\dim X = 1$, we have $\mathrm{Hilb}^n(X) = \mathrm{Sym}^n(X)$ and

$$\sum_{n=0}^{+\infty} \chi(\mathrm{Hilb}^n(X)) q^n = \frac{1}{(1-q)^{\chi(X)}}. \tag{1.2}$$

Since $\mathrm{Hilb}^n(\mathcal{P}^1) \cong \mathcal{P}^n$, we see that (1.1) is a special case of (1.2).

When $\dim X = 2$, $\mathrm{Hilb}^n(X)$ is far more complicated than $\mathrm{Sym}^n(X)$. In fact, $\mathrm{Hilb}^n(X)$ is a desingularization of $\mathrm{Sym}^n(X)$ [**5**]. Göttsche [**10**] proved the following

1991 *Mathematics Subject Classification.* Primary 14J60, 57R55; Secondary 14D20, 14F05, 14J27.

Partially supported by the grant HKUST631/95P.
Partially supported by an NSF grant and an Alfred P. Sloan Research Fellowship.

beautiful formula:

$$\sum_{n=0}^{+\infty} \chi(\mathrm{Hilb}^n(X))q^n = q^{\frac{\chi(X)}{24}} \cdot \frac{1}{\eta(q)^{\chi(X)}}$$

where $\eta(q) = q^{\frac{1}{24}} \prod_{n\geq 1} (1-q^n)$ is the Dedekind η-function, i.e.,

$$q^{-\frac{\chi(X)}{24}} \cdot \sum_{n=0}^{+\infty} \chi(\mathrm{Hilb}^n(X))q^n = \frac{1}{\eta(q)^{\chi(X)}}. \tag{1.3}$$

One can check that, unlike the case of $\chi(\mathcal{P}^n)$, the formula for each individual $\chi(\mathrm{Hilb}^n(X))$ is quite complicated. However, (1.3) is a very simple formula. It is not hard to believe that some beautiful but deep mathematics is hidden behind the formula. Let $\mathcal{H}$ be the upper-half plane of the complex plane $\mathcal{C}$, and $\tau \in \mathcal{H}$. Then it is well-known that $\eta(q)$ is a modular form for $SL(2, \mathcal{Z})$ of weight 1/2 with a suitable multiplier system. So a profound implication of formula (1.3), which only became clear after the work of Vafa and Witten [**24**], is that over a complex algebraic surface, the generating function of the Euler numbers of certain moduli spaces has modularity property.

2. Hilbert schemes, Quot-schemes, and vertex algebras

Throughout the rest of this article, X will stand for a complex algebraic surface. We now discuss the results of Nakajima and Grojnowski [**22, 13, 23**] on Hilbert schemes $\mathrm{Hilb}^n(X)$ (and more generally, the Quot-schemes $\mathrm{Quot}^n_{\mathcal{O}_X^{\oplus 2}}$).

In [**22**], Nakajima found an interesting connection between the infinite dimensional Heisenberg algebra and the Hilbert schemes $\mathrm{Hilb}^n(X)$, and thus provided a geometric interpretation for Göttsche's formula (1.3). More precisely, let L be the lattice $H^*(X; \mathcal{Z})$/torsion endowed with the intersection pairing. Nakajima proved that the direct sum of the homology groups of all the Hilbert schemes $\mathrm{Hilb}^n(X)$:

$$S \overset{\text{def}}{=} \bigoplus_{n\geq 0} H_*(\mathrm{Hilb}^n(X); \mathcal{C})$$

admits a natural representation of the Heisenberg/Clifford algebra modeled on the integral lattice L. By comparing the character of this representation with Göttsche's formula (1.3), Nakajima concluded that S is the Fock space modeled on the integral lattice L.

The results in [**22**] also appeared in the announcement [**13**]. In fact, Grojnowski partially generalized these results to the Quot-schemes $\mathrm{Quot}^n_{\mathcal{O}_X^{\oplus 2}}$. Moreover, a connection with vertex algebras can be found in [**13**]. This connection was further worked out in Chapter 9 of [**23**]. Let NS(X) $\subset$ $\mathrm{H}^2(\mathrm{X}; \mathcal{Z})$/torsion be the Néron-Severi group of X endowed with the restriction of the intersection pairing. By identifying the moduli space of rank-1 torsion free sheaves with the Hilbert schemes, Nakajima [**23**] gave an algebro-geometric construction of the simple lattice vertex algebra associated to the integral lattice NS(X). In particular, all the vertex operators correspond to certain geometric objects.

The recent work of Baranovsky [**1**] generalized the results of Nakajima and Grojnowski [**22, 13**] to the moduli space of Gieseker semistable sheaves over algebraic surfaces.

3. Vector bundles and the S-duality conjecture formulated by Vafa-Witten

Vector bundles over algebraic varieties have been studied extensively for a long time. In order to construct a "good" space to parameterize the set of algebro-geometric objects (for example, the set of curves and the set of vector bundles over curves), Mumford [**6**] introduced the concept of stability. Later on, Gieseker [**9**] applied Mumford's geometric invariant theory to moduli problem for vector bundles over algebraic surfaces and proved the existence of a coarse moduli space for Gieseker semistable torsion free sheaves over algebraic surfaces. The study of stable bundles over algebraic surfaces acquired a new facet thanks to Donaldson's fundamental work [**3, 4**] on smooth invariants of algebraic surfaces and on the equivalence between instantons and stable bundles. Since then, significant contributions have been made by Friedman, Gieseker, Jun Li, O'Grady, and many others.

The Hilbert schemes may be viewed as some special types of the Gieseker moduli spaces, while the Quot-schemes are related to the boundary of the Gieseker moduli spaces. From this point of view, it would be natural to speculate that the generating functions of the Euler numbers of the Gieseker moduli spaces over algebraic surfaces might satisfy certain modularity property. As we pointed out earlier, this only became clear after the work of Vafa-Witten [**24**] on S-duality which is a conjecture in physics. In [**24**], Vafa and Witten tried to test the S-duality. Mathematically, their test boils down to a conjecture concerning certain modularity property for the generating functions of the Euler numbers of the Gieseker moduli spaces over algebraic surfaces. To state the conjecture, we introduce some notations. For simplicity, we restrict ourselves to rank-2 case, and also assume that our algebraic surface is simply connected. Let H be an ample divisor on a simply connected algebraic surface X, and c_1 be a divisor on X. For an integer n, let $\mathfrak{M}_H(c_1, n)$ be the moduli space of Mumford-Takemoto H-stable rank-2 bundles with Chern classes c_1 and n, and $\mathfrak{M}_H^G(c_1, n)$ be the moduli space of Gieseker H-semistable rank-2 torsion-free sheaves with Chern classes c_1 and n. It is well-known that the Gieseker moduli space is projective and $\mathfrak{M}_H(c_1, n)$ is an open subset of $\mathfrak{M}_H^G(c_1, n)$. Let $q = e^{2\pi i\tau}$ with $\tau \in \mathcal{H}$. Form two generating functions:

$$Z_X(\tau, SU(2)) = \tfrac{1}{2} \cdot q^{-\frac{\chi(X)}{12}} \cdot \textstyle\sum_n \chi(M_H^G(0, n))q^n \tag{3.1}$$

$$Z_X(\tau, SO(3)) = q^{-\frac{\chi(X)}{12}} \cdot \textstyle\sum_{c_1 \in H^2(X, \mathcal{Z}_2)} \sum_n \chi(M_H^G(c_1, n))q^{n - \frac{c_1^2}{4}}. \tag{3.2}$$

Let $\Gamma_0(m)$ be the subgroup of $SL(2, \mathcal{Z})$ consisting of matrices whose lower left entry is a multiple of m. Vafa and Witten [**24**] proposed the following.

Conjecture 3.3. (The S-duality conjecture formulated by Vafa and Witten [**24**]) *The generating functions $Z_X(\tau, SU(2))$ and $Z_X(\tau, SO(3))$ in (3.1) and (3.2) are invariant under the action of $\Gamma_0(4) \subset SL(2, \mathcal{Z})$ (or $\Gamma_0(2) \subset SL(2, \mathcal{Z})$ when the underlying 4-manifold of the algebraic surface X is spin). Moreover, there exists a transformation law between $Z_X(\tau, SU(2))$ and $Z_X(\tau, SO(3))$:*

$$Z_X(-1/\tau, SU(2)) = \pm 2^{-\frac{\chi(X)}{2}} \cdot \left(\frac{\tau}{i}\right)^{\frac{\chi(X)}{2}} \cdot Z_X(\tau, SO(3)). \tag{3.4}$$

This conjecture remains open, although it has been verified and discussed for certain special algebraic surfaces such as $\mathcal{P}^2$, $K3$-surfaces, and regular elliptic surfaces [**11, 12, 24, 26, 27**].

4. The blowup formula and vertex operator algebras

Vafa and Witten [**24**] noticed that an interesting consequence of the transformation law (3.4) is something called the blowup formula. Let $\pi : \widetilde{X} \to X$ be the blowing-up of a simply connected algebraic surface X at a point $x_0 \in X$, and E be the exceptional divisor. Fix a divisor c_1 on X, $\tilde{c}_1 = \pi^* c_1 - aE$ with $a = 0$ or 1, and an ample divisor H on X with odd $(H \cdot c_1)$. For $r \gg 0$, the divisors $H_r = r \cdot \pi^* H - E$ on $\widetilde{X}$ are ample; moreover, all the moduli spaces $\mathfrak{M}_{H_r}(\tilde{c}_1, n)$ (resp. $\mathfrak{M}^G_{H_r}(\tilde{c}_1, n)$) can be naturally identified. So we shall use $\mathfrak{M}_{H_\infty}(\tilde{c}_1, n)$ (resp. $\mathfrak{M}^G_{H_\infty}(\tilde{c}_1, n)$) to denote the moduli space $\mathfrak{M}_{H_r}(\tilde{c}_1, n)$ (resp. $\mathfrak{M}^G_{H_r}(\tilde{c}_1, n)$) with $r \gg 0$. Let $\theta_a(q) = \sum_{n \in \mathcal{Z}} q^{(n+\frac{a}{2})^2}$ $(a = 0, 1)$ be the theta functions.

Theorem 4.1 *(The blowup formula for the S-duality conjecture of Vafa and Witten [**24**]) Let notations be as in the preceding paragraph. Then,*

$$\begin{aligned}&\left(q^{-\frac{\chi(\widetilde{X})}{12}} \cdot \sum_n \chi(\mathfrak{M}^G_{H_\infty}(\tilde{c}_1, n)) q^{n - \frac{\tilde{c}_1^2}{4}}\right) \\ &= \frac{\theta_a(q)}{\eta(q)^2} \cdot \left(q^{-\frac{\chi(X)}{12}} \cdot \sum_n \chi(\mathfrak{M}^G_H(c_1, n)) q^{n - \frac{c_1^2}{4}}\right). \qquad (4.2)\end{aligned}$$

Formula (4.2) was conjectured by Vafa and Witten, partially based on a blowup formula over finite fields proved by Yoshioka [**25**]. In [**16**] and [**17**], a blowup formula for the generating functions of the virtual Hodge numbers of the Gieseker moduli spaces was found, which implies formula (4.2). Göttsche [**11**] gave a much simpler proof of (4.2) by using the Weil conjecture (a theorem due to Deligne) and Yoshioka's blowup formula over finite fields (this simpler proof was first observed by Jun Li).

In [**19**] (see also the announcement [**18**]), some interesting relations between the blowup formula (4.2) with $a = 0$ and vertex operator algebras were found. The starting point is the modular invariance of the universal function $\dfrac{\theta_0(q)}{\eta(q)^2}$ in (4.2) (the theta function $\theta_0(q)$ is a modular form for $\Gamma_0(4)$ of weight 1/2 with a suitable multiplier system). Vertex algebras came from the study [**21**] of 2-dimensional conformal field theory in physics and the study [**2, 7, 8**] of certain simple finite group known as the Monster group and its representations known as the Moonshine modules. Borcherds [**2**] gave the first axioms of vertex algebras. Vertex operator algebras are a special type of vertex algebras and their characters have certain modular invariance [**28**]. Let $L = \mathcal{Z}^\ell$ be a lattice with a $\mathcal{Z}$-valued symmetric bilinear form $< \cdot | \cdot >$, and Δ be a sublattice of L such that the restriction $< \cdot | \cdot > |_\Delta$ is even and positive-definite. By the constructions in [**2, 8, 15**], one obtains a vertex operator algebra $V_{\Delta \subset L}$ with the character

$$\mathrm{ch}(V_{\Delta \subset L}) = \frac{\sum_{\alpha \in \Delta} q^{\frac{1}{2} < \alpha | \alpha >}}{\eta(q)^\ell}. \qquad (4.3)$$

In particular, let $L = \mathcal{Z}^2$ with the standard bilinear form $< \cdot | \cdot >$, and let $\Delta = \{(n, n) |\ n \in \mathcal{Z}\}$ be the diagonal in $\mathcal{Z}^2$. Then, $< (n, n) | (n, n) >= 2n^2$. So $< \cdot | \cdot > |_\Delta$

is even and positive-definite. Therefore we obtain a vertex operator algebra $V_{\Delta\subset\mathcal{Z}^2}$ with the character

$$\mathrm{ch}(V_{\Delta\subset\mathcal{Z}^2}) = \frac{\sum_{\alpha\in\Delta} q^{\frac{1}{2}<\alpha|\alpha>}}{\eta(q)^2} = \frac{\sum_{n\in\mathcal{Z}} q^{n^2}}{\eta(q)^2} = \frac{\theta_0(q)}{\eta(q)^2}. \tag{4.4}$$

This character is precisely the universal function appeared in the blowup formula (4.2) for $a = 0$. Moreover, the Lie algebra associated to $V_{\Delta\subset\mathcal{Z}^2}$ is isomorphic to the Lie algebra $\mathfrak{gl}_2$ with the invariant symmetric bilinear form $(\cdot|\cdot)$ defined by $(A|B) = \mathrm{tr}(AB)$. For a complex scheme Y, let $\mathcal{C}(Y)$ be the space of all $\mathcal{C}$-valued constructible functions over Y [**20**]. The following has been proved in [**19**].

Theorem 4.5 *Assume that $(H \cdot c_1)$ is odd and the moduli space $\mathfrak{M}_H(c_1, m)$ is nonempty. Then there exists a linear embedding:*

$$\phi_m : V_{\Delta\subset\mathcal{Z}^2} \to \bigoplus_{n,p\in\mathcal{Z}} \mathcal{C}(\mathfrak{M}^G_{H_\infty}(\pi^*c_1, n+p^2))$$

of the vertex operator algebra $V_{\Delta\subset\mathcal{Z}^2}$ in $\bigoplus_{n,p\in\mathcal{Z}} \mathcal{C}(\mathfrak{M}^G_{H_\infty}(\pi^*c_1, n+p^2))$ such that

(i) *the image of the embedding ϕ_m can be explicitly identified;*

(ii) *for every $a \in V_{\Delta\subset\mathcal{Z}^2}$ with the field $Y(a, z) = \sum_{n\in\mathcal{Z}} a_{(n)} z^{-n-1}$, the operators*

$$\phi_m \circ a_{(n)} \circ (\phi_m)^{-1} \quad (n \in \mathcal{Z})$$

can be expressed in terms of certain geometric operators;

(iii) *ϕ_m induces a representation of $\hat{\mathfrak{g}}$ in $\bigoplus_{n,p\in\mathcal{Z}} \mathcal{C}(\mathfrak{M}^G_{H_\infty}(\pi^*c_1, n+p^2))$,*

where $\hat{\mathfrak{g}}$ is the affinization of the Lie algebra $\mathfrak{g} = \mathfrak{gl}_2$ with the invariant symmetric bilinear form $(\cdot|\cdot)$ defined by $(A|B) = \mathrm{tr}(AB)$.

We refer the details of Theorem 4.5 to [**18, 19**]. The main idea in the proof is that the embeddings ϕ_m and the operators $\phi_m \circ a_{(n)} \circ (\phi_m)^{-1}$ come from the geometric constructions in our proof of the blowup formula (4.2) in [**16**]. The importance of the geometric realization described in Theorem 4.5 is that the existence of a vertex operator algebra structure on the Gieseker moduli spaces should indicate the presence of some very rich mathematical structures on these moduli spaces such as the modularity of the generating functions of their Euler numbers. This is precisely what Conjecture 3.3 of Vafa and Witten amounts to.

References

[1] V. Baranovsky, *Moduli of sheaves on surfaces and action of the oscillator algebra,* **Preprint.**

[2] R. Borcherds, *Vertex algebras, Kac-Moody algebras and the monster,* **Proc. Natl. Acad. Sci. 83** (1986), 3068-3071.

[3] S.K. Donaldson, *Anti-self-dual Yang-Mills connections over complex algebraic surfaces and stable vector bundles* **Proc. Lond. Math. Soc. 50** (1985), 1–26.

[4] ——— *Polynomial invariants for smooth four-manifolds,* **Topology 29** (1990), 257–315.

[5] J. Fogarty, *Algebraic families on an algebraic surface,* **Amer. J. Math. 90** (1968), 511-521.

[6] J. Fogarty, D. Mumford, **Geometric Invariant Theory** (Berlin-New York: Springer-Verlag, 1982.)

[7] I. Frenkel, J. Lepowsky, A. Meurman, *Vertex operator algebra and the monster* **Pure and Applied Math.** (Academic Press), **134** (1988).

[8] I. Frenkel, J. Lepowsky, A. Meurman, *A natural representation of the Fischer-Griess Monster with the modular function J as a character* **Proc. Nat. Acad. Sci. USA 81** (1984) 3256-3260.

[9] D. Gieseker, *On the moduli of vector bundles on an algebraic surface* **Ann. of Math. 106** (1977), 45-60.

[10] L. Göttsche, **Hilbert schemes of zero-dimensional subschemes of smooth varieties** Lecture Notes in Mathematics **1572** (Berlin: Springer-Verlag 1994).

[11] L. Göttsche, *Theta functions and Hodge numbers of moduli spaces of sheaves on rational surfaces,* **Preprint.**

[12] L. Göttsche, and D. Huybrechts, *Hodge numbers of moduli spaces of stable bundles on K3 surfaces* **Internat. J. Math. 7** (1996), 359-372.

[13] I. Grojnowski, *Instantons and affine algebras I: the Hilbert scheme and vertex operators,* **Math. Res. Lett. 3** (1996), 275-291.

[14] A. Grothendieck, *Techniques de construction et théorèmes d'existence en géométrie algébrique IV,* **Sem. Bourbaki 221** (1960-1961), 13.

[15] V. Kac, **Vertex Algebras for Beginners,** University Lecture Series **10** (Providence, Rhode Island: AMS, 1997.)

[16] W.-P. Li, Z. Qin, *On blowup formulae for the S-duality conjecture of Vafa and Witten* **To appear in Invent. Math.**

[17] ——— *On blowup formulae for the S-duality conjecture of Vafa and Witten II: the universal functions,* **Math. Res. Letters 5** (1998), 439-453.

[18] ——— *Vertex operator algebras and the blowup formula for the S-duality conjecture of Vafa and Witten,* **To appear in Math. Res. Letters.**

[19] ——— *On blowup formulae for the S-duality conjecture of Vafa and Witten III: relations with vertex operator algebras,* **In preparation.**

[20] R. MacPherson, *Chern classes of singular varieties* , **Ann. Math. 100** (1974), 423-432.

[21] G. Moore, N. Seiberg, *Classical and quantum conformal field theory,* **Comm. Math. Phys. 123** (1989), 177-254.

[22] H. Nakajima, *Heisenberg algebra and Hilbert schemes of points on projective surfaces,* **Ann. Math. 145** (1997), 379-388.

[23] ——— *Lectures on Hilbert schemes of points on surfaces,* **Preprint.**

[24] C. Vafa, E. Witten, *A strong coupling test of S-duality,* **Nucl. Phys. 431** (1994), 3-77.

[25] K. Yoshioka, *The Betti numbers of the moduli space of stable sheaves of rank 2 on* $\mathcal{P}^2$ **J. reine angew. Math. 453** (1994), 193-220.

[26] ——— *Numbers of* $\mathcal{F}_q$*-rational points of moduli of stable sheaves on elliptic surfaces* **Moduli of vector bundles, Lect. Notes in Pure and Applied Math. 179** (1994), 297-305

[27] ——— *Euler characteristics of* $SU(2)$ *instanton moduli spaces on rational elliptic surfaces,* **Preprint.**

[28] Y. Zhu, *Modular invariance of characters of vertex operator algebras,* **J. of Amer. Math. Soc. 9** (1996), 237-302.

Department of Mathematics, HKUST, Clear Water Bay, Kowloon, Hong Kong
E-mail address: `mawpli@uxmail.ust.hk`

Department of Mathematics, Oklahoma State University, Stillwater, OK 74078, USA
E-mail address: `zq@math.okstate.edu`

AMS/IP Studies in Advanced Mathematics
Volume 20, 2001

Cayley-Bacharach Property and k-Very Ampleness

Sheng-Li Tan

Abstract

In this article we shall describe the equivalence between the Cayley-Bacharach property of a smooth complex projective variety and the k-very ampleness of the adjoint linear system, both studies have received much attention in recent years. We use the Cayley-Bacharach type techniques to give a canonical construction of some cohomology classes, and to study vector bundles and the geometry of codimension 2 subschemes. The main results are from [**23**] and [**24**].

1. Cayley-Bacharach Theorem and Generalizations

Cayley-Bacharach Theorem ([**2**], [**7**]) is a generalization of some classical theorems in plane projective geometry, for example, Pappus Theorem, Pascal Theorem and Chasles Theorem.

Cayley-Bacharach Theorem. *Let A, B and C be three plane curves in $\mathbb{P}^2$ of degree a, b and $a+b-3$, respectively. Assume that A and B meet in ab points. If C passes through all but one of the ab points, then it passes through the remaining one.*

See also ([**21**], p.89). In the last decades, this theorem was generalized to projective space by using various methods. We refer to a very nice survey [**9**] for the history and the modern generalizations to $\mathbb{P}^n$ from the point of view of duality theory.

Griffiths and Harris are the first who generalized the theorem to n dimensional complex manifold X [**13**]. Let $\pi : E \to X$ be a rank n vector bundle and $s \in H^0(X, E)$ be a global section whose zero set $Z = Z(s)$ is of zero dimension. We denote by K_X the canonical divisor of X.

Griffiths-Harris Theorem. *If Z consists of distinct simple points, then each $F \in |K_X + \det E|$ that passes through all but one point of Z necessarily contains that remaining point.*

1991 *Mathematics Subject Classification.* Primary 14C20, 14F05; Secondary 14J60, 14C07.

Key words and phrases. Cayley-Bacharach property, k-very ample, vector bundle, cohomology.

This work is supported by NSF for Outstanding Youths, the 973 Project Foundation and the Doctoral Program Foundation of EMC.

When $E = \bigoplus_{i=1}^{n} \mathcal{O}_X(F_i)$ and $Z = F_1 \cdots F_n$ is the intersection subscheme of the hypersurfaces F_i, the above theorem implies the Cayley-Bacharach Theorem. For example, if $X = \mathbb{P}^2$ and $E = \mathcal{O}_X(a) \oplus \mathcal{O}_X(b)$, then one gets exactly the Cayley-Bacharach Theorem.

In order to get a more general extension, we need to consider hypersurfaces F in the linear system $|\det E - L|$ for any divisor L. We use the notion "k points" for any zero-dimensional subscheme of length k, not requiring the points to be distinct. The degree of an object is defined with respect to a fixed ample divisor A on X, hence the degree of a codimension r subscheme Y of X is defined by $\deg Y = A^{n-r}Y$, although A is not mentioned in the statements. The author and E. Viehweg ([**23**], [**24**]) prove that the general Cayley-Bacharach theorem is equivalent to the k-very ampleness of some linear systems. Let us first recall the definition of *k-very ampleness* (cf. [**3**], [**4**] or [**15**]).

Definition 1.1. A linear system $|D|$ on X is called k-very ample if for any zero-dimensional subscheme Y of degree $k+1$, the restriction map

$$\rho_Y : H^0(\mathcal{O}(D)) \to H^0(\mathcal{O}_Y(D))$$

is surjective, which is equivalent to the injectivity of

$$\beta_Y : H^1(\mathcal{I}_Y(D)) \to H^1(\mathcal{O}(D)) \to 0.$$

Note that "0-very ample" is equivalent to "base point free", and "1-very ample" is equivalent to "very ample".

Theorem 1.2. *For a fixed divisor L on X and a positive integer k, the following conditions are equivalent.*

1. *Let E be a rank n vector bundle with a non zero global section s such that $Z = Z(s)$ is a zero-dimensional subscheme, and let $F \in |\det E - L|$. If F passes through a subscheme $Z' \subset Z$ whose degree $\geq \deg Z - k$, then F passes through Z.*
2. *Let $F_1, \cdots, F_n$ be hypersurfaces such that the intersection $Z = F_1 \cdots F_n$ is a zero-dimensional subscheme, and let $F \in |F_1 + \cdots + F_n - L|$. If F passes through a subscheme $Z' \subset Z$ whose degree $\geq \deg Z - k$, then F passes through Z.*
3. *$|K_X + L|$ is $(k-1)$-very ample.*

If $L = -K_X$, then $|K_X + L| = |0|$ is obviously base point free (0-very ample). Hence the first and second parts of Theorem 1.2 are true for $k = 1$. This is the Griffiths-Harris Theorem without the assumption that $Z(s)$ is reduced (cf. [**13**], p.677).

Since $\mathcal{O}_{\mathbb{P}^n}(k)$ is k-very ample, one obtains a Cayley-Bacharach theorem on $\mathbb{P}^n$. In fact, this theorem is sharp.

The following conjecture is equivalent to Fujita's conjecture [**11**].

Conjecture 1.3. *Suppose $F_1, \cdots, F_n$ are n hypersurfaces such that the intersection $Z = F_1 \cdots F_n$ is a zero-dimensional subscheme. Let H be an ample divisor on X, and let*

$$F \in |F_1 + \cdots + F_n - \ell H|.$$

If F passes through all but $k \leq \ell - n$ points of Z, then F passes through Z.

We can see from Theorem 1.2 that the conjecture for $k = 1$ (resp. 2) is equivalent to the base point freeness (resp. very ampleness) part of Fujita's conjecture.

We know that this conjecture is true for surfaces ($n = 2$) because $|K_X + \ell H|$ is $(\ell - 3)$-very ample (cf. [**19**], [**3**], [**4**]).

Thanks to the study of Fujita's conjecture, the 0- and 1-very ampleness are recently known for some linear systems on higher dimensional varieties, thus Conjecture 1.3 is true in the corresponding cases. We refer to [**1**], [**8**], [**10**], [**12**], [**14**], [**22**], $\cdots$ for the precise results.

2. A Representation of Cohomology Classes

In this section, we shall present a canonical construction of non zero cohomology classes in $H^{q+1}(X, \mathcal{I}_\Delta(K_X + L))$ for any L, where Δ is a q dimensional subscheme of X with ideal sheaf $\mathcal{I}_\Delta$. The k-very ampleness of $|K_X + L|$ is about this cohomology group ($q = 0$).

Let X be a smooth projective variety over $\mathbb{C}$ of dimension $n \geq 2$, and let Z be a subscheme of X of pure codimension $r \geq 2$.

Given a subscheme $Z' \subset Z$, the "complement" Z'' of Z' in Z is the canonical closed subscheme $Z'' \subset Z$ with sheaf of ideals $\mathcal{I}_{Z''} = [\mathcal{I}_Z : \mathcal{I}_{Z'}]$, i.e., for any open set $U \subset X$, we define

$$\mathcal{I}_{Z''}(U) := \{g \in \mathcal{O}_X(U) \mid g\mathcal{I}_{Z'}(U) \subset \mathcal{I}_Z(U)\},$$

or equivalently,

$$\mathcal{I}_{Z''}/\mathcal{I}_Z = \mathcal{H}om_{\mathcal{O}_X}(\mathcal{O}_{Z'}, \mathcal{O}_Z).$$

The second description implies that $Z'' = Z$ if the support of Z' does not contain any irreducible component of Z. Moreover, if Z is reduced, then Z'' is the closure of $Z - Z'$. We call Z'' the *residual subscheme* of Z' in Z and denote it by

$$Z'' = Z - Z'.$$

Let E be a vector bundle on X of rank $r \geq 2$, let s be a global section of E and let $Z = Z(s) \subset X$ be its zero scheme. As above we shall assume that Z is of pure codimension r, hence it is a local complete intersection. For a divisor L and a subscheme $\Delta \subset Z(s)$, we want to study hypersurfaces F in X satisfying the equations

$$\begin{cases} \Delta = Z(s) - Z(s)F, \\ L \equiv \det E - F. \end{cases} \tag{2.1}$$

Given Δ and L we shall call (E, s, F) a solution of (2.1) if $Z(s)$ is of pure codimension $r = \operatorname{rank}(E)$ and if the equation (2.1) holds true.

Here and throughout this article $Z(s)F$ denotes the intersection subscheme of $Z(s)$ and a hypersurface (or effective divisor) F in X. If a hypersurface F satisfies the first equation in (2.1) we shall say that F does not pass through Δ. In a similar way, if Z' is a subscheme of F, we shall say that F passes through Z'.

If $Z(s)$ is a reduced subscheme of X, then F satisfies the first equation in (2.1) if Δ is the union of all irreducible components of $Z(s)$ which are not contained in F.

Theorem 2.1. *Let E, s and Z be as above, let $Z'' \subset Z'$ be two codimension r subschemes of Z, and let L be a divisor. Then there exists a complex of vector spaces*

$$\begin{aligned} 0 \to H^0(\mathcal{I}_{Z-Z''}(\det E - L)) &\xrightarrow{\alpha} H^0(\mathcal{I}_{Z-Z'}(\det E - L)) \\ \xrightarrow{\mu} H^{n-r+1}(\mathcal{I}_{Z'}(K_X + L)) &\xrightarrow{\beta} H^{n-r+1}(\mathcal{I}_{Z''}(K_X + L)) \to 0, \end{aligned}$$

exact except at $H^{n-r+1}(\mathcal{I}_{Z'}(K_X + L))$. If E is sufficiently ample, then the complex is exact everywhere.

Remark 2.2. The condition "E is sufficiently ample" we used in the theorem stands for the following vanishing conditions:

$$H^j(X, \wedge^i E^\vee(\det E - L)) = 0, \quad \text{for } i,\ j = 1, \cdots, r-1. \tag{2.2}$$

If $X = \mathbb{P}^n$ and E splits (the hypersurface case), then (2.2) is always true. In general (2.2) can be enforced by replacing E by $E \otimes H$, for a sufficiently ample line bundle H, which keeps the Cayley-Bacharach property unchanged (cf. [**24**], Lemma 4).

The connecting map μ is not "natural", but there is a natural map to the dual of $\ker \beta$.

Theorem 2.3. *With the same assumptions as in the above theorem. For any subscheme $Z' \subset Z = Z(s)$, we have a connecting homomorphism*

$$\mu_{Z'} : H^0(X, \mathcal{I}_{Z-Z'}(\det E - L)) \to H^{n-r+1}(X, I_{Z'}(K_X + L))^\vee$$

such that

1. *$\mu_{Z'}$ is natural. Namely, if $Z'' \subset Z'$ is another subscheme, then we have the following commutative diagram:*

$$\begin{array}{ccc} H^0(\mathcal{I}_{Z-Z'}(\det E - L)) & \xrightarrow{\mu_{Z'}} & H^{n-r+1}(\mathcal{I}_{Z'}(K_X + L))^\vee \\ \uparrow & & \uparrow \\ H^0(\mathcal{I}_{Z-Z''}(\det E - L)) & \xrightarrow[\mu_{Z''}]{} & H^{n-r+1}(\mathcal{I}_{Z''}(K_X + L))^\vee \end{array}$$

2. *If E is sufficiently ample, i.e., (2.2) holds true, then $\mu_{Z'}$ is surjective for any Z'.*
3. *$\mu_{Z'}(f) = 0$ iff f is in the image of the following map*

$$s : H^0(E^\vee(\det E - L)) \to H^0(\mathcal{I}_Z(\det E - L)) \subset H^0(\mathcal{I}_{Z-Z'}(\det E - L)).$$

   So $\ker \mu_{Z'}$ is independent of Z', and $\mu_{Z'}(f) = 0$ implies that f vanishes on Z, the converse is also true provided $H^{n-r+1}(\mathcal{O}(K_X + L)) = 0$.

Note that if $s = (f_1, \cdots, f_r) \in H^0(E)$ (here $E = \oplus_{i=1}^r \mathcal{O}(F_i)$), then $\mu_{Z'}(f) = 0$ if and only if there are global sections $g_i \in H^0(\det E - L - F_i)$ such that

$$f = g_1 f_1 + \cdots + g_r f_r.$$

Corollary 2.4. *Let Δ be a subscheme of X of pure codimension r and let L be a divisor on X. Then the following conditions are equivalent.*

1. *There are a hypersurface F, a rank r vector bundle E and a non zero global section s of E whose zero set $Z = Z(s)$ is an $n-r$ dimensional subscheme such that*

$$\begin{cases} \Delta = Z(s) - Z(s)F, \\ L \equiv \det E - F. \end{cases} \tag{2.3}$$

2. *There are $r+1$ hypersurfaces $F, F_1, \cdots, F_r$ such that $Z = F_1 \cdots F_r$ is an $n-r$ dimensional subscheme and*

$$\begin{cases} \Delta = F_1 \cdots F_r - F_1 \cdots F_r F, \\ L \equiv F_1 + \cdots + F_r - F. \end{cases} \tag{2.4}$$

3. *There is an element η in $H^{n-r+1}(\mathcal{I}_\Delta(K_X+L))^\vee$ such that for any proper codimension r closed subscheme $\Delta' \not\subseteq \Delta$, η is not in the image of the following natural inclusion map:*

$$H^{n-r+1}(\mathcal{I}_{\Delta'}(K_X+L))^\vee \to H^{n-r+1}(\mathcal{I}_\Delta(K_X+L))^\vee.$$

We shall give some applications of Corollary 2.4 in Sect. 4.

3. On the Vector Bundle Method

It is well-known that vector bundle method is a very powerful tool in the study of k-very ampleness of linear systems on curves and surfaces. Beltrametti and Sommes [**3**], [**4**] solve the problem in surface case by using Reider's method [**19**], which is a beautiful application of Bogomolov's inequality [**5**] for rank two semistable vector bundles. Unfortunately, this method does not work for higher dimensional varieties.

Equivalently, we can attack the k-very ampleness problem by using the Cayley-Bacharach property for hypersurfaces F_i (Theorem 2.1, 2), and we can also assume that F_i is sufficiently ample .

The Cayley-Bacharach property for hypersurfaces is a problem of finding the minimal degree of the syzygies of $F = (F_0, F_1, \cdots, F_n)$ or $f = (f_0, f_1, \cdots, f_n)$, where F_0 is the hypersurface F in Sect. 2 and f_i is a global section of $\mathcal{O}(F_i)$ whose zero scheme is F_i.

Let $\mathcal{M}$ be an invertible sheaf on X. We consider sections $g_i \in H^0(X, \mathcal{M} \otimes \mathcal{O}(-F_i))$ for $i = 0, 1, \cdots, n$.

Definition 3.1. $g = (g_0, g_1, \cdots, g_n)$ is called an $\mathcal{M}$-*syzygy* (or simply *syzygy*) of f if in $H^0(\mathcal{M})$ we have

$$f_0 g_0 + f_1 g_1 + \cdots + f_n g_n = 0. \tag{3.1}$$

We denote by G_i the zero subscheme of g_i if it is non zero.

If $g \neq (0, \cdots, 0)$ and H is an ample divisor on X, we shall call $c_1(\mathcal{M})H^{n-1}$ the degree of the syzygy g. Cayley-Bacharach property means that if f has a big common zero set, then it admits a non zero syzygy g of small degree. From the definition of Δ, we can see that G_0 passes through Δ if $g_0 \neq 0$. When F_0 passes through all of the intersection points $F_1 \cdots F_n$, then the minimal degree of the non zero syzygies is $\deg F_0$, i.e., g_0 is a non zero constant (cf. [**23**]).

In order to use the vector bundle method, we consider the reflexive syzygy sheaf $\mathcal{F}$ of F:

$$0 \to \mathcal{F} \to \bigoplus_{i=0}^{n} \mathcal{O}(-F_i) \xrightarrow{f} \mathcal{I}_{Z'} \to 0, \tag{3.2}$$

where $Z' = F_0 F_1 \cdots F_n$. If X is a surface, then $\mathcal{F}$ is a vector bundle.

Problem 3.2. *Find numerical conditions on the Chern classes of a rank n sheaf $\mathcal{F}$ such that $\mathcal{F}$ admits an invertible subsheaf $\mathcal{M}^\vee$ of higher degree (e.g., $\deg \mathcal{M}^\vee > \deg \mathcal{F}/\mathrm{rk}\,\mathcal{F}$).*

For rank 2 vector bundles, this is just a problem on unstability, and the Bogomolov's inequality gives the solution. When $n > 2$, what we need is the rank 1 stability, instead of the usual stability.

Definition 3.3. Rank r vector bundle E is called *rank* 1 *unstable* if there is an invertible subsheaf $\mathcal{L}$ of E such that for any rank ≥ 2 subsheaf $\mathcal{E}' \subset E$, we have $\deg \mathcal{L} > \deg \mathcal{E}'/\mathrm{rk}\,\mathcal{E}'$.

Some results on stability are still true for rank 1 stability, for example,

Proposition 3.4. *E is rank 1 unstable if and only if there is an m_0 such that for any $m > m_0$ and for any general $Y \in |mH|$, $E|_Y$ is rank 1 unstable.*

PROOF. This follows from Mehta-Ramanathan's extension theorem (cf. [**16**] or [**17**]). □

If $X = C$ is a curve and

$$0 \subset \mathcal{E}_1 \subset \mathcal{E}_2 \subset \cdots \subset \mathcal{E}_n = \mathcal{E}$$

is the Harder-Narasimhan filtration of $\mathcal{E}$, then $\mathcal{E}$ is rank 1 unstable iff $\mathrm{rk}\,\mathcal{E}_1 = 1$.

If $\pi : \widetilde{C} \to C$ is a finite covering of degree d, then $\mathcal{E}$ is rank 1 unstable iff $\pi^*\mathcal{E}$ is.

The numerical conditions in the problem should be invariant up to tensoring a line bundle L, because E is rank 1 unstable iff $E \otimes L$ is. Bogomolov's polynomial

$$P_2(E) = c_2 - \frac{r-1}{2r} c_1^2$$

is the unique invariant quadratic homogeneous polynomial on Chern classes. There is only one invariant cubic homogeneous polynomial:

$$P_3(E) = -c_3 + \frac{r-2}{r} c_2 c_1 - \frac{(r-2)(r-1)}{3r^2} c_1^3.$$

In general, for any $k \leq r$, the polynomial

$$P_k(E) = -\frac{k-1}{r^k}\binom{r}{k} c_1^k + \sum_{j=0}^{k-2} \frac{(-1)^{k-j}}{r^j}\binom{r-k+j}{j} c_{k-j} c_1^j.$$

is invariant. In fact, P_k is the k-th elementary symmetric polynomial of $\dfrac{c_1}{r} - t_i$, $i = 1, \cdots, r$, where $t_1, \cdots, t_r$ are the Chern roots of E. Note that from the definition, $P_0 = 1$, $P_1 = 0$, and $P_k = 0$ if $k > r$. But for $k \geq 4$, there is at least one different way to construct such polynomials, for example, let $p : P(E) \to X$ be the projective bundle of E, and let H be the divisor of $\mathcal{O}(1)$ on $P(E)$. Then we have

$$H^r - p^*(c_1(E))H^{r-1} + \cdots + (-1)^r p^*(c_r(E)) = 0.$$

Thus

$$Q_r(E) = -\left(H - \frac{1}{r}p^*(c_1(E))\right)^{n+r-1}$$

is also an invariant homogeneous polynomials on the Chern classes. For example, if $r = 4$, then

$$Q_4(E) = \frac{13}{16}c_2c_1^2 - \frac{1}{4}c_3c_1 - \frac{39}{256}c_1^4 + c_4 - c_2^2,$$

thus $Q_4 = P_4 - P_2^2$.

There is no new homogeneous inequality for semistable vector bundles on a 3-fold [**20**], but $-P_3(E)$ may be bounded from above by a polynomial of $P_2(E)$, which depends only on X (cf. [**6**]).

4. Applications

In codimension 2 case, we can apply directly Bogomolov's inequality to our problems. A divisor L is called numerically effective (nef) if the intersection number LC is non negative, for all irreducible curves C on X.

Theorem 4.1. *Let L be a nef divisor and $r = 2$. Assume that one of the three equivalent conditions in Corollary 2.4 is true. If $\deg \Delta < \deg L^2/4$, then either Δ is empty or there exists an effective divisor D passing through Δ such that*

$$\deg DL - \deg\Delta \le \deg D^2 < \tfrac{1}{2}\deg DL \le \\ \tfrac{1}{4}\left(\deg L^2 - \sqrt{\deg L^2}\sqrt{\deg L^2 - 4\deg\Delta}\right) < \deg\Delta.$$

PROOF. This is just an application of Bogomolov's inequality to the rank 2 syzygy sheaf $\mathcal{F}$ of (F, F_1, F_2) (cf. [**23**], [**24**]). □

The following corollary is a slight generalization of Paoletti's theorem in [**18**].

Corollary 4.2. *As in the above theorem, assume that $X = \mathbb{P}^n$, $L = \mathcal{O}(\ell)$ and $d = \deg\Delta$. Denote by s the minimal degree of the hypersurfaces containing Δ. If $d \le \ell^2/4$, then*

$$s \le \frac{1}{2}\ell - \frac{1}{2}\sqrt{\ell^2 - 4d},$$

with equality iff Δ is a complete intersection of type $(s, \ell - s)$.

The invariant

$$e = \max\{\, m \mid H^{n-1}(\mathcal{I}_Y(m)) \ne 0 \,\}$$

is called the specialty of Y. When $\Delta = Y$ is reduced and irreducible, and $\ell = e+n+1$, then Y satisfies obviously the third condition in Corollary 2.4, so Corollary 4.2 is true. This is Paoletti's theorem on the specialty.

The following corollary is a generalization of the classical Cayley-Bacharach Theorem.

Corollary 4.3. *Assume that E is a rank 2 vector bundle with a global section s whose zero scheme has codimension 2. Let H be an ample divisor on X, let ℓ be a positive integral, and let $F \in |\det E - \ell H|$. If F passes through an $(n-2)$-dimensional subscheme of $Z(s)$ whose degree is larger than or equal to $\deg Z(s) - \ell + 2$, then F passes through $Z(s)$.*

PROOF. We only need to prove that under the second condition of Corollary 2.4, $\deg \Delta \geq \ell - 1$. This follows from the inequalities in Theorem 4.1. □

Corollary 2.4 can be used to give an explicit construction of rank 2 vector bundles from codimension 2 subschemes (cf. [**24**], Theorem 10).

Finally, one can also use Corollary 2.4 and Corollary 4.2 to classify the pure codimension r subschemes Z such that $H^{n-r+1}(\mathcal{I}_Z(K_X+L)) \neq 0$. In fact, we only need to classify the minimal Z in the following sense: for any closed $Z' \not\subseteq Z$ we have $H^{n-r+1}(\mathcal{I}_{Z'}(K_X+L)) = 0$. Because $H^{n-r+1}(\mathcal{I}_Z(K_X+L)) \neq 0$ iff Z contains a minimal one. For simplicity, we just give some examples in the case $X = \mathbb{P}^2$. We always assume that Z is minimal, and $\deg Z < \ell^2/4$, $(\ell = d+3)$.

Example 4.4. Assume $\deg Z \leq 12$. If $H^1(\mathcal{I}_Z(4)) \neq 0$, then Z can be classified as follows.

1. C^3C^4 is a complete intersection of type $(3,4)$;
2. 10 points lie on a conic;
3. 6 points lie on a line.

PROOF. From Corollary 4.2, we have $s \leq 3$.
If $s = 3$, then $\deg Z \geq \ell s - s^2 = 12$, so $\deg Z = 12$ and $Z = C^3C^4$.
If $s = 2$, then $\deg Z \geq 10$, so 10 points are in a conic.
If $s = 1$, then $\deg Z \geq 6$, so 6 points are collinear. □

Example 4.5. $\deg Z \leq 15$, $H^1(\mathcal{I}_Z(5)) \neq 0$:

1. $Z = C^3C^5$;
2. 12 points lie on a conic;
3. 7 points lie on a line.

Example 4.6. $\deg Z \leq 20$, $H^1(\mathcal{I}_Z(6)) \neq 0$:

1. C^4C^5;
2. C^3C^6;
3. 19 points lie on a cubic;
4. 14 points lie on a conic;
5. 8 points lie on a line.

Example 4.7. $\deg Z \leq 24$, $H^1(\mathcal{I}_Z(7)) \neq 0$:

1. C^4C^6;
2. C^3C^7;
3. 22 points lie on a cubic;
4. 16 points lie on a conic;
5. 9 points lie on a line.

Example 4.8. $\deg Z \leq 30$, $H^1(\mathcal{I}_Z(8)) \neq 0$:

1. C^5C^6;
2. $C^4C^8 - \{$ 2 points $\}$;
3. $C^3C^7 + \{$ 9 collinear points $\}$;
4. $C^4C^8 - \{$ 3 collinear points $\}$;
5. C^4C^7;
6. 25 points lie on a cubic;
7. C^3C^8;
8. 18 points lie on a conic;
9. 10 points lie on a line.

References

[1] U. Angehrn, Y.-T. Siu, *Effective freeness and point separation for adjoint bundles,* Invent. Math. **122** (1995), 291–308

[2] I. Bacharach, *Uber den Cayley'schen Schnittpunktsatz,* Math. Ann. **26** (1886), 275–299

[3] M. C. Beltrametti, A. J. Sommese, *Zero cycles and k-th order embeddings of smooth projective surfaces,* (with an appendix by L. Göttsche), in: Problems in the Theory of Surfaces and Their Classification, (eds F. Catanese, C. Ciliberto and M. Cornalba), Sympos. Math. vol. 32 (1991), Academic Press, London, 33–48

[4] M. C. Beltrametti, A. J. Sommese, The Adjunction Theory of Complex Projective Varieties, De Gruyter Expositions in Mathematics, vol. 16, Walter de Gruyter, Berlin, New York (1995),

[5] F. A. Bogomolov, *Holomorphic tensors and vector bundles on projective varieties,* Math. USSR Izv. **13** (1978), 499–555

[6] F. Catanese, M. Schneider, *Bounds for stable bundles and degrees of Weierstrass schemes,* Math. Ann. **293** (1992), 579–594

[7] A. Cayley, *On the intersection of curves,* Cambridge Math. J. **3** (1843), 211–213; Collected math. papers I, vols. 25–27, Cambridge Univ. Press, 1889

[8] J.-P. Demailly, *Effective bounds for very ample line bundles,* Invent. Math. **124** (1996), 243–261

[9] D. Eisenbud, M. Green, J. Harris, *Cayley-Bacharach theorems and conjectures,* Bull. of the A.M.S. **33** (1996), 295–324

[10] L. Ein, R. Lazarsfeld, *Global generation of pluricanonical and adjoint linear series on smooth projective threefolds,* J.A.M.S. **6** (1993), 875–903

[11] T. Fujita, *On polarized manifolds whose adjoint bundles are not semipositive,* in: Algebraic Geometry, Sendai, Advanced Studies in Pure Math., vol. 10, (1987), 167–178

[12] T. Fujita, *Remarks on Ein-Lazarsfeld criterion of spannedness of adjoint bundle,* of polarized threefolds, preprint

[13] P. Griffiths, J. Harris, Principles of Algebraic Geometry, Wiley-Interscience (1978)

[14] Y. Kawamata, *On Fujita's freeness conjecture for 3-folds and 4-folds,* preprint

[15] R. Lazarsfeld, *Lectures on Linear Series,* in: Complex Algebraic Geometry, (ed. J. Kollar) Park-City IAS series, vol. 3 (1997), 163–219

[16] V. B. Mehta, A. Ramanathan, *Semi-stable sheaves on projective varieties and their restriction to curves,* Math. Ann. **258** (1982), 213–224

[17] Y. Miyaoka, *The Chern classes and Kodaira dimension of a minimal variety,* in: Algebraic geometry, Sendai 1985, Adv. Stud. Pure Math. **10** (1987), 449–476

[18] R. Paoletti, *On Halphen's specialty theorem,* in: Higher-Dimensional Complex Varieties (Trento, 1994), de Gruyter, Berlin (1996), 341–355

[19] I. Reider, *Vector bundles of rank 2 and linear systems on algebraic surfaces,* Ann. of Math. **127** (1988), 309–316

[20] M. Schneider, *Vector bundles and low-codimensional submanifolds of projective space: a problem list,* in: Topics in Algebra, Banach Center Publications, vol. 26, Part 2, PWN-Polish Scientific Publishers, Warsaw (1990),

[21] J. G. Semple, L. Roth, Introduction to Algebraic Geometry, Clarendon Press, Oxford (1949)

[22] Y.-T. Siu, *Effective very ampleness,* Invent. Math. **124** (1996), 563–571

[23] S.-L. Tan, *Cayley-Bacharach property of an algebraic variety and Fujita's conjecture,* J. of Algebraic Geometry **9** (2000), 201–222

[24] S.-L. Tan, E. Viehweg, *A note on Cayley-Bacharach property for vector bundles,* in: Complex Analysis and Algebraic Geometry, (eds Thomas Peternell and Frank-Olaf Schreyer), de Gruyter, Berlin (2000), 361–373

DEPARTMENT OF MATHEMATICS, EAST CHINA NORMAL UNIVERSITY, SHANGHAI 200062, P. R. OF CHINA

E-mail address: sltan@math.ecnu.edu.cn

AMS/IP Studies in Advanced Mathematics
Volume 20, 2001

Sharp polynomial upper estimate of number of positive integral points in tetrahedron and coordinate free characterization of homogeneous polynomials

KE-PAO LIN and STEPHEN S.-T. YAU

1. Introduction

Let Δ_n be a n-dimensional tetrahedron defined by $\frac{x_1}{a_1}+\frac{x_2}{a_2}+\cdots+\frac{x_n}{a_n}\leq 1$ and $x_1 \geq 0$, $\cdots$, $x_n \geq 0$ where $a_1, a_2 \cdots, a_n$ are positive integer. Let $P_n = \sharp\{(x_1,\cdots,x_n)\in \mathbf{Z}_+^n : \frac{x_1}{a_1}+\frac{x_2}{a_2}+\cdots+\frac{x_n}{a_n}\leq 1\}$=number of positive integral points in Δ_n. The general problem of counting the number P_n has been a challenge problem for many years. It has been discussed by many authors including Lehmer, Lochs and Ehrhart. There are extensive references in Review in Number Theory 1940-1972 and 1973-1983, especially in the sections H40, Vol.3 and P28, Vol.4. One tries to estimate P_n in terms of $a_1,\ a_2,\cdots,\ a_n$. Classically P_n is estimated by rational function and more complicated functions of $a_1,\ a_2,\cdots,\ a_n$ whose values are not easy to be estimated. Motivated from the number theory and singularities theory, we are interested in counting P_n for positive real numbers $a_1,\ a_2,\cdots,\ a_n$. This latter problem is much harder than the former one because the technique of toric variety from algebraic geometry is no longer applicable. Beside the applications in number theory and singularities theory, the latter problem has an interesting application in coordinate free characterization of homogeneous polynomial. In this paper, we shall discuss some of the recent results in this direction.

2. Estimation of P_n with integral vertices

We first begin with a simple example.

Example. Consider the 3-dimensional tetrahedron Δ_3:
$\frac{x}{5}+\frac{y}{4}+\frac{z}{3}\leq 1$, $x\geq 0,\ y\geq 0,\ z\geq 0$. It is clear that there are only two positive integral points, namely (1,1,1) and (2,1,1), in Δ_3. Hence $P_3 = 2$ in this case.

In order to compute P_n, in 1967 Ehrhart introduced the following function:
$\ell_{\Delta_n}(k) := \sharp(k\Delta_n \cap \mathbf{Z}^n)\quad , k\in \mathbf{Z}$
$:=$ number of integral points in Δ dilated by a factor of the integer k

Research supported by National Science Foundation.

Observe that $\ell_{\Delta_n}(1) = P_n$.

Ehrhart proved the following beautiful theorem.

Theorem 2.1[Eh]. $\ell_{\Delta_n}(k)$ is a polynomial function in k of degree n: $b_n k^n + b_{n-1}k^{n-1} + \cdots + b_0$ with rational coefficients and

b_n = volume of Δ_n

b_{n-1} = half the sum of the volumes of (n-1)-dimensional faces of Δ_n (measured with respect to the (n-1)-dimensional lattice in the (n-1)-plane)

b_0=1

Remark. The above polynomial is now called Ehrhart polynomial. For $n = 2$, Ehrhart's Theorem above reduces to the classical Pick's Theorem.

Theorem 2.2 [Pick]. $P_2 = vol(\Delta_2) + \frac{1}{2}S(\Delta_2) + 1$ where $S(\Delta_2)$ is the sum of the lattice lengths of the edges of Δ_2.

For $n = 3$, it was Mordell [Mo] who first noticed the connection between counting lattice points in a tetrahedra and certain Dedekind sums. In fact, he gave a formula for P_3 which is expressed in terms of a rational function and three Dedekind sums in case a_1, a_2, a_3 are pairwise relatively prime.

Definition. Let p, q be two relatively prime integers, then the clasical Dedekind sum is

$$s(p,q) = \sum_{i=1}^{q}((\frac{i}{q}))((\frac{pi}{q}))$$

where

$$((x)) = \begin{cases} x - [x] - \frac{1}{2} & x \notin \mathbf{Z} \\ 0 & x \in \mathbf{Z} \end{cases}$$

In 1993, Pommersheim [Po] gave a more general formula than Mordell.

Theorem 2.3 [Po]. For 3-dimensional tetrahedron Δ_3:

$$\frac{x_1}{a_1} + \frac{x_2}{a_2} + \frac{x_3}{a_3} \leq 1, \quad and \quad x_1 \geq 0,\ x_2 \geq 0,\ x_3 \geq 0,$$

if $gcd(a_1, a_2, a_3) = 1$,then

$$\ell_{\Delta_3}(k) = \frac{a_1a_2a_3}{6}k^3 + \frac{a_1a_2 + a_1a_3 + a_2a_3 + d}{4}k^2 + [\frac{1}{12}(\frac{a_1a_3}{a_2} + \frac{a_2a_3}{a_1} + \frac{a_1a_2}{a_3}$$

$$+\frac{d^2}{a_1a_2a_3}) + \frac{a_1 + a_2 + a_3 + A + B + C}{4} - As(\frac{a_2a_3}{d}, \frac{a_1A}{d})$$

$$-Bs(\frac{a_1a_3}{d}, \frac{a_2B}{d}) - Cs(\frac{a_1a_2}{d}, \frac{a_3C}{d})]k + 1$$

where $A = gcd(a_2, a_3)$, $B = gcd(a_1, a_3)$, $C = gcd(a_1, a_2)$, and $d = ABC$.

In fact for general integral Δ_3, Ehrhart's result says that

$$\ell_{\Delta_3}(k) = vol(\Delta_3)k^3 + \frac{1}{2}S(\Delta_3)k^2 + b_1 k + 1$$

where $S(\Delta_3)$ = sum of the lattice volumes of the 2-dimensional faces of Δ_3. Pommersheim's main contribution in his paper [Po] is the explicit computation of b_1.

Later Kantor and Khovanskii [Ka-Kh] succeeded in computing b_{n-2} in general. In fact they gave a general formula for the number of integral points in any integral polytope in $\mathbf{R}^4$. Most recently Capppell and Shaneson [Ca-Sh] have announced a fantastic result with which they can compute all the coefficients b_i in the Ehrhart polynomial. Unfortunately, from Pommersheim's result to the result of Cappell and Shaneson, although we have an exact formula for P_n, we still don't know how big P_n is because their formulas involve complicated functions which do not have sharp estimates.

3. Polynomial estimation of P_n with non-integral vertices

In this section, we shall consider n-dimensional tetrahedra Δ_n:

$$\frac{x_1}{a_1} + \frac{x_2}{a_2} + \cdots + \frac{x_n}{a_n} \leq 1, \quad and \quad x_1 \geq 0,\ x_2 \geq 0,\ \cdots,\ x_n \geq 0,$$

with non-integral vertices i.e. $a_1, a_2 \cdots a_n$ are positive real numbers. There are at least two reasons to count the number P_n of positive integral points in this Δ_n. Firstly, we are told by Professor Granville that this is an extremely important question in number theory; it would have many applications to current problems in analytic number theory, primality testing and in factoring. Given a set $\mathcal{P}$ of primes $p_1 < p_2 < \cdots < p_n \leq y$. Number theorists are interested in counting the number of integers $m \leq x$ where $m = p_1^{\ell_1} p_2^{\ell_2} \cdots p_n^{\ell_n}$ is composed only of primes from the $\mathcal{P}$, and $x = y^u$,for all $u > 2$. Thus they wish to count the number of $(\ell_1, \cdots, \ell_n) \in (\mathbf{Z}_+ \cup \{0\})^n$ such that

$$\frac{\ell_1}{a_1} + \frac{\ell_2}{a_2} + \cdots + \frac{\ell_n}{a_n} \leq 1 \quad where \quad a_i = \frac{\log x}{\log p_i} \geq u$$

Observe that a_i's are real numbers. This is of course the problem that we consider above.

Secondly, the problem has interesting application in singularity theory. Let $f : (\mathbf{C}^n, 0) \to (\mathbf{C}, 0)$ be the germ of a complex analytic function with isolated critical point at the origin. Let $V = \{z \in \mathbf{C}^n : f(z) = 0\}$.

Definition 3.1.
(a) The Milnor number of the singularity $(V, 0)$ is

$$\mu = dim\mathbf{C}\{z_1, z_2, \cdots, z_n\}/(f_{z_1}, f_{z_2}, \cdots, f_{z_n}).$$

(b) The geometric genus of the singularity $(V, 0)$ is

$$p_g = dim\Gamma(V - \{0\}, \Omega^n)/L^2(V - \{0\}, \Omega^n),$$

where Ω^n denotes the sheaf of germs of holomorphic n-forms on $V - \{0\}$ and $L^2(V - \{0\}, \Omega^n) = \{\omega \in \Gamma(V - \{0\},\ \Omega^n) :\ \int_{V-\{0\}} \omega \wedge \bar{\omega} < \infty\}$.

In 1978 Durfee [Du] made the following conjecture.

Durfee Conjecture(1978). $n!\ p_g \leq \mu$ with equality only when $\mu = 0$.

The connection between Durfee conjecture and the proposed problem is as follows.

Definition 3.2. Let $f(z_1, \cdots, z_n) = \sum a_\nu z^\nu$ be the Tayler expansion of a complex analytic function f, where $z^\nu = z_1^{\nu_1} \cdots z_n^{\nu_n}$. The Newton boundary $\Gamma(f)$ is the union of the compact faces of $\Gamma_+(f)$ where $\Gamma_+(f)$ is the convex hull of the union of the subsets $\nu + (\mathbf{R}_+)^n$ of $\mathbf{R}^n$ for ν such that $a_\nu \neq 0$. Newton polyhedron $\Gamma_-(f)$ of f is a cone over $\Gamma(f)$ with cone point at the origin. For any closed face Δ of $\Gamma(f)$, we associate a polynomial $f_\Delta(z) = \sum_{\nu \in \Delta} a_\nu z^\nu$. f is said to be nondegenerate on Δ if $\dfrac{\partial f_\Delta}{\partial z_1} = \cdots = \dfrac{\partial f_\Delta}{\partial z_n} = 0$ has no solution in $(\mathbf{C}^*)^{\mathbf{n}}$. f is said to be nondegenerate if f is nondegenerate on any face Δ of $\Gamma(f)$.

Example 3.1. $f(x, y, z) = x^5 + x^2y^2 + y^7$,

$\Gamma(f) = \Delta_1 \cup \Delta_2$

$f_{\Delta_1} = x^2y^2 + y^7$

$$\frac{\partial f_{\Delta_1}}{\partial x} = 2xy^2, \quad \frac{\partial f_{\Delta_1}}{\partial y} = 2x^2y + 7y^6 = y(2x^2 + 7y^5)$$

It is easy to see that f_{Δ_1} has critical points on x-axis.
$f_{\Delta_2} = x^5 + x^2y^2$

$$\frac{\partial f_{\Delta_2}}{\partial x} = 5x^4 + 2xy^2 = x(5x^3 + 2y^2), \quad \frac{\partial f_{\Delta_2}}{\partial y} = 2x^2y$$

It is easy to see that f_{Δ_2} has critical points on y-axis. Hence f is a nondegenerate polynomial.

We first recall a theorem of Merle-Teissier.

Theorem 3.1 [Me-Te]. Let $(V, 0)$ be an isolated hypersurface singularity defined by a nondegenerate holomorphic function $f : (\mathbf{C}^n, 0) \to (\mathbf{C}, 0)$. Then the geometric genus p_g is equal to $\sharp\{p \in \mathbf{Z}_+^n \cap \Gamma_-(f)\}$, the number of positive integral points in the Newton Polyhedron $\Gamma_-(f)$ of f.

Definition 3.3. A polynomial $f(z_1, \cdots, z_n)$ is weighted homogeneous of type $(w_1, \cdots, w_n)$, where $w_1, \cdots, w_n$ are fixed positive rational numbers, if f can be expressed as a linear combination of monomials $z_1^{i_1} \cdots z_n^{i_n}$ for which $i_1/w_1 + \cdots + i_n/w_n = 1$

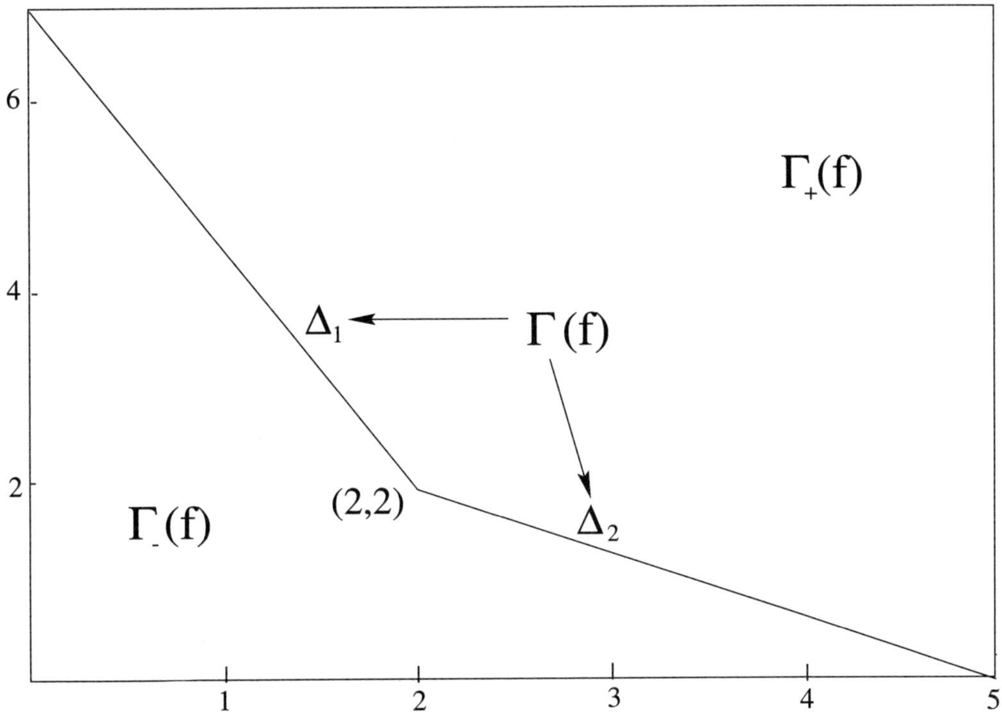

FIGURE 1. Figure of Example 3.1

Remark. If f is a weighted homogeneous polynomial, then the Newton polyhedron $\Gamma_-(f)$ is a n-dimensional tetrahetra. Therefore the theorem of Merle-Teissier implies that computing the geometric genus of isolated singularity defined by a weighted homogeneous polynomial is equivalent to counting the number of positive integral points in a tetrahedron Δ_n.

We next recall a theorem of Milnor and Orlik.

Theorem 3.2 [Mi-Or]. Let $f(z_1, \cdots, z_n)$ be a weighted homogeneous polynomial of type $(w_1, \cdots, w_n)$ with isolated singularity at the origin. Then the Milnor number is $\mu = (w_1 - 1)(w_2 - 1) \cdots (w_n - 1)$

Durfee conjecture together with Theorem 3.1 and Theorem 3.2 suggest the following polynomial estimate of P_n for Δ_n with non-integral vertices.

Conjecture. Let Δ_n be the n-dimensional tetrahedra:

$$\frac{x_1}{a_1} + \frac{x_2}{a_2} + \cdots + \frac{x_n}{a_n} \leq 1 \quad \textit{and} \quad x_1 \geq 0,\ x_2 \geq 0,\ \cdots, x_n \geq 0,$$

where $a_1, a_2, \cdots, a_n$ are positive real numbers greater than or equal to 2. Let P_n be the numberof positive integral points in Δ_n. Then

$$n! P_n \leq (a_1 - 1)(a_2 - 1) \cdots (a_n - 1)$$

For $n = 3$, Xu and Yau [Xu-Ya 1] in 1992 established the following polynomial estimate.

Theorem 3.3 [Xu-Ya 1] Let $a \geq b \geq c \geq 2$ be real number. Let $P_3 =$ number of positive integral solutions of $\frac{x}{a} + \frac{y}{b} + \frac{z}{c} \leq 1$. Then

$$6P_3 \leq (a-1)(b-1)(c-1) - c + 1$$

and the equality is attained if and only if $a = b = c = \textit{integer}$.

The sharp estimate in Theorem 3.3 can be used to give lower bound of the Dedekind sums in Theorem 2.3.

Given a function f with an isolated singularity at the origin, it is an important question to know whether f is a weighted homogeneous polynomial or a homogeneous polynomial after a biholomorphic change of variables. The former question was answered by a celebrated paper [Sa] of Saito in 1973. However the latter question has remained open ever since. To illustrate the difficulty of the latter problem, let us consider the following example.

Example 3.2. Let $f(x,y,z) = x^2 + y^3 + z^4$. Take a nonsingular change of coordinate

$$\begin{pmatrix} x \\ y \\ z \end{pmatrix} = \begin{pmatrix} a_{11} & a_{12} & a_{13} \\ a_{21} & a_{22} & a_{23} \\ a_{31} & a_{32} & a_{33} \end{pmatrix} \begin{pmatrix} x_1 \\ y_1 \\ z_1 \end{pmatrix}$$

Then we get a new polynomial

$$\begin{aligned} \tilde{f}(x_1,y_1,z_1) &= (a_{11}x_1 + a_{12}y_1 + a_{13}z_1)^2 + (a_{21}x_1 + a_{22}y_1 + a_{23}z_1)^3 \\ &\quad + (a_{31}x_1 + a_{32}y_1 + a_{33}z_1)^4 \end{aligned}$$

On the other hand, let $g(x,y,z) = x^2 + y^2 + z^2$. Take a nonsingular change of coordinate

$$\begin{pmatrix} x \\ y \\ z \end{pmatrix} = \begin{pmatrix} b_{11} & b_{12} & b_{13} \\ b_{21} & b_{22} & b_{23} \\ b_{31} & b_{32} & b_{33} \end{pmatrix} \begin{pmatrix} x_1 \\ y_1 \\ z_1 \end{pmatrix} + \begin{pmatrix} q_1 \\ q_2 \\ q_3 \end{pmatrix}$$

where q_1, q_2, q_3 are homogeneous polynomials of degree two in x_1, y_1, z_1. Then we get a new polynomial

$$\begin{aligned} \tilde{g}(x_1,y_1,z_1) &= (b_{11}x_1 + b_{12}y_1 + b_{13}z_1 + q_1(x_1,y_1,z_1))^2 \\ &\quad + (b_{21}x_1 + b_{22}y_1 + b_{23}z_1 + q_2(x_1,y_1,z_1))^2 \\ &\quad + (b_{31}x_1 + b_{32}y_1 + b_{33}z_1 + q_3(x_1,y_1,z_1))^2 \end{aligned}$$

We can see that although $\tilde{f}$ and $\tilde{g}$ have the same format, $\tilde{g}$ can be changed to homogeneous polynomial while $\tilde{f}$ cannot be changed to a homogeneous polynomial.

In 1993, Xu and Yau [Xu-Ya 2] sharpened their Theorem 3.3 and proved the following theorem.

Theorem 3.4 [Xu-Ya 2]. Let $(V, 0)$ be a two dimensional isolated singularity defined by a weighted homogeneous polynomial $f(z_1, z_2, z_3) = 0$. Let μ, p_g, and ν be the Milnor number, geometric genus, and multiplicity of the singularity respectively. Then

$$\mu - \nu + 1 \geq 6p_g$$

and equality holds if and only if $(V, 0)$ is defined by the homogeneous polynomial.

As a corollary of Theorem 3.4, Xu and Yau [Xu-Ya 2] obtained the following coordinate free characterization of homogeneous polynomial of three variables.

Corollary 3.5 [Xu-Ya 2]. Suppose that $f : (\mathbf{C}^3, 0) \rightarrow (\mathbf{C}, 0)$ has an isolated critical point at the origin. Let μ , p_g,and ν as in Theorem 3.4. Then f is a homogeneous polynomial after a change of coordinate if and only if $\mu - \nu + 1 = 6p_g$ and $\mu = \tau$, where $\tau = dim\mathbf{C}\{z_1, z_2, z_3\}/(f, f_{z_1}, f_{z_2}, f_{z_3})$.

In view of Corollary 3.5, in 1995, Yau made the following conjecture.

Durfee-Yau Conjecture(1995). Let $f : (\mathbf{C}^n, 0) \rightarrow (\mathbf{C}, \mathbf{0})$ be a weighted homogeneous polynomial with an isolated critical point at the origin. Then

$$\mu - h(\nu) \geq n!\, p_g$$

and equality holds if and only if f is a homogeneous polynomial. Here $h(\nu)$ is a polynomial function on the multiplicity ν with the properties $h(\nu) \geq 0$ and $h(\nu) = 0$ if and only if $\nu = 1$.

Remark.
For $n = 3$, Corollary 3.5 implies that Durfee-Yau Conjecture holds with $h(\nu) = \nu - 1$.

For $n = 4$, Xu and Yau [Xu-Ya 3] get the following polynomial estimate for P_4 in 1996.

Theorem 3.6 [Xu-Yau 3]. Let $a \geq b \geq c \geq d \geq 2$, and $P_4 = \sharp\{(x, y, z, w) \in \mathbf{Z}^4_+ : \frac{x}{a} + \frac{y}{b} + \frac{z}{c} + \frac{w}{d} \leq 1\}$. If $P_4 > 0$, then

$$24P_4 \leq abcd - \frac{3}{2}(abc + abd + acd + bcd) + \frac{11}{3}(ab + ac + bc) - 2(a + b + c)$$

and the equality is attained if and only if $a = b = c = d = integer$.

If $a \geq b \geq c \geq d \geq 3$, then the same conclusion hold without the assumption $P_4 > 0$

The purpose of this paper is to announce the following result which implies a coordinate free characterization of homogeneous polynomials in 4 variables.

Main Theorem. Let $(V, 0)$ be a three dimensional isolated singularity defined by a weighted homogeneous polynomial $f(x, y, z, w) = 0$. Let μ , p_g,and ν be the Milnor number, geometric genus, and multiplicity of the singularity. Then we have

$$\mu - (2\nu^3 - 5\nu^2 + 2\nu + 1) \geq 4!\, p_g$$

and the equality is attained if and only if $(V, 0)$ is defined by an homogeneous polynomial.

Corollary. (Coordinate free characterization of homogeneous polynomial of 4 variables) Let $f : (\mathbf{C}^4, 0) \to (\mathbf{C}, \mathbf{0})$ be a complex analytic function with an isolated critical point at the origin. Let μ , p_g,and ν be as above in the Main Theorem. Then f is a homogeneous polynomial after change of coordinate if and only if $\mu - (2\nu^3 - 5\nu^2 + 2\nu + 1) = 4!p_g$ and $\mu = \tau$ where $\tau = dim\mathbf{C}\{z_1, z_2, z_3, z_4\}/(f, f_{z_1}, f_{z_2}, f_{z_3}, f_{z_4})$.

The following is a sketch of the proof of the Main Theorem. We first prove the following theorem.

Theorem A. Let $a \geq b \geq c \geq d \geq 3$ or $a \geq b \geq c \geq d = 2$ be real numbers. Let $P_4 = \sharp\{(x, y, z, w) \in \mathbf{Z}_+^4 : \frac{x}{a} + \frac{y}{b} + \frac{z}{c} + \frac{w}{d} \leq 1\}$. Define $\mu = (a-1)(b-1)(c-1)(d-1)$. Then

$$\mu \geq 24P_4 + 2d^3 - 5d^2 + 2d + 1$$

and the equality is attained if and only if $a = b = c = d = integer$.

Recall that the multiplicity of f is the smallest integer bigger than or equal to the smallest weight of f. Hence Theorem A is exactly our Main Theorem in case the smallest weight of f is an integer.

We next prove the following theorem.

Theorem B. Let $a \geq b \geq c \geq d > 3$ be real number and $P_4 = \sharp\{(x, y, z, w) \in \mathbf{Z}_+^4 : \frac{x}{a} + \frac{y}{b} + \frac{z}{c} + \frac{w}{d} \leq 1\}$. Suppose d is not an integer and $d = [d] + \beta$ where β is either $\frac{d}{c}$ or $\frac{d}{b}$ or $\frac{d}{a}$. Define $\mu = (a-1)(b-1)(c-1)(d-1)$. Then

$$\begin{aligned} 24P_4 < {} & \mu - (2\nu^3 - 5\nu^2 + 2\nu + 1) \mid_{\nu = d - \beta + 1} \\ = {} & abcd - (abc + abd + acd + bcd) + (ab + ac + ad + bc + bd + cd) \\ & -(a + b + c) - (2d^3 + d^2 - d - 1) + 2\beta^3 - \beta^2(6d + 1) + \beta(6d^2 + 2d - 2) \end{aligned}$$

We observe that if $w_1 \geq w_2 \geq w_3 \geq w_4$, $w_4 = [w_4] + \beta$ with $0 < \beta < 1$, where (w_1, w_2, w_3, w_4) is a weight type of a weighted homogeneous polynomial, then β is either $\frac{w_4}{w_1}$ or $\frac{w_4}{w_2}$ or $\frac{w_4}{w_3}$. Hence Theorem B implies the Main Theorem if the smallest weight is a non-integer bigger than 3.

We next prove the following theorem.

Theorem C. Let $a \geq b \geq c \geq d > 2$ be real numbers and $P_4 = \sharp\{(x, y, z, w) \in \mathbf{Z}_+^4 : \frac{x}{a} + \frac{y}{b} + \frac{z}{c} + \frac{w}{d} \leq 1\}$. Suppose d is not an integer and $d = [d] + \beta$ with $0 < \beta < 1$.

(i)If β is either $\frac{d}{c}$ or $\frac{d}{b}$, then the inequality in Theorem B holds.

(ii)If β is $\frac{d}{a}$ and $P_4 > 0$. Then the inequality in Theorem B holds.

Now our Main Theorem follows from Theorem A, Theorem B, and Theorem C except in the case that the smallest weight w_4 is of the form $2 + \frac{w_4}{w_1}$ and $p_g = 0$, in which case the Main Theorem can be proved case by case checking because we have classification of weighted homogeneous polynomials with $p_g = 0$.

References

[Ca-Sh] S.E. Cappell and J.L. Shaneson, Genera of algebraic varieties and counting lattice points, Bull. A.M.S.(New series) 30 (1994), 62-69.

[Di-Ro] R. Diaz and S. Robins, The ehrhart polynomial of a lattice polytope, Annals of mathematics, 145 (1997), 503-518.

[Du] A.H. Durfee, The signature of smoothings of complex surface singularities, Math. Ann. 232 (1978) 85-98.

[Eh] E.Ehrhart, Sur un problèm de géométrie diophantienne linéaire II, J. reine angew. Math. 227 (1967), 25-49.

[Ka-Kh] J.M. Kantor and A. Khovanskii, Une application du Théorème de Riemann-Roch combinatoire au polynôme d'Ehrhart des polytopes entier de $\mathbf{R}^d$, C.R. Acad. Sci. Paris 317 (I) (1993),501-507.

[Me-Te] M. Merle and B. Teissier, Conditions d'adjonction d'aprěs Du Val, Sěminaire sur les singularités des surfaces (Center de Math. de l'Ecole Polytechniqe, 1976-1977), Lecture Notes in Math., Vol. 777, Springer, Berlin, 1980, 229-245.

[Mi-Or] J. Milnor & P. Orlik, Isolated singularities defined by weighted homogeneous polynomials, Topology 9 (1970) 385-393.

[Mo] L.J. Mordell, Lattice points in tetrahedron and generalized Dedekind sums, J. Indian Math. 15 (1951), 41-46.

[Po] J. Pommershim, Toric variety, lattice points and Dedekind sums, Math. Ann. 295 (1993), 1-24.

[Sa] K. Saito, Quasihomogene isolierte singularitäten von Hyperflachen, Invent. Math. 14 (1971) 123-142.

[Xu-Ya 1] Y.-J. Xu and S.S.-T. Yau, Sharp estimate of number of integral points in a tetrahedron, J. reine angew. Math. 423 (1992),199-219.

[Xu-Ya 2] Y.-J. Xu and S.S.-T. Yau, Durfee conjecture and coordinate free characterization of homogeneous singularities, J. Diff. Geom. 37 (1993), 375-396.

[Xu-Ya 3] Y.-J. Xu and S.S.-T. Yau, Sharp estimate of number of integral points in a 4-dimensional tetrahedron, J. reine angew. Math. 473 (1996),1-23.

Department of MSCS (M/C 249), University of Illinois at Chicago, 851 South Morgan Street, Chicago, Illinois 60607-7045

E-mail address: klin@math.uic.edu; yau@uic.edu

AMS/IP Studies in Advanced Mathematics
Volume 20, 2001

Spectral theory and geometry of locally symmetric spaces

Lizhen Ji

ABSTRACT. We discuss several results about the spectrum of locally symmetric spaces and their connection with the geometry at infinity. Specifically, we present a solution of the trace class conjecture in the theory of the Selberg trace formula, the Weyl type upper bound on the counting function of the discrete spectrum of locally symmetric spaces, and new compactifications of quotients by arithmetic subgroups whose boundary components play a crucial role in the proof of the trace class conjecture.

1. Introduction

For a close, compact Riemannian manifold M, the spectrum Spec(Δ) of the Laplace-Beltrami operator Δ acting on $L^2(M)$ is discrete: $0 = \lambda_0 \leq \lambda_1 \leq \lambda_2 \leq \cdots$, where the eigenvalues are listed with multiplicity. In the following, the spectrum ia also denoted by Spec(M).

Though the eigenvalues λ_k do not determine M up to isometry, they are important invariants of the manifold. Denote the counting function of Spec(M) by $N(\lambda) = |\{\lambda_i \in \mathrm{Spec}(M) \mid \lambda_k \leq \lambda\}|$. Then $N(\lambda)$ satisfies the Weyl law:

$$\lim_{\lambda\to+\infty} \frac{N(\lambda)}{\lambda^{\frac{n}{2}}} = (4\pi)^{-\frac{n}{2}} \frac{\mathrm{vol}(M)}{\Gamma(\frac{n}{2}+1)}, \quad n = \dim M.$$

From this Weyl law, it is clear that Spec(M) determines the dimension and volume of M. In general, it is very difficult to compute explicitly the eigenvalues of M, but the Weyl law says that λ_k is asymptotic to $4\pi(\Gamma(\frac{n}{2}+1)/\mathrm{vol}(M))^{\frac{2}{n}} k^{\frac{2}{n}}$ as $k \to +\infty$, in particular, that M has infinitely many eigenvalues.

On the other hand, if the manifold M is noncompact, Spec(M) is not discrete in general and consists of a continuous part $\mathrm{Spec}_{con}(M)$, a discrete part $\mathrm{Spec}_{dis}(M)$ (or the pure point spectrum), and possibly some accumulation points of the discrete part. Each part could be empty, and the discrete spectrum could be nonempty but finite. For example, when $M = \mathbb{R}^n$, $\mathrm{Spec}(M) = [0, +\infty)$ is absolutely continuous, and there is no discrete spectrum, i.e., no L^2-eigenfunction.

For a general noncompact Riemannian manifold M, very little is known about Spec(M), in particular its discrete part, and its connection with the geometry at infinity of M. In the following, we will concentrate on an important class of

Partially Supported by NSF grants and an Alfred P. Sloan Research Fellowship.

complete, noncompact Riemannian manifolds: locally symmetric spaces $\Gamma\backslash X = \Gamma\backslash G/K$ of finite volume. We will study the following problems:

1. The discrete spectrum and generalization of the Weyl law for compact manifolds.
2. The continuous spectrum of $\Gamma\backslash X$ and its generalized eigenfunctions.
3. The geometry at infinity and compactifications of $\Gamma\backslash X$, and their connection to the spectrum of $\Gamma\backslash X$.

For locally symmetric spaces $\Gamma\backslash X$, study of the discrete spectrum is closely related to study of the continuous spectrum. For example, the analogue of the Weyl law involves counting of both the discrete and continuous spectra, and has been obtained through the Selberg trace formula in special cases. The main result in this paper is a complete solution to the trace class conjecture in the theory of the Selberg trace formula for general arithmetic subgroups. An important point that will become clear is that the spectral theory of $\Gamma\backslash X$ can be better understood through the right regular representation of G on $L^2(\Gamma\backslash G)$.

The rest of this paper is organized as follows. In §2, we first discuss the spectral theory of the simplest locally symmetric spaces $\Gamma\backslash SL(2,\mathbb{R})/SO(2)$, and then explain how to reduce the study of the spectrum of $\Gamma\backslash SL(2,\mathbb{R})/SO(2)$ to understanding irreducible subrepresentations of the right regular representation of $SL(2,\mathbb{R})$ in $L^2(\Gamma\backslash SL(2,\mathbb{R}))$. In §3, we state the trace class conjecture for general semisimple Lie groups and explain its relation to bounds on the counting function of the discrete spectrum of the associated locally symmetric spaces. In §4, we explain the key steps in the proof of polynomial upper bounds on the counting function of the discrete spectrum of $\Gamma\backslash X$, and discuss several results on compactifications of $\Gamma\backslash G$ and $\Gamma\backslash X$.

2. Riemann Surfaces

In this section, we recall several results from the spectral theory of arithmetic Riemann surfaces to motivate the general problems to be studied in later sections.

Let $\mathbb{H} = \{x + iy \mid x \in \mathbb{R}, y > 0\}$ be the upper half plane with the Poincare metric $ds^2 = y^{-2}(dx^2 + dy^2)$. Then $SL(2,\mathbb{R})$ acts isometrically and transitively on $\mathbb{H}$, and the isotropy subgroup of the basepoint $x_0 = i$ is $SO(2)$. Hence $\mathbb{H} = SL(2,\mathbb{R})/SO(2)$. Let Γ be equal to $SL(2,\mathbb{Z})$ or a subgroup of finite index. Then $\Gamma\backslash\mathbb{H}$ is the simplest example of locally symmetric spaces of finite volume. When Γ is a congruence subgroup, $\Gamma\backslash\mathbb{H}$ is a modular curve.

The continuous spectrum of $\Gamma\backslash\mathbb{H}$ is equal to $[\frac{1}{4}, +\infty)$ with multiplicity equal to the number of cusps of $\Gamma\backslash\mathbb{H}$, and the generalized eigenfunctions are given by the Eisenstein series. Briefly, $\mathrm{Spec}(\mathbb{H}) = [\frac{1}{4}, +\infty)$, and its generalized eigenfunctions are analogues of the exponential functions for $\mathbb{R}^n$; and the Eisenstein series for $\Gamma\backslash\mathbb{H}$ are averages of the exponential functions of $\mathbb{H}$ so that they are Γ invariant.

More precisely, there is an Eisenstein series for every cusp. If $i\infty$ is a cusp, then the associated Eisenstein series is

$$E_\infty(s,z) = \sum_{\gamma \in \Gamma_\infty\backslash\Gamma} \mathrm{Im}(\gamma z)^s, \quad \mathrm{Re}(s) > 1,$$

where Γ_∞ is the stabilizer in Γ of $i\infty$. The Eisenstein series $E_\infty(s,z)$ admits a meromorphic continuation to the whole s-complex plane and satisfies the eigen-equation $\Delta E_\infty(s,z) = s(1-s)E_\infty(s,z)$. When $\mathrm{Re}(s) = \frac{1}{2}$, $s(1-s)$ is real and belongs to $[\frac{1}{4}, +\infty)$, and $E_\infty(s, z)$ gives a generalized eigenfunction of the continuous spectrum $[\frac{1}{4}, +\infty)$. The Eisenstein series for other cusps can be defined similarly. Together, they form all the generalized eigenfunctions of $\mathrm{Spec}(\Gamma\backslash\mathbb{H})$. (See [**8**].)

On the other hand, less is known about the discrete spectrum of $\Gamma\backslash\mathbb{H}$. It is clear that $\lambda_0(\Gamma\backslash\mathbb{H}) = 0$, but bounds on $\lambda_1(\Gamma\backslash\mathbb{H})$ are very subtle. When $\Gamma = SL(2,\mathbb{Z})$, $\lambda_1(\Gamma\backslash\mathbb{H}) > \frac{1}{4}$, i.e., all λ_i, $i \geq 1$, are embedded in the continuous spectrum. If Γ is a congruence subgroup of $SL(2,\mathbb{Z})$, Selberg made the following famous conjecture:

$$\lambda_1(\Gamma\backslash\mathbb{H}) \geq \frac{1}{4}.$$

This conjecture is still open and has important applications to analytic number theory (see [**19**] for a nice summary and details about this conjecture).

Another problem concerns the distribution of the discrete spectrum $\mathrm{Spec}_{dis}(\Gamma\backslash\mathbb{H})$, in particular, whether it is infinite. Let $N_d(\lambda)$ be the counting function of $\mathrm{Spec}_{dis}(\Gamma\backslash\mathbb{H})$. A natural question is whether $N_d(\lambda)$ satisfies the Weyl law as in the compact case. It turns out that the continuous spectrum comes in naturally, and the answer depends on the size of the continuous spectrum. The growth $N_c(X)$ of the continuous spectrum can be measured in terms of the determinant φ of the scattering matrix:

$$N_c(\lambda) = -\frac{1}{4\pi}\int_{-\sqrt{\lambda-\frac{1}{4}}}^{\sqrt{\lambda-\frac{1}{4}}} \frac{\varphi'}{\varphi}(\frac{1}{2} + ir)dr, \quad \lambda \geq \frac{1}{4}$$

where the scattering matrix is determined by the constant terms of the Eisenstein series (see [**8**]). Then the analogue of the Weyl law for the noncompact space $\Gamma\backslash\mathbb{H}$ is the following Weyl–Selberg law:

$$N_d(\lambda) + N_c(\lambda) \sim \frac{\mathrm{area}(\Gamma\backslash\mathbb{H})}{4\pi}\lambda, \qquad \text{as } \lambda \to +\infty,$$

which follows from the Selberg trace formula.

In general, it is difficult to determine the relative sizes of $N_d(\lambda)$ and $N_c(\lambda)$. When Γ is a congruence subgroup, it can be shown that $N_c(\lambda) = o(\lambda)$ as $\lambda \to +\infty$, and hence $N_d(\lambda)$ satisfies the Weyl law [**21**] [**8**] [**9**]:

$$N_d(\lambda) \sim \frac{\mathrm{area}(\Gamma\backslash\mathbb{H})}{4\pi}\lambda, \qquad \text{as } \lambda \to +\infty.$$

On the other hand, if Γ is a non-congruence arithmetic subgroup, the Weyl law is not known, though $N_d(\lambda) \to +\infty$ as $\lambda \to +\infty$. In fact, it can be shown that $\lim_{\lambda\to+\infty} \inf N_d(\lambda)/\lambda > 0$ (see [**23**]). For a non-arithmetic subgroup Γ, it is conjectured by Phillips and Sarnak (see [**20**]) that $\mathrm{Spec}_{dis}(\Gamma\backslash\mathbb{H})$ is finite.

The eigenfunctions u_k, $\Delta u_k = \lambda_k u_k$, of $\Gamma\backslash\mathbb{H}$ are Maass forms. They are real analytic but not holomorphic. A (classical) modular form of weight k on $\Gamma\backslash\mathbb{H}$ is a holomorphic function f on $\mathbb{H}$ satisfying the condition

$$f(\gamma \cdot z) = f(\frac{az+b}{cz+d}) = (cz+d)^k f(z)$$

for all $\gamma \in \Gamma$, $\gamma = \begin{pmatrix} a & b \\ c & d \end{pmatrix}$, and suitable growth condition at infinity.

Such a modular form is a holomorphic section of a line bundle over $\Gamma\backslash\mathbb{H}$ and can be lifted to a function $\tilde{f}$ on $\Gamma\backslash SL(2,\mathbb{R})$ by

$$\tilde{f}(g) = (ci+d)^{-k} f(g\cdot i),$$

where $g = \begin{pmatrix} a & b \\ c & d \end{pmatrix}$. Similarly, Maass forms on $\Gamma\backslash\mathbb{H}$ can also be lifted to functions on $\Gamma\backslash SL(2,\mathbb{R})$. To study simultaneously the Maass forms and modular forms of all weights together with more general eigensections of the line bundles on $\Gamma\backslash\mathbb{H}$, we study functions on $\Gamma\backslash SL(2,\mathbb{R})$.

The space $\Gamma\backslash SL(2,\mathbb{R})$ is a principal $SO(2)$-bundle over $\Gamma\backslash\mathbb{H}$, and different irreducible (1-dimensional) representations of $SO(2)$ give rise to holomorphic line bundles corresponding to modular forms of different weights. One advantage of $\Gamma\backslash SL(2,\mathbb{R})$ over $\Gamma\backslash\mathbb{H}$ is that $\Gamma\backslash SL(2,\mathbb{R})$ is a homogeneous space admitting a right G-action, and one can study Maass and modular forms using representation theory.

Denote the right regular representation of $SL(2,\mathbb{R})$ on $L^2(\Gamma\backslash SL(2,\mathbb{R}))$ by R: $R(g)f(x) = f(xg)$, $g \in SL(2,\mathbb{R})$.

PROPOSITION 2.1. *Lift a L^2-Maass or modular form f to a function $\tilde{f}$ in $L^2(\Gamma\backslash SL(2,\mathbb{R}))$ as above. Then the subspace spanned by the right translates $R(g)\tilde{f}$ of $\tilde{f}$, $g \in SL(2,\mathbb{R})$, is an irreducible subrepresentation of the regular representation R.*

Such irreducible subrepresentations of R are special. In general, an irreducible subrepresentation of R is generated by an automorphic function, which is an eigensection of a holomorphic line bundle for modular forms. From this description, it is clear that the Maass and modular forms are special automorphic forms. See **[6]** for more discussions about the relation between automorphic forms and subrepresentations of R.

Let $L^2_{dis}(\Gamma\backslash SL(2,\mathbb{R}))$ be the direct sum of all irreducible subrepresentations π of R: $L^2_{dis}(\Gamma\backslash SL(2,\mathbb{R})) = \oplus_\pi m(\pi)\pi$, where $m(\pi)$ is the multiplicity of π. Then the study of the discrete spectrum of $\Gamma\backslash\mathbb{H}$ and the modular forms on it is reduced to understanding the discrete subspace $L^2_{dis}(\Gamma\backslash SL(2,\mathbb{R}))$ and its irreducible subspaces.

3. The Trace Class Conjecture

In this section, we set up the framework of the Selberg trace formula and state the trace class conjecture.

Let $\mathbf{G}$ be a semisimple linear algebraic group defined over $\mathbb{Q}$, $\Gamma \subset \mathbf{G}(\mathbb{Q})$ an arithmetic subgroup, and $G = \mathbf{G}(\mathbb{R})$. Since G is semisimple, it admits a bi-invariant Haar measure. The discrete subgroup Γ acts properly discontinuously on G, and the quotient $\Gamma\backslash G$ has finite volume. In the following, we assume that $\Gamma\backslash G$ is noncompact.

The right regular representation of G on $L^2(\Gamma\backslash G)$ is denoted by R. The direct sum of all the irreducible subrepresentations of R gives the discrete subspace $L^2_{dis}(\Gamma\backslash G)$:

$$L^2_{dis}(\Gamma\backslash G) = \oplus_{\pi\in\hat{G}} m(\pi)\pi.$$

The othrogonal completemt of $L^2_{dis}(\Gamma\backslash G)$ in $L^2(\Gamma\backslash G)$ is nonempty, called the continuous subspace and denoted by $L^2_{con}(\Gamma\backslash G)$.

The irreducible subrepresentations π in $L^2_{dis}(\Gamma\backslash G)$ are automorphic representations of G and play an important role in Langlands' program. As explained above near the end of §2, study of the discrete spectrum of the locally symmetric space $\Gamma\backslash X = \Gamma\backslash G/K$, where K is a maximal compact subgroup of G, can be reduced to understanding these automorphic forms. Therefore, it is a fundamental problem to understand these automorphic representations and determine their multiplicities. A powerful tool for this purpose is the Selberg trace formula.

Briefly, the Selberg trace formula can be explained as follows. For every function $\alpha \in C_o^\infty(G)$, the regular representation R defines an operator $R(\alpha)$ on $L^2(\Gamma\backslash G)$:

$$R(\alpha)f(x) = \int_G \alpha(g)R(g)f(x)dg.$$

Then $R(\alpha)$ preserves the subspace $L^2_{dis}(\Gamma\backslash G)$ and its restriction is denoted by $R_{dis}(\alpha)$.

Basically, the Selberg trace formula, as envisioned by Selberg, computes the trace $\mathrm{tr}R_{dis}(\alpha) = \sum_{\pi\in\hat{G}} m(\pi)trR(\alpha)|_\pi$ in terms of geometry of $\Gamma\backslash G$ (orbital integrals parametrized by conjugacy classes in Γ), and continuous spectrum (scattering matrices or intertwining operators), which is related to the geometry at infinity.

The first natural problem is that $\mathrm{tr}R_{dis}(\alpha)$ needs to be well-defined.

CONJECTURE 3.1 (Trace Class Conjecture). *For all $\alpha \in C_0^\infty(G)$, $R_{dis}(\alpha)$ is of trace class.*

Recall that an operator A on a Hilbert space is of trace class if sum of all the eigenvalues of the positive operator $|A| = \sqrt{A^*A}$ is finite.

THEOREM 3.2. *The trace class conjecture is true.*

This theorem is proved independently in [**10**] and [**17**]. In fact, for more general functions α belonging to the space $\mathcal{C}^1(G)$ of integrable Schwartz functions on G, $R_{dis}(\alpha)$ is of trace class. Consequently, $trR_{dis} : \mathcal{C}^1(G) \to \mathbb{C}$ defines a distribution, and the Selberg trace formula is a useful method to describe this distribution.

Partial results on this conjecture have been obtained by various people in the following chronological order. Selberg [**21**] proved it when $\mathbf{G} = SL(2)$ and α is $SO(2)$-invariant; Warner [**24**] proved the conjecture when $\mathbf{G}$ has $\mathbb{R}$-rank 1, i.e., when X has strictly negative curvature, and α is K-finite; Donnelly [**4**] proved it when $\mathbf{G}$ has $\mathbb{Q}$-rank 1, i.e., the ends of $\Gamma\backslash G$ are topological cylinders, and α is K-finite; Müller proved it for general $\mathbf{G}$ and K-finite α; and Langlands proved the conjecture for all α, i.e., without the K-finiteness assumption, when $\mathbf{G}$ has $\mathbb{Q}$-rank 1.

The removal of the K-finiteness is interesting for several reasons. From an intrinsic point of view, it is natural to remove this extra assumption. More importantly, some problems require us to treat all automorphic forms simultaneously, for example, all modular forms of different weights on $\Gamma\backslash\mathbb{H}$, and it is important for such purposes not to restrict the K-types of α. For example, in [**15**], asymptotic properties of a family of modular forms are studied as the weight goes to infinity.

It seems to be natural to add a remark on relation between the Arthur–Selberg trace formula and the trace class conjecture. As known, Arthur (see [**1**] and the references there) has established a trace formula useful for comparison of automorphic forms on different groups before the trace class conjecture was proved. In fact, the trace he computed is the trace of $R(\alpha)$ restricted to a smaller subspace than $L^2_{dis}(\Gamma\backslash G)$. More precisely, the discrete subspace $L^2_{dis}(\Gamma\backslash G)$ contains a proper subspace $L^2_{cus}(\Gamma\backslash G)$, called the cuspidal subspace. Functions in $L^2_{cus}(\Gamma\backslash G)$ have vanishing constant terms along all rational parabolic subgroups, and automorphic forms in $L^2_{cus}(\Gamma\backslash G)$ have fast decay at infinity. The complement of $L^2_{cus}(\Gamma\backslash G)$ in $L^2_{dis}(\Gamma\backslash G)$ is called the residual subspace, denoted by $L^2_{res}(\Gamma\backslash G)$, since it is spanned by square integrable residues of Eisenstein series. In contrast to cuspidal automorphic functions, automorphic functions in $L^2_{res}(\Gamma\backslash G)$ decay very slowly at infinity, for example, the constant function belongs to $L^2_{res}(\Gamma\backslash G)$ but does not decay at all.

Gelfand and Piatetski-Shapiro [**7**] proved that the restriction $R_{cus}(\alpha)$ of $R(\alpha)$ to $L^2_{cus}(\Gamma\backslash G)$ is of trace class, and Arthur computed the trace of $R_{cus}(\alpha)$. The original Selberg trace formula computing $R_{dis}(\alpha)$ as envisioned by Selberg in [**22**] has not been established for general **G**, though it is a very interesting problem.

The proof of the trace class conjecture is related to upper bounds on the counting function of the discrete spectrum of locally symmetric spaces $\Gamma\backslash X$ and $\Gamma\backslash G$.

Let K be a maximal compact subgroup of G. Then $X = G/K$ is a Riemannian symmetric space of noncompact type, and $\Gamma\backslash X$ is a locally symmetric space of finite volume. For each finite dimensional representation $\sigma : K \to GL(E)$, we get a bundle E_σ over $\Gamma\backslash X$. The space $L^2(\Gamma\backslash X, E_\sigma)$ has a decomposition

$$L^2(\Gamma\backslash X, E_\sigma) = L^2_{dis}(\Gamma\backslash X, E_\sigma) + L^2_{con}(\Gamma\backslash X, E_\sigma)$$

such that the Bochner-Laplacian Δ_σ has a discrete spectrum $\mathrm{Spec}_{dis}(\Gamma\backslash X, E_\sigma)$ on $L^2_{dis}(\Gamma\backslash X, E_\sigma)$ and a continuous spectrum on $L^2(\Gamma\backslash X, E_\sigma)$. Denote the counting function of $\mathrm{Spec}_{dis}(\Gamma\backslash X, E_\sigma)$ by $N_{d,\sigma}(\lambda)$.

PROPOSITION 3.3 ([**18**]). *If for every irreducible representation σ of K, there exist an integer m and a positive constant c_σ such that $N_{d,\sigma}(\lambda) \leq c_\sigma(1 + \lambda^m)$ for $\lambda > 0$, then the trace class conjecture is true for all K-finite convolution functions α.*

By a K-finite function, we mean a function whose translates under elements in K span a finite dimensional space. In [**18**], Müller proved such a bound and hence proved the trace class conjecture for K-finite functions α.

To remove the K-finiteness condition, Müller showed in [**17**] that c_σ can be chosen to depend polynomially on the norm of the highest weight parametrizing the irreducible representation σ of K. Such an uniform bound on $N_{d,\sigma}(\lambda)$ allows him to prove the trace class conjecture in complete generality. On the other hand, I avoid this problem by working on $\Gamma\backslash G$ and hence treating all representations σ of K simultaneously. This point of view was emphasized in the earlier sections.

To explain the novelty in this approach, we start with some comments on differences between G and X. Let $\mathfrak{g} = \mathfrak{k} + \mathfrak{p}$ be the Cartan decomposition of the Lie algebra $\mathfrak{g}$ of G. Let x_0 be the basepoint in X corresponding to K. Then the tangent space $T_{x_0}X$ can be identified with $\mathfrak{p}$. Since the Killing form $\langle , \rangle$ is positive definite on $\mathfrak{p}$, it defines an inner product on $T_{x_0}X$ and hence an invariant Riemannian metric on X. On the other hand, the Killing form $\langle , \rangle$ is negative on $\mathfrak{k}$ and hence defines an indefinite bilinear form on $\mathfrak{g}$ and an invariant pseudo-Riemannian metric on G having both positive and negative signatures.

This difference is also reflected in the action of the Casimir element. It defines an elliptic operator on X, but a non-elliptic operator on G. As is well-known, ellipticity is crucial to spectral analysis.

The basic idea to overcome this difficulty is to change the sign of the Killing form $\langle , \rangle$ on $\mathfrak{k}$ to get a positive definite inner product $\langle , \rangle'$ on $\mathfrak{g}$. Then $\langle , \rangle'$ defines a left G-invariant Riemannian metric on G. Note that this metric is not right G-invariant, but is right K-invariant. This right K-invariance plays an important role in the proof of the trace class conjecture.

Let Δ be the Beltrami-Laplacian of $\Gamma\backslash X$ with respect to this metric. Then $L^2(\Gamma\backslash G) = L^2_{dis}(\Gamma\backslash G) + L^2_{con}(\Gamma\backslash G)$ and Δ has a discrete spectrum on $L^2_{dis}(\Gamma\backslash G)$. Denote the counting function of the discrete spectrum of Δ by $N_d(\lambda)$.

PROPOSITION 3.4 ([**10**, §7]). *If there exist an integer m and a constant c such that $N_d(\lambda) \leq c(1+\lambda^m)$, then the trace class conjecture holds for all $\alpha \in C_0^\infty(G)$.*

PROPOSITION 3.5 ([**10**, Theorem 1.1.3]). *There exist an integer and a constant c such that $N_d(\lambda) \leq c(1+\lambda^m)$.*

Such a bound implies the uniform bound on $N_{d,\sigma}(\lambda)$ mentioned earlier. In fact, the right K-invariance of the metric of G shows that Δ preserves the following decomposition

$$L^2(\Gamma\backslash G) = \oplus_{\sigma\in\hat{K}} \dim\sigma\, L^2(\Gamma\backslash X, E_\sigma).$$

Since the restriction of Δ to each $L^2(\Gamma\backslash X, E_\sigma)$ is a shift of the Bochner-Laplacian Δ_σ by a constant determined by the value of the Casimir element of $\mathfrak{k}$ on the highest weight of σ, a polynomial bound on $N_d(\lambda)$ implies a polynomial bound on $N_{d,\sigma}(\lambda)$ with polynomial dependence on σ.

4. The Weyl upper bounds

In this section, we explain how to obtain the polynomial upper bound on $N_d(\lambda)$ in Proposition 3.5 and the sharp Weyl bound on $N_d(\lambda)$ in several cases.

As explained in §1, bounds on $N_d(\lambda)$ are difficult, though interesting in themselves. One difficulty is the presence of the continuous spectrum caused by the non-compactness of $\Gamma\backslash G$.

The basic idea to overcome this difficulty is as follows:

1. Modify $L^2(\Gamma\backslash G)$ to remove the continuous spectrum.
2. Understand how the discrete spectrum is perturbed after the modification.

The removal of the continuous spectrum is based on two observations.

Observation 1. The continuous spectrum is caused by noncompactness of $\Gamma\backslash G$. In fact, by the decomposition principle [**5**], the essential spectrum, which is the complement of the discrete spectrum, does not change under any compact perturbation. The infinity of $\Gamma\backslash G$ is described by Siegel sets of proper rational parabolic subgroups. Roughly, one Γ-conjugacy class of parabolic subgroups corresponds to a "cusp" of $\Gamma\backslash G$. Naturally, on the subspace of functions in $L^2(\Gamma\backslash G)$ which vanish at all "cusps" the Laplacian clearly has only a discrete spectrum. However, this space is too small since it does not include the cuspidal subspace $L^2_{cus}(\Gamma\backslash G)$, on which the Laplacian has a discrete spectrum as mentioned earlier.

Observation 2. The continuous spectrum is caused by the constant terms along all proper rational parabolic subgroups. This is the philosophy of cusp forms due to Gelfand and Piatetski-Shapiro [**7**].

These two observations imply that the continuous spectrum is caused by the constant terms at all "cusps" at infinity. Therefore, by removing the constant terms at all the cusps, we get a subspace of $L^2(\Gamma\backslash G)$ which has only a discrete spectrum.

More precisely, choose a sufficiently large height T and define a subspace $L^2_T(\Gamma\backslash G)$ to consist of functions in $L^2(\Gamma\backslash G)$ whose constant term along every proper rational parabolic subgroup vanishes above the height T. A particular nice example to illustrate this subspace is a Riemann surface $\Gamma\backslash\mathbb{H}$ with only one cusp at $i\infty$. The constant term is

$$f_\infty(y) = \int_{\Gamma_\infty \backslash N} f(n(x+iy))dn,$$

where $N = \{\begin{pmatrix} 1 & b \\ 0 & 1 \end{pmatrix} \mid b \in \mathbb{R}\}$, and $\Gamma_\infty = \Gamma \cap N$; and $f \in L^2_T(\Gamma\backslash G)$ if and only if $f_\infty(y) = 0$ for $\ln y \geq T$.

The Dirichlet quadratic form $D(f,f) = \int_{\Gamma\backslash G} |\nabla f|^2$ defines a self-adjoint operator Δ_T on $L^2_T(\Gamma\backslash G)$, called the pseudo-Laplacian at the truncation height T.

PROPOSITION 4.1 ([**10**, Theorem 1.2.1]). *The pseudo-Laplacian Δ_T has a discrete spectrum, and its counting function $N_T(\lambda)$ satisfies the Weyl upper bound:*

$$\lim_{\lambda\to+\infty} \sup \frac{N_T(\lambda)}{\lambda^{\frac{n}{2}}} \leq (4\pi)^{-\frac{n}{2}} \frac{vol(\Gamma\backslash G)}{\Gamma(\frac{n}{2}+1)}, \qquad n = \dim \Gamma\backslash G.$$

If the discrete spectrum of Δ were not perturbed during this truncation of the constant term and hence equal to a part of the spectrum of Δ_T, this proposition immediately implies the Weyl upper bound on $N_d(\lambda)$. A crucial problem is to understand the relation between the discrete spectrum of Δ and the spectrum of Δ_T.

As mentioned earlier, $L^2_{dis}(\Gamma\backslash G) = L^2_{cus}(\Gamma\backslash G) \oplus L^2_{res}(\Gamma\backslash G)$. If $f \in L^2_{cus}(\Gamma\backslash G)$, the constant terms of f vanish everywhere, and hence $f \in L^2_T(\Gamma\backslash G)$. Therefore, $\mathrm{Spec}_{cus}(\Delta) \subset \mathrm{Spec}(\Delta_T)$. On the other hand, functions in $L^2_{res}(\Gamma\backslash G)$ are residues of Eisenstein series and have non-zero constant terms, and hence the truncation procedure does affect them. To explain the relation between $\mathrm{Spec}_{res}(\Delta)$ and $\mathrm{Spec}(\Delta_T)$, we need to decompose $L^2(\Gamma\backslash G)$ and $L^2_d(\Gamma\backslash G)$ according to association classes of rational parabolic subgroups of $\mathbf{G}$.

Two rational parabolic subgroups of $\mathbf{G}$ are defined to be associated if their split components in the Langlands decomposition are conjugate under G. For each association class $\mathcal{C}$ of rational parabolic subgroups, there is a subspace $L^2_{\mathcal{C}}(\Gamma\backslash G)$ of $L^2(\Gamma\backslash G)$. Briefly speaking, this subspace is spanned by functions having only nonzero constant terms along parabolic subgroups in $\mathcal{C}$.

When $\mathcal{C} = \{\mathbf{G}\}$, $L^2_{\mathcal{C}}(\Gamma\backslash G)$ is the cuspidal subspace $L^2_{cus}(\Gamma\backslash G)$. Hence the residual subspace $L^2_{res}(\Gamma\backslash X, \sigma)$ can be decomposed as follows:

$$L^2_{res}(\Gamma\backslash G) = \sum_{\mathcal{C}\neq\{\mathbf{G}\}} L^2_d(\Gamma\backslash G) \cap L^2_{\mathcal{C}}(\Gamma\backslash G),$$

where the sum over $\mathcal{C}$ is over association classes of proper parabolic subgroups. Define $L^2_{d,\mathcal{C}}(\Gamma\backslash G) = L^2_{\mathcal{C}}(\Gamma\backslash G) \cap L^2_d(\Gamma\backslash G)$.

Our basic results concerning the relation between the spectra of Δ_σ and $\Delta_{T,\sigma}$ are as follows:

1. When $\mathcal{C}$ is of rank one, i.e., the parabolic subgroups in $\mathcal{C}$ are of rank one, the eigenvalues of Δ in $L^2_{d,\mathcal{C}}(\Gamma\backslash X)$ can be uniformly approximated by the eigenvalues of Δ_T on the corresponding subspace except for a subset of smaller order.
2. When $\mathcal{C}$ is of rank greater than one, the counting function of the discrete spectrum in $L^2_{d,\mathcal{C}}(\Gamma\backslash G)$ can be bounded in terms of a power of $N_T(\lambda)$ of some $\mathbb{Q}$-rank one spaces of more general than $\Gamma\backslash G$ but of smaller dimension.

Together with the bound on $N_T(\lambda)$ in Proposition 4.1, these two results imply the desired polynomial bound on $N_d(\lambda)$ in Proposition 3.5, and hence the trace class conjecture in Theorem 3.2.

The spaces appearing in (2) can be explained as follows. For each rational parabolic subgroup $\mathbf{P}$, there is a Langlands decomposition of $P = \mathbf{P}(\mathbb{R})$:

$$P = N_P A_P M_P,$$

where N_P is the unipotent radical of $\mathbf{P}$, A_P is the split component, and M_P is a reductive group. The arithmetic subgroup Γ defines an arithmetic subgroup Γ_{M_P} of M_P. Let $X_P = M_P/K_P$, where $K_P = K \cap M_P$. Then the space appearing in (2) is a principal K-bundle over $\Gamma_{M_P}\backslash X_P$. Since $\Gamma_{M_P}\backslash M_P$ is a principal K_P-bundle over $\Gamma_{M_P}\backslash X_P$ and K_P is a proper subgroup of K, this space is a fiber bundle over $\Gamma_{M_P}\backslash M_P$ with fiber equal to K/K_P. These spaces are also crucial to bounding the error terms in (1), and hence play an important role in the proof of the trace class conjecture. They are also boundary components of a natural compactification of $\Gamma\backslash G$ as shown in [**3**].

THEOREM 4.2. *$\Gamma\backslash G$ has a compactification $\overline{\Gamma\backslash G}^{RBS}$ satisfying the properties:*

1. *The boundary components of parabolic subgroups are of the type described above.*
2. *The right G action on $\Gamma\backslash G$ extends to $\overline{\Gamma\backslash G}^{RBS}$.*
3. *It is a K-principal bundle over $\overline{\Gamma\backslash X}^{RBS}$, the reductive Borel-Serre compactification of $\Gamma\backslash X$, when Γ is neat.*

This compactification has an interesting realization when Γ and all the induced subgroups Γ_{M_P} are maximal. Specifically, let $\mathcal{S}(G)$ be the space of closed subgroups of G. Then the map $g\Gamma \to g\Gamma g^{-1}$ defines an embedding $\Gamma\backslash G$ into $\mathcal{S}(G)$. The closure of $\Gamma\backslash G$ in $\mathcal{S}(G)$ defines a compactification and is equal to the compactification $\overline{\Gamma\backslash G}^{RBS}$ under the above maximality assumption. For other points of view on the connection between the spectral theory and compactifications of $\Gamma\backslash X$, see [**14**].

As mentioned earlier, it is a natural and important problem to understand asymptotic behaviors of the counting function of the discrete spectrum, in particular, to improve the above polynomial upper bounds on the discrete spectrum. In [**2**, §4.7], Borel and Garland raised the following problem.

PROBLEM 4.3. *For every irreducible representation σ of K, decide whether the counting function $N_{d,\sigma}(\lambda)$ of the discrete spectrum of $L^2(\Gamma\backslash X, E_\sigma)$ satisfies the Weyl upper bound:*

$$\lim_{\lambda\to+\infty} \sup \frac{N_{d,\sigma}(\lambda)}{\lambda^{\frac{n}{2}}} \le (4\pi)^{-\frac{n}{2}} \frac{\mathrm{vol}(\Gamma\backslash X)}{\Gamma(\frac{n}{2}+1)} \dim E_\sigma.$$

THEOREM 4.4 ([**11**, Theorem 1.1.2]). *When the $\mathbb{Q}$-rank of $\mathbf{G}$ is equal to 1 or the $\mathbb{R}$-rank is less than or equal to 2, $N_{d,\sigma}(\lambda)$ satisfies the Weyl upper bound.*

This result certainly suggests that the Weyl upper bound on $N_{d,\sigma}(\lambda)$ holds in general. A more difficult problem is to determine when the Weyl law holds for $N_{d,\sigma}(\lambda)$.

References

[1] J.Arthur, *The trace formula and Hecke operators*, in *Number theory, trace formulas and discrete groups*, ed. by K.Aubert et al., Academic Press, 1989, pp. 11–27.

[2] A.Borel, H.Garland, *Laplacian and the discrete spectrum of an arithmetic group*, Amer. J. Math. 105 (1983), 309–335.
[3] A.Borel, L.Ji, *Compactifications of locally spaces*, preprint, 1999.
[4] H.Donnelly, *Eigenvalue estimates for certain noncompact manifolds*, Michigan Math. J. 31 (1984), 349–357.
[5] H.Donnelly & P.Li, *Pure point spectrum and negative curvature for noncompact manifolds*, Duke Math. J. 46 (1979), 497–503.
[6] S.Gelbart, *Automorphic forms on adele groups*, Princeton University Press, 1975.
[7] I.M.Gelfand, I.I.Piatetski–Shapiro, *Unitary representations in homogeneous spaces with discrete stationary groups*, Sov. Math., Dokl. 3 (1962), 1528–1531.
[8] D.Hejhal, *The Selberg trace formula*, vol. 2, Lecture Notes in Math. 1001, Springer–Verlag, 1983.
[9] M.Huxley, *Scattering matrices for congruence subgroups*, in *Modular forms*, ed. by R.A.Rankin, Horwood, Chichester, 1984, pp. 141–156.
[10] L.Ji, *The trace class conjecture for arithmetic groups*, J. Diff. Geom. 48 (1988), 165–203.
[11] L.Ji, *The Weyl upper bound on the discrete spectrum of locally symmetric spaces*, J. Diff. Geom. 51 (1999), 97–147.
[12] L.Ji, *The greatest common quotient of Borel–Serre and the toroidal compactifications of locally symmetric spaces*, GAFA 8 (1998), 978–1015.
[13] L.Ji, *Metric compactifications of locally symmetric spaces*, International J. of Math. 9 (1998), 465–491.
[14] L.Ji & R.MacPherson, *Geometry of compactifications of locally symmetric spaces*, preprint 1997.
[15] N.Katz & P.Sarnak, *Zeroes of Zeta function and symmetry*, Bull. A.M.S. 36 (1999), 1–26.
[16] R.Langlands, *Rank–one residues of Eisenstein series*, in *Festschrift in honor of I. I. Piatetski–Shapiro on the occasion of his sixtieth birthday*, Part II, Israel Math. Conf. Proc. vol. 3, Weizmann, Jerusalem, 1990, pp. 111–125.
[17] W.Müller, *The trace class conjecture in the theory of automorphic forms II*, Geom. Funct. Anal. 8 (1998), 315–355.
[18] W.Müller, *The trace class conjecture in the theory of automorphic forms*, Ann. of Math. 130 (1989), 473–529.
[19] P.Sarnak, *Selberg's eigenvalue conjecture*, Notices. A.M.S. 42 (1995), 1272–1277.
[20] P.Sarnak, *On cusp forms*, in *The Selberg trace formula and related topics*, ed. by D.Hejhal et al., Contemporary Math. vol 53, Amer. Math. Soc., 1984, pp. 393–407.
[21] A.Selberg, *Harmonic analysis*, in *Collected papers of A. Selberg*, vol. 1, Springer–Verlag, 1989, pp. 626–674.
[22] A.Selberg, *Discontinuous groups and harmonic analysis*, Proc. of ICM, 1962.
[23] A.Venkov, *Spectral theory of automorphic forms*, Kluwer Academic Publishers, 1990.
[24] G.Warner, *Selberg's trace formula for nonuniform lattices: The $\mathbb{R}$–rank 1 case*, Adv. in Math. Suppl. Stud., vol.6, Academic Press, New York, 1979, pp. 1–142.

DEPARTMENT OF MATHEMATICS, UNIVERSITY OF MICHIGAN, ANN ARBOR, MI 48109
E-mail address: `lji@math.lsa.umich.edu`

AMS/IP Studies in Advanced Mathematics
Volume 20, 2001

On the First Eigenvalue of Laplacian for Locally Symmetric Manifolds

Jian-shu Li

1. Introduction

Let G be a simple Lie group. For the present purpose we may assume G is a closed subgroup of $GL(n, \mathbb{C})$ invariant under the involution $\theta : A \mapsto {}^t\bar{A}^{-1}$. Let $K = G^\theta = G \cap U(n)$ be the subgroup of fixed points of θ. Then K is a maximal compact subgroup of G, and $X = G/K$ is a globally symmetric Riemannian manifold. A *lattice* in G is a discrete subgroup $\Gamma \subset G$ such that the volume of $\Gamma\backslash G$ (or equivalently, of $\Gamma\backslash G$) is finite. Fix a normalization of the invariant metric on X, and therefore one on every $\Gamma\backslash X$. Let Δ be the corresponding Laplace operator. In this note we shall be concerned with the first positive eigenvalue of Δ on $\Gamma\backslash X$. We denote this by $\lambda_1(\Gamma)$.

EXAMPLE 1.1. Let $G = SL(2, \mathbb{R})$. Then X is identified with the Poincaré upper half plane H in such way that the natural action of G on $X = G/K$ becomes the usual action of $SL(2, \mathbb{R})$ on H via linear fractional transformations. We may take

$$\Delta = -y^2(\frac{\partial^2}{\partial x^2} + \frac{\partial^2}{\partial y^2})$$

Let Γ be a *congruence subgroup* of $SL(2, \mathbb{Z})$. A famous conjecture of Selberg asserts that $\lambda_1(\Gamma) \geq 1/4$, and it can be shown that this is best possible. Selberg himself proved $\lambda_1(\Gamma) \geq 3/16$, and for many years this was the best estimate available. It was in a relatively recent paper that Luo, Rudnick and Sarnak [**LRSa**] decisively improved this lower bound.

EXAMPLE 1.2. Let $G = SO(n, 1)$ or $SU(n, 1)$. For such G there is an analogue of Selberg's "3/16 estimate" [**EGM**] [**CLPS**] [**Lia**] [**BS**]. Combination of the methods in [**LRSa**] and [**BS**] gives rise to substantial improvements of these estimates [**LRSb**]

Up to coverings, $SO(n, 1)$ and $SU(n, 1)$ are the only simple Lie groups without Kazhdan's property T. So suppose G does have Kazhdan's property T. Then every lattice in G is automatically a congruence subgroup with respect to some rational

Supported in part by NSF Grant DMS-9801713, HKUST DAG Grant DAG97/98.SC16 and RGC-CERG grant HKUST6169/99P..

form of G. Our goal here is to establish an explicit and optimal lower bound for $\lambda_1(\Gamma)$, for all lattices Γ in such G. Set

$$\gamma(G) = \inf\{\lambda_1(\Gamma) \,|\, \Gamma \text{ any lattice in } G\} \tag{1.1}$$

We will relate $\gamma(G)$ to another invariant $\lambda_1(G)$ (see §2) which is defined in terms of the unitary representation theory of G and has apparently nothing to do with lattices. Kazhdan's property T is essentially equivalent to $\lambda_1(G) > 0$. From the definitions it will be clear that $\gamma(G) \geq \lambda_1(G)$. The constant $\lambda_1(G)$ will be explicitly calculated. The present note is devoted to a brief discussion for the proof of the following result. Details will appear elsewhere.

THEOREM 1.3. *For most classical groups and some exceptional groups G, we have $\gamma(G) = \lambda_1(G)$.*

Some explicit examples appear in §4. This result immediately suggest the conjecture that $\gamma(G) = \lambda_1(G)$ for *all* G. However this is not the case: there is a strong indication that one should have $\gamma(G) > \lambda_1(G)$ for $G = SO(2n+1, \mathbb{C})$ $(n \geq 4)$.

2. $\lambda_1(G)$ and Matrix Coefficients

Let $\mathfrak{g}_0$ be the Lie algebra of G with Cartan decomposition $\mathfrak{g}_0 = \mathfrak{k}_0 \oplus \mathfrak{p}_0$. As is customary, we drop the subscript "0" to denote complexification. Thus $\mathfrak{g} = \mathfrak{g}_0 \otimes \mathbb{C}$, etc. Fix an invariant bilinear form $<,>$ on $\mathfrak{g}_0$ and extend it to $\mathfrak{g}$. We assume that $<,>$ is positive definite on $\mathfrak{p}_0$, so that it is a positive multiple of the Killing form. Let $\{x_j\}$ be a basis of $\mathfrak{g}$, and $\{x_j^*\}$ the dual basis with respect to $<,>$. That is $< x_i, x_j^* >= \delta_{ij}$. Define the Casimir element $\mathcal{C}$ by

$$\mathcal{C} = \sum x_j x_j^* \tag{2.1}$$

Then $\mathcal{C}$ is in the center of the universal enveloping algebra of $\mathfrak{g}$. Assume that the Riemannian metric of $X = G/K$ is induced by $<,>$; let Δ be the corresponding Laplace operator. Equivalently, we define Δ to be the operator on X induced by $-\mathcal{C}$. Thus $\mathcal{C}$ acts on the space of (say smooth) functions on G, and Δ is the restriction of $\mathcal{C}$ to the subspace of K-invariant functions.

Let π be an irreducible admissible representation of G on a Hilbert space H. By Diximir's Schur lemma, we know $\mathcal{C}$ acts on (the space of K-finite vectors in) H as a scalar operator. Thus

$$\pi(\mathcal{C}) = -\lambda_\pi \cdot I \tag{2.2}$$

with $\lambda_\pi \in \mathbb{C}$. If π is unitary then $\lambda_\pi \in \mathbb{R}$. Set

$$\lambda_1(G) = \inf\{\, \lambda_\pi \,|\, \pi \text{ non-trivial, spherical, irreducible, and unitary}\} \tag{2.3}$$

From well known facts in representation theory, we have $\lambda_1(G) \geq 0$. Kazhdan's property T for G is more or less equivalent to $\lambda_1(G) > 0$.

Now let Γ be a lattice in G. The right regular representation of G on $L^2(\Gamma\backslash G)$ decomposes into a direct integral of irreducible unitary representations of G. This gives rise to a spectral decomposition for the action of $\mathcal{C}$ on $L^2(\Gamma\backslash G)$. On the other hand, we have a spectral decomposition of the self-adjoint operator Δ acting on $L^2(\Gamma\backslash X)$. Since $L^2(\Gamma\backslash X)$ is precisely the space of K-fixed functions in $L^2(\Gamma\backslash G)$, we have $\lambda_1(\Gamma) \geq \lambda_1(G)$ for all Γ.

Our first task is to determine $\lambda_1(G)$ explicitly for each G. This is by no means trivial. In fact, in this note we will only discuss the calculation of $\lambda_1(G)$, leaving the construction of Γ with $\lambda_1(\Gamma) = \lambda_1(G)$ to some parenthetical remarks.

Let $B = MAN$ be a minimal parabolic subgroup of G. The restriction of $<,>$ to $\mathfrak{a}_0$ is positive definite, and we may identify $\mathfrak{a}$ with its linear dual $\mathfrak{a}^*$ by means of $<,>$. The exponential map from $\mathfrak{g}_0$ to G restricts to an isomorphism of abelian groups

$$\exp : \mathfrak{a}_0 \longrightarrow A$$

For any $\nu \in \mathfrak{a}$ we let $\mathrm{Ind}_B^G(\nu)$ be the representation of G induced from the quasi-character

$$m \exp(x) n \mapsto e^{<x,\nu>} \qquad (m \in M, x \in \mathfrak{a}_0, n \in N)$$

Here we use normalized induction. The family of induced representations $\mathrm{Ind}_B^G(\nu)$ constitute the so-called spherical principal series of G. Let $\pi(\nu)$ be the unique irreducible spherical subquotient of $\mathrm{Ind}_B^G(\nu)$. Every irreducible spherical representation of G is of the form $\pi(\nu)$ for some ν. Let ρ be the half-sum of positive roots of $\mathfrak{a}$ in $\mathfrak{n}$:

$$< \rho, x >= \text{trace of } \frac{1}{2} ad(x)|_{\mathfrak{n}} \qquad (x \in \mathfrak{a})$$

If π is any subquotient of $\mathrm{Ind}_B^G(\nu)$ then

$$\lambda_\pi =< \rho, \rho > - < \nu, \nu > . \tag{2.4}$$

Let now π be any admissible representation of G on a vector space V. Let (π^*, V^*) be the contragredient of (π, V). A *matrix coefficient* of π is a function on G of the form

$$g \mapsto (\pi(g)v, v^*), \qquad (v \in V, v^* \in V^*)$$

where $(,)$ is the canonical pairing between V and V^*. Let r be a positive real number. We say π is *strongly* $L^{r+\varepsilon}$ if $(\pi(g)v, v^*) \in L^{r'}(G)$ for any $r' > r$ and any K-finite vectors $v \in V, v^* \in V^*$.

It is known that the spherical representation $\pi(\nu)$ is strongly $L^{r+\varepsilon}$ if and only if the real part of ν is dominated by $(1 - \frac{2}{r})\rho$ (cf. [**Kn**]). By an elementary argument, this implies

$$< \mathrm{Re}(\nu), \mathrm{Re}(\nu) > \leq (1 - \frac{2}{r})^2 < \rho, \rho > .$$

Now if $\pi(\nu)$ is unitary then $< \nu, \nu > \leq < \mathrm{Re}(\nu), \mathrm{Re}(\nu) >$, and we have

$$< \rho, \rho > - < \nu, \nu > \geq (1 - (1 - \frac{2}{r})^2) < \rho, \rho >= \frac{4}{r}(1 - \frac{1}{r}) < \rho, \rho >$$

We therefore obtain

LEMMA 2.1. *Let π be an irreducible spherical unitary representation of G. Suppose π is strongly $L^{r+\varepsilon}$. Then*

$$\lambda_\pi \geq \frac{4}{r}(1 - \frac{1}{r}) < \rho, \rho > \tag{2.5}$$

3. The Estimates

We would like to calculate $\lambda_1(G)$. Our strategy is as follows. First we propose a candidate $c(G)$ for the value of $\lambda_1(G)$. It is not hard to see that the set of π's with λ_π greater than or equal to a given constant is closed with respect to the Fell topology. To prove $c(G) = \lambda_1(G)$ we need to show $\lambda_\pi \geq c(G)$ for all (non-trivial, irreducible, unitary) π and $\lambda_\pi = c(G)$ for some π.

Next we divide the set of irreducible unitary representations of G into two subsets consisting of "large" and "small" representations respectively. Technically, for classical groups "small" means "of low rank" in the sense of [**H**] and [**Lib**]. For exceptional groups we use a suitable generalization of this notion. Then of course "large" simply means not small.

In principle, if π is large then λ_π should be *much* bigger than $\lambda_1(G)$, and a reasonable estimate would prove that π is strongly $L^{r+\varepsilon}$ with (cf. (2.5))

$$\frac{4}{r}(1-\frac{1}{r}) < \rho, \rho > \; > \; c(G).$$

(cf. 2.5).

On the other hand, if π is small then (by [**Lib**] and [**Lie**]) it is a "local theta lift" from a group G' much smaller than G. By making use of this fact we may write π as the Langlands quotient of some standard representation $\mathrm{Ind}_P^G(\sigma \otimes \nu_0)$, where $P = MAN$, σ is large for M, and ν_0 is a parameter on A which is explicitly determined. The representation σ is also spherical. Let ν_1 be its parameter. Then $\nu = (\nu_1, \nu_0)$ is the parameter for π. Since σ is large (for M) we have some control of the size of ν_1. This will yield the desired estimate $\lambda_\pi \geq \lambda_1(G)$ by (2.4).

This strategy of separating representations into large and small ones is effective (and good enough for our purpose) when the split rank of G is big enough. For groups with small split rank, we must make a direct analysis of the unitary dual, or appeal to the known results on unitary dual for some of these groups.

4. Examples

In this section we give some of the main examples covered by Theorem 1.3. From the very beginning we have assumed G to be a simple Lie group: this means G is connected, and its Lie algebra $\mathfrak{g}_0$ is simple. Suppose $G \longrightarrow G'$ is a surjective homomorphism with central, discrete kernel. We may identify $\mathfrak{g}_0$ with $\mathfrak{g}_0'$ via this homomorphism (which induces an isomorphism of Lie algebras). Then, as long as we take the same normalization of the invariant form $<,>$ on $\mathfrak{g}_0$ and $\mathfrak{g}_0'$, we would have $\lambda_1(G) = \lambda_1(G')$ and $\gamma(G) = \gamma(G')$. Thus it suffices to take one representative from each "isogeny class" of simple Lie groups.

4.1. Classical groups. Let D be one of the division algebras $\mathbb{R}, \mathbb{C}$ or $\mathbb{H}$ (the quarternion algebra) over $\mathbb{R}$. If $G = SL(n, D)$ then it is a trivial computation to obtain $\lambda_1(G)$ from Vogan's classification of the unitary dual of $GL(n, D)$ [**Va**]. Similarly for the complex classical groups $SO(n, \mathbb{C})$ and $Sp_{2m}(\mathbb{C})$, we may calculate $\lambda_1(G)$ from the classification in [**B**].

The remaining classical groups are essentially isometry groups of various sesquilinear forms. Let D be as above. We have the standard involution $x \mapsto \bar{x}$ on D: it is trivial for $\mathbb{R}$, complex conjugation for $\mathbb{C}$, and quaternionic conjugation for $\mathbb{H}$. Let V be a vector space over D with a non-degenerate hermitian or skew-hermitian form $(,)$. We denote by U the group of isometries of U. Let G be the connected component of the identity in the derived group of U. By choosing a basis for V, we may identify elements of G and $\mathfrak{g}_0$ with $n \times n$ matrices with entries in D, where n is the dimension of V over D. Set

$$< x, y >= \text{trace}(xy) \qquad (x, y \in \mathfrak{g}_0) \tag{4.1}$$

Here the trace of a matrix A with diagonals $a_{11}, \cdots, a_{nn}$ is defined to be

$$tr_{D/\mathbb{R}}(a_{11} + \cdots + a_{nn}).$$

For $G = SL(n, D), SO(n, \mathbb{C})$ or $Sp_{2m}(\mathbb{C})$ we define $<,>$ by the same formula (4.1). With this normalization of $<,>$, the constant $\lambda_1(G)$ is as given in Table 1. The second column of this table gives the conditions for G to have Kazhdan's property T.

TABLE 1. $\lambda_1(G)$ for classical groups

G	Condition for "T"	$\lambda_1(G)$
$SL(n, \mathbb{R})$	$n \geq 3$	$\frac{1}{4}n(n-1)$
$SL(n, \mathbb{H})$	$n \geq 3$	$2n(n-1)$
$SU(p, q)$	$p \geq q \geq 2$	$(p-1)q$
$SO_0(p, q)$	$p \geq q \geq 2, p+q \geq 5$	$\frac{1}{2}(p-2)q$
$SO^*(2n)$	$n \geq 4$	$n(n-3)$
$Sp_{2n}(\mathbb{R})$	$n \geq 2$	$\frac{1}{2}n^2$
$Sp(p, q)$	$p \geq q \geq 1,\ p \geq 2$	$2pq$
$SL(n, \mathbb{C})$	$n \geq 3$	$\frac{1}{4}n(n-1)$
$SO(n, \mathbb{C})$	$n \geq 5$	$\frac{1}{4}n(n-4)$
$Sp_{2n}(\mathbb{C})$	$n \geq 2$	$\frac{1}{4}n(2n+1)$

Here is a more precise version of Theorem 1.3 for classical groups.

THEOREM 4.1. *Let G be a real or complex classical group different from $SL(n, \mathbb{H})$ or $SO(2n+1, \mathbb{C})$. Then $\gamma(G) = \lambda_1(G)$*

Given the explicit calculation of $\lambda_1(G)$ outlined in the previous section, this theorem is about the construction of eigenforms of the Laplacian, or equivalently the construction of automorphic forms with a prescribed type at the archimedean places. The main tools for the proof of this result were developed in [**Lid**].

4.2. Exceptional groups. Here we give a few examples of exceptional groups G with $\gamma(G) = \lambda_1(G)$.

EXAMPLE 4.2. Consider $G = F_{4(-20)}$. This is the exceptional simple Lie group of type F_4 and split rank 1. The spherical complementary series of G was determined in a well known paper of Kostant [**Ko**]. It can be parameterized by a subset of the interval $0 \leq \nu \leq 11$, as depicted in Fig. 1. Here $\nu = \rho = 11$ corresponds to

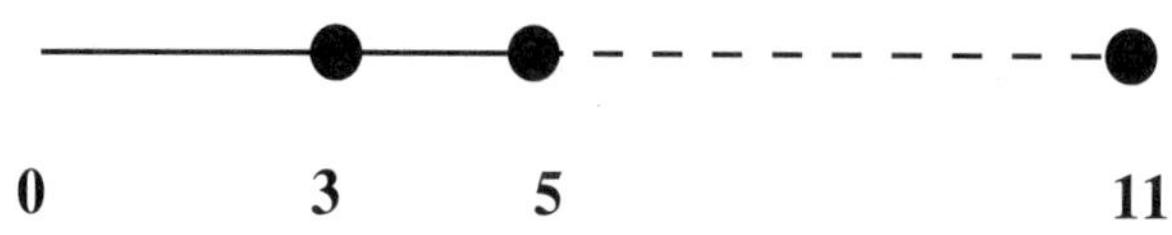

FIGURE 1. Unitary and automorphic points for $F_{4(-20)}$

the trivial representation. The subinterval $0 \leq \nu \leq 5$ parameterizes the rest of the unitary points.

The point $\nu = 5$ is at the end of the complementary series. It corresponds to the unique spherical irreducible unitary representation $\pi = \pi_5$ such that $\lambda_\pi = \lambda_1(G)$. Using the calculations in [**Lic**], it can be shown that there is a lattice $\Gamma \subset G$, such that $L^2(\Gamma\backslash G)$ contains π_5 as an irreducible summand. Thus we have $\gamma(G) = \lambda_1(G)$ for $G = F_{4(-20)}$.

In [**BLS**] it was shown that the point $\nu = 3$ in Fig. 1 corresponds to a limit of automorphic representations. Concretely this means the following: for any $\varepsilon > 0$ there is a spherical unitary representation $\pi(\nu)$ with parameter ν, such that $|\nu - 3| < \varepsilon$, and $\pi(\nu)$ occurs in $L^2(\Gamma\backslash G)$ for some lattice Γ. Of course one expects the point $\nu = 3$ itself to be automorphic. It is natural to wonder whether the point $\nu = 1$ also corresponds to an automorphic representation, and that $\nu = 1, 3, 5$ are the only automorphic points contained in the spherical complementary series of $G = F_{4(-20)}$.

EXAMPLE 4.3. Let G be a real split simple Lie group of type G_2. The unitary dual of this group has been determined by Vogan [**Vb**]. Fig. 2, which is reproduced from [**Vb**], p.742, describes the spherical unitary representations with real infinitesimal characters. The picture is centered around the origin. The trivial representation corresponds to some isolated points which are not shown in this picture. The non-trivial ones correspond to the shaded regions, including their boundaries. Thus the unitary points consists of everything inside the larger hexagon, with six open triangular regions deleted.

Of course, points which are conjugate under the Weyl group correspond to the same representation. In particular, the six vertices of the larger hexagon define the same representation which we denote by $\pi(\beta)$, where β is a short root for G. (We use this notation because the six vertices are the end points of the six short roots of G). Clearly, $\pi = \pi(\beta)$ is the unique spherical unitary representation with $\lambda_\pi = \lambda_1(G)$. Langlands [**L**] has shown that $\pi(\beta)$ occurs in the residue spectrum of $L^2(\Gamma\backslash G)$ for some lattice Γ. In [**LS**] it was further shown that $\pi(\beta)$ occurs in the cuspidal spectrum of $L^2(\Gamma\backslash G)$ for any lattice Γ with deep enough level. Thus we have $\gamma(G) = \lambda_1(G)$ for $G = G_2$.

EXAMPLE 4.4. Finally we consider $G = E_{6(-14)}$, a simple Lie group of type E_6 which gives rise to a hermitian symmetric domain. In this case the unitary dual is not known. Yet it is possible to calculate $\lambda_1(G)$. The (unique) spherical unitary representation π_0 with $\lambda_{\pi_0} = \lambda_1(G)$ can be constructed by restricting the minimal representation of $E_{7(-5)}$ (the rank 4 real form of E_7) to G. The restriction is a direct sum of irreducible unitary representations. Exactly one of them is spherical, and it is isomorphic to π_0. By "globalizing" this construction we can prove the existence of a lattice Γ with $\lambda_1(\Gamma) = \lambda_1(G)$. So once again we have $\gamma(G) = \lambda_1(G)$.

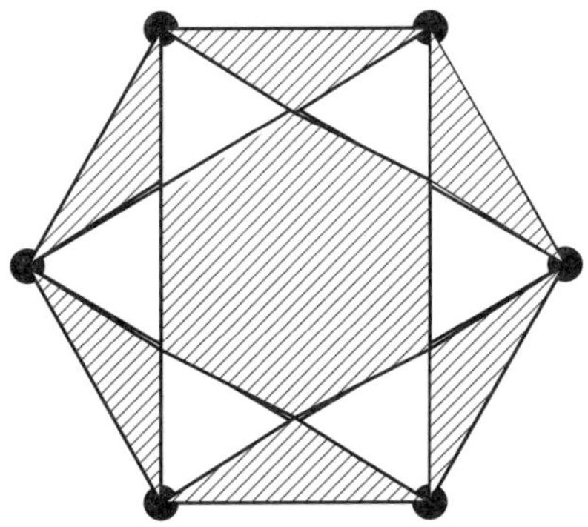

FIGURE 2. Unitary and automorphic points for G_2

References

[B] D. Barbasch, *The Unitary Dual for Complex Classical Lie Groups*, Invent. math. **96** (1989), 103-176.

[BLS] M. Burger, J-S. Li, and P. Sarnak, *Ramanujan Duals and automorphic spectrum*, Bulletin of Amer. Math. Soc. **26** (1992), 253-257.

[BS] M. Burger, and P. Sarnak, *Ramanujan Duals* II, Invent. Math. **106** (1991), 1-11.

[CLPS] J. Cogdell, J-S. Li, Piatetski-Shapiro, and P. Sarnak, *Poincare series for $SO(n,1)$*, Acta Math. **167** (1991), 229-285.

[EGM] J. Elstrodt, F. Grunewald, and J. Mennicke, *Kloosterman sums for Clifford algebras and a lower bound for the positive eigenvalues of the Laplacian for congruence subgroups acting on hyperbolic spaces*, Inven. Math. **101** (1990), 641-685.

[H] R. Howe, *On a notion of rank for unitary representations of classical groups*, in: *Harmonic analysis and group representations:* II Circlo 1980, Palazzone della Scuola normale superiore, 223-331.

[Kn] A. W. Knapp, *Representation Theory of Semisimple Groups: an overview based on examples*, Princeton Univ. Press, Princeton, N.J. 1986

[Ko] B. Kostant, *On the existence and irreducibility of certain series of representations*, Bulletin of AMS **75** (1969), 627-642.

[L] R. Langlands, *On the functional equations satisfied by Eisenstein series*, Lecture Note in Math. **544**, Springer-Verlag, Berlin (1976).

[Lia] J-S. Li, *Kloosterman-Selberg zeta functions on complex hyperbolic spaces*, Amer. Jour. of Math. **113** (1991), 653-731.

[Lib] ———, *Singular Unitary Representations of Classical Groups*, Inven. Math. **97** (1989), 237-255.

[Lic] ———, *Two reductive dual pairs in groups of type E*, Manuscripta Math. **91** (1996), 163-177.

[Lid] ———, *Non-vanishing Theorems for the Cohomology of Certain Arithmetic Quotients*, J. reine angew. Math. **428** (1992), 177-217.

[Lie] ———, *On the classification of irreducible low rank unitary representations of classical groups*, Compositio Math. **71**, (1989), 29-48.

[LS] ———, and J. Schwermer, *Constructions of automorphic forms and related cohomology classes for arithmetic subgroups of G_2*, Compositio Math. *87* (1993), 45-78.

[LRSa] W. Luo, Z. Rudnick, and P. Sarnak, *On Selberg's eigenvalue conjecture*, Geom. Func. Anal. **5** (1995), 387-401.

[LRSb] ———, *On the generalized Ramanujan conjecture for* GL(n), in: Automorphic forms, automorphic representations, and arithmetic (Fort Worth, TX, 1996), 301–310, Proc. Sympos. Pure Math., 66, Part 2, Amer. Math. Soc., Providence, RI, 1999.

[Va] D. Vogan, *The Unitary Dual of GL(n) Over an Archimedean Field*, Inven. Math. **83** (1986), 449-505.

[Vb] ———, *The unitary dual of G_2*, Invent. Math. **116** (1994), 677-791.

Department of Mathematics, HKUST and Univ. of Maryland, Clear Water Bay, Kowloon, Hong Kong
E-mail address: jsl@math.umd.edu

AMS/IP Studies in Advanced Mathematics
Volume 20, 2001

DGBV Algebras and Mirror Symmetry

Huai-Dong Cao, Jian Zhou

ABSTRACT. We describe some recent development on the theory of formal Frobenius manifolds via a construction from differential Gerstenhaber-Batalin-Vilkovisk (DGBV) algebras and formulate a version of mirror symmetry conjecture: the extended deformation problems of the complex structure and the Poisson structure are described by two DGBV algebras; mirror symmetry is interpreted in term of the invariance of the formal Frobenius manifold structures under quasi-isomorphism.

1. Introduction

According to Getzler [**15**], topological string theory is conformal field theorist's algebraic topology. Indeed, ideas from cohomology theory has also been used extensively by physicists, especially in the two widely used quantization schemes: the BRST formalism and the BV formalism. On the other hand, infinite algebra structures and the notion of operad originally developed in homotopy theory, have also shown up in many places in string theory. A well-known connection between the cohomology theory and the homotopy theory is provided by the rational homotopy theory. One naturally speculates that such a connection should have its counterpart in string theory. Here we report some recent work on mirror symmetry which reflects this connection.

The mirror symmetry is one of the mysteries in string theory. For its history and backgrounds, see Yau [**31**] and Greene-Yau [**18**]. In physicist's language, given a Calabi-Yau manifold, one can define two kinds of superconformal field theories on it: the A-type theories and the B-type theories. The mirror symmetry conjecture says for certain Calabi-Yau manifolds M, there exist Calabi-Yau manifolds $\widehat{M}$, such that a certain A-type theory on M can be identified with a certain B-type theory on $\widehat{M}$. There are three issues involved in this conjecture:

(a) the construction of the mirror manifolds;
(b) the mathematical formulations of the A and B-type theories;
(c) the identifications of the relevant theories.

Most of researches so far has focused on (a) and (b). We present here an approach which deals with (b) and (c) in the same framework.

[1]The first author's research was supported in part by NSF grants # DMS-9803549. The second author's research was supported in part by NSF grant # DMS-9632028.

According to physicists, an A-type theory should be sensitive to the deformation of the Kähler structure, but is independent of the complex structure, while a B-type theory should be sensitive to the deformation of the complex structure, but is independent of the Kähler structure. So it is reasonable to speculate that such theories are related to the deformation theories of the Kähler structure and the complex structure respectively. It will be clear from below that it is more natural to consider the deformation of the Poisson structure defined by the Kähler structure.

In deformation theory, the following principle due to Deligne is well-known (see e.g. Goldman-Millson [**16**]): "In characteristic zero a deformation problem is controlled by a differential graded Lie algebra with the property that quasi-isomorphic differential graded Lie algebras give the same deformation theory." In this principle, one finds many objects in the rational homotopy theory. The notion of quasi-isomorphism was used in Sullivan's minimal model theory approach to rational homotopy theory [**27**]. As suggested by the BV quantization scheme, the consideration of the extended deformation problem is necessary. As a modification of Deligne's principle, one notices that such problems are usually governed by a differential Gerstenhaber (or Poisson) algebra. For example, the extended deformation problems of the complex structure and the Poisson structure of a Calabi-Yau manifold are controlled by two differential Gerstenhaber-Batalin-Vilkovisky (DGBV) algebras in the title. Such algebras are combinations of differential graded algebras (DGA's) and differential graded Lie algebras (DGLA's). Chen [**10**] developed a theory for DGA's to compute the (co)homology of loop spaces. Hain [**21**] generalized it to DGLA's and showed that Chen's theory is an alternative to Sullivan's theory. Our approach to (b) uses a construction of formal Frobenius manifolds from DGBV algebras in which the formal power series connection on a DGLA is used. We have defined in [**9**] a notion of quasi-isomorphisms of DGBV algebras, and have shown that formal Frobenius manifolds obtained from quasi-isomorphic DGBV algebras can be identified with each other. This is our approach to (c).

2. Frobenius algebras and formal Frobenius manifolds

2.1. Frobenius algebras. Throughout this paper, $\mathbf{k}$ will be a commutative $\mathbb{Q}$-algebra. An *invariant metric* on a commutative algebra $(H, \wedge)$ over $\mathbf{k}$ is a non-degenerate bilinear map $\eta : H \times H \to \mathbf{k}$ such that

$$\eta(X, Y) = \eta(Y, X), \tag{1}$$

$$\eta(X \wedge Y, Z) = \eta(X, Y \wedge Z), \tag{2}$$

for $X, Y, Z \in H$. The triple $(H, \wedge, \eta)$ is called a *Frobenius algebra.* Suppose that H is free as $\mathbf{k}$-module, and fix a basis $\{e_a\}$ of H. Then there are elements $\phi^c_{ab} \in \mathbf{k}$, such that

$$e_a \wedge e_b = \phi^c_{ab} e_c.$$

Let $\eta_{ab} = \eta(e_a, e_b)$ and $\phi_{abc} = \eta(e_a \wedge e_b, e_c) = \phi^p_{ab}\eta_{pc}$. From (1) and (2), one easily sees that ϕ_{abc} is symmetric in all three indices. Denote by (η^{ab}) the inverse matrix of (η_{ab}), then $\phi^c_{ab} = \phi_{abp}\eta^{pc}$. The associativity of the multiplication is equivalent to the following system of equations

$$\phi_{abp}\eta^{pq}\phi_{qcd} = \phi_{bcp}\eta^{pq}\phi_{aqd}. \tag{3}$$

When $(H, \wedge)$ has an identity 1, only ϕ is needed. In fact, one can take $e_0 = 1$. Then from (2), one gets $\eta_{ab} = \phi_{0ab}$. To summarize, the structure of a Frobenius algebra with identity is determined by the symmetric 3-tensor ϕ such that $\eta_{ab} = \phi_{0ab}$. One can easily generalize this discussion to the graded case.

EXAMPLE 2.1. Let M be an oriented connected closed n-dimensional smooth manifold, then the de Rham cohomology $H^*(M)$ with the wedge product $\wedge$ is a graded commutative algebra over $\mathbb{R}$. Let $\int_M : H^*(M) \to \mathbb{R}$ be defined by integrations of the top degree components. Then Poincaré duality implies that $\int_M$ induces a Frobenius algebra structure on $(H^*(M), \wedge)$.

2.2. WDVV equations and formal Frobenius manifold structures. Denote by $\{x^a\}$ the linear coordinates in the basis $\{e_a\}$. Consider a family of commutative associative multiplications $\wedge_x$ on H, one for each $x \in H$, such that

$$\eta(X \wedge_x Y, Z) = \eta(X, Y \wedge_x Z),$$

for all $X, Y, Z, x \in H$. We then have a family of 3-tensors $\phi_{abc}(x)$. Such families with the property that

$$\frac{\partial}{\partial x^d}\phi_{abc} = \frac{\partial}{\partial x^c}\phi_{abd}$$

are particularly interesting to physicists (see e.g. Dijkgraaf-Verlinde-Verlinde [**12**]). Given such a family, one can find a function $\Phi : H \to \mathbf{k}$, such that

$$\phi_{abc} = \frac{\partial^3 \Phi}{\partial x^a \partial x^b \partial x^c}.$$

By (3), Φ satisfies the Witten-Dijkgraaf-E. Verlinde-H. Verlinde (WDVV) equations:

$$\frac{\partial^3 \Phi}{\partial x^a \partial x^b \partial x^p}\eta^{pq}\frac{\partial^3 \Phi}{\partial x^q \partial x^c \partial x^d} = \frac{\partial^3 \Phi}{\partial x^b \partial x^c \partial x^p}\eta^{pq}\frac{\partial^3 \Phi}{\partial x^a \partial x^q \partial x^d}. \tag{4}$$

The function Φ is called the *potential function* for the family. A *formal Frobenius manifold structure* on $(H, \wedge, \eta)$ is a formal power series Φ which satisfies the WDVV equations (Manin [**25**]). We refer to $(H, \wedge)$ as the initial data for the WDVV equations. If $(H, \wedge)$ has an identify $1 = e_0$ which is also an identity for all of $\wedge_x$, then we have

$$\eta_{ab} = \frac{\partial^3 \Phi}{\partial x^0 \partial x^a \partial x^b}. \tag{5}$$

If a formal Frobenius manifold structure Φ satisfies (5), it is called a structure of formal Frobenius manifold with identity. Again, it is straightforward to generalize the above discussions to the graded case.

3. Some notions from rational homotopy theory

3.1. Quasi-isomorphisms. A *quasi-isomorphism* between two DGA's $\mathcal{A}$ and $\mathcal{B}$ is a series of DGAs $\mathcal{A}_0, \cdots, \mathcal{A}_n$, and DGA-homomorphisms either $f_i : \mathcal{A}_i \to \mathcal{A}_{i+1}$, or $f_i : \mathcal{A}_{i+1} \to \mathcal{A}_i$ for $0 \le i \le n-1$, such that $\mathcal{A}_0 = \mathcal{A}$, $\mathcal{A}_n = \mathcal{B}$ and each f_i induces isomorphism on cohomology. A DGA $\mathcal{A}$ is called *formal* if it is quasi-isomorphic to its cohomology algebra (regarded as a DGA with zero differential). Every simply connected DGA has a minimal model and quasi-isomorphic DGAs have the same minimal model (see e.g. Giffiths-Morgan [**20**]).

3.2. Formal power series connections. Given a *differential graded Lie algebra* (DGLA) $(\mathcal{L}, [\cdot,\cdot], \mathfrak{d})$, fix a decomposition $\mathcal{L} = \mathcal{H} \oplus \mathfrak{d}M \oplus M$, such that $\mathcal{H} \subset \operatorname{Ker}\mathfrak{d}$, the natural map $\mathcal{H} \to H(\mathcal{L}, \mathfrak{d})$ is an isomorphism and $\mathfrak{d}|_M$ is injective. Such a decomposition is called a *cohomological decomposition.* Fix a homogeneous basis $\{\alpha_j \in \mathcal{H}\}$, denote by $\{X^j\}$ the dual basis. Each X^j is given the degree $-|\alpha_j| + 1$. The dual space of $\mathcal{H}$ with such a grading is denoted by $s^{-1}\mathcal{H}^t$. Let $\overline{S}(s^{-1}\mathcal{H}^t) = \sum S^n(s^{-1}\mathcal{H}^t)$. Modifying the method in Chen [**10**], Hain [**21**] inductively constructed a differential b on $\overline{S}(s^{-1}\mathcal{H}^t)$ and a *formal power series connection* of the form

$$\omega = \sum \alpha_i X^i + \cdots + \sum \alpha_{i_1\cdots i_n} X^{i_1} \odot \cdots \odot X^{i_n} + \cdots,$$

where $\alpha_{i_1\cdots i_n} \in \mathcal{L}$ has degree $|X^{i_1}| - \cdots - |X^{i_n}| + 1$, such that

$$b\omega + \mathfrak{d}\omega + \frac{1}{2}[\omega, \omega] = 0.$$

Here we have naturally extended b, $\mathfrak{d}$ and $[\cdot,\cdot]$ on $\mathcal{L} \otimes \overline{S}(s^{-1}\mathcal{H}^t)$.

4. Formal Frobenius manifold structures from DGBV algebras

4.1. DGBV algebras. A *Gerstenhaber algebra* (G-algebra) is a graded commutative algebra $(\mathcal{A}, \wedge)$ over $\mathbf{k}$ together with a bilinear map $[\cdot \bullet \cdot] : \mathcal{A} \otimes \mathcal{A} \to \mathcal{A}$ of degree -1, such that

$$\begin{aligned}
&[a \bullet b] = -(-1)^{(|a|-1)(|b|-1)}[b \bullet a],\\
&[a \bullet [b \bullet c]] = [[a \bullet b] \bullet c] + (-1)^{(|a|-1|)(|b|-1)}[b \bullet [a \bullet c]],\\
&[a \bullet (b \wedge c)] = [a \bullet b] \wedge c + (-1)^{(|a|-1|)|b|} b \wedge [a \bullet c],
\end{aligned}$$

for homogeneous $a, b, c \in \mathcal{A}$. For any linear operator $\Delta : \mathcal{A} \to \mathcal{A}$ of degree -1, define

$$[a \bullet b]_\Delta = (-1)^{|a|}(\Delta(a \wedge b) - (\Delta a) \wedge b - (-1)^{|a|} a \wedge \Delta b),$$

for homogeneous elements $a, b \in \mathcal{A}$. If $\Delta^2 = 0$ and

$$[a \bullet (b \wedge c)]_\Delta = [a \bullet b]_\Delta \wedge c + (-1)^{(|a|-1)|b|} b \wedge [a \bullet c]_\Delta,$$

for all homogeneous $a, b, c \in \mathcal{A}$, then $(\mathcal{A}, \wedge, \Delta, [\cdot \bullet \cdot]_\Delta)$ is a *Gerstenhaber-Batalin-Vilkovisky (GBV) algebra.* When there is no confusion, we will simply write $[\cdot \bullet \cdot]$ for $[\cdot \bullet \cdot]_\Delta$. It can be checked that a GBV algebra is a G-algebra (cf. Koszul [**23**], Getzler [**15**] and Manin [**26**]). A *DGBV algebra* is a GBV algebra with a differential δ with respect to $\wedge$, such that $\delta\Delta + \Delta\delta = 0$ (hence δ is also a differential of $[\cdot \bullet \cdot]_\Delta$).

4.2. Nice integrals. A $\mathbf{k}$-linear functional $\int : \mathcal{A} \to \mathbf{k}$ on a DGBV algebra $(\mathcal{A}, \wedge, \delta, \Delta, [\cdot \bullet \cdot])$ is called *an integral* if for all homogeneous $a, b \in \mathcal{A}$,

$$\int (\delta a) \wedge b = (-1)^{|a|+1} \int a \wedge \delta b, \tag{6}$$

$$\int (\Delta a) \wedge b = (-1)^{|a|} \int a \wedge \Delta b. \tag{7}$$

Given an integral, $\eta(\alpha, \beta) = \int \alpha \wedge \beta$ defines a graded symmetric bilinear form η on $H(\mathcal{A}, \delta)$. If η is non-degenerate, we say that the integral is *nice.* Therefore, if $\mathcal{A}$ has a nice integral, then $(H(\mathcal{A}, \delta), \wedge, \eta)$ is a Frobenius algebra.

4.3. Formal Frobenius manifold structure on the cohomology of a DGBV algebra. Let $(\mathcal{A}, \wedge, \delta, \Delta, [\cdot \bullet \cdot])$ be a DGBV algebra satisfying the following conditions:

(i) The cohomology algebra $H(\mathcal{A}, \delta) = \operatorname{Ker} \delta / \operatorname{Im} \delta$ is free and of finite rank as a $\mathbf{k}$-module.
(ii) There is a nice integral on $\mathcal{A}$.
(iii) The inclusions $(Ker\Delta, \delta) \hookrightarrow (\mathcal{A}, \delta)$ and $(Ker\delta, \Delta) \hookrightarrow (\mathcal{A}, \Delta)$ induce isomorphisms of cohomology.

Then there is a canonical construction of a formal Frobenius manifold structure with identity on $H(\mathcal{A}, \delta)$.

This construction was implicit in Bershadsky-Cecotti-Ooguri-Vafa [**2**] in the special case of the extended deformation theory Calabi-Yau manifolds based on the work of Tian [**28**] and Todorov [**29**]. It was mathematically formulated by Barannikov and Kontsevich [**1**]. The details for general DGBV algebras can be found in Manin [**26**]. Here we give a description in terms of Chen's construction of formal power series connections. Since $(s\mathcal{A}, \delta, [\cdot \bullet \cdot])$ is a DGLA, by §3.2 one gets a universal formal power series connection Γ and a derivation ∂ on $K = \overline{S}(H(\mathcal{A}, \delta)^t)$, such that

$$\partial\Gamma + \delta\Gamma + \frac{1}{2}[\Gamma \bullet \Gamma] = 0.$$

However, the condition (iii) above implies that $\partial = 0$. Such a method was first used in Tian [**28**] and Todorov [**29**]. Therefore, Γ satisfies

$$\delta\Gamma + \frac{1}{2}[\Gamma \bullet \Gamma] = 0, \tag{8}$$

Furthermore, one can take $\Gamma = \Gamma_1 + \Delta B$, where $\Gamma_1 = \sum \alpha_j X^j$. Set $\delta_\Gamma = \delta + [\Gamma \bullet \cdot]$, then δ_Γ is a derivation of $\mathcal{A}_K = (\mathcal{A} \otimes K, \wedge)$. Now

$$\delta_\Gamma^2 = [(\delta\Gamma + \frac{1}{2}[\Gamma \bullet \Gamma]) \bullet \cdot] = 0.$$

It can be proved that $H(\mathcal{A}_K, \delta_\Gamma) \cong H(\mathcal{A}, \delta) \otimes K$, hence the multiplication in $H(\mathcal{A}_K, \delta_\Gamma)$ induces a deformation of the multiplication in $H(\mathcal{A}, \delta)$. Given $X \in H(\mathcal{A}, \delta)$, the contraction with X from the right induces a right derivation of degree $|X|$ on K and hence also on $\mathcal{A}_K$. For $\alpha \in \mathcal{A}_K$, denote by $X\alpha$ the contraction by X of α. Now applying contraction by X on both sides of (8), one gets

$$\delta(X\Gamma) + [\Gamma \bullet (X\Gamma)] = 0.$$

So $X\Gamma$ represents a class in $H(\mathcal{A}_K, \delta_\Gamma)$. Following Bershadsky *et al* [**2**], set

$$\Phi = \int \frac{1}{6}\Gamma^3 - \frac{1}{2}\delta B \Delta B. \tag{9}$$

A calculation in Barannikov-Kontsevich [**1**] and Manin [**26**] shows that

$$X^3\Phi = \int (X\Gamma) \wedge (X\Gamma) \wedge (X\Gamma),$$

i.e. Φ is the potential function of the deformation.

4.4. Gauge invariance. We recall some results proved in Cao-Zhou [**9**]. Let $\mathfrak{m}$ be the maximal ideal of K. Consider the group

$$\mathcal{G} = \exp(s\mathcal{A} \otimes \mathfrak{m})^e,$$

with the multiplication $e^A e^B = e^C$ defined by the Campbell-Baker-Hausdorff formula. Assume $(\mathcal{A}, \wedge, \delta, \Delta, [\cdot \bullet \cdot])$ is a DGBV algebra which satisfies the conditions (i)-(iii) in §4.3. Then it is shown in [**9**] that given any two universal normalized solutions Γ and $\overline{\Gamma}$, there exists an odd element $A \in (\operatorname{Im}\Delta \otimes \mathfrak{m})$, such that $e^A \cdot \Gamma = \overline{\Gamma}$. Furthermore, the potential function Φ is gauge invariant: $\Phi(e^A \cdot \Gamma) = \Phi(\Gamma)$.

4.5. Invariance under quasi-isomorphisms. A homomorphism between two DGBV algebras with nice integrals over $\mathbf{k}$, $(\mathcal{A}_i, \wedge_i, \delta_i, \Delta_i, [\cdot \bullet \cdot]_i, \int_i)$, $i = 1, 2$, is a homomorphism of graded algebras $f : \mathcal{A}_1 \to \mathcal{A}_2$ such that $f\delta_1 = \delta_2 f$, $f\Delta_1 = \Delta_2 f$, and $\int_2 f(\alpha) = \int_1 \alpha$ for all $\alpha \in \mathcal{A}$. It is a quasi-isomorphism if f induced isomorphisms $H(\mathcal{A}_1, \delta_1) \to H(\mathcal{A}_2, \delta_2)$ and $H(\mathcal{A}_2, \Delta_2) \to H(\mathcal{A}_2, \Delta_2)$. We prove in [**9**] the following result: If there is a quasi-isomorphism between two DGBV algebras with nice integrals which satisfy the conditions (i)-(iii) in §4.3, then the formal Frobenius manifolds constructed from them as in §4.3 can be naturally identified with each other.

5. DGBV algebras from symplectic and complex geometries

5.1. GBV algebra structure on the space of polyvector fields. Denote by $\Omega^{-*}(M) = \Gamma(M, \Lambda^* TM)$ the space of polyvector fields on M. The Lie bracket $[\cdot, \cdot]$ of vector fields can be extended to the *Schouten-Nijenhuis bracket* $[\cdot, \cdot]_S : \Omega^{-*}(M) \times \Omega^{-*}(M) \to \Omega^{-*}(M)$ of degree -1 given locally by

$$\begin{aligned}&[X_1 \wedge \cdots \wedge X_p, Y_1 \wedge \cdots \wedge Y_q]_S \\ =&\sum_{i=1}^{p}\sum_{j=1}^{q}(-1)^{i+j+p+l}[X_i, Y_j] \wedge X_1 \wedge \cdots \wedge \widehat{X}_i \wedge \cdots \wedge X_p \wedge Y_1 \wedge \cdots \wedge \widehat{Y}_j \wedge \cdots \wedge Y_q,\end{aligned}$$

for local vector fields $X_1, \cdots, X_p, Y_1, \cdots, Y_q$. Then $(\Omega^{-*}(M), \wedge, [\cdot, \cdot]_S)$ becomes a G-algebra. There are two methods to make it a GBV algebra. The first method due to Koszul [**23**] uses the generalized divergence operator for any torsion free connection, while the second method due to Witten [**30**] uses any volume form in a way similar to earlier construction in Tian [**28**] and Todorov [**29**] (see §5.5).

5.2. DGBV algebras from Poisson geometry. A *Poisson structure* on a manifold M is an element $w \in \Omega^{-2}(M)$ such that $[w, w]_S = 0$. On a Poisson manifold (M, w), define $\sigma : \Omega^{-*}(M) \to \Omega^{-(*+1)}(M)$ by $\sigma(P) = [w, P]$. Then $(\Omega^{-*}(M), \wedge, \sigma, [\cdot, \cdot]_S)$ is a differential G-algebra. If a Poisson manifold (M, w) is *regular*, i.e., w has constant rank, then there is a (not unique) torsion free connection for which the Poisson structure w is parallel. If Δ is the generalized divergence operator for such a connection, then $[\Delta, \sigma] = 0$ and $(\Omega^{-*}(M), \wedge, \sigma, \Delta, [\cdot, \cdot]_S)$ is a DGBV algebra.

There is another DGBV algebra associated with a Poisson manifold (M, w). Koszul defined an operator $\Delta = [i(w), d] : \Omega^*(M) \to \Omega^{*-1}(M)$, where $i(w)$ denotes the contraction by w. He proved $[d, \Delta] = \Delta^2 = 0$ and $[\cdot \bullet \cdot]_\Delta$ satisfies

$$[\alpha \bullet (\beta \wedge \gamma)]_\Delta = [[\alpha \bullet \beta]_\Delta \wedge \gamma + (-1)^{(|\alpha|-1)|\beta|} \beta \wedge [\alpha \bullet \gamma]_\Delta,$$

hence $(\Omega^*(M), \wedge, d, \Delta, [\cdot \bullet \cdot]_\Delta)$ is a DGBV algebra.

5.3. DGBV algebras from Symplectic manifolds. A *symplectic manifold* M is automatically a Poisson manifold, hence there are associated DGBV algebras from it. Indeed, the symplectic form ω induces isomorphisms $\sharp : \Omega^*(M) \to \Omega^{-*}(M)$ and $\flat : \Omega^{-*}(M) \to \Omega^*(M)$. Using the Darboux theorem, it can be easily seen that the bivector filed $w = \omega^\sharp$ is a Poisson structure. Furthermore, it can be seen that $d = [\omega \bullet \cdot]_\Delta$. The symplectic form induces an isomorphism of DGBV algebras $(\Omega^{-*}(M), \wedge, \sigma, \Delta, [\cdot,\cdot]_S) \cong (\Omega^*(M), \wedge, d, \Delta, [\cdot \bullet \cdot]_\Delta)$.

5.4. Differential G-algebras from complex manifolds. Given a complex n-manifold M, let

$$\Omega^{-*,-*}(M) = \sum_{p,q\geq 0} \Omega^{-p,-q}(M) = \sum_{p,q\geq 0} \Gamma(M, \Lambda^p T^{1,0}M \otimes \Lambda^q T^{0,1}M).$$

Similarly define $\Omega^{-*,*}(M)$ and $\Omega^{*,*}(M)$. Since $\Omega^{-*,-*}(M) = \Omega^{-*}(M) \otimes \mathbb{C}$, it is a complex G-algebra. From $[\Omega^{-1,0}(M), \Omega^{-1,0}(M)] \subset \Omega^{-1,0}(M)$, one sees that $\Omega^{-*,0}(M)$ is G-subalgebra of $\Omega^{-*,-*}(M)$. Using local coordinates, it is easy to see that the Schouten-Nijenhuis bracket on $\Omega^{-*,0}(M)$ can be extended to a G-algebra structure on $\Omega^{-*,*}(M)$. Since $\Omega^{-*,*}(M)$ is the space of $(0,*)$-form with values in the holomorphic bundle $\Lambda^* T^{1,0}M$, there is an operator $\bar\partial$ acting on it. The tuple $(\Omega^{-*,*}(M), \wedge, [\cdot,\cdot]_S, \bar\partial)$ is a differential G-algebra.

5.5. DGBV algebras from Calabi-Yau manifolds. Recall that a Kähler manifold is called a *Calabi-Yau manifold* if it admits a parallel holomorphic volume form. The terminology comes from Yau's solution to Calabi's conjecture [**32**]. Such manifolds are very important in string theory. Given a Calabi-Yau n-fold M, the holomorphic volume form Ω defines an isomorphism $\Omega^{-*,*}(M) \to \Omega^{n-*,*}(M)$. Let $\Delta : \Omega^{-*,*}(M) \to \Omega^{-(*-1),*}(M)$ be the conjugation of ∂ by this isomorphism. A formula in Tian [**28**] shows that $[\cdot \bullet \cdot]_\Delta = [\cdot,\cdot]_S$, hence Δ gives a GBV algebra structure on $\Omega^{-*,*}(M)$. From $\bar\partial\Omega = 0$, one sees that $\bar\partial\Delta + \Delta\bar\partial = 0$. Hence $(\Omega^{-*,*}(M), \wedge, \bar\partial, \Delta, [\cdot,\cdot])$ is a DGBV algebra.

6. Deformations of complex and Poisson structures

6.1. Deformation of complex structures. Given a complex manifold M, the small deformations of the complex structure can be studied via the Maurer-Cartan equation

$$\bar\partial\omega + \frac{1}{2}[\omega,\omega] = 0, \tag{10}$$

where $\omega \in \Omega^{-1,1}(M)$. Fix a Hermitian metric on M, let $\bar\partial^*$ be the formal adjoint of $\bar\partial$ and $\Box_{\bar\partial} = \bar\partial\bar\partial^* + \bar\partial^*\bar\partial$. As a consequence of the Hodge decomposition

$$\Omega^{-*,*}(M) = \mathcal{H}_{\bar\partial}^{-*,*}(M) \oplus \operatorname{Im}\bar\partial \oplus \operatorname{Im}\bar\partial^*,$$

we have $H^{-*,*}(M) \cong \mathcal{H}_{\bar\partial}^{-*,*}(M)$. Take a basis $\{\alpha_j\}$ of $\mathcal{H}_{\bar\partial}^{-1,1}$, and let $\{t^j\}$ be coordinates in this basis. Let $\omega(t) = \omega_1(t) + \cdots + \omega_n(t) + \cdots$ be a formal power series, such that $\omega_1(t) = \sum_j \alpha_j t^j$, and ω_n is homogeneous of degree n in t^j's. Then (10) is equivalent to a sequence of equations

$$\bar\partial\omega_n = -\frac{1}{2}\sum_{p+q=n} [\omega_p, \omega_q].$$

Suppose that $\omega_1, \cdots, \omega_n$ have been defined, a calculation shows that

$$\bar{\partial}(-\frac{1}{2} \sum_{p+q=n+1} [\omega_p, \omega_q]) = 0.$$

Let $Q = G_{\bar{\partial}} \bar{\partial}^*$, where $G_{\bar{\partial}}$ is the Green's operator of $\Box_{\bar{\partial}}$. Take

$$\omega_{n+1} = -\frac{1}{2} Q \sum_{p+q=n+1} [\omega_p, \omega_q].$$

This is equivalent to

$$\omega + \frac{1}{2} Q[\omega, \omega] = \omega_1. \tag{11}$$

Following Kodaira-Spencer [**22**], one can show that for small $||t||$, $\omega(t)$ is convergent. Taking $\bar{\partial}$ on both sides of (11), one gets

$$\bar{\partial}\omega + \frac{1}{2}[\omega, \omega] = -\frac{1}{2}[\omega, \omega]^H,$$

where $[\omega, \omega]^H$ is the harmonic part of $[\omega, \omega]$. When $H^{-1,2}(M) = 0$, a neighborhood of 0 in $H^{-1,1}(M)$ then parameterizes the small deformations. This is the construction by Kodaira-Spencer [**22**] of a complete family when $H^{-1,2}(M) = 0$. When $H^{-1,2}(M) \neq 0$, one gets a map from $U \subset \mathcal{H}_{\bar{\partial}}^{-1,1}(M)$ to $\mathcal{H}_{\bar{\partial}}^{-1,2}(M)$ given by $t = (t^j) \mapsto [\omega(t), \omega(t)]^H$, its zero set gives solutions ω to the Maurer-Cartan equation. This is due to Kuranishi (see e.g. [**24**]).

The extended deformation problem in this case is to find solutions to the Maurer-Cartan equation with $\omega \in \Omega^{-*,*}(M)$. The Hodge decomposition gives the cohomological splitting, and the above iterative method of ω corresponds to Chen's construction of formal power series connection modified by Hain [**21**] for DGLA's. The formal power series $-\frac{1}{2}[\omega(t), \omega(t)]^H$ corresponds to the differential b (see §3.2), where $\{t^j\}$ are the coordinates of $\mathcal{H}^{-*,*}(M)$ in a homogeneous basis.

6.2. Formal Frobenius manifolds from Calabi-Yau manifolds. Motivated by a result of Bogomolov [**4**], Tian [**28**] (see also Todorov [**29**]) introduced an ingenious method to prove that the deformations of complex structures on Calabi-Yau manifolds are unobstructed. This includes the introduction of the operator Δ in §5.5, showing $[\cdot \bullet \cdot]_\Delta = [\cdot, \cdot]_S$, using Hodge theory to find power series solution $\omega(t)$ such that $\omega(t) = \omega_1(t) + \Delta B(t)$, and then showing that $[\omega(t), \omega(t)]^H = 0$. This method has been used by Bershadsky-Cecotti-Ooguri-Vafa [**2**] in what they called Kodaira-Spencer theory of gravity. Fixing a holomorphic volume form Ω on a Calabi-Yau n-manifold M, define $\int : \Omega^{-*,*}(M) \to \mathbb{C}$ by

$$\int \alpha = \int_M \alpha^\flat \wedge \Omega.$$

It is easy to see that $\int$ is a nice integral for $(\Omega^{-*,*}(M), \wedge, \bar{\partial}, \Delta, [\cdot, \cdot]_\Delta)$. Bershadsky *et al.* found a Lagrangian invariant under the action of diffeomorphisms.They showed that its critical points correspond to the complex structures nearby. They discussed the quantization via BV formalism in which the extended deformation problem suggested by Witten arises naturally. They also argued how the values of the Lagrangian at the critical points give the potential for the deformed multiplicative structure. Barannikov and Kontsevich [**1**] formulated such results in terms of Frobenius manifolds introduced by Dubrovin [**13**]. They also remarked that such construction can be carried out for DGBV algebras satisfying conditions in §4.3. The details of this construction of formal Frobenius manifold can be found in Manin [**26**]. Anyway, the moral is that the extended deformation problem leads to the DGBV algebra $(\Omega^{-*,*}(M), \wedge, \bar{\partial}, \delta, [\cdot,\cdot]_S)$, which further leads to a canonical construction of formal Frobenius manifold structure on $H^{-*,*}(M)$.

6.3. Deformations of Poisson structures. Given a Poisson structure w_0 on M, any Poisson structure close to w_0 can be written as $w = w_0 + \gamma$. Let $\sigma = [w_0, \cdot]_S : \Omega^{-*}(M) \to \Omega^{-(*+1)}(M)$, then $[w, w]_S = 0$ can be rewritten as

$$\sigma\gamma + \frac{1}{2}[\gamma,\gamma]_s = 0. \tag{12}$$

Now assume that w_0 comes from a symplectic form ω_0 on M. The deformation theory of symplectic structure is "trivial". It is well-known that any closed 2-form close to a symplectic 2-form in C^0-norm is also symplectic. However, the deformation of the Poisson structure is nonlinear: equation (12) corresponds to

$$d\omega + \frac{1}{2}[\omega \bullet \omega] = 0. \tag{13}$$

Given a Riemannian metric on M, then one can consider the formal adjoint d^* of d and the Laplacian operator $\Box_d : \Omega^*(M) \to \Omega^*(M)$. The power series method can be carried out similar to the complex case. Also one can consider the extended deformation problem of finding solutions of (13) in $\Omega^*(M)$.

6.4. Kähler gravity. The mirror analogue of the Kodaira-Spencer theory of gravity is the theory of Kähler gravity studied in Bershadsky-Sadov [**3**]. Let M be a closed Kähler manifold with Kähler form ω_0. By Stokes theorem and Poincaré duality, the integral of top degree forms on M is a nice integral of the DGBV algebra $(\Omega^*(M), \wedge, d, \Delta, [\cdot \bullet \cdot])$. Kähler identities shows that $\Delta = (d^c)^*$. By imposing the gauge fixing condition $\Gamma \in \operatorname{Ker}\Delta$, Bershadsky and Sadov wrote a Lagrangian similar to that used in Kodaira-Spencer gravity and showed that the Euler-Lagrangian equation of it on $\operatorname{Ker}\Delta$ is

$$d\Gamma + \frac{1}{2}\Delta(\Gamma \wedge \Gamma) = 0,$$

or equivalently,

$$d\Gamma + \frac{1}{2}[\Gamma \bullet \Gamma] = 0.$$

According to §5.3, it describes the deformations of the Poisson structure corresponding to ω_0. Given a solution K_0 with $(d^c)^*K_0 = 0$, Bershadsky and Sadov pointed out the operator $D = d + [(d^c)^*, K_0]$ squares to zero. Notice that $D = d + [K_0 \bullet \cdot] = [(\omega_0 + K_0) \bullet \cdot]$, so deforming d to D corresponds to deforming $[w_0, \cdot]_S$ to $[w, \cdot]_S$ on $\Omega^{-*}(M)$.

Similarly, consideration of BV quantization leads to the extended deformation problem, which is described by the DGBV algebra $(\Omega^*(M), \wedge, d, \Delta, [\cdot \bullet \cdot])$. Again, Hodge theory of Kähler manifold can be used to show that this DGBV algebra satisfies all the conditions in §4.3. Hence there is a canonical construction of formal Frobenius manifold structure on the de Rham cohomology $H^*(M)$ for any Kähler manifold. This was done in Cao-Zhou [**6**]. Notice that we can complexify this DGBV algebra. Then we have $d = \bar{\partial} + \bar{\partial}$, $(d^c)^* = \sqrt{-1}\bar{\partial}^* - \sqrt{-1}\partial^*$. In Cao-Zhou [**5**], we proved that $(\Omega^{*,*}(M), \wedge, \bar{\partial}, -\sqrt{-1}\partial^*, [\cdot \bullet \cdot]_{-\sqrt{-1}\partial^*})$ is a DGBV algebra which satisfies the conditions in §4.3. Hence there is a natural construction of the

Frobenius manifold structure on the Dolbeault cohomology $H^{*,*}(M)$. Furthermore, we showed in [**6**] that the formal Frobenius manifold structures on $H^*(M, \mathbb{C})$ and $H^{*,*}(M)$ given above can be identified with each other. This generalizes of the well-known isomorphism $H^*(M, \mathbb{C}) \cong H^{*,*}(M)$ (as complex vector spaces) to the highly nonlinear formal Frobenius manifold structures on these spaces.

The above results have been generalized in various directions. When M is a hyperkähler manifold, we showed in [**7**] there are many different ways to construct DGBV algebras for which the construction in §4.3 applies. Also, when a Kähler manifold admits a Hamiltonian action by holomorphic isometries, we showed in [**8**] that the Cartan model admits a structure of DGBV algebra for which the method of §4.3 yields a formal Frobenius manifold structure.

7. Concluding remarks

In the above we have discussed the relevance of rational homotopy theory in an A-type theory, the theory of Kähler gravity, and a B-type theory, the Kodaira-Spencer theory of gravity. The strategy is to obtain DGBV algebra structures by considering extended deformation problems of the relevant geometric objects. Ideas from rational homotopy theory can then be used to construct and identify formal Frobenius manifold structures on the the cohomology of such algebras. As a result, we propose the following version of mirror symmetry conjecture: For a Calabi-Yau manifold M and a suitably defined mirror manifold $\widehat{M}$, the DGBV algebras

$$(\Omega^{*,*}(M), \wedge, d, (d^c)^*, [\cdot \bullet \cdot]_{(d^c)^*})$$

and

$$(\Omega^{-*,*}(\widehat{M}), \wedge, \bar{\partial}, \Delta, [\cdot,\cdot]_S)$$

are quasi-isomorphic, hence by [**9**] the formal Frobenius manifold structures on their cohomology can be identified with each other. The examples of mirror manifold constructed by Greene and Plesser [**17**] involves the quotients of Calabi-Yau manifold by finite automorphism groups. The extension of the results described above to the quotient case is in progress. Finally we want to mention that the above version of mirror symmetry suggests an application being developed of Quillen's closed model category theory to the study of mirror symmetry.

References

[1] S. Barannikov, M. Kontsevich, *Frobenius Manifolds and Formality of Lie Algebras of Polyvector Fields*, **Internat. Math. Res. Notices** (1998), no. 4, 201–215, alg-geom/9710032.

[2] M. Bershadsky, S. Cecotti, H. Ooguri, C. Vafa, *Kodaira-Spencer theory of gravity and exact results for quantum string amplitudes* **Comm. Math. Phys. 165** (1994), no. 2, 311–427, hep-th/9309140.

[3] M. Bershadsky, V. Sadov, *Theory of Kähler gravity*, **Internat. J. Modern Phys. A 11** (1996), no. 26, 4689–4730, hep-th/9410011.

[4] F.A. Bogomolov, *Hamiltonian Kählerian manifolds*, **Soviet Math. Dokl. 19** (1978), no. 6, 1462 - 1465 (1979).

[5] H-D. Cao, J. Zhou, *Frobenius Manifold Structure on Dolbeault Cohomology and Mirror Symmetry*, to appear in **Comm. Anal. Geom.**, math.DG/9805094.

[6] H-D. Cao, J. Zhou, *Identification of Two Frobenius Manifolds*, **Math. Res. Lett. 6** (1999), no. 1, 17–29, math.DG/9805095.

[7] H-D. Cao, J. Zhou, *Frobenius manifolds from hyperkähler manifolds*, preprint.

[8] H-D. Cao, J. Zhou, *Formal Frobenius manifold structure on equivariant cohomology*, **Commun. Contemp. Math. 1** (1999), no. 4, 535–552, math.DG/9903024.

[9] H-D. Cao, J. Zhou, *On quasi-isomorphic DGBV algebras*, math.DG/9904168.

[10] K.-T. Chen, *Extension of C^∞ function algebra by integrals and Malcev completion of π_1*, **Advances in Math. 23** (1977), no. 2, 181–210.
[11] P. Deligne, P. Griffiths, J. Morgan, D. Sullivan, *Real homotopy theory of Kähler manifolds*, **Invent. Math. 29** (1975), no. 3, 245–274.
[12] R. Dijkgraaf, H. Verlinde, E. Verlinde, *Notes on topological string theory and 2D quantum gravity*, in **String theory and quantum gravity** (Trieste, 1990), 91–156, World Sci. Publishing, River Edge, NJ, 1991.
[13] B. Dubrovin, *Geometry of 2D topological field theories*, in **Integrable systems and quantum groups** (Montecatini Terme, 1993), 120–348, Lecture Notes in Math., 1620, Springer, Berlin, 1996.
[14] M. Gerstenhaber, *The cohomology structure of an associative ring*, **Ann. of Math. (2) 78** (1963), 267–288.
[15] E. Getzler, *Two-dimensional topological gravity and equivariant cohomology*, **Comm. Math. Phys. 163** (1994), no. 3, 473–489.
[16] W.M. Goldman, J.J. Millson, *The homotopy invariance of the Kuranishi space*, **Illinois J. Math. 34** (1990), no. 2, 337–367.
[17] B.R. Greene, M.R. Plesser, *Duality in Calabi-Yau moduli space*, **Nuclear Phys. B 338** (1990), no. 1, 15–37.
[18] B. Greene, S.-T. Yau ed., **Mirror symmetry. II.**, AMS/IP Studies in Advanced Mathematics, 1. American Mathematical Society, Providence, RI; International Press, Cambridge, MA, 1997.
[19] P. Griffiths, J. Harris, **Principles of algebraic geometry**. Pure and Applied Mathematics. Wiley-Interscience [John Wiley & Sons], New York, 1978.
[20] P. Griffiths, J. Morgan, **Rational homotopy theory and differential forms**, *Progress in Mathematics*, 16. Birkhuser, Boston, Mass., 1981.
[21] R. Hain, *Twisting cochains and duality between minimal algebras and minimal Lie algebras*, **Trans. Amer. Math. Soc. 277** (1983), no. 1, 397–411.
[22] K. Kodaira, D.C. Spencer, *Existence of complex structure on a differentiable family of deformations of compact complex manifolds*, **Ann. of Math. (2) 70** (1959) 145–166.
[23] J.-L. Koszul, *Crochet de Schouten-Nijenhuis et cohomologie.* **The mathematical heritage of Élie Cartan** (Lyon, 1984). Astérisque 1985, Numero Hors Serie, 257–271.
[24] M. Kuranishi, **Deformations of compact complex manifolds**. Séminaire de Mathématiques Supérieures, No. 39 (Été 1969). Les Presses de l'Université de Montréal, Montreal, Que., 1971.
[25] Yu. Manin, **Frobenius manifolds, quantum cohomology, and moduli spaces**. American Mathematical Society Colloquium Publications, 47. American Mathematical Society, Providence, RI, 1999.
[26] Y. Manin, *Three constructions of Frobenius manifolds: a comparative study*, Sir Michael Atiyah: a great mathematician of the twentieth century. **Asian J. Math. 3** (1999), no. 1, 179–220, math.QA/9801006.
[27] D. Sullivan, *Infinitesimal computations in topology*, **Inst. Hautes Études Sci. Publ. Math. No. 47** (1977), 269–331 (1978).
[28] G. Tian, *Smoothness of the universal deformation space of compact Calabi-Yau manifolds and its Petersson-Weil metric*, in **Mathematical aspects of string theory** (San Diego, Calif., 1986), 629–646, Adv. Ser. Math. Phys., 1, World Sci. Publishing, Singapore, 1987.
[29] A.N. Todorov, *The Weil-Petersson geometry of the moduli space of* $\mathrm{SU}(n \geq 3)$ *(Calabi-Yau) manifolds. I.*, **Comm. Math. Phys. 126** (1989), no. 2, 325–346.
[30] E. Witten, *A note on the antibracket formalism*, **Modern Phys. Lett. A 5** (1990), no. 7, 487–494.
[31] S.-T. Yau ed., **Mirror symmetry. I.**, AMS/IP Studies in Advanced Mathematics, 9. American Mathematical Society, Providence, RI; International Press, Cambridge, MA, 1998.
[32] S.-T. Yau, *On the Ricci curvature of a compact Kähler manifold and the complex Monge-Ampère equation I*, **Comm. Pure Appl. Math. 31** (1978), no. 3, 413–414.
[33] J. Zhou, *Rational homotopy types of the mirror manifolds*, preprint, math.DG/9910027.
[34] J. Zhou, *Homological perturbation theory and mirror symmetry*, preprint math.DG/9906096.

DEPARTMENT OF MATHEMATICS, TEXAS A&M UNIVERSITY, COLLEGE STATION, TX 77843
E-mail address: cao@math.tamu.edu, zhou@math.tamu.edu

AMS/IP Studies in Advanced Mathematics
Volume 20, 2001

Harmonic maps on noncompact manifolds

Luen-fai Tam

§1. In this note, we will give a brief survey of some results on harmonic maps between noncompact manifolds.

Let us begin with bounded harmonic maps. A well-known result of Yau [**29**] says that there is no nonconstant bounded harmonic functions on a complete noncompact manifold with nonnegative Ricci curvature. Using similar methods in [**29, 7**], Cheng [**4**] proved that there is no nonconstant bounded harmonic maps from a manifold with nonnegative Ricci curvature into a Cartan-Hadamard manifold. Recall that a Cartan-Hadamard manifold is a complete simply connected manifold with nonpositive sectional curvature. It is interesting that Cheng's theorem is still true in a much more general setting. One can prove that if M is a complete noncompact manifold which supports no nonconstant bounded harmonic functions, then there is no nonconstant bounded harmonic map from M into a Cartan-Hadamard manifold. This result was proved by Kendall [**15**], see also [**24**]. We should mention that unlike [**15, 24**] Cheng's work contains useful estimates for the energy density of a harmonic map.

In view of the works [**15, 24**], it is natural to ask what can be said if the dimension of the space of bounded harmonic functions is finite but greater than one. In this case, we have the following result by Sung, Wang and the author [**25**]:

Theorem 1 *Let M be a complete noncompact manifold and let $\mathcal{H}^0(M)$ be the space of bounded harmonic functions on M. Suppose* $\dim\left(\mathcal{H}^0(M)\right) = \ell < \infty$, *then given ℓ points q_i in a Cartan-Hadamard manifold N, there is a harmonic map u from M to N such that $q_i \in \overline{u(M)}$, and $u(M)$ is contained in the convex hull of the q_i's.*

The harmonic map in the theorem is unique in the following sense. Any bounded continuous map from M into N can be extended to be bounded and continuous on the Stone-Cech compactification $\hat{M}$ of M. There exist ℓ points $\hat{x}_i \in \hat{M}$, $1 \le i \le \ell$, such that if u and v are two bounded harmonic maps from M into N with $u(\hat{x}_i) = v(\hat{x}_i)$, then $u \equiv v$. In the work of Li and Wang [**18**], the structure of the image of a harmonic map from a complete noncompact manifold into a Cartan-Hadamard manifold have been studied thoroughly. One of their results is the foilowing: If M and N are as in Theorem 1, then the image of a bounded

Research partially supported by Earmarked Grant of Hong Kong #CUHK325/96P.

harmonic map u from M into N is contained in the convex hull of at most ℓ points in N. Theorem 1 is complement to this result.

There are many examples of manifolds M with $\dim\left(\mathcal{H}^0(M)\right) = \ell < \infty$. In [**16**], it was proved that the dimension of the space of bounded harmonic functions on a complete noncompact manifold with nonnegative sectional curvature outside a compact set is equal to the number of ends with large volume growth. Since the number of ends of such a manifold is finite, this dimension may be larger than one but finite. In fact, one can show that the space of bounded harmonic functions on a complete noncompact manifold with nonnegative Ricci curvature outside a compact set is finite and can be estimated by the geometric data on the manifold, see [**9, 5, 17**].

In case $\ell = \infty$, Aviles, Choi and Micallef [**3**] obtained the following result:

Theorem 2 *Let $\tilde{M}$ be a compactification of M, such that ∂M is regular with respect to the Laplacian equation, then for any Cartan-Hadamard manifold N every $f \in C^0(\partial M, N)$ can be extended to a harmonic map u from M to N.*

It was proved by Anderson [**2**] and Sullivan [**23**] independently that any continuous function on the geometric boundary of a Cartan-Hadamard manifold M with sectional curvature pinched between two negative constants can be extended to a harmonic function on M. The result was generalized by Ancona [**1**] to complete manifolds which are quasi-isometric to Cartan-Hadamard manifolds with sectional curvature pinched between two negative constants and by Cheng [**6**] to Cartan-Hadamard manifolds with positive first eigenvalues and 'locally pinched' sectional curvatures. In all these cases the dimension of $\mathcal{H}^0(M)$ is no smaller than the dimension of $C^0(\partial M)$ which is obviously infinite.

§2. Let us now consider harmonic maps whose images may not be bounded. As we mention before, there is a thorough study in the general nature of the image set of a harmonic map from a complete noncompact manifold into a Cartan-Hadamard manifold by Li and Wang [**18**]. Here we only concentrate on the case that M is the Euclidean space $\mathbb{R}^m$ and N is the hyperbolic space $\mathbb{H}^n$. We will first consider harmonic maps from $\mathbb{R}^2 = \mathbb{C}$ into $\mathbb{H}^2$. In [**20**], Schoen asked whether Riemann surfaces which are related by harmonic diffeomorphisms are necessarily quasi-conformally related. Since there is no quasi-conformal homeomphism between $\mathbb{C}$ and $\mathbb{H}^2$, one would like to know whether there is a harmonic diffeomorphism between them. It was proved by Heinz in [**13**] that there is no harmonic diffeomorphism from $\mathbb{H}^2$ onto $\mathbb{C}$. In fact, there is no harmonic diffeomorphism from $\mathbb{H}^2$ onto a complete surface with nonnegative Gaussian curvature [**20**]. Note that unlike quasi-conformal map, the inverse of a harmonic map may not be harmonic and it is still unknown whether there is a harmonic diffeomorphism from $\mathbb{C}$ onto $\mathbb{H}^2$. However, there are some partial results.

Let us first review some well-known facts on harmonic maps between surfaces. Let (M, g), (N, h) be two oriented surfaces with conformal metrics: $g = \lambda^2(z)|dz|^2$, $h = \sigma^2(u)|du|^2$. Suppose u is a harmonic map from M into N. Define $|\partial u|^2 = \lambda^{-2}(z)\sigma^2(u(z))\left|\frac{\partial u}{\partial z}\right|^2$, $|\bar{\partial} u|^2 = \lambda^{-2}(z)\sigma^2(u(z))\left|\frac{\partial u}{\partial \bar{z}}\right|^2$, $\Phi = \sigma^2(u(z))\frac{\partial u}{\partial z}\frac{\partial \bar{u}}{\partial z}dz^2$. Φ is holomorphic and is called the Hopf differential of u. The Jacobian of u is given by $\mathrm{Jac}(u) = |\partial u|^2 - |\bar{\partial} u|^2$. We always assume that $\mathrm{Jac}(u) > 0$. That is, u is a orientation preserving local diffeomorphism. In case $M = \mathbb{C}$, $\Phi = \phi(z)dz^2$ with ϕ to be an entire function. In [**27**], Wan proved the following:

Theorem 3 *Let u be a harmonic diffeomorphism from $\mathbb{C}$ into $\mathbb{H}^2$ with Hopf differential ϕdz^2. Then ϕ is a polynomial if and only if u is of polynomial growth.*

In [**11**], Han proved the following:

Theorem 4 *Let u be a harmonic diffeomorphism from $\mathbb{C}$ into $\mathbb{H}^2$ with Hopf differential ϕdz^2 such that $|\partial u|^2 dz^2$ is a complete metric on $\mathbb{C}$. Suppose ϕ is a polynomial, then the image of the harmonic map is the convex hull of a set L in $\partial\mathbb{H}^2$, which is totally disconnected. In particular, the map is not onto.*

Note that if u is an orientation preserving harmonic diffeomorphism from $\mathbb{C}$ onto $\mathbb{H}^2$, then $|\partial u|^2 dz^2$ is complete, see [**20**]. Therefore Han's theorem implies that there is no polynomial growth harmonic diffeomorphism from $\mathbb{C}$ onto $\mathbb{H}^2$. Actually, Theorem 4 can be improved significantly. In fact, we have

Theorem 5 [**12**] *Let u be a harmonic diffeomorphism from $\mathbb{C}$ into $\mathbb{H}^2$ with Hopf differential ϕdz^2. Suppose $|\partial u|^2|dz|^2$ is complete. Then the image of u is an ideal polygon with $n+2$ vertices at the geometric boundary of $\mathbb{H}^2$ if and only if ϕ is a polynomial of degree n.*

Li and Wang [**18**] proved that one direction of Theorem 5 is true in a much more general setting. For example, they proved that:

Theorem 6 *Let M^n be a complete noncompact manifold with nonnegative Ricci curvature and N be a Cartan-Hadamard manifold with sectional curvature bounded between two negative constants. Let u be a harmonic map with polynomial growth of degree at most ℓ, then $u(M)$ is contained in the convex hull of k points on the geometric boundary of N with $k \leq C\ell^{n-1}$ where C is a constant depending only on n.*

A map u is said to be of polynomial growth of degree at most ℓ if there is a constant C such that $d_N(u(x), o') \leq C\left(d_M(x, o) + 1\right)^{\ell}$ for some fixed points $o \in M$ and $o' \in N$.

§3. Theorem 5 is a characterization of polynomial growth harmonic diffeomorphisms from $\mathbb{C}$ into $\mathbb{H}^2$. In this section, we will discuss methods to construct harmonic maps. The first method of construction is due to Wan and Au [**28**], see also [**26**].

Theorem 7 *Given a holomorphic quadratic differential ϕdz^2 on $\mathbb{C}$, there is a harmonic diffeomorphism u from $\mathbb{C}$ into $\mathbb{H}^2$ such that the Hopf differential of u is ϕdz^2 and that $|\partial u|^2|dz|^2$ is complete. The harmonic diffeomorphism is unique up to isometries of $\mathbb{H}^2$.*

In this construction, it is unclear what is the image of the harmonic map. We are interested in the following 'boundary value' problem: Given $m \geq 3$ points on $\mathbb{S}^1$, try to construct a harmonic map from $\mathbb{C}$ into $\mathbb{H}^2$ such that its image is the ideal polygon with the given m points as the vertices. This problem was solved by Choi and Treibergs [**8**]. In our setting, what they proved is the following:

Theorem 8 *Given $m \geq 3$ points on $\mathbb{S}^1$, there exists an entire closed spacelike constant mean curvature $H > 0$ surface in the Minkowski three space, such that the image of the Gauss map is the ideal polygon with the given m points as vertices. Moreover, the surface is conformally equivalent to $\mathbb{C}$.*

The constant mean curvature surface in the theorem is called a constant mean curvature cut. The Gauss map is a diffeomorphism from the surface into $\mathbb{H}^2$. Moreover, since the Gauss map of the constant mean curvature surface is a harmonic map into $\mathbb{H}^2$ [**19, 14, 8**], by composing a conformal map from $\mathbb{C}$ onto the surface we obtain a harmonic map from $\mathbb{C}$ into $\mathbb{H}^2$ satisfying the required condition. Actually they have proved more. But let us first mention the following question by Schoen [**20**]:*Is a harmonic diffeomorphism from $\mathbb{C}$ onto an ideal polygon of $\mathbb{H}^2$ unique up to composition with a conformal automorphism of $\mathbb{C}$?* Base on Theorem 5 and 7, we have the following observation. Consider the space $\{\phi dz^2$ with ϕ to be a polynomial of degree $n\}/\{$conformal diffeomorphisms of $\mathbb{C}\}$. This set has real dimension $2(n-1)$. On the other hand, consider the class of sets of $n+2$ points on $\mathbb{S}^1$. Two sets of points are said to be equivalent if there is a linear fractional transformation on the unit disk which takes one set to another. This family of equivalent classes has $n-1$ real parameters. In view of Theorem 5 and 7, the answer to Schoen question should be negative. One could guess that that there is a $n-1$ real parameters family of harmonic diffeomorphisms with the same ideal polygon with $n+2$ vertices as image. This was confirmed by the works [**8, 12**]. Before we state the results, let us see what are the parameters in the family. A constant mean curvature cut in the Minkowski space $\mathbb{E}^{2,1}$ can be represented as the graph of a function f defined on the whole plane. We have the following two kinds of boundary values for f:

- *Projective boundary value*:
$$V_f(x) = \lim_{r\to\infty} \frac{f(rx)}{r}.$$
The limit exists and there is a closed set $L_f \subset \mathbb{S}^1$ so that
$$V_f(x) = \sup\{\xi \bullet x : \xi \in L_f\}$$
where "$\bullet$" is the usual $\mathbb{R}^2$ inner product;
- *Asymptotic boundary value*:
$$\Lambda_f(\theta) := \lim_{r\to\infty} (f(r\theta) - r)$$
which decreases either to a finite limit or to $-\infty$. See [**8**] for example.

In our case, let $\theta_1, \ldots, \theta_{n+2} \in \mathbb{S}^1$, $n \geq 1$. Then by [**8**] it is known that given the boundary conditions

$$(*) \qquad \begin{cases} V_f(x) &= \max_{k=1,\ldots,m} x \bullet \theta_k \\ \Lambda_f(\theta) &= \begin{cases} t_k, & \text{if } \theta = \theta_k, \\ -\infty, & \text{otherwise,} \end{cases} \end{cases}$$

there is a unique f satisfying (*) whose graph is a constant mean curvature cut. In this case, the image of the Gauss map of the constant mean curvature cut is the ideal polygon with vertices $\theta_1, \ldots, \theta_{n+2}$. The parameters mentioned above come from the t_i's. More precisely, let
$$N = \{s \in \mathbb{R}^{n+2} | \ s_k = (a,b) \bullet \theta_k + c, \text{ for some } (a,b,c) \in \mathbb{R}^3\}.$$

Then we have [**12**]:

Theorem 9 *With the above notations, for each class* $[t] \in \mathbb{R}^{n+2}/N$*, there is a harmonic diffeomorphism* $u_{[t]}$ *from* $\mathbb{C}$ *into* $\mathbb{H}^2$ *whose image is the ideal polygon with vertices* $\theta_1, \ldots, \theta_{n+2}$. $u_{[t]}$ *is obtained by composing the Gauss map of the constant mean curvature cut satisfying (*) with a conformal diffeomorphism from* $\mathbb{C}$ *onto the constant mean curvature cut. If* $[t] \neq [s]$*, then* $u_{[t]}$ *and* $u_{[s]}$ *are distinct in the sense that they do not differ by a conformal diffeomorphism on* $\mathbb{C}$*. Moreover, all orientation preserving harmonic diffeomorphism* u *from* $\mathbb{C}$ *into an ideal polygon* $\mathbb{P}$ *with* $|\partial u|^2 dz^2$ *to be complete on* $\mathbb{C}$ *is equal to* $u_{[t]}$ *for some* $[t]$.

However, in this construction we do not know the relation between the coefficients of the Hopf differential of a harmonic diffeomorphism and its image. In this regard, we have the following partial result by Shi and Tam [**22**]:

Proposition 1 *Let* $\phi(z) = z^{2m} + az^{m-1}$*, where* a *is a real number. Suppose* u *is an orientation preserving harmonic diffeomorphism from* $\mathbb{C}$ *to* $\mathbb{H}^2$ *with Hopf differential* ϕdz^2 *such that* $|\partial u|^2 |dz|^2$ *is complete. Then by composing an isometry of* $\mathbb{H}^2$ *if necessary, the image of* u *is a regular ideal polygon.*

The methods in Theorem 5 and 9 cannot be generalized to higher dimensions. In fact, there are very few examples of nontrivial harmonic maps in higher dimensions. Recently, Shi and the author [**22**] are able to use a more direct method to construct a harmonic diffeomorphism from $\mathbb{C}$ to $\mathbb{H}^2$ so that its image is a regular ideal polygon. Using similar methods and induction, one can construct nontrivial harmonic maps from $\mathbb{R}^2$ into $\mathbb{H}^m$ for $m \geq 2$ so that the intersection of the closure of such a harmonic map with the geometric boundary of $\mathbb{H}^m$ consists of finitely many points. One can also construct nontrivial harmonic maps from $\mathbb{R}^m$ into $\mathbb{H}^m$. A harmonic map u from $\mathbb{R}^n$ to $\mathbb{H}^m$ is said to be nontrivial if its image is not contained in any totally geodesic submanifold of $\mathbb{H}^m$ and u cannot be decomposed as $u = F \circ G$ such that F is an isometry of $\mathbb{R}^m$ and $G = G(y^1, \ldots, y^{n-1})$ is independent of y^n.

References

[1] A. Ancona *Negatively curved manifolds, elliptic operators, and the Martin boundary,* **Annals Math. 125** (1987), 495–536.

[2] M. Anderson, *The Dirichlet problem at infinity for manifolds of negative curvature,* **J. Differential Geom. 18** (1983), 701–721.

[3] P. Avilés, H.I. Choi, and M. Micallef, *Boundary behavior of harmonic maps on non-smooth domains and complete negatively curved manifolds,* **J. Funct. Anal. 99** (1991), 293–331.

[4] S. Y. Cheng, *Liouville theorem for harmonic maps,* **Proc. Symp. Pure Math. 36** (1980), 147–151.

[5] S.-Y. Cheng, *Finite dimensionality of the spaces of positive and bounded harmonic functions,* preprint.

[6] S.-Y. Cheng, *The Dirichelet problem at infinity for nonpositively curved manifolds,* **Comm. Anal. Geom. 1** (1993), 101–112.

[7] S.-Y. Cheng and S.-T. Yau, *Differential equations on Riemannian manifolds and their geometric applications,* **Comm. Pure Appl. Math. 28** (1975), 333-354.

[8] H. I. Choi and A. Treibergs, *Gauss map of spacelike constant mean curvature hypersurface of Minkowski Space,* **J. Diff. Geom. 32** (1990), 775–817.

[9] H. Donnelly, *Bounded harmonic functions amd positive Ricci curvature,* **Math. Z. 191** (1986), 559–565.

[10] A. Grigor'yan, *Dimension of spaces of harmonic functions,* **Matem. Zametki 48** (1990), 55–61.

[11] Z. C. Han, *Some Remarks Concerning the Geometric Behavior of Harmonic Maps between Surfaces,* Preprint, (1994).

[12] Z. Han, L. F. Tam, A. Treibergs and T. Wan, *Harmonic maps from the complex plane into surfaces with nonpositive curvature,* **Comm. Anal. Geom. 3** (1995) 85–114.

[13] E. Heinz, *Über die Losungen der Minimalflächengleichung,* **Nach. Akad. Wiss.,Göttingen Math. Phys. Kl. II** (1952), 51–56.

[14] T. Ishihara *The harmonic Gauss map in a generalized sense,* **J. London Math. Soc. 26** (1982) 104–112.

[15] W. S. Kendall, *Probability, convexity, and harmonic maps with small image I: uniqueness and fine existence,* **Proc. London Math. Soc. 61** (1990), 371-406.

[16] P. Li and L.-F. Tam, *Positive harmonic functions on complete manifolds with nonnegative curvature outside a compact set,* **Ann. Math. 125** (1987), 171-207.

[17] P. Li and L.-F. Tam, *Harmonic functions and the structure of complete manifolds,* **J. Differential Geom. 35** (1992), 359–383.

[18] P. Li and J.-P. Wang, *Convex hull properties of harmonic maps,* preprint.

[19] E. A. Ruh and J. Vilms, *The tension field of the Gauss map,* **Trans. Amer. Math. Soc. 271** (1982), 639–652.

[20] R. Schoen, *The role of harmonic mappings in rigidity and deformation problems,* in Collection: Complex Geometry (Osaka, 1990), Lecture Notes in Pure and Applied Mathematics Dekker, New York (1993), **143** 179–200.

[21] R. Schoen & S. T. Yau, *On Univalent Harmonic Maps between Surfaces,* **Invent. Math. 44** (1978), 265-278.

[22] Y. Shi and L.-F. Tam, *Harmonic maps from* $\mathbb{R}^n$ *to* $\mathbb{H}^m$ *with symmetry,* preprint.

[23] D. Sullivan, *The Dirichlet problem at infinity for a negatively curved manifold,* **J. Differential Geom. 18** (1983), 723–732.

[24] S. J. Sung, L.-F. Tam and J. P. Wang, *Bounded harmonic maps on a class of manifolds,* **Proc. AMS 124** 2241–2248.

[25] S. J. Sung, L.-F. Tam and J. P. Wang, *Spaces of harmonic functions,* preprint.

[26] L. F. Tam and T. Y.-H. Wan, *Harmonic diffeomorphisms into Cartan-Hadamard surfaces with prescribed Hopf differentials,* **Comm. Anal. Geom., 2** (1995), 593–625.

[27] T. Y.-H. Wan, *Constant mean curvature surface, harmonic maps, and universal Teichmller space,* **J. Differential Geom. 35** (1992), 643–657.

[28] T. Y.-H. Wan and T. Au, *Parabolic constant mean curvature spacelike surfaces,* **Proc. Amer. Math. Soc. 120** (1992), 559–564.

[29] S.-T. Yau, *Harmonic fucntions on complete Riemannian manifolds,* **Comm. Pure Appl. Math. 28** (1975), 201–228.

Department of Mathematics, The Chinese University of Hong Kong, Shatin, Hong Kong, China,

E-mail address: lftam@math.cuhk.edu.hk

AMS/IP Studies in Advanced Mathematics
Volume 20, 2001

Curvature, Bounded cohomology and Path integrals

Jyh-Yang Wu

1. Introduction

This note was based on the lecture that the author gave in ICCM'98 Beijing, China. The purpose of this paper is twofold. On one hand, we report some new developements in Riemannian geometry along the traditional road. This is related to the theory of Hausdorff convergence, initiated by Gromov ([**Gr1**] [**GLP**]) around 1980. On the other hand, we introduce a path integral approach to Riemannian geometry. The path integral approach to quantum mechanics was developed by Dirac ([**D**]) and Feynman ([**F**] [**FH**]). We would like to emphasis this approach to Riemannian geometry by extending the Feynman formula for the Schrödinger equation and the Feynman-Kac formula for the diffusion equation on Euclidean spaces to Riemannian manifolds.

It is a fundamental problem in Riemannian geometry to understand the interaction between the topology and the geometry of a given class $\mathcal{M}$ of Riemannian manifolds (or geometric objects). This fundamental problem can be subdivided into three questions:

1. find suitable obstructions or estimates about manifolds in $\mathcal{M}$,
2. yield a criterion to recognize certain manifolds, and
3. give a calssification of manifolds in $\mathcal{M}$.

We give some well-known facts to illustrate roughly these questions.

1. If M is in the class $\mathcal{M}_1$ of Riemannian n-manifolds with Ricci curvature $\geq n-1$, then its fundamental group $\pi_1(M)$ is finite ([**CE**]). Furthermore, if its diameter d_M is π, then it is isometric to the standard sphere $\mathbb{S}^n$ ([**Ch**]).
2. So far, people only know that the fundamental group of a Riemannian manifold in $\mathcal{M}_1$ is finite. However, there is no known obstruction to these finite groups. We conjectured in [**Wu1**] that there is a prime p_n such that the first Betti number $b_1(M, \mathbb{Z}_p)$ is at most $n-1$ for every $M \in \mathcal{M}_1$ when the prime $p \geq p_n$.
3. The eigenvalues of the Laplace operator can be estimated in terms of Ricci curvature and diameter the the underlying Riemannian manifold ([**Ch**] [**Cl**]).
4. If a compact spin Riemnnian manifold M has positive scalar curvature, then its $\hat{A}$-genus vanishes ([**Pt**]).

The author is partially supported by a Taiwan NSC grant.

To understand the relation between the geometry and the topology of a Riemannian manifold, one tries to relate its geometric invariants with its topological invariants. Here is the geometric invariants of a Riemannnian manifolds that we concern most:

Curvature, Diameter, Volume, Spectrum,
Heat kernel, Curvature decay, Entropy, Systoles, $\cdots$.

The topological invariants that people consider most are:

Homotopy groups, Homology, Cohomology, Euler characteristic,
Special topological structure, Characteristic classes, Group Actions, $\cdots$.

The rest of this note is organized as follows. In section two, we report some recent developements of the study of certain classes of Riemannian manifolds under curvature bounds. In section three, we introduce the bounded cohomology and present some recent results of the joint work [**SW**] with Z. Shen (at IUPUI, USA). In section four, we shall describe the path integral approach to Riemannian geometry by deveolping a path integral formula (4.4) for the fundamental solutions of a family of time evolution equations, which is obtained in [**Wu2**]. Some important applications [**Wu5, Wu7, Wu8**] of these results will be announced.

2. The Hausdorff Convergence Theory of Curvature

The most natural class of study is the class $\mathcal{N}_1$ of all closed Riemannian n-manifolds M with Ricci curvature $Ric_M \geq (n-1)k$ and diameter $d(M) \leq D$ since it is precompact with respect to the Gromov-Hausdorff topology ([**Gr1**] [**Wu3**]). Around 1980, Gromov extended the Hausdorff distance of compact metric spaces in a amibent metric space to two abstract metric spaces. It can be described briefly as follows: Consider two compact metric spaces X and Y. If there is a metric on the disjoint union $X \amalg Y$, extending the metrics on X and Y in which the classical Hausdorff distance between X and Y in $X \amalg Y$ is less than ϵ, then we shall say that the Gromov-Hausdorff distance, $d_{GH}(X,Y)$, between X and Y is less than ϵ. See also [**Wu4**] for a definition of the parametrized Gromov-Hausdorff distance between fibered spaces and its applications.

The theory of Hausdorff convergence has provided a systemetic way to investigate the class $\mathcal{N}_1$ and its subclasses. The general strategy of this technique is to consider a sequence of Riemannian n-manifolds M_i (or Riemannian metrics on a given Riemannian manifold) in a precompact class that we want to study, which converges to a metric space X. Note that the limit space X is, in general, non-smooth even in the topological sense. There are, in usual, three major problems or their variants that we need to consider:

1. What topological and geometric properties of the M_i's can be carried over to the limit space X?
2. Does the limit space have the special properties that we expect? Is it good enough? And in what sense?
3. To what extent can these nice properties of X be pulled back to the manifold M_i for large i? Or, what can we say about M_i from the information on the limit space X?

In order to manage these problems well, one always need to tackle them from many different viewpoints in algebraic, analytic, metric and topological respects.

Moreover, some comparison techniques also play important roles along this line of research. To name some, we just mention the followings: Rauch Comparison Theorem, Bochner Formula, Toponogov Comparison Theorem, Relative Volume Comaprison Theorem, Abresch-Gromoll Inequality, and Splitting Theorems ([**BC**], [**CE**], [**Pt**]).

Next we shall report some of the recent developments of the class $\mathcal{N}_1$ and its subclasses. We start from the story of sectional curvature. Let $\mathcal{M}_{k,d,v}^{K,D,V}(n)$ denotes the class of closed Riemannian n-manifolds M with sectional curvature $k \leq K_M \leq K$, diameter $d \leq d(M) \leq D$ and volume $v \leq vol(M) \leq V$. Here we list some of the main results and useful properties of the limit spaces of the class $\mathcal{M}_{k,d,v}^{K,D,V}(n)$.

(A-I) The class $\mathcal{M}_{k,\cdot,v}^{K,D,\cdot}(n)$ and $M_i \in \mathcal{M}_{k,\cdot,v}^{K,D,\cdot}(n)$ with $X = \lim M_i$:

1. The Cheeger finiteness theorem ([**Cd**]),
2. The $C^{1,\alpha}$-convergence theorem ([**GLP**]),
3. X is a smooth manifold with a $C^{1,\alpha}$-metric,
4. M_i is diffeomorphic to X for large i and the convergence takes place in the $C^{1,\alpha}$ topology.

(A-II) The class $\mathcal{M}_{k,\cdot,v}^{\cdot,D,\cdot}(n)$ and $M_i \in \mathcal{M}_{k,\cdot,v}^{\cdot,D,\cdot}(n)$ with $X = \lim M_i$:

1. The Grove-Petersen-Wu-Perelman finiteness theorem ([**GPW**][**Pe**]),
2. The C^0-convergence theorem ([**Pe**]),
3. X is a topological manifold,
4. M_i is homeomorphic to X for large i and the convergence takes place in the uniform topology.

(A-III) The class $\mathcal{M}_{k,\cdot}^{K,D}(n)$ and $M_i \in \mathcal{M}_{k,\cdot}^{K,D}(n)$ with $X = \lim M_i$:

1. Cheeger-Fukaya-Gromov's theory of nilpotent structures ([**CFG**]),
2. X is isometric to a quotient space $N/O(n)$ with a $C^{1,\alpha}$-Riemannian manifold N, and the orthogonal group $O(n)$ acting isometrically on N.
3. Rong's almost-cyclic-fundamental-group theorem for $k > 0$ ([**R**]).

(A-IV) The class $\mathcal{M}_{k,\cdot}^{\cdot,D}(n)$ and $M_i \in \mathcal{M}_{k,\cdot}^{\cdot,D}(n)$ with $X = \lim M_i$:

1. Gromov's Betti number theorem ([**Gr3**])
2. Finiteness of isomorphism types of bounded cohomology ([**SW**]),
3. X is an Alexandrov space with curvature $\geq k$ ([**GP**][**GLP**]),
4. The group $Isom(X)$ of isometries of X is a Lie group ([**FY1**]),
5. If X is a smooth Riemannian manifold, then M_i fibers over X for large i ([**FY2**][**Y**]).

Next we shall report the story of Ricci curvature. Let $\mathcal{N}_{k,d,v}^{K,D,V}(n)$ denotes the class of closed Riemannian n-manifolds M with Ricci curvature $(n-1)k \leq K_M \leq (n-1)K$, diameter $d \leq d(M) \leq D$ and volume $v \leq vol(M) \leq V$. Again, we list some of the main results and useful properties of the limit spaces of the class $\mathcal{N}_{k,d,v}^{K,D,V}(n)$.

(B-I) The class $\mathcal{N}_{k,\cdot,v}^{K,D,\cdot}(n)$ and $M_i \in \mathcal{N}_{k,\cdot,v}^{K,D,\cdot}(n)$ with $X = \lim M_i$:

1. It is conjectured that there is a finiteness theorem of diffeomorphism types concerning this class,
2. If X is a smooth Riemannian n-manifold, then M_i is diffeomorphic to X for large i ([**CC1, CC2**]).

(B-II) The class $\mathcal{N}_{k,\cdot,v}^{\cdot,D,\cdot}(n)$ and $M_i \in \mathcal{N}_{k,\cdot,v}^{\cdot,D,\cdot}(n)$ with $X = \lim M_i$:

1. Anderson's finiteness theorem of fundamental groups ([**A**]).
2. The group $Isom(X)$ of isometries is a Lie group ([**CC1, CC2**]).
3. X contains some homology manifold points.

(B-III) The class $\mathcal{N}_{k,\cdot}^{K,D}(n)$ and $M_i \in \mathcal{N}_{k,\cdot}^{K,D}(n)$ with $X = \lim M_i$:

1. If X is a smooth Riemannian n-manifold, then M_i is $C^{1,\alpha}$ convergent to X ([**CC1, CC2**]),
2. Not much about the singularities of X is known.

(B-IV) The class $\mathcal{N}_{k,\cdot}^{\cdot,D}(n)$ and $M_i \in \mathcal{N}_{k,\cdot}^{\cdot,D}(n)$ with $X = \lim M_i$:

1. Gromov's precompactness theorem ([**GLP**]),
2. Eigenvalues estimates ([**Ch**][**Cl**]),
3. First Betti number estimates of Gromov ([**GLP**]) and Gallot ([**Ga**]).
4. The bounded Betti number estimate of Shen-Wu ([**SW**])
5. Cheeger and Colding's analysis on singular/regular points of X ([**CC1, CC2**]).
6. The bounded Betti number $\tilde{\beta}^j(M_i)$ vanishes for large i when $j > \dim X$ ([**SW**]).

For applications of these results and the theory of Hausdorff convergence, we refer to [**Gr1**] [**GLP**], [**Pt**] and [**Wu1, Wu3, Wu6**].

3. Bounded Betti numbers

In this section, we briefly discuss the bounded cohomology of a topological manifold and the joint work with Z. Shen, i.e.,(A-IV)-(2) and (B-IV)-(4)(6), since the theory of bounded cohomology is not widely known to Riemannian geometers.

Let M be a connected topological space. Denote by Σ the set of all singular simplices $\sigma : \Delta \to M$. The standard pseudo-L^∞ norm on $C^*(M)$ is defined by

$$\|c\|_\infty := \sup_{\sigma\in\Sigma} |c(\sigma)|.$$

Consider the subcomplex $\hat{C}^*(M)$ of bounded singular cochains of M. The homology groups $\hat{\mathrm{H}}^*(M)$ of $\hat{C}^*(M)$ is called the *bounded cohomology* of M. Let $\|\cdot\|_\infty^b$ denote the induced norm on $\hat{\mathrm{H}}^*(M)$. Then $(\hat{\mathrm{H}}^*(M), \|\cdot\|_\infty^b)$ becomes a normed space. The natural inclusion $I : \hat{C}^*(M) \to C^*(M)$ induces a homomorphism

$$\iota : \hat{\mathrm{H}}^*(M) \to \mathrm{H}^*(M).$$

Put

$$\tilde{\mathrm{H}}^*(M) := \iota\Big[\hat{\mathrm{H}}^*(M)\Big].$$

$\tilde{\mathrm{H}}^*(M)$ is called the *bounded part* of $\mathrm{H}^*(M)$ (see [**BC**]). Cohomology classes in $\tilde{\mathrm{H}}^*(M)$ are called *bounded classes* of $\mathrm{H}^*(M)$. Put

$$\|\alpha\|_\infty := \inf_{\beta\in I^{-1}(\alpha)} \|\beta\|_\infty^b.$$

Then $(\tilde{\mathrm{H}}^*(M), \|\cdot\|_\infty)$ is a normed space.

The inclusion $I : \hat{C}^*(M) \to C^*(M)$ induces a homomorphism $\iota : \hat{\mathrm{H}}^*(M) \to \mathrm{H}^*(M)$ with image $\tilde{\mathrm{H}}^*(M) := \iota\left[\hat{\mathrm{H}}^*(M)\right]$. We set

$$\begin{aligned} \tilde{\beta}^i(M) : &= \dim \tilde{\mathrm{H}}^i(M), \\ \hat{\beta}^i(M) : &= \dim \hat{\mathrm{H}}^i(M). \end{aligned}$$

Obviously,

$$\tilde{\beta}^*(M) \leq \beta^*(M), \quad \tilde{\beta}^*(M) \leq \hat{\beta}^*(M).$$

But there is no direct relationship between $\beta^*(M)$ and $\hat{\beta}^*(M)$. Both $\tilde{\beta}^i(M)$ and $\hat{\beta}^i(M)$ are called the ith *bounded Betti numbers* of M.

The bounded Betti numbers behave quite differently from the ordinary Betti numbers. In particular, the Poincaré duality for $\beta^*(M)$ does not hold for $\tilde{\beta}^*(M)$ and $\hat{\beta}^*(M)$. By [**Gr3**], we know that the bounded Betti numbers $\hat{\beta}^*(M)$ are completely determined by $\pi_1(M)$. However, the bounded Betti numbers $\tilde{\beta}^*(M)$ depend not only on $\pi_1(M)$, but also on the higher dimensional topological structure of M, when $\pi_1(M)$ is not amenable.

Here are some basic properties of the bounded cohomology:

1. A continous map $f : X \to Y$ induces a homomorphism $\hat{f} : \hat{C}^*(Y) \to \hat{C}^*(X)$ which is bounded relative to l_∞-norm, i.e., $\|\hat{f}(\beta)\|_\infty \leq \|\beta\|_\infty$ for all $\beta \in \hat{C}^*(Y)$.
2. A homotopy h between two maps $f, g : X \to Y$ gives a chain homotopy $\hat{h}$ between $\hat{f}$ and $\hat{g}$ such that $\hat{h}$ is bounded in each dimension relative to the l_∞-norm and hence $\hat{f}^*$ and $\hat{g}^* : \hat{H}^*(Y) \to \hat{H}^*(X)$ are equal.
3. $\hat{H}^i(X)$ may not be zero even when $i > \dim X$.
4. $\hat{H}^1(X)$ is always zero.
5. Thurston's theorem says that the homomorphisms $\hat{H}^i(X) \to H^i(X, \mathbb{R})$ is surjective for $i \geq 2$ if X is a closed manifold of negative curvature.
6. The groups $\hat{H}^i, i = 2, 3$ do not vanish for the Hawaii ring.

Recall that a group Π is called amenable if its action on the space of bounded functions $\Pi \to \mathbb{R}$ has a left invariant mean (average), that is a Π-invariant projection P of norm one from the space $L^\infty(\Pi)$ onto the one dimensional subspace of constant functions. Note that abelian groups, finite froups, nilpotent groups and solvable groups are amenable. If a normal subgroup $G \subset \Pi$ as well as the quotient group Π/G is amenable, then Π is amenable. the converse is also true. Direct sums of amenable groups are amenable, while direct products may be not amenable. The simplest examples of non-amenable groups are free non-abelian groups. If V is a non-flat manifold with non-positive curvature, then its fundamental group $\pi_1(V)$ is non-amenable. Let $f : Y \to X$ be a regular covering with an amenable Galois group Π. Then the induced map $\hat{f}^* : \hat{H}^*(X) \to \hat{H}^*(Y)$ is injective and isometric relative to the l_∞-norms.

The importance of amenable groups in the theory of bounded cohomology lies in the theorem of Hirsch-Thurston: If a group Π is amenable, then the bounded cohomology vanishes:

$$\hat{H}^i(\Pi) := \hat{H}^i(K(\Pi, 1)) = 0, \quad i > 0.$$

Gromov proves in [**Gr1**] that for any Riemannian n-manifold (M,g) in $\mathcal{M}_{k,\cdot}^{\cdot,D}(n)$, the total Betti number (with respect to any field) is bounded, namely,

$$\sum_{i=0}^{n} \beta^i(M) \le C(n,D).$$

However, Sha-Yang's examples [**SY**] show that this Gromov Betti Number Theorem is not true for Riemannian n-manifolds in $\mathcal{N}_{k,\cdot}^{\cdot,D}(n)$. Nevertheless, the first Betti number is still bounded, i.e., $\beta^1(M) \le C(n,D)$ ([**Ga**], [**GLP**]).

Thus it is natural to ask that besides the ordinary Betti number, what topological invariants are still controlled for manifolds in $\mathcal{N}_{k,\cdot}^{\cdot,D}(n)$?

To answer this question, Shen and the author prove in [**SW**] the following results.

Theorem A *There is a constant $C(n,D)$ depending only on n, D such that for closed Riemannian n-manifold (M,g) satisfying $Ric_M \ge -(n-1)$ and $d(M) \le D$, the total bounded Betti numberis bounded*

$$\sum_{i=0}^{n} \tilde{\beta}^i(M) \le C(n,D).$$

This result generalizes, in some sense, Gromov's Betti nember theorem to the case of Ricci curvature. The statements (B-IV)-(4) and (6) can now be formulated as

Theorem B *Let X be a compact metric space. There is a small constant $\epsilon = \epsilon(n,X) > 0$ such that for any closed n-manifold M satisfying $Ric_M \ge -(n-1)$, if the Gromov-Hausdorff distance $d_{GH}(M,X) < \epsilon$, then $\tilde{\beta}^j(M) = 0$ for $j > \dim X$.*

This result provides an obstruction to the collapsing under a lower Ricci curvature bound in the theory of Hausdorff convergence.

Theorem C *There are only finitely many isometric isomorphism types of bounded cohomology groups $(\hat{H}^*(M), \|\cdot\|_\infty)$ among closed n-manifolds satisfying $K_M \ge k$ and $d(M) \le D$.*

This extends Gromov's Betti number theorem to the finiteness of the bounded cohomology. For more results and techniques about bounded cohomology, we refer it to the paper [**Gr3**] by Gromov.

4. Path integral approach to Riemannian geometry

In this section we discuss some results in [**Wu2**] and try to introduce the approach of path integrals in Physics to Riemannian geometry. Especially, this will allows one to handle the eigenvalue estimates and the heat kernel estimates of the Laplace operator on a Riemannian n-manifold. One of the motivation to introduce this new approach is the search of new methods to obtain such estimate like Cheeger-Yau's estimate ([**CY**], [**Cl**]) of heat kernel under a lower Ricci curvature bound. Since most of Riemannian geometers are not familiar with the path integral formulation of the heat kernel, we shall give here a brief discussion about this approach by presenting a result in [**Wu2**]. However, for the physical motivation, we refer to the book of Feynman and Hibbs ([**FH**]).

We consider the time evolution equation on a Riemannian manifold (M, g):

$$\alpha \frac{\partial \Phi(x,t)}{\partial t} = \Delta \Phi(x,t) + W(x)\Phi(x,t) \tag{4.1}$$

where α is a nonzero complex number with $Re(\alpha) \geq 0$. The underlying space M will be required to have uniform properties and the function $W(x)$ a Lipschtz differentiable real-valued function on M. The reason that the nonzero complex number α is required to have $Re(\alpha) \geq 0$ is due to the Fresnel integral formula: for nonzero real number a and $\alpha = ai$,

$$\int_{-\infty}^{\infty} \frac{1}{\sqrt{2\pi}} e^{-\frac{ai}{2}x^2} dx = \frac{(\sqrt{i})^{-a/|a|}}{\sqrt{|a|}}$$

and the Gauss formula: for $Re(\alpha) > 0$,

$$\int_{-\infty}^{\infty} \frac{1}{\sqrt{2\pi}} e^{-\frac{\alpha}{2}x^2} dx = \frac{1}{\sqrt{\alpha}}.$$

However, for $Re(\alpha) < 0$ the integral

$$\int_{-\infty}^{\infty} \frac{1}{\sqrt{2\pi}} e^{-\frac{\alpha}{2}x^2} dx$$

does not exist! Note also that Fresnel integral formula is an analytic extension of the Gauss formula when α with $Re(\alpha) > 0$ approaches purely imaginary number α with $Re(\alpha) = 0$.

Given any nonzero complex number α with $Re(\alpha) \geq 0$, we define the α-action, denoted by $\omega_\alpha(\gamma)$, for a curve $\gamma : [t_0, t_1] \to M$ in the Sobolev space $H^{1,2}([t_0, t_1], M)$, by

$$\omega_\alpha(\gamma) = \int_{t_0}^{t_1} -\frac{\alpha}{4} < \gamma'(t), \gamma'(t) > +\frac{1}{\alpha} W(\gamma(t)) + \frac{1}{6\alpha} R(\gamma(t)) dt \tag{4.2}$$

where $R(\gamma(t))$ denotes the scalar curvature of (M, g) at the point $\gamma(t)$. Note that the Sobolev embedding theorem implies that a curve γ in $H^{1,2}([t_0, t_1], M)$ is continuous since $\dim[t_0, t_1] = 1$ and thus $\omega_\alpha(\gamma)$ is well-defined.

When $\alpha = 1$ and $W = 0$, (4.1) is the heat equation on a Riemannian manifold (M, g) and (4.2) takes the form

$$\omega_1(\gamma) = \int_{t_0}^{t_1} -\frac{1}{4} < \gamma'(t), \gamma'(t) > +\frac{1}{6} R(\gamma(t)) dt. \tag{4.3}$$

Next we shall discuss some sutiable path subspaces of $H^{1,2}([t_0, t_1], M)$ that will be used to define the path integrals.

DEFINITION 1. *Definition (Quantum-Path Spaces $\mathbb{D}_k(a, b)$) For $k \in \mathbb{N}$, $t_1 > t_0 \geq 0$ and $x, y \in M$ we define the path space $\mathbb{D}_k(x, t_0, y, t_1)$ to be the space of all broken geodesics $\gamma : [t_0, t_1] \to M$ from x to y with possible broken points at $y_j := \gamma(t_0 + \frac{j(t_1 - t_0)}{k})$, $j = 1, 2, \ldots, k-1$. Note that $x = y_0$ and $y = y_k$. Another way to describe γ is that a broken geodesic passing the point y_j at time $t_0 + \frac{j(t_1-t_0)}{k}$. The curve $\gamma|_{[t_0 + \frac{(j-1)(t_1-t_0)}{k}, t_0 + \frac{j(t_1-t_0)}{k}]}$ is a smooth minimal geodesic from y_{j-1} to y_j and thus it has the speed $\|\gamma'(s)\| = d(y_{j-1}, y_j)k/(t_1 - t_0)$ when s is in $(t_0 +$*

*$\frac{(j-1)(t_1-t_0)}{k}, t_0 + \frac{j(t_1-t_0)}{k})$. From this description, we know that a broken geodesic γ in $\mathbb{D}_k(x, t_0, y, t_1)$ corresponds canonically to a ordered set of points $\{y_j\}_{j=1}^{k-1}$ when the point y_{j-1} is not in the cut locus of y_j (c.f. [**BC**]). As long as the integration is concerned, we can always assume this for all curves $\gamma \in \mathbb{D}_k(x, t_0, y, t_1)$. Under this assumption, $\mathbb{D}_k(x, t_0, y, t_1)$ has a natrual manifold structure of dimension $n(k-1)$ since it can be identified with the open submanifold of all points $(y_1, y_2, \cdots, y_{k-1})$ in M^{k-1} such that y_j is not a cut point of y_{j-1} for all $j = 1, 2, \ldots, k$. Note that in this way, $\mathbb{D}_k(x, t_0, y, t_1)$ can be viewed as an open submanifold of M^{k-1} and indeed, it is obtained by taking away a measure zero subset from M^{k-1}. We can also set the starting event point $a = (x, t_0)$ and the ending event point $b = (y, t_1)$ and denote the path spaces $\mathbb{D}_k(x, t_0, y, t_1)$ briefly by $\mathbb{D}_k(a, b)$.*

Thus, the Quantum-path space $\mathbb{D}_k(a, b)$ can be endowed with a canonical measure, denoted by $\mathcal{D}_k\gamma$, from the product manifold M^{k-1} via the correspondence of γ and the ordered set $\{y_j\}_{j=1}^{k-1}$ viewed as the point $(y_1, y_2, \cdots, y_{k-1})$ in M^{k-1}. Let $K_\alpha(a, b)$ denote the fundamental solution of the time evolution equation (4.1) at a. In [**Wu2**], we provide a mathematical theory of the path integral formulation for the family of time evolution equations (4.1). More specifically, we prove

Theorem D *If the Riemannian n-manifold (M, g) satisfies the uniformality condition:*

1. *the injectivity radius is poisitve, say $inj(M) \geq 10\epsilon > 0$, and*
2. *the Riemannian curvature and its covariant derivative have an absolute bound, say b^2 for some $b \geq 0$,*

and W is a C^2-Lipschtz function on M, then one has

$$\mathrm{K}_\alpha(\mathrm{a}, \mathrm{b}) = \lim_{\mathrm{k}\to\infty} \int_{\mathbb{D}_\mathrm{k}(\mathrm{a},\mathrm{b})} (\frac{\mathrm{k}\alpha}{4\pi \mathrm{t}})^{\mathrm{kn}/2} \mathrm{e}^{\omega_\alpha(\gamma)} \mathcal{D}_\mathrm{k}\gamma. \tag{4.4}$$

We can also put the formula (4.4) in a physical notation as

$$K_\alpha(a, b) = \int_a^b e^{\omega_\alpha(\gamma)} \mathcal{D}\gamma.$$

Note also that when (M, g) is a compact manifold or a homogeneous space, the uniformality condition holds automatically. When α is a positive real number, for example 2, we obtain the diffusion (heat) equation. Hence, the formula (4.4) generalizes the Feynman-Kac formula to curved spaces and allows one to study Brownian motions on those spaces ([**CW**]).

In [**CY**], Cheeger and Yau give a lower bound on the heat kernel of the Laplace operator for a Riemmian n-manifold with a lower Ricci curvature bound $(n-1)k$. Indeed, the heat kernel is bounded below by the corresponding heat kernel on the simply connected space n-form $\mathbb{M}_k^n$ with constant sectional curvature k. In view of the relative volume comparison theorem ([**BC**]) under a lower Ricci curvature bound, the formula (4.4) provides a heuristic and intrinsic understanding of the heat kernel comparison theorem of Cheeger and Yau.

In order to get some feeling why the formula (4.4) holds, we discuss briefly its proof when the potential W vanishes and the underlying manifold M is closed. That is, we consider the time evolution equation on a closed Riemannian n- manifold

(M, g):

$$\alpha \frac{\partial \Phi(x,t)}{\partial t} = \Delta \Phi(x,t) \tag{4.5}$$

and let $p_\alpha(x, y, t)$ denote its fundamental solution for $t_0 = 0$ and $t = t_1$. Without loss of generality, to prove Theorem D, we only need to establish the formula (4.4) for $p_\alpha(x, y, t)$. It is well known that the fundamental solution $p_\alpha(x, y, t)$ has the semigroup property:

$$p_\alpha(x, z, t + s) = \int_M p_\alpha(x, y, t) p_\alpha(y, z, s) dy. \tag{4.6}$$

Note that this is the most important common property of the time evolution equation (4.5) for all nonzero complex number α with $Re(\alpha) \geq 0$. To prove the formula (4.4), we basically use the formula (4.6) and the following asymptotic behavior of the heat kernel:

$$p_\alpha(x, y, t) = \omega(x, y, t/\alpha)[\rho(d(x, y))(u_0(x, y) + \frac{t}{\alpha} u_1(x, y)) + o(t)] \tag{4.7}$$

where the smooth functions $\rho(d(x, y))$, $u_0(x, y)$ and $u_1(x, y)$ have the properties:

1. the cut-off function $\rho : [0, \infty] \to [0, 1] \in C^\infty$, such that $\rho|[0, \epsilon] = 1$ and $\rho|(2\epsilon, \infty) = 0$; (Recall that one has $inj(M) \geq 10\epsilon > 0$ and the sectional curvature is absolutely bounded by b^2 with $b \geq 0$.)
2. one has $u_0(y, y) = 1$ for all $y \in M$ and for $y_r = \exp_y(rv)$ with v a unit vector in TM_y, a direct computation (cf. [**BGM**]) shows

$$u_0(y_r, y) = 1 + \frac{Ric(v, v)}{12} r^2 + o(r^2) \tag{4.8}$$

where $r = d(y_r, y)$ and $o(r^2)$ is controlled by n and b, and

3. one also has $u_1(x, y) = R(x)/6 + O(d(x, y))$ for all $y \in M$.
4. the function $\omega(x, y, t/\alpha)$ is given by

$$\omega(x, y, t/\alpha) = (4\pi t/\alpha)^{-n/2} e^{-\frac{\alpha d^2(x,y)}{4t}}. \tag{4.9}$$

Next we use (4.6-9) to develope the path integral formulation of the heat kernel. Using $k - 1$ times of the formula (4.6) to get, for fixed $x, y \in M$ and $t > 0$, the formula

$$p_\alpha(x, y, t) = \int_{M^{k-1}} \Pi_{j=1}^k p_\alpha(y_{j-1}, y_j, t/k) dvol_{M^{k-1}}$$

where we set $x = y_0$ and $y = y_k$. In particular, we have

$$p_\alpha(x, y, t) = \lim_{k \to \infty} \int_{M^{k-1}} \Pi_{j=1}^k p_\alpha(y_{j-1}, y_j, t/k) dvol_{M^{k-1}}. \tag{4.10}$$

To investigate the behavior of $\Pi_{j=1}^k p_\alpha(y_{j-1}, y_j, t/k)$ for large k, we only need to keep track of the leading term when k goes to ∞. We have

$$\Pi_{j=1}^{k} p_\alpha(y_{j-1}, y_j, t/k) = \Pi_{j=1}^{k}\omega(y_{j-1}, y_j, \tfrac{t}{k\alpha}) \times \Pi_{j=1}^{k}[\rho(d(y_{j-1}, y_j))(u_0(x,y) + \tfrac{t}{k\alpha}u_1(x,y)) + O(\tfrac{1}{k^2})]. \quad 4.11$$

There are four terms that we need to handle:

1. $\Pi_{j=1}^{k}\omega(y_{j-1}, y_j, \frac{t}{k\alpha})$,
2. $\Pi_{j=1}^{k}\rho(d(y_{j-1}, y_j))$,
3. $\Pi_{j=1}^{k}u_0(y_{j-1}, y_j)$, and
4. $\Pi_{j=1}^{k}(1 + \frac{t}{k\alpha}\frac{u_1(y_{j-1},y_j)}{u_0(y_{j-1},y_j)} + O(1/k^2))$.

Now use (4.7-9) and the definition of the path space $\mathbb{D}_k(a,b)$, one can obtain

$$\Pi_{j=1}^{k} p_\alpha(y_{j-1}, y_j, \frac{t}{k}) = (\frac{k\alpha}{4\pi t})^{\frac{kn}{2}} e^{(-\frac{\alpha}{4}+0(k^{-1}))\int_0^t <\gamma'(s),\gamma'(s)>ds + \int_0^t \frac{R(\gamma(s))}{6\alpha}ds + o(1)}$$

for large k. Indeed, Term (1) gives the kinetic energy and Trem (4) gives the scalar curvature term, while Term (2,3) make small contribution $o(1)$ when compared with Term (1,4) as k is large.

Now the canonical identification between $(M^{k-1}, dvol_{M^{k-1}})$ and $(\mathbb{D}_k(a,b), \mathcal{D}_k\gamma)$ gives

$$p_\alpha(x,y,t) = \lim_{k\to\infty}\int_{\mathbb{D}_k(a,b)} (\frac{k\alpha}{4\pi t})^{kn/2} e^{\int_0^t (-\frac{\alpha}{4})<\gamma'(s),\gamma'(s)> + \frac{R(\gamma(s))}{6\alpha}ds}\mathcal{D}_k\gamma$$

and this proves Theorem D.

To extend Theorem D to the general case when the potentiel $W(x)$ does not vanish, one only need to establish the short-time asymptotic expansion: as $t \to 0$,

$$K_\alpha(x,y,t) = p_\alpha(x,y,t)[1 + \frac{t}{\alpha}a_1(x,y) + o(t)]$$

where p_α is the function as in (4.7) and $o(t)$ depends only on M and W. Moreover, one also has $a_1(x,y) = W(x) + O(d(x,y))$ for all $y \in M$, provided that W is a C^2-Lipschtz function on M. We refer to [**Wu2**] for the details.

Finally, we would like to point out that when $\alpha = -\frac{2im_o}{\hbar}$ and $W = -\frac{2m_o}{\hbar^2}V$, (4.1) is the Schrödinger equation on curved spaces:

$$i\hbar\frac{\partial\Phi}{\partial t} = -\frac{\hbar^2}{2m_o}\Delta\Phi + V\Phi \quad 4.12$$

where m_o denotes the mass of the particle and $\hbar$ Planck's constant. Thus, Theorem D also gives a non-relativistic path integral formulation of quantum mechanics on curves spaces. To apply this result to motion in a space-time, we propose in [**Wu5, Wu7**] the following duality:

Classical-Quantum Duality *The equation of motion of a free particle on a curved space is the Euler-Lagrange equation of the action $S[\gamma(t)]$ which appears in the path integral formula of the fundamental solution of the Schrödinger equation (4.12):*

$$K(b,a) = \int_a^b e^{\frac{i}{\hbar}S[\gamma(t)]}\mathcal{D}\gamma(t).$$

The Euler-Lagrangian equation of the action $\omega_{-\frac{2im_o}{\hbar}}$ gives a generalized Newton's equation of motion:

$$F = m_o \nabla_{\gamma'(\tau)} \gamma'(\tau) - \frac{\hbar^2}{12m_o} \nabla R$$

4.13

where ∇R is the gradient of the scalar curvature R and F the external force $-\nabla V$. Thus, the vector ∇R represents, in some sense, an intrinsic force caused by the curvedness of the Riemannian manifold.

For massless particles, such as photos, , m_o is replaced by its mass of equivalence via Einstein's formula $E = m_o c^2$ with c the speed of light. Hence when there is no external force, (4.13) becomes

$$\nabla_{\gamma'} \gamma' = \frac{\lambda^2 c^2}{48\pi^2} \nabla R$$

4.14

where λ stands for the wavelength of the massless particle. In the paper [**Wu5**], This generalized Newton equation of motion (4.13-14) in spacetime will play an important role in the problem of quantum effect in general relativity. It will describe a new phenomenon of Grativational Rainbow. One of Einstein's prediction from his theory of general relativity is the bending of light when it passing a gravitational field. This was confirmed in 1919. Now again if the space-time does not have constant scalar curvature in some region, then when a white light passing throught such a region, it will bend and be dispersed into its colored rays since ,for instance, red light has larger wavelenghth than blue light, so the intrinsic force $\frac{\lambda^2 c^2}{48\pi^2} \nabla R$ will have more impact on red light than blue one. Hence, this gives us a new phenomenon of "Gravitational Rainbow". The equation (4.14) is a dispersive equation and it also provides us a new explanation of the evaporation of a black hole rather than that given by Hawking. See [**Wu7, Wu8**] for detailed discussions.

Since the importance of the path integral approach to quantum mechanics is now becoming clear, we also expect that the path integral approach should also have some impact on the theory of Riemannian geometry. Thus, we would like to take this opportunity to provoke people's interest in this approach from quantum mechanics to geometry.

References

[A] M. Anderson, *Short geodesics and gravitational instantons,* **J. Diff. Geom. 31** (1990), 265–275.

[BC] R. Bishop and R.L. Crittendon, **Geometry of manifolds, (New York:** Academic Press, 1974.)

[BGM]] M. Berger, P. Gauduchon and E. Mazet, **Le spectre d'une variété riemanniennes, (New York:** Springer-Verlag, 1971.)

[BGP] Yu. Burago, M. Gromov and G. Perelman, *A. D. Alexandrov's spaces with curvatures bounded from below,* **Usp. Mat. Nauk 47** (1992), 3–51.

[Cd] J. Cheeger, *Finiteness theorems for Riemannian manifolds,* **Amer. J. Math. 92** (1970), 61–74.

[Ch] S. Y. Cheng, *Eigenvalues comparison theorems and its geometric applications,* **Math. Zeit. 143** (1975), 289–297.

[Cl] I. Chavel, **Eigenvalues in Riemannian geometry, (New York:** Academic Press, 1984.)

[CC1] J. Cheeger and T. H. Colding, *Lower bounds on Ricci curvature and almost rigidity of warped products,* **Ann. of Math. 144** (1996), 189–237.

[CC2] J. Cheeger and T. H. Colding, *On the structure of spaces with Ricci curvature bounded below, I* **J. Differ. Geom. 46** (1997), 406–480.

[CE] J. Cheeger and D. Ebin, **Comparison theorems in Riemannian geometry, (Amsterdam and New York:** North-Holland, 1975.)

[CFG] J. Cheeger, K. Fukaya and M. Gromov, *Nilpotent structures and invariant metrics on collapsed manifolds,* **Jour. Amer. Math. Soc., 5** (1992), 327–372.

[CW] M.-H. Chi and J.-Y. Wu, *Brownian motion on curved spaces,* **in preparation.**

[CY] J. Cheeger and S.-T. Yau, *A lower bound for the heat kernel,* **Commun. Pure Appl. Math. 34** (1981), 465–480.

[D] P.A.M. Dirac, *The Lagrangian in Quantum Mechanics,* **Physikalische Zeitschrift der Sowjetunion, Band 3, Heft 1** (1933), 64–70.

[ET] K.D. Elworthy and A. Truman, *Classical mechanics, the diffusion (heat) equation and the Schrödinger equation,* **J. Math. Phys. 22-10** (1981), 2144–2166.

[F] R.P. Feynman, *Space-time approach to non-relativistic quantum mechanics,* **Rev. Med. Pyhs. 20** (1948), 267.

[FH] R.P. Feynman and A.R. Hibbs, **Quantum Mechanics and Path integrals, (New York:** McGraw-Hill Publishing Company, 1965.)

[FY1] K. Fukaya and T. Yamaguchi, *Isometry groups of singular spaces,* **Math. Zeit. 216** (1994), 31–44.

[FY2] K. Fukaya and T. Yamaguchi, *The fundamental group of almost nonnegatively curved manifolds,* **Annals of Math. 136** (1992), 253–333.

[Ga] S. Gallot, *Inégalités isopérimétriques, courbure de Ricci et invariants géométriques, I,* **C. R. Acad. Sci. Sér. I Math. 296** (1983), 333-336

[Gl] Y. Gliklikh, **Global Analysis in Mathematical Physics: Geometric and Stochastic Methods,** (Springer 1997.)

[Gr1] M. Gromov, *Groups of polynomial growth and expanding maps,* **IHES Publ. Math. 53** (1981), 183–215.

[Gr1] M. Gromov, *Curvature, diameter and Betti numbers,* **Comment. Math. Helv. 56** (1981), 179–195.

[Gr3] M. Gromov, *Volume and bounded cohomology,* **Publ. Math. IHES 56** (1981), 213–307.

[GP] K. Grove and P. Petersen V, *Manifolds near the boundary of existence,* **J. Differ. Geom. 33** (1991), 379–394.

[GLP] M. Gromov, J. Lafontaine and P. Pansu, **Structure metrique pour les varites riemanniennes, (Paris:** Cedic / Fernand Nathan, 1981.)

[GPW] K. Grove, P. Petersen V and J.-Y. Wu, *Geometric finiteness theorems via controlled topology,* **Invent. Math. 99** (1990) 205–213.

[Pe] G. Perelman, *Alexandrov's spaces with curvatures bounded from below, II,* **preprint.**

[Pt] P. Petersen, **Riemannian Geometry, GTM 171 (New York:** Springer, 1998.)

[Pt1] P. Petersen V, *Gromov-Hausdorff convergence of metric spaces,* in **Proc. Symp. Pure Math. ed. by S.-T. Yau and R. Greene, Vol 54, Part 3** (1993), 489–504.

[R] X. Rong, *The almost cyclicity of the fundamental groups of positively curved manifolds,* **Invent. Math. 126** (1996), 47–64.

[SW] Z. Shen and J.-Y. Wu, *Curvature, diameter and bounded Betti numbers,* **preprint.**

[SY] J. Sha and D. Yang, *Examples of manifolds of positive Ricci curvature,* **J. Differ. Geom. 29** (1989), 95–103.

[Wu1] J.-Y. Wu, *An obstruction to fundamental groups of positively Ricci curved manifolds,* **Annals of Global Analysis and Geometry 16** (1998), 371–382.

[Wu2] J.-Y. Wu *A mathematical backgroud for path integrals* **Journal of Physics A: Mathematical and General, to appear.**

NCTS Preprints in Mathematics 1999-2, National Center for Theoretical Sciences, Taiwan, Hsin-Chu 1-14.

[Wu3] J.-Y. Wu, *Hausdorff convergence and sphere theorems,* **Proc. Symp. Pure Math., ed. by R. Greene and S.-T. Yau 54 part 3** (1993), 685–692.

[Wu4] J.-Y. Wu, *A parametrized geometric finiteness theorem,* **Indiana University Math. Jour. 45** (1996), 511–528.

[Wu5] J.-Y. Wu, *Quantum Effect in General Relativity,* **NCTS Preprints in Mathematics 1999-6, National Center for Theoretical Sciences, Taiwan, Hsin-Chu** 1–9.

[Wu6] J.-Y. Wu, *The topological spectrum of a smooth closed manifold,* **Indiana University Math. Jour. 44** (1995), 511–534.

[Wu7] J.-Y. Wu, *Gravitational Rainbow,* **NCCU Math. Technical report No. JYWU 1999-4. Taiwan, Chia-Yi.**

[Wu8] J.-Y. Wu, *The Evaporation of Black Holes,* **NCCU Math technical report No. JYWU 1999-5, Taiwan, Chia-Yi.**

[Y] T. Yamaguchi, *Collaping and pinching under a lower curvature bound,* **Ann. Math. 133** (1991), 317–357.

Department of Mathematics, National Chung Cheng University, Ming-Hsiung, Chia-Yi 621, Taiwan

E-mail address: `jywu@math.ccu.edu.tw`

AMS/IP Studies in Advanced Mathematics
Volume 20, 2001

Vanishing Theorems for Loop Spaces

Kefeng Liu

ABSTRACT. We briefly discuss several vanishing theorems for the indices of elliptic operators on loop space.

1. Introduction:

Let M be a smooth connected compact spin manifold of dimension $2k$. Let $\{\pm x_j\}_{j=1}^k$ be the Chern roots of the complexified tangent bundle $TM \otimes_{\mathbf{R}} \mathbf{C}$. Recall that the $\hat{A}$-genus of M, denoted by $\hat{A}(M)$, is the intgeral over M of the following differential form

$$\hat{A}(TM) = \prod_{j=1}^{k} \frac{x_j/2}{\sinh x_j/2}.$$

Given any vector bundle E on M, we can consider the twisted Dirac operator $D \otimes E$. The Atiyah-Singer index theorem tells us that

$$\operatorname{Ind} D \otimes E = \int_M \hat{A}(TM) \operatorname{ch} E$$

where $\operatorname{ch} E$ is the Chern character of E. In fact, let $\{y_j\}$ denote the Chern root of E, or $E \otimes_{\mathbf{R}} \mathbf{C}$ if E is a real bundle, then

$$\operatorname{ch} E = \sum_j e^{y_j}.$$

Then we have the following famous vanishing theorem, which was a surprising application of the Atiyah-Bott-Segal-Singer fixed point formula,

Theorem 1.1: (Atiyah-Hirzrbruch)*Assume that there is a nontrivial circle action on M, then $Ind D = \hat{A}(M) = 0$.*

The above Atiyah-Hirzebruch vanishing theorem is for spin manifold of dimension $4k$. In [LY], Lawson-Yau proved a much more refined version for spin manifold with $SU(2)$-action.

Theorem 1.2: (Lawson-Yau) *Let M be a spin manifold with an $SU(2)$-action. Then $Ind D = 0$.*

This theorem implies the mod 2 index for a dimension $8k+1$ or $8k+2$ spin manifold with $SU(2)$-action. Note that the mod 2 index does not have a fixed point formula, which is crucial for the proof of the Atiyah-Hirzebruch vanishing theorem. Since any compact nonabelian Lie group contains $SU(2)$ as a subgroup, so the above theorem holds for any compact nonabelian Lie group action.

The method of the proof of Lawson-Yau is to construct a metric with positive scalar curvature by scaling the metric on the orbits. We know that for Dirac operator D on M, there is the Lichnerovicz formula

$$D^2 = \nabla^*\nabla + \frac{k}{4}.$$

From this one easily derives the vanishing of the index of D for spin manifold with positive scalar curvature.

In the following sections, we will describe several loop space analogues of the above vanishing theorems, as well as certain vanishing theorems that do not have any finite dimensional analogues.

2. Loop space vanishing theorems

In [W], by applying formally the Atiyah-Bott-Segal-Singer fixed point formula to the Dirac operator on LM, the free loop space of M, Witten derived the following twisted Dirac operator,

$$\mathcal{D}_L = D \otimes \otimes_{j=1}^{\infty} S_{q^j} TM.$$

Here $q = e^{2\pi i \tau}$ with τ in the upper half plane is a parameter, and

$$S_{q^j} TM = 1 + q^j TM + q^{2j} S^2(TM) + \cdots$$

is the symmetric operation.

By expanding into power series in q, we have

$$\otimes_{j=1}^{\infty} S_{q^j} TM = \sum_{l=0}^{\infty} A_l q^l$$

where each A_l is a linear combinations of the symmetric products of TM. Obviously $A_0 = 1$. Therefore

$$D \otimes \otimes_{j=1}^{\infty} S_{q^j} TM = \sum_{l=0}^{\infty} D \otimes A_l q^l$$

should be understood as infinite number of twisted Dirac operators $D \otimes A_l$, $0 = 1, 2, \cdots$. Up to a normalization, the index of $\mathcal{D}_L$ is called the Witten genus.

Note that, formally by the fixed point principle, the operator $\mathcal{D}_L$ captures all of the non-torsion information of the Dirac operator on LM. Our first theorem can be viewed as a loop space generalization of the above $\hat{A}$-vanishing theorem of Atiyah and Hirzebruch. To state it, we first recall the basic definition of equivariant characteristic classes.

Recall the universal construction

$$\pi : \ ES^1 \times_{S^1} M \to BS^1.$$

The cohomology group of $ES^1 \times_{S^1} M$ is defined to be the equivariant cohomology of M, denoted by $H^*_{S^1}(M)$. Here we consider cohomology with rational coefficient.

Note that $H^*(BS^1) = \mathbf{Q}[[u]]$ with u a degree two cohomology class. We have the pull-back map

$$\pi^* : H^*(BS^1) \to H^*_{S^1}(M).$$

Given any equivariant vector bundle E on M, we know that $ES^1 \times_{S^1} E$ is a vector bundle on $ES^1 \times_{S^1} M$. The characteristic classes of $ES^1 \times_{S^1} E$ is callled the corresponding equivariant characteristic classes of E. We use $p_1(M)_{S^1}$ to denote the first equivariant Pontryagin class of the tangent bundle of M.

Theorem 2.1: *If the S^1-action is nontrivial, and $p_1(M)_{S^1} = n\,\pi^* u^2$ for an integer n, then $Ind\mathcal{D}_L = 0$.*

Since, modulo torsion, the condition $p_1(M) = 0$ can be viewed as the spin condition on LM, so roughly speaking, the condition $p_1(M)_{S^1} = n\,\pi^* u^2$ is equivalent to saying that the S^1-action preserves the spin structure of LM.

Dessai showed that for $G = SU(2)$ acting on M, one has the following exact sequence

$$H^4(BG) \to H^4(EG \times_G M) \to H^4(M)$$

which implies that, if the S^1-action is induced from an $SU(2)$-action, then the condition $p_1(M)_{S^1} = n\,\pi^* u^2$ is equivalent to $p_1(M) = 0$.

In fact, one may see this more easily, from degree counting we may actually think

$$p_1(M)_{S^1} = p_1(M) + A\,u + n\,\pi^* u^2$$

for some class $A \in H^2(M)$. If the S^1-action is induced from an $SU(2)$-action on M, then the $SU(2)$-equivariant cohomology is the Weyl invariant part in the S^1 equivariant cohomology. One has $H^*_{SU(2)}(M) = H^*_{S^1}(M)^W$. The action of the Weyl group W of $SU(2)$ is $u \to -u$, so $p_1(M)_{S^1}$, if invariant under the Weyl group, must satisfy the condition $A = 0$.

Corollary 2.2: *Assume there is an $SU(2)$-action on M. If $p_1(M) = 0$, then $Ind\mathcal{D}_L = 0$.*

Motivated by the above results and the Lawson-Yau vanishing theorem, Hoehn and Stolz made the following conjecture [S].

Conjecture 2.3: *Let M be a spin manifold with positive Ricci curvature. If $\frac{1}{2}p_1(M) = 0$, then $Ind\mathcal{D}_L = 0$.*

The idea of this conjecture is to find a loop space analogue of the Lichnerovicz formula. The Ricci curvature on M may be considered as the scalar curvature on the loop space of M. Note that the condition on p_1, as well as the condition on the Ricci curvature both are crucial for the conjecture. There are counter-examples if we drop any one of the conditions.

The puzzle for this conjecture is the relationship between the Ricci curvature and the modular property of $q^{-\frac{k}{12}}\mathrm{Ind}\,\mathcal{D}_L = 0$. The modular invariance is the key point for the proof of the above vanishing theorems for group actions. The condition $\frac{1}{2}p_1(M) = 0$ is exactly the spin condition for LM. See [S] for some details about this conjecture.

We remark that, at this point, we do not have a vanishing theorem for the mod 2 index $\mathrm{Ind}\,\mathcal{D}_L = 0$, except for $8k+1$ dimensional manifolds with odd involution [Liu2].

Other vanishing theorems proved in [Liu2] do not have finite dimensional analogues. Here we only describe one of them. More general theorems involving loop group representations can be found in [Liu2].

For an equivariant spin vector bundle V on M, let

$$d_L^S = \mathcal{D}_L \otimes \otimes_{j=1}^{\infty} \wedge_{q^j} V$$

be a twisted Dirac operator on the loop space LM, where

$$\Lambda_t V = 1 + t\Lambda V + t^2 \Lambda^2 V + \cdots$$

denotes the wedge operattion on V. The index $\mathrm{Ind}\, d_L^S$ is called the universal elliptic genus of Ochanine-Landweber-Stong. Then we have the following theorem:

Theorem 2.4: *If $p_1(V)_{S^1} - p_1(M)_{S^1} = n\,\pi^* u^2$ with $n < 0$, then $Ind\, d_L^S = 0$.*

If $n = 0$, then the equivariant index of d_L^S is a constant, which is precisely the famous Witten rigidity theorem.

As in the above, one may derive some easy corollaries for compact nonabelian Lie group action. In particular, if the S^1-action is induced from an $SU(2)$-action on M, $p_1(V) = p_1(M)$ and $\sum_i n_i^2 - \sum_j m_j^2 < 0$ for some fixed component F, then $\mathrm{Ind}\, d_L^S = 0$. Here $\{n_i\}$ and $\{m_j\}$ denote the corresponding weights of V and TM restricted to F with respect to the S^1-action.

To prove the above loop space vanishing theorems, we first used the Atiyah-Bott-Segal-Singer fixed point formula to express the equivariant indices of the twised operators in terms of the Jacobi theta-functions. We then combine the modular property of the theta-functions with the index theory to show these equivariant indices are actually holomorphic Jacobi forms of negative index, which therefore must be zero. For the details of the proof and more general rigidity and vanishing theorem involving loop group representations, see [Liu2].

3. Family vanishing theorems for loop spaces.

Recently in [LiuM1], [LiuM2], [LiuMaZ1] and [LiuMaZ2], we have proved several results which can be considered as vanishing theorems for families of elliptic operators on loop space.

To be more precise, let W, B be two compact smooth manifolds, and $\pi : W \to B$ be a smooth fibration with compact fiber X. Assume a compact Lie group G acts fiberwise on W. We always assume that W is connected. Let P be a family of G-equivariant elliptic operators along the fiber, then the family index of P is defined to be

$$\mathrm{Ind}(P) = \mathrm{Ker}P - \mathrm{Coker}P \in K_G(B)$$

In many situations in geometry and topology, it is rather natural to generalize the vanishing theorems to family case. The vanishing theorems in the family case contain more information, encoded in the higher degree terms of the indices.

Let TX be the tangent bundle of X which is a vector bundle on W. Given a Riemannian metric on TX, one can construct the Clifford module and spinor bundles of TX. From these data we construct the family Dirac operator D^X by using the metric connection on TX.

The following theorem was proved in [LiuMa1].

Theorem 3.1: *Assume there is a nontrivial fiberwise S^1-action on W, if $p_1(TX)_{S^1} = n\,\bar{\pi}^*u^2$, then the equivariant Chern character of the index bundle of $\mathcal{D}_L^X$, $ch(Ind\,\mathcal{D}_L^X)$, is zero in $H^*_{S^1}(B)$.*

Here $\bar{\pi}^*$ denotes the pull-back homomorphism

$$\bar{\pi}^* : \ H^*(BS^1) \to H^*_{S^1}(W)$$

and the family operator,

$$\mathcal{D}_L^X = D^X \otimes \otimes_{j=1}^{\infty} S_{q^j} TX$$

can be considered as the family of loop space Dirac operator on the fibration $\pi : W \to B$.

Similar to the discussion in last section, we have the following simple corollary:

Theorem 3.2 *Assume W is connected and admits an fiberwise $SU(2)$-action. If $p_1(X) = 0$, then $ch(Ind\,\mathcal{D}_L^X)$ vanishes in $H^*_{S^1}(B)$.*

We may also consider the family of twisted Dirac operators on loop space. Let

$$d_L^X = \mathcal{D}_L^X \otimes \otimes_{j=1}^{\infty} \Lambda_{q^j} V$$

for some spin equivariant vector bundle V on W, we then have the family analogue of the vanishing theorem:

Theorem 3.3: *If $p_1(V)_{S^1} - p_1(TX)_{S^1} = n\,\bar{\pi}^*u^2$ with $n < 0$, then $ch(Ind\,d_L^X)$ vanishes in $H^*_{S^1}(B)$.*

Again this theorem does not have a finite dimensional analogue. There are much more general family rigidity and vanishing theorems proved in [LiuMa2] involving loop group representations. Similar theorems for Spin^c manifold were proved in [LiuMa2].

If the S^1-action is induced from an $SU(2)$-action on W, then we have a simple corollary under the condition on the nonequivariant first Pontryagin class and the weights of V and TX on the fixed points.

Note that, the above theorems implies the vanishing of the many higher degree terms in $\mathrm{ch}(\mathrm{Ind}\,\mathcal{D}_L^X)$ and $\mathrm{ch}(\mathrm{Ind}\,d_L^X)$. These higher degree terms contain important information about the family.

To prove these theorems, we used the modular invariance argument as described in last section, together with a scaling trick and some basic properties of Jacobi forms. For the details of the proofs, see [LiuMa1] and [LiuMa2].

4. Vanishing theorems in K-theory for loop spaces

In last section, we see the vanishing theorems for the Chern characters of the index bundles. In many cases, it is possible that, the index bundle may involve torsion elements, which are killed by taking Chern charater.

In [LiuMaZ1] and [LiuMaZ2], we start the program to understand the possible torsions. As a first step, we prove a refined K-theory version of the fixed point formula for circle action. This fixed point formula is more subtle than the one in [AS], since the usual localization process may kill the torsion elements.

By using this refined fixed point formula for S^1-action, we were able to prove several rigidity and vanishing theory for the index bundle. Here we only state two vanishing theorems in [LiuMaZ1] for elliptic genera.

Theorem 4.1: *Assmue there is a nontrivial fiberwise S^1-action on W, if $\frac{1}{2}p_1(TX)_{S^1} = n\,\bar{\pi}^*u^2$, then $Ind\mathcal{D}_L^X$ is zero in $K_{S^1}(B)$.*

In particular if the S^1-action is induced from an $SU(2)$-action, then we have

Corollary 4.2: *Assmue there is a nontrivial fiberwise $SU(2)$-action on W, if $\frac{1}{2}p_1(TX) = 0$, then $Ind\mathcal{D}_L^X$ is zero in $K_{S^1}(B)$.*

Similarly given any spin vector bundle V on W, we have the following family vanishing theorem in K-group:

Theorem 4.3: *If $\frac{1}{2}(p_1(V)_{S^1} - p_1(TX)_{S^1}) = n\,\bar{\pi}^*u^2$ with $n < 0$, then $Ind\,d_L^X$ vanishes in $K_{S^1}(B)$.*

The proofs of the above theorems used the very subtle shifting operation as first explored in [T] and [BT]. Here it is interesting to see that, only the elliptic property of the elliptic genera is essentially used. This shifting operation involves some interesting operations on families of vector bundles, which may be related to the Hecke operators of modular forms and Jacobi forms. A clear understanding of the relationship between the shifting operation and the modular property of the elliptic genera is extremally interesting.

For the details of the proofs of the above theorems, see [LiuMaZ1] and [LiuMaZ2].

5. Concluding remarks

As one can see, essentially the vanishing properties of the loop space elliptic operators can be viewed as the combination of modular invariance or elliptic property with index theory. Though the index theory is for finite dimensional manifolds, it is impossible to discover and prove such vanishing results without using the infinite dimensional symmetry: the modular invariance or the elliptic property.

There are some other situations that one can study the vanishing properties of the elliptic operators appeared there by using our methods. The first is the orbifold elliptic genus, in which we should use delocalized index of the elliptic operators as proposed by physicists [Zhou]. In this case one may try to get vanishing theorem by using the modular invariance method. The second is the elliptic genus for foliated manifolds. Again one may derive vanishing results by using the modular invariance of the corresponding equivariant index.

I would like to than Xiaonan Ma and Weiping Zhang for valuable discussions.

References

[AH] Atiyah M.F. and Hirzebruch F., Spin manifolds and groups actions, *Essays on topology and Related Topics, Memoires dédié à Georges de Rham* (ed. A. Haefliger and R. Narasimhan), Springer-Verlag, New York-Berlin (1970), 18-28.

[AS] Atiyah M.F. and Segal G., The index of elliptic operators II. *Ann. of Math.*87 (1968), 531-545

[BT] Bott R. and Taubes C., On the rigidity theorems of Witten, *J.A.M.S.* 2 (1989), 137-186.

[L] Landweber P.S., *Elliptic Curves and Modular forms in Algebraic Topology*, Landweber P.S., SLNM 1326, Springer, Berlin.

[LY] Lawson B., Yau S.-T., Scalar curvature, non-abelian group actions, and the degree of symmetry of exotic spheres. Comment. Math. Helv. 49 (1974), 232–244.

[Liu1] Liu K., On elliptic genera and theta-functions, *Topology* 35 (1996), 617-640.

[Liu2] Liu K., On modular invariance and rigidity theorems, *J. Diff. Geom.* 41 (1995), 343-396.

[LiuMa1] Liu K. and Ma X., On family rigidity theorems I. *Duke Math. J.*, to appear.

[LiuMa2] Liu K. and Ma X., On family rigidity theorems II. *Preprint*, 1999.

[LiuMaZ1] Liu K., Ma X. and Zhang W., Rigidity and Vanishing Theorems in K-Theory I. *Preprint*, 1999.

[LiuMaZ2] Liu K., Ma X. and Zhang W., Rigidity and Vanishing Theorems in K-Theory II. *Preprint*, 1999.

[S] Stolz, S., A conjecture concerning positive Ricci curvature and the Witten genus. Math. Ann. 304 (1996), no. 4, 785–800

[T] Taubes C., S^1 actions and elliptic genera, *Comm. Math. Phys.* 122 (1989), 455-526.

[W] Witten E., The index of the Dirac operator in loop space, in [L], pp. 161-186.

[Zhou] Zhou, J., Calculations of the delocalized Hirzebruch χ_y-genera of symmetric products by holomorphic Lefschetz formula, preprint 1999.

Kefeng Liu, Department of Mathematics, Stanford University, Stanford, CA 94305, USA.

E-mail address: kefeng@math.stanford.edu

AMS/IP Studies in Advanced Mathematics
Volume 20, 2001

SYMPLECTIC FORMS, 2+1 SOLITON EQUATIONS, AND SUPERSYMMETRIC GAUGE THEORIES

D.H. Phong

ABSTRACT. We review some recent developments in Seiberg-Witten theory and in the Hamiltonian formulation of soliton theory, with particular emphasis on the symplectic forms common to both.

1. INTRODUCTION

One of the most important problems in current theoretical physics is the understanding of physical theories at strong coupling. Of considerable interest are highly supersymmetric theories, where exact solutions may be amenable thanks to the strong holomorphicity constraints on the low energy effective action resulting from supersymmetry. For $\mathcal{N} = 2$ supersymmetric gauge theories in four dimensions, there has been rapid progress since the 1994 solution of the $SU(2)$ model by Seiberg and Witten [47,48]. In particular, a deep correspondence has emerged between three seemingly different classes of objects, namely

- (1) $\mathcal{N} = 2$ gauge theories, indexed by their gauge group G and the representation R of their matter hypermultiplets (subject to the constraints of asymptotic freedom or of vanishing beta function);
- (2) solitons and integrable models;
- (3) and fibrations of Riemann surfaces equipped with a suitable meromorphic 1-form.

We shall not be able to describe fully in this short survey the many facets of this correspondence, which is still far from being completely understood. Rather, we shall concentrate on one common theme underlying all three classes, namely symplectic forms, and take the opportunity to discuss some joint work of the author with E. D'Hoker and I.M. Krichever.

Symplectic structures have of course long played a central role in the theory of solitons and integrable models. And while the Seiberg-Witten Ansatz is usually formulated in terms of a 1-form $d\lambda$ on a fibration of Riemann surfaces (c.f. (4.4) below), it was recognized very early on [48,18] that the intrinsic notion for $\mathcal{N} = 2$

Research supported in part by the National Science Foundation under NSF Grant DMS-98-00783.

gauge theories was actually the corresponding symplectic form $\omega = \delta(\sum_{i=1}^{g} d\lambda(z_i))$ on the fibration of Jacobians. However, this finite-dimensional symplectic form seemed quite different from the usual infinite-dimensional forms in soliton theory. It turned out that, by extending an earlier construction of Novikov and Veselov [43] for hyperelliptic Riemann surfaces, both types of symplectic forms can be obtained from the same fundamental formula [32,33]

$$\omega_{m_0} = \mathrm{Res}_\infty < \Psi_0^*(A_{m_0}^{(1)}\delta L - L^{(1)}\delta A_{m_0}) \wedge \delta\Psi_0 >_0 dk \tag{1.1}$$

(c.f. Theorems 1-4 below). Surprisingly, in the process, we also shed some new light on a number of still puzzling issues in soliton theory: the existence of an infinite number of symplectic structures for KdV, including the Adler-Magri structure; integro-differential constraints and a proper Hamiltonian formulation for 2+1 soliton equations; and universal formulas for the many disparate symplectic forms known so far.

This survey is organized as follows. In Section II, we present the new Hamiltonian formalism of [33], for soliton equations of the form $[\partial_y - L, \partial_t - A] = 0$. The key ingredient is the correct Bloch function Ψ_0, which determines both the integro-differential constraints on L and the universal, infinite-dimensional, symplectic form (1.1). The symplectic forms of Seiberg-Witten theory turn out to be the pull-backs of this universal symplectic form to various fibrations of Jacobians by the Krichever map. Sections III and IV provide a precise version of this last statement. First, we explain in Section III how fibrations of Riemann surfaces equipped with a meromorphic 1-form $d\lambda$, and hence a symplectic structure, arise naturally from the leaves of a foliation on the moduli space of Riemann surfaces with two Abelian integrals E and Q. If the leaves of this foliation are imbedded into the space of operators L by the Krichever map, then this symplectic form reduces to (1.1). Next, in Section IV, we explain how to view the quantum moduli space of vacua of $SU(N)$ gauge theories with matter in the fundamental representation as a leaf of the foliation in Section III. Which leaf is determined by the physical parameters of the gauge theory such as the masses of the matter hypermultiplets, and the quantum moduli parameters turn out to be just the action variables for the symplectic structure on the fibration of Jacobians above the leaf. We conclude with a brief mention of some of the most recent related developments.

2. HAMILTONIAN FORMALISM FOR 2+1 SOLITON EQUATIONS

Hamiltonian Formalism for 1+1 Soliton Equations

A fundamental property of many 1+1 soliton equations is their interpretation as Hamiltonian systems (see e.g. [22,50,51] and references therein). For example, consider the KdV equation

$$u_t - \frac{3}{2}uu_x - \frac{1}{4}u_{xxx} = 0. \tag{2.1}$$

Recall that a Hamiltonian system on a manifold M with Hamiltonian H is a flow on M of the form

$$\dot{u}^i = \sum_{j=1}^{dim\,M} \omega^{ij}(u)\frac{\partial H}{\partial u^j}. \tag{2.2}$$

Here u^i are local coordinates on M, $\omega = \sum_{i,j=1}^{dim\,M} \omega_{ij} du^i \wedge du^j$ is a symplectic form, i.e., a closed and non-degenerate 2-form, and ω^{ij} is the inverse of ω_{ij}, $\omega_{ij}\omega^{jk} = \delta_i^j$.

In the case of KdV, let M be the (infinite-dimensional) space of functions $u(x)$, and let the Hamiltonian $H(u)$ be given by

$$H(u) = \int_{-\infty}^{\infty} (\frac{1}{4}u^3 - \frac{1}{8}u_x^2)dx. \tag{2.3}$$

Then $\frac{\delta H}{\delta u(x)} = \frac{3}{4}u^2 + \frac{1}{4}u_{xx}$, so that the KdV equation (2.1) can be rewritten as

$$\dot{u}(x) = \frac{\partial}{\partial x}\frac{\delta H}{\delta u(x)} = -\int_{-\infty}^{\infty} \delta'(x-x')\frac{\delta H}{\delta u(x')}dx'. \tag{2.4}$$

This is an equation of the form (2.2), with the inverse ω^{ij} of the symplectic form given by $\omega^{xx'} = -\delta'(x-x')$. Solving the equation $\int \omega^{xz}\omega_{zy}dz = \delta(x-y)$ for the symplectic form $\omega = \int\int \omega_{xy}dx \wedge dy$, we find the Gardner-Faddeev-Zakharov symplectic form for the KdV equation

$$\omega = -\int dx\big(\delta u(x) \wedge \int^x \delta u(y)dy\big). \tag{2.5}$$

Hamiltonian formulations also exist for other soliton equations such as Toda, Boussinesq, and Calogero-Moser systems. Remarkably, in some cases, several distinct Hamiltonian formulations may exist. For example, the KdV equation can also be rewritten as

$$\dot{u}(x) = \big(\partial_x^3 + 2(u\partial_x + \partial_x u)\big)\frac{\delta H'}{\delta u}, \quad H'(u) = \int_{-\infty}^{\infty} \frac{1}{8}u^2 dx, \tag{2.6}$$

where $H'(u)$ is the new Hamiltonian, and the kernel of the antisymmetric operator $\partial_x^3 + 2(u\partial_x + \partial_x u)$ is the inverse of the new Adler-Magri symplectic form.

2+1 Soliton Equations

By contrast, Hamiltonian formulations for 2+1 soliton equations are not as well understood. A 2+1 soliton equation is an equation of the form

$$[\partial_y - L, \partial_t - A] = 0, \tag{2.7}$$

where the unknown functions are the coefficients $u = (u_0(x,y,t), \cdots, u_{n-2}(x,y,t))$ of the ordinary differential operator

$$L = \partial_x^n + \sum_{i=0}^{n-2} u_i\partial_x^i, \tag{2.8}$$

and $A = \partial_x^m + \sum_{i=0}^{m-2} v_i\partial_x^i$ is an ordinary differential operator of a fixed order m. The main issue in a Hamiltonian formulation is the operator A. For the equation (2.7) to be a closed system for the unknown functions u, A should be determined by u, in which case A provides a Hamiltonian on the phase space of all operators L. Now the operator $[L, A]$ is of order $\leq n+m-1$, and the equation (2.7) implies only that it is actually of order $\leq \max(n-2, m-2)$. This is in marked contrast to 1+1 soliton equations, which correspond to the y-independent case $[L, \partial_t - A] = 0$. In this case, $[L, A]$ must be of order $\leq n-2$, and this does determines A in terms of L, as shown by Gelfand and Dickey [24]. No such result is available for 2+1 equations, even in the simplest case of the Kadomtsev-Petviashvili hierarchy, originating with$\frac{3}{4}u_{yy} = \partial_x(u_t - \frac{3}{2}u\partial_x u - \frac{1}{4}\partial_x^3 u)$.

We shall see that the key to a Hamiltonian formulation of 2+1 soliton equations is the identification of suitable integro-differential constraints on the dynamical variables u, thanks to which the operator A in the equation $L_t - A_y = [L, A]$ is well-defined on the space of operators L. More precisely [33],

Theorem 1. *Let n be a fixed integer ≥ 2, and let $\mathcal{L}(b)$ be the space of all operators L of the form (2.8) with C^∞ coefficients $u_i(x,y)$ of two variables (x,y) satisfying the periodicity conditions $u_i(x+1,y) = u_i(x,y+1) = u_i(x,y)$ and the constraints (2.14) below. Then*

- *The expression $\omega = \mathrm{Res}_\infty < \Psi_0^* \delta L \wedge \delta \Psi_0 > dk$, where Ψ_0, Ψ_0^* are the Bloch and dual Bloch functions below and $< \cdot >$ denotes averaging in x and y, is a symplectic form on $\mathcal{L}(b)$;*
- *For each integer $m \geq 1$, the flows given by (2.17) below are Hamiltonian with respect to the symplectic form ω, with Hamiltonian $2nH_{m+n}$, where H_s is defined by (2.13) below. Furthermore, they all commute pairwise.*

The Bloch Function Ψ_0

There exists a unique asymptotic series $\Psi_0(x,y,k)$ of the form

$$\Psi_0(x,y,k) = (1 + \sum_{s=1}^{\infty} \xi_s(x,y) k^{-s}) \exp(kx + k^n y + \sum_{i=0}^{n-2} B_i(y) k^i), \quad \Psi_0(0,0,k) = 1, \tag{2.9}$$

which satisfies the conditions

$$\Psi_0(x+1,y,k) = e^k \Psi_0(x,y,k), \quad (\partial_y - L)\Psi_0(x,y,k) = 0. \tag{2.10}$$

Here the functions $B_i(y)$, $0 \leq i \leq n-2$, are integro-differential expressions in the $u_i(x,y)$. The asymptotic series $\Psi_0(x,y,k)$ is then Bloch with respect to y, i.e.,

$$\Psi_0(x,y+1,k) = w_2(k)\Psi_0(x,y,k), \tag{2.11}$$

where the *quasi-energy* $w_2(k)$ can be expressed as

$$w_2(k) = (1 + \sum_{s=1}^{\infty} J_s k^{-s}) \exp(k^n + \sum_{i=0}^{n-2} B_i(1) k^i), \tag{2.12}$$

for suitable coefficients J_s.

Quasi-Momentum and Quasi-Energy

In the above formulation, the quasi-energy $w_2(k)$ appeared as an asymptotic series in the quasi-momentum k. This can be reversed by setting $K^n = \log w_2(k)$ with $K = k + O(k^{-1})$, so that for suitable coefficients H_s,

$$\left\{ \begin{array}{c} w_1 = e^k \\ w_2 = (1 + \sum_{s=1}^{\infty} J_s k^{-s}) e^{k^n + \sum_{i=0}^{n-2} B_i(1) k^i} \end{array} \right\} \Leftrightarrow \left\{ \begin{array}{c} w_1 = e^{K + \sum_{s=1}^{\infty} H_s K^{-s}} \\ w_2 = e^{K^n} \end{array} \right\} \tag{2.13}$$

Integro-differential Constraints

We restrict ourselves henceforth to the space $\mathcal{L}(b)$ of operators satisfying the following integro-differential constraints

$$\frac{\partial B_i}{\partial y}(y) = b_i, \quad 0 \leq i \leq n-2, \tag{2.14}$$

where the b_i are $(n-1)$ fixed constants. It is not difficult to see that this is equivalent to fixing the values of the first $n-1$ functionals H_i, $1 \leq i \leq n-1$, in (2.13).

The Dual Bloch Function Ψ_0^*

The dual Bloch function $\Psi_0^*(x,y,k)$ is the asymptotic series of the form $\Psi_0^*(x,y,k)=(1+\sum_{s=1}^{\infty}\xi_s^* k^{-s})\exp(-kx-(k^n+\sum_{i=0}^{n-2}b_ik^i)y))$ satisfying the condition

$$\mathrm{Res}_\infty(\Psi_0^*(x,y,k)\partial_x^m\Psi_0(x,y,k)dk)=0 \tag{2.15}$$

for all $m\in\mathbf{N}$. It satisfies the equation $\Psi_0^*(\partial_y-L)=0$ (where the action on the left of a differential operator ∂_x^i is defined by $g\partial_x^i=(-\partial_x)^ig)$, and is a Bloch function with Bloch multipliers w_1^{-1} and w_2^{-1}

$$\Psi_0^*(x+1,y,k)=w_1^{-1}\Psi_0^*(x,y,k),\quad \Psi_0^*(x,y+1,k)=w_2^{-1}\Psi_0^*(x,y,k). \tag{2.16}$$

The Operator A_m *and the* m*-th Hamiltonian Flow*

Again, solving recursively, we can find for any $m\geq 1$ a unique operator $A_m=\partial_x^m+\sum_{i=0}^{m-2}v_{im}\partial_x^i$ satisfying the condition $(K^m-A_m)\Psi_0(x,y,k)=O(k^{-1})\Psi_0(x,k)$. The operator A_m can be viewed as an operator-valued function defined on the phase space $\mathcal{L}(b)$ of operators L satisfying the constraints (2.14). We define the mth-flow of the hierarchy to be the flow with time t_m given by

$$\frac{\partial L}{\partial t_m}=\partial_yL+[A_m,L]. \tag{2.17}$$

Higher Symplectic Forms

It follows from the uniqueness of the Bloch function Ψ_0 that we must have

$$(\partial_m-A_m)\Psi_0=-\tilde{\Omega}_m\Psi_0, \tag{2.18}$$

for some suitable eigenvalue $\tilde{\Omega}_m$. The eigenvalues $\tilde{\Omega}_m$ are asymptotic series in k (or equivalently in K), with coefficients which are integro-differential expressions in u. They are of course independent of (x,y), but may depend on t if L varies according to any of the flows (2.17). Fix $m_0\geq 1$, and let $\mathcal{D}_{m_0}^{erg}$ be the space of operators for which the m_0-th flow (2.17) with initial data $L(t_{m_0})_{|t_{m_0}=0}=L$ exists for all time t_{m_0} with uniformly bounded coefficients. Introduce $\Omega_{m_0}(K)$ and the Hamiltonians Ω_{s,m_0} by

$$\Omega_{m_0}(K)=\mathrm{lim}_{T\to\infty}\frac{1}{T}\int_0^T\tilde{\Omega}_m(K,t)dt_m=K^{m_0}+\sum_{s=-n+2}^{\infty}\Omega_{s,m_0}K^{-s} \tag{2.19}$$

We note that although the averaging in t_m seems to eliminate only the dependence of $\Omega_{m_0}(K)$ on t_{m_0}, Ω_{m_0} is actually independent of all the times t_m if L evolves according to any of the flows (2.17).

Theorem 2. *Let* $I=(I_0,\cdots,I_{n-2})$ *be* $(n-1)$ *fixed constants. Then the expression*

$$\omega_{m_0}=\mathrm{Res}_\infty<\Psi_0^*(A_{m_0}^{(1)}\delta L-L^{(1)}\delta A_{m_0})\wedge\delta\Psi_0>_0 dk$$

where $A^{(1)}=\sum_{i=0}^m iv_i\partial_x^{i-1}$ *and* $<f(x,y,t)>_0=\mathrm{lim}_{T\to\infty}\frac{1}{T}\int_0^T<f(\cdot,\cdot,t)>dt$, *defines a symplectic form on the subspace of* $\mathcal{D}_{m_0}^{erg}$ *satisfying the constraints* $\Omega_{-s,m_0}=I_s$, $0\leq s\leq n-2$. *The flows (2.17) are Hamiltonian with respect to* ω_{m_0}, *with Hamiltonian given by* $2n\Omega_{n+m,m_0}$.

Theorem 2 is proved in [33]. We note that for $m_0=1$, $A_{m_0}=\partial_x$, and the form ω_{m_0} reduces to the symplectic form of Theorem 1.

Further Results

• We illustrate explicitly the above concepts in the simplest case $n = 2$, corresponding to the Kadomtsev-Petviashvili hierarchy. In this case $L = \partial_x^2 + u(x,y)$, $B_0(y) = \int_0^y \int_0^1 u(\alpha,\beta)d\alpha d\beta$, and the coefficients ξ_s of the Bloch function Ψ_0 are given recursively by $\xi_0 = 1$, $\xi_{s+1} = c_{s+1} + \xi_{s+1}^0$, $\xi_{s+1}^0(x,y) = \frac{1}{2}\int_0^x [\partial_y \xi_s - \xi_s'' + (\partial_y B_0 - u)\xi_s](\alpha,y)d\alpha$, $c_{s+1}(y) = -\int_0^y \int_0^1 [\partial_y \xi_{s+1}^0 + (\partial_\beta B_0 - u)\xi_{s+1}^0](\alpha,\beta)d\alpha d\beta$. The constraints (2.14) are $\int_0^1 u(x,y)dx = constant$, and the symplectic form is $\omega = - < \delta u \wedge \int_{x_0}^x \delta u dx' >$. When $u(x,y)$ is independent of y, the KP equation reduces to the KdV equation, and ω reduces to the Gardner-Faddeev-Zakharov form.

• The Boussinesq hierarchy corresponds to $n = 3$, $L = \partial_x^3 + u\partial_x + v$. The constraints defining the space $\mathcal{L}(b)$ are of the form $\int_0^1 u(x,y) = constant$, $\int_0^1 v(x,y)dx = constant$, and the basic symplectic form of Theorem 1 is given by

$$\omega = -\frac{2}{3} < \delta u \wedge \int_{x_0}^x \delta v dx' + \delta v \wedge \int_{x_0}^x \delta u dx > .$$

On the other hand, if we consider the higher symplectic form ω_2 of Theorem 2 corresponding to $m = 2$, then the constraints are of the form $\int_0^1 u(x,y)dx = constant$, $\int_0^1 v(x,y)dx = constant$, $\int_0^1 u^2(x,y)dx = constant$, $L^{(1)} = 3\partial_x^2 + u$, $A^{(1)} = 2\partial_x$, and

$$\omega_2 = 2 < \frac{3}{4} u\delta u \wedge \int_{x_0}^x \delta u dx' + 2\delta v \wedge \int_{x_0}^x \delta v dx' + 2\delta u \wedge \delta v - \frac{3}{2}\delta u_x \wedge \delta u > .$$

• We can now explain the existence of an infinite number of symplectic structures for 1+1 soliton equations. The distinguishing feature of 1+1 solitons is the fact that $E(k) = \log w_2(k)$ is a genuine function. Thus in the case of KdV, the basic symplectic structure of Theorem 1 reduces to the Gardner-Faddeev-Zakharov symplectic form (2.5), but an infinite number of other symplectic structures can be obtained similarly by setting

$$\omega^{(l)} = \text{Res}_\infty (E(k)^l < \Psi_0^* \delta L \wedge \delta \Psi_0 > dk).$$

Formally, the Adler-Magri form should correspond to $l = -1$, with the residue at ∞ replaced by the residue at 0. More precisely, let $\Psi_i(x,E)$ be n linearly independent solutions of $(L - E)\Psi_i(x,E) = 0$, with distinct Bloch multipliers $\Psi_i(x+1,E) = w_i(E)\Psi_i(x,E)$, and let $\Psi_i^*(x,E)$ be their duals, defined by $\Psi_i^*(L - E) = 0$, $< \Psi_i^*(x,E)\Psi_j(x,E) >= \delta_{ij}$. Then the Adler-Magri form is given by

$$\omega^{(-1)} = \sum_{i=1}^n \frac{< \Psi_i^*(x,0)\delta L \wedge \delta \Psi_i(x,0) >}{< \Psi_i^* L^{(1)} \Psi_i >},$$

which defines a symplectic form with respect to which the flows of the KdV hierarchy are Hamiltonian.

• For simplicity, we have formulated our formalism only in the case of a scalar operator L and for continuous time (which corresponds to the KdV hierarchy in the case of 1+1 equations and to the Kadomtsev-Petviashvili hierarchy in the case of 2+1 equations). However, our formalism extends readily to other settings, including matrix equations and discrete times (as in the case of Toda chains).

3. FINITE-DIMENSIONAL SYMPLECTIC FORMS

So far, the Bloch function Ψ_0 is only an asymptotic series in k, and the construction of the infinite-dimensional forms of Theorems 1 and 2 on a space of operators L required only the formal properties of Ψ_0. To construct finite-dimensional symplectic forms, we exploit the geometric theory of solitons. Its key feature is the Krichever correspondence, which produces genuine functions on a Riemann surface Γ for which k is a local coordinate. More precisely, the Krichever correspondence is the map

$$(\Gamma, k^{-1}, \gamma_1, \cdots, \gamma_g) \longrightarrow \Psi(z,t). \tag{3.1}$$

Here Γ is a fixed Riemann surface of genus g, $z = k^{-1}$ is a local coordinate system in a neighborhood of a marked point $P \in \Gamma$, $(z_1, \cdots, z_g)$ are g points of Γ in general position, $t = (t_1, t_2, t_3, \cdots)$ is a set of times (only a finite number of which are non zero), and $\Psi(z,t)$ is the Baker-Akhiezer function, i.e. the unique function on Γ which is meromorphic away from P, has at most simple poles at $z_1, \cdots, z_g$, and admits the following essential singularity near P

$$\Psi(z,t) = (1 + \sum_{i=1}^{\infty} \xi_s z^s)\exp(\sum_{i=1}^{\infty} t_i k^i). \tag{3.2}$$

The Baker-Akhiezer function $\Psi(z,t)$ yields a doubly infinite set of 2+1 soliton equations $[\frac{\partial}{\partial t_m} - L_m, \frac{\partial}{\partial t_n} - L_n] = 0$, where L_n is the differential operator in $x = t_1$ defined by the condition $(\frac{\partial}{\partial t_n} - L_n)\Psi(z,t) = 0$.

Since we shall focus only on a single equation, the following modified correspondence is more appropriate for our considerations. Let $N \geq 1$, $n = (n_1, \cdots, n_N)$, $m = (m_1, \cdots, m_N)$ be fixed integers. Let $t \in \mathbf{R}$ be a single time, and consider the map

$$(\Gamma, E, Q, z_1, \cdots, z_g) \to \Psi(z; x, y, t) \to (L, A), \tag{3.3}$$

where E and Q are two Abelian integrals on Γ, with poles at most of order n_α and m_α at N marked points $P_1, \cdots, P_N$. (We always assume that the Riemann surface Γ has been cut apart, so that E and Q are well-defined on the cut surface. Our main results do not depend on the choice of cuts). The equation $E = z^{-n} + (\mathrm{Res}_{P_1} E)\log z$ (or $E = (\mathrm{Res}_{P_1} E)\log z$ when $n = 0$) defines a particular coordinate z near P. If we define the polynomial $Q_+(k)$ by the requirement that $Q - Q_+(k)$ be regular near P, the Baker-Akhiezer function of $\Psi(z,t)$ of (3.2) can now be characterized by the same conditions as before, but with the essential singularity (3.2) replaced now by

$$\Psi(z; x, y, t) = (1 + \sum_{s=1}^{\infty} \xi_s z^s)\exp(xk + yk^n + Q_+(k)t). \tag{3.4}$$

The operators L and A are again defined by $(\partial_y - L)\Psi = O(k^{-1})$ and $(\partial_t - A)\Psi = O(k^{-1})$. It follows that the right hand sides of these equations are actually 0, and that $[\partial_y - L, \partial_t - A] = 0$.

Let $\mathcal{N}_g^k(n,m)$ be the space of data of the form $(\Gamma, E, Q, z_1, \cdots, z_k)$. It can be viewed as a fibration over the space $\mathcal{M}_g(n,m) = (\Gamma, E, Q)$, the fiber of which is the symmetric product $S^k(\Gamma)$ of k copies of the Riemann surface Γ. For $k = g$, as in (3.3), $S^k(\Gamma)$ can be identified with the Jacobian of Γ by the Abel map. It follows readily from the Riemann-Roch theorem that the dimensions of $\mathcal{M}_g(n,m)$ and $\mathcal{N}_g^k(n,m)$ are given respectively by $dim\, \mathcal{M}_g(n,m) = 5g - 3 + 3N + \sum_{\alpha=1}^{N}(n_\alpha + m_\alpha)$

and $dim\,\mathcal{N}_g^k(n,m) = k+5g-3+3N+\sum_{\alpha=1}^N(n_\alpha+m_\alpha)$. In general, the coefficients of the operators L and A are quasi-periodic. But they become periodic if all the periods of the Abelian differentials dE and dQ are integer multiples of some fixed real number.

The main properties of the space $\mathcal{M}_g(n,m)$ are the existence of

(a) a natural (multiple-valued) meromorphic 1-form $d\lambda = QdE$ on each curve Γ,

(b) and of a natural foliation, the leaves of which are equipped with the finite-dimensional symplectic forms we seek.

To describe it, we note that the following provide local coordinates for $\mathcal{M}_g(n,m)$ for the open set where the divisors $\{z; dE=0\}$ and $\{z; dQ=0\}$ do not intersect:

- the $\sum_{\alpha=1}^N(n_\alpha+m_\alpha)+N-1$ pole coefficients in the Laurent expansions of $d\lambda$ at each of the punctures P_α (for residues, the coefficients at only $N-1$ punctures suffice);
- the $2N-2$ residues of dE and dQ;
- the $4g$ periods of dE and dQ;
- the A-periods $a_I = \oint_{A_I} d\lambda$ of the differential $d\lambda$.

Here A_I are half of a homology basis (A_I, B_I), $1\leq I\leq g$, $\#(A_I\cap A_J) = \#(B_I\cap B_J) = 0$, $\#(A_I\cap B_J) = \delta_{IJ}$, which we fix locally on $\mathcal{M}_g(n,m)$. Let ω_I, $1\leq I\leq g$, be the dual basis of holomorphic Abelian differentials $\oint_{A_I} d\omega_J = \delta_{IJ}$. A foliation of $\mathcal{M}_g(n,m)$ is provided then by fixing the values of all parameters except a_I. We note that this foliation does not depend on the choices made in cutting the Riemann surface Γ. Let $\mathcal{M}$ be a fixed leaf of this foliation, and let $\mathcal{N}^k$ the fibration $\mathcal{L}^k(n,m)$ restricted to $\mathcal{M}$. Now initially, the Abelian integrals E and Q are multiple valued meromorphic functions on Γ, and their differentials dE, dQ should be viewed as sections of the vertical cotangent bundle of the fibration $\mathcal{N}_g^{k=1}(n,m)\to\mathcal{M}_g(n,m)$. However, the differentials δE and δQ of E and Q over the total space $\mathcal{N}^{k=1}$ over the leaf $\mathcal{M}$ are genuine 1-forms over $\mathcal{N}^{k=1}$. This is because the ambiguities in both E and Q are multiples of the residues and periods of dE and dQ, and thus disappear upon differentiation along $\mathcal{M}$. Similarly, the full differential $\delta(QdE)$ is well-defined as a 2-form on $\mathcal{N}$. Then we have [32]

Theorem 3. *Let $\mathcal{N}^{k=g}$ be the fibration of Jacobians $S^g(\Gamma)$ over a leaf $\mathcal{M}$ of the fibration described above. Then the expression*

$$\omega_{\mathcal{M}} = \delta(\sum_{I=1}^g Q(z_I)dE(z_I)) = \sum_{I=1}^g da_I\wedge d\omega_I$$

defines a symplectic form on $\mathcal{N}^g$. For leaves $\mathcal{M}$ for which the periods of dE and dQ are all multiples of a fixed real number, the map (3.3) provides an imbedding of $\mathcal{N}^g$ into the space of operators L with periodic coefficients. We have then $\omega_{\mathcal{M}} = \mathcal{G}^*(\omega_m)$.

Further Results

• As Donagi and Witten have pointed out ([18], see also [17,23]), a symplectic form arising as in Theorem 3 from a fibration of Riemann surfaces leads conversely to an integrable model. Indeed, by construction, $\omega_{\mathcal{M}}$ vanishes along the fibers, so that a maximum set of commuting Hamiltonians is provided by any system u_i of local coordinates on the base with respect to which $\omega_{\mathcal{M}}$ can be written as $\omega_{\mathcal{M}} = \sum_{i=1}^N dz_i \wedge du_i$.

• The elliptic Calogero-Moser system is the system $\ddot{q}_i = m^2 \sum_{i\neq j} \wp'(q_i - q_j)$, $1 \leq i,j \leq N$, where $\wp(z)$ is the Weierstrass $\wp$-function defined by a fixed associated lattice $\mathcal{L} = 2\omega_1\mathbf{Z} + 2\omega_2\mathbf{Z}$ and m is a fixed mass parameter. It admits a Lax pair $L(z), A(z)$ with spectral parameter z, with $L_{ij}(z) = p_i\delta_{ij} - m(1 - \delta_{ij})\frac{\sigma(z-q_i+q_j)}{\sigma(z)\sigma(q_i-q_j)}e^{\zeta(z)(q_i-q_j)}$, $1 \leq i,j \leq N$ [28] (see also the exposition in [52]). Its spectral curves $\Gamma = \{(k,z); det(kI - L(z)) = 0\}$ can also be expressed as a leaf $\mathcal{M}$ in the moduli space $\mathcal{M}_g(n,m)$ of surfaces of genus $g = N$ with N punctures P_α, and Abelian integrals $E = z$, $Q = k$ with poles of order $n_\alpha = -1$, $m_\alpha = 1$ at each P_α. The dimension of $\mathcal{M}_g(n,m)$ is then $8N - 3$, and the desired Calogero-Moser leaf $\mathcal{M}$ is obtained by imposing the following $7N - 3$ constraints [32,8]

$$\begin{array}{lll} \mathrm{Res}_{P_\alpha} dQ = 0 & 1 \leq \alpha \leq N, & \oint_C dQ = 0 \\ \int_{P_1}^{P_\alpha} dE \in \mathcal{L} & 2 \leq \alpha \leq N & \oint_C dE \in \mathcal{L} \\ \mathrm{Res}_{P_\alpha}(QdE) = -m, & 2 \leq \alpha \leq N, & \mathrm{Res}_{P_1}(QdE) = m(N-1). \end{array}$$

The construction of Theorem 3 provides a symplectic form $\omega_{\mathcal{M}} = \delta(\sum_{I=1}^N k(z_I)\delta z_I)$. Applying the same arguments as for Theorem 3, we can also write

$$\omega = \sum_{\alpha=1}^N \mathrm{Res}_{P_\alpha} \frac{< \delta\tilde{C}^\dagger \wedge \delta\tilde{L}\,\tilde{C} >}{< \tilde{C}^\dagger\tilde{C} >} dE. \tag{3.5}$$

Here $\tilde{L}_{ij} = e^{k(q_i-q_j)}L_{ij}(z)$, and $\tilde{C} = e^{kq_i}C_{ij}$, $(2k + L(z))C = 0$, is the analogue of the Bloch function. As noted earlier, the construction of Theorem 3 identifies the parameters $a_I = \oint_{A_I} kdz$ as the action variables for the angle variables ω_I on the Jacobian of Γ, while it follows readily from the expression in terms of $\tilde{L}(z)$ and $\tilde{C}$ that ω coincides with the canonical symplectic form $\omega = \frac{1}{2}\sum_{I=1}^N \delta p_I \wedge \delta q_I$. This gives a new proof of the fact [26] that (a_I, ω_I) are action-angle variables for the Calogero-Moser system with respect to the canonical symplectic form.

• The preceding can be extended in several directions, including spin Calogero-Moser systems and Ruijsenaars-Schneider models. The spin Calogero-Moser system is the following extension of Calogero-Moser systems $\ddot{q}_i = m^2 \sum_{j\neq i}(b_i^\dagger a_j)(b_j^\dagger a_i) \wp'(q_i - q_j)$, $\dot{a}_i = -m^2 \sum_{j\neq i} a_j(b_j^\dagger a_i)\wp(q_i - q_j)$, $\dot{b}_i = m^2 \sum_{j\neq i} b_j^\dagger(b_j^\dagger a_i)\wp(q_i - q_j)$, where for each $1 \leq i \leq N$, a_i is an l-dimensional vector and b_i is an l-dimensional covector. It admits a Lax pair with $L_{ij}(z) = p_i\delta_{ij} - m^2(1 - \delta_{ij})(b_i^\dagger a_j)\frac{\sigma(z-q_i+q_j)}{\sigma(z)\sigma(q_i-q_j)}e^{\zeta(z)(q_i-q_j)}$, and spectral curves $\Gamma = \{(k,z); det(kI - L(z)) = 0\}$ of genus $g = Nl - \frac{1}{2}l(l-1)$ [31]. The leaf $\mathcal{M}$ of spin Calogero-Moser spectral curves is of dimension $g + l - 1$. It is shown in [30] that the symplectic form $\omega = \delta(\sum_{i=1}^{g+l-1} k(z_i)dz_i)$ associated by Theorem 3 to the fibration $\mathcal{L}_g^{k=g+l-1}$ with fiber $S^{g+l-1}(\Gamma)$ can be expressed in the form (3.5) for a suitable vector $\tilde{C}$, and consequently as $\omega = -\sum_{i=1}^N(\delta p_i \wedge \delta x_i + \sum_{\alpha=1}^l \delta B_i^\alpha \wedge \delta A_{i\alpha})$, where $A_i = W_0^{-1}a_i$, $B_i^\dagger = b_i^\dagger W_0$, for a suitable constant matrix W_0.

• The elliptic Ruijsenaars-Schneider system is the relativistic version of the Calogero-Moser system, given by $\ddot{q}_i = \sum_{j\neq i} \dot{q}_i\dot{q}_j(\zeta(q_j - q_i + \eta) - \zeta(q_i - q_j + \eta) + 2\zeta(q_i - q_j))$, where η is a parameter [46,6,7,34,5]. The operator $L(z)$ in the Lax pair is $L_{ij}(z) = [\frac{\sigma(z-\eta)}{\sigma(z+\eta)}]^{\frac{q_i-q_j-\eta}{2\eta}} \tilde{L}_{ij}(z)$, $\tilde{L}_{ij}(z) = \dot{q}_i \frac{\sigma(z+q_i-q_j)}{\sigma(z+\eta)\sigma(q_i-q_j-\eta)}$. The analogue of the symplectic form of Theorem 3 and of the Calogero-Moser system is now $\omega = 2\sum_{i=1}^N \delta z_i \wedge \delta \log k(z_i) = \sum_{\alpha=1}^N \mathrm{Res}_{P_\alpha} < \tilde{C}^* \delta \tilde{L}\, \tilde{L}^{-1} \wedge \delta \tilde{C} > dz$. This attractive formula as well as generalizations to spin Ruijsenaars-Schneider systems can be found in [30].

4. SUPERSYMMETRIC $SU(N)$ GAUGE THEORIES

We shall now explain how the symplectic structures constructed in the previous sections are precisely the ones arising from exact solutions to N=2 supersymmetric $SU(N)$ gauge theories. Recall that such gauge theories admit a moduli space of vacua of dimension $N-1$, corresponding to the constant configurations satisfying $[\phi, \phi^\dagger] = 0$ of the scalar field ϕ (valued in the adjoint representation) of the N=2 gauge field multiplet. Generically, the $SU(N)$ gauge symmetry is broken down to $U(1)^{N-1}$. Since the N=2 supersymmetry is unbroken, the effective action must be the N=2 supersymmetric action of $N-1$ interacting $U(1)$ gauge fields A_j

$$I_{eff} = \frac{1}{4}\mathrm{Im}(\tau_{ij})F_i \wedge *F_j + \frac{1}{4}\mathrm{Re}(\tau_{ij})F_i \wedge F_j + d\bar{\phi}^j \wedge d\phi_{Dj} + \text{fermions}, \tag{4.1}$$

where F_j is the curvature of A_j, ϕ_j is the scalar field in the multiplet of the $U(1)$ gauge field A_j, and ϕ_{Dj} is the dual gauge scalar. Both ϕ_{Dj} and the effective gauge coupling are given by a single holomorphic function $\mathcal{F}(\phi, \Lambda)$, called the prepotential

$$\tau_{ij} = \frac{\partial^2 \mathcal{F}}{\partial\phi_i \partial\phi_j}(\phi, \Lambda), \quad \phi_{Dj} = \frac{\partial \mathcal{F}}{\partial \phi_j}(\phi, \Lambda). \tag{4.2}$$

Here Λ is the renormalization scale. From perturbative analysis, it is known that for an $SU(N)$ theory with N_f hypermultiplets of bare masses m_j in the fundamental representation, $\mathcal{F}$ admits the following expansion for Λ small

$$\begin{aligned}\mathcal{F}(a,\Lambda) = \quad & \mathcal{F}^{\text{class}} - \tfrac{1}{8\pi i}[\textstyle\sum_{i\neq j}(a_i - a_j)^2 \ln \frac{(a_i-a_j)^2}{\Lambda^2} \\ & - \textstyle\sum_{j=1}^{N_f}\sum_{i=1}^N (a_i + m_j)^2 \ln \frac{(a_i+m_j)^2}{\Lambda^2}] \quad + \textstyle\sum_{d=1}^\infty \mathcal{F}_d \Lambda^{(2N-N_f)d},\end{aligned} \tag{4.3}$$

where the logarithmic terms are due to one-loop quantum corrections, and the Λ^d terms are the contributions of d-instanton processes. In general, it is not possible to evaluate $\mathcal{F}$ directly. The Seiberg-Witten Ansatz asserts that to each N=2 supersymmetric gauge theory corresponds a fibration of Riemann surfaces Γ (or spectral curves) over the moduli space of vacua equipped with a meromorphic form $d\lambda$, and that $\mathcal{F}$ is then given by

$$a_I = \oint_{A_I} d\lambda, \quad a_{DI} = \oint_{B_I} d\lambda, \quad a_{DI} = \frac{\partial \mathcal{F}}{\partial a_I}. \tag{4.4}$$

Based on several consistency checks (see [1,35] and references therein) and on an explicit evaluation of their logarithmic singularities [14,15], we know now the spectral curves for the SU(N) gauge theory with an arbitrary number of hypermultiplets N_f ($N_f \leq 2N$). Our goal here is to describe how the same spectral curves and meromorphic form $d\lambda$ (and hence the same symplectic form on the fibration of Jacobians) arise readily from the soliton set-up of Section III. More precisely, let

$g = N - 1$, and consider Riemann surfaces Γ of genus g with $2 + N_f$ marked points $(P_\alpha) = (P_+, P_-; P_1, \cdots, P_{N_f})$. Then [32]

Theorem 4. *For each fixed $N_f < 2N$, $m_1, \cdots, m_{N_f}$, let $\mathcal{M}$ be the leaf characterized by*

(a) dE has simple poles with residues $-N$, $N - N_f$, 1 at the points P_+, P_-, P_j, $1 \leq j \leq N_f$ respectively, and all the periods of dE are integer multiples of $2\pi i$;

(b) Q is a single-valued function with simple poles at P_+ and P_-, and $Q(P_i) = -m_i$, $1 \leq i \leq N_f$;

(c) $\mathrm{Res}_{P_-}(zd\lambda) = (N - N_f)(\frac{\Lambda^{2N-N_f}}{2})^{1/(N-N_f)}$;

(d) $E = N\log Q + \log 2 + O(Q^{-1})$ near P_+,

where $z = E^{-1/N}$ or $z = E^{1/(N-N_f)}$ is the local coordinate near the marked point P_+ or P_-. (The fourth condition (d) is just a normalization of the additive constant in the Abelian integral E). Then the corresponding fibration of Riemann surfaces and meromorphic form $d\lambda = QdE$ are precisely the Seiberg-Witten spectral curves for the $SU(N)$ gauge theory with N_f hypermultiplets in the fundamental representation.

Further Results

• It has been verified explicitly that the prepotential $\mathcal{F}(a, \Lambda)$ resulting from the spectral curves in Theorem 4 has the required logarithmic singularities [14,15]. The methods of residues and analytic continuation of [14,15] can also determine the instanton corrections $\mathcal{F}_d$ to arbitrary order d. More recently, they have been applied to $SU(N)$ gauge theories with matter in the symmetric or the anti-symmetric representations [21,41].

• The effective prepotential $\mathcal{F}(a, \Lambda)$ has a lot of structure. In particular, its dependence on the renormalization scale is determined by a renormalization group equation, the beta function of which is a modular form (see [16] and references therein). In the theory of solitons, this equation is the first of a whole "Whitham hierarchy" of equations (see [49,39,20,25] and references therein). Instanton corrections for $SU(N)$ gauge theories with matter in the fundamental representation have recently been also evaluated using Whitham equations [19].

• The effective prepotential $\mathcal{F}$ also exhibits several features reminiscent of the free energy of topological field theories. As in topological field theories (see [29,33] and references therein), it is known that, at least for $SU(N)$ theories with matter in the fundamental representation, the third order derivatives of $\mathcal{F}$ are given by residues $\frac{\partial^3 \mathcal{F}}{\partial a_I \partial a_J \partial_K} = \mathrm{Res}_{dE=0}(\frac{d\omega_I d\omega_J d\omega_K}{dQ\, dE})$, and they satisfy WDVV equations (see [37] and references therein).

• Although we lack space to discuss it in any detail, there has been also a lot of progress in gauge theories with a general gauge group G as well as matter in different representations from the fundamental. Of particular interest is the gauge theory with a matter hypermultiplet of mass m in the adjoint representation of G. This theory admits a microscopic gauge coupling τ, acquires an $N = 4$ supersymmetry when $m = 0$, and can reduce to other important theories in suitable decoupling limits [8,54]. When $G = SU(N)$, Donagi and Witten [18] had proposed that the gauge theory mith matter in the adjoint representation should correspond to the $SU(N)$ Hitchin system [27,17,36]. Nekrasov [42], Martinec [40], and Krichever subsequently recognized that the $SU(N)$ Hitchin spectral curves and the elliptic Calogero-Moser spectral curves are identical. The precise correspondence between the $SU(N)$ gauge theory with matter in the adjoint representation and the elliptic

Calogero-Moser system was established in [8]. A noteworthy feature is that, unlike in the case of the fundamental representation, the moduli of the gauge theory are not apparent from previously known parametrizations for the Calogero-Moser spectral curves. They can however be readily identified as the zeroes of the monic polynomial $H(k)$ in the following new parametrization [8]

$$det(\lambda I - L(z)) = \frac{\theta_1(\frac{1}{2\omega_1}(z - m\frac{\partial}{\partial k})|\tau)}{\theta_1(\frac{z}{2\omega_1}|\tau)} H(k)|_{k=\lambda+m\partial_z \log\theta_1(\frac{z}{2\omega_1}|\tau)} \tag{4.5}$$

The polynomial $H(k)$ also encodes conservation laws for the Calogero-Moser system [12], and has been useful in the study of certain thermodynamic limits [53]. Another important feature of the analysis in [8] is the emergence of a new renormalization group equation for the effective prepotential, $\frac{\partial \mathcal{F}}{\partial \tau} = \frac{1}{2} Tr L^2$, where the variable is now the microscopic gauge coupling τ, and the beta function is the Hamiltonian of the Calogero-Moser system. This relation between the Hamiltonian of the integrable model and the beta function of a renormalization group type equation for the four-dimensional gauge theory, is a fundamental feature of the correspondence between the two theories, going far beyond the original coincidence of spectral curves.

- For G a general simple Lie group of rank r, it is shown in [9-11] that the integrable model corresponding to the gauge theory with gauge group G and matter in the adjoint representation is a new model, the twisted elliptic Calogero-Moser, with Hamiltonian

$$H_G^{twisted}(q,p) = \frac{1}{2}\sum_{i=1}^{r} p_i^2 - \frac{1}{2}\sum_{\alpha\in\mathcal{G}} m_{|\alpha|}^2 \wp_{\nu(\alpha)}(\alpha\cdot q), \quad \wp_\nu(z) = \sum_{\sigma=0}^{\nu-1} \wp(z + 2\omega_a \frac{\sigma}{\nu}) \tag{4.6}$$

Here $\mathcal{R}(G)$ is the set of roots of G, $m_{|\alpha|}$ is a mass parameter which may depend on the length of the root α, and we have set $\nu(\alpha)$ for all long roots α, $\nu(\alpha) = 2$ for the short roots of B_n, C_n, F_4, and $\nu(\alpha) = 3$ for the short roots of G_2. The system (4.6) reduces of course to the (untwisted) systems introduced long ago by Olshanetsky and Perelomov [44] when G is simply-laced. One of its main properties is that it scales to the affine dual Toda system for $(G^{(1)})^\vee$, as the gauge coupling τ tends to $i\infty$ according to the renormalization group. Lax pairs with spectral parameter were obtained for both untwisted and twisted elliptic Calogero-Moser systems for all Lie algebras in [9]. Another construction has been subsequently proposed in [2-4].

- The correspondence between integrable models and supersymmetric gauge theories is still far from complete. For example, we do not know whether the spin Calogero-Moser model corresponds to any gauge theory. It is suggestive that its $Nl - \frac{1}{2}l(l-1)$ dimensional moduli of spectral curves admit a natural parametrization generalizing (4.5)

$$det(\lambda I - L(z)) = \sum_{p=1}^{l} \partial_z^{p-1}\left(\frac{\theta_1(\frac{z}{2\omega_1}(z - m\frac{\partial}{\partial k})|\tau)}{\theta_1(\frac{1}{2\omega_1}|\tau)} H_p(k)\right)|_{k=\lambda+m\partial_z \log\theta_1(\frac{z}{2\omega_1}|\tau)}, \tag{4.7}$$

where $H_p(k)$ is a polynomial of degree $N - p + 1$, $1 \leq p \leq l$, $H_1(k)$ is monic, and $H_p(k)$ has no term of order k^0 [13]. In another direction, it has been argued that the relativistic Ruijsenaars-Schneider system should correspond to the Seiberg-Witten solution of supersymmetric gauge theories in dimensions > 4, with the relativization due to the tower of Kaluza-Klein modes [38,5]. This seems to be a promising area for further investigation.

5. ACKNOWLEDGEMENTS

The author would like to thank Eric D'Hoker and Igor Krichever for collaboration on the work described here, and for many helpful conversations. He would like to express his appreciation to the organizers of the ICCM 98 in Beijing, especially to Professors Yang Lo, Gang Tian, and Shing-Tung Yau, for the opportunity to participate and for their very warm welcome and hospitality there.

References

[1] Argyres, P., R. Plesser, and A. Shapere, "The Coulomb phase of N=2 supersymmetric QCD", Phys. Rev. Lett. **75** (1995) 169
[2] Bordner, A., E. Corrigan, and R. Sasaki, "Calogero-Moser systems: a new formulation", hep-th/9805106.
[3] Bordner, A., R. Sasaki, and K. Takasaki, "Calogero-Moser systems II: symmetries and foldings", hep-th/9809068.
[4] Bordner, A. and R. Sasaki, "Calogero-Moser systems III: symmetries and twisting", hep-th/9812232.
[5] Braden, H., A. Marshakov, A. Mironov, and A. Morozov, "The Ruijsenaars-Schneider model in the context of Seiberg-Witten theory", hep-th/9902205.
[6] Bruschi, M. and F. Calogero, "The Lax representation for an integrable class of relativistic dynamical systems", Comm. Math. Phys. **109** (1987) 481-492.
[7] Bruschi, M. and O. Ragnisco, "Lax representation and complete integrability for the periodic relativistic Toda lattice", Phys. Lett. **A 134** (1989) 365-370.
[8] D'Hoker, E. and D.H. Phong, "Calogero-Moser systems in SU(N) Seiberg-Witten theory", Nucl. Phys. **B 513** (1998) 405-444.
[9] D'Hoker, E. and D.H. Phong, "Calogero-Moser Lax pairs with spectral parameter for general Lie algebras", Nucl. Phys. **B 530** (1998) 537-610, hep-th/9804124.
[10] D'Hoker, E. and D.H. Phong, "Calogero-Moser and Toda systems for twisted and untwisted affine Lie algebras", Nucl. Phys. **530** (1998) 611-640, hep-th/9804125.
[11] D'Hoker, E. and D.H. Phong, "Spectral curves for super Yang-Mills with adjoint hypermultiplet for general Lie algebras", Nucl. Phys. **B 534** (1998) 697-719, hep-th/9804126.
[12] D'Hoker, E. and D.H. Phong, "Order parameters, free fermions, and conservation laws for Calogero-Moser systems, Asian J. Math. **2** (1998), 655-665, hep-th/9808156.
[13] D'Hoker, E. and D.H. Phong, "Lax pairs and spectral curves for Calogero-Moser and spin Calogero-Moser systems", **3** (1998) 27-39 in *Regular and Chaotic Dynamics.*
[14] D'Hoker, E., I.M. Krichever, and D.H. Phong, "The effective prepotential for N=2 supersymmetric SU(N) gauge theories", Nucl. Phys. **B 489** (1997) 179-210.
[15] D'Hoker, E., I.M. Krichever, and D.H. Phong, "The effective prepotential for N=2 supersymmetric SO(N) and Sp(2N) gauge theories", Nucl. Phys. **B 489** (1997) 211-222.
[16] D'Hoker, E., I.M. Krichever, and D.H. Phong, "The renormalization group equation for N=2 supersymmetric gauge theories", Nucl. Phys. **B 494** (1997) 89-104.
[17] Donagi, R., "Seiberg-Witten integrable systems", alg-geom/9705010.
[18] Donagi, R. and E. Witten, "Supersymmetric Yang-Mills and integrable systems", Nucl. Phys. **B 460** (1996) 288-334.
[19] Edelstein, J.D., M. Gomez-Reina, and J. Mas, "Instanton corrections in N=2 supersymmetric theories with classical gauge groups, and fundamental matter hypermultiplets", hep-th/9904087.
[20] Eguchi, T. and S.K. Yang, "Prepotentials of N=2 supersymmetric gauge theories and soliton equations", Mod. Phys. Lett. **A 11** (1996) 131-138, hep-th/9510183.
[21] Ennes, I, S. Naculich, H. Rhedin, and H. Schnitzer, "One-instanton predictions of a Seiberg-Witten curve from M-theory: the symmetric representation of SU(N)", Int. J. Mod. Phys. A **14** (1999), 301, hep-th/9804141.
[22] Faddeev, L. and L. Takhtajan, "*Hamiltonian methods in the theory of solitons*", Springer-Verlag, 1987.
[23] Freed, D.S., "Special Kähler manifolds", hep-th/9712042.

[24] Gelfand, I.M. and L. Dickey, "Fractional powers of operators and Hamiltonian systems", Funct. Anal. and Appl. **10** (1976) 13-29.

[25] Gorsky, A., A. Marshakov, A. Mironov, and A. Morozov, "RG equations from Whitham hierarchy", Nucl. Phys. **B 527** (1998) 690, hep-th/9802007.

[26] Gorsky, A. and N. Nekrasov, "Elliptic Calogero-Moser system from two-dimensional current algebra", preprint ITEP-NG/1-94, hep-th/9401121.

[27] Hitchin, N., "Stable bundles and integrable systems", Duke Math. J. **54** (1987) 91-114.

[28] Krichever, I.M., "Elliptic solutions of the Kadomtsev-Petviashvili equation and integrable systems of particles", Funct. Anal. Appl. **14** (1980) 282-290.

[29] Krichever, I.M., "The τ-function of the universal Whitham hierarchy, matrix models, and topological field theories", Comm. Pure Appl. Math. **47** (1994) 437-475.

[30] Krichever, I.M., "Elliptic solutions to difference non-linear equations and nested Bethe ansatz equations", solv-int/9804016.

[31] Krichever, I.M., O. Babelon, E. Billey, and M. Talon, "Spin generalization of the Calogero-Moser system and the matrix KP equation", Amer. Math. Soc. Transl. **170** (1995) 83-119.

[32] Krichever, I.M. and D.H. Phong, "On the integrable geometry of soliton equations and N=2 supersymmetric gauge theories", J. Differential Geometry **45** (1997) 349-389.

[33] Krichever, I.M. and D.H. Phong, "Symplectic forms in the theory of solitons", *Surveys in Differential Geometry* **IV**, edited by C.L. Terng and K. Uhlenbeck, 239-313.

[34] Krichever, I.M. and A. Zabrodin, "Spin generalizations of the Ruijsenaars-Schneider model, non-abelian 2D Toda chain, and representations of the Sklyanin algebra", Uspekhi Mat. Nauk **50** (1995) 3-56, hep-th/9505039.

[35] Lerche, W., "Introduction to Seiberg-Witten theory and its stringy origins", Proceedings of the *Spring School and Workshop on String Theory*, ICTP, Trieste (1996), hep-th/9611190, Nucl. Phys. Proc. Suppl. **B 55** (1997) 83.

[36] Markman, E., "Spectral curves and integrable systems", Comp. Math. **93** (1994) 255-290.

[37] Marshakov, A., A. Mironov, and A. Morozov, "WDVV-like equations in N=2 Susy gauge theory", Phys. Lett. **B 357** (1995) 342-348.

[38] Marshakov, A. and A. Mironov, "5d and 6d supersymmetric gauge theories: prepotentials from integrable systems", hep-th/9711156.

[39] Martinec, E. and N. Warner, "Integrable systems and supersymmetric gauge theories", Nucl. Phys. **B 459** (1996) 97-112.

[40] Martinec, E., "Integrable structures in supersymmetric gauge and string theory", Phys. Lett. **B 367** (1996) 91-96, hep-th/9510204.

[41] Naculich, S., H. Rhedin, and H. Schnitzer, "One-instanton test of a Seiberg-Witten curve from M-Theory: the antisymmetric representation of SU(N)", Nucl. Phys. **B 533** (1998), 275, hep-th/9804105.

[42] Nekrasov, N., "Holomorphic bundles and many-body systems", Comm. Math. Phys. **180** (1996) 587-604, hep-th/9503157.

[43] Novikov, S. and A. Veselov, "On Poisson brackets compatible with algebraic geometry and Korteweg-de Vries dynamics on the space of finite-zone potentials", Sov. Math. Doklady **26** (1982) 357-362.

[44] Olshanetsky, M.A. and A.M. Perelomov, "Completely integrable Hamiltonian systems connected with semisimple Lie algebras", Inventiones Math. **37** (1976) 93-108.

[45] Polychronakos, A., "Generalized Calogero-Sutherland systems from many matrix models", Nucl. Phys. **B 546** (1999) 495-502.

[46] Ruijsenaars, S.N.M. and H. Schneider, "A new class of integrable systems and its relation to soliton equations", Ann. Phys. (NY) **170** (1986) 370-405.

[47] Seiberg, N. and E. Witten, "Electro-magnetic duality, monopole condensation, and confinement in N=2 supersymmetric Yang-Mills theory", Nucl. Phys. **B 426** (1994) 19-53.

[48] Seiberg, N. and E. Witten, "Monopoles, duality, and chiral symmetry breaking in N=2 supersymmetric QCD", Nucl. Phys. **B 431** (1994) 494.

[49] Sonnenschein, J., S. Theisen, and S. Yankielowicz, "On the relations between the holomorphic prepotential and the quantum moduli in SUSY gauge theories", Phys. Lett. **B 367** (1996) 145-150.

[50] Terng, C.L., "Soliton equations and differential geometry", J. Differential Geometry **45** (1997) 407-445.

[51] Terng, C.L. and K. Uhlenbeck, *editors*, *Surveys in Differential Geometry*, Vol. III, International Press (1998).
[52] Treibich, A. and J.L. Verdier, "Solitons elliptiques", The Grothendieck Festschrift, Vol. III, 437-480 (1990), Birkhauser, Boston.
[53] Vaninsky, K., "On explicit parametrization of spectral curves for Moser-Calogero particles and its applications", December 1998 preprint, to appear in Int. Math. Res. Notices.
[54] Witten, E., "Solutions of four-dimensional field theories via M theory", Nucl. Phys. **B 500** (1997) 3, hep-th/9703166.

Department of Mathematics, Columbia University, New York, NY 10027
E-mail address: phong@math.columbia.edu

AMS/IP Studies in Advanced Mathematics
Volume 20, 2001

Quotients, invariant version of Cartan's lemma, and the minimum principle

Xiang-Yu Zhou

In [Zh1], [Zh2], the author gave a proof of the extended future tube conjecture which states that the extended future tube is a domain of holomorphy. The conjecture was put in a general setting of the reductive Lie group actions on Stein manifolds. The author found a general idea dealing with the conjecture in this general setting and relating the minimum principle and the slice theory to the conjecture. This idea is different from the previous ones trying to deal with the conjecture, and had been ignored. As an application of the general approach together with some key techniques, the proof was given. Essentially and simply speaking, the previous ideas are to try to prove the holomorphical convexity of the involved domain, *i.e.*, to construct a holomorphic function which tends to $+\infty$ near to the boundary. For example, a nice idea due to Sergeev is based on the assumption of the invariant version of the Cartan's lemma. These previous ideas are not finally successful, probably because constructing nontrivial holomorphic functions in several complex variables is harder. Our idea is to prove the pseudoconvexity of the domain, *i.e.*, to construct a plurisubharmonic function which tends to $+\infty$ near to the boundary. In the present paper, we'll discuss both ideas and their relation.

The present work was partially supported by the Chinese NSF and Qiu Shi foundation.

The present paper is grew out of my talk at ICCM held in Beijing during Dec. 12–Dec. 16, 1998. The author would like to thank the organizer for the invitation.

1. Some examples

Let M^{n+1} be the $n+1$ dimensional Minkowski space, *i.e.*, M^{n+1} is $\mathbb{R}^{n+1}$ equipped with the Lorentz metric. More precisely, for a pair of vectors $x = (x_0, x_1, \cdots, x_n), y = (y_0, y_1, \cdots, y_n)$, the metric is defined by $xy = x_0y_0 - x_1y_1 - \cdots - x_ny_n$. The future light cone is $V^+ = \{x \in M^{n+1} : x_0 > 0, x^2 > 0\}$; the past cone is $V^- = -V^+ = \{x \in M^{n+1} : x_0 < 0, x^2 > 0\}$.

The Lorentz group L of all linear transformations preserviing the metric acts on M^{n+1}. We denote by $L_+^{\uparrow}$ the identity component of L, and by $L_+(\mathbb{C})$ the identity component of the complex Lorentz group $L(\mathbb{C})$ of all complex linear transformations of $\mathbb{C}^{n+1}$ preserving the bilinear form $z\xi = z_0\xi_0 - z_1\xi_1 - \cdots - z_n\xi_n$ for $z = (z_0, z_1, \cdots, z_n), \xi = (\xi_0, \xi_1, \cdots, \xi_n) \in \mathbb{C}^{n+1}$. In other words, $L(\mathbb{C}) = O(1, n; \mathbb{C}), L = O(1, n)$. It's clear that $L_+^{\uparrow}$ is the real form of $L_+(\mathbb{C})$.

We denote by $x = (x_1, \cdots, x_N)$ a point of $M^{N(n+1)} = M^{n+1} \times \cdots \times M^{n+1}$ (N factors) with components $x_j = (x_{0j}, x_{1j}, \cdots, x_{nj}) \in M^{n+1}$. The N–point future (past) cone $V_N^+(V_N^-)$ is the products of N cones $V^+(V^-)$. The N–point future (past) tube domain with base $V_N^+(V_N^-)$ is given by $T_N^+ = T^{V^+} \times \cdots \times T^{V^+} \subset \mathbb{C}^{N(n+1)} (T_N^- = T^{V^-} \times \cdots \times T^{V^-})$. The action of the group $L(\mathbb{C})$ on $\mathbb{C}^{N(n+1)}$ is given by the rule $z = (z_1, \cdots, z_N) \longmapsto (\Lambda z_1, \cdots, \Lambda z_N)$, where $z \in \mathbb{C}^{N(n+1)}, \Lambda \in L(\mathbb{C})$.

The "extended future tube domain" is given by

$$T_N' = L_+(\mathbb{C}) \cdot T_N^+ = \cup_{\Lambda \in L_+(\mathbb{C})} \Lambda(T_N^+),$$

where $\Lambda(T_N^+)$ is the image of the set $T_N^+ \subset \mathbb{C}^{N(n+1)}$ under the transformation $\Lambda \in L_+(\mathbb{C})$.

A question asks if the extended future tube domain is a domain of holomorphy (extended future tube conjecture).

For the case $n = 1$ an affirmative answer was given by R. Jost ([J]).

For the case $n = 3$ which is an interesting case in physics, an affirmative answer was given by Zhou ([Zh1,Zh2]).

A matrix analogue of the above question can be formulated as follows.

Let $\mathbb{C}[l \times l]$ be the set of all $l \times l$ complex matrices, $\mathbb{C}^N[l \times l]$ be the space consisting of all N $l \times l$ complex matrices.

Denote by $H = \{Z \in \mathbb{C}\ [l \times l] : \frac{Z - \overline{Z}'}{2i} > 0\}$, where $\overline{Z}'$ is the conjugate transpose of Z; by $H_N = H \times \cdots \times H \subset \mathbb{C}^N[l \times l]$.

Let $SL(l, \mathbb{C}) \times SL(l, \mathbb{C})$ act on $\mathbb{C}^N[l \times l]$ by the rule

$$(A, B) \cdot Z = (AZ_1B^{-1}, \cdots, AZ_NB^{-1}),$$

where $A, B \in SL(l, \mathbb{C}), Z = (Z_1, \cdots, Z_N), Z_i \in \mathbb{C}[l \times l]$.

The induced real action of $SL(2, \mathbb{C})$ on $\mathbb{C}^N[l \times l]$ is given by

$$A \cdot Z = (A, (\overline{A}')^{-1}) \cdot Z = (AZ_1\overline{A}', \cdots, AZ_N\overline{A}').$$

Denote by $H_N' = (SL(l, \mathbb{C}) \times SL(l, \mathbb{C})) \cdot H_N$.

A question asks if H_N' is a domain of holomorphy.

It's known that the question when $l = 2$ is equivalent to the extended future tube conjecture when $n = 3$ in terms of Pauli mapping.

2. A general problem

Now we consider the above questions in a general setting of group actions.

Let X be a connected Stein manifold, $G_{\mathbb{C}}$ be a connected complex (reductive) Lie group holomorphically acting on X. Let $G_{\mathbb{R}}$ be a connected closed real form of $G_{\mathbb{C}}$. Suppose $D \subset X$ is a $G_{\mathbb{R}}$–invariant Stein domain, one may ask if $G_{\mathbb{C}} \cdot D$ is a Stein domain (under some mild assumptions).

The questions in §1 are special cases of the above question.

In the first example, $X = \mathbb{C}^{N(n+1)}, G_{\mathbb{C}} = L_+(\mathbb{C}), G_{\mathbb{R}} = L_+^{\uparrow}, D = T_N^+$ and $G_{\mathbb{C}} \cdot D = T'_N$.

In the second example,

$$X = \mathbb{C}^N[l \times l], G_{\mathbb{C}} = SL(l, \mathbb{C}) \times SL(l, \mathbb{C}), G_{\mathbb{R}} = SL(l, \mathbb{C}), D = H_N$$

and

$$G_{\mathbb{C}} \cdot D = H'_N.$$

3. Categorical quotient, slice theory

Definition. Let X be a $G_{\mathbb{C}}$ –space, a $G_{\mathbb{R}}$–domain D is called orbit connected if $b_x^{-1}(D)$ is connected for any $x \in D$ where b_x is the orbit map: $G_{\mathbb{C}} \to X, g \mapsto g \cdot x$.

Example. N–point future tube T_N^+ is orbit connected.

The fact for $n = 3$ is given in the proof of the Bargman–Hall–Wightman theorem (cf. Jost [J], Streater–Wightman[SW]), for general n we refer to Bros et al ([BEG]).

A modern form of BHW theorem. Let X be a complex space, $G_{\mathbb{C}}$ be a connected complex Lie group holomorphically acting on X, $G_{\mathbb{R}}$ be a connected real form of $G_{\mathbb{C}}$. Suppose D is a $G_{\mathbb{R}}$–invariant orbit connected domain, f is a $G_{\mathbb{R}}$–invariant (or equivariant holomorphic mapping) holomorphic function on D, then there exists a holomorphic extension of f on $G_{\mathbb{C}} \cdot D$ which is $G_{\mathbb{C}}$–invariant (or equivariant).

The proof of the fact is the same as the proof of BHW theorem given in [SW].

A natural question asks if the above extension can be further extended.

In the above question of §2, this is equivalent to ask if $G_{\mathbb{C}} \cdot D$ is Stein when D is π–saturated , where $\pi : X \to X//G_{\mathbb{C}}$ is the categorical quotient. Now we explain the concept of the categorical quotient.

Definition. A triple (X, Y, π), where X is a complex space with a holomorphic action of $G_{\mathbb{C}}, Y$ a complex space, $\pi : X \to Y$ a holomorphic mapping, is called a categorical quotient if given an holomorphic mapping $\varphi : X \to Z$ which is constant on any $G_{\mathbb{C}}$ –orbit , there exists a unique holomorphic mapping $\psi : Y \to Z$ *s.t.*,

$$\begin{array}{ccc} X & \xrightarrow{\pi} & Y \\ \varphi\searrow & & \swarrow\psi \\ & Z & \end{array}$$

commutes.

If such a categorical quotient exists, we denote by $Y = X//G_{\mathbb{C}}$.

Example 1. When X is an affine varienty and redutive linear algebraic group $G_{\mathbb{C}}$ over $\mathbb{C}$ acts on X morphically, then $X//G_{\mathbb{C}} = Spec\mathbb{C}[X]^{G_{\mathbb{C}}}$ which is affine again. When X is a projective variety, the categorical quotient exists in terms of semi–stable points by a result of Mumford's (cf. Mumford–Fogarty–Kirwan [MFK]).

Example 2. When X is a Stein space, $X//G_{\mathbb{C}}$ can be given by the quotient X/R where the equivalence relation is given by $R = \{(x_1, x_2) \in X \times X : f(x_1) = f(x_2) \text{ for all } f \in \mathcal{O}(X)^{G_{\mathbb{C}}}\}$. The structure sheaf is determined by the following presheaf. Associated to an open set of the quotient the presheaf consists of continuous functions whose pull backs are holomorphic on the inverse open set of X.

Definition. A $G_{\mathbb{C}}$–set $F \subset X$ is called π–saturated if $\pi^{-1}(\pi(F)) = F$. A $G_{\mathbb{R}}$–set D is called π–saturated if $G_{\mathbb{C}} \cdot D$ is π–saturated.

Keeping the notations of §2.

Set

$k = \min\{\dim(G_{\mathbb{C}})_x : G_{\mathbb{C}} \cdot x \text{ is closed}\}$ and

$$l = \min\{\# \text{ of components of } (G_{\mathbb{C}})_x : G_{\mathbb{C}} \cdot x \text{ is closed and } \dim(G_{\mathbb{C}})_x = k\}.$$

The orbit through $x \in X$ *s.t.*, $G_{\mathbb{C}} \cdot x$ is closed , $\dim(G_{\mathbb{C}})_x = k$ and $(G_{\mathbb{C}})_x$ has l components is principal , the set consisting all such orbits are denoted by X'. The isotropy groups of the principal orbits are called principal.

By slice theory due to Luna–Snow, a closed orbit $G_{\mathbb{C}} \cdot X$ is principal $\Longleftrightarrow \pi(x)$ is a regular point of $X//G_{\mathbb{C}}$ and $\pi : X \longrightarrow X//G_{\mathbb{C}}$ is nondegenerate at x; put $X_{pr} = \pi^{-1}(\pi(X'))$, then X_{pr} is Zariski spen in X, *i.e.,*, $X \backslash X_{pr}$ is an analytic subset of X; all principal isotropy groups are conjugate.

If X has FPIG (*i.e.*, all principal isotropy groups are finite), then $X' = X_{pr}$, adn $\pi^{-1}(\pi(x)) = G_{\mathbb{C}} \cdot x$ for $x \in X'$.

We observe the following relationship between the FPIG action and the π–saturated domain.

Proposition ([Zh3]). Let X be a connected Stein manifold, $G_{\mathbb{C}}$ be a connected reductive complex Lie group holomorphically acting on X. Suppose that the action is of FPIG, then any $G_{\mathbb{C}}$–invariant domain is π–saturated.

We refer to [Zh3] for a detailed proof.

Let $x \in X', H = (G_{\mathbb{C}})_x$. Denote by $F = X^H$ the fixed point set of H, $N = N_{G_{\mathbb{C}}}(H)/H$, where $N_{G_{\mathbb{C}}}(H)$ is the normalizer of H in $G_{\mathbb{C}}$. A result due to D. Luna -R.W. Richardson [LR] means (although they state their result only in algebraic setting) that if $X^H//N$ is irreducible, then the unique mapping ψ satisfying $\psi \circ \pi_F = \pi|_F$ is biholomorphic. Let A be an irreducible component of X^H *s.t.* $\pi(A)$ meets

$\pi(X')$ and let $L = \{g \in G_{\mathbb{C}} | g \cdot A = A\}$; one has $N_{G_{\mathbb{C}}}(H)^\circ \subset L \subset N_{G_{\mathbb{C}}}(H)$. Let $W = L/H$. Then the inclusion map $A \to X$ induces an isomorphism of varieties $A//W \to X//G_{\mathbb{C}}$.

Let W_0 be the identity component of W, then the natural mapping $A//W_0 \to A//W$ is finite. Denote by $\pi' : A \to A//W_0$, Then $\pi(D)$ is Stein if and only if $\pi'(D \cap A)$ is Stein. For considering the problem of §2, one can reduce the general case of large PIG to the special case of trivial PIG.

4. Fundamentals of several complex variables

The concept of a domain of holomorphy is one of the most fundamental concepts in several complex variables. In higher dimensional case, a domain is not always a domain of holomorphy.

For example, the punctured domain is not a domain of holomorphy. A domain is called a domain of holomorphy, if there exists a holomorphic function on D which can't be extended further.

Holomorphic convexity and pseudoconvexity play a very important role in understanding of domains of holomorphy.

A domain is called holomorphically convex if given a discret point sequence there exists a holomorphic function which goes to $+\infty$ while restrict to the sequence.

It's known that a domain is a domain of holomorphy if and only if it's holomorphically convex.

A domain is called pseudoconvex, if there is a plurisubharmonic function f *s.t.* $f(x) \to +\infty$ when $x \to a$ (a is any boundary point) (we call it weak exhaustion).

A fundamental theorem in several complex variable asserts that a domain is a domain of holomorphy if and only if it's pseudoconvex.

Our idea approaching the extended future tube conjecture is to prove the pseudoconvexity of the domain, *i.e.*, to construct a special plurisubharmonic function. Constructing plurisubharmonic functions seems to be not so hard as constructing holomorphic functions. As Lelong said , plurisubharmonic functions are among "les objets souples de l'analyse complexe" (cf.[K1]).

5. The minimum principle.

In the present section, we recall the minimum principle and observe its simple consequence. Kiselman ([K2]) first studied the minimum principle and coined the term "minimum principle". The following general case was due to Loeb ([L2]).

Let X be a complex manifold, $G_{\mathbb{C}}$ be a connected complex Lie group, $G_{\mathbb{R}}$ be a connected closed real form of $G_{\mathbb{C}}$. Denote by $\psi : G_{\mathbb{C}} \to G_{\mathbb{C}}/G_{\mathbb{R}}$, $p : X \times G_{\mathbb{C}} \to X$ the natural projections.

Suppose $G_{\mathbb{C}}$ acts on $X \times G_{\mathbb{C}}$ on the right as follows:

$$(X \times G_{\mathbb{C}}) \times G_{\mathbb{C}} \longrightarrow X \times G_{\mathbb{C}}$$

$$((x, g), h) \longmapsto (x, gh)$$

Let $\Omega \subset X \times G_{\mathbb{C}}$ be a right $G_{\mathbb{R}}$–invariant domain *s.t.* all fibres of p are connected; $u \in C^\infty(\Omega)$ be a right $G_{\mathbb{R}}$–invariant function (then u naturally induces a smooth function $\dot{u}(x, \psi(g))$ on $\dot{\Omega} := (id_X, \psi)(\Omega)$.

Assume that

1) u is *p.s.h.* on Ω;

2) $\forall x \in p(\Omega), u(x, \cdot)$ is strictly *p.s.h.* on $\Omega_x = \Omega \cap p^{-1}(x) \subset G_{\mathbb{C}}$;

3) $\dot{u}(x,\cdot)$ is exhaustive on $\dot{\Omega}_x = \psi(\Omega_x)$.

Then the minimum principle asserts that $v(x) := \inf_{g\in\Omega_x} u(x,g)$ is C^∞ *p.s.h.* on $p(\Omega)$.

From Loeb's results ([L1], [L2]), one can easily deduce the following:

Proposition. Let X be a connected Stein manifold. $G_{\mathbb{C}}$ be a connected complex Lie group holomorphically acting on X, $G_{\mathbb{R}}$ be a connected closed real form of $G_{\mathbb{C}}$. Let $D\subset X$ be a $G_{\mathbb{R}}$ –invariant orbit connected Stein domain. Suppose that $(G_{\mathbb{C}}, G_{\mathbb{R}})$ is a pseudoconvex pair (*i.e.*, $\exists$ a $G_{\mathbb{R}}$–invariant *p.s.h.* function on $G_{\mathbb{C}}$ which is exhaustive on $G_{\mathbb{C}}/G_{\mathbb{R}}$). Then $G_{\mathbb{C}}\cdot D$ is Stein.

Proof. Consider the mapping:

$$\begin{aligned} X\times G_{\mathbb{C}} &\xrightarrow{\ \rho\ } X \\ (x,g) &\longmapsto g^{-1}\cdot x. \end{aligned}$$

Set $\Omega := \rho^{-1}(D)$.

Note that Ω has connected fibres of $p \Longleftrightarrow \Omega$ is orbit connected.

By the definition of pseudoconvex pair and a result of Loeb's ([L1]), one can get a smooth $G_{\mathbb{R}}$–invariant *p.s.h.* function $u(x,g)$ on Ω satisfying the assumptions of the minimum principle, and $\dot{u}$ is exhaustion on $\Omega/G_{\mathbb{R}}$, therefore $v(x) = \inf_{g\in\Omega_x} u(x,g)$ is smooth *p.s.h.* exhaustion on $p(\Omega) = G_{\mathbb{C}}\cdot D$. so $G_{\mathbb{C}}\cdot D$ is Stein. □

Remark. When $G_{\mathbb{R}}$ is compact or nilpotent, $(G_{\mathbb{C}}, G_{\mathbb{R}})$ is a pseudoconvex pair ([L1], [L2]). Actually, this obervation of ours was one of the starting points of our proof of the extended future tube conjecture.

6. Main idea of the proof: based on the minimum principle

Let X be a connected Stein manifold, $G_{\mathbb{C}}$ be a connected reductive complex Lie group holomorphically acting on X, $G_{\mathbb{R}}$ be a connected real form of $G_{\mathbb{C}}$.

In terms of Luna–Richardson's theorem (cf. §4), one can suppose that X has FPIG for simplicity.

Fact ([Zh1, 2]): In the case of the extended future tube conjecture, $\mathbb{C}^{4N}$ has FPIG *w.r.t.* $L_+(\mathbb{C})$–action for $N \geq 3$.

Put $X_0 := \{\xi \in X : G_{\mathbb{C}} \cdot \xi \text{ is closed}, (G_{\mathbb{C}})_\xi \text{ is finite and principal } \}$. Then $A = X \backslash X_0$ is closed analytic subset of X and $\pi^{-1}(\pi(X_0)) = X_0$.

Let D be a $G_{\mathbb{R}}$–invariant, orbit connected, Stein domain. Then $D_0 := D \cap X_0 = D \backslash A$ is also orbit connected.

Set

$$X \times G_{\mathbb{C}} \xrightarrow{\rho} X$$

$$(\xi, g) \longmapsto g^{-1} \cdot \xi$$

Consider $\Omega_0 := \rho^{-1}(D_0), \Omega := \rho^{-1}(D)$.

Note that Ω has connected fibres of $p \iff D$ is orbit connected.

Suppose that there is a $G_{\mathbb{R}}$–invariant C^∞ *s.p.s.h.* and weak exhaustion φ on D. For $G_{\mathbb{R}}$ compact, such a fuction (even exhaustion) obviously exists by "averaging" technique. This technique fails for noncompact Lie groups. However, for $G_{\mathbb{R}}$ noncompact semisimple Lie group, we have the following observation.

Fact: Let $G_{\mathbb{R}}$ be a real semisimple Lie group linearly act on $D \subset \mathbb{C}^n$, then the Bergman kernel satisfies: $K(g \cdot z, g \cdot w) = K(z, w)$, for $g \in G_{\mathbb{R}}, z, w \in D$.

Since $G_{\mathbb{R}}$ is semisimple, then every element of $G_{\mathbb{R}}$ is generated by elements like $f \circ g \circ f^{-1} \circ g^{-1}$. Note that its Jacobian is 1. The result follows from the transformation formular of the Bergman kernel. □

Define

$$v(\xi) := \inf_{g \in \Omega_\xi} \varphi(g^{-1} \cdot \xi) = \inf_{y \in D \cap G_{\mathbb{C}} \cdot \xi} \varphi(y).$$

Note that $v(\xi)$ is $G_{\mathbb{C}}$–invariant, defined and upper semicontinuous on $p(\Omega) = G_{\mathbb{C}} \cdot D$.

Although one cann't directly apply the minimum principle to the whole Ω, one can apply the principle to Ω_0.

It's easy to check that the assumption 2) of the principle is satisfied.

If assumption 3) of the principle is also satisfied, *i.e.*, the function $\dot{u}(\xi, \psi(g))$ induced by $u(\xi, g)$ is exhaustive on $\dot{\Omega}_\xi$, then applying the minimum principle to Ω_0, we see that $v(\xi)$ is C^∞ *p.s.h.* on $p(\Omega_0) = G_{\mathbb{C}} \cdot D_0 = G_{\mathbb{C}} \cdot D \backslash A$.

Since $v(\xi)$ is upper semicontinuous on $G_{\mathbb{C}} \cdot D$, we see by the extension theorem for *p.s.h.* functions, that $\exists$ a *p.s.h.* function $\tilde{v}(\xi)$ on $G_{\mathbb{C}} \cdot D$ *s.t.* $\tilde{v}(\xi) = v(\xi)$ on $G_{\mathbb{C}} \cdot D \backslash A$ and

$$\tilde{v}(\xi_0) = \overline{\lim_{\substack{\xi \to \xi_0 \\ \xi \in G_{\mathbb{C}} \cdot D_0}}} \, v(\xi)$$

for $\xi_0 \in G_{\mathbb{C}} \cdot D \cap A$.

If $\tilde{v}(\xi)$ is a weak exhaustion function, then $G_{\mathbb{C}} \cdot D$ is Stein.

This is the main idea of the proof. In order to make the idea work, we are left to check two assumptions: assumption 3) of the principle and the weak exhaustion of the constructed *p.s.h.* function.

In the case of the extended future tube conjecture, we were successful to check the above two assumptions, and then finished a proof of the conjecture.

7. Another idea of the proof: based on the invariant version of Cartan's lemma.

We mean the Cartan's lemma by the following: a holomorphic function on an analytic subset of a Stein domain can be extended to a holomorphic function on the Stein domain.

An invariant version of the Cartan's lemma means the following: an invariant holomorphic function on an invariant analytic subset of an invariant Stein domain can be extended to an invariant holomorphic function on the domain.

Remark. In general , the above invariant version of Cartan's lemma is not always true (cf. [SZ]). However, if the group $G_{\mathbb{R}}$ is compact, then it's true , by using the Cartan's lemma and the "averaging" technique (integration over a compact group).

Let X be a connected Stein manifold. $G_{\mathbb{C}}$ be a connected redutive complex Lie group holomorphically acting on X, $G_{\mathbb{R}}$ be a connected closed real form of $G_{\mathbb{C}}$. Let $D \subset X$ be a $G_{\mathbb{R}}$–invariant orbit connected, π–saturated, Stein domain, one has the following:

Proposition. Assume that the invariant version of the Cartan's lemma holds, then $G_{\mathbb{C}} \cdot D$ is Stein.

This is another idea due to Sergeev tackling the extended future tube conjecture. The essential point of the idea is to prove the holomorphical convexity of the domain $G_{\mathbb{C}} \cdot D$. Up to now one is still not able to check directly the invariant version of the Cartan's lemma so that one can get $G_{\mathbb{C}} \cdot D$ is Stein.

Actually, we have the following:

Theorem ([Zh3]). The converse of the above proposition is also valid, *i.e.*, $G_{\mathbb{C}} \cdot D$ is Stein if and only if the invariant version of the Cartan's lemma on D holds.

We refer to [Zh3] for the detailed proof.

Corrollary. Under the above assumption, then $G_{\mathbb{C}} \cdot D$ is Stein if and only if $G_{\mathbb{C}} \cdot D$ is holomorphically convex *w.r.t.* $\mathcal{O}(D)^{G_{\mathbb{R}}} = \mathcal{O}(G_{\mathbb{C}} \cdot D)^{G_{\mathbb{C}}}$ (when $\mathcal{O}(D)^{G_{\mathbb{R}}} \neq \mathbb{C}$).

References

[BEG] J. Bros, H. Epstein, V. Glaser: On the connection between analyticity and Lorentz covariance of Wightman functions. Comm. Math. Phys. 6(1967), 77-100.

[G] R. Gunning: Introduction to holomorphic fuctions of several variables, vol.1, Brooks-Cole Publishing Company, Pasific Grove, CA 1990.

[J] R. Jost: The general theory of quantized fields. AMS, Providence, R.I., 1965.

[K1] C.O. Kiselman: Plurisubharmonic functions and potential theory in several complex variables. in "Développment des mathématiques au cours de la seconde moitié du XXe siècle (Development of Mathematics 1950-2000)", ed. by Jean-Paul Pier. Forthcoming.

[K2] C.O. Kiselman: The partial Legendre transformation for plurisubharmonic functions. Invent. Math. 49(1978), 137-148.

[L1] J.J. Loeb: Pseudo-convexité des ouverts invariants et convexite geodesique dans certains symetriques. Sém. Lelong-Skoda, Lect. Notes in Math. 1198(1983-1984), 172-190.

[L2] J.J. Loeb: Action d'une forme réele d'un groupe de Lie complexe sur les fonctions plurisousharmoniques. Ann. L'Inst. Fourier, fasc.4, t.35 (1985), 59-97.

[LR] D. Luna, R.W. Richardson: A generalization of the Chevalley restriction theorem. Duke math. J. 46(1979), no.3, 487-496.

[MFK] D. Mumford, J.C. Forgarty, F. Kirwan: Geometric invariant theory. Third edition. Ergebnisse der Mathematik und ihrer Grenzgebiete(2). Springer-Verlag, Berlin, 1994.

[SV] A.G. Sergeev, V.S. Vladimirov: Complex analysis in the future tube. in "Encyclopaedic of Mathematical Sciences", Vol. 8, Several complex Variables II, 1992.

[SW] R.F. Streater, A.S. Wightman: PCT, Spin and statistics, and all that. Benjamin Inc., New York–Amsterdam, 1964.

[SZ] A. Sergeev, Xiang–Yu Zhou: On invariant domains of holomorphy. Trudy Mat. Inst. Steklov. 203(1994), 159-172.

[V1] V.S. Vladimirov: Analytic functions of several complex variables and quantum field theory. Proc. of the Steklov Math. Inst., 135(1975) 68–80.

[V2] V.S. Vladimirov: Several complex variables in Mathematical physics. Sém. Lelong–skoda, Lect. Notes in Math. 919(1982), 358–386.

[V3] V.S. Vladimirov: Analytic functions of several complex variables and axiowatic quantum field theory. Actes, Congrés intern. Math., 1970, Tome 3, p.21 à 26.

[Zh1] Xiang–Yu Zhou: The extended future tube is a domain of holomorphy. Math. Res. Letters, 5(1998). 185–190.

[Zh2] Xiang–Yu Zhou: A proof of the extended future tube conjecture (in Russian). Izvestiya, Russian Academy of Sciences, Ser. Math., T.62, N.1, 1998, 211–224 = English version, Izvestiya: Mathematics 62:1 (1998) 201-213.

[Zh3] Xiang–Yu Zhou: Invariant version of Cartan's lemma and invariant domains (in Russian). To appear in Dokl. Russian Academy of Sciences.

[Zh4] Xiang–Yu Zhou: On orbit connectedness, orbit convexity, and envelopes of holomorphy. Izvestiya, Russian Academy of Sciences, Ser. Math., T.58, N.2, 1994, 195-205.

INSTITUTE OF MATHEMATICS, ACADEMIA SINICA, BEIJING, CHINA
E-mail address: xyzhou@math08.math.ac.cn

AMS/IP Studies in Advanced Mathematics
Volume 20, 2001

An Elementary Exposé on Equisingularities

Tzee-Char Kuo and Laurentiu Paunescu

ABSTRACT. Take two convergent power series in n variables. When should they be declared "equivalent"? This is the equisingularity problem. For the case of n real variables the notion of blow-analytic equivalence has led to interesting results. The main ideas and results can be fully explained in the two variables case. This we do in §2. The notion of blow-up plays a vital rôle, hence we give an elementary description of it in §1. This notes can be understood by non-specialists.

1. Blow-up

Take a Möbius band, $\mathbf{M}$, as obtained by rotating a line around the centre circle. A natural map $\beta : \mathbf{M} \to \mathbb{R}$ can be visualized as follows.

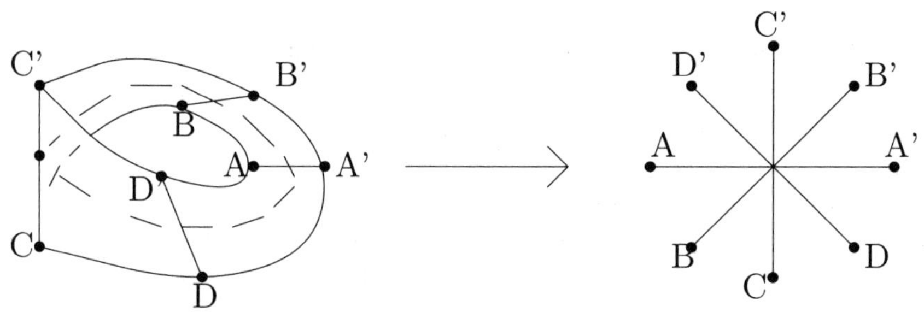

Every generating line is mapped to a line through the origin O. The entire centre circle collapses to the origin; elsewhere, β is bijective.

We call β (or more properly β^{-1}) the blow-up of $\mathbb{R}^2$ at the origin O, or with centre O. A neighbourhood of O is magnified, infinitely many times, to a neighbourhood of the centre circle. The precise definition is given below.

Consider the canonical line bundle over the real projective line

---

1991 *Mathematics Subject Classification.* 32S05, 32S15, 14H20.

The authors are supported by ARC Large Grant A699045676.
We like to thank Craig Jin for his help in computer graphing.

$$\pi_1 : \mathbf{E} \to \mathbb{R}P^1 \quad (l,x) \to l, \qquad \mathbf{E} := \{(l,x) | x \in l \in \mathbb{R}P^1\}.$$

Recall that a point, l, of $\mathbb{R}P^1$ is just a line in $\mathbb{R}^2$ through O.

Topologically, $\mathbf{E}$ is a Möbius band, $\mathbb{R}P^1$ is a circle. The restriction to $\mathbf{E}$ of the second projection

$$\beta := \pi_2|\mathbf{E} : \mathbf{E} \to \mathbb{R}^2, \qquad (l,x) \to x,$$

is called the blow-up of $\mathbb{R}^2$ at O.

Note that $\beta^{-1}(O) = \mathbb{R}P^1 \times \{O\}$ is the zero cross-section of the bundle, outside of which β is a bianalytic isomorphism.

Intuitively, $\mathbf{E}$ is obtained by separating lines in $\mathbb{R}^2$ through O and spread them over $\mathbb{R}P^1$ in a natural manner; what β does is just putting each line back to where it used to be in $\mathbb{R}^2$, as indicated earlier.

The above readily generalises to any 2-dimensional real analytic manifold $\mathbb{M}$. Take a point $P \in \mathbb{M}$. A neighbourhood of P is a copy of $\mathbb{R}^2$ with $P = O$, whence we can define the blow-up of $\mathbb{M}$ at P:

$$\beta : (\mathbf{M}, \beta^{-1}(P)) \to (\mathbb{M}, P),$$

which is an analytic isomorphism outside $\beta^{-1}(P)$.

We can also compose them. Take $\beta_n : \mathbf{M}_n \to \mathbf{M}_{n-1}, 1 \le n \le N, \mathbf{M}_0 := \mathbb{M}$, and consider the composition:

$$\tilde{\beta} = \beta_1 \circ \beta_2 \circ \cdots \circ \beta_N : \mathbf{M}_N \to \mathbb{M}.$$

We shall discuss the generalisations to higher dimensional, and the complex cases in §3. Now consider the map

$$\beta_1^* : \mathbb{R}^2 \to \mathbb{R}^2, \quad (X,Y) \to (x,y) = (X, XY).$$

Each horizontal line $Y = c$, c a constant, has image $y = cx$. The Y-axis ($X = 0$), collapses to the origin, while $X = c, c \neq 0$, has image $x = c$. The rectangle in the (X,Y)-plane is mapped to a sector around the x-axis, as indicated below.

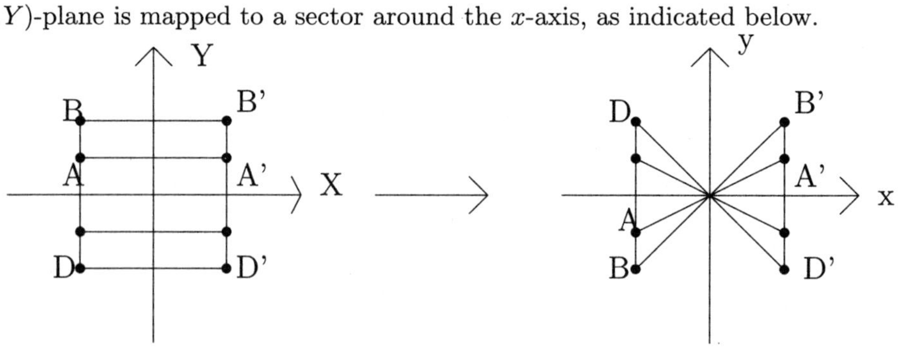

Philosophy: When an object lies in the sector, $(\beta_1^*)^{-1}$ would magnify it into an object in the rectangle, often making it simpler. For example, the cusp $y^2 = x^3$ is magnified to $X^2(Y^2 - X) = 0$, consisting of a parabola plus the Y-axis, as indicated below.

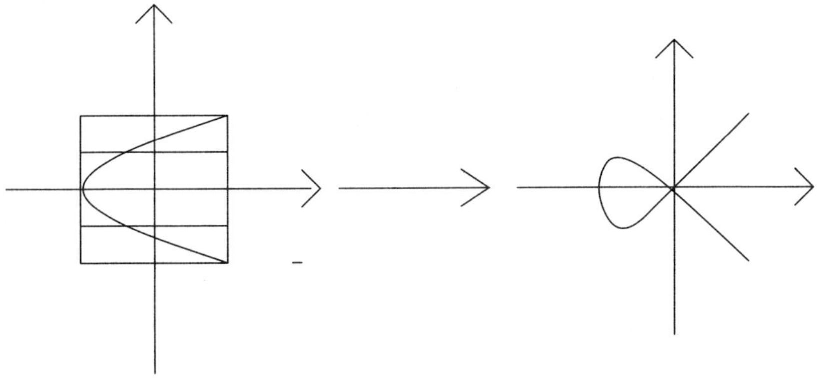

Objects that are tangent to the y-axis, however, need a different magnification:

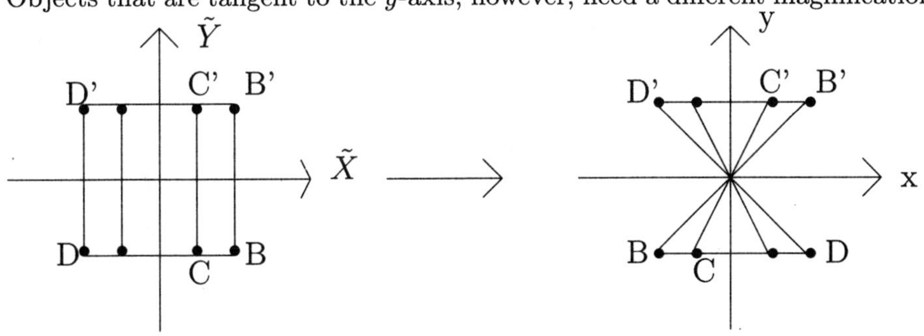

$$\beta_2^* : \mathbb{R}^2 \to \mathbb{R}^2, \quad (\tilde{X}, \tilde{Y}) \to (x, y) = (\tilde{X}\tilde{Y}, \tilde{Y}).$$

Gluing up the above two rectangles yields a neighbourhood of the centre circle of **E**. Using the language of Differential Topology, β_1^*, β_2^*, are two coordinate charts for the blow-up map β.

In order to study a given function $f(x, y)$, we often study $f \circ \beta_1^*$ and $f \circ \beta_2^*$, which can be simpler.

2. Blow-analytic equisingularity

Consider the familiar function of two real variables:

$$m(x, y) = \frac{x^2 y}{x^2 + y^2}, \qquad m(0, 0) = 0,$$

which is continuous, but not C^1.

Take the blow-up $\beta : \mathbf{M} \to \mathbb{R}^2$ introduced earlier, and then consider the composition $m \circ \beta$.

In either charts,

$$(m \circ \beta_1^*)(X, Y) = \frac{XY}{1 + Y^2}, \quad (m \circ \beta_2^*)(\tilde{X}, \tilde{Y}) = \frac{\tilde{X}^2 \tilde{Y}}{1 + \tilde{X}^2},$$

both are analytic functions.

Thus $m \circ \beta$ is an analytic function on $\mathbf{M}$.

A function $q : \mathbb{R}^2 \to \mathbb{R}$ is called *blow-analytic* if there exists a succession of blow-ups:

$$\tilde{\beta} := \beta_1 \circ \cdots \circ \beta_N : \mathbf{M_N} \to \mathbf{M_{N-1}} \to \cdots \to \mathbf{M_1} \to \mathbb{R}^2,$$

such that the composition $q \circ \tilde{\beta}$ is analytic on $\mathbf{M_N}$.

Let us also localize this notion.

Take a germ $q : (\mathbb{R}^2, 0) \to (\mathbb{R}, 0)$. We say q is *blow-analytic* if there exists a composite (as above) $\tilde{\beta}$, for which each β_i is the blow-up of $\mathbf{M_{i-1}}$ with centre a point in $(\beta_1 \circ \beta_2 ... \circ \beta_{i-1})^{-1}(0)$, such that $q \circ \tilde{\beta}$ is analytic in a neighbourhood of $\tilde{\beta}^{-1}(0)$.

An important example of blow-analytic function, found in [**12**], is

$$Q(x, y) = \frac{xy^5}{x^4 + y^6}$$

The readers are invited to find out for themselves how many blow-ups are needed to make Q analytic.

We introduce a more effective way in the following.

According to [**11**], we say that q is *arc analytic* if for any two convergent power series $x(t), y(t)$,we have also that $q(x(t), y(t))$ is a convergent power series.

The readers are invited to check, quite easily, that Q is arc analytic.

Consider a meromorphic function, i.e. a quotient

$$m(x, y) = \frac{n(x, y)}{d(x, y)},$$

where $n(x, y), d(x, y)$ are analytic.

An important fact is that $m(x, y)$ is arc-analytic if and only if it is blow-analytic.

This follows immediately from the celebrated Desingularization Theorem of Hironaka. It is also a special case of the Bierstone-Milman theorem on arc-analytic functions [**1**].

Following [**9**], [**10**], we call two germs of analytic functions $f, g : (\mathbb{R}^2, 0) \to (\mathbb{R}, 0)$ *blow-analytically equivalent* if there exists a local homeomorphism $h : (\mathbb{R}^2, 0) \to (\mathbb{R}^2, 0)$, such that $g \circ h = f$, and the components of h and h^{-1} are (germs of) blow-analytic functions.

In 1965, Whitney [**13**] discovered a family of germs,

$$w_t(x, y) = xy(x + y)(x - ty), \quad 0 < t < \infty,$$

parametrized by t, which consists of infinitely many different C^1-types.

This family, however, consists of only one blow-analytic type [**9**].

In 1998, Fukui [**2**], proved that if f and g are blow-analytically equivalent, then they have the same multiplicity. His proof is very short and elegant, making use of the Fukui invariant set $A(f)$. An effective method of computing the Fukui set has recently been found in [**6**].

In 1999, Fukui and Paunescu [**5**], have proved that a t-parametrised family of weighted homogeneous forms $w_t(x, y)$, having an isolated singularity at $(x, y) =$

$(0,0)$, consists of only one blow-analytic equivalence class. A typical example is:

$$x^2 + txy^2 + y^4, \quad t > 2.$$

3. Higher dimensional case

It is rather straightforward to generalize the notion of blow-up.

Take $\mathbb{K} = \mathbb{C}$ or $\mathbb{R}$. Consider $\mathbb{K}^n$ and the canonical $\mathbb{K}$-line bundle over the $\mathbb{K}$-projective space $\mathbb{K}P^{n-1}$:

$$\pi : \mathbf{E} \to \mathbb{K}P^{n-1}, \quad \pi(l, x) = l, \quad \mathbf{E} := \{(l, x) \in \mathbb{K}P^{n-1} \times \mathbb{K}^n | x \in l\}.$$

The restriction to $\mathbf{E}$ of the second projection

$$\beta := \pi_2 | \mathbf{E} : \mathbf{E} \to \mathbb{K}^n, \qquad \beta(l, x) = x,$$

is called the blow-up of $\mathbb{K}^n$ at the origin O.

Now, take a $\mathbb{K}$-analytic manifold, $\mathbf{M}$, say of dimension m, and a closed, smooth submanifold, $\mathbf{C}$, say of codimension n. Consider a tubular neighbourhood of $\mathbf{C}$, its normal sections can be considered as copies of $\mathbb{K}^n$. We can blow-up each copy at its origin to yield

$$\beta : (\tilde{\mathbf{M}}, \beta^{-1}(\mathbf{C})) \to (\mathbf{M}, \mathbf{C}),$$

which we call the blow-up of $\mathbf{M}$ with centre $\mathbf{C}$.

The blow-analytic equivalence readily generalises to the higher dimensional real analytic case ($\mathbb{K} = \mathbb{R}$), but not to the complex case.

For further details see [**3**].

A number of puzzling examples have been discovered in recent years.

The first was by Koike in [**8**]. He discovered a blow-analytic homeomorphism on $\mathbb{R}^3$ which is not Lipschitz.

In [**12**] the following simpler example is constructed. Consider

$$h : \mathbb{R}^3 \to \mathbb{R}^3, \qquad h(X, Y, Z) = (X, Y, Z - 2\frac{XY^5}{X^4 + Y^6}),$$

whose inverse is

$$h^{-1} : \mathbb{R}^3 \to \mathbb{R}^3, h^{-1}(x, y, z) = (x, y, z + 2\frac{xy^5}{x^4 + y^6}).$$

The component functions are arc-analytic whence blow-analytic.

Take a smooth arc $\lambda(t) = (t^3, t^2, t)$. Then $h \circ \lambda(t) = (t^3, t^2, 0)$ is a cusp. Hence h cannot be Lipschitz.

In [**7**], some blow-analytic homeomorphisms on $\mathbb{R}^2$ are constructed which can transform singular curves to smooth ones; in particular, they are not Lipschitz.

In a forthcoming paper [**4**], a more systematic construction of these kind of homeomorphisms is developed.

Much research remains to be accomplished in this dirrection.

References

[1] E.Bierstone and P.D.Milman, *Arc-analytic functions*, Invent. math., **101** (1990), 411-424.

[2] T. Fukui, *Seeking invariants for blow-analytic equivalence*, Compositio Math. **105** (1997), 95–107.

[3] T. Fukui,S. Koike, T.-C.Kuo, *Blow-analytic equisingularities, properties, problems and progress*, in "Real analytic and algebraic singularities", Pitman Research Notes in Mathematics Series, **381**, 1997, Longman, 8–29.

[4] T. Fukui, T.-C.Kuo, L. Paunescu, *Remarks on blow-analytic homeomorphisms*, in preparation.

[5] T. Fukui, L. Paunescu, *Modified analytic trivialization for weighted homogeneous function-germs*, J. Math. Soc. Japan (to appear).

[6] S. Izumi,S. Koike, T.-C.Kuo, *A formula for the Fukui invariant and stability*, in preparation.

[7] M. Kobayashi, T.-C.Kuo, *On Blow-analytic equivalence of embedded curve singularities*, in "Real analytic and algebraic singularities", Pitman Research Notes in Mathematics Series, **381**, 1997, Longman, 30–37.

[8] S. Koike, *On strong C^0- equivalence of real analytic functions* J. Math. Soc. Japan **45** (1993), 313–320.

[9] T.-C.Kuo, *The modified analytic trivialization of singularities*, J. Math. Soc. Japan **32** (1980), 605–614.

[10] T.-C.Kuo, *On classification of real singularities*, Invent. math., **82** (1985), 257–262.

[11] K. Kurdyka, *Ensembles semi-algébriques symétriques par arcs*, Math. Ann., **282** (1988), 445-462.

[12] L.Paunescu, *An example of blow-analytic homeomorphism*, in "Real analytic and algebraic singularities", Pitman Research Notes in Mathematics Series, **381**, 1997, Longman, 62–63.

[13] H.Whitney, *Local properties of analytic varieties*, in "A Symposium in Honor of M.Morse", (Ed. by S.S.Cairns), Princeton Univ. Press. 1965, 205–244.

T.C. Kuo: School of Mathematics and Statistics, University of Sydney, Sydney, NSW, 2006, Australia.
E-mail: tck@maths.usyd.edu.au.

Laurentiu Paunescu: School of Mathematics and Statistics, University of Sydney, Sydney, NSW, 2006, Australia.
E-mail: laurent@maths.usyd.edu.au.

AMS/IP Studies in Advanced Mathematics
Volume 20, 2001

A proof of the strong Kervaire invariant in dimension 62

Wen-Hsiung Lin

1. Introduction

Let $\pi_* = {}_2\pi_*^s$ denote the 2-primary stable homotopy groups of spheres. Whether or not there are elements $\theta_i \in \pi_{2^{i+1}-2}$ with non-zero Kervaire invariant is an important problem in the study of π_* ([3], [5]). The first three cases of this problem are trivial: $\theta_1 = \eta^2, \theta_2 = \nu^2, \theta_3 = \sigma^2$ where η, ν and σ are the three Hopf classes. Mahowald and Tangora prove in [8] that θ_4 exists. In fact, they have shown that $\pi_{30} \cong \mathbf{Z}/2$, generated by θ_4. Barratt, Jones and Mahowald prove in [4] that θ_5 also exists by using a lot of information about π_k for k up to 48. For $i \geq 6$, whether there are θ_i is still open.

The strong Kervaire invariant in the title referes to any θ_j which has order 2 (see [3]). The cases $\theta_1, \theta_2, \theta_3, \theta_4$ above all have order 2. The θ_5 constructed in [4] has not been shown yet to satisfy $2\theta_5 = 0$. Kochman recorded in his book [7] a proof, due to Mahowald, of the following result.

Theorem 1.1. *There exists a θ_5 with $2\theta_5 = 0$.*

In [7] this is deduced from the following result.

Theorem 1.2. $\theta_4^2 = 0$.

That (1.2) implies (1.1) follows from the following general result due to Barratt, Jones and Mahowald [3]. Throughout this paper all cohomology groups have mod 2 coefficients.

Theorem 1.3. *For $i \geq 5$ if θ_{i-1} exists and satisfies $2\theta_{i-1} = 0$ then there exists a stable complex $X = S^n \cup_{\theta_{i-1}^2} e^{n+2^{i+1}-3} \cup_\eta e^{n+2^{i+1}-1} \cup_{2\iota} e^{n+2^{i+1}}$ such that*

$$Sq^{2^{i+1}} : H^n(X) = \mathbf{Z}/2 \longrightarrow H^{n+2^{i+1}}(X) = \mathbf{Z}/2 \quad \text{is non-zero.}$$

Thus if θ_{i-1} further satisfies $\theta_{i-1}^2 = 0$ then this implies the existence of a stable complex $Y = S^n \cup e^{n+2^{i+1}-1} \cup_{2\iota} e^{n+2^{i+1}}$ such that

$$Sq^{2^{i+1}} : H^n(Y) = \mathbf{Z}/2 \longrightarrow H^{n+2^{i+1}}(Y) = \mathbf{Z}/2 \quad \text{is non-zero}$$

and this is equivalent to the existence of a θ_i with $2\theta_i = 0$.

1991 *Mathematics Subject Classification.* 55Q45, 55T15.

Supported by N.S.C. of Taiwan and the Foundation for Advancement of Outstanding Scholarship.

(1.2) is obtained in [7] by a lot of information about π_k, particularly for k from 48 to 60, and the proof is non-trivial in the sense that it is not easy to keep track the details of each step.

In this paper we will give a proof of Theorem 1.1 which perhaps is simpler. This proof avoids showing $\theta_4^2 = 0$ but will depend on the first conclusion of Theorem 1.3 for $i = 5$. Except for this, the new proof is based entirely only upon

(∗) the knowledge of π_k for k up to 48

before Kochman's gigantic calculations in [7]. (∗) refers to the knowledge provided by the authors Barratt, Jones, Mahowald, May, Mimura, Tangora and Toda in [2], [4], [8], [11], [12], [13], [17], [18]. These publications together with Bruner's work [6] give a rather complete picture of π_k for k up to 48 at least for the groups structure. By (∗) we mean the knowledge of these groups structure together with some well known composition relations. Among these relations those that we need in this paper will be given in the beginning of §2.

The main result in the new proof is the following

Theorem 1.4. $0 \in < \theta_4, 2\iota, \sigma^2 > \subseteq \pi_{45}$.

Here $< \ , \ , \ >$ denotes Toda bracket.

(1.4) will be proved in §2. The remainder of this section is devoted to deducing (1.1) from (1.4). Before deducing (1.1) we make some conventions about cell diagrams that will also be used in the proof of (1.4).

All spaces to be considered are stable objects and will be localized at prime 2. Thus, for example, S^n will denote sphere spectrum in stable dimension n. For $\beta \in \pi_*$ any stable map representing β will also be denoted by β. In addition to S^n, we will also consider stable complexes of the form $X = S^n \cup_\beta e^{n+\ell}$ for $\beta \in \pi_{\ell-1}$. To simplify, following [4], we will use nodes and edges to denote S^n, X by $\underset{n}{\circ}$, $\underset{n}{\circ}\overset{\beta}{\text{——}}\underset{n+\ell}{\circ}$ as their cell diagrams. In introducing further diagrams we will omit indicating stable dimensions $n, n+\ell$ on these nodes. For $\alpha, \beta \in \pi_*$ with $\alpha\beta = 0$, $\begin{array}{l}\circ \\ \uparrow \alpha \\ \circ \overset{\beta}{\text{——}} \circ\end{array}$ will mean an extension of α where α is a map with source space the bottom sphere of $\circ\overset{\beta}{\text{——}}\circ$. We will use $\bar{\alpha}$ or $\tilde{\alpha}$ to denote such an extension and indicate it by the diagram $\uparrow \bar{\alpha} \quad \begin{array}{l}\circ \\ \uparrow \alpha \\ \circ \overset{\beta}{\text{——}} \circ\end{array}$. Thus the source space of $\bar{\alpha}$ is $\circ\overset{\beta}{\text{——}}\circ$. Dually, if $\begin{array}{r}\circ \overset{\alpha}{\text{——}} \circ \\ \uparrow \beta \\ \circ\end{array}$ is a coextension of β then we also use $\bar{\beta}$ or $\tilde{\beta}$ to denote such a coextension and indicate it by the diagram $\uparrow \bar{\beta} \quad \begin{array}{r}\circ \overset{\alpha}{\text{——}} \circ \\ \uparrow \beta \\ \circ\end{array}$. If, furthermore, $\gamma \in \pi_*$ is some other element satisfying $\beta\gamma = 0$ and $0 \in < \alpha, \beta, \gamma >$ so that one can form $\begin{array}{r}\circ \overset{\alpha}{\text{——}} \circ \\ \uparrow \beta \\ \circ \overset{\gamma}{\text{——}} \circ\end{array}$

then this map is denoted by $\bar{\bar{\beta}}$. Thus the restriction of $\bar{\bar{\beta}}$ to the bottom sphere of $\circ \xrightarrow{\gamma} \circ$ is some coextension $\bar{\beta}$ of β and the portion $\begin{array}{c} \circ \\ \uparrow \beta \\ \circ \xrightarrow{\gamma} \circ \end{array}$ (obtained by pinching the bottom sphere of $\circ \xrightarrow{\alpha} \circ$) is some extension $\tilde{\beta}$ of β.

Theorem 1.1 is proved as follows. Recall ([8]) that the composite

$$\begin{array}{ccccc} & 30 \circ \xrightarrow{2\iota} \circ\, 31 & & & \\ \bar{\bar{\sigma}}^2 \uparrow & \uparrow \sigma^2 & & & \\ & 45 \quad \circ \xrightarrow{2\iota} \circ\, 46 & & & \\ \bar{\sigma}^2 \uparrow & \uparrow \sigma^2 & & & \\ & \circ\, 60 & & & \end{array}$$

is the composite $\underset{60}{\circ} \xrightarrow{\theta 4} \underset{30}{\circ} \hookrightarrow \underset{30}{\circ} \xrightarrow{2\iota} \underset{31}{\circ}$ no matter what $\bar{\bar{\sigma}}^2, \bar{\sigma}^2$ are chosen. Here one can form $\begin{array}{c} \circ \xrightarrow{2\iota} \circ \\ \uparrow \sigma^2 \\ \circ \xrightarrow{2\iota} \circ \end{array}$ because $\sigma^2\eta \in < 2\iota, \sigma^2, 2\iota >$ is zero ([18]) and also note that $< \sigma^2, 2\iota, \sigma^2 > \subset \pi_{29} = 0$. So the composite

$$(**) \qquad \begin{array}{ccc} & 0 \quad \circ & \\ \bar{\theta}_4 \uparrow & \uparrow \theta_4 & \\ & 30 \quad \circ \xrightarrow{2\iota} \circ\, 31 & \\ \bar{\bar{\sigma}}^2 \uparrow & \uparrow \sigma^2 & \\ & 45 \quad \circ \xrightarrow{2\iota} \circ\, 46 & \\ \bar{\sigma}^2 \uparrow & \uparrow \sigma^2 & \\ & \circ\, 60 & \end{array}$$

is θ_4^2 for any choice of $\bar{\theta}_4, \bar{\bar{\sigma}}^2$ and $\bar{\sigma}^2$. Since the composite

$$\begin{array}{cc} & 30 \circ \xrightarrow{2\iota} \circ\, 31 \\ \tilde{\sigma}^2 \uparrow & \uparrow \sigma^2 \\ & \circ\, 45 \\ 2\iota \uparrow & \uparrow 2\iota \\ & \circ\, 45 \end{array}$$

is always zero for any choice of $\tilde{\sigma}^2$, from Theorem 1.4 we see $\bar{\bar{\sigma}}^2$ and $\bar{\theta}_4$ can be chosen so that $\bar{\theta}_4\, \bar{\bar{\sigma}}^2\, |S^{45}$ is zero. Thus the composite in $(**)$ for these $\bar{\bar{\sigma}}^2$ and $\bar{\theta}_4$ is the composite $\underset{60}{\circ} \xrightarrow{\sigma^2} \underset{46}{\circ} \xrightarrow{\xi} \underset{0}{\circ}$ for some $\xi \in \pi_{46}$, that is, $\theta_4^2 = \xi\sigma^2$. We recall from [18] that $0 \in < \sigma^2, \eta, 2\iota >$. This implies that there a map $\underset{60}{\circ} \xrightarrow{\eta} \underset{62}{\circ} \xrightarrow{2\iota} \underset{63}{\circ} \xrightarrow{\hat{\sigma}^2} \underset{46}{\circ}$ such that $\hat{\sigma}^2|S^{60} = \sigma^2$. Let $\bar{h}_6$ be the composite $\underset{60}{\circ} \xrightarrow{\eta} \underset{62}{\circ} \xrightarrow{2\iota} \underset{63}{\circ} \xrightarrow{\hat{\sigma}^2} \underset{46}{\circ} \xrightarrow{\xi} \underset{0}{\circ}$. Then $\bar{h}_6|S^{60} = \theta_4^2$. Let Y be the mapping cone of $\bar{h}_6$. The cell structure of Y is of the form $\underset{0}{\circ}\overset{\theta_4^2}{\text{---}}\underset{61}{\circ}\overset{\eta}{\text{---}}\underset{63}{\circ}\overset{2\iota}{\text{---}}\underset{64}{\circ}$. It is clear that $\bar{h}_6$ does not carry the functional Sq^{64}, that is,

$$Sq^{64}: H^0(Y) = \mathbf{Z}/2 \longrightarrow H^{64}(Y) = \mathbf{Z}/2 \quad \text{is zero.}$$

On the other hand, Theorem 1.3 shows that there is a complex X of the same form $\underset{0}{\circ}\overset{\theta_4^2}{\text{---}}\underset{61}{\circ}\overset{\eta}{\text{---}}\underset{63}{\circ}\overset{2\iota}{\text{---}}\underset{64}{\circ}$ such that

$$Sq^{64}: H^0(X) = \mathbf{Z}/2 \longrightarrow H^{64}(X) = \mathbf{Z}/2 \quad \text{is non-zero.}$$

Since the first two cells of X and Y are the same it follows that there is a complex Z of the form $\underset{0}{\circ}\text{---}\underset{63}{\circ}\overset{2\iota}{\text{---}}\underset{64}{\circ}$ such that

$$Sq^{64}: H^0(Z) = \mathbf{Z}/2 \longrightarrow H^{64}(Z) = \mathbf{Z}/2 \quad \text{is non-zero.}$$

This implies the existence of a θ_5 with $2\theta_5 = 0$ (see [3]). This proves Theorem 1.1.

The key point of the proof above is to deduce $\theta_4^2 = \xi\sigma^2$ from (1.4). Kochman proves in [7] the following non-trivial result.

Theorem 1.5. $\eta\pi_{45} = \pi_{46}$.

If we use this result then, since $\xi \in \pi_{46}$, say $\xi = \xi'\eta$, it follows that $\theta_4^2 = \xi\sigma^2 = \xi'\eta\sigma^2 = 0$, and so (1.1) follows. The proof of (1.5) in [7], however, is also based on a lot of non-trivial calculations, and our purpose here is to show the existence of a θ_5 with $2\theta_5 = 0$ without results from such calculations.

2. Proof of Theorem 1.4

We will look at π_* from the viewpoint of the Adams spectral sequence which has E_2 term $E_2^{s,t} = Ext_A^{s,t}(\mathbf{Z}/2, \mathbf{Z}/2)$, where A denotes the mod 2 Steenrod algebra, and will freely use the knowledge of $Ext_A^{s,t}(\mathbf{Z}/2, \mathbf{Z}/2)$ for $t - s \leq 49$ together with the knowledge of the differentials of the spectral sequence in this range as described in the chart of [15] which are summarized from the works [2], [4], [6], [8], [11], [16] and [17]. If $a \in E_2$ is an infinite cycle then we use the usual notation that $\{a\} \subseteq \pi_*$ denotes the set of homotopy elements detected by a.

Besides elementary composition relations such as $2\eta = 0$, $2\nu^2 = 0$ we will need the following relations.

$$\text{(2.1)} \qquad \begin{array}{l} (1)\ \ \sigma^2\eta = 0, \quad (2)\ \ \bar{\epsilon}\sigma = 0, \quad (3)\ \ \eta^2\bar{\kappa} = \bar{\epsilon}\kappa \neq 0, \\ (4)\ \ \theta_4\kappa = 0, \quad (5)\ \ \theta_4\eta^2 = 0. \end{array}$$

Here $\kappa, \bar{\epsilon}$ are Toda's elements ([18]) ($\bar{\epsilon}$ here is ϵ in [18]), $\bar{\kappa}$ is Mimura and Toda's element ([12]) which are detected respectively by d_0, c_0 and g_1. (1) and (2) are proved in [18]. (4) is proved in [4]. To see (5) we first note that $\pi_{32} \cong \mathbf{Z}/2 \oplus \mathbf{Z}/2 \oplus \mathbf{Z}/2 \oplus \mathbf{Z}/2$ on $\gamma_1 \in \{h_5 h_1\}$, $\gamma_2 \in \{d_1\}$, $\gamma_3 \in \{q\}$, $\gamma_4 \in \{P^3 c_0\}$ and that $\eta\gamma_3 \neq 0$, $\eta\gamma_4 \neq 0$ are linearly independent in π_{33} ([2]). Let $\theta_4\eta^2 = \sum_{i=1}^{4} \delta_i\gamma_i$ where $\delta_i = 0$ or 1. It is clear that $AF(\theta_4\eta^2) \geq 5$ where AF stands for "*Adams filtration*". So $\delta_1 = \delta_2 = 0$. Since $\theta_4\eta^3 = 4\nu\theta_4 = 0$ and $\eta\gamma_3 \neq 0$, $\eta\gamma_4 \neq 0$ are linearly independent it follows that $\delta_3 = \delta_4 = 0$ also. Thus $\theta_4\eta^2 = 0$. This proves (5). For (3) we note that $\bar{\epsilon}\kappa \neq 0$ is proved in [13]. It is easy to see that this element is detected by $P^1 d_0$. The relation $\eta^2\bar{\kappa} = \bar{\epsilon}\kappa$ follows from the computations on π_k for $k \leq 22$ in [12], [13], [18]. A proof of it can be found in [8].

What we need from (2.1)(2), (3) is to prove the following where the cell diagram is as described in §1.

Lemma 2.2. *(1) $< \nu^2, 2\iota, \sigma^2 >$ consists of σ^3 and 0.*
(2) Consider the composite

$$\begin{array}{ccccc} & & \circ & & \\ \bar{\nu^2} & \uparrow & \uparrow \nu^2 & & \\ & & \circ & \xrightarrow{2\iota} & \circ \\ \bar{\sigma^2} & \uparrow & & & \uparrow \sigma^2 \\ & & & & \circ . \end{array}$$

Then $\bar{\nu^2}\bar{\sigma^2}$ is σ^3 or zero depending on the mapping cone $C_{\bar{\nu^2}}$ has zero Sq^8 or not in cohomology no matter what coextension $\bar{\sigma^2}$ is chosen.

PROOF: $\pi_{21} \cong \mathbf{Z}/2 \oplus \mathbf{Z}/2$ on σ^3 and $\eta\bar{\kappa}$ ([13]). Given any $\psi \in < \nu^2, 2\eta, \sigma^2 > \subset \pi_{21}$. Then $\psi = \delta_1\sigma^3 + \delta_2\eta\bar{\kappa}$ where $\delta_i = 0$ or 1. The set $< \eta, \nu^2, 2\iota >$ is the set $\{\bar{\epsilon}, \bar{\epsilon} + \sigma\eta\}$. So $\eta < \nu^2, 2\iota, \sigma^2 > = < \eta, \nu^2, 2\iota > \sigma^2 = 0$ by (2.1)(1), (2). Thus $\eta\psi = 0$. This implies $\delta_2\eta^2\bar{\kappa} = \delta_2\bar{\epsilon}\kappa = 0$. Since $\bar{\epsilon}\kappa \neq 0$ it follows that $\delta_2 = 0$. So $\psi = \delta_1\sigma^3$. This shows $< \nu^2, 2\iota, \sigma^2 >$ consists of σ^3 and 0 since σ^3 lies in the indeterminacy of $< \nu^2, 2\iota, \sigma^2 >$. This proves (1). To show (2) let $\psi = \bar{\nu^2}\bar{\sigma^2}$ be as in the diagram above. Since the mapping cone $C_{\bar{\sigma^2}}$ always has non-zero Sq^{16} in cohomology for any coextension $\bar{\sigma^2}$ (by Adams' Theorem [1]) it follows from Moss' Theorem ([14]) that $\bar{\nu^2}\bar{\sigma^2}$ is detected by $h_3^3 = h_4h_2^2$ if $C_{\bar{\nu^2}}$ has zero Sq^8 in cohomology, and so must be σ^3. Otherwise, $\psi = \bar{\nu^2}\bar{\sigma^2}$ is zero. This proves Lemma 2.2.

Now we prove Theorem 1.4. We begin by recalling ([4], [17]) that

$$\pi_{45} \cong \mathbf{Z}/16 \oplus \mathbf{Z}/2 \oplus \mathbf{Z}/2 \oplus \mathbf{Z}/2$$

and that one can choose $\alpha \in \{h_4^3\}, \beta \in < 2\iota, \theta_4, \kappa > \cap \{h_5d_0\}, \gamma \in \{g_2\} \subset \pi_{44}$, $\delta \in \{w\}$ such that $\alpha, \beta - 2\alpha, \eta\gamma, \delta$ are the generators of the cyclic groups in the decomposition in that order. Note that $4\alpha \in \{h_0h_5d_0\}$. The following is clear.

(2.3) For $\xi \in \pi_{45}$ let $\xi = \lambda_1\alpha + \lambda_2(\beta - 2\alpha) + \lambda_3\eta\gamma + \lambda_4\delta$ where $0 \leq \lambda_1 \leq 15$, $\lambda_i = 0$ or 1 for $2 \leq i \leq 4$. Then
$\xi \in \{h_4^3\}$ for $\lambda_1 \equiv 1 \pmod 2$,
$\xi \in \{h_4^3h_0\}$ for $\lambda_1 \equiv 2 \pmod 4$, $\lambda_2 = 0$ or $\lambda_1 \equiv 0$ (mod 4), $\lambda_2 = 1$,
$\xi \in \{\lambda_2h_5d_0 + \lambda_3g_2h_1\}$ for $\lambda_1 \equiv 2 \pmod 4$, $\lambda_2 = 1$ or $\lambda_1 \equiv 0 \pmod 4$, $\lambda_2 = 0$,
$\xi \in \{h_0h_5d_0\}$ for $\lambda_1 \equiv 4 \pmod 8$, $\lambda_2 = \lambda_3 = 0$,
$\xi \in \{h_0^2h_5d_0\}$ for $\lambda_1 = 8$, $\lambda_2 = \lambda_3 = 0$,
$\xi \in \{\lambda_4w\}$ for $\lambda_1 = \lambda_2 = \lambda_3 = 0$.

Now given any $\phi \in < \theta_4, 2\iota, \sigma^2 > \subseteq \pi_{45}$. It is easy to see that $AF(\phi) \geq 4$. So, by (2.3),
(1) $\phi = 2\epsilon_1\alpha + \epsilon_2(\beta - 2\alpha) + \epsilon_3\eta\gamma + \epsilon_4\delta$ where $0 \leq \epsilon_1 \leq 7$, $\epsilon_i = 0$ or 1 for $2 \leq i \leq 4$.
We will show

(1.4)′ (1) $2\epsilon_1 = 8\epsilon_1'$ where $\epsilon_1' = 0$ or 1,
(2) $\epsilon_2 = 0$,
(3) $\epsilon_3 = 0$,
(4) $\epsilon_4 = 0$.

Granting this we conclude $\phi = 8\epsilon_1'\alpha$.

Lemma 2.4. *$\rho\theta_4 = 8\alpha$ where $\rho = \beta_{15}$ is the generator of $imJ_{15} \cong \pi_{15} \cong \mathbf{Z}/32$.*

Thus $\phi = 8\alpha = \rho\theta_4$ or zero according as $\epsilon_1' = 1$ or 0. This will prove Theorem 1.4 since $\rho\theta_4$ lies in the indeterminacy of $< \theta_4, 2\iota, \sigma^2 >$.

We recall the following proved in [17].

Lemma 2.5. *(1)* $\rho\theta_4 \in \{h_0^2 h_5 d_0\}$. *(2)* $\eta^2\delta \in \{P^1 u\}$; *so* $\eta^2\delta \neq 0$.

PROOF OF LEMMA 2.4: Let $\rho\theta_4 = \lambda_1\alpha + \lambda_2(\beta - 2\alpha) + \lambda_3\eta\gamma + \lambda_4\delta$ where $0 \leq \lambda_1 \leq 15$, $\lambda_i = 0$ or 1 for $2 \leq i \leq 4$. By (2.5)(1), $\rho\theta_4 \in \{h_0^2 h_5 d_0\}$. So $\lambda_1 = 8, \lambda_2 = \lambda_3 = 0$ by (2.3). Thus $\rho\theta_4 = 8\alpha + \lambda_4\delta$. By (2.5)(2), $\eta^2\delta \neq 0$. Since $\eta^2\rho\theta_4 = 0$, $\eta^2 8\alpha = 0$ it follows that $\lambda_4 = 0$. This proves Lemma 2.4.

We proceed to prove (1.4)′. (1.4)′(1) follows from (2.3) and the fact $2 < \theta_4, 2\iota, \sigma^2 >= \theta_4 < 2\iota, \sigma^2, 2\iota >= \theta_4\left(\sigma^2\eta + 2\pi_{15}\right) = 0$. Note this shows $2\phi = 0$ for any $\phi \in < \theta_4, 2\iota, \sigma^2 >$. (1) thus becomes

$$(2) \qquad \phi = 8\epsilon_1'\alpha + \epsilon_2(\beta - 2\alpha) + \epsilon_3\eta\gamma + \epsilon_4\delta.$$

We want to show for (1.4)′(2) that $\epsilon_2 = 0$. We prove this by contradiction. Assume $\epsilon_2 = 1$. So $\phi = 8\epsilon_1'\alpha + (\beta - 2\alpha) + \epsilon_3\eta\gamma + \epsilon_4\delta$ which implies $\nu\phi = \nu\beta - 2\nu\alpha + \epsilon_4\nu\delta$. Recall $\alpha \in \{h_4^3\}$. In the Adams spectral sequence for π_* we have $h_4^3 h_2 = 0$ and the differentials $d_2(h_3 c_2) = h_0 h_2 g_2 \neq 0$, $d_2(h_5 e_0) = h_1^2 h_5 d_0 \neq 0$ ([4]). So $AF(\nu\alpha) \geq 6$ which implies $AF(2\nu\alpha) \geq 7$. Clearly $AF(\nu\delta) \geq 10$. $\beta \in \{h_5 d_0\}$ implies $\nu\beta$ is detected by $h_2 h_5 d_0 \neq 0$. Thus $\nu\phi = \nu\beta - 2\nu\alpha + \epsilon_4\nu\delta$ is detected by $h_2 h_5 d_0 \neq 0$. We also have the differential $d_2(h_5 f_0) = h_0 h_2 h_5 d_0 \neq 0$ ([4]). Since $2\nu\phi = 0$, this implies that any extension

$$\begin{array}{ccccc} & 0 & \circ & & \\ \overline{\nu\phi}\uparrow & & \uparrow \nu\phi & & \\ & 48 & \circ & \xrightarrow{2\iota} & \circ\ 49 \end{array}$$

of $\nu\phi$ must be detected by $h_5 f_0$ on the top cell e^{49}. But $2\phi = 0$ which shows that the composite

$$\begin{array}{ccccc} & 0 & \circ & & \\ \uparrow \nu & & \uparrow \nu & & \\ & 3 & \circ & & \\ \uparrow \bar{\phi} & & \uparrow \phi & & \\ & 48 & \circ & \xrightarrow{2\iota} & \circ\ 49 \end{array}$$

is also an extension of $\nu\phi$ where $\bar{\phi}$ is any extension of ϕ. Clearly such an extension is not detected by $h_5 f_0$ on the cell e^{49}. This contradiction proves $\epsilon_2 = 0$.

Thus (2) becomes

$$(3) \qquad \phi = 8\epsilon_1'\alpha + \epsilon_3\eta\gamma + \epsilon_4\delta.$$

We want to show for (1.4)′(3) that $\epsilon_3 = 0$. Again we prove this by contradiction. Assume $\epsilon_3 = 1$. So $\phi = 8\epsilon_1'\alpha + \eta\gamma + \epsilon_4\delta$. Then, by (2.3), $\phi \in \{g_2 h_1\}$ which implies $\sigma\phi$ is detected by $h_3 g_2 h_1 = h_5 g_1 h_1 \neq 0$. This is the only situation in the whole proof that we consider $Ext_A^{s,t}(\mathbf{Z}/2, \mathbf{Z}/2)$ for $t - s$ beyond 49. Thus $AF(\sigma\phi) = 6$. Note that $\sigma\phi \in < \theta_4, 2\iota, \sigma^2 > \sigma \subseteq < \theta_4, 2\iota, \sigma^3 >$. A contradiction will arise if we can show

(*) any element in $< \theta_4, 2\iota, \sigma^3 >$ has $AF \geq 7$.

Consider the composite

$$(**)\qquad \begin{array}{lccccc}
 & 0\circ & & & & \\
\bar{\nu^2}\ \uparrow & \uparrow \nu^2 & & & & \\
 & 6\circ & \xrightarrow{2\iota} & \circ\, 7 & & \\
\overline{\overline{\sigma^2}}\ \uparrow & & & \uparrow \sigma^2 & & \\
 & & 21 & \circ & \xrightarrow{2\iota} & \circ\, 22 \\
\bar{\theta}_4\ \uparrow & & & & & \uparrow \theta_4 \\
 & & & & & \circ\, 52
\end{array}$$

for some $\bar{\theta}_4, \overline{\overline{\sigma^2}}$ and $\bar{\nu^2}$ where $\bar{\nu^2}$ is not detected by h_3 on the cell e^7. That we can choose such a $\bar{\nu^2}$ is clear. By (2.2) the composite

$$\begin{array}{lccc}
 & 0\circ & & \\
\bar{\nu^2}\ \uparrow & \uparrow \nu^2 & & \\
 & 6\circ & \xrightarrow{2\iota} & \circ\, 7 \\
\bar{\sigma^2}\ \uparrow & & & \uparrow \sigma^2 \\
 & & & \circ\, 21
\end{array}$$

is σ^3 where $\bar{\sigma^2} = \overline{\overline{\sigma^2}}\ |S^{21}$. So the composite $\bar{\nu^2}\ \overline{\overline{\sigma^2}}\ \bar{\theta}_4$ in (**) is an element of $< \theta_4, 2\iota, \sigma^3 >$. The composite

$$\begin{array}{lccc}
 & 7\circ & & \\
\bar{\sigma^2}\ \uparrow & \uparrow \sigma^2 & & \\
 & 21\circ & \xrightarrow{2\iota} & \circ\, 22 \\
\bar{\theta}_4\ \uparrow & & & \uparrow \theta_4 \\
 & & & \circ\, 52
\end{array}$$

which is a portion of (**) is some element ϕ' in $< \theta_4, 2\iota, \sigma^2 >$. ϕ' may not be equal to ϕ, but must be of the form as in (3) since what we have proved up to (3) is that any element in $< \theta_4, 2\iota, \sigma^2 >$ must be of the form in (3). So $AF(\phi') \geq 5$. The composite in (**) is the composite

$$\begin{array}{lccc}
 & 0\circ & & \\
\bar{\nu^2}\ \uparrow & \uparrow \nu^2 & & \\
 & 6\circ & \xrightarrow{2\iota} & \circ\, 7 \\
\bar{\phi}'\ \uparrow & & & \uparrow \phi' \\
 & & & \circ\, 52
\end{array}$$

for some $\bar{\phi}'$. Since $AF(\phi') \geq 5$ and $Ext_A^{s,s+46}(\mathbf{Z}/2, \mathbf{Z}/2) = 0$ for $s \leq 5$ it follows that $\bar{\phi}'$ has $AF \geq 5$ in the ASS for $\pi_*^s\left(S^6 \cup_{2\iota} e^7\right)$. Since $\bar{\nu^2}$ does not carry h_3 on e^7, it has $AF \geq 2$ in the ASS for the stable groups $\left[S^6 \cup_{2\iota} e^7, S^0\right]_*$. These imply that
$\bar{\nu^2}\ \overline{\overline{\sigma^2}}\ \bar{\theta}_4 = \bar{\nu^2}\bar{\phi}'$, which is an element of $< \theta_4, 2\iota, \sigma^3 >$, has $AF \geq 7$. The indeterminacy of $< \theta_4, 2\iota, \sigma^3 >$ is $\sigma^3\pi_{31} + \theta_4\pi_{22}$. Elements of π_{31} are of the form $\delta_1\eta\theta_4 + \zeta$ where $\delta_1 = 0$ or 1 and ζ has $AF \geq 7$. So $AF\left(\sigma^3\pi_{31}\right) \geq 8$. $\pi_{22} \cong \mathbf{Z}/2 \oplus \mathbf{Z}/2$ on $\nu\bar{\sigma}$ and $\bar{\epsilon}\kappa$ where $\bar{\sigma} \in \{c_1\}$. We have $\bar{\epsilon}\kappa \in \left\{P^1 d_0\right\}$. So $AF\left(\bar{\epsilon}\kappa\right) = 8$. $\nu\bar{\sigma}$ is detected by $h_2 c_1$. Since $h_4^2 h_2 c_1 = 0$, it follows that $AF\left(\theta_4\nu\bar{\sigma}\right) \geq 7$. Thus elements in $\theta_4\pi_{22}$ all have $AF \geq 7$. We conclude that any element in the indeterminacy $\sigma^3\pi_{31} + \theta_4\pi_{22}$ has $AF \geq 7$ also. This proves (*). This contradiction proves $\epsilon_3 = 0$.

Thus (3) becomes

(4) $\phi = 8\epsilon_1'\alpha + \epsilon_4\delta$.
We want to show for $(1.4)'(4)$ that $\epsilon_4 = 0$. We have

$$\eta^2\phi \in \eta^2 < \theta_4, 2\iota, \sigma^2 > = < \eta^2, \theta_4, 2\iota > \sigma^2 \subseteq \sigma^2\pi_{33}.$$

$\pi_{33} \cong \mathbf{Z}/2 \oplus \mathbf{Z}/2 \oplus \mathbf{Z}/2 \oplus \mathbf{Z}/2$ on $\eta\eta_5, \nu\theta_4, \eta^2\beta_{31}, \mu_4$ ([2]) where η_5 is Mahowald's element ([9]), β_{31} is the generator of $imJ_{31} \subseteq \pi_{31}$ and μ_4 is Barratt's element which is detected by P^4h_1; so $AF(\mu_4) = 21$. Since $\sigma^2\eta = 0, \sigma^2\nu = 0$, it follows that $\sigma^2\pi_{33}$ is the set $\{\sigma^2\mu_4, 0\}$ with $AF(\sigma^2\mu_4) \geq 23$ (it is easy to see $\sigma^2\mu_4 = 0$, but we don't need this). So $AF(\eta^2\phi) \geq 23$. If $\epsilon_4 = 1$, so that $\phi = 8\epsilon_1'\alpha + \delta$, then $\eta^2\phi = \eta^2\delta$ which, by (2.5)(2), is detected by P^1u and therefore has $AF = 13$. This contradiction proves $\epsilon_4 = 0$.

This completes the proof of $(1.4)'$ and therefore Theorem 1.4.

References

[1] J.F. Adams, On the non-existence of elements of Hopf invariant one, Annals of Math. 72 (1960), 20-104.
[2] M.G. Barratt, M.E. Mahowald and M.C. Tangora, Some differentials in the Adams spectral sequence-II, Topology 9 (1970), 309-316.
[3] M.G. Barratt, J.D.S. Jones and M.E. Mahowald, The Kervaire invariant problem, Contemporary Math., Vol.19 (AMS 1983), 9-23.
[4] M.G. Barratt, J.D.S. Jones and M.E. Mahowald, Relations amongest Toda brackets and the Kervaire invariant in dimension 62, J. London Math. Soc. (2), 30 (1984), 533-550.
[5] W. Browder, The Kervaire invariant of framed manifolds and its generalizations, Annals of Math. 90 (1969), 157-186.
[6] R. Bruner, A new differential in the Adams spectral sequence, Topology 23 (1954), 271-276.
[7] S. Kochman, Stable Homotopy Groups of Spheres, Lecture Notes in Math. No.1423, Springer-Verlag (1990).
[8] M.E. Mahowald and M.C. Tangora, Some differentials in the Adams spectral sequence, Topology 6 (1967), 349-369.
[9] M.E. Mahowald, A new infinite family in ${}_2\pi_*^s$, Topology 16 (1977), 249-256.
[10] M.E. Mahowald, Reviwer of AMS MR 91 # 55016.
[11] J.P. May, The cohomology of restricted Lie algebras and of Hopf algebras; applications to the Steenrod algebras, Thesis, Princeton University, 1964.
[12] M. Mimura and H. Toda, The $(n+20)^{th}$ homotopy groups of n-spheres, J. Math. Kyoto Univ. 3 (1963), 37-58.
[13] M. Mimura, On the generalized Hopf homomorphism and the higher composition, Part II, J. Math. Kyoto Univ. 4 (1964-65), 301-326.
[14] R.M.F. Moss, Secondary compositions and the Adams spectral sequence, Math. Zeit. 115 (1970), 283-310.
[15] Paul Shick, Adams spectral sequence chart, Contemparary Math., Vol.146 (1991), 479-481.
[16] M.C. Tangora, On the cohomology of the Steenrod algebra, Math. Zeit. 116 (1970), 18-64.
[17] M.C. Tangora, Some extension questions in the Adams spectral sequence, Proceedings of the Advanced Study on algebraic topology, August 1970, Vol.III (Aarhus University, Aarhus, 1970).
[18] H. Toda, Composition mehtods in homotopy groups of spheres, Ann. of Math. Studies No.49, Princeton Univ. Press, Princeton, N.J. (1962).

Department of Mathematics
National Tsing Hua University
Hsinchu, Taiwan 30043

AMS/IP Studies in Advanced Mathematics
Volume 20, 2001

Melvin-Morton Conjecture: A Survey

Xiao-Song Lin

1. The Melvin-Morton conjecture

The Alexander knot polynomial and the Jones knot polynomial are characterized by similar skein relations. For the Alexander polynomial $A(K,t)$, the skein relation is

$$A(K_+,t) - A(K_-,t) = (t^{-1/2} - t^{1/2})\, A(K_0,t), \qquad A(\text{unknot},t) = 1$$

and for the Jones polynomial $J(K,t)$, the skein relation is

$$t^{-1}\, J(K_+,t) - t\, J(K_-,t) = (t^{-1/2} - t^{1/2})\, J(K_0,t), \qquad J(\text{unknot},t) = 1.$$

Here, as usual, $K_+ =$ , $K_- =$  and $K_0 =$ are links given by diagrams which differ only at a single crossing as indicated.

Note that if K is a knot, $A(K,t)$ is a polynomial of $(t^{-1/2} - t^{1/2})^2$. So, it is a symmetric Laurent polynomial in t, which is the variable in the topological definition of the Alexander polynomial.

A relationship between the Alexander polynomial and the Jones polynomial was known almost immediately after the discovery of the Jones polynomial in 1984. They are both specializations of a two variable knot polynomial known as the HOMFLY polynomial. In the language of quantum groups, the HOMFLY polynomial $H(K,t) = H(K,t,N)$ comes from the fundamental representation of $U_h\mathfrak{sl}(N)$, where the variables t and h are related by $t = e^h$. It is determined by the skein relation

$$t^{-N/2}\, H(K_+,t) - t^{N/2}\, H(K_-,t) = (t^{-1/2} - t^{1/2})\, H(K_0,t), \qquad H(\text{unknot},t) = 1.$$

We get the Jones polynomial if $N = 2$, and get the Alexander polynomial by taking the limit when $N \to 0$.

The Melvin-Morton conjecture is about another relationship between the Alexander polynomial and the Jones polynomial, which is made possible not by extending the quantum groups from $U_h\mathfrak{sl}(2)$ to $U_h\mathfrak{sl}(N)$ but by extending the fundamental representation of $U_h\mathfrak{sl}(2)$ to its higher dimensional irreducible representations.

The set of finite dimensional irreducible representations of the quantum group $U_h\mathfrak{sl}(2)$ can be listed as $V_1, V_2, V_3, \ldots$, where V_d is d-dimensional. The fundamental representation is V_2, which is the one used to construct the Jones polynomial $J(K)$.

Based on a lecture delivered at ICCM 98. The author's work is supported in part by an N.S.F. grant.

Other representations can also be used to construct knot polynomials. The knot polynomial obtained by "coloring" the (zero framed) knot K with the irreducible representation V_d is called the colored Jones polynomial $J(K, V_d)$ [**14**]. We have

$$J(\text{unknot}, V_d) = [d] = \frac{t^{d/2} - t^{-d/2}}{t^{1/2} - t^{-1/2}},$$

and $J(K, V_2) = (t^{1/2} + t^{-1/2})J(K, t)$.

We may also color K by non-irreducible representations, for example, by $V_2^{\otimes d}$. Such a colored Jones polynomial can be interpreted in two ways:

1. Assume that K has zero framing, let K^d be the link obtained by replacing K with d parallel copies (this is the zero framing cabling operation), then $J(K, V_2^{\otimes d}) = J(K^d, V_2)$.
2. We have the following relation in the representation ring of $U_h\mathfrak{sl}(2)$: $V_2 \otimes V_d = V_{d+1} \oplus V_{d-1}$. Thus, $V_2^{\otimes d}$ is a linear combination of the irreducible modules $V_{d+1}, V_{d-1}, V_{d-3}, \ldots$ and $J(K, V_2^{\otimes d})$ is the same linear combination of $J(K, V_{d+1}), J(K, V_{d-1}), J(K, V_{d-3}), \ldots$.

These two interpretations can be used to establish a precise relation between the colored Jones polynomials and the cablings of the Jones polynomial. We quote from [**8**] such a relation in the case considered here:

$$J(K, V_{d+1}) = \sum_{j=0}^{d/2} (-1)^j \binom{d-j}{j} J(K^{d-2j}, V_2).$$

Now let $t = e^h$ and consider the power series expansion

$$\frac{J(K, V_{d+1})}{[d+1]}(t) = \sum_{n=0}^{\infty} P_n(K, d)h^n.$$

THEOREM 1.1. (Melvin-Morton conjecture) *The coefficient $P_n(K, d)$ is a degree n polynomial in d:*

$$P_n(K, d) = a_{n,0}(K) + a_{n,1}(K)d + a_{n,2}(K)d^2 + \cdots + a_{n,n}(K)d^n. \tag{1}$$

Moreover, the power series formed by the diagonal of the infinite lower triangular matrix $\{a_{n,m}(K)\}$ is equal to the reciprocal of the Alexander polynomial:

$$\sum_{n=0}^{\infty} a_{n,n}(K)h^n = \frac{1}{A(K, e^h)}. \tag{2}$$

In Theorem 1.1, Equation (1) comes more or less directly from the construction of the colored Jones polynomial. Equation (2), on the other hand, seemed to be rather unexpected when Melvin and Morton first proposed their conjecture [**12**]. They observed that the two sides of Equation (2) behave in the same way when acted on by the connected cabling operation on knots. Later, Morton established the conjecture in the case of torus knots [**13**].

Notice that assuming Equation (1) in Theorem 1.1, Equation (2) is equivalent to

$$\lim_{d\to\infty} \frac{J(K, V_{d+1})}{[d+1]}(e^{h/d}) = \frac{1}{A(K, e^h)}. \tag{3}$$

Rozansky arrived at a non-rigorous proof of Theorem 1.1 using the path integral interpretation of the Jones polynomial given by Witten [**15**]. It should be remarked that Rozansky's work went beyond the particular simple Lie algebra $\mathfrak{sl}(2)$ and extended the Melvin-Morton conjecture to its full generality. We recommend that the reader consults [**2**] for a clear outline of Rozansky's argument.

The first rigorous proof of Theorem 1.1 was given by Bar-Natan and Garoufalidis [**2**]. Variations of the Bar-Natan-Garoufalidis proof were later given by Vaintrob [**16, 17**], Chmutov [**4, 5**], and Kricker [**9**] independently. These proofs all rely on Kontsevich's construction of universal Vassiliev invariants (the Kontsevich integral). Proofs using the state model of the colored Jones polynomial were obtained independently by Kurpita and Murasugi [**10**] and by Lin and Wang [**11**]. We will outline the proofs of the Melvin-Morton conjecture by Bar-Natan and Garoufalidis and by Lin and Wang in the next two sections of this article. These two proofs all reduce Equation (2) to some general combinatorial identities. On the other hand, the proof by Lin and Wang is based on calculating the limit of the left hand side of Equation (3), whereas the proof by Bar-Natan and Garoufalidis establishes Equation (2) directly.

Other proofs of the Melvin-Morton conjecture all have their own interest and perspectives. We will leave it to the reader to explore and find treasures in these different approaches by himself. It is a remarkable fact to us that the Melvin-Morton conjecture can be seen in so many different directions. We think that this reflects the true nature of knot theory as a unifying thread of many different fields.

2. Bar-Natan and Garoufalidis' proof

The theory of knot invariants of finite type (Vassiliev invariants) was developed by introducing the notion of singular knots, and the combinatorial feature of singular knots is captured by the notion of chord diagrams. A singular knot is an immersion of the circle S^1 into $\mathbb{R}^3$ whose only singularities are finitely many transverse double points. The preimage of these double points may be characterized combinatorially as a pairing of even number of points on S^1, and this paring is seen as a chord diagram where a pair of points are connected by a chord. A knot invariant of finite type will then induce a function on chord diagrams called a weight system. A weight system is characterized by certain constrains and the fundamental theorem of Kontsevich concludes that any rational weight system is induced by a knot invariant of finite type. See [**1**] for an introduction to all these said here and more.

The key notion in the Bar-Natan-Garoufalidis proof of the Melvin-Morton conjecture is that of canonical Vassiliev power series. Let

$$\tilde{A}(K,h) = \frac{h}{e^{h/2} - e^{-h/2}} A(K, e^h),$$

thought of as a power series in h. This is a canonical Vassiliev power series in the sense that it is the Kontsevich integral of a weight system W_A defined as follow:

For a chord diagram D, the intersection graph of D, $IG(D)$, is given such that its vertices are chords of D, and there is an edge between two vertices whenever the corresponding chords intersect. Label the vertices of $\mathrm{IG}(D)$ (or the chord of D) and define the intersection matrix, $\mathrm{IM}(D)$, by

$$\mathrm{IM}(D)_{ij} = \begin{cases} \mathrm{sign}(i-j) & \text{if there is an edge } (ij) \text{ in } IG(D); \\ 0 & \text{otherwise.} \end{cases}$$

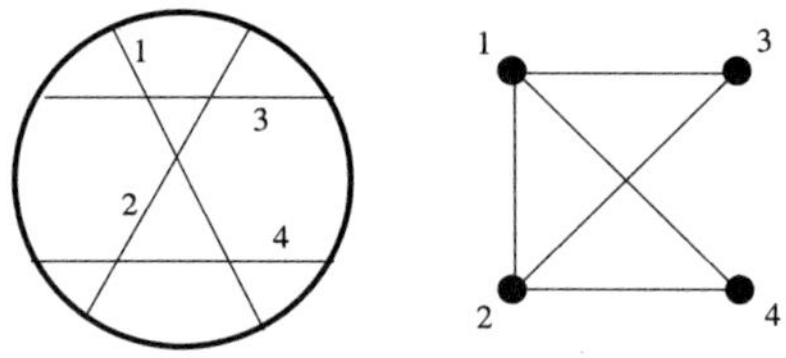

FIGURE 1

Then

$$W_A(D) = \det \mathrm{IM}(D). \tag{4}$$

EXAMPLE 2.1. We look at the chord diagram D and its intersection graph in Figure 1.

We have

$$W_A(D) = \det \begin{pmatrix} 0 & -1 & -1 & -1 \\ 1 & 0 & -1 & -1 \\ 1 & 1 & 0 & 0 \\ 1 & 1 & 0 & 0 \end{pmatrix} = 0.$$

Another canonical Vassiliev power series comes from the colored Jones polynomial. Assuming that the first part of the Melvin-Morton conjecture holds, we write

$$\frac{J(K, V_{d+1})}{d+1}(e^h) = \sum_{n \geq m \geq 0} b_{n,m}(K)\, d^m h^n.$$

Much of the effort in [2] is spent on proving that the power series

$$\tilde{M}(K,h) = \sum_{n=0}^{\infty} b_{n,n}(K) h^n$$

is a canonical Vassiliev power series of the weight system

$$W_M(D) = \mathrm{per}\, \mathrm{IM}(D). \tag{5}$$

Here per is the permanent of matrices: If $M = (m_{ij})_{n \times n}$ is a matrix, then

$$\det M = \sum_{\sigma \in S_n} (-1)^\sigma m_{1\sigma(1)} m_{2\sigma(2)} \cdots m_{n\sigma(n)}$$

$$\mathrm{per}\, M = \sum_{\sigma \in S_n} m_{1\sigma(1)} m_{2\sigma(2)} \cdots m_{n\sigma(n)}.$$

The advantage of considering canonical Vassiliev power series is that the product of two canonical Vassiliev power series with weight system W_1 and W_2 respectively is again a canonical Vassiliev power series with the weight system $W_1 \star W_2$ (see the definition of the convolution $\star$ below). It is easy to check that Equation (2) in Theorem 1.1 is equivalent to

$$\tilde{M}(K,h)\, \tilde{A}(K,h) = 1.$$

Therefore, all we need to show is that

$$W_M \star W_A = 0 \tag{6}$$

Let us recall the definition of the convolution of two weight systems W_1 and W_2 [1]. Let D be a chord diagram. A sub-diagram D' of D is obtained by dropping off some chords from D, and the complement diagram D'' of D' is obtained by dropping off chords in D' from D. With this definition, we have

$$W_1 \star W_2\,(D) = \sum_{D'} W_1(D')W_2(D'')$$

summing over all sub-diagrams D'.

For a matrix $M = (m_{ij})_{n\times n}$, a sub-matrix M' of M is obtained by dropping off some rows and the corresponding columns, and the complement matrix M'' of M' is obtained by dropping off the rows and columns that M' occupies. Then, Equation (6) is clearly implied by the following theorem. (Our statement here is more general than what was actually used and proved in [**2**]. It is hard to believe that one could not find in the literature such an elementary fact and its equally elementary proof, but that is the case for us.)

THEOREM 2.2. *Let M be a $n \times n$ anti-symmetric matrix, we have*

$$\sum_{M'} \det M' \cdot \operatorname{per} M'' = 0,$$

summing over all sub-matrices M' of M.

PROOF. If $\sigma \in S_n$ is a cycle, $\sigma = (i_1 i_2 \dots i_s)$, we write

$$m_\sigma = m_{i_1i_2}m_{i_2i_3}\dots m_{i_{s-1}i_s}m_{i_si_1}.$$

In general, if $\sigma \in S_n$ decomposes into cycles

$$\sigma = \sigma_1\sigma_2\cdots\sigma_k,$$

we have

$$m_{1\sigma(1)}m_{2\sigma(2)}\cdots m_{n\sigma(n)} = m_{\sigma_1}m_{\sigma_2}\cdots m_{\sigma_k}.$$

With this notation, we have

$$\sum_{M'} \det M' \cdot \operatorname{per} M'' = \sum_{\sigma=\sigma_1\cdots\sigma_k} \prod_{r=1}^{k} \left((-1)^{\sigma_r} m_{\sigma_r} + m_{\sigma_r}\right),$$

summing over all $\sigma \in S_n$.

Notice that if σ_r is an even cycle, $(-1)^{\sigma_i} = -1$; and if $\sigma_r = (i_1 i_2 \dots i_s)$ is an odd cycle, let $\bar{\sigma}_r = (i_s \dots i_2 i_1)$, we have $m_{\sigma_r} = -m_{\bar{\sigma}_r}$. These facts are enough to establish the theorem. □

So Equation (6) holds and this finishes our outline of the Bar-Natan-Garoufalidis proof of the Melvin-Morton conjecture.

3. Lin and Wang's proof

Let us describe a model of random walks on knot diagrams first. This model was hinted originally by a remark of V.F.R. Jones on a probabilistic interpretation of the Burau representation of braid groups. In this model, imagine that a knotted track is laid out in the space and a ball travels on the track in a given direction. When it comes to an ϵ-crossing ($\epsilon = \pm 1$ is the sign of the crossing) from the under-crossed segment, it may either jump up with probability $1 - t^\epsilon$ and continue to travel on the over-crossed segment, or continue to travel with probability t^ϵ on the under-crossed segment. This is of course a realistic probability model only when all the crossings are of the same sign.

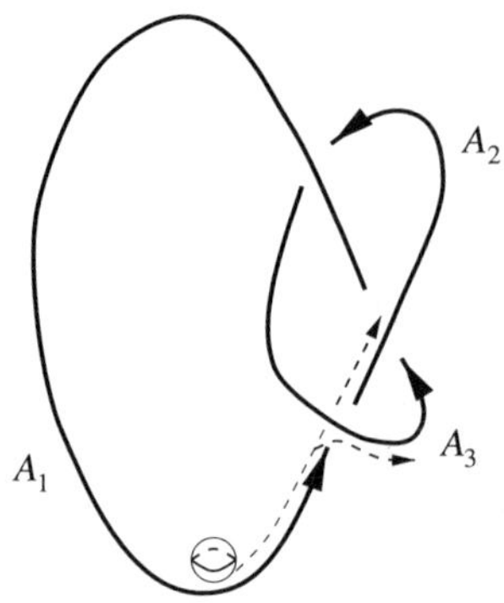

FIGURE 2

We draw a knot diagram on the plane and list the arcs in the knot diagram by $\{A_1, A_2, \ldots, A_n\}$. This set will be our space of states. The transition matrix $\tilde{\mathcal{B}}$ is defined as follows. The j-th row of $\tilde{\mathcal{B}}$ has at most two non-zero entries: If the ball is on the arc A_j and it comes to an ϵ-crossing with A_i as the over-crossed arc and A_k the arc on the other side, when $k \neq i$, the (j,k)-entry is t^ϵ and (j,i)-entry is $1-t^\epsilon$; when $k = i$, only the (j,k)-entry in the j-th row is non-zero, which is equal to 1.

Let $\mathcal{B}$ be the $(n-1)\times(n-1)$ matrix obtain from $\tilde{\mathcal{B}}$ by deleting the first row and the first column. Then $\det(I-\mathcal{B})$ and the Alexander polynomial $A(K,t)$ differ only by a factor of a power of t. Let us look at an example.

EXAMPLE 3.1. The trefoil knot K is shown in Figure 2. We have three arcs $\{A_1, A_2, A_3\}$ and the transition matrix is

$$\tilde{\mathcal{B}} = \begin{pmatrix} 0 & t & 1-t \\ 1-t & 0 & t \\ t & 1-t & 0 \end{pmatrix}.$$

So

$$\det(I-\mathcal{B}) = t^2 - t + 1.$$

On the other hand, we have

$$A(K,t) = 1 + (t^{-1/2} - t^{1/2})^2 = t - 1 + t^{-1}.$$

So we have

$$A(K,t) = t^{-1}\det(I-\mathcal{B}).$$

In general, it is a little bit subtle to determine what exactly the difference between $\det(I-\mathcal{B})$ and $A(K,t)$ is. Here is one way: Realize the knot K as the closure of a $(1,1)$-tangle diagram T whose writhe, i.e. the sum of signs of all crossings, is zero. If we smooth each crossing of T in accordance with the orientation, we get a collection of oriented circles. We denote by $\mathrm{rot}(T)$ the sum of these circles, 1 for each counter-clock-wisely oriented ones and -1 for each clock-wisely oriented ones. Mark the arc that closes T to K by A_1, then

$$A(K,t) = t^{\frac{\mathrm{rot}(T)}{2}}\det(I-\mathcal{B}). \tag{7}$$

The tangle diagram T will also be useful in our application of the beautiful Foata-Zeilberger formula [**6**]. A cycle in our model of random walks is a sequence of transition of states:

$$A_{i_1} \to A_{i_2} \to \cdots \to A_{i_m} \to A_{i_1}.$$

We will not distinguish a cycle and its cyclic permutations. We have a weight $W(c)$ for each cycle c, which is the product of transition probabilities along the cycle. We will consider cycles on K containing no A_1. All these cycles are carried by T. We denote by $\mathcal{Q}$ the set of all such cycles.

THEOREM 3.2. *We have*

$$(8) \qquad \begin{aligned} (\det(I-\mathcal{B}))^{-1} &= \prod_{c\in\mathcal{Q}} (1-W(c))^{-1} \\ &= 1 + \sum_{k=1}^{\infty} \sum_{(c_1,\ldots,c_k)\in\mathcal{Q}^k} W(c_1)\cdots W(c_k). \end{aligned}$$

Let us recall the celebrated MacMahon Master Theorem [**3, 7**]. We will use Cartier's formulation: For a matrix A,

$$(9) \qquad (\det(I-tA))^{-1} = \sum_{k=0}^{\infty} t^k \operatorname{Tr}(S^k A),$$

where $S^k A$ is the k-th symmetric tensor of A, see [**3**].

Suppose that $A = (a_{ij})_{n\times n}$ is the matrix of a linear transformation on the vector space V under the basis $\{e_1, e_2, \ldots, e_n\}$. The monomials $e_1^{k_1} e_2^{k_2} \cdots e_n^{k_n}$ with $k_1 + k_2 + \cdots + k_n = k$ form a basis of $S^k V$. We have

$$S^k A\,(e_1^{k_1} e_2^{k_2} \cdots e_n^{k_n}) = \prod_{i=1}^{n} (a_{i1}e_1 + a_{i2}e_2 + \cdots + a_{in}e_n)^{k_i}.$$

In order to reorganize the expansion of the right side of Equation (9), we consider $e_1, e_2, \ldots, e_n$ as ordered non-commuting variables. A Lyndon word in $e_1, e_2, \ldots, e_n$ is defined to be a primitive word, i.e. not the power of any other word, and is minimal among all its cyclic permutations within the lexicographic order of words. Then a result of Lyndon says that any word can be uniquely written as a non-increasing juxtaposition of Lyndon words.

Given a Lyndon word $l = e_{i_1} e_{i_2} \cdots e_{i_p}$, define

$$\beta_{\text{circ}}(l) = a_{i_1 i_2} a_{i_2 i_3} \cdots a_{i_{p-1} i_p} a_{i_p i_1}.$$

For an arbitrary word w in $e_1, e_2, \ldots, e_n$, write it as a non-increasing juxtaposition of Lyndon words

$$w = l_1 l_2 \cdots l_m, \qquad \text{each } l_i \text{ is a Lyndon word, } l_1 \geq l_2 \geq \cdots \geq l_m,$$

and define

$$\beta(w) = \beta_{\text{circ}}(l_1)\beta_{\text{circ}}(l_2)\cdots\beta_{\text{circ}}(l_m).$$

Then the Foata-Zeilberger formula is

$$(10) \qquad \operatorname{Tr}(S^k A) = \sum_{w} \beta(w)$$

summing over all word w in $e_1, e_2, \ldots, e_n$ where each e_i appears k_i times. It is easy to see that Equation (8) is a consequence of Equation (9) and (10).

FIGURE 3

We consider now the colored Jones polynomial. As before, let T be a $(1,1)$-tangle diagram with zero framing whose closure is the knot K that we are looking at. A family of cycles in the model of random walks avoiding the arc A_1 is called *simple* if none of the arcs A_2, A_3, ..., A_n appear twice in these cycles, so they can be seen as a generic family of closed plane curves. We denote by $\mathcal{Q}^t$ the set of all simple families of cycles. For $c \in \mathcal{Q}^t$, let $\mathrm{rot}(c)$ be the sum of rotation numbers of cycles in c thought of as closed plane curves, $\alpha(c)$ be the number of double points in c, and $W(c)$ the product of weights of cycles in c.

THEOREM 3.3. *We have*

$$J(K,t) = t^{\frac{-\mathrm{rot}(T)}{2}} \left(1 + \sum_{c \in \mathcal{Q}^t} t^{\mathrm{rot}(c)-\alpha(c)} W(c)\right).$$

EXAMPLE 3.4. We look again at the trefoil knot, now realized by a zero framing tangle as shown in Figure 3. We have only one element $c \in \mathcal{Q}^t$ shown in Figure 3 as the dashed curve. We have $W(c) = t(1-t)$, $\mathrm{rot}(c) = 1$, and $\alpha(c) = 0$. Also, $\mathrm{rot}(T) = -2$. Therefore

$$J(K,t) = t(1 + t^2(1-t)) = t + t^3 - t^4$$

as expected.

In general, for the colored Jones polynomial $J(K, V_{d+1})$, we look at the d parallel copies of T, the (d,d)-tangle T^d. We consider the set $\mathcal{Q}^t_d$ of simple families of cycles on T^d which avoid the parallel copies of the arc A_1. We will define $\mathrm{rot}(c)$, $\alpha(c)$ and $W(c)$ for $c \in \mathcal{Q}^t_d$ similarly as before for $c \in \mathcal{Q}^t$.

THEOREM 3.5. *We have*

$$\frac{J(K, V_{d+1})}{[d+1]}(t) = t^{\frac{-\mathrm{rot}(T^d)}{2}} \left(1 + \sum_{c \in \mathcal{Q}^t_d} t^{\mathrm{rot}(c)-\alpha(c)} W(c)\right). \tag{11}$$

With Equations (7, 8, 11) at hand, we consider lifting families of cycles on T with $t = e^h$ to simple families of cycles on T^d with $t = e^{h/d}$. A quite straightforward calculation of limit gives us

$$\lim_{d\to\infty} \frac{J(K, V_{d+1})}{[d+1]}(e^{h/d}) = t^{\frac{-\mathrm{rot}(T)}{2}} \left(1 + \sum_{k=1}^{\infty} \sum_{(c_1, c_2, \ldots, c_k) \in \mathcal{Q}^k} W(c_1)W(c_2)\cdots W(c_k)\right)$$
$$= \frac{1}{A(K, e^h)}.$$

This finishes our outline of the Lin-Wang proof of the Melvin-Morton conjecture.

References

[1] D. Bar-Natan, *On the Vassiliev knot invariants*, Topology **34**(1995), 423–472.

[2] D. Bar-Natan and S. Garoufalidis, *On the Melvin-Morton-Rozansky conjecture*, Invent. Math. **1**(1996), 103-133.

[3] P. Cartier, *A course on determinants*, Conformal Invariance and String Theory, ed. P. Dita and V. Georgescu, Academic Press, 1989, pp. 443–557.

[4] S. Chmutov, *A proof of Melvin-Morton conjecture and Feynman diagrams*, preprint.

[5] S. Chmutov, *Combinatorial analog of the Melvin-Morton conjecture*, preprint.

[6] D. Foata and D. Zeilberger, *A combinatorial proof of Bass' evaluations of the Ihara-Selberg zeta function for graphs*, Trans. AMS, to appear; `math.CO/9806037`.

[7] I.P. Goulden and D.M. Jackson, Combinatorial Enumeration, Wiley-Interscience Series in Discrete Mathematics, John Wiley & Sons, 1983.

[8] R. Kirby and P. Melvin, *The 3-manifold invariants of Witten and Reshetikhin-Turaev for* $\mathfrak{sl}(2)$, Invent. Math. **105**(1991), 473-545.

[9] A. Kricker, *Alexander-Conway limits of many Vassiliev weight systems*, preprint, `math.QA/9603011`.

[10] B.I. Kurpita and K. Murasugi, *A graphical approach to the Melvin-Morton conjecture, Part I*, Topo. Its Appl. **82**(1998), 297–316.

[11] X.-S. Lin and Z. Wang, *Random walk on knot diagrams, the colored Jones polynomial and the Ihara-Selberg zeta function*, preprint, `math.GT/9812039`.

[12] P. Melvin and H. Morton, *The coloured Jones function*, Comm. Math. Phys. **169**(1995), 501-520.

[13] H. Morton, *The coloured Jones function and Alexander polynomial for torus knots*, Math. Proc. Camb. Philos. Soc. **117**(1995), 129–136.

[14] N. Yu. Reshetikhin and V. G. Turaev, *Ribbon graphs and their invariants derived from quantum groups,* Comm. Math. Phys. **127**(1990), 1–26.

[15] L. Rozansky, *A contribution of the trivial connection to the Jones polynomial and Witten's invariant of 3-manifolds,* Comm. Math. Phys. **175**(1996), 275-318.

[16] A. Vaintrob, *Melvin-Morton conjecture and primitive Feynman diagrams,* Internat. J. Math. **8** (1997), 537–553.

[17] A. Vaintrob, *Universal weight systems and the Melvin-Morton conjecture*, preprint, `math.QA/9605003`.

Department of Mathematics, University of California, Riverside, CA 92521
E-mail address: `xl@math.ucr.edu`

AMS/IP Studies in Advanced Mathematics
Volume 20, 2001

The Seiberg-Witten Equations and The Milnor Torsion

G. Meng

This talk is about a joint work with C. H. Taubes [**6**]. Throughout this talk the followings are assumed.

1. X is an oriented, connected, compact 3-manifold without boundary or with toroidal boundaries, i.e., X is the compact compliment of a link (could be empty) in an oriented, connected compact 3-manifold without boundary; moreover, a triangulation on X is given. This is assumed when we discuss Milnor Torsion. Or

2. X is an oriented, connected 3-manifold without boundary, with or without ends, and the compact compliment of the ends is a manifold described above; moreover a Riemannian metric on X which is the product of a flat metric on torus with the standard metric on $(0, \infty)$ on each end, is given. This is assumed when we discuss the Seiberg-Witten Equations. We denote the end of X by $End\ X$.

3. X is homologically oriented and its first Betti number $b_1(X)$ is nonzero.

Under the above assumptions, two topological invariants are defined, one is the new Seiberg-Witten invariant $\underline{SW}(X)$ and the other is the old Milnor Torsion $\tau(X)$. The main theorem in [**6**] states that

$$\underline{SW}(X) = \tau(X).$$

1. Relative $Spin^c$-structure on X

We start with the exact sequence:

$$1 \to U(1) \to U(2) \overset{Ad}{\to} SO(3) \to 1$$

Let $P \to X$ be the oriented orthonormal tangent frame bundle over X, this is a principal $SO(3)$ bundle over X. The restriction of this bundle to the ends of X has a canonical trivialization defined by the flat structure of the ends of X.

Roughly speaking, a relative $Spin^c$-structure on X is a lifting of $P \to X$ to a principal $U(2)$ bundle $Q \to X$ compatible with the group homomorphism Ad : $U(2) \to SO(3)$ together with a lifting of the canonical trivialization of P on $End\ X$ to a trivialization of Q on $End\ X$, modulo natural equivalences.

More precisely, a relative $Spin^c$-structure on X is an isomorphism class of triple (Q, α, A_0), where Q is a principal $U(2)$ bundle over X which is trivial on $End\ X$, α is a bundle isomorphism: $Q \times_{Ad} SO(3) \to P$, A_0 is a smooth connection on $\det Q := Q \times_{\det} U(1)$ which is trivial on $End\ X$.

We let $Spin^c(X, \partial X)$ denote the set of relative $Spin^c$-structures on X. By obstruction theory, the obstruction to the existence of relative $Spin^c$-structures is the Euler number of the pair $(X, \partial X)$, hence is zero; moreover, $Spin^c(X, \partial X)$ is a principal $H^2(X, \partial X; \mathbb{Z})$-space over a point.

Next we will introduce Seiberg-Witten Equations for each relative $Spin^c$-structure on X.

2. The Seiberg-Witten Equations

Given a relative $Spin^c$-structure (Q, α, A_0), one can introduce the Seiberg-Witten Equations and the associated boundary condition at infinity.

The equations are the following equations for pairs (A, ψ) where A is a smooth connection on $\det Q$, ψ is a smooth section of the associated spinor bundle $S = Q \times_F \mathbb{C}^2$ (F is the fundamental representation of $U(2)$):

$$\begin{aligned} F_A &= \psi\bar{\psi} + i\mu, \\ D_A\psi &= 0. \end{aligned} \tag{1}$$

where D_A is the Dirac operator determined by A and the Riemannian metric on X, F_A is the curvature of A, μ is a real smooth differential two-form with compact support on X. $\psi\bar{\psi}$ is a quadratic term in ψ that can be best understood in representation theory.

In the case where X is not compact, the equations in (1) are augmented with the "boundary condition": There exists a smooth gauge transform of A on X, say A', such that, $A' - A_0$ is an one-form without dt component on $\partial X \times [T, \infty)$ for some $T > 0$, and as $t \to +\infty$,

$$(A' - A_0, \psi)|_{\partial X \times \{t\}} \longrightarrow (0, 0) \tag{2}$$

in the C^∞ topology.

Remark that this boundary condition implies the "finite energy" condition:

$$\int_X |F_A|^2 < \infty. \tag{3}$$

3. The solution space

The set of smooth solutions to (1) and (2) should be topologized by its imbedding in the Frechét space $i\Omega^1(X) \oplus C^\infty(S) \oplus \mathbb{R}$ which sends (A, ψ) to $(A - A_0, \psi, \int_X |F_A|^2)$. (Here, the vector space $\Omega^1(X)$ and $C^\infty(S)$ are topologized by the weak C^∞ topology.)

If (A, ψ) is a smooth solution to (1) and (2), then so is $(A + 2udu^{-1}, u \cdot \psi)$ if u is a smooth map from X to $U(1)$. Let $\mathfrak{M}(X; Q, \alpha, A_0; g, \mu)$ denote the space of irreducible smooth solutions to (1) and (2) modulo this smooth free action of gauge group $C^\infty(X; S^1)$ (also given the weak C^∞ topology). (A solution (A, ψ) is irreducible means ψ is not identically zero.)

Here are some important properties of $\mathfrak{M}(X; Q, \alpha, A_0; g, \mu)$:

PROPOSITION 3.1. *$\mathfrak{M}(X; Q, \alpha, A_0; g, \mu)$ is a zero dimensional, compact, oriented manifold when μ is chosen from a certain Baire set of smooth compactly supported real 2-forms. The orientation of $\mathfrak{M}(X; Q, \alpha, A_0; g, \mu)$ is induced by the homology orientation of X. Furthermore, $\mathfrak{M}(X; Q, \alpha, A_0; g, \mu)$ is non-empty only* *when* $[(Q, \alpha, A_0)]$ *belongs to some finite set of isomorphism classes of relative* $Spin^c$*-structures.*

4. The invariant SW

DEFINITION 4.1. *: In the case where* $b_1 > 1$*, define*

$$SW: \ Spin^c(X, \partial X) \to \mathbb{Z}$$

to be the map that sends (Q, α, A_0) *to the sum of signs of points in* $\mathfrak{M}(X; Q, \alpha, A_0; g, \mu)$*. Here, it is assumed that* $\mathfrak{M}(X; Q, \alpha, A_0; g, \mu)$ *is defined by* μ *from the Baire set mentioned in (3.1).*

In the case $b_1 = 1$*, then* SW *is defined on the set of triples* (Q, α, A_0, o) *where* o *is a generator of* $H^1(X; \mathbb{Z}) \cong \mathbb{Z}$*. The definition is the same as above, but with the following added restriction: The cohomology class* $[\mu]$ *of* μ *must obey* $\langle [\mu] \cup o, [X] \rangle > 2\pi \langle c_1(\det Q, A_0) \cup o, [X] \rangle$*. Here,* $[X]$ *denotes the fundamental class of* X*,* $c_1(\det Q, \alpha)$ *is the first Chern class of the relative* $U(1)$ *bundle* (Q, A_0)*, and* $\langle , \rangle$ *denotes the natural pairing between* H^3 *and* H_3*.*

(Note that in the case $b_1 = 1$ where X is noncompact, o determines a homology orientation and vice versa.)

The following theorem asserts, in part, that SW is well defined:

THEOREM 4.2. *: When* $b_1 > 1$*, the map* SW *from Definition 4.1 depends only on* $(X; Q, \alpha, A_0)$ *up to natural equivalence of relative spin*^{c} *structures under change of metric, and up to proper, orientation preserving diffeomorphisms which also preserve the homology orientation. When* $b_1 = 1$*, the map* SW *from Definition 4.1 depends only on* $(X; Q, \alpha, A_0; o)$ *up to natural equivalence of relative* $Spin^c$*-structures under change of metric and up to proper orientation preserving diffeomorphism which also preserve* o *and the homology orientation.*

Remark that in the case where X is compact, the preceding theorem can be deduced from the properties of the 4-dimensional Seiberg-Witten invariant for the manifold $S^1 \times X$. (See, e.g. [**11**] for a detailed description of the 4-dimensional invariant. A number of authors have also discussed this invariant in the compact case from a strictly 3-dimensional point of view. See, e.g. [**3**], [**10**], [**1**], [**5**].)

Here are some properties of SW:

- SW is finitely supported if $b_1(X) > 1$.
- SW is identically zero if X is the connected sum of two 3-manifolds each of which has nonzero first Betti number.
- $SW(c) = (-1)^n SW(\bar{c})$ where $\bar{c}$ is the complex conjugation of c. (Here n is the number of components of $End\ X$)

When X is the compliment of a link, these three properties implies that X is essentially a polynomial invariant of links which is trivial for split links and cannot distinguish a link from its mirror.

Now anyone who is slightly familiar with the classical knot theory would surely make the following conclusion:

The Seiberg-Witten Invariant = The Alexander Polynomial.

A more professional way of leading to this observation comes from the application of the Main Theorem in [**12**] to those 3-manifolds which fiber over the circle.

And this brings us to the Alexander polynomial/Milnor Torsion/Turaev Torsion in the next.

5. The Milnor Torsion

Let $C_2 \xrightarrow{\phi} C_1$ be an one-step finite dimensional based chain complex over F with zero Euler characteristic number ("based" means that a basis is chosen for each chain group), then $\det\phi$ is defined and is an element of F. It is clear that $\det\phi$ is nonzero if and only if ϕ is an isomorphism. The determinant has a natural generalization to an n-step finite dimensional based chain-complex over F with zero Euler characteristic: $C_* = (C_n \to C_{n-1} \to \cdots \to C_1 \to C_0)$, and is called the Reidemeister Torsion, denoted by $\tau(C_*)$. $\tau(C_*) \in F$ is nonzero if and only if C_* has no homology.

Let $\tilde{X}$ be the universal free abelian cover of X, the triangulation of X lifts to a H-equivariant triangulation of $\tilde{X}$, here H is the quotient of the first integral homology group of X modulo its torsion. Let $C_*(\tilde{X})$ be the simplicial chain complex of $\tilde{X}$ over $\mathbb{Z}$, $Q(X)$ be the quotient field of $\mathbb{Z}[H]$. Define

$$\Delta_*(X) := C_*(\tilde{X}) \otimes_H Q(H),$$

then $\Delta_*(X)$ is a finite dimensional chain-complex over field $Q(X)$. Let e_* be a basis for $\Delta(X)$ consisting of oriented cells on $\tilde{X}$, then the Reidemeister torsion of $\Delta(X)$ with basis e_* is defined, and is denoted by $\tau(X, e_*)$. The collection of $\tau(X, e_*)$ with e_* running over the all possible choices can be seen to be a coset of $\pm H$ in $Q(H)$. If we require all e_* to be compatible with the homology orientation on X, we get a coset of H in $Q(X)$ (actually in $\mathbb{Z}[H]$ if $b_1(X) > 1$). This coset can be shown to be independent of the triangulation of X and is denoted by $\tau(X)$.

Milnor [**7**] introduced $\tau(X)$ (up to sign, the sign improved version was introduced by Turaev [**13**]) and observed that it is essentially the Alexander polynomial of X when X is the compact compliment of a classical link.

6. The Main Theorem

$Spin^c(X, \partial X)$ is in one-to-one correspondence with $H^2(X, \partial X; \mathbb{Z})$, or $H_1(X; \mathbb{Z})$, however, there is no canonical correspondence. Fixing a relative $Spin^c$-structure c on X will fix an one-to-one correspondence, which in turn fixes a projection

$$P_c: \; Spin^c(X, \partial X) \to H.$$

Combining this projection with SW: $Spin^c(X, \partial X) \to \mathbb{Z}$, we get a map

$$SW_c: \; H \to \mathbb{Z}.$$

Here, by definition $SW_c(h) = \sum_{c' \in P_c^{-1}(h)} SW(c')$.

The collection of all SW_c with c running over all possible relative $Spin^c$-structures on X, is a coset of H in the space of $\mathbb{Z}$-valued function on H: $\mathbb{Z}[[H]]$ (actually in $\mathbb{Z}[H]$ when $b_1(X) > 1$). Let us denote it by $\underline{SW}(X)$. Now we can state the

THEOREM 6.1 (Main Theorem). *: With the assumptions given in the beginning of this talk, we have*

$$\underline{SW}(X) = \tau(X).$$

Remark that in the case when $b_1(X) = 1$, modulo H, $\underline{SW}(X)$ is a Laurent series in t and $\tau(X)$ is a rational function in t whose denominator is $(1-t)^2$ if X is compact or $(1-t)^{-1}$ if X is noncompact. The equality in the main theorem means that, modulo H, the Laurent expansion of $\tau(X)$ about $t = 0$ is $\underline{SW}(X)$.

An (essentially) axiomatic proof is outlined in [**6**].

7. Some related works

Shortly after [**6**] appeared, quite a few interesting works appeared. I list three here:

• In [**2**], M. Hutching and Y.-J. Lee gave a Morse theoretical definition for Milnor Torsion and that could lead to another proof of the main theorem.

• In [**4**], Stern-Fitushel provided a method to construct infinitely many topologically equivalent but smoothly inequivalent smooth 4-manifolds by using knots or links. These manifolds are smoothly inequivalent because their Seiberg-Witten invariants can be explicitly computed based on the main theorem and the gluing formula [**9**].

• In [**15**], Turaev observed that his further improved version of torsion [**14**] in dimension 3 is equivalent to a map

$$T: \ Spin^c(X, \partial X) \to \mathbb{Z}$$

and suggested that the improved version of the main theorem should be

$$SW = T.$$

References

[1] A. L. Carey, J. McCarthy, B. L. Wang, R. B. Zhang, *Topological quantum field theory and Seiberg-Witten monopole*, preprint.

[2] M. Hutchings and Y-J Lee, *Circle-valued Morse theory, Reidemeister torsion, and Seiberg-Witten invariants of 3-manifolds*, xxx.lanl.gov e-Print archive, dg-ga/9612004.

[3] P. B. Kronheimer and T. S. Mrowka, *The genus of embedded surfaces in the projective plane*, Math. Res. Letters **1**(1994),797-808.

[4] R. Fintushel and R. Stern, *Knots, links, and 4-manifolds*, Invent. Math. 134 (1998), 363–400.

[5] Y. Lim, *Seiberg-Witten Theory of 3-manifolds*, Preprint 1996

[6] G.Meng and C.H. Taubes, *$\underline{SW}$ = Milnor Torsion*, Math. Res. Lett. **3**(1996), 661-674.

[7] J. W. Milnor, *A duality theorem for Reidemeister torsion*, Ann. of Math. (2)**76**(1962), 137-147.

[8] J. W. Morgan and Z. Szabo, *Lectures at Harvard.*

[9] J. W. Morgan, T. S. Mrowka, and Z. Szabo, *Product formula along T^3 for the Seiberg-Witten invariants*, Math. Res. Lett. **4**(1997), 915-929.

[10] J. W. Morgan, Z. Szabo and C. H. Taubes, *A product formula for the Seiberg-Witten invariants and the generalized Thom Conjecture*, Jour. Diff Geom. **44**(1996), 706-788.

[11] J. W. Morgan, *The Seiberg-Witten Equations and Applications to the Topology of Smooth 4-manifolds*, Mathematical Notes **44**, Princeton University press, 1996.

[12] C. Taubes, *The Seiberg-Witten and the Gromov Invariants*, Math. Res. Lett., **2**, 221 (1995); *From Seiberg-Witten to Gromov*, Journal of AMS **9**, 845 (1996); *From Gromov to Seiberg-Witten*, preprint 1997; *Counting Connections and Curves*, preprint 1997; *Singular Symplectic Forms and p-holomorphic Curves*, lectures given at Harvard and IP Lectures Irvine 1996; *The Seiberg-Witten Invariants and Symplectic Froms*, Math. Res. Lett. **1**, 809 (1994); *Counting Pseudo-holomorphic Submanifolds in Dimension 4*, J. Diff. Geom., **44**, 818 (1996).

[13] V. G. Turaev, *Reidemeister torsion in knot theory*, Russian Math. Surveys 41.1 (1986), 119-182.

[14] V. G. Turaev, *Euler structures, nonsingular vector fields, and torsions of Reidemeister type*, Math USSR Izvestia 34: 3(1990), 627-662.

[15] V. G. Turaev, *A combinatorial formulation for the Seiberg-Witten invariants of 3-manifolds*, Math. Res. Lett. **5** (1998), 583–598.

Guowu Meng, Department of Mathematics, Hong Kong University of Science and Technology, Clear Water Bay, Kowloon, Hong Kong

E-mail address: mameng@uxmail.ust.hk

AMS/IP Studies in Advanced Mathematics
Volume 20, 2001

The Bergman Kernel Function of Some Reinhardt Domains

Sheng Gong and Xuean Zheng

The Bergman kernel function $K_\Omega(z,\bar{w}), z, w \in \Omega$ for a domain $\Omega \subset \mathbf{C^n}$ is the kernel of Bergman projection operator,the operator projecting $L^2(\Omega)$ onto its holomorphic subspace $A^2(\Omega)$.

The Bergman kernel can almost never be calculated explicitly, unless the domain Ω has a great deal of symmetry, so that a useful orthonormal basis for $A^2(\Omega)$ can be determined, for example, the classical domains(L.K.Hua[1]). To study the boundary behavior of $K_\Omega(z,\bar{w})$ for a domain $\Omega \subset \mathbf{C^n}$ is quite important. This problem is quite old, going back to the original inquire of Bergman in his famous paper(S. Bergman[2]).

In 1974, C. Fefferman introduced a new technique for obtaining an asymptotic expansion for the Bergman kernel on strictly pseudo-convex domains(C. Fefferman[3]), (An alternate approach, see L. Boutet deMonvel and J. Sjöstrand[4]). This work enable rather explicit estimations of the Bergman metric and opened up an entire branch of analysis on domains in $\mathbf{C^n}$. (This is one of the main contributions for C. Fefferman to get Fields medal). He proved that:

Let $\Omega = \{z \in \mathbf{C^n} | \psi(z) > 0\}$ be a strictly pseudo-convex domain, where $\psi(z) \in C^\infty(\overline{\Omega})$ satisfies $|grad\psi| \neq 0$ on $\partial\Omega$, then

$$K(z,\bar{w}) = (const.)|grad\psi(w)|^2 det L(w) X^{-(n+1)}(z,w)$$
$$+\tilde{K}(z,w) \tag{1}$$

where $z, w \in \Omega, L(w)$ denote the Levi form, and $X(z,w) = \psi(w) + \sum \frac{\partial\psi}{\partial w_j}(z_j - w_j) + \frac{1}{2}\sum \frac{\partial^2\psi}{\partial w_j \partial w_k}(z_j - w_j)(z_k - w_k)$, $\tilde{K}(z,w)$ is the lower degree term.

In particular,

$$K(z,\bar{z}) = \Phi(z)\psi^{-(n+1)}(z) + \Psi(z) log\psi(z) \tag{2}$$

where $\Phi, \Psi \in C^\infty$ and $\Phi \neq 0$ near $\partial\Omega$.

He used a basic classical fact: A biholomorphic mapping $F : D_1 \to D_2$ is not merely a homeomorphism, but an isometry when both domains are eqipped with Bergman metric, to prove the following

Fefferman Theorem Let $\Omega_1, \Omega_2 \subset \mathbf{C^n}$ be strictly pseudo-convex domains with C^∞ boundary and $\phi : \Omega_1 \to \Omega_2$ is a biholomorphic mapping, then ϕ extends uniquely to a C^∞ diffemorphism of $\overline{\Omega_1}$ onto $\overline{\Omega_2}$.

After that, there are lots important works, e.g. Chern-Moser invariant, S. Bell[5], S. Bell and E.Ligocka[6] introduced "condition R" to simplify the proof of Fefferman Theorem.

How about the boundary behavior of the Bergman kernel K_Ω of other kind of domains, especially, for weakly pseudo-convex domains. There are lots important works.

For example, for domains in $\mathbf{C^2}$, Catlin[7], has given a precise description of the behavior of $K_\Omega(z, \bar{z})$ near finite-type points in $\partial\Omega$. Sharp estimates for $K_\Omega(z, \bar{w})$ for this class of domains were obtained by McNeal [8] and Negal, Rosay, Stein and Wainger[9]. For a "decoupled" class of $\Omega \subset \mathbf{C^n}$, McNeal[10] described the exact estimates on $K_\Omega(z, \bar{w})$ for z near a point of finite type in $\partial\Omega$.

For weakly pseudo-convex domains, the boundary behavior of the Bergman kernel function of the domain is more complicated. As we know, "locally", everystrictly pseudo-convex domain looks much like the unit ball. More precisely, given $p \in \partial\Omega$, we may osculate $\partial\Omega$ at p by a ball, i.e., we can find a region $\tilde{\Omega}$ internally tangent to Ω to third order at p, and an explicit biholomorphic change of coordinates F mapping a neighborhood of p in $\tilde{\Omega}$ to a neighborhood of $F(p)$ in the unit ball. The Bergman kernel of the unit ball is known already. We can pull back the Bergman kernel from the unit ball to obtain an explicit formula for K_Ω. This is the main idea of the Fefferman's paper.

In the case of weakly pseudo-convex domains, it is never holds true. We need to treat the boundary behavior of Bergman kernel for weakly pseudo-convex domains by different method for different kind of weakly pseudo-convex domains.

One kind of pseudo-convex domains is more interesting for us. There are many examples to show the surprising effectiveness of the explicit or asymptotic formula of the Bergman kernel function of this kind weakly pseudo-convex domains.

For example, in 1980, Mostow and Y.T.Siu[11] gave a counter-example to the important conjecture that the universal covering of a compact Kähler manifold of negative sectional curvature should be biholomorphic to the ball. In their counter-example, the explicit calculation of the Bergman kernel function and metric of the weakly pseudo-convex domain $\{z = (z_1, z_2) \in \mathbf{C^n} | |z_1|^2 + |z_2|^{14} < 1\}$ played an essential role.

We are particularly interesting in to study the boundary behavior of the Bergman kernel of the following kind of the weakly pseudo-convex domains:

$$D = \{z = (z_1, z_2, \cdots, z_n) \in \mathbf{C^n} | \|z\|_\alpha = \sum_{j=1}^{n} |z_j|^{\frac{2}{\alpha_j}} < 1\} \tag{3}$$

where $\alpha_j > 0, j = 1, 2, \cdots, n$, are real numbers.

Perhaps D'Angelo was the first mathematician who studied the Bergman kernel function of such kind Reinhardt domains. In 1978, D'Angelo [12] studied the Reinhardt domain

$$D_1 = \{z \in \mathbf{C^n} | |z_1|^2 + \cdots + |z_{n-1}|^2 + |z_n|^{2p} < 1\}$$

where p is a positive **integer**. He Proved that

$$K_{D_1}(z,\bar{w}) = \sum_{k=1}^{n} c_k \frac{(1-\sum_{j=1}^{n-1} z_j\overline{w_j})^{-n+\frac{k}{p}}}{[(1-\sum_{j=1}^{n-1} z_j\overline{w_j})^{\frac{1}{p}} - z_n\overline{w_n}]^{k+1}} \tag{4}$$

where $c_1, \cdots, c_n$ are constants which are independent on $z = (z_1, \cdots, z_n)$ and depend on p and n only. He also proved that for domain (3), the series expansion of its Bergman kernel function is as follows:

$$K_D(z,\bar{w}) = \pi^{-n} \prod_{j=1}^{n} \frac{1}{\alpha_j} \sum_{m\geq 0} \frac{\Gamma(\sum_{j=1}^{n} \alpha_j(m_j+1)+1)}{\Gamma(\alpha_1 m_1+\alpha_1)\cdots\Gamma(\alpha_n m_n+\alpha_n)} \times (z_1\overline{w_1})^{m_1}\cdots(z_n\overline{w_n})^{m_n}.$$

where $m = (m_1, m_2, \cdots, m_n)$.

In 1996, by using the Appell's multivariable hypergeometric functions, G. Francsics and N. Hanges[13] considered the Reinhardt domain

$$D_2 = \{z \in \mathbf{C^n} | |z_1|^{2p_1} + |z_2|^{2p_2} + \cdots + |z_n|^{2p_n} < 1\}$$

where all the $p_1, p_2, \ldots, p_n$ are positive **integers**. They proved that in this case the Bergman kernel function of D_2 is

$$K_{D_2}(z,\bar{\omega}) = \pi^{-n} \prod_{j=1}^{n} p_j \sum_{q_1=0}^{p_1-1} \cdots \sum_{q_n=0}^{p_n-1} \frac{\Gamma(1+\sum_{j=1}^{n} \frac{q_j+1}{p_j})}{\prod_{j=1}^{n}\Gamma(\frac{q_j+1}{p_j})} (z\bar{w})^q$$

$$\times F_A^{(n)}(1+\sum_{j=1}^{n}\frac{q_j+1}{p_j}, \mathbf{1}, \frac{\mathbf{q}+\mathbf{1}}{\mathbf{p}}, (z\bar{w})^p) \tag{5}$$

for all $(z,w) \in D_2 \times D_2$, where $(z\bar{w})^q = (z_1\overline{w_1})^{q_1}\cdots(z_n\overline{w_n})^{q_n}, q = (q_1, \cdots, q_n), z = (z_1, \cdots, z_n), w = (w_1, \cdots, w_n)$, $\mathbf{1} = (1, \cdots, 1)$, $\frac{\mathbf{q}+\mathbf{1}}{\mathbf{p}} = (\frac{q_1+1}{p_1}, \cdots, \frac{q_n+1}{p_n})$ and $F_A^{(n)}$ is the Appell's multivariable hypergeometric function

$$F_A^{(n)}(\alpha,\beta,\gamma,x) = \frac{1}{\Gamma(\alpha)} \prod_{j=1}^{n} \frac{\Gamma(\gamma_j)}{\Gamma(\beta_j)} \sum_{m\geq 0} \Gamma(\alpha+\sigma(m)) \times (\prod_{j=1}^{n} \frac{\Gamma(\beta_j+m_j)}{\Gamma(\gamma_j+m_j)m_j!} x_j^{m_j})$$

where $m = (m_1, \cdots, m_n) \in \mathbf{R^n}_+, \sigma(m) = \sum_{j=1}^{n} m_j, \alpha \in \mathbf{R}, \beta = (\beta_1, \cdots, \beta_n) \in \mathbf{R^n}, \gamma = (\gamma_1, \cdots, \gamma_n) \in \mathbf{R^n}$.

They also considered the Reinhardt domain

$$D_3 = \{z = (z_1, \cdots, z_n) \in \mathbf{C^n} | |z_1|^{2p_1} + \cdots + |z_{n-1}|^{2p_{n-1}} + |z_n|^{2p_n} < 1\}$$

where all the $p_1, \cdots, p_{n-1}$ are positive **integers** and $p_n > 0$ is an arbitrary positive number. They gave the Bergman kernel function of D_3 as

$$K_{D_3}(z, \bar{w}) = \pi^{-n}(\prod_{j=1}^{n} p_j)h(z\bar{w}) \tag{6}$$

where the function h is given by

$$h(x) = \frac{\partial^{n-1}}{\partial x_1 \cdots \partial x_{n-1}} \left(\sum_{j_1=0}^{p_1-1} \cdots \sum_{j_{n-1}=0}^{p_{n-1}-1} \frac{(1 - \sum_{l=1}^{n-1} \omega_l^{j_l} x_l^{p_l})^{\frac{1}{p_l}-1}}{[(1 - \sum_{l=1}^{n-1} \omega_l^{j_l} x_l^{p_l})^{\frac{1}{p_l}} - x_n]^2} \right)$$

and $\omega_j = e^{2\pi p_j i}, j = 1, 2, \cdots, n-1$.

Of course, D_1, D_2, D_3 are special cases of domain (3), but the kernel functions (4), (5), (6) are complicated. It can not explain the boundary behavior of the Bergman kernel functions of these domains.

In 1996, we (Gong and Zheng[14]) proved that for any positive **real numbers** $\alpha_j > 0, j = 1, 2, \cdots, n$, their exists two positive numbers m and M and a function F_0 such that

$$mF_0(z, \bar{z}) \leq K_D(z, \bar{z}) \leq MF_0(z, \bar{z}) \tag{7}$$

holds for every $z \in \Omega$. Here

$$F_0(z, \bar{z}) = (-r(z))^{-n-1} \prod_{j=1}^{n} (-r(z) + |z_j|^{\frac{2}{\alpha_j}})^{1-\alpha_j} \tag{8}$$

and $r(z) = \|z\|_\alpha - 1$ is the defining function for D, the constants m and M depend only on $\alpha = (\alpha_1, \alpha_2, \cdots, \alpha_n)$ and n, not on z.

D (defined by (3)) is weakly pseudo-convex domain if at least one of $\alpha_j, j = 1, 2, \cdots, n$ is less than 1. The Bergman kernel functions (4), (5), (6) are complicated, but the boundary behavior (7) and (8) of the Bergman kernel function is very simple. It completely characterized the behavior on the boundary. Compare with Fefferman's result (2), we may find that at some strictly pseudo-convex points of D, e.g. at $z = (z_1, z_2, \cdots, z_n)$, all $z_j \neq 0, j = 1, 2, \cdots, n$, the boundary behavior are same as (2) when $r(z) \to 0$, it is $O((-r(z))^{-n-1})$. But for some other points, e.g. $z_1 = 0.z_j \neq 0, j = 2, 3, \cdots, n$, the boundary behavior are $O((-r(z))^{-n-\alpha_1})$ when $r(z) \to 0$. It is different with (2). Actually, (7) and (8) are clearly described the boundary behavior of the Bergman kernel function of D.

The proof of this result is at some length. It needed 13 lemmas, most of the lemmas are combinatorial.

In 1994, D'Angelo[15] considered another kind Reinhardt domains

$$\Omega_0 = \{z = (z_1, z_2, \cdots, z_N) \in \mathbf{C}^\mathbf{N} | \sum_{j=1}^{l} |z_j|^2 + (\sum_{j=l+1}^{N} |z_j|^2)^{\frac{1}{\mathrm{p}}} < 1\} \tag{9}$$

where $1 \leq l \leq N-1$ and $p > 0$ is a positive real number. When $N > l+1$, the domain is not "decoupled". He gave the explicit formula of the Bergman kernel functiom of Ω_0 as

$$K_{\Omega_0}(z,\bar{z}) = \sum_{k=0}^{l+1} c_k(1-\sum_{j=1}^{l}|z_j|^2)^{-l-1+pk}$$

$$\times((1-\sum_{j=1}^{l}|z_j|^2)^p - \sum_{j=l+1}^{N}|z_j|^2)^{-N+l-k} \tag{10}$$

where $c_k, k = 1,2,\cdots,l+1$ are constants which only depend on k,l,N and p. May be this is the first example for not " decoupled" domain where the Bergman kernel is known explicitely.

Domain (3) and (9) are not included to each other. So (7) and (10) are not included to each other. Of course, to study the boundary behavior of the Bergman kernel function of a not "decoupled" domain is more difficult than the "decoupled domain.

In 1996, Francsics and Hanges[13] considered a more general Reinhardt domain

$$\Omega_1 = \{Z = (Z_1, Z_2, \cdots, Z_n) = (z_1, z_2, \cdots, z_N) \in \mathbf{C^N},$$

$$Z_j \in \mathbf{C^{N_j}}, j = 1,2,\cdots,n | \|Z\|_p = \sum_{j=1}^{n}\|Z_j\|^{2p_j} < 1\} \tag{11}$$

where $N_1, N_2, \cdots, N_n, p_1, p_2, \cdots, p_n$ are positive **integers**, $N = N_1 + N_2 + \cdots + N_n$, $\|Z_j\|$ is the standard Euclidean norm of $\mathbf{C^{N_j}}, j = 1,2,\cdots,n$. They gave the explicit formula of the Bergman kernel function of Ω_1 as

$$K_{\Omega_1}(Z,\bar{W}) = \pi^{-n}(\prod_{j=1}^{n} p_j)h(Z_1\overline{W_1'},\cdots,Z_n\overline{W_n'}) \tag{12}$$

when $(Z,W) \in \Omega_1 \times \Omega_1$, where

$$h(y) = \frac{\partial^{N_1-1}}{\partial y_1^{N_1-1}}\cdots\frac{\partial^{N_n-1}}{\partial y_n^{N_n-1}}(\sum_{q_1=0}^{p_1-1}\cdots\sum_{q_n=0}^{p_n-1}$$

$$\frac{\Gamma(1+\sum_{j=1}^{n}\frac{q_j+1}{p_j})}{\prod_{j=1}^{n}\Gamma(\frac{q_j+1}{p_j})}F_A(1+\sum_{j=1}^{n}\frac{q_j+1}{p_j},(\mathbf{1},\frac{\mathbf{q}+\mathbf{1}}{\mathbf{p}},\mathbf{y^P})$$

and $y = (y_1,\cdots,y_n), y^p = y_1^{p_1}y_2^{p_2}\cdots y_n^{p_n}$.

This expression is too complicated, and nearly impossible to understand the boundary behavior of the Bergman kernel function of Ω_1.

Recently we(Gong and Zheng) studied a very general kind Reinhardt domains

$$\Omega = \{Z = (Z_1, Z_2, \cdots, Z_n) = (z_1, z_2, \cdots, z_N) \in \mathbf{C^N},$$

$$Z_j \in \mathbf{C_j^N}, j = 1,2,\cdots,n | \|Z\|_\alpha = \sum_{j=1}^{n}\|Z_j\|^{\frac{2}{\alpha_j}} < 1\} \tag{13}$$

where $N_1, N_2, \cdots, N_n$ are positive integers, $N = N_1 + N_2 + \cdots + N_n$, $\alpha_j > 0, j = 1,2,\cdots,n$ are positive **real numbers**, $\|Z_j\|$ is the standard Euclidean norm in $\mathbf{C^{N_j}}, j = 1,2,\cdots,n$.

When all $N_j = 1, j = 1, 2, \cdots, n$, then $\Omega = D$. When all $\frac{1}{\alpha_j}, j = 1, 2, \cdots, n$ are positive integers, then $\Omega = \Omega_1$. If at least one of $N_j, j = 1, 2, \cdots, n$ is greater than 1, then Ω is not "decoupled " domain. We have the following

Theorem 1 There exist two positive constants m and M which depend only on $N_j, \alpha_j, j = 1, 2, \cdots, n$, such that

$$mF(Z, \bar{Z}) \leq K_\Omega(Z, \bar{Z}) \leq MF(Z, \bar{Z})$$

holds for every $Z \in \Omega$, where

$$F(Z, \bar{Z}) = (-r(Z))^{-N-1} \prod_{j=1}^{n} (-r(z) + \|Z_j\|^{\frac{2}{\alpha_j}})^{(1-\alpha_j)N_j}$$

and $r(Z) = \|Z\|_\alpha - 1$ is the defining function of Ω.

Obviously, Theorem 1 extend our previous result, the boundary behavior of Bergman kernel function of the domain D. In this case, $F = F_0$. Theorem 1 is coincided the result (10) of D'Angelo, when $n = 2, N = N_1 + N_2, N_1 = l, N_2 = N - l, \alpha_1 = 1, \alpha_2 = p$. The proof of Theorem 1 is at some length again. It develops the method which we used in the previous paper(TAMS 348(1996)1771-1803).

Moreover, we give the estimates of the Bergman metric of domain Ω.

Theorem 2 Let $ds^2 = dZH\overline{dZ'}$ be the Bergman metric of domain Ω, then we may express H as

$$H(Z) = (-r(Z))^{-1}\theta(Z)\Lambda(Z) + (-r(Z))^{-2}\Lambda(Z)\xi(Z)\Lambda(Z) \tag{14}$$

where

$$\begin{aligned}
\theta(Z) &= diag[\theta_1(Z)I_{N_1}, \theta_2(Z)I_{N_2}, \cdots, \theta_n(Z)I_{N_n}] \in \mathbf{C}^{\mathbf{NxN}}, \\
\Lambda(Z) &= diag[(-r(Z) + \|Z_1\|^{\frac{2}{\alpha_1}})^{1-\alpha_1} I_{N_1}, \\
&\quad (-r(Z) + \|Z_2\|^{\frac{2}{\alpha_2}})^{1-\alpha_2} I_{N_2}, \cdots, \\
&\quad (-r(Z) + \|Z_n\|^{\frac{2}{\alpha_n}})^{1-\alpha_n} I_{N_n}] \in \mathbf{C}^{\mathbf{NxN}}, \\
\xi(Z) &= \overline{\xi(Z)'} = (\xi_{ij}(Z)\overline{z_i}z_j)_{1 \leq i,j \leq N} \in \mathbf{C}^{\mathbf{NxN}}.
\end{aligned}$$

Moreover, $\theta_l(Z), \xi_{ij}(Z), 1 \leq l \leq n, 1 \leq i, j \leq N$ have the following properties:

1. $\xi_{ij}(Z) = \eta_{rs}(Z)$ when $P(i) = r, P(j) = s$ where $P(i) = r, P(j) = s$ means $z_i \in \mathbf{C}^{\mathbf{N_r}}, z_j \in \mathbf{C}^{\mathbf{N_s}}$;

2. there exist two positive constants m and M which are depend on $N_1, N_2, \cdots, N_n$ and $\alpha = (\alpha_1, \alpha_2, \cdots, \alpha_n)$ only and not on Z, such that

$$m \leq \theta_l(Z) \leq M, m \leq \eta_{ll}(Z) \leq M$$

and $m \leq |\eta_{rs}(Z)| \leq M$ when $r \neq s$.

That means, (14) is clearly to describe the boundary behavior of $H(Z)$. It is decided by $-r(Z)$ and $\Lambda(Z)$ only.

References

[1] L. K. Hua, Harmonic analysis of functions of several complex variables in classical domains, Transl. Amer. Math. Soc. Vol. 6., 1963.

[2] S. Bergman, Über die Kernfunction eines Bereiches und ihr Verhalten am Rande, J. Reine Angew. Math. 169 (1933) 1-65.

[3] C. Fefferman, The Bergman kernel and biholomorphic mappings of strictly pseudoconvex domains, Invent. Math. 26 (1974) 1-63.

[4] L. Boutet de Monvel and J. Sjöstrand, Sur la singularite des noyaux de Bergman et de Szegö, Astérique 34-45, Soc. Math. de Paris, France, 1976, 123-164.

[5] S. Bell, Biholomorphic mappings and the $\bar{\partial}$ problem, Ann. Math., 114 (1981) 103-113.
[6] S. Bell and E. Ligocka, A simplication and extension of Fefferman's theorem on biholomorphic mappings, Invent. Math. 57 (1980) 283-289.
[7] D. Catlin, Estimates of invariant metrics on pseudoconvex domains of dimension two, Math. Z. 200 (1989) 429-466.
[8] J. McNeal, Boundary behavior of the Bergman kernel function in $\mathbf{C}^2$, Duke Math. J. 58 (1989) 499-512.
[9] A. Negal, J. P. Rosay, E. M. Stein and S. Wainger, Estimates for the Bergman and Szegö kernels in $\mathbf{C}^2$, Ann. Math. 129 (1989) 113-149.
[10] J. McNeal, Local geometry of decouple pseudoconvex domain, Aspekte der Math. E17 (1990) 223-230.
[11] G. D. Mostow and Y. T. Siu, A compact Kähler surface of negative curvature and covered by the ball, Ann. Math. 112 (1980) 321-360.
[12] J. P. D'Angelo, A note on the Bergman kernel, Duke Math. J. 45 (1978) 259-266.
[13] G. Francsics and N. Hanges, The Bergman kernel of complex ovals and multivariable hypergeometric functions, J. of Funct. Anal. 142 (1996) 494-510.
[14] S. Gong and X. A. Zheng, The Bergman kernel function of some Reinhardt domains, Tran. Amer. Math. Soc. 348 (1996) 1771-1803.
[15] J. P. D'Angelo, An explicit computation of the Bergman kernel function, J. Geom. Anal. 4 (1994) 23-34.

DEPARTMENT OF MATHEMATICS, UNIVERSITY OF SCIENCE AND TECHNOLOGY OF CHINA HEFEI, ANHUI, 230026, CHINA

DEPARTMENT OF MATHEMATICS, BEIJING NORMAL UNIVERSITY BEIJING, 100875, CHINA

AMS/IP Studies in Advanced Mathematics
Volume 20, 2001

On Some Problems in Several Complex Variables and CR Geometry

Xiaojun Huang

1. Introduction

Geometric analysis in several complex variables and Cauchy Riemann geometry has been a quite active subject in pure and applied mathematics for decades. The classical work of Poincaré, Hartogs, Levi, E. Cartan, etc at the beginning of the century laid down part of the frame work of such a study. Since then, more and more attention has be paid to this field, as people discover its deep relationship with many fundamental problems in classical analysis and geometry, and its importance for partial differential equations, algebraic and symplectic geometry, topology, mathematical physics. In this note, we would like to present some results and problems along the lines of this research. Limited to the page numbers, we will focus our discussion on some of the problems, which the author's recent work has touched upon. We will also have to leave out many important references, which can be found, say, in the survey paper of Forstneric [Fr1], the author's papers [Hu1] [Hu2] [Hu3] [Hu6], and the recent research monographs of Baouendi-Ebenfelt-Rothschild [BER], D'Angelo [DA1].

2. Rigidity problem of holomorphic maps between balls

In an important development of several complex variables, Poincaré discovered in 1907 that any biholomorphic map between two pieces of the boundary of the unit ball $\mathbf{B}^2$ in $\mathbf{C}^2$ is the restrict of a certain automorphism of $\mathbf{B}^2$. This phenomenon fails obviously in the setting of one complex variable and reveals for the first time a strong super-rigidity property of holomorphic mappings in several complex variables. Poincaré's result was later extended to higher dimensions by Tanaka and Chern-Moser. In the 1970's, Alexander [Alx] and Webster [We1] provided two more beautiful rigidity theorems for mappings between balls and algebraic domains in complex spaces of the same dimension. Their work has motivated many of the recent exploitations of rigidity phenomena of holomorphic objects. Here, we will first

1991 *Mathematics Subject Classification.* 32, 53.

Key words and phrases. Rigidity problems, elliptic complex tangents, Levi flat submanifolds, reflection principles, Kobayashi extremal mappings, iterates of holomoprhic mappings.

Supported in part by NSF-9970439 and an NSF postdoctoral fellowship.

constraint ourselves to problems for mappings between balls in complex spaces of different dimensions. (For more recent development on mappings between algebraic sets, we refer the reader to [BER] and the references therein.)

In 1978, Webster in [We3] first took up the problem of considering proper holomorphic mappings from $\mathbf{B}^n$ into $\mathbf{B}^{n+1}$ with $n \geq 3$ by using the Cartan-Chern-Moser CR projective bundle theory. He proved that any such a map must be linear whenever it is in $C^3(\overline{\mathbf{B}^n})$. Here, $\mathbf{B}^n$ denotes the unit ball in $\mathbf{C}^n$ centered at the origin. (Also, we recall that two maps $f, g : \mathbf{B}^n \to \mathbf{B}^N$ are called equivalent if there are elements $\sigma \in \mathrm{Aut}(\mathbf{B}^n)$ and $\tau \in \mathrm{Aut}(\mathbf{B}^N)$ such that $f = \tau \circ g \circ \sigma$. In particular, we call f linear or a totally geodesic embedding if f is equivalent to the standard 'big circle' embedding sending $(z_1, \cdots, z_n)$ to $(z_1, \cdots, z_n, 0, \cdots, 0)$.)

In [Fa1], Faran classified proper maps from $\mathbf{B}^2$ into $\mathbf{B}^3$, which are in the smooth class $C^3(\overline{\mathbf{B}^n})$. Different from the linearity phenomenon of Webster, he surprisingly showed that up to automorphisms, there are four different maps. These results were proved for mappings which are only C^2-regular up to the boundary in 1983 by Cima-Suffridge [CS]. In the same paper, they first conjectured that any proper holomorphic map from $\mathbf{B}^n$ into $\mathbf{B}^N$ with $N \leq 2n-2$ must be linear once it is in $C^2(\overline{\mathbf{B}^n})$. Notice that there are non-linear proper rational maps from $\mathbf{B}^n$ into $\mathbf{B}^{2n-1}$. Cima-Suffridge's conjecture was verified in [Fa2] when the maps are holomorphic up to the boundary, by reducing it to the problem concerning mappings between projective spaces. Forstneric [Fo2] then in his work carried out a systematic study on the reflection principle for mappings between balls in different complex spaces [Fr2]. His work together with Faran's result gives the super-rigidity property for mappings from $\mathbf{B}^n$ into $\mathbf{B}^N$ with $N \leq 2n-2$, when the map is C^{N-n+1}-smooth up to the boundary. In a different direction, due to the discovery of inner functions over the ball, it is known that there are many proper holomorphic mappings from $\mathbf{B}^n$ into $\mathbf{B}^{n+1}$ which are continuous up to the boundary but not rational. (See [Fr1], [Theorem 2, Dr]). This opens up a very interesting but also a difficult subject to clarify the minimal boundary regularity required for mappings between balls to guarantee the rigidity phenomenon. It particular, it has been wondered for years if there is a fixed number t independent of the codimension such that the above described rigidity holds for mappings whose boundary regularity is C^t up to the boundary. In [Hu1], we introduced a new approach for such a study and were able to obtain the following:

Theorem 2.1 *Let M_1 and M_2 be two open connected pieces of the boundaries of the unit balls $\mathbf{B}^n$ and $\mathbf{B}^N$, respectively. Let F be a twice continuously differentiable CR mapping from M_1 and M_2. If $N \leq 2n-2$, then F extends to a linear map from $\mathbf{B}^n$ into $\mathbf{B}^N$, unless it is constant.*

Here, we mention that a bounded function defined over a real analytic hypersurface M is a CR function when it is annihilated by CR vector fields along M. In particular, when M does not contain any germs of complex hypersurface, then g is CR if and only if g extends holomorphically to a certain side of M by a result of Baouendi-Treves [BT]. A mapping from M is called a CR map if all of its components are CR functions.

Ideas and results developed in [Hu1] have some other applications. For instance, the following comes as an immediate Corollary of the work in [Hu1]:

Corollary 2.1 *Let M_1 and M_2 be two open connected pieces of the boundaries of the unit balls $\mathbf{B}^n$ and $\mathbf{B}^{2n-1}$, respectively ($n > 1$). Let F be a twice continuously differentiable CR mapping from M_1 and M_2. Then F must be rational.*

The proof of Corollary 2.1 can be seen as follows: As in [Hu1], we may assume without loss of generality that M_1 and M_2 are open pieces of the Heisenberg hypersurfaces containing the origin, and $F = (f, \phi, g)$ satisfies the normalization condition as in [(3.1), Hu1]. By [Lemma 5.3, Hu1], we have that $f = z + \frac{iw}{2} a^{(1)}(z) + o_{wt}(3)$ and $g = w + o_{wt}(4)$ with $< a^{(1)}(z), \overline{z} > |z|^2 = \sum_{j=1}^{n-1} |\phi_j^{(2)}(z)|^2$. For each $(z_0, w_0) \in M_1$, define $F^{**}_{(z_0,w_0)}$ as in [Hu1]. (See notation following [Lemma 4.1, Hu1]. Also, we notice that $F^{**}_{p_0}$ only depends on p_0 and the choice of $(C_l)_{p_0}$). Then we claim that if $F \not\equiv (z, 0, w)$, after replacing F by a certain $F^{**}_{(z_0,w_0)}$ with $(z_0, w_0) \approx 0$ (if necessary), it must hold that $\{\phi_j^{(2)}\}_{j=1}^{n-1}$ is a linearly independent family over $\mathbf{C}$. Indeed, suppose not. We can then assume that $\phi_{n-1}^{(2)} \equiv \sum_{j=1}^{n-2} \mu_j \phi_j^{(2)}$ for some complex numbers μ_j, without loss of generality. Notice that $\sum_{j=1}^{n-1} |\phi_j^{(2)}|^2$ can now be written as $\sum_{j=1}^{n-2} \chi_j(z) \overline{\phi_j^{(2)}}(z)$ for certain holomorphic polynomials $\chi_j(z)$. Applying [Lemma 3.2, Hu1] to the equation $< \overline{z}, a^{(1)}(z) > |z|^2 = \sum_{j=1}^{n-2} \chi_j(z) \overline{\phi_j^{(2)}}(z)$, it follows that $a^{(1)}(z), \phi_j^{(2)}(z) \equiv 0$ for each j. Since this phenomenon also holds for $F^{**}_{p_0}$ for each p_0 and $(C_l)_{p_0}$, it follows from [Proposition 4.2, Lemma 5.3, Hu1] that $F \equiv (z, 0, w)$. This is a contradiction. Next, write $\phi_j^{(2)} = \sum_{1 \le k \le l \le n-1} a_{kl}^j z_k z_l$. Write $\xi_{kl} = (a_{kl}^1, \cdots, a_{kl}^{n-1})$. By the above, the family of vectors $\{\xi_{kl}\}_{kl}$ has maximum rank $n-1$. Hence, we can find $\{k_j\}$ and $\{l_j\}$ such that $\det(a_{k_i l_i}^j)_{1 \le i,j \le n-1} \neq 0$. Together with the normalization condition, we thus conclude that the following is also linearly independent: $\{L_1 \widetilde{f}, \cdots, L_{n-1} \widetilde{f}, L_{k_1} L_{l_1} \widetilde{f}, \cdots, L_{k_{n-1}} L_{l_{n-1}} \widetilde{f}\}|_0$. (Here $L_j = 2\sqrt{-1}\overline{z_j} \frac{\partial}{\partial w} + \frac{\partial}{\partial z_j}$). Now, applying [Theorem 1, Fa3] and a result of Forstneric [Fo2], we get the proof of Corollary 2.1.

The main idea introduced in [Hu1] depends on the following two steps (a): Derive a jet relation for the normalized map (b): Push, by using the huge automorphisms, the normalized equation into a partial differential equation, from which one tries to dig out the possible rigidity property of the maps. This idea might be also useful to attack the following two long standing open questions:

Problem 2.2 *Let $M_1 \subset \mathbf{C}^n$ and $M_2 \subset \mathbf{C}^N$ be open pieces of spheres and let f be a CR mapping from M_1 into M_2 ($1 < n < N$). Can one find a fixed constant t, independent of $N - n$, such that whenever f is C^t over M_1, f is then rational? Furthermore, can one take $t = 2$ or even 1?*

Problem 2.3 ([Fr1] [DA1]) *Let M_1, M_2 and f be as above. Suppose that f is rational. What is the possible maximum degree of f? When $n = 2$, is then the degree of f bounded by $2N - 3$?*

Problem 2.2 was solved by Forstneric [Fr2] when f is C^{N-n+1}-regular, by a reflection principle argument. He also obtained a rough bound estimate for the degree of F in terms of n and N. The key point of Problem 2.2 here is whether one can get a number t which is independent of the codimension.

3. Rigidity of holomorphic mappings with group actions

This problem has been motivated by the well-known super-rigidity problem in differential geometry and ergodic theory: Let G_1 and G_2 be two non-compact semi-simple Lie groups with $dim(G_2) \geq dim(G_1)$. Suppose $\Gamma_j \subset G_j$ are lattices (with certain density or co-compactness properties) and there is a injective homomorphism $\phi : \Gamma_1 \to \Gamma_2$. Can one then extend ϕ to a global homomorphism from G_1 into G_2? This question has been quite well understood when the real ranks of the groups are at least 2 by the work of Mostow, Margulis and Mok-Siu-Yeung (see [MSY], for example). In the rank one case, there are essentially four cases to be considered: The automorphism groups of real, complex and quaternionic hyperbolic spaces, together with that of the hyperbolic Cayley plane. By the work of many people (see [Cor] for related references) including Mostow, Gromov-Piastetski-Shaprio, Johnson-Milnor, Corlette, the important remaining open case is when $G_1 = Aut(\mathbf{B}^n)$ and $G_2 = Aut(\mathbf{B}^N)$ with $1 < n < N$, Γ_1 co-compact in G_1 and $\phi(\Gamma_1)$ convex-cocompact. (Notice that when $n = 1$, certain kind of counter-examples have been constructed by Mostow). This problem, after applying the harmonic mapping theory, is reduced to the following (see [Yue] or [Cor]):

Problem 3.1 *Let $\mathbf{B}^n \subset \mathbf{C}^n$ and $\mathbf{B}^N \subset \mathbf{C}^N$ be the unit balls, and let f be a proper holomorphic embedding from $\mathbf{B}^n$ into $\mathbf{B}^N$ $(n, N > 1)$. Suppose that there is a discrete subgroup $\Gamma \subset Aut(\mathbf{B}^N)$ such that Γ fixes $M = f(\mathbf{B}^n)$ and acts co-compactly over M. Is f then a linear embedding?*

The result of Gromov, etc, says that the supper-rigidity fails in case the groups come from the isometry groups of real hyperbolic spaces; and the result of Corlette states the supper-rigidity indeed holds for the isometry groups of the hyperbolic balls in the quaternionic space $\mathbf{Q}^n$ with $n \geq 2$ or hyperbolic Cayley plane (see [Cor] for a more precise and detailed account on this matter). Hence, it is very interesting to find out what happens for the automorphism groups $Aut(\mathbf{B}^n)$ of balls in complex spaces with $n > 1$? Does the quaternionic space $\mathbf{Q}^n$ with $n \geq 2$ enjoy more rigidity than $\mathbf{C}^n$ with $n \geq 2$ in this regard? A solution to Problem 3.1 is desirable, not just only from the point view of complex analysis, but also from the ergodic theory, dynamics and Lie group theory point of view.

We mention that this question has been answered in the affirmative in the work of Cao-Mok [CaoM] in the small codimensional case ($N \leq 2n - 1$) by using some pure Kähler Geometry arguments. Notice that the map f in problem 3.1 is always very nice along the orbit. Hence, it may be possible to consider some kind of the Moser formal theory along each orbit and then derive a certain kind of difference equation as in [Hu1] so that the argument we developed in [Hu1] can be made into prospective. To make such a process workable, one needs a very good mastering of the dynamic property of hyperbolic elements of $Aut(\mathbf{B}^n)$.

4. Rigidity of algebraic domains and their mappings

This study was initiated in a celebrated paper of S. Webster [We1], who first proved that any biholomorphic map between two algebraic Levi non-degenerate hypersurfaces must be an algebraic map. Recent developments of his theorem can be found in the work of [BER], [Zat], etc. In [HJ], we proved a sup-rigidity theorem concerning algebraic spheric domains, which claims all bounded algebraic spherical domains in $\mathbf{C}^n$ with $n > 1$ must be biholomorphically equivalent to the ball.

The motivation for the work [HJ] with Ji can be described as follows: After the fundamental work of Chern-Moser [CM] and Fefferman, people started to attack the complex structures of domains by using the boundary CR geometry. A very natural question is then to ask if one can characterize the ball by its boundary CR geometry. More precisely, is any bounded spherical domain biholomorphic to the ball? Indeed, the work of Krushilin-Pinchuk-Vitsushili in the 1970's shed the light toward the positive side of such a problem. Surprisingly, in 1980, Burns-Schnider [BS], provided a counter-example to this. They showed that there are many bounded real analytic spherical domains in $\mathbf{C}^n$, which are topologically equivalent to the torus—hence, can not be the ball. A naive observation is as follows: If the boundary is spherical, it should be flat in a certain sense. Hence, to make it compact, one has to use a periodic function to define it. (The defining functions of Burns-Schnider's examples are trigonometric functions). The domain then can not be algebraic. It is indeed such a (non-rigorous) intuitive which led us to the work in [HJ].

Theorem 2 of [HJ] also tells the set of all bounded strongly pseudoconvex algebraic domains indeed forms a strictly smaller subset of all real analytic domains; and indicates that there are many rigidities for algebraic domains which analytic domains do not share. It seems to us our understanding to the specific holomorphic properties of algebraic domains is still at a very early stage, comparing to the role of complex algebraic geometry in complex analytic geometry (see [We4]). An interesting question arising naturally is first to develop a systematic invariant theory to distinguish strongly pseudoconvex algebraic hypersurfaces from strongly pseudoconvex real analytic hypersurfaces. Such an investigation might motivate out new interesting problems, which can lead to a further understanding of rigidities for algebraic domains. Perhaps, a good start is to find a motivated proof whether $M_1 = \{(z, u + iv) \in \mathbf{C}^2 : v = e^{|z|^2} - 1\}$ is holomorphically equivalent to an algebraic hypersurface. Peter Ebenfelt recently communicated to us that by the result in [BHR1] (see Theorem 7.5 in the next section), his example in [E] on a Levi non-flat real analytic hypersurface M in $\mathbf{C}^2$ can not be biholomorphic to any algebraic hypersurface, due to the presence of non-trivial holomorphic curves and non-degenerate smooth but not analytic CR mappings from his M to other Levi non-flat algebraic hypersurfaces (see [E]).

When one deals with strongly pseudoconvex hypersurfaces, a totally different argument is needed. An idea to carry out this study might be to use the Cartan-Chern-Moser connections and curvatures, which and whose functional relations are algebraic for algebraic strongly pseudoconvex hypersurfaces. It is very interesting to see what such an implication would really be.

5. Smooth and analytic structures of a Bishop submanifold in $\mathbf{C}^n$ and local complex Plateau problems

Let M be a real k-manifold in $\mathbf{C}^n$ with $p \in M$. One of the fundamental problems in complex analysis is to understand the holomorphic invariants of M near p. When p is a totally real point, namely, $T_p^{(1,0)} = CT_pM \cap \sqrt{-1}CT_pM = \{0\}$, M near p basically has the character of the standard $\mathbf{R}^k$ in $\mathbf{C}^n$. When the complex tangent space $T_p^{(1,0)}M \neq 0$, the situation can be much more complicated. Thus far, only sufficiently non-degenerate complex tangents have been studied. Recall that p is called a point with an elliptic complex tangent if the Levi-form of M in a certain sense is positive at p ([CM], [Bis]). In this subject, there are two cases which are of particular interest– namely, $k = 2n-1$ and $k = n$. When M is a hypersurface ($k = 2n-1$), p is a point with an elliptic complex tangent if and only if M is strongly pseudoconvex near p. In this case, the geometry of M near p has been relatively quite well understood and has been successfully used for the study of the equivalence problem as in the work of Cartan and Chern-Moser [CM]. When $k = n$ with $p \in M$ an elliptic complex tangent, one has the least possible complex structure near p, and away from a thin set, M is totally real. In this situation, the existence of 'good geometry', which can be used for the equivalence problem, is still not completely understood in some important situation.

Bishop [Bis] in 1965 first introduced the second order biholomorphic invariant, now called the Bishop invariant for the case $n = k$, which corresponds to the Levi-eigenvalue in the hypersurface case. He showed that M near such a p bounds many complex curves and has a non-trivial local hull of holomorphy $\widetilde{M}$. Bishop conjectured that the local hull $\widetilde{M}$ is a Levi flat submanifold which is foliated by pairwise disjoint embedded complex analytic disks attached to M. Also, he proposed the study of the fine structure of $\widetilde{M}$ near p. In the case of $n = 2$, Kenig-Webster later showed, in their paper [KW1], that $\widetilde{M}$ is indeed a smooth-up-to-the boundary Levi-flat hypersurface in $\mathbf{C}^2$ foliated by a family of analytic disks. (See also the work done in [BG] [BK] for the global case.) When $n > 2$, Kenig-Webster [KW2] showed that for each l, M bounds a Levi flat submanifold $\widetilde{M_l}$, which is C^l up to M. But they left open whether all these $\widetilde{M_l}$ are the same and give the local hull of holomorphy. In the work of Moser-Webster [MW], it was shown that $\widetilde{M}$ is real analytic across M in case M is a real analytic submanifold and when the Bishop invariant attached at p does not vanish. But Moser-Webster's argument does not apply in the case of the vanishing Bishop invariant. (See also the work done in [Mos] where the case some special case in the vanishing Bishop situation invariant was studied.) In the author's joint paper with Krantz [HK], in case $M \subset \mathbf{C}^2$, we studied the real analytic structure of $\widetilde{M}$ in the vanishing Bishop case, settling an open question of Moser in [Mos]. Recently, In [Hu2], with a more subtle analysis of the singular Bishop equation and formal theory argument, we established the above mentioned results in any dimensions, and showed that the $\widetilde{M_l}$'s constructed by Kenig-Webster in [KW2] are indeed the same. The effort in [Hu2] provides, together with the work of the above mentioned mathematicians, a complete solution to an old problem initiated in the work of Bishop [Bis]:

Theorem 5.1 *Let $M \subset \mathbf{C}^n$ be a smooth submanifold of real dimension n. Let $p \in M$ be an elliptic complex tangent point. Then M bounds, near p, a unique Levi-flat CR submanifold $\widetilde{M}$ of real dimension $n+1$. Moreover, $\widetilde{M}$ is foliated by embedded complex analytic disks and $\widetilde{M}$ extends smoothly across the boundary M. Also, $\widetilde{M}$ gives the local hull of holomorphy of M. When M is real analytic, $\widetilde{M}$ is real analytical across M and M can be flattened (namely, M can be biholomorphically transformed into $\mathbf{R}^{n-1} \times \mathbf{C}$).*

Notice that in the case of dimension two, when p is elliptic, after a holomorphic change of coordinates, we can always assume that $p = 0$ and M near p is defined by an equation of the form: $w = z\overline{z} + \lambda(z^2 + \overline{z}^2) + o(|z|^2)$, $\lambda \in [0, 1/2)$. Here λ is the Bishop invariant. The nice property of the local hull of holomorphy $\widetilde{M}$

provides some important information for the holomorphic structure of M. However, a further understanding of the whole set of invariants of surfaces of this type is of considerable interest and desirable because of the following reasons: First, from the point view of complex analysis, they can be viewed as the simplest higher codimensional analogy of strongly pseudoconvex hypersurfaces; secondly, they have a rich complex structure at the elliptic complex tangent and have trivial complex structure elsewhere, namely they can also be viewed as the simplest models where one sees the CR singularity; thirdly, from the celebrated work of Moser-Webster [MW], one sees here a tremendous interaction of complex analysis with the classical dynamics problems in Celestial Mechanics [SM]– An understanding of such a problem may provide information and motivation to some convergence problems in Mechanics.

In [MW], the normal form was established for the elliptic complex tangent with non-vanishing Bishop invariant, and M is showed to posses only 'two and a half' biholomorphic invariants near p. However, the situation with vanishing Bishop invariant is quite different and M may support infinitely many biholomorphic invariants, as conjectured by Moser and Webster [MW] [Mos]. As is known, one of the major difficulties to the study of the normal form theory is to provide the convergence proof for certain formal power series. In the hypersurface case, Chern-Moser [CM] circumvented this in their work by applying the intrinsic CR geometry of the manifolds (namely, CR structure bundles, the Cartan-Chern-Moser chains, etc). In the non-vanishing Bishop invariant case with $k = n = 2$, Moser-Webster [MW] obtained this by ingeniously using the geometry of a pair of involutions (intertwined by the natural conjugation operator) defined over the complexification of M, and then reducing the problem to the normalization problems of these involutions. The convergence proofs can then be done by a very clever use of the majorant methods as in the work, dealing with the classical mechanics, of Birkhoff, Siegel, Moser [SM]. In the vanishing Bishop invariant case, namely, $\lambda = 0$, the aforementioned Moser-Webster involutions do not exist anymore and other geometry has to been employed for the equivalence problem. In [HK] and [Hu2], we showed that M can be mapped into the Levi-flat analytic hypersurface $\mathbf{R} \times \mathbf{C}$ and bounds a family of holomorphic disks depending real analytically on a parameter. This may give some very useful information for the convergence proof of the normal-form related formal series. For instance, in [Hu2] using this geometry, we were able to handle the convergence proof of the normalized formal solution of an important functional equation as in Theorem 5.2:

Theorem 5.2 *Let M be an elliptic Bishop n-manifold in $\mathbf{C}^n$ near the origin. For any real analytic function $G(z, \overline{z})$ defined near 0, there is a holomorphic solution F to the functional equation: $Im(F(z)) = G(z, \overline{z})$, where $z \in M$.*

We think this should also be one of the basic tools to answer the following fundamental question of Moser-Webster [MW] and Moser [Mos] in this subject of several complex variables:

Problem 5.3 *Let $E \subset \mathbf{C}^2$ be a real analytic submanifold of real dimension 2 defined by $w = |z|^2 + o(|z|^2)$. (That is, $0 \in E$ is an elliptic complex tangent point with Bishop invariant $\lambda = 0$). What are the biholomorphic invariants of E at 0? Is E biholomorphic to a real algebraic surface?*

6. Topology of pseudoconvex domains in $\mathbf{C}^2$

We now mention some other related global problems, which are related to our work described above. Probably, the simplest non-trivial case of the above local consideration is the filling problem of a general 2-submanifold M in $\mathbf{C}^2$ by holomorphic disks. Through the work of Bedford-Gaveau [BG], Bedford-Klingenberg [BK], Gromov [Gr] and Eliashberg [Eli], etc, certain positive results have been obtained in case M is a topological 2-sphere in generic position with a certain convexity assumption. However, when M does not have these properties, the problem is far from being understood. A natural question is to understand the geometric and topological obstructions to the filling process. Another interesting problem in this field is to ask what kind of Levi-flat hypersurface $\widetilde{M}$ a totally real 2-torus M in $\mathbf{C}^2$ can bound. We notice that Alexander's recent solution to the Bennequin problem reveals that $\widetilde{M}$ is usually not smooth. We also remark that, in case M is the graph of a torus in $\mathbf{R}^3$, the problem is equivalent to studying the regularity problem of the Dirichlet problem of a certain totally degenerate non-linear elliptic differential equation.

The combining effort of the two dimensional filling result of Bedford-Gaveau [BG] and the regularity result of Kenig-Webster[KW1] gives the following Theorem (BGKW): Any generic two-sphere $S \subset \mathbf{C}^2$ with only two elliptic complex tangents and contained in the boundary of a certain strongly pseudoconvex domain bounds a unique smooth Levi-flat holomorphically convex hypersurface M which is smooth up to S. This is a deep achievement in the subject of several complex variables, whose proof is quite non-trivial. Hence, it will not be surprised to us if one finds some deep and important applications. For us, it is desirable to see if they can be applied to work on various problems concerning the topology of pseudoconvex domains. For instance, we believe that together with the Morse Theory and a subtle use of Theorem (BGKW), one should be able to get a complete understanding of the following problem (see [Fr3] for certain studies related to this problem):

Problem 6.1 *Let D be a simply-connected smoothly bounded pseudoconvex domain in $\mathbf{C}^2$. What can we say about its second fundamental group? In particular, can D be homotopic to the unit 2-sphere? What can we say about the topology of ∂D? Is it also simply connected?*

Problems of this type are crucial for the understanding of many problems in both Geometry and Topology. (See [Za])

One idea to work on the last part of Problem 6.1 is to use the Morse theory and Theorem (BGKW). Namely, one will try to deform any 2-sphere embedded in $\overline{D}$, touching a certain fixed boundary point p to ∂D, relative to p. Use the Morse theorem to understand the character of the critical points of the Morse defining function and use Theorem (BGKW) to pave a way for the the 2-sphere to get rid of the critical points with index 2. When $n > 2$, this process goes through very easily. (See §5 of [HJ].)

7. Regularity of CR mappings between real hypersurfaces in $\mathbf{C}^n$

One of the important properties for holomorphic functions in one complex variable is the Schwarz reflection principle, which states that any holomorphic function f defined over a certain side of a real analytic curve γ and continuous up to γ must extend holomorphically across γ once $f(\gamma)$ is also contained in a real analytic curve. A natural consideration of this phenomenon in higher dimensions seems to replace γ by real analytic hypersurfaces. However, a moment thought indicates that the situation is not that simple anymore: Let $M_1 = M_2 = \{(z,w) \in \mathbf{C}^2 : Im(w) = \mathrm{Re}(w)|z|^2\}$ and $F = (e^{-1/w^{1/3}}, 0)$. Then $F(M_1) \subset M_2$, F extends holomrophically to a certain side of M_1 and smooth up to M_1, but F does not extend holomorphically across M_1. An easy investigation shows that the failure of the reflection principle in this example is due to the existence of non-trivial holomorphic curves inside M_1 and the degeneracy of the map. Hence, this suggests one to formulate the following problems. (Here we recall a real analytic hypersurface is of D'Angelo finite type if it does not contain any non-trivial germs of holomorphic curves):

Problem 7.1 *Let M_1 and M_2 be two real analytic hypersurfaces in $\mathbf{C}^n$ of finite D'Angelo type. Let f be a continuous CR mapping from M_1 into M_2. Is f then real analytic over M_1?*

Problem 7.2 *Let M_1 and M_2 be sufficiently non-degenerate hypersurfaces but of infinite type. Let f be a sufficiently smooth CR map from M_1 into M_2, which is also sufficiently non-degenerate. Is then f real analytic?*

Indeed, although a lot of attention has been paid by many mathematicians in the past twenty-five years, whether the above holds or not still remains to be unsolved questions in the subject.

The most important contribution along this direction is contained in the the celebrated work of C. Fefferman, which, in particular, indicates that any CR homeomorphism between strongly pseudoconvex real analytic hypersurfaces is a smooth CR diffeomorphism. The work of Chern-Moser [CM] [Jac] and Lewy [Le], Pinchuk [Pi] can then be used to give the analyticity for such a map. An extensive survey on this subject with a fairly complete list of references, after the work of Fefferman and Chern-Moser, can be found in the survey paper of Forstneric [Fr1]. Hence, we will only focus on the development after the survey paper [Fr1].

Regarding Problem 7.1 and before [Hu3] and [BHR1], there had been many studies for maps which either have some kind of Hopf lemma property or extend properly to a certain side through the work of many people (See [Fr1] for a detailed survey and related references). However, very little had been known about the regularity of general CR mappings between non-strongly pseudoconvex hypersurfaces. Indeed, the following theorem proved in [Hu3] was, at the time, new under the additional assumption that the map is smooth or the hypersurfaces are pseudoconvex. (This result is due to Baouendi-Bell-Rothschild [BBR] when f is smooth and does not vanish to infinite order at any point, and is due to Diederich-Fornaess-Ye [DFY] when the map is a CR homoemorphism.)

Theorem 7.3 *Let M_1 and M_2 be two (connected) real analytic hypersurfaces in $\mathbf{C}^2$ which do not contain any non-trivial holomorphic curves inside. Let f be a non-constant continuous CR mapping from M_1 into M_2. Then f is real analytic along M_1. Moreover, f extends locally properly to a neighborhood of M_1.*

For the general dimensions, we state here a result obtained in the joint work with Pan [HP] and Baouendi, Rothschild [BHR2], which, together with the theorem of Baouendi-Rothschild [BR1] and Diederich-Fornaess [DF1], gives the following:

Theorem 7.4 *Let M_1 and M_2 be two real analytic hypersurfaces of finite D'Angelo type in $\mathbf{C}^n$ $(n > 1)$. Suppose that f is a smooth CR mapping from M_1 into M_2. If f is real analytic at most away from a certain compactly supported subset of M_1, then f is actually real analytic over M_1. In particular, any proper holomorphic map between two bounded real analytic domains in $\mathbf{C}^n$ admits a holomorphic extension across the boundary of the source domain if it is C^∞-smooth up to the boundary of the source domain.*

In regard to Problem 7.2, using the strengthened version of the Artin approximation theorem, we were able to obtain in a joint work with Baouendi-Rothschild [BHR1] the following, which essentially gives a fairly nice solution of Problem 7.2 in the algebraic category: (See [Sta] [BHR1] for the definition of related concepts).

Theorem 7.5 *Let M_1 and M_2 be two connected algebraic hypersurfaces in $\mathbf{C}^n$. Assume that M_1 is holomorphically non-degenerate. Then there is an integer k depending only on the degrees of M_1 and M_2 such that any CR mapping f from M_1 into M_2, which is C^k- smooth and whose Jacobian is not identically zero, must be real analytic over M_1.*

Concerning Problem 7.1, by Baouendi-Treves [BT], one knows that the map there extends holomorphically to a certain side D. However, it is usually very difficult to know whether a certain type of the Hopf lemma still holds for the normal component of the map or whether the map extends properly to D so that many results developed for the global case can be applied. In fact, it is often the case that one has to get the regularity before understanding the Hopf lemma property and the local properness. For example, the following seems to be a non-trivial open question: Is any continuous CR mapping from an open piece of the sphere in $\mathbf{C}^3$ near the origin into $M_2 = \{(z_1, z_2, w = u + iv) \approx 0 : v = |z_1|^2 - |z_2|^2 + 100|z_2|^4\}$ a constant map? (The difficulty lies in that we do not know if such an f extends properly to a certain side.)

Theorem 7.3 gives a complete solution to Problem 7.1 in the two dimensional case. Since it fails for hypersurfaces of infinity type and fails for non-continuous bounded CR maps, it can be regarded as a natural extension of the classical Schwarz reflection principle in $\mathbf{C}^2$. In the following, we give a brief account on the proof of Theorem 7.3 (more detailed account on the closely related history and references can be found in [Hu3], in particular, the third paragraph on page 1785 of [Hu3]):

By [BT], the map in Theorem 7.3 is known to extend holomorphically to a certain side D of M_1 with $f \in C(D \cup M_1) \cap Hol(D)$. Hence, one needs to construct the extension of the map to another side D^c. Since it is difficult to get the direct holomorphic extension of the map to the D^c-side (unless the target hypersurface is strongly pseudoconvex), one wishes to split the proof into two parts, as suggested in many papers (see [BJT], [DF1], [DFY], for example): *Part (a)*: Construct the multiple valued extension of the map f to the D^c-side; and *Part (b)*: Prove the single value property assuming the multiple-valued extension.

Part (b) follows from the Malgrange theorem as in the work of [BJT] [BBR] [DF1] when the map is assumed to be smooth. Apriori to our work, there had been the work done by Diederich-Fornaess [DF2], which gave a solution for (b) in the CR homeomorphism case (or, globally, biholomorphic case.) Their approach, however, is difficult to extend to proper and CR mappings. We circulated a proof for this step, namely, Part (b) for proper and general CR mappings in the earlier version [Hu7] of [Hu3], by a quite different approach from [DF2]. (In November 1995,

Diederich-Pinchuk announced at MSRI that they also had a proof of this result for proper holomorphic mappings.)

Concerning Part (a), at the time, there had been the work done by Diederich-Fornaess-Ye in [DFY]. Meanwhile, Diederich-Pinchuk had also distributed a lecture note on the construction of the multiple valued extension for biholomorphic maps, by using the above mentioned Part (b) result in [DF2]. (See [Hu3, pp 1785, Note added February, 1996] for some discussion on this matter) Their note, referenced as [23] in [Hu3] (see also [DP] in [Hu7]), was later called by Diederich-Pinchuk in the Remark on Page 1091 of [DiPi] as the preliminary notes. In passing, we would like to make a clarification of this remark, simply by mentioning to the reader that the later published paper [DiPi], comparing with their previously circulated note mentioned above, contains a fundamentally new and different part , namely, the proof of the above mentioned Part (b) for proper holomorphic maps, (which had been previously obatined in [Hu7, August 1995], and annouced by Diederich-Pinchuk at MSRI in November 1995.)

It had been clear that once (b) was settled, the proof of Part (a) for mappings which are assumed apriori to be proper from D could be achieved by slightly modifying the above mentioned work for Part (a) in the biholomorphic case. However, the situation for the study of Part (a) for general CR mappings was different, partially because the previously established global results for the pseudoconvex domains by Baouendi-Bell-Rothschild, etc, (see [BC]) and the pseudoconcave preserving property of proper mappings were no longer applicable. We succeeded in writing up a proof of Part (a) for general CR mappings by using a continuity method, which was also used in the papers of Webster [We2] and Diederich-Fornaess-Ye [DFY]. The major difficulty in the continuity method, as usual, was to get the closeness. The key point which enabled us to get (a) for general CR mappings was a way to use the the geometric size of the Segre varieties to control the rate of blowing-up of the multiple-valued extension of f to the D^c-side, from which the closeness could be concluded. (See Lemma 3.1 of [Hu3], for example). This idea also works for the n-dimensional pseudoconvex case and was further exploited in [Hu4].

[Hu4] is concerned with the problem of Bell and Bell-Catlin [BC]:

Problem 7.6 *Let M_1 and M_2 be two pseudoconvex real analytic hypersurfaces of finite D'Angelo type in $\mathbf{C}^n$. Let f be a continuous CR mapping from M_1 into M_2. Is then f real analytic over M_1.*

Regarding Problem 7.6, Bell-Catlin [BC] showed that f is real analytic if for each $p \in M_1$, $F_p = f^{-1}(f(p))$ is a compact subset of M_1. On the other hand, if one would have already known that f is non-constant and real analytic, then $Q_q = Q_p$ for each $q \in F_p$ and F_p would be discrete. (Here Q_q is the Segre variety of M_1 associated to q). Indeed, the work in [Hu4] essentially tells that we need only to verify the Bell-Catlin condition on $E_p = Q_p \cap F_p$. This, in particular, gives the following corollary: The answer to Problem 6 is affirmative if M_1 has the following locally finite stratification $M_1 = \cup_j M_1^j$, where for each $q \in M_1^j$, $(Q_q \setminus \{q\}) \cap O(q) \subset \cup_{l<j} M_1^l$.

The implication of the work in [Hu3] [BHR1] may also give some hints to work on the following problem of Alinhac-Baouenid-Rothschild:

Problem 7.7 *Let f be a smooth non-constant CR-mapping from M_1 into M_2, two smooth real hypersurfaces of finite D'Angelo-type in $\mathbf{C}^n$ $(n > 1)$. Does the* *normal component of f have non-vanishing normal derivative at each point of M_1?*

References

[Alx] H. Alexander *Proper holomorphic maps in* $\mathbf{C}^n$, **Ind. Univ. Math. Journal 26,** (1977), 137–146

[BBR] S. Baouendi, S. Bell and L. Rothschild; *Mappings of three-dimensional CR manifolds and their holomorphic extension,* **Duke Math. J. 56** (1988), 503–530.

[BER] S. Baouendi, P. Ebenfelt and L. Rothschild, *Real submanifolds in complex spaces and their mappings,* **Princeton Mathematical Series, 47** (Princeton, New Jersey: 1999.)

[BHR1] S. Baouendi, X. Huang and L. Rothschild, *Regularity of CR mappings between algebraic hypersurfaces* **Invent. Math. 125** (1996), 13–36.

[BHR2] S. Baouendi, X. Huang and L. Rothschild *Nonvanishing of the differential of holomorphic mappings at boundary points* **Math. Res. Lett. 2** (1995), 737–750.

[BJT] S. Baouendi, H. Jacobowitz and F. Treves, *On the analyticity of CR mappings* **Ann. of Math. 122** (1985), 365-400

[BR1] S. Baouendi and L. Rothschild, *Germs of CR maps between real analytic hypersurfaces,* **Invent. Math. 93** (1988), 481-500.

[BT] S. Baouendi and F. Treves, *About the holomorphic extension of CR functions on real hypersurfaces in complex space,* **Duke Math. J. 51** (1985), 77–107.

[BG] E. Bedford and B. Gaveau, *Envelopes of holomorphy of certain 2-spheres in* $\mathbf{C}^2$ **Amer. J. Math. 105** (1983), 975-1009.

[BK] *E. Bedford and W. Klingenberg, holomorphy of a 2-sphere in* $\mathbf{C}^2$ **Journal of AMS 4** (1991), 623-655.

[Be] S. Bell *Analytic hypoellipticity of the* $\overline{\partial}$*-Neumann problem and extendibility of holomorphic mappings,* **Acta Math. 147** (1981), 109-116.

[BC] S. Bell and D. Catlin, *Regularity of CR mappings,* **Math. Z.** (1988) , 357-368.

[Bis] E. Bishop, *Differentiable manifolds in complex Euclidean space,* **Duke Math. J. 32** (1965), 1-21.

[BS] D. Burns and S. Shnider, *Spherical hypersurfaces in complex manifolds,* **Invent. Math. 33** (1976), 223-246.

[CaM] H. Cao and N. Mok, *Holomorphic immersions between compact hyperbolic space forms,* **Invent. Math. 100** (1990), 49–61.

[Ca] D. Catlin, *Boundary invariant of pseudoconvex domains,* **Ann of Math 120** (1984), 529-586.

[CS] J. Cima and T. J. Suffrige, *A reflection principle with applications to proper holomorphic mappings,* **Math Ann. 265** (1983), 489-500.

[CY] S. Y. Cheng and S. T. Yau, *On the existence of a complex Kähler metric on non-compact complex manifolds and the regularity of Fefferman's equation,* **Comm. Pure Appl. Math. 33** (1980), 507–544.

[CM] S. S. Chern and J. K. Moser, *Real hypersurfaces in complex manifolds,* **Acta Math. 133** (1974), 219-271.

[Cor] K. Corlette, *Archimedean superrigidity and hyperbolic geometry,* **Ann. of Math. (2) 135** (1992), 165–182.

[DA1] J. P. D'Angelo, *Several Complex Variables and the Geometry of Real Hypersurfaces,* **Boca Raton: CRC Press,** 1993).

[DA2] J. D'Angelo, *Real hypersurfaces, orders of contact, and applications,* **Ann. Math. 115** (1982), 615-637.

[DA3] J. P. D'Angelo, *Polynomial proper holomorphic mappings between balls,* **Duke Math J. 57** (1988), 211-219.

[DF1] K. Diederich and J.E. Fornaess, *Proper holomorphic mappings between real analytic pseudoconvex domains in* $\mathbf{C}^n$, **Math. Ann. 282** (1988), 681-700.

[DF2] K. Diederich and E. Fornaess, *Applications holomorphes propres entre domaines a bord analytique reel,* **C-R-Paris-Ser.I-Math. 307** (1988), 321- 324.

[DFY] K. Diederich, J. Fornaess, and Z. Ye, *Biholomorphisms in Dimension 2,* **J. Geom. Anal. 4** (1994), 539-552.

[DiPi] K. Diederich and S. Pinchuk, *Proper holomorphic maps in Dimension 2 extend,* **Indiana Univ. Math. J. 44** (1995), 1089-1126.

[Dr] A. Dor, Proper holomorphic maps between balls in one co-dimension, **Ark. Math.28** (1990), 49-100.

[E] P. Ebenfelt, *on the unique continuation problem for CR mappings into nonminimal hypersurfaces,* **J. Geom. Anal. 6** (1996) 385–405.

[Eli] Y. Eliashberg *Filling by holomorphic discs and its applications, Geometry of Low-Dimensional Manifolds* **London Math. Soc. Lecture Notes 151** 1997

[Fa1] J. Faran *Maps from the two ball to the three ball* **Invent Math. 68** (1982), 441-475.

[Fa2] J. Faran, *The linearity of proper holomorphic maps between balls in the low codimension case,* **J. Diff. Geom. 24** (1986), 15-17.

[Fa3] J. Faran, *A reflection principle for proper holomorphic mappings and geometric invariants,* **Math. Z. 203** (1990) 363-377.

[Fr1] F. Forstneric, *A survey on proper holomorphic mappings,* **Proceeding of Year in SCVs at Mittag-Leffler Institute, Math. Notes, (Princeton, NJ: Princeton University Press** 1992) **38.**

[Fr2] F.Forstneric, *Extending proper holomorphic mappings of positive codimension,* **Invent. Math. 95** (1989), 31-62.

[Fr3] F. Forstneric, *Complex tangents of real surfaces in complex surfaces,* **Duke Math. J. 67** (1992), 353–376.

[GR] M. Gromov, *Pseudo holomorphic curves in symplectic geometry,* **Invent Math. 82** (1985), 307-347.

[HS] M. Hakim and N. Sibony, *Fonctions holomorphes bornes sur la boule unite de* $\mathbf{C}^n$, **Invent. Math. 67** (1982), 213–222.

[Hu1] X. Huang, *On a linearity problem of proper holomorphic mappings between balls in complex spaces of different dimensions,* **Jour of Diff. Geom. 51 (No. 1)** (1999), 13-33.

[Hu2] X. Huang, *On an n-manifold in* $\mathbf{C}^n$ *near an elliptic complex tangent,* **J. Amer. Math. Soc. 11** (1998), 669–692.

[Hu3] X. Huang, *Schwarz reflection principle in complex spaces of dimension two,* **Comm. in PDE 21** (1996), 1781-1828.

[Hu4] X. Huang, *A Removable singularity property of CR mappings between real analytic hypersurfaces in* $\mathbf{C}^n$, **Comm. in PDE 25** (2000), 299-317.

[Hu5] X. Huang, *On the mapping problem for algebraic real hypersurfaces in complex spaces of different dimensions,* **Annales de L'Institut Fourier44** (1994), 433-463.

[Hu6] X. Hunag, *A non-degeneracy property of extremal mappings and iterates of holomorphic mappings,* **Annali Scoula Norm. Sup. Pisa, Series IV. Vol. XXI** (1994), 399-419.

[Hu7] X. Huang, *Regularity of holomorphic correspondences and applications to the mapping problem,* **Preprint circulated in August 1995.**

[HJ] X. Huang and S. Ji, *Global extension of local holomorphic maps and the Riemann mapping theorem for algebraic domains,* **Math. Res. Lett. 5** (1998), 247–260.

[HP] X. Huang and Y. Pan, *Proper holomorphic mappings between real analytic domains in* $\mathbf{C}^n$, **Duke Math. J. 82** (1996), 437-446.

[HK] X. Huang and S. Krantz, *On a problem of Moser,* **Duke Math. J. 78** 213-228.

[Jac] H. Jacobowitz, *An introduction to CR structures Mathematical Surveys and Monographs,* **American Mathematical Society, 32** (1990).

[KW1] C. Kenig and S. Webster, *The local hull of holomorphy of a surface in the space of two complex variables,* **Invent. Math. 67** (1982), 1-21.

[KW2] C. Kenig and S. Webster, *On the hull of holomorphy of an n-manifold in* C^n, **Annali Scoula Norm. Sup. de Pisa IV 11 No. 2** (1984), 261-280.

[Lem1] L. Lempert, *La metrique de Kobayashi et la representation des domain sul la boule,* **Bull. Soc. Math. France 109** (1981), 427-474.

[Lem2] L. Lempert, *A precise result on the boundary regularity of biholomorphic mappings,* **Math Z. 193** (1986), 559-579.

[Le] H. Lewy, *On the boundary behavior of holomorphic mappings,* **Acad. Zaz. Lincei. 35** (1977), 1-8.

[MSY] N. Mok, Y. T. Siu and S. K. Yeung, *Geometric superrigidity,* **Invent. Math. 113** (1993), 57-83.

[Mos] J. Moser, *Analytic surfaces in* $\mathbf{C}^2$ *and their local hull of holomorphy,* **Annales AcademiæFennicae Series A.I. Mathematica 10** (1985), 397-410.

[MW] J. Moser and S. Webster, *Normal forms for real surfaces in* $\mathbf{C}^2$ *near complex tangents and hyperbolic surface transformations,* **Acta Math. 150** (1983), 255-296.

[Pi] S. Pinchuk, *On the proper holomorphic mappings of real analytic pseudoconvex domains,* **Siberian Math. 15** (1974), 909-917.

[PS] S. Pinchuk and Tsyganov, *Maps between strongly pseudoconvex hypersurfaces,* **Akad. Nauk. SSSR 53** (1989), 1120-1129.

[Sta] N. Stanton, *Infinite CR automorphism of real heypersurfaces,* **Amer. Jour. of Math. 118** (1996), 209-233.

[SM] C. L. Siegle and J. Moser, *Lectures on Celestial Mechanics,* **Classics in Mathematics, Springer-verlag** (1995).

[TU] A. Tumanov, *Extension of CR functions into a wedge from a manifold of finite type,* **Math. USSR-Sb. 64** (1989), 129-140.

[We1] S. Webster, *On the mapping problem for algebraic real hypersurfaces,* **Invent. Math. 43** (1977), 53-68

[We2] S. Webster, *Biholomorphic mappings and the Bergmann kernels off diagonal,* **Invent. Math. 51** (1979) 155-169.

[We3] S. Webster, *On mappings an (n+1)-ball in the complex space,* **Pac. J. Math. 81** (1979), 267-272.

[We4] S. Webster, Surveys on mappings bewteen analytic and algebraic hypersurfaces, **Preprint** (1995).

[Yue] C. Yue, *Dimension and rigidity of quasi-Fuchsian representations,* **Ann. of Math. 143** (1996), 331–355.

[Za] M. Zaidenberg, *Open problems on open algebraic varieties,* **Montreal** (1994).

[Zat] D. Zaitsev, *Mappings between algebaric sets,* **Preprint** (1998).

Department of Mathematics, Rutgers University, New Brunswick, NJ 08903, USA

AMS/IP Studies in Advanced Mathematics
Volume 20, 2001

A KAM Theory for Resonant Tori and A Generalization of Poincaré-Birkhoff Fixed Point Theorem

Chong-Qing Cheng

Consider a nearly integrable Hamiltonian system, it has n degrees of freedom $(n \geq 3)$:

$$\dot{q} = \frac{\partial H}{\partial p}, \qquad \dot{p} = -\frac{\partial H}{\partial q}, \qquad\qquad q \in \mathbb{T}^n,\ p \in \mathbb{R}^n. \qquad 1$$

The Hamiltonian H has the KAM type, i.e.

$$H = N(p) + P(p, q),$$

where $N(p)$ is the main part and is assumed satisfying Kolmogorov nondegenerate condition: the determinent of its Hessian matrix is not zero, and P is a small term, representing small perturbation. If P vanishes, the system is completely integrable and the phase space has a foliation, every leaf is an n-dimensional invariant torus $p = p_o$, the flow generated by the vector field is quasi-periodic, its frequency Ω is determined by

$$\Omega = \frac{\partial N}{\partial p}(p_o).$$

If Ω satisfies some nonresonant condition called diophantine condition

$$|\langle k, \Omega \rangle| \geq D|k|^{-\mu}, \qquad \forall k \in \mathbb{Z}\backslash\{0\},$$

by the KAM theory, the torus has some persistency subject to the small perturbation ($|P| \ll 1$), and is still invariant but undergoes small deformation only. Such invariant n-dimensional tori occupy a set possessing nearly whole Lebesgue measure, but is nowhere dense. The complementary set, being dense set, is called instability region.

If Ω does not satisfy nonresonant conditions, it is believed that any typical small perturbation will destroy the corresponding invariant torus of the integrable system. Let's consider such a case, i.e. we assume that the Ω satisfies exactly k $(k < n)$ resonant conditions

$$\langle \Omega, I_j \rangle = 0, \qquad I_j \in \mathbb{Z}^n\backslash\{0\},\ n-k < j \leq n$$

where all I_j are integer-valued vectors, linearly independent. By the abelian group theory there exist other $(n-k)$integer-valued vectors $I_1, I_2, \ldots, I_{n-k}$ so that they

Supported by NNSF of China and Qiu Shi Foundation.

can construct a unimodular matrix, i.e.

$$\det I = \det|(I_1, I_2, \ldots, I_n)| = 1.$$

With this matrix we can construct a canonical transformation $(p, q) \to (x, y)$

$$x = qI, \qquad p = yI^t$$

which keep x be angle variable still. In the new coordinates we have

$$\frac{\partial N}{\partial y}(y_o) = (\omega_1, \omega_2, \ldots, \omega_{n-k}, 0, \ldots, 0),$$

from it we can see clearly that the torus $y = y_o$ has a foliation of $(n-k)$-dimensional tori, each of which is invariant for the flow $x(t) = x_o + \omega t$, $y(t) = y_o$. Such an n-torus has no any persistency for small perturbation, it is usually destroyed. Nevertheless, the disintegration does not imply the non-existence of lower dimensional invariant tori. The study in this aspect can trace back to the Poincaré-Birkhoff fixed point theorem for area-preserving twist diffeomorphisms of annulus or cylinder, where the Ω has the form $(1, 0)$ since a result of Moser [Mo] proves that every twist diffeomorphism is the time one map associated to a suitable positive definite Lagrange system. Of course, the Poincaré-Birkhoff fixed point theorem is established not only for nearly integrable systems but for arbitrary twist map. For the Hamiltonian systems with arbitrary n degrees of freedom the first study of such a problem is own to Berstein and Katok [BK], they proved such a result:

Theorem 1 *If the main part N is convex in y, and $\Omega = (\omega, 0, \ldots, 0)$, if P is sufficient small, then there are at least n periodic orbits, its frequency is $\omega \in \mathbb{R}$.*

Such a result is proved by using variational method, associating the proof with the existence of critical points of a function defined on an n-torus. Since objects we look for is periodic orbits, there is no problem of small divisor. It is no longer the case when we consider the case when $k < n - 1$, where the small divisor problem is involved and some KAM-type techniques need to be used.

Let $\omega = (\omega_1, \omega_2, \ldots, \omega_{n-k})$ and assume that ω satisfies some diophantin condition. A step of KAM iteration can be always done since the main part N is independent of angle variable. After such a step of canonical coordinate change, the new main part depends also on some of the angle variables in general, $N = N(y, \hat{x})$, $\hat{x} = (x_{n-k+1}, \ldots, x_n)$. Such N consists of the original main part plus some average part of P over $\mathbb{T}^k$. Since the Kolmogorov non-degenerate condition is assumed, the equation

$$\frac{\partial N}{\partial y}(y, \hat{x}) = \Omega$$

has unique solution $y = y(\hat{x})$ close to $y = y_o$. With such a function $y(\hat{x})$ we can construct a Lagrangian type function

$$\mathcal{L}(\hat{x}) = \langle \Omega, y(\hat{x}) \rangle - N(y(\hat{x}, \hat{x}),$$

which is defined on k-torus. So it has at least $k + 1$ critical point by the Lusternik-Schnirelman catergry theory. It is easy to show that each critical point of $\mathcal{L}(\hat{x})$ corresponds to a $(n-k)$-dimensional invariant torus of the Eq.(1) when $H = N$, the frequency of the flow on these $\mathbb{T}^{n-k}$ tori is ω. What we need to show is that such tori remain invariant under small perturbations. We have not yet obtained the existence of such $(k+1)$ $(n-k)$-dim invariant tori for arbitrary k, which still remains as a conjecture, but when $k = 1$ we have the following theorem [Cg2]:

Theorem 2 *Let the Hamiltonian be of KAM type H(p,q)=N(p) +P(p,q) and be real analytic in a complex neighbourhood*

$$\Sigma_{\sigma,\tau}: \quad |Imq| \le \tau, \qquad |p - p_o| \le \sigma, \quad \forall p_o \in \mathbb{D}$$

of the set $\mathbb{D} \times \mathbb{T}^n$, *where* $|\cdot| = \max_{1\le j\le n} |\cdot_j|$ *and assume the following conditions:*

1, Nondegeneracy: the main part N satisfies Kolmogorov's nondegeneracy condition, i.e. the determinant of its Hessian

$$\left| det\left(\frac{\partial^2 N}{\partial p_i \partial p_j}\right)\right| \ge \lambda > 0, \qquad \forall p \in \mathbb{D}.$$

2, Resonance: for all $p \in \mathbb{M}$ *the following holds:*

$$\frac{\partial N}{\partial p_n}\bigg|_{p\in\mathbb{M}} = 0, \qquad \frac{\partial^2 N}{\partial p_n^2}\bigg|_{p\in\mathbb{M}} \ne 0.$$

Then there is a positive $d = d(N,n,\sigma,\tau) > 0$ *such that if* $|P| < d$ *in* $\Sigma_{\sigma,\tau}$ *there is a set* $\mathbb{S} \subset \mathbb{M}$ *with positive* $(n-1)$ *dimensional Lebesgue measure, such that if* $p_o \in \mathbb{S}$, *the Hamiltonian flow defined by Eq.(1.1) admits at least two* $(n-1)$*-dimensional invariant submanifolds of the form*

$$\begin{aligned} p &= p_o + \Gamma_j(\phi_1, \phi_2, \cdots, \phi_{n-1}, q_{nj}), \qquad j = 1,2; \\ q &= (\phi_1, \phi_2, \cdots, \phi_{n-1}, q_{nj}) + \Theta_j(\phi_1, \phi_2, \cdots, \phi_{n-1}, q_{nj}), \end{aligned}$$

with Γ_j, Θ_j *being real analytic functions of period* 2π *in the complex domain* $|Im\phi| \le \frac{\tau}{2}$, *and* q_{nj} *is some constant. The parametrization is chosen so that the induced flow on* $\mathbb{T}^{n-1}$ *is still given by*

$$\phi = \phi_o + \frac{\partial N}{\partial p}(p_o)t.$$

Moreover, for each $\epsilon > 0$ *there exists a positive* $d' = d'(N,n,\sigma,\tau,\epsilon) \le d$ *such that if in addition* $|P| < d'$ *in* $\Sigma_{\sigma,\tau}$, *then the functions* Γ_j *and* Θ_j *satisfy*

$$|\Gamma_j| + |\Theta_j| < \epsilon.$$

As $d \to 0$ *the Lebesgue* $meas(\mathbb{S}) \to meas(\mathbb{M})$.

It is worthy to point out that the number d is independent of the especial form of the perturbation, that is to say, we have a positive lower bound, which is uniform for perturbations, for the existence of 2 $(n-1)$-dim. invariant tori.

The proof of such a theorem contains two key steps. The first one is how to handle the problem that the critical point of $\mathcal{L}$ may be degenerate, the new critical point may move far away from the original one when $\mathcal{L}$ is subject to a small perturbation, so we have to choose the nested neighborhood of the critical point in an implicit way, not like the usual KAM techniques. In our iteration scheme, such domain of normal coordinates may not shrink to a point, which needs substantial new ideas, see [Cg1]. The second key step is how to handle the problem of possible resonance between normal and tangent frequencies when we are going to prove the existence of elliptical tori. Obviously, the normal frequency comes from the perturbation, therefore, any hypothesis should not be assumed on the normal frequency, see [Cg2] for the details of the proof. Such a result can be thought as the extension of the KAM theory to resonant tori as well as a generalization of Poincaré-Birkhoff fixed theorem.

When k is an arbitrary number between 1 and $n-1$ there is no such a existence theorem yet, but under some generic assumptions we have the following result [CW]:

Theorem 3 *If N is convex in $\hat{y}$, there exist $d_o = d_o(N, P, n, k, r) > 0$ and an open dense set $\mathcal{G}$ in $C^r(\mathbb{T}^n \times \mathbb{D}, \mathbb{R})$ such that for $P \in \mathcal{G}$ and $\|P\|_{C^r} \leq d_o$, there is a Cantor set $\mathcal{S} \subset \mathcal{M}$ such that the Eq.(1) admits at least 2^k smooth families (in Whitney's sense) of (n-k)-dim. invariant tori. The (n-k)-dim. Lebesgue measure of $\mathcal{S}$ tends to that of $\mathcal{M}$ as $d_o \to 0$, where $\mathcal{M} = \{y \in D|\omega(y) = (\omega_1(y), \ldots, \omega_{n-k}(y), 0, \ldots, 0)\}$.*

It should be pointed out that the constant d_o in this theorem depends on the form of perturbation P, so we do not get the same conclusion as we get in the Theorem 2, i.e. uniformity for perturbations in $\mathcal{P}$.

¿From the proof of the Theorem 3 we can see that among these 2^k invariant tori there is at least one hyperbolic torus and one elliptic torus. If we drop the condition that N is convex in $\hat{y}$, it is no longer the case. Consider the following example:

$$H = \frac{1}{2}(y_1^2 - y_2^2) + f \circ g(x_1, x_2) + \frac{1}{2}\sum_{i=3}^{n} y_i^2 + \sum_{i=3}^{n} \omega_i y_i,$$

obviously $(y = 0, x_1 = x_1^o, x_2 = x_2^o)$ is the $(n-2)$-dimensional invariant torus of Eq.(1) if (x_1^o, x_2^o) is the critical point of $f \circ g$. We choose

$$\begin{aligned} f(x_1, x_2) &= \epsilon(\cos x_1 + \cos x_2) \\ g(x_1, x_2) &= \begin{cases} (x_1, x_2) & \text{if}(x_1, x_2) \in B_{\frac{\pi}{2}}(x_1^o, x_2^o) \\ (x_1^o, x_2^o) + T^{-1}(r, \theta + \alpha(r)) & \text{elsewhere} \end{cases} \end{aligned}$$

where T is the standard transformation from Cartesian coordinates to polar coordinates, α is such a function that has the following properties:

$$\begin{cases} \alpha(r) \equiv \frac{\pi}{2} & \text{when } r < \frac{\pi}{8} \\ \alpha'(r) < 0 & \text{when } \frac{\pi}{8} < r < \frac{\pi}{2} \\ \alpha(r) \equiv 0 & \text{when } r > \frac{\pi}{2}. \end{cases}$$

Obviously, there are 4 distinct critical point of $f \circ g$ $A_1 = (0, 0)$, $A_2 = (0, \pi)$, $A_3 = (\pi, 0)$, $A_4 = (\pi, \pi)$. Both A_1 and A_4 represent a mixed type $(n-2)$-torus if we choose $(x_1^o, x_2^o) = (\pi, 0)$. The other two points A_2 and A_3 represent a hyperbolic torus. If we fix $(x_1^o, x_2^o) = (0, \pi)$ then both A_2 and A_3 represent an elliptic torus.

If we drop the generic condition the situation turns more complicated even if we impose on some positive definite condition on the main part N. The minimizer of the Lagrange like function does not necessarily correspond to a hyperbolic torus if it is nondegenerate. Here is an example to show it.

$$\begin{aligned} H = \; & \left\{\tfrac{1}{2}(y_1^2 + y_2^2) + a(y_1 \cos x_2 + y_2 \cos x_1) + b(\cos x_1 + \cos x_2)\right\} \\ & \times (1 + \delta(x_1, x_2)) + \tfrac{1}{2}\textstyle\sum_{i=3}^{n} y_i^2 + \sum_{i=3}^{n} \omega_i y_i, \end{aligned}$$

where both a and b are small positive numbers, satisfying the condition $a^2 > b$,$\delta(x_1, x_2) = \delta(\|(x_1, x_2) - (x_1^o, x_2^o)\|)$, $x_1^o = \cos^{-1}(\frac{b}{a^2})$, $x_2^o = -\cos^{-1}(\frac{b}{a^2})$, $\delta(r) \in C^\infty$, $\delta(r) \geq 0$, $\delta'(r) \leq 0$, $\delta(0) = 0$, supp$\delta \subset B_\epsilon(0)$, ϵ is sufficiently small. It is easy to check that the point (x_1^o, x_2^o) represents such an ellpitical $(n-2)$-torus $\mathbb{T}^{n-2}$

$$\begin{cases} x_1 = \cos^{-1}(\frac{b}{a^2}) & x_2 = -\cos^{-1}(\frac{b}{a^2}) \\ y_1 = -\frac{b}{a} & y_2 = -\frac{b}{a} \\ y_i = \omega_i & i = 3, 4, \ldots, n \end{cases}$$

on which the frequency of the linear flow is $(\omega_3, \dots, \omega_n)$. Note the function $\mathcal{L}(x_1, x_2)$ takes its minimum at (x_1^o, x_2^o), the only global minimum.

Recall a Mather's result [M1]. Consider a positive definite Lagrangian, defined on a tangent bundle of a compact manifold M and periodic in t, i.e.

$$L(\xi, t+1) = L(\xi, t), \qquad \xi \in TM, \ t \in \mathbb{R}.$$

It has fiberwise superlinear growth:

$$L(\xi, t)/\|\xi\| \to +\infty \qquad \text{as } \|\xi\| \to +\infty \quad \text{for } \xi \in TM, \ t \in \mathbb{R}.$$

Every trajectory of its Euler-Lagrange flow E_L is defined for all time. Let μ denote E_L-invariant probability measure on $TM \times \mathbb{T}$, the average action is defined by

$$A(\mu) = \int L d\mu.$$

To every invariant measure we can define a rotation vector $\rho(\mu) \in H_1(M, \mathbb{R})$ [M2]. It is shown in [M1] that for any rotation vector $\rho(\mu)$ there exist an E_L-invariant probability measure μ which minimizes the action $A(\mu)$. A KAM torus is clearly the support of such a minimal measure. It is believed that the support of minimal measure is of hyperbolic when it is nondegenerate in the sense of dynamics. Consider this example where the rotation vector $\rho = (0, 0, \omega_3, \dots, \omega_n) \in H_1(M, \mathbb{R})$. The $(n-2)$-torus, corresponding the minimun of the Lagrange like function $\mathcal{L}$, is the support of some invariant measure whose rotation vector is exactly ρ, but is obviously not the support of the minimal measure since it is ellptic type.

In the case when there is exactly only one resonant condition, the minimum of the Lagrange like function $\mathcal{L}(x)$ always correspondes to a hyperbolic $(n-1)$ torus, is it the support of an minimal measure? If it is the support of minimal measure, these $(n-1)$ tori can be connected by some trajectory [M2].

References

[BK] Berstein, D. & Katok, A. *Birkhoff periodic orbits for small perturbations of completely integrable Hamiltonian system with convex Hamiltonians.* Invent. Math. **88** (1987) 224-241.

[Cg1] Cheng, C.-Q. *Birkhoff-Kolmogorov-Arnold-Moser tori in convex Hamiltonian systems.* Commun. Math. Phys. **177** (1996) 529-559.

[Cg2] Cheng, C.-Q. *Lower dimensional invaraint tori in the regions of instability for nearly integrable Hamiltonian systems.* Commun. Math. Phys.**203**1999 385-419.

[CW] Cheng, C.-Q. & Wang, S. *The surviving of lower dimensional tori from a resonant torus of Hamiltonian systems* J. Diff. Eqns. **155** (1999) 311-326.

[M1] Mather, J.N. Action minimizing invariant measures for positive definite Lagrangian systems, *Math. Z.* **207** (1991) 169-207.

[M2] Mather, J.N. Variational construction of connecting orbits. *Ann. Inst. Fourier* **43**(1993) 1349-1386.

[Mo] Moser, J.K. Monotone twist mappings and the calculus of variations. *Ergodic Th. Dynam. Syst.* **6** (1986) 401-413.

DEPARTMENT OF MATHEMATICS, NANJING UNIVERSITY, NANJING 210093, CHINA
E-mail address: chengcqnju.edu.cn

AMS/IP Studies in Advanced Mathematics
Volume 20, 2001

PERSISTENCE OF C^k NORMALLY HYPERBOLIC INVARIANT MANIFOLDS FOR INFINITE DIMENSIONAL DYNAMICAL SYSTEMS

Peter W. Bates, Kening Lu, and Chongchun Zeng

ABSTRACT. In this paper, we establish a general theory for the persistence of C^k $(k \geq 1)$ normally hyperbolic invariant manifolds (not necessarily compact, or finite dimensional) and the existence of the C^k center-stable and center-unstable manifolds of the normally hyperbolic invariant manifolds.

1. Introduction

This paper, which is a sequel to [**5**] by the same authors, is devoted to the persistence of C^k normally hyperbolic Invariant manifolds for infinite dimensional dynamical systems in a Banach space.

We consider a C^1 semiflow defined on a Banach space X, that is, it is continuous on $[0, \infty) \times X$, and for each $t \geq 0, T^t : X \to X$ is C^1, and

$$T^t \circ T^s(x) = T^{t+s}(x)$$

for all $t, s \geq 0$ and $x \in X$. A typical example is the solution operator for a differential equation.

In [**5**] we proved that a compact normally hyperbolic invariant manifold M persists under small C^1 perturbations in the semiflow. We also showed that in a neighborhood of M, there exist a center-stable manifold and a center-unstable manifold which intersect in the manifold M. In [**5**], the compactness and invariance of the manifold M were important assumption.

In the present paper, we establish a general theory for the persistence of C^k $(k \geq 1)$ normally hyperbolic invariant manifolds (not necessarily compact, or finite dimensional) and the existence of the C^k center-stable and center-unstable manifolds of the normally hyperbolic invariant manifolds for C^k semiflows in a Banach space. The main ideas for the proofs are the same as in [**5**]. However, here we have to overcome the difficulties caused by the noncompactness of the manifold such as the non-uniform tubular neighborhood. Some ideas developed in [**6**] are also used.

The theory of invariant manifolds for discrete and continuous dynamical systems has a long and rich history. For the case of M consisting of a single equilibrium point, Hadamard [**18**] constructed the unstable manifold of a hyperbolic fixed point of a diffeomorphism of the plane by iterating the mapping applied to a curve in the plane, thereby obtaining a convergent sequence of curves. The limit of this sequence of curves gives the unstable manifold. The convergence makes use of certain "invariant cone" arguments. This is precisely the approach we take but with the obvious difficulties resulting from the lack of local compactness of the space and the lack of an inverse of our time-t map.

Lyapunov [**38**] and Perron [**43**], [**44**], [**45**] constructed the unstable manifold of an equilibrium point by formulating the problem in terms of an integral equation. This method involves capturing invariant manifolds (stable, unstable, center, etc.) as the set of initial data for which solutions exist globally in forward or/and backward time with appropriate bounds on their growth. This method is analytic rather than geometric. This method does not appear to be available in our general case because of the lack of a trivial normal bundle for the manifold.

There is an extensive literature on the stable, unstable, center, center-stable, and center-unstable manifolds of equilibrium points for both ordinary differential equations and partial differential equations. See, for example, [**1**], [**2**], [**3**], [**9**], [**10**], [**20**], [**21**], [**22**], [**25**], [**28**], [**42**], [**48**], [**46**], [**53**], [**54**], [**57**], [**58**], and [**59**]. A good treatment of center manifold theory using the Lyapunov-Perron approach for the finite dimensional case may be found in the monograph by Carr [**9**], where many applications are also set forth. Certain infinite dimensional settings are also treated. The Lyapunov-Perron method was used by Ball [**1**] to obtain local stable, unstable and center manifolds for dynamical systems in Banach space, with application to the beam equation. Henry [**20**] developed the theory in the case when the linearized operator generates an analytic semigroup. Later, Chow and Lu [**10**] used this approach to prove the existence of smooth center-unstable manifolds.

The local theory for dynamical and semidynamical systems in Banach space using Hadamard's approach may be found in the work of Bates and Jones [**3**], where applications are given demonstrating the stability of a pulse solution to the FitzHugh-Nagumo equations and the instability of stationary solutions to the nonlinear Klein-Gordon equation.

The above citations refer to *local* theory, that is, near a critical point, in finite and infinite dimension systems. Our theorem, given below, refers to the persistence of a *global* invariant manifold for a semiflow in a Banach space.

The theory of the persistence of invariant manifolds for finite dimensional dynamical systems and its applications were developed by many authors including Krylov and Bogoliubov [**30**], Bogoliubov and Mitropolski [**8**], Hale [**19**], Kyner [**29**], Levinson [**34**], Diliberto [**12**], Hufford [**24**], Kurzweil [**33**], Marcus [**40**], McCarthy [**41**], Sacker [**52**], Pliss and Sell [**50**], Yi [**60**], and others.

In the seventies, Fenichel [**13**], [**14**], [**15**], and Hirsch, Pugh and Shub [**22**] (independently) further developed this theory, clearly defining normal hyperbolicity and showing that this condition was sufficient for persistence. Mañé [**39**] proved that normal hyperbolicity is a necessary condition for the persistence of an invariant manifold for finite dimensional dynamical systems, therefore showing that it is *exactly* the right condition. Since then, the applications of this theory to problems from science and engineering have flourished, especially to singular perturbation problems (see, for example, [**11**], [**16**], [**17**], [**23**], [**26**], [**27**], [**31**], [**35**], [**55**], and [**56**]).

The theory of integral manifolds developed by Hale has been generalized by Henry in [**20**], to semilinear parabolic equations where the associated linearized operator generates an analytic semigroup and when a global coordinate system is available.

Recently, Li, McLaughlin, Shatah, and Wiggins [**36**] proved some results on the persistence of invariant manifolds for flows (having forward and backward orbits) arising from the nonlinear Schroedinger equation. In that situation, the Lyapunov-Perron method was available because of the triviality of the normal bundle of the manifold involved.

Considering the semiflow generated by a parabolic equation, Lu in [**37**] constructed infinitely many invariant manifolds as perturbations of eigenspaces of the operator obtained by linearizing at an equilibrium point. With these and corresponding invariant foliations, a new coordinate system was constructed in a neighborhood of the equilibrium point. This facilitated a proof of a Hartman-Grobman theorem for scalar parabolic PDE's, which yields that the flow near a hyperbolic equilibrium point is structurally stable. In [**4**] a more general theorem on the existence of invariant manifolds and foliations was proved. This theorem was then used to obtain a Hartman-Grobman result for both the phase-field system and the Cahn-Hilliard equation.

Very recently, Zeng [**61**] proved the existence of homoclinic orbits for perturbed nonlinear Schrödinger equations with unbounded perturbations (diffusion term) by using invariant manifolds and invariant foliations.

2. Main Results

In this section, we introduce basic notations and hypotheses, then state our main results for maps and semiflows.

Let X be a Banach space. Consider $M \subset X$ a C^1 connected submanifold of X and $T \in C^1(B(M,r),X)$, where $B(M,r) = \{x \in X : \operatorname{dist}(x,M) < r\}$ for some $r > 0$.

Definition 2.1 M is said to be a normally hyperbolic invariant manifold if the following conditions hold

(H1) M is invariant, $T(M) = M$.

(H2) M is normally hyperbolic, that is, for each $m \in M$ there is a decomposition

$$X = X_m^c \oplus X_m^u \oplus X_m^s$$

of closed subspaces with X_m^u, X_m^s being transversal to $T_m M$, the tangent space of M at m. Furthermore, for any $m_0 \in M$,

$$\Pi_{m_1}^\alpha DT(m_0) : X_{m_0}^\alpha \to X_{m_1}^\alpha$$

are isomorphisms for $\alpha = c, u$, where $m_1 = T(m_0) \in M$ and Π_m^α is the projection onto X_m^α with kernel being the sum of the other subspaces. Also there exists $\lambda \in (0,1)$ such that

$$||\Pi_{m_1}^s DT(m_0)|_{X_{m_0}^s}|| < \lambda \min\{1, \inf\{|\Pi_{m_1}^c DT(m_0)x^c| \ : \ x^c \in X_{m_0}^c, |x^c| = 1\}\},$$

$$\lambda \inf\{|\Pi_{m_1}^u DT(m_0)x^u| : x^u \in X_{m_0}^u, |x^u| = 1\} > \max\{1, ||\Pi_{m_1}^u DT(m_0)|_{X_{m_0}^c}||\}.$$

In order to establish local tubular neighborhoods with a uniform size, in the following we shall assume that the projections Π_m^α for $\alpha = c, u, s$ are Lipschitz in m and the manifold M does not 'twist' too much:

(H3) There is an $r > 0$ such that for any $m_0 \in M$, $m_1, m_2 \in B(m_0, r)$, $m_1 \neq m_2$, and $\alpha = c, u, s$,

$$||\Pi^\alpha_{m_1} - \Pi^\alpha_{m_2}|| \leq L|m_1 - m_2|$$

and

$$\frac{|m_1 - m_2 - \Pi^c_{m_0}(m_1 - m_2)|}{|m_1 - m_2|} \leq \epsilon_1 < 1,$$

where $1 \leq L < \frac{\sqrt{2}-1}{r}$ and ϵ_1 are constants.

As an example, when M is a C^2 compact manifold imbedded in a Hilbert space, the hypothesis (H3) holds if Π^c_m is the orthogonal projection onto the tangent space of M and r is chosen to be small enough. When M is a C^1 manifold, this particular projection is only C^0 in m. However, if M is finite dimensional and has a countable basis, one may construct a C^1 approximation of any C^0 family of projections. This issue is addressed in [**7**]. We do not know if such approximations exists for general infinite dimensional manifolds.

Since we do not assume that M is compact or finite dimensional, for technical reasons, we need to assume that T, Π^c_m, Π^u_m, and Π^s_m have some uniform properties:

(H4) (1) There is a constant $B > 0$ such that $||\Pi^\alpha_m|| \leq B$ for all $m \in M$, and $\alpha = c, u, s$;

(2) There exists $\mu_0 > 0$ such that for any $m \in M$ and $\alpha = u, s$,

$$||\Lambda^\alpha_m|| \leq \mu_0,$$

where $\Lambda^u_m \in L(X^c_m, X^u_m)$ and $\Lambda^s_m \in L(X^c_m, X^s_m)$ are determined by

$$T_m M = (I + \Lambda^u_m + \Lambda^s_m)X^c_m;$$

(3) For any $\eta > 0$, there exists $\epsilon > 0$, such that for any $x_1, x_2 \in B(M, \epsilon)$ with $|x_1 - x_2| < \epsilon$,

$$||DT(x_1) - DT(x_2)|| \leq \eta;$$

(4) There are constants $a > 0$, $B_1 > 0$, and $b_1 > 0$ such that

$$\inf\{|\Pi^c_{m_1} DT(m_0)x^c| \ : \ x^c \in X^c_{m_0}, |x^c| = 1\} \geq a,$$

$$||DT|_{B(M,r)}|| \leq B_1,$$

$$||\Pi^\alpha_m DT(m_0)|_{X^u_{m_0}}|| \leq b_1,$$

for $\alpha = c, s$, and

$$||\Pi^\alpha_m DT(m_0)|_{X^s_{m_0}}|| \leq b_1,$$

for $\alpha = c, u$.

Condition (2) implies that the space X^c_m is an approximation of the tangent space of M at m with an error bounded by μ_0. Condition (3) automatically holds when M is contained in a compact set. Roughly, it is an assumption on the uniform continuity of DT, but weaker than that. The reason for this assumption is that the graph transform is a global transform and some uniform estimates are needed.

We shall see that if X^c_m is a good approximation of the tangent space of M, then we have the existence and the persistence of unstable manifolds of the overflowing manifolds.

To get higher smoothness of the manifolds, we assume that the following "spectral gap" condition holds, which is necessary even when M is an equilibrium point,

(H5)

$$\lambda \inf\{|\Pi^u_{m_1} DT(m_0)x^u| : x^u \in X^u_{m_0}, |x^u| = 1\} > (\|\Pi^c_{m_1} DT(m_0)|_{X^c_{m_0}}\|)^i$$

and

$$\|\Pi^s_{m_1} DT(m_0)|_{X^s_{m_0}}\| < \lambda(\inf\{|\Pi^c_{m_1} DT(m_0)x^c| \ : \ x^c \in X^c_{m_0}, |x^c| = 1\})^i,$$

for all $m_0 \in M$, $m_1 = T(m_0)$, $i = 1, 2, \cdots, k$, and some $0 < \lambda < 1$.

THEOREM 2.1. *Assume that (H1) – (H4) hold. Then, there exist* $\epsilon_1^* > 0$, *and* $\mu_0^* > 0$, $b_1^* > 0$, *and* $\sigma > 0$ *such that if* $\epsilon_1 \le \epsilon_1^*$, $\mu_0 \le \mu_0^*$, $b_1 < b_1^*$, *and* $\|\tilde{T} - T\|_{C^1(B(M,r),X)} \le \sigma$, *then*

(a) *Persistence:* $\tilde{T}$ *has a unique* C^1 *connected normally hyperbolic invariant manifold* $\tilde{M}$ *near* M.
(b) *Convergence:* $\tilde{M}$ *converges to* M *in the* C^1 *topology as* $\|\tilde{T}-T\|_{C^1}$ *approaches zero;*
(e) *Smoothness:* $\tilde{M}$ *is* C^k *if* $\tilde{T} \in C^k(B(M,r),X)$ *and* $(H5)$ *holds and* $\tilde{M}$ *is* $C^{k,1}$ *when* $T \in C^{k,1}(B(M,r),X)$ *and* $(H5)$ *holds.*
(c) *Existence: For the map* $\tilde{T}$, *there are unique* C^1 *center-stable and center-unstable manifolds,* $\tilde{W}^{cs}$ *and* $\tilde{W}^{cu}$, *in the tubular neighborhood.*
(d) *Characterization:*

$$\tilde{W}^{cs} = \left\{x_0 \in N : \tilde{T}^k(x_0) \in N, k \ge 1, \ \tilde{T}^k(x_0) \to \tilde{M}\right\}$$

$$\tilde{W}^{cu}(\epsilon) = \Big\{ \ x_0 \in N : \ \exists \ \{x_k\}_{k>0} \subset N, \text{ satisfying } \tilde{T}(x_k) = x_{k-1}, \text{ for } k \ge 1, \text{ and } x_k \to \tilde{M} \text{ as } k \to \infty\Big\},$$

and $\tilde{M} = \tilde{W}^s(\epsilon) \cap \tilde{W}^u(\epsilon)$, *where* N *is a tubular neighborhood of* $\tilde{M}$.

We now consider the manifolds for semiflows. Let a C^1 map T and M satisfy (H1)-(H3). Let $\tilde{T} \in C([0, +\infty) \times B(M,r), X)$ be a semiflow, i.e.,

$$\tilde{T}^0 = I, \qquad \tilde{T}^{t+s} = \tilde{T}^t \circ \tilde{T}^s, \quad \text{for } t, s \ge 0.$$

We assume that for all $t \ge 0$, $\tilde{T}^t \in C^1(B(M,r),X)$. Suppose there exists $t_0 > 0$ such that $\|\tilde{T}^{t_0} - T\|_{C^1(B(M,r),X)} < \sigma$. From Theorem 2.1, when σ is sufficiently small, there exists a C^1 normally hyperbolic invariant manifold $\tilde{M}$ for $\tilde{T}^{t_0}$.

Furthermore, if we assume

(H6) For any $\eta > 0$, there exists $\zeta > 0$, such that for any $x \in B(M,r)$, $t \in [0, \zeta]$, we have

$$|\tilde{T}^t(x) - x| < \eta.$$

Then we have

THEOREM 2.2. *The normally hyperbolic invariant manifold* $\tilde{M}$ *for* $\tilde{T}^{t_0}$ *is the normally hyperbolic invariant manifold for the semiflow* $\tilde{T}^t$. *Moreover,* $\tilde{M}$ *is* C^k *if* $\tilde{T}^{t_0} \in C^k(B(M,r),X)$ *and* $(H5)$ *holds and* $\tilde{M}$ *is* $C^{k,1}$ *when* $T^{t_0} \in C^{k,1}(B(M,r),X)$ *and* $(H5)$ *holds.*

The basic ideas of the proof of Theorem 2.1 are the same as those we used in [**5**]. However, here we have to overcome the difficulties caused by the noncompactness of the manifold and extend the idea of proving the smoothness of the manifold to establish C^k smoothness. The complete proof of these results will appear in [**7**].

References

[1] J. M. Ball, Saddle point analysis for an ordinary differential equation in a Banach space and an application to dynamic buckling of a beam, *Nonlinear Elasticity*, edited by R. W. Dickey, Academic Press, NewYork, (1973), 93–160.

[2] A. Burchard, B. Deng and K. Lu, Smooth Conjugacy of Center Manifolds, *Proceedings of the Royal Society of Edingburgh*, **120A** (1992), 61-77.

[3] P. Bates and C. K. R. T. Jones, Invariant manifold theorems for semilinear partial differential equations, *Dynamics Reported* **2** (1988), 1-38.

[4] P. W. Bates and K. Lu, A Hartman-Grobman theorem for the Cahn-Hilliard and phase-field equations, *J. Dynamics and Differential Equations* **6** (1994) 101–145.

[5] P. W. Bates, K. Lu and C. Zeng, Existence and persistence of invariant manifolds for semiflows in Banach spaces, *Memoirs of the AMS.* **645** (1998).

[6] P. W. Bates, K. Lu and C. Zeng, Persistence of Overflowing Manifold for Semiflows, *Communications in Pure and Applied Mathematics*, **52** (1999), 983-1046.

[7] P. W. Bates, K. Lu and C. Zeng, Invariant Manifolds and Foliations for Infinite Dimensional Dynamical Systems, *in preparation.*

[8] N. Bogoliubov and Y. Mitroposky, Asymptotic methods in the theory of nonlinear oscillations, *Gordon and Breach, New York,* (1961).

[9] J. Carr, Applications of Centre Manifold Theory, *Springer-Verlag, New York,* 1981.

[10] S-N. Chow and K. Lu, Invariant manifolds for flows in Banach spaces, *J. Differential Equations* **74** (1988), 285-317.

[11] B. Deng, The existence of infinitely many traveling front and back waves in the FitzHugh-Nagumo equations, *SIAM J. Math. Anal* **22** (1991), 1631-1650

[12] S. P. Diliberto, Perturbation theorems for periodic surfaces, *rend. Circ. Mat. Palermo Ser.* 2, bf 9 (1960), 265–299.

[13] N. Fenichel, Asymptotic stability with rate conditions, *Indiana Univ. Math. Journal* **23** (1974) 1109–1137.

[14] N. Fenichel, Asymptotic stability with rate conditions II, *Indiana Univ. Math. Journal* **26** (1977), 81–93.

[15] N. Fenichel, Geometric singular perturbation theory for ordinary differential equations, *J. Differential Equations* **31** (1979), 53–98.

[16] R. A. Gardner, An invariant manifold analysis of electrophoretic traveling waves, *J. Dynamics and Differential Equations,* **5** (1993), 599-606.

[17] I. Gasser and P. Szmolyan, A geometric singular perturbation analysis of detonation and deflagration waves, *SIAM J. Math. Anal.* **24** (1993), 968-986.

[18] J. Hadamard, Sur l'iteration et les solutions asymptotiques des equations differentielles, *Bull. Soc. Math. France,* **29** (1901), 224–228.

[19] J. K. Hale, Integral manifolds of perturbed differential systems, *Ann. Math.,* **73** (1961),

[20] D. Henry, Geometric Theory of Semilinear Parabolic Equations, *Lect. Notes in Math.* **840**, Springer-Verlag, 1981.

[21] M. W. Hirsch, and C. C. Pugh, Stable manifolds and hyperbolic sets, *Global Analysis, Proc. Sympos. Pure Math.,* **14**, Amer. Math. Soc., Providence, (1970), 133-163.

[22] M. W. Hirsch, C. C. Pugh and M. Shub, Invariant manifolds, *Lecture Notes in Mathematics,* **583** *Springer-Verlag, New York* (1977).

[23] G. Haller and S. Wiggins , N-pulse homoclinic orbits in perturbation of resonant Hamiltonian systems, *Arch. Rat. Mech. Anal.* **130** (1995) 25-101.

[24] G. Hufford, Banach spaces and the perturbation of ordinary differential equations, *Contributions to the theory of nonlinear oscillations,* **III** Ann. of Math. Studies, **36**, Princeton University Press, Princeton, (1956), 173–195.

[25] M. C. Irwin, On the stable manifold theorem, *Bull. London Math. Soc.* **2** (1970), 196.

[26] C. K. R. T. Jones, Geometric singular perturbation theory, *C.I.M.E. Lectures* (1994).

[27] C. Jones and N. Kopell, Tracking invariant manifolds with differential forms in singularly perturbed systems, *J. Diff. Eq.* **108** (1994), 64-88.

[28] A. Kelley, The stable, center-stable, center, center-unstable, and unstable manifolds, *J. Differential Equations,* **3** (1967), 546-570.

[29] W. T. Kyner, Invariant manifolds, *Rend. Circ. Mat. Palermo, Ser.* **10** (1961), 98–110

[30] N. Krylov and N. Bogoliubov, The application of methods of nonlinear mechanics to the theory of stationary oscillations, *Publication 8 of the Ukrainian Academy of Science, Kiev,* 1934.

[31] G. Kovacic and S. Wiggins, Orbits homoclinic to resonances with applications to chaos in a model of the forced and damped sine-Gordon equation, *Physica-D* **57** (1992), 185-225.

[32] U. Kirchgraber and K. J. Palmer, Geometry in the neighborhood of invariant manifolds of maps and flows and linearization, *Pitman Research Notes in Mathematics Series, Longman Scientific and Technical, published in the United States with John Wiley and Sons, Inc. New York* (1990).

[33] J. Kurzweil, Invariant manifolds for flows, *Proc. Sym. Diff. Equations and Dynamical Systems.* Academic Press, New York, 1967.

[34] N. Levinson, Small periodic perturbations of an autonomous system with a stable orbit, *Ann. of Math.,* **5** (1950), 727–738.

[35] X-B. Lin, Shadowing lemma and singularly perturbed boundary value problems, *SIAM J. Appl. Math.* **49** (1989), 26-54.

[36] Y. Li, D. W. McLaughlin, J. Shatah and S. Wiggins, Persistent homoclinic orbits for perturbed nonlinear Schrödinger equations, *Comm. Pure Appl. Math* **49** (1995), 1175–1255.

[37] K. Lu, A Hartman-Grobman theorem for scalar reaction-diffusion equations, *J. Differential Equations* **93** (1991), 364-394.

[38] A. M. Liapunov, Problème géneral de la stabilité du mouvement, *Annals Math. Studies,* **17**, 1947, Princeton, N.J. (originally published in Russian, 1892).

[39] R. Mañé, Liapunov exponents and stable manifolds for compact transformations, *Geometrical Dynamics Lecture Notes in Math.* **1007** Springer Verlag, New York (1985), 522–577.

[40] M. Marcis, An variant surface theorem for a non-degenerate system, *Contribution to the theory of nonlinear oscillations,* **III**, Ann. of Math. Studies, Number 36, Princeton University Press, Princeton, (1956), 243–256.

[41] J. McCarthy, Stability of invariant manifolds, *Bull. Amer. Math. Soc.* **61** (1955), 149–150.

[42] J.E. Marsden and J. Scheurle, The construction and smoothness of invariant manifolds by the deformation method, *SIAM J. Math. Anal.* **18**, 1987.

[43] O. Perron, Über Stabilität and asymptotisches Verhalten der Integrale von Differentialgleichungssystemen, *Math. Z.***29** (1928), 129–160.

[44] O. Perron, Über Stabilität und asymptotisches Verhalten der Lösungen eines systems endlicher Differenzengleichungen, *J. Reine Angew. Math.* **161** (1929), 41–64.

[45] O. Perron, Die Stabilitätsfrage bei Differentialgleichungen, *Math. Z.* **32** (1930), 703–728.

[46] V. A. Pliss, A reduction principle in the theory of the stability of motion, *Izv. Akad. Nauk SSSR, Mat. Ser,* **28** (1964), 1297-1324.

[47] J. Palis and S. Smale, Structural stability theorem, Global Analysis, *Proc. Sympos. Pure Math.* **14** (1970), 223-232.

[48] Palis and F. Takens, Topological equivalence of normally hyperbolic dynamical systems, *Topology,* **16** (1977), 335–345.

[49] M.M. Peixoto, Structural stability on two dimensional manifolds, *Topology* **1** (1962), 102–120.

[50] V. A. Pliss and G. R. Sell, Perturbations of attractors of differential equations, *J. Differential Equations,* 92 (1991), no. 1, 100–124.

[51] D. Ruelle, Characteristic exponents and invariant manifolds in Hilbert space, *Annals of Math* **115** (1982), 243–290.

[52] R. Sacker, A perturbation theorem for invariant manifolds and Hölder continuity, *J. Math. Mech.* **18** (1969), 705–762.

[53] Sijbrand, Properties of center manifolds, *Trans. A.M.S.,* **289** (1985), 431–469.

[54] S. Smale, Differentiable dynamical systems, *Bull. Amer. Math. Soc.* **73** (1967), 747-817.

[55] P. Szmolyan, Transversal heterclinic and homoclinic orbits in singular perturbed problems, *J. Differential Equations* **92** (1991), 252-281.

[56] D. Terman, The transition fromm bursting to continuous spiking in exitable membrane models, *J. Nonlinear Sci.* **2** (1992), 135-182.

[57] A. Vanderbauwhede, Center manifolds, normal forms and elementary bifurcations, *Dynamics Reported* **2** (1989), 89-169.

[58] A. Vanderbauwhede and G. Iooss, Center manifold theory in infinite dimensions, 1990.

[59] A. van Gils and A. Vanderbauwhede, Center manifolds and contractions on a scale of Banach spaces, *J. Funct. Anal.* **72** (1987), 209–224.
[60] Y. Yi, A generalized integral manifold theorem, *J. Differential Equations,* **102** (1993), 153-187.
[61] C. Zeng, Homoclinic orbits for Perturbed Nonlinear Schrödinger Equation, *preprint*, 1999.

Department of Mathematics, Brigham Young University, Provo, Utah 84602

Department of Mathematics, Brigham Young University, Provo, Utah 84602

Courant Institute of Mathematical Sciences, New York University, New York, NY 10012

AMS/IP Studies in Advanced Mathematics
Volume 20, 2001

Fine Analysis of Blow up and Applications

YanYan Li

0. Introduction

Blow up analysis has been very useful in the study of nonlinear partial differential equations. Usually blow up analysis is done with respect to energy norms, namely, giving asymptotic description in Sobolev norms to a sequence of solutions (or approximate solutions) of nonlinear partial differential equations. In a number of applications, pointwise asymptotic analysis is needed. This is a finer blow up analysis. In this note we briefly describe some of the problems we have been working on and many of our studies rely on such finer analysis of blow up. These problems include the Yamabe problem on manifolds with boundary, the problem of prescribing scalar curvature on $\mathbb{S}^n$, and the best Sobolev inequality on Riemannian manifolds.

1. The Yamabe problem on manifolds with boundary

Let (M, g) be an n-dimensional compact, smooth, Riemannian manifold without boundary. For $n = 2$, the Uniformization Theorem of Poincaré says that there exist metrics on M which are pointwise conformal to g and have constant Gauss curvature. For $n \geq 3$, the well known Yamabe conjecture states that there exist metrics on M which are pointwise conformal to g and have constant scalar curvature. The answer to the Yamabe conjecture is proved to be affirmative through the work of Yamabe [**Y**], Trudinger [**T1**], Aubin [**A1**], and Schoen [**S1**]. See Lee and Parker [**LP**] for a survey. See also Bahri and Brezis [**BB**], Bahri [**B1**], and Schoen [**S2, S3**] for works on the problem and related ones. More recently, Schoen has obtained compactness results for the Yamabe problem. He proved in [**S3**] that when (M, g) is locally conformally flat but not conformally equivalent to the standard sphere, then all solutions to the Yamabe problem stay in a compact set of $C^2(M)$ and the total degree of all solutions is equal to -1. In the same paper he also announced, with indication of the proof, the same result for general manifolds. This is much stronger than the existence results. Analogues of the Yamabe problem for compact Riemannian manifolds with boundary have been studied by Cherrier, Escobar, and others. In particular, Escobar proved that a large class of compact Riemannian manifolds with boundary are conformally equivalent to one

1991 *Mathematics Subject Classification.* AMS subject classification: 58, 53, 35.
Key words and phrases. Yamabe problem, Scalar curvature, Best Sobolev inequality.

with constant scalar curvature and zero mean curvature on the boundary. In the following, (M, g) denotes some smooth, compact, n-dimensional Riemannian manifold with boundary, ν denotes the outward unit normal on ∂M, h_g denotes the mean curvature of ∂M with respect to ν. We restrict ourselves to manifolds of positive type. This means that the sign of the first eigenvalue λ_1 of the following eigenvalue problem is positive:

$$\begin{cases} -\Delta_g \varphi + \frac{n-2}{4(n-1)} R_g \varphi = \lambda \varphi, & M^\circ, \\ \frac{\partial \varphi}{\partial \nu} + \frac{n-2}{2} h_g \varphi = 0, & \partial M. \end{cases}$$

Escobar proved ([**E2**]) that for a large class of manifolds, there exists $\tilde{g}$, conformally equivalent to g, such that $R_{\tilde{g}} = 1$ in M and $h_{\tilde{g}} = 0$ on ∂M. He continued to show ([**E3**]) that for the same class of manifolds there exist $c_+ > 0$ and $c_- < 0$ and two conformal metrics $\tilde{g}_\pm$ such that $R_{\tilde{g}_\pm} = 1$ in M and $h_{\tilde{g}_\pm} = c_\pm$ on ∂M. Together with Z.C. Han we proposed in [**HL2**]

Conjecture 1 *For all $c \in \mathbb{R}$, there exists $\tilde{g}$ conformal to g such that*

$$R_{\tilde{g}} = 1 \text{ in } M, \qquad h_{\tilde{g}} = c \text{ on } \partial M.$$

Conjecture 2 *All solutions to the above stay in a compact set of $C^3(M)$, unless (M, g) is conformally equivalent to standard half spheres.*

Theorem 1.1 ([**HL2**]) *Conjecture 1 and Conjecture 2 hold when (M, g) is locally conformally flat with umbilic boundary.*

Theorem 1.2 ([**HL3**]) *Conjecture 1 holds when $n \geq 5$ and the boundary of M has at least one non-umbilic point.*

The boundary of M is called umbilic if the traceless part of the second fundamental form is identically zero on the boundary. A boundary point of M is called non-umbilic if the traceless part of the second fundamental form is nonzero at the point. One ingredient in the proof of Theorem 1.1 is the following Liouville type theorem in $\mathbb{R}^n_+$

Theorem 1.3 ([**LZ1**]) *Solutions to*

$$\begin{cases} -\Delta u = n(n-2) u^{\frac{n+2}{n-2}}, \quad u > 0, & \mathbb{R}^n_+, \\ \frac{\partial u}{\partial x_n} = c u^{\frac{n}{n-2}}, & \partial \mathbb{R}^n_+, \end{cases}$$

are of form

$$u_{\lambda,c}(x', x_n) = \left(\frac{\lambda}{1 + \lambda^2 |(x', x_n) - (\overline{x}', \overline{x}_n)|^2}\right)^2,$$

where $\lambda > 0, \overline{x}' \in \mathbb{R}^{n-1}$, and $\overline{x}_n = \frac{c}{(n-2)\lambda}$.

Such Liouville type theorem in $\mathbb{R}^n$ is due to Caffarelli, Gidas and Spruck ([**CGS**]), and, under an additional hypothesis $u(x) = O(|x|^{2-n})$ for large $|x|$, is due to Obata ([**O**]) and Gidas, Ni and Nirenberg ([**GNN**]). Under the same additional decay hypothesis the result in $\mathbb{R}^n_+$ is due to Escobar ([**E1**]).

In [**HL4**] we have also proved Conjecture 1 for $n = 3, 4, 5$. For the Yamabe problem, the solutions of Schoen in low dimensions rely on a consequence of the Positive Mass theorem of Schoen and Yau ([**SY1**], [**SY2**]): In appropriate local coordinates, the expansion of the Green's function is of the form $|x|^{2-n} + A + O(|x|^\alpha)$ with constant $A \geq 0$ and $A = 0$ if and only if the manifold is conformally equivalent to standard spheres. For the Yamabe problem with boundary, new difficulties occur in low dimensions when the boundary has no umbilic points. In any local coordinates with $g_{ij}(0) = \delta_{ij}$ near a non-umbilic boundary point (identified as $x = 0$), the term after $|x|^{2-n}$ in the expansion of the Green's function is singular. For instance, in dimension $n = 3$, the expansion of the Green's function is of the form $|x|^{-1} + A(x) + O(|x|^\alpha)$, where $A(x)$ is homogeneous of degree 0, non-constant, with no definite sign. Nevertheless we obtain some kind of average sign condition on $A(x)$ and prove Conjecture 1.

2. Prescribing scalar curvature on S^n

Let $(\mathbb{S}^n, g_0)$ be the standard n–sphere. The following question was raised by L. Nirenberg. Which function $K(x)$ on $\mathbb{S}^2$ is the Gauss curvature of a metric g on $\mathbb{S}^2$ conformally equivalent to g_0? Naturally one may ask a similar question in higher dimensional case, namely, which function $K(x)$ on $\mathbb{S}^n$ is the scalar curvature of a metric g on $\mathbb{S}^n$ conformally equivalent to g_0?

For $n \geq 3$, we write $g = u^{\frac{4}{n-2}} g_0$, the problem is equivalent to finding a solution of

$$-\Delta_{g_0} u + \frac{n(n-2)}{4} u = \frac{n-2}{4(n-1)} K(x) u^{\frac{n+2}{n-2}}, \quad u > 0, \quad \text{on } \mathbb{S}^n, \tag{2.1}$$

where Δ_{g_0} denotes the Laplace-Beltrami operator associated with the metric g_0. For $n = 2$, we write $g = e^{2u} g_0$, and the equation takes the form

$$-\Delta_{g_0} u + 1 = K(x) e^{2u}, \quad \text{on } \mathbb{S}^2. \tag{2.2}$$

A necessary condition for solving (2.1) or (2.2) is that K being positive somewhere. A more subtle and important necessary condition was given by Kazdan and Warner in [**KW**] (see also [**BE**] for a generalization). These necessary conditions are not sufficient for the existence, see [**CLi1**] and [**HL1**]. Much effort has been made in finding various sufficient conditions. In particular, the sufficient condition on $\mathbb{S}^2$ given by Moser ([**M1**]) is that $K(-x) \equiv K(x)$ and $K(x) > 0$ somewhere. This was extended by Escobar and Schoen ([**ES**]) to higher dimensions: $K(-x) \equiv K(x)$ and there exists a point $\bar{x}$ such that $K(\bar{x}) = \max K > 0$ and all derivatives of K up to $n-2$ order at $\bar{x}$ vanish. In dimension $n = 3$ this is a complete generalization of the result of Moser. Such complete generalization is no longer valid in dimension $n \geq 4$, see Bianchi and Engnell ([**BiE**]) and Bianchi ([**Bi**]). Further extensions were made by Chen ([**Ch**]) and Hebey ([**H1**]) to functions K which are invariant under more general groups of isometries. Another type of sufficient condition, discovered by Chang and Yang ([**CY1**] and [**CY2**])) on $\mathbb{S}^2$ and Bahri and Coron ([**BC**]) on $\mathbb{S}^3$ is as follows: K is a C^2 positive Morse function having nonzero ΔK at critical points of K and satisfying

$$\sum_{\nabla K(x)=0, \Delta K(x)<0} (-1)^{i(x)} - (-1)^n \neq 0. \tag{2.3}$$

Due to some more recent work of Schoen ([**S4**]), Chang, Gursky and Yang ([**CGY**]), and Schoen and Zhang ([**SZ**]), C^3 norms of all solutions are bounded and the number on the left hand side of (2.3) is the Leray-Schauder degree of all solutions. Chang and Yang showed in [**CY3**] that (2.3) is sufficient in all dimensions if we further assume that K is close to 1 in $L^\infty(\mathbb{S}^n)$ norm. Without this closeness hypothesis (2.3) is not sufficient for the existence on $\mathbb{S}^4$, a fact which can be deduced, by elementary consideration, from the results of Bianchi and Engnell ([**BiE**] and [**Bi**]) and Li ([**Li2**]). It is reasonable to believe that (2.3) is not sufficient for $n \geq 5$ either. A third type of sufficient conditions (Mountain Path type) was given by Chen and Ding ([**CD**]). There are more existence results for dimension $n = 2, 3$. For higher dimensions, see for instance [**ES**], [**CY3**], [**BCH**], [**Li1**], [**Li2**], [**B2**], [**AB**], [**CL4**], [**AGP**], and the references therein. In the following we present our results in [**Li2**] for $n = 4$. Let $\mathcal{M}_K$ denote the set of positive solutions of (2.1) and set $C^2(\mathbb{S}^4)^+ = \{K \in C^2(\mathbb{S}^4) \mid K(x) > 0 \ \forall \ x \in \mathbb{S}^4\}$. We have defined explicitly a subset $\mathcal{A} \subset C^2(\mathbb{S}^4)^+$, which is open and dense with respect to the C^2 topology, and on which we have defined an integer valued, continuous function Index : $\mathcal{A} \to \mathbb{Z}$. Though there is an explicit formula of Index(K) for Morse functions K in $\mathcal{A}$, it is by no means obvious that the Index mapping can be extended as a continuous function on $\mathcal{A}$. The next theorem gives optimal compactness results, as well as existence results and a degree counting formula of solutions.

Theorem 2.1 ([**Li2**]) *(a) For any $K \in \mathcal{A}$, $K_i \to K$ in $C^2(\mathbb{S}^4)$, and any $u_i \in \mathcal{M}_{K_i}$, we have*

$$0 < \liminf_{i\to\infty}(\min_{\mathbb{S}^4} u_i) \leq \limsup_{i\to\infty}(\max_{\mathbb{S}^4} u_i) < \infty.$$

(b) For any $K \in C^2(\mathbb{S}^4)^+ \setminus \mathcal{A} = \partial\mathcal{A}$, there exists $K_i \to K$ in $C^2(\mathbb{S}^4)$, and $u_i \in \mathcal{M}_{K_i}$ such that

$$\lim_{i\to\infty}(\max_{\mathbb{S}^4} u_i) = \infty, \qquad \lim_{i\to\infty}(\min_{\mathbb{S}^4} u_i) = 0.$$

(c) For any $K \in \mathcal{A}$, $0 < \alpha < 1$, there exists constant $C = C(K)$ such that for all $R \geq C$,

$$\deg\left(u - \frac{1}{6}(-\Delta_{g_0} + 2)^{-1}(Ku^3), \mathcal{O}_R, 0\right) = \textit{Index }(K), \tag{2.4}$$

where $\mathcal{O}_R = \{u \in C^{2,\alpha}(\mathbb{S}^4) \mid 1/R < u < R, \|u\|_{C^{2,\alpha}(\mathbb{S}^4)} < R \}$, and deg *denotes the Leray-Schauder degree in $C^{2,\alpha}(\mathbb{S}^4)$. As a consequence, $\mathcal{M}_K \neq \phi$ provided Index $(K) \neq 0$.*

We only briefly describe what is needed in the proof of Part (c). For any Morse function $K \in \mathcal{A}$, let $K_\lambda = (1-\lambda) + \lambda K$ for $0 \le \lambda \le 1$. Because of Part (a), there exist $0 < \lambda_1 < \cdots < \lambda_l < 1$ such that for all $\lambda \in (0,1] \setminus \{\lambda_1, \cdots, \lambda_l\}$ the total Leray-Schauder degree of all possible solutions with respect to K_λ, denoted as d_λ, is well defined and is a constant function of λ in each component of $(0,1] \setminus \{\lambda_1, \cdots, \lambda_l\}$. The left hand side of (2.4) is d_1, which is what we want to calculate. The value of d_λ for small λ is known due to the work of Chang and Yang [**CY3**]. So it is sufficient to calculate the jump-values of d_λ at λ_m for $1 \le m \le l$. Indeed we know these jump-values from [**Li2**] which yields (2.4) for a Morse function K in $\mathcal{A}$. The validity of (2.4) for all K in $\mathcal{A}$ then follows from Part (a). This is different from the situation in dimension $n = 2, 3$ where d_λ is a constant in the whole interval $(0,1]$, no jumps occur.

We have also given a complete characterization of blow up solutions of (2.1) in dimension $n = 4$. For $K \in C^2(\mathbb{S}^4)^+$, we have defined explicitly a set $\mathcal{S}(K)$ consisting of points of the form $(q^{(1)}, \cdots, q^{(k)})$ with $k \ge 1$, $\nabla K(q^{(j)}) = 0, \Delta K(q^{(j)}) \le 0$ for each j, and some additional properties. This set satisfies the property that $\mathcal{S}(K) = \phi$ if and only if $K \in \mathcal{A}$.

Theorem 2.2 ([**Li2**]) *(a) Let $K \in C^2(\mathbb{S}^4)^+, K_i \to K$ in C^2, $u_i \in \mathcal{M}_{K_i}$ with $\max_{\mathbb{S}^4} u_i \to \infty$. Then $K \in C^2(\mathbb{S}^4)^+ \setminus \mathcal{A}$ and for some $(q^{(1)}, \cdots, q^{(k)}) \in \mathcal{S}(K)$, $\{u_i\}$, after passing to a subsequence, blows up at precisely the k points.*

(b) Let $K \in C^2(\mathbb{S}^4)^+ \setminus \mathcal{A}, (q^{(1)}, \cdots, q^{(k)}) \in \mathcal{S}(K)$. Then there exists $K_i \to K$ in C^2, $u_i \in \mathcal{M}_{K_i}$ such that $\{u_i\}$ blows up at precisely the k points.

The above compactness results are new and optimal. The existence result extends that of Bahri and Coron [**BC**], Benayed, Chen, Chtioui, and Hammami [**BCH**], and Zhang [**Zh**]. It follows from our results that solutions to the problem may have more than one point of blow up, a new phenomena which is very different from the situation in dimension $n = 2, 3$. The phenomena of more than one point of blow up has also been exhibited for $n \ge 5$ in [**Li2**].

The main ingredient in establishing Theorem 2.1-2.2 is some fine analysis of blow up to solutions of (2.1). Such analysis, local in nature, was carried out by Schoen ([**S4**]) and Schoen and Zhang ([**SZ**]) in dimension $n = 3$ for any positive C^2 functions K, and in dimension $n \ge 4$ for constant K. This was extended by the author to dimension $n = 4$ for any positive C^2 functions K, and in dimension $n \ge 5$ for those K which further satisfy suitable $(n-2)$-flatness hypothesis near critical points of K. Such analysis in particular implies, under the above hypothesis, the following Harnack type inequality for solutions u of (2.1)

$$(\sup_{B_\epsilon} u)(\inf_{B_{2\epsilon}} u) \le C\epsilon^{2-n},$$

where B_ϵ and $B_{2\epsilon}$ are concentric balls and C is independent of u. The flatness order $n-2$ is the borderline for various existence and compactness results to hold. The relevance to the existence of the flatness order $n-2$ to the existence was exhibited in [**ES**]. Our notion of the flatness order, introduced in [**Li1**], is stronger than merely requiring derivatives up to $n-2$ order vanish at critical points, as in [**ES**]. As an application, existence and degree counting formula are obtained for such K. For instance, one result, an extension of known results in dimension $n = 2, 3$, is as follows.

Theorem 2.3 (**[Li1]**) *For $n \geq 3$, we suppose that $K \in C^1(\mathbb{S}^n)$ is some positive function such that for any critical point q_0 of K, there exists some real number $\beta = \beta(q_0) \in (n-2, n)$ such that in some geodesic normal coordinate system centered at q_0, $K(y) = K(0) + \sum_{j=1}^n a_j |y_j|^\beta + R(y)$, where $a_j = a_j(q_0) \neq 0, \sum_{j=1}^n a_j \neq 0$, $R(y)$ is $C^{[\beta]-1,1}$ near 0 and satisfies $\lim_{|y|\to 0} \sum_{0\leq|\alpha|\leq[\beta]} |\partial^\alpha R(y)||y|^{-\beta+|\alpha|} = 0$. Then all solutions of (2.1) are uniformly bounded in $C^2(\mathbb{S}^n)$ norm and the Leray-Schauder degree of all solutions is equal to $\sum_{\nabla_{g_0} K(q_0)=0, \sum_{j=1}^n a_j(q_0)<0} (-1)^{i(q_0)} - (-1)^n$, where $i(q_0) = \#\{a_j(q_0) \mid a_j(q_0) < 0, 1 \leq j \leq n\}$. In particular if the number is not zero, (2.1) has at least one solution.*

We remark that $i(q_0)$ is the Morse index of K at q_0 if $\beta = 2$. When $\beta = n-2$ the existence part no longer holds, and, following from [**CL4**], the Leray-Schauder degree of all solutions is equal to zero if $\beta(q_0) \in ((n-2)/2, n-2)$ for all the q_0's, which is completely different from the case $\beta(q_0) \in (n-2, n)$. A consequence of the results in [**Li1**] and [**Li2**] is the following density result: C^∞ scalar curvature functions are dense in $C^{1,\alpha}$ $(0 < \alpha < 1)$ norms among positive functions. This is not true in the C^2 norm.

In a series of papers Chen and Lin have made further investigation on the scalar curvature equations. Among other things, they used the method of moving planes to establish such Harnack type inequalities in [**CL1**] under suitable $(n-2)$-flatness hypothesis. Moreover they have obtained delicate estimates when the flatness order is $\beta \in (1, n-2)$. In particular they exhibited that $(n-2)/2$ is the borderline flatness order for the existence of a sequence of solutions with energy tending to infinity. Their analysis should be useful in establishing strong existence results in high dimensions for those K with flatness order $\beta \in (1, n-2)$. Indeed they have already established in [**CL4**] a degree counting formula in dimension $n = 5$ for such functions.

In the above discussion we have required that K be positive everywhere. If K changes signs, what was missing is the control of solutions u near $K^{-1}(0)$. This has been settled for $n \geq 3$ by Chen and Li in [**CLi2**] where they showed that if 0 is a regular value of K, the L^∞ norms of u near $K^{-1}(0)$ is uniformly bounded for all solutions u. Though the estimates were established on $\mathbb{S}^n$, the proof applies to locally conformally flat Riemannian manifolds. With the help of this result, one can extend a number of previously known results for positive K to include those K which change signs. For instance, Theorem 2.3 can be extended to allow K to change signs, as long as 0 is a regular value of K. More specifically, in the statement of Theorem 2.3, q_0's are being understood as critical points with positive values $K(q_0)$ and the number is therefore changed to $\sum_{\nabla_{g_0} K(q_0)=0, K(q_0)>0, \sum_{j=1}^n a_j(q_0)<0} (-1)^{i(q_0)} - (-1)^n$. One can also extend the previously mentioned density result in [**Li1**] to its full generality: C^∞ scalar curvature functions are dense in $C^{1,\alpha}$ $(0 < \alpha < 1)$ norms among functions which are positive somewhere. This is a generalization of the L^p density result in [**BE**] and the C^0 density result in [**Li1**].

Another way to obtain the estimates of Chen and Li is to show that there is no positive solution to

$$-\Delta u = x_n u^{\frac{n+2}{n-2}}, \qquad \text{in } \mathbb{R}^n.$$

This was established independently by Zhu ([**Z1**]) for even n, and by Lin ([**L**]) for all n.

3. The best Sobolev inequality on Riemannian manifolds

It is well-known that sharp Sobolev inequalities are important in the study of partial differential equations, especially in the study of those arising from geometry and physics. There has been much work on such inequalities and their applications. See, for example, Trudinger [**T2**], Moser [**M2**], Aubin [**A2, A3**], Talenti [**Ta**], Lieb [**L1, L2**], Brezis-Nirenberg [**BN**], Cherrier [**C**], Brezis-Lieb [**BL**], Carleson-Chang [**CC**], Struwe [**St**], Escobar [**E1**], Carlen and Loss [**CaL**], Beckner [**Be**], Adimurthi and Yadava [**AY**], Hebey and Vaugon [**HV1, HV2**], Hebey [**H2, H3**], Li and Zhu [**LZ2, LZ3**], Zhu [**Z2, Z3**], Druet [**D**], Aubin, Druet and Hebey [**ADH**], and the references therein.

For $n \geq 2$, $1 \leq p < n$, and $p^* = np/(n-p)$, let

$$\frac{1}{K(n,p)} = \inf\left\{ \frac{\|\nabla u\|_{L^p(\mathbb{R}^n)}}{\|u\|_{L^{p^*}(\mathbb{R}^n)}} \,\middle|\, u \in L^{p^*}(\mathbb{R}^n)\setminus\{0\},\ \nabla u \in L^p(\mathbb{R}^n)\right\}.$$

Aubin [**A2**] and Talenti [**Ta**] calculated the value of the above Sobolev constant $K(n,p)$ and found the extremal functions.

Let (M^n, g) be a compact Riemannian manifold, it was shown [**A2**] that for any $\epsilon > 0$ there exists a constant $A_p(\epsilon)$ (depending also on the manifold (M^n, g)) such that

$$\|\varphi\|_{L^{p^*}} \leq [K(n,p)+\epsilon]\|\nabla\varphi\|_{L^p} + A_p(\epsilon)\|\varphi\|_{L^p}, \qquad \forall\ \varphi \in W^{1,p}(M^n),$$

and $K(n,p)$ is the smallest constant having this property. A natural question arises: Is the best constant achieved? i.e. does there exist $A_p(0)$? Aubin made a conjecture in [**A3**] concerning the following inequalities:

Conjecture *There exist constants $A(p)$ such that any $\varphi \in W^{1,p}(M_n)$ satisfies*

$$\|\varphi\|^p_{L^{p^*}} \leq K(n,p)^p\|\nabla\varphi\|^p_{L^p} + A(p)\|\varphi\|^p_{L^p} \qquad \text{if } 1 \leq p \leq 2, \tag{3.1}$$

and

$$\|\varphi\|^{\frac{p}{p-1}}_{L^{p^*}} \leq K(n,p)^{\frac{p}{p-1}}\|\nabla\varphi\|^{\frac{p}{p-1}}_{L^p} + A(p)\|\varphi\|^{\frac{p}{p-1}}_{L^p} \qquad \text{if } 2 < p < n. \tag{3.2}$$

A stronger form of (3.2) is

$$\|\varphi\|^2_{L^{p^*}} \leq K(n,p)^2\|\nabla\varphi\|^2_{L^p} + A(p)\|\varphi\|^2_{L^p} \qquad \text{if } 2 < p < n. \tag{3.3}$$

The above conjecture was made because he proved these inequalities when the manifold is the standard $n-$sphere $\mathbb{S}^n$. He also proved that the best constant is achieved for manifolds of dimension two, and for manifolds of constant sectional curvature. Related problems on domains of $\mathbb{R}^n$ were studied by Brezis-Nirenberg [**BN**], Brezis-Lieb [**BL**], and Adimurthi and Yadava [**AY**]. Hebey and Vaugon proved in [**HV1**] and [**HV2**] inequality (3.1) for $p = 2$ under the following condition:

$$\left\{\begin{array}{l} (M_n, g) \text{ has a positive injectivity radius } d > 0, \\ \qquad |R_{ijkl}| \text{ and } |\nabla_m R_{ijkl}| \text{ are bounded by } k. \end{array}\right. \tag{H}$$

Results on compact manifolds with boundaries, also for $p = 2$, were obtained by Li and Zhu in [**LZ2**] and [**LZ3**]. Further results were given by Zhu in [**Z2**] and [**Z3**]. Recently Druet has shown in [**D**] that inequality (3.1) is false for $4 < p^2 < n$ if the scalar curvature is positive somewhere. Then Aubin, Druet, and Hebey proved in [**ADH**] that inequality (3.1) holds for all $p \in (1, n)$ on compact manifolds of dimension $2, 3$ or 4 with non-positive sectional curvature. Together with Aubin we established

Theorem 3.1 (**[AL]**) *Let (M_n, g) be a C^∞ complete Riemannian manifold satisfying (H). Then there exist constants $A(p)$, depending also on n, d and k, such that for all $\varphi \in W^{1,p}(M_n, g)$, inequality (3.1) holds for all $1 < p \leq 2$, and inequality (3.3) holds for all $2 < p < n$.*

The larger the exponent of the norms is, the stronger is the inequality, so the conjecture for $1 < p < n$ follows from the above theorem. In fact we established stronger inequalities where the norms are to some exponent which seems optimal, see [**AL**] for details. Theorem 3.1 was also independently obtained by Druet.

References

[AY] Adimurthi and S.L. Yadava, *Some Remarks on Sobolev Type Inequalities,* **Calc. Var. and PDEs 2** (1994), 427-442.

[AGP] A. Ambrosetti, J. Garcia Azorero and I. Peral, *Perturbation of $\Delta u + u^{\frac{N+2}{N-2}} = 0$, the scalar curvature problem in $\mathbb{R}^N$ and related topics,* **J. Funct. Analysis,** to appear.

[A1] T. Aubin, *Équations différentielles non linéaires et probléme de Yamabe concernant la courbure scalaire,* **J. Math. Pures Appl.55** (1976), 269-296.

[A2] T. Aubin, *Espaces de Sobolev sur les variétés Riemanniennes,* **Bull. Sc. Math. 100** (1976), 149-173.

[A3] T. Aubin, *Problèmes isopérimétriques et espaces de Sobolev,* **J. Diff. Geom. 11** (1976), 573–598.

[ADH] T. Aubin, O. Druet and E. Hebey, *Best constants in Sobolev inequalities for compact manifolds of nonpositive curvature,* **C. R. Acad. Sci. Paris,** to appear.

[AB] T. Aubin and A. Bahri, *Méthodes de topologie algébrique pour le problème de la courbure scalaire prescrite,* **J. Math. Pures Appl. 76** (1997), 525-549.

[AL] T. Aubin and Y.Y. Li, *On the best Sobolev inequality, J. Math. Pures Appl.* **78** (1999), 353-387.

[B1] A. Bahri, *Proof of the Yamabe conjecture without the positive mass conjecture for locally conformally flat manifolds,* **Nonlinear variational problems and partial differential equations** (Isola d'Elba, 1990), Pitman Res. Notes Math. Ser., 320, Longman Sci. Tech., Harlow (1995), 13–43.

[B2] A. Bahri *An invariant for Yamabe-type flows with applications to scalar-curvature problems in high dimension,* **Duke Math. J., 81** (1996), 323-466.

[BB] A. Bahri and H. Brezis, *Non-linear elliptic equations on Riemannian manifolds with the Sobolev critical exponent,* **Topics in geometry, Progr. Nonlinear Differential Equations Appl., 20, Birkhauser Boston, Boston, MA** (1996), 1–100.

[BC] A. Bahri and J.M. Coron, *The scalar-curvature problem on standard three-dimensional sphere,* **J. of Func. Anal. 95** (1991), 106-172.

[Be] W. Beckner, *Sharp Sobolev inequalities on the sphere and the Moser-Trudinger inequality,* **Ann. of Math. 138** (1993), 213-242.

[BCH] M. Benayed, Y. Chen, H. Chtioui, and M. Hammami, *On the prescribed scalar curvature problem on 4-manifolds,* **Duke Math. J., 85** (1996), 633-677.

[Bi] G. Bianchi, *Non-existence and symmetry of solutions to the scalar curvature equation,* **Comm. Partial Differential Equations 21** (1996), 229-234.

[BiE] G. Bianchi and H. Egnell,*An ODE approach to the equation $\Delta u + K u^{\frac{n+2}{n-2}} = 0$ in R^n,* **Math. Z. 210** (1992), 137-166.

[BE] J.P. Bourguignon and J.P. Ezin, *Scalar curvature functions in a conformal class of metric and conformal transformations,* **Tran. Amer. Math. Soc. 301** (1987), 723-736.

[BL] H. Brezis and E.H. Lieb, *Sobolev inequalities with remainder terms,* **J. Funct. Anal. 62** (1985), 73-86.

[BN] H. Brezis and L. Nirenberg, *Positive solutions of nonlinear elliptic equations involving critical Sobolev exponents,* **Comm. Pure Appl. Math. 36** (1983) 437-477.

[CGS] L. Caffarelli, B. Gidas and J. Spruck, *Asymptotic symmetry and local behavior of semilinear elliptic equations with critical Sobolev growth,* **Comm. Pure Appl. Math. 42** (1989) 271-297.

[CaL] E. Carlen and M. Loss *Extremals of functionals with competing symmetry* **J. Funct. Anal. 88** (1990), 437-456.

[CC] L. Carleson and S. Y. A. Chang, *On the existence of an extremal function for an inequality of J. Moser,* **Bull. Sci. Math. 110** (1986), 113-127.

[CL] K.C. Chang and J. Liu, *On Nirenberg's problem,* **International J. of Math. 4** (1993), 35-58.

[CY1] S.Y. Chang and P. Yang, *Conformal deformations of metrics, on* $\mathbb{S}^2$ **J. Diff. Geom. 27** (1988), 256-296.

[CY2] S.Y. Chang and P. Yang, *Prescribing Gaussian curvature on* $\mathbb{S}^2$, **Acta Math. 159** (1987), 215-259.

[CY3] S.Y. Chang and P. Yang, *A perturbation result in prescribing scalar curvature on* $\mathbb{S}^n$, **Duke Math. J. 64** (1991), 27-69.

[CGY] S.Y. Chang, M.J. Gursky and P. Yang, *The scalar curvature equation on 2- and 3-sphere,* **Calc. Var. Partial Differential Equations 1** (1993), 205-229.

[CL1] C.C. Chen and C.S. Lin, *Estimates of the conformal scalar curvature equation via the method of moving planes,* **Comm. Pure Appl. Math. 50** (1997) 971-1017.

[CL2] C.C. Chen and C.S. Lin, *Blowing up with infinite energy of conformal metrics on* $\mathbb{S}^n$, **Comm. in PDEs 24** (1999), 785-799.

[CL3] C.C. Chen and C.S. Lin, *On axisymmetric solutions of the conformal scalar curvature equations on* $\mathbb{S}^n$, **preprint.**

[CL4] C.C. Chen and C.S. Lin, *On prescribing scalar curvature on* $\mathbb{S}^n$, *part 2: Counting the Leray-Schauder degree,* **in preparation.**

[Ch] W. Chen, *Scalar curvatures on* $\mathbb{S}^n$, **Math. Ann. 283** (1989) 353-365.

[CD] W. Chen and W. Ding, *Scalar curvature on* $\mathbb{S}^2$, **Trans. Amer. Math. Soc. 303** (1987), 365-382.

[CLi1] W. Chen and C. Li, *A necessary and sufficient condition for the Nirenberg problem,* **Comm. Pure Appl. Math. 48** (1995), 657-667.

[CLi2] W. Chen and C. Li, *A priori estimates for prescribing scalar curvature equations,* **Ann. of Math. 145** (1997), 547-564.

[C] P. Cherrier, *Meilleures constantes dans inégalités relatives aux espace de Sobolev,* **Bull. Sci. Math. 108** (1984), 225-262.

[D] O. Druet, *Optimal Sobolev inequalities of arbitrary order on compact Riemannian manifolds,* **J. Functional Analysis 159** (1998), 217-242.

[E1] J.F. Escobar, *Uniqueness theorems on conformal deformation of metric, Sobolev inequalities, and an eigenvalue estimate,* **Comm. Pure Appl. Math. 43** (1990), 857-883.

[E2] J.F. Escobar, *The Yamabe problem on manifolds with boundary,* **J. Diff. Geom. 35** (1992), 21-84.

[E3] J.F. Escobar, *Conformal deformation of a Riemannian metric to a constant scalar curvature metric with constant mean curvature on the boundary,* **Indiana Univ. Math. J. 45** (1996), 917-943.

[ES] J.F. Escobar and R. Schoen, *Conformal metrics with prescribed scalar curvature,* **Invent. Math. 86** (1986), 243-254.

[GNN] B. Gidas, W.M. Ni and L. Nirenberg, *Symmetry and related properties via the maximum principle,* **Comm. Math. Phys. 68** (1979), 209-243.

[Ha] Z.C. Han, *Prescribing Gaussian curvature on* $\mathbb{S}^2$, **Duke Math. J. 61** (1990), 679-703.

[HL1] Z.C. Han and Y.Y. Li, *A note on the Kazdan-Warner type condition,* **Ann. Inst. H. Poincaré Anal. Non Linéaire 13** (1996), 541-597.

[HL2] Z.C. Han and Y.Y. Li, *The Yamabe problem on manifolds with boundaries: Existence and compactness results,* **Duke Math. J., 99** (1999), 489-542.

[HL3] Z.C. Han and Y.Y. Li, *The existence of conformal metrics with constant scalar curvature and constant boundary mean curvature,* **Comm. Anal. Geom.** to appear.

[HL4] Z.C. Han and Y.Y. Li, *Further results on the Yamabe problem with boundary,* in preparation.

[H1] E. Hebey, *Changements de métriques conformes sur la sphère. Le problème de Nirenberg,* **Bull. Sci. Math. 114** (1990), 215-242.

[H2] E. Hebey, *Optimal Sobolev inequalities on complete Riemannian manifolds with Ricci curvature bounded below and positive injectivity radius,* **Amer. J. Math. 118** (1996), 291-300.

[H3] E. Hebey, *Sobolev Spaces on Riemannian Manifolds,* **Lecture Notes in Mathematics 1635, Springer-Verlag,**(1996).

[HV1] E. Hebey and M. Vaugon, *Meilleures constantes dans le theoreme d'inclusion de Sobolev,* **Annales de l'Institut Henri Poincare, Analyse nonlineaire 13** (1996), 57-93.

[HV2] E. Hebey and M. Vaugon, *The best constant problem in the Sobolev embedding theorem for complete Riemannian manifolds,* **Duke Math. J. 79** (1995), 235-279.

[KW] J. Kazdan and F. Warner, *Scalar curvature and conformal deformations of Riemannian structure,* **J. Diff. Geom. 10** (1975), 113-134.

[LP] J. Lee and T. Parker, *The Yamabe problem,* **Bull. Amer. Math. Soc. 17** (1987), 37-91.

[Li1] Y.Y. Li, *Prescribing scalar curvature on S^n and related problems, Part I,* **J. Diff. Equ.120** (1995), 319-410.

[Li2] Y.Y. Li, *Prescribing scalar curvature on S^n and related problems, Part II,* **Comm. Pure Appl. Math. 49** (1996), 541-597.

[LZ1] Y.Y. Li and M. Zhu, *Uniqueness theorems through the method of moving spheres,* **Duke Math. J. 80** (1995), 383-417.

[LZ2] Y.Y. Li and M. Zhu, *Sharp Sobolev trace inequalities on Riemannian manifolds with boundaries,* **Comm. Pure Appl. Math. 50** (1997), 449-487.

[LZ3] Y.Y. Li and M. Zhu, *Sharp Sobolev inequalities involving boundary terms,* **Geometric And Functional Analysis 8** (1998), 59-87.

[L1] E.H. Lieb, *Existence and uniqueness of the minimizing solution of Choquard's nonlinear equation,* **Studies in Appl. Math. 57** (1976/77), 93-105.

[L2] E.H. Lieb, *Sharp constants in the Hardy-Littlewood-Sobolev and related inequalities,* **Ann. of Math. 118** (1983), 349-374.

[L] C.S. Lin, *On Liouville theorem and apriori estimates for the scalar curvature equations,* **Ann. Scuola Norm. Sup. Pisa Cl. Sci. 27** (1998), 107-130.

[M1] J. Moser, *On a nonlinear problem in differential geometry,* in *Dynamical systems (M. Peixoto, ed.) New York: Academic Press, 1973)*

[M2] J. Moser, *A sharp form of an inequality by N. Trudinger,* **Indiana Univ. Math. J. 20** (1970/71), 1077-1092.

[O] M. Obata, *The conjecture on conformal transformations of Riemannian manifolds,* **J. Diff. Geom. 6** (1971), 247-258.

[S1] R. Schoen, *Conformal deformation of a Riemannian metric to constant scalar curvature,* **J. Diff. Geom. 20** (1984), 479-495.

[S2] R. Schoen, *Variational theory for the total scalar curvature functional for Riemannian metrics and related topics* **Topics in Calculus of Variations, Lecture notes in mathematics, No. 1365, edited by M. Giaquinta, Springer-Verlag** (1989), 120-154.

[S3] R. Schoen, *On the number of constant scalar curvature metrics in a conformal class,* **Differential Geometry: A symposium in honor of Manfredo Do Carmo (H.B. Lawson and K. Tenenblat, eds), Wiley** (1991), 311-320.

[S4] R. Schoen, *Courses at Stanford University (1988) and Courant Institute (1989).*

[SY1] R. Schoen and S.T. Yau, *On the proof of the positive mass conjecture in General Relativity,* **Comm. Math. Phys. 65** (1979), 45-76.

[SY2] R. Schoen and S.T. Yau, *Conformally flat manifolds, Kleinian groups, and scalar curvature,* **Invent. Math. 92** (1988), 47-71.

[SZ] R. Schoen and D. Zhang, *Prescribed scalar curvature on the $n-$sphere,* **Calc. Var. Partial Differential Equations 4** (1996), 1-25.

[St] M. Struwe, *Critical points of embeddings of $H_0^{1,n}$ into Orlicz spaces,* **Ann. Inst. H. Poincaré Anal. Non Linéaire 5** (1988), 425-464.

[Ta] G. Talenti, *Best constant in Sobolev inequality,* **Ann. Mat. Pura Appl. 110** (1976), 353-372.

[T1] N. Trudinger, *Remarks concerning the conformal deformation of Riemannian structures on compact manifolds,* **Ann. Scuola Norm. Sup. Cl. Sci. 22** (1968), 265-274.

[T2] N. Trudinger, *On imbeddings into Orlicz spaces and some applications,* **J. Math. Mech. 17** (1967), 473-483.

[Y] H. Yamabe, *On a deformation of Riemannian structures on compact manifolds,* **Osaka Math. J. 12** (1960), 21-37.

[Zh] D. Zhang, *New results on geometric variational problems, thesis, Stanford University, 1990.*

[Z1] M. Zhu, *Liouville theorems on some indefinite equations,* **Proc. Roy. Soc. Edingburgh Sect. A.** to appear.

[Z2] M. Zhu, *Some general forms of sharp Sobolev inequalities,* **J. Functional Analysis 156** (1998) 75-120.

[Z3] M. Zhu, *Sharp Sobolev inequalities with interior norms,* **Calc. Var. Partial Differential Equations . 8** (1999), 27–43.

DEPARTMENT OF MATHEMATICS, RUTGERS UNIVERSITY, NEW BRUNSWICK, NJ 08903, USA

AMS/IP Studies in Advanced Mathematics
Volume 20, 2001

Varifold Type Theory for Sobolev Mappings

Fanghua Lin

0. Introduction

The purpose of this article is to describe some recent works concerning harmonic, approximate harmonic maps and their associated flows. A general picture seems to emerge from these works is what the title of the present paper indicated, that is, a varifold type theory for Sobolev mappings between Riemannian manifolds. Though such a theory has never been explicitly explored in any existing literature known to the author, nor has it been developed in these works [**1**], [**2**], [**3**] and [**5**], which I hope to put them in this general context, the principle feature of it has, however, already presented in these cited works. Let us recall (one of) the original motivations of the theory of varifolds (see [**6**] and [**7**]).

Let $\{T_i\}$ be a sequence of n-dimensional integer multiplicity rectifiable currents such that the corresponding integer multiplicity rectifiable varifolds (see [**7**]) are stationary in an open set U of $\mathbb{R}^{n+k}$. Suppose that $\partial T_i = 0$ in U and there is a uniform local mass bound $\sup_{i\geq 1} M(T_i \llcorner W) < \infty$, for every open subset W with compact closure in U. By the compactness theorem of Federer-Fleming [**8**] for integral rectifiable currents we can assert that there is a subsequence $\{T_{i'}\}$ of $\{T_i\}$ such that $T_{i'}$ converges to T weakly as currents. Moreover, T is also an integral rectifiable current. However, it is not clear (in fact, it is not true in general) that T is stationary. The chief difficult is that it is not generally true that the corresponding sequence of measures $\mu_{T_{i'}}$ (see [**7**]) converges to μ_T. Indeed, if $\mu_{T_{i'}}$ converges to μ_T (as they would be by [**7**, 34.5] in case T_i are area-minimizing in U) then it is not hard to prove that T is stationary in U. This leads one to consider measure theoretic convergence rather than weak convergence of the currents. However, if we take a limit (in the sense of Radon measures) of some subsequence $\{\mu_{T_{i'}}\}$ of the $\{\mu_{T_i}\}$, then we get merely an abstract Radon measure on U, and first variation of this does not make sense.

To resolve these difficulties, the general notions of varifolds were introduced by Almgren (see [**6**]). General varifolds in U are simply Radon measures on $U \times GL(n, n+k)$. Here $GL(n, n+k)$ denotes the set of all n-dimensional subspaces in $\mathbb{R}^{n+k}$. We associate each T_i a Radon measure V_i on $U \times G\ell(n, n+k)$,

$$V_i(A) = \mu_{T_i}(\pi_i(A)), \quad A \subset U \times G\ell(n, n+k),$$

where $\pi_i(A) = \{x \in U; \langle x, \vec{T}_i(x)\rangle \in A\}$, and $\vec{T}_i(x)$ denotes the n-dimensional tangent space of T_i at $x \in \operatorname{spt} T_i$. One then uses compactness theorem for measures on $U \times G\ell(n, n+k)$ to obtain $V_{i'} \to V$ for some subsequence $\{V_{i'}\}$ of $\{V_i\}$. V is again a Radon measure on $U \times G\ell(n, n+k)$. It turns out to be possible to define a notion of stationarity of such Radon measures (i.e. varifolds) V on $U \times G\ell(n, n+k)$, and in the circumstances above V turns out to be a stationary rectifiable varifold, see [**6**] and [**7**] for the details.

We now come back to mappings between Riemannian manifolds. First, let us consider a special case. Let Ω be a bounded domain in $\mathbb{R}^n$, and K be a compact subset of $\mathbb{R}^{n+k}$. Giving a weakly converging sequence of maps $\{u_i\}$ from Ω into K such that $u_i \rightharpoonup u$ weakly in $L^2(\Omega)$. Then the well-known theorem of L. C. Young (see [**9**], [**10**]) asserts that there is a family of probability measures $\{\nu_x\}$, $x \in \Omega$, on K such that for every $f \in C(K)$, one has

$$f(u_i)(x) \rightharpoonup \overline{f}(x) = \int_K f(y)\, d\nu_x(y).$$

Moreover, $u_i(x) \to u(x)$ strongly in $L^2(\Omega)$ if and only if $\nu_x = \delta_{u(x)}$

Viewing from the above Young-Measure theorem, it is easy to see why one defines the "generalized surfaces" (integral rectifiable currents or varifolds) as Radon measures on the Grassmanian manifold $U \times G\ell(n, u+k)$. However, the Young measures described above do not involve derivatives of mappings. In the study of harmonic mappings, one has the additional control on L^2-norms of the gradients of mappings. This naturally leads to the strong convergence of mappings and the weak convergence of derivatives of mappings. Therefore, one should study the Young measures associated with gradients of mappings instead. The gradient Young measures are, however, much hard to characterize and to understand. Despite many efforts, our knowledges about them are still very limited (see the recent survey article [**11**]). Moreover, from the form of partial differential equations for harmonic maps, one ought to study Young measures associated with weakly converging sequences of functions of the form $\sum_{i,j=1}^n C_{ij}(U_k)\, \frac{\partial U_k}{\partial x_i} \cdot \frac{\partial U_k}{\partial x_j}$, where $U_k : \Omega \subseteq \mathbb{R}^n \to N \hookrightarrow \mathbb{R}^p$ (isometrically) is a sequence of maps weakly converging in $H^1(\Omega, N)$. As far as I know, the general studies for these Young measures have not yet been carried out.

1. Generalized Varifolds

The following notion of the generalized varifolds were already remarked in the earlier unpublished notes of Almgren (see [**6**] and reference therein). It has been recently adopted in the study of the Ginzburg-Landau equations by Ambrosio-Soner [**12**] and, in the study of mapping problems by the author in [**2**].

DEFINITION 1. A k-dimensional generalized varifold V is simply a Radon measure on $\Omega \times \mathbb{S}^k_{\lambda,\wedge}$. Where Ω is an open subset of $\mathbb{R}^n$ and

$$\begin{aligned}\mathbb{S}^k_{\lambda,\wedge} =& \{A : A \text{ is an } n \times n \text{ symmetric matrix with} \\ & tr\, A = k \quad \text{and} \quad -\lambda I \le A \le \wedge I\}.\end{aligned}$$

for positive constants λ, $\wedge$. The set of all such V will be denoted by $V^*(k, \lambda, \wedge)$. We shall simply denote $\mathbb{S}^{n-2}_{1,1}$ by $\mathbb{S}$, and $V^*(n-2, 1, 1)$ by V^*. The associated important notion is the so-called first variations of $V \in V^*(n-2, \lambda, \wedge)$.

DEFINITION 2. Let $V \in V^*(n-2,\lambda,\wedge)$. Then the first variation of V, denoted by δV, is a distribution on $C_0^\infty(\Omega,\mathbb{R}^n)$ such that

$$\delta V\,(X) = -\int_{\Omega\times S^{n-2}_{\lambda,\wedge}} DX(x) : A\,dV(x,A),\ X \in C_0^\infty(\Omega,\mathbb{R}^n). \tag{1}$$

Here $A : B = tr(AB)$ for $n\times n$ matrices A and B. V is called *stationary* if $\delta V = 0$.

EXAMPLE a. Let $u : \Omega \to N$ be a map in the space $H^1(\Omega, N)$. Define $V_u = \delta_{A(u)}|\nabla u|^2(x)\,dx$, $A(u) = I_n - \frac{2\nabla u\otimes\nabla u}{|\nabla u|^2}$. Here $A(u)(x)$ is the $n\times n$ matrix such that

$$A(u)(x) = \begin{cases} I_n - 2\left(\frac{u_{x_i}\cdot u_{x_j}}{|\nabla u|^2}\right)(x), & \text{whenever } |\nabla u|(x)\neq 0 \\ I_{n-2} & \text{if } |\nabla u(x)| = 0. \end{cases} \tag{2}$$

The V_u defines an element in V^* by the following

$$V_u(B) = \int_{\pi(B)} |\nabla u|^2(x)\,dx. \text{ Here } B\subset \Omega\times S \text{ and} \tag{3}$$
$$\pi(B) = \{x\in\Omega : (x, A_u(x)\in B\}.$$

The first variation of V_u is given by

$$\delta V_u = \operatorname{div}\ \left[|\nabla u|^2 - 2\nabla u\otimes\nabla u\right] \text{ in the sense of distributions.} \tag{4}$$

A consequence of the above definitions is that: let u be a stationary harmonic map (cf. [**1**]), then V_u is a stationary generalized varifold in V^*. Conversely, given a weakly harmonic map $u : \Omega\to N$. Then u is a stationary harmonic map whenever V_u is a stationary generalized varifold.

We also note that if $\{V_i\}\subset V^*(n-2,\lambda,\wedge)$, is a sequence of stationary generalized varifolds, and if $V_i \rightharpoonup V$ weakly as Radon measures on $S^{n-2}_{\lambda,\wedge}$, then V is also a stationary generalized varifold in $V^*(n-2,\lambda,\wedge)$. Suppose $\{u_i\}$ is a sequence of stationary harmonic maps from Ω into N that converges weakly in $H^1(\Omega,N)$. Let $\{V_{u_i}\}$ be the associated sequence of generalized varifolds in V^* as above. Then $V_{u_i}\rightharpoonup V$ weakly in V^*, and V is also a stationary generalized varifold. One of the main results of [**1**] implies that V is a sum $V_U + V_R$, where $V_u\in V^*$ as above, and V_R is a rectifiable varifold in the classical sense of [**7**]. That is $V_R = (M,\Theta)$, here M is an $(n-2)$-rectifiable subset of Ω (in fact, [**1**] says that M is of locally finite mass and relatively closed in Ω), and $\Theta(x)\in L_{\mathrm{loc}}(\Omega)$ is the multiplicity function which satisfies $\Theta(x)\geq\epsilon_0(n,N) > 0$ for all $x\in M$.

EXAMPLE b. Let $u_i \in C^0(\Omega,N)\cap H^1(\Omega,N)$ be a minimizing sequence for the usual energy. Let $V_{u_i}\in V^*$ be the associated sequence of generalized varifolds. By taking subsequence if necessary, one has $V_{u_i}\rightharpoonup V$ in V^*. Moreover V is a stationary generalized varifold (see [**2**]). In other words, div $\left[|\nabla u_i|^2 - 2\nabla u_i\otimes\nabla u_i\right]\to 0$ in the sense of distributions.

One of the principle result of [**2**] is that V is a sum, $V = V_U + V_R$. Where u is a weakly harmonic map which is smooth away from a relatively closed, rectifiable subset M of Ω with $\mathcal{H}^{n-2}$-measure of $M\cap\Omega'$ bounded for every $\Omega'\Subset\Omega$. Here V_R is an integral rectifiable varifold in Ω as defined in [**7**]. In fact, $V_R = (M,\Theta)$ with $\Theta(x) = \sum_{j=1}^{\ell_x} E(\phi_j)$ for $\mathcal{H}^{n-2}$-a.e. $x\in M$. Here $\{\phi_j\} : \mathbb{S}^2\to N$ are smooth harmonic maps.

EXAMPLE c. For $0 < \epsilon \ll 1$, consider a family of critical points of $I_\epsilon(\cdot)$,

$$I_\epsilon(u_\epsilon) = \int_\Omega \left[\frac{1}{2}|\nabla u_\epsilon|^2 + \frac{1}{\epsilon^2} F(u_\epsilon)\right] dx \leq C_0 \tag{5}$$

$$\Delta\, u_\epsilon + \frac{1}{\epsilon^2} f(u_\epsilon) = 0 \quad \text{in} \quad \Omega. \tag{6}$$

Here $f = -\nabla F$,

$$F(u) = \begin{cases} \text{dist}^2(u, N), & \text{if dist } (u, N) \leq \delta_N, \\ 4\delta_N^2, & \text{if dist } (u, N) \geq 2\delta_N, \\ \text{smooth monotone} & \text{for } \delta_N \leq \text{dist } (u, N) \leq 2\delta_N. \end{cases} \tag{7}$$

Define V_{u_ϵ} as before. Then there are sequences $\{\epsilon_i\}$, $\epsilon_i \downarrow 0$ such that $V_i = V_{u_{\epsilon_i}} \rightharpoonup V$ in V^*. Moreover, V is stationary (see [**3**]).

In fact, $\delta V_{u_\epsilon}(X) = -\int_\Omega \text{div}\, X \frac{F(u_\epsilon)}{\epsilon^2}\, dx \to 0$ as $\epsilon \to 0$, for every $X \in C_0^1(\Omega, \mathbb{R}^n)$. Again, it is shown in [**3**] that V is a sum $V_U + V_R$ with u a weakly harmonic and smooth away from $M \subset \Omega$, $V_R = (M, \Theta)$.

2. Some Basic Results for the Generalized Varifolds

For $V \in V^*(k, \lambda, \wedge)$, we let $\|V\|(E) = V(E \times S^k_{\lambda,\wedge})$ for $E \subset \Omega$, and

$$d(V, a, r) = \frac{\|V\|\,(B(a, r))}{\alpha(k)\, r^k} \tag{8}$$

whenever $a \in \Omega$, $0 < r < \infty$, $B(a, r) = \{x \in \mathbb{R}^n : |x - a| < r\} \subseteq \Omega$. Here $\alpha(k)$ is the Lebesgue measure of the k-dimensional unit ball.

Define $d(V, a) = \lim_{r\downarrow 0} d(V, a, r)$ provided the limit exists. Note that $d^*(V, a) = \overline{\lim}_{r\downarrow 0}\, d(V, a, r)$ always exists.

We shall state below a few results their proofs are quite standard.

LEMMA 1. *Suppose $V \in V^*(k, \lambda, \wedge)$ and δV is a measure. Let $f : \Omega \to R$ be a smooth function such that spt $\|V\| \subseteq \{x \in \Omega : f(x) = 0\}$. Then $A \subset \ker Df(x)$ for V almost all $(x, A) \in \Omega \times S^k_{\lambda,\wedge}$.*

We note that if δV is a measure, that is, if

$$\begin{aligned} \|\delta V\|(G) &= \sup\{|\delta V(X)| : X \in C_0^\infty(\Omega, \mathbb{R}^n),\ \|X\|_{L^\infty} \leq 1, \text{spt } X \subset G\} \\ &\leq C(G) < \infty \quad \text{for every} \quad G \Subset \Omega, \end{aligned} \tag{9}$$

then the Rietz Representation Theorem implies that

$$\delta V(X) = \int X(x)\, \eta(x)\, d\|\delta V\|(x), \quad \text{for} \quad X \in C_0^\infty(\Omega, \mathbb{R}^n). \tag{10}$$

Here η is a $\|\delta V\|$-measure, $\mathbb{S}^{n-1}$-valued function.

LEMMA 2. *Suppose $V \in V^*(k, \lambda, \wedge)$ is a stationary. Then*

$$\begin{aligned} &\alpha(k)\,[d\,(V, a, s) - d\,(V, a.r,)] = \\ &\int_{A(r,s)} |x - a|^{-k-2} \left[1 - \left\langle A\, \tfrac{x-a}{|x-a|}, \tfrac{x-a}{(x-a)} \right\rangle\right] dV(x, A). \end{aligned} \tag{11}$$

where $A(r, s) = \{x : r \leq |x - a| \leq s\} \times S^k_{\lambda,\wedge}$.

It is interesting to remark that if $V = V_u \in V^*$, and u is a stationary harmonic map, then Lemma 2 is exactly the usual energy monotonicity formula for stationary harmonic maps. Suppose $V \in V^*$ is a limit of some weakly converging sequence V_{u_i}, where u_i are stationary harmonic maps. Then Lemma 2 gives the energy monotonicity formula for the defect measures introduced in [**1**], [**2**]. Also we note that when $V \in V^*$, $d(V, a, r)$ is increasing in r.

The following basic rectifiability lemma is due to Ambrosio-Soner.

LEMMA 3. *Let $V \in V^*(k, \lambda, \wedge)$, and suppose δV is a Radon measure on Ω. Suppose that*

$$\Theta^*_\alpha(\|V\|, x) > 0 \quad \textit{for} \quad \|V\| - \textit{a.e.} \quad x \in \Omega,$$

$$\Theta^*_\alpha(\|V\|, x) = \lim_{r \downarrow 0} \frac{\|V\|(B(x,r))}{r^\alpha}, \quad \textit{for some} \quad \alpha > 0.$$

Let $\{V_x\}_{x\in\Omega}$ be a family of the probability measures which is obtained from slicing V by the projection map $\pi : \Omega \times S^k_{\lambda,\wedge} \to \Omega$, and let $\bar{A}(x) = \int_{S^k_{\lambda,\wedge}} A \, dV_x(A)$. Then

(i) *if $\alpha < k+2$, then $\bar{A}(x) \geq 0$ for $\|V\|$-a.e. $x \in \Omega$;*
(ii) *if $\alpha < k+1$, then $\bar{A}(x) \in G\ell(k, n+k)$ for $\|V\|$-a.e. $x \in \Omega$, herence $\widetilde{V} = \delta_{\bar{A}(x)}\|V\|$ is a classical varifold;*
(iii) *if $\alpha = k$, then $\widetilde{V}$ is a rectifiable varifold, i.e., $\|V\| = \|\widetilde{V}\|$ is supported on a $\mathcal{H}^k$-rectifiable set.*

Note that the last statement (iii) of the above lemma follows from [**6**]. An improvement of the above statement (iii), which follows from arguments in [**1**] and [**2**], is the following:

PROPOSITION 1. *Let $V \in V^*(k, \lambda, \wedge)$ be stationary. Then the set $\{x \in \Omega : d(V, x) > 0\}$ is a $\mathcal{H}^k$-rectifiable. In general, if $V \in V^*(k, \lambda, \wedge)$ and if $\|V\|(\Omega) < \infty$, $\|\delta V\|$ is a Radon measure on Ω. Then, for any $\epsilon > 0$, there are k-dimensional Lipschitz graphs $M_1, \ldots, M_{j(\epsilon)}$ in Ω and positive real numbers $C_1, \ldots, C_{j(\epsilon)}$ such that the $\|V\|\left(\{x \in \Omega : d(V, x) > 0\} \setminus \bigcup_{i=1}^{j(\epsilon)} M_i\right) < \epsilon$ and that the total variation of the signed measure*

$$\widetilde{V} - \sum_{i=1}^{j(\epsilon)} V(M_i, C_i)$$

is also bounded by ϵ. Here $V \lfloor \{x \in \Omega : d(V, x) > 0\}$.

Finally, we note that the monotonicity Lemma 2 above can also be easily generalized to obtain the following:

PROPOSITION 2. *Suppose $V \in V^*$, δV is a measure, $a \in spt \|V\|$, $0 < r < s$ and $B(a, s) \subset \Omega$. Then*

$$d(V, a, s) \exp \int_r^s \frac{\|\delta V\|(B(a, \rho))}{\|V\|(B(a, \rho))} \, d\rho \geq d(V, a, r).$$

Though we have discussed so far examples related to the harmonic maps, the general p-harmonic mapping theory fits in very well too. Indeed, let $u : \Omega \to N$ be a stationary m-harmonic map. We define $V_u = \delta_{A(u)}\|\nabla u|^m(x)\, dx$, here $A(u) =$

$I_n - m\,\frac{\nabla u\otimes\nabla u}{|\nabla u|^2}$. Then V_u is a $(n-m)$-dimensional stationary generalized varifold. Moreover the monotonicity Lemma 2 implies that

$$\begin{aligned}&\frac{1}{r^{n-m}}\int_{B_r(a)}|\nabla u|^m(x)\,dx - \frac{1}{s^{n-m}}\int_{B_s(a)}|\nabla u|^m(x)\,dx\\ &= \int_{B_r\setminus B_s(a)} m|\nabla u|^{m-2}\,u_\rho^2\,\frac{dx}{\rho^{n-m+2}} \geq 0.\end{aligned}$$

3. Flow of Generalized Varifolds

Let μ be the total variation measure of a k-dimensional rectifiable varifold in $\mathbb{R}^n$, and let ϕ be in $C_0^2(\mathbb{R}^n\times[0,\infty))$. We define

$$\mathcal{B}(\mu,\phi) = -\int_{\mathbb{R}^n}\phi\,|H|^2 - \langle T_x^\perp\,\mu\,\nabla\phi, H\rangle\,d\mu, \tag{12}$$

Here $H\,d\mu = \delta V$, $H\in L^2(d\mu)$, $T_x^\perp\,\mu$ means the orthogonal projection onto the Normal space of the tangent space $T_x\,\mu$ (tangent space of M at x, if $V=(M,\Theta)$). Then Brakke [**13**] introduced the so-called Brakke-flow:

Let $\{\mu_t\}_{t\geq 0}$ be a Radon measures on $\mathbb{R}^n$. We say that $\{\mu_t\}$ is a Brakke-flow if

$$\limsup_{s\to t}\frac{\mu_s(\phi)-\mu_t(\phi)}{s-t}\leq\mathcal{B}(\mu_t,\phi)$$

for any $t\geq 0$ and $\phi\in C_0^2(\mathbb{R}^n\times[0,\infty))$. Here $\mu_t = \|V_t\|$, and $\mathcal{B}(\mu_t,\phi)$ is a define by [**13**]. We note the point of the Brakke's definition is the identity

$$\frac{d}{dt}\int_{\overline{Z}_t}\phi d\mathcal{H}^k \equiv -\int_{\overline{Z}_t}\phi|\mathcal{H}_t|^2 - \langle\nabla\phi,\mathcal{H}_t\rangle\,d\mathcal{H}^k \tag{13}$$

which holds for any smooth flow $\{\overline{Z}_t\}$ of k-dimensional surfaces by mean curvature and $\phi\in C_0^2(\mathbb{R}^n)$.

For the generalized varifolds one may introduce the similar notion as flow of these generalized varifolds (see also [**12**]). In [**5**], we consider the Chen-Struwe [**14**] approach for the heat-flow of harmonic maps. Thus one consider (see also [**7**] above)

$$\begin{cases}\frac{\partial u_\epsilon}{\partial t} = \Delta u_\epsilon + \frac{1}{\epsilon^2}f(u_\epsilon) & \text{in } M\times R_+\\ u_\epsilon(x,0) - = u_0(x),\end{cases} \tag{14}$$

for maps from $M\times R_+\hookrightarrow\mathbb{R}^P$. As $\epsilon\to 0^+$ $u_\epsilon(x,t)\rightharpoonup u(x,t)$ weakly in $H^1_{\text{loc}}(M\times R_+)$. $u(x,t)$ satisfies weak heat-flow equation for maps from M into N and, $u(x,t)$ is smooth away from a relatively closed subset, $\sum$ in $M\times R_+$ such that the $n=\dim M$ dimensional parabolic Hausdorff measure of $\sum$ is locally finite (see [**14**]). In [**5**] we show if there is no harmonic $\mathbb{S}^2$ in N, then $P^n(\sum)=0$ and $u(x,t)$ is stationary (see [**3**] and [**5**]). Moreover, $u_\epsilon(x,t)\to u(x,t)$ in $H^1_{\text{loc}}(M\times R_+)$. In general, one has

$$\begin{aligned}&V_{u_\epsilon}\rightharpoonup V \text{ is generalized varifolds in } M\times R_+\times S_-\\ &\mu_t^\epsilon = \|V_{u_\epsilon}\|(t)\rightharpoonup\mu_t = \|V\|(t)\quad\text{for all } t\geq 0.\\ &\text{Note}\quad\mu_t^\epsilon\,dt = \|V_{u_\epsilon}\|,\qquad\|V\| = \mu_t\,dt.\end{aligned}$$

Let $V = \nu_{t,x}\,\mu_t(x)\,dt$. Here $\nu_{t,x}$ is a probability measure on $\mathbb{S}$.

Next, we let $\sigma_\epsilon = -\nabla_t\,u^\epsilon\,\nabla u^\epsilon\,dx\,dt$, then $\sigma_\epsilon\rightharpoonup\sigma = H_t\,\mu_t\,dt$.

$$\int_0^\infty\int_M|\mathcal{H}_t|^2\,d\mu_t\,dt\leq 2\,\underline{\lim}_\epsilon\int_0^\infty\int_M\left|\frac{\partial u^\epsilon}{\partial t}\right|^2\,dxdt.$$

It turns out $\delta V_t = H_t\,\mu_t$ for L^1-a.e. $t > 0$. Again, this V can be written as a sum $V = V_u + V_R$, V_u is the generalized varifold associated with the limiting map $u(x,t)$ and V_R turns out to be a rectifiable varifold in $M \times R_+$. $\|V\| = \frac{|\nabla u|^2}{2}\,dxdt + \nu_t\,dt$. Finally, one has the energy inequality

$$
\begin{aligned}
(15) \qquad & \frac{d}{dt}\int_M \phi\,\frac{|\nabla u|^2}{2}\,dx + \int_M \phi\,d\mu_t \\
& \le -\int_M (\phi\,u_t^2 - \nabla u\,u_t) \\
& \quad - \int_{\overline{Z}_t} (\phi|H_t|^2 - \langle T_x^{\perp}\,\overline{Z}_t\,\nabla\phi,\, H_t\rangle)\,d\mu_t
\end{aligned}
$$

$\overline{Z}_t = \operatorname{spt}\,\nu_t$, and $u(x,t)$ is smooth away from $\overline{Z}_t$, $t > 0$. For the details, we refer to [**5**]. Inequality (15) is certainly the right definition also for flows of the generalized varifolds.

4. Concluding Remarks

In above discussions, we have put the theory of harmonic maps and its associated gradient flows in a rather general content. It is obviously useful to have such a global picture. Moreover, this general theory has put the mapping theory and the classical theory of the minimal surfaces in an essentially the same content as one may naturally believes so. However, a more serious question one may ask, is that why one needs such general theory anyway, besides whatever has been described or said. It will be very proper to mention here a very nice general theory of the so-called Cartensian currents which has been systematically developed by Gianquinta-Modica-Soucek [**15**]. The Cartensian currents have many interesting applications in calculus of variations. A particular example is [**16**] which concerns with least energy maps among continuous maps which have the same boundary condition. Indeed, before [**2**], [**16**] is probably the best result with regards to this question. Unfortunately, the Cartensian current theory when apply to the harmonic maps, it works well only when either domain or target being two-dimensional. This is a serious restriction. The theory we have explained here (see [**2**]) applies without any restrictions. One should view the theory developed in [**15**] as current-type theoyr for Sobolev mappings. Thus certain advantage we have for the varifold type theory of mappings is obvious as in the case of minimal surface theory.

References

[1] Fanghua Lin, Gradient estimates and blow-up analysis for stationary harmonic maps, *Annals of Math.* **149** (1999), 785–829.

[2] Fanghua Lin, Mapping problems, fundamental groups and defect measures. *Acta Math. Sin.* (English Ser.) **15** No. 1, (1999), 25–52.

[3] Fanghua Lin and Chang You Wang, Harmonic and quasi harmonic spheres, *Comm. Anal. Geom.* **7** No. 2, (1999), 397–429.

[4] Fanghua Lin and Chang You Wang, Harmonic and quasi harmonic spheres, Part II and Part III, preprint, 2000.

[5] Fanghua Lin and Chang You Wang, Rectifiability of defect measures in the harmonic map heat flow, preprint.

[6] W. K. Allard, First variation of a varifold, *Annals of Math.* **95** (1972), 417–491.

[7] L. Simon, Lectures on Geometric Measure Theory, *Proc. of CMA of ANU* vol 3, 1983.

[8] H. Federer and W. Fleming, Normal and integral currents, *Annals of Math.* **72** (1960), 458–520.

[9] J. M. Ball, A version of the fundamental theorem for Young measures, P.D.E.'s and continuum models of phase transition, *Lecture Notes in Physics* **344** Springer, 207–215.

[10] L. C. Young, Generalized surfaces in the calculus of variatoins, I and II, *Annals of Math.* **43**, 84–103 and 530–544.

[11] Pablo Pedregal, Optimization, Relaxation and Young Measures, *Bull. A.M.S.* **36** No. 1 (1999), 27–58.

[12] L. Ambrosio and M. Soner, A measure theoretic approach to higher dimension mean curvature flows, *Annali Scuola Norm. Sup. Pisa* (1998), 27–49.

[13] K. Brakke, The motion of a surface by its mean curvature, *Math. Notes* Princeton Univ. Press, 1978.

[14] Y. M. Chen and M. Struwe, Existence and partial regularity results for the heat flow of harmonic maps, *Math Z.* **201** (1989), 83–103.

[15] M. Giaquinta, L. Modica and J. Soucek, *Cartensian currents in the Calculus of Variations*, I and II, Springer-Verlag, Berlin (1998), 697 and 711.

[16] M. Giaquinta, L. Modica and J. Soucek, The Dirichlet energy of mappings with values into the sphere, *Manuscripta Math.* **65** No. 4, (1989), 489–507.

Courant Institute, 251 Mercer Street, New York, NY 10012
E-mail address, Fanghua Lin: `linf@courant.nyu.edu`

AMS/IP Studies in Advanced Mathematics
Volume 20, 2001

Cellular Neural Networks: Pattern and Waves

Song-Sun Lin

ABSTRACT. We consider 2-d CNN and compute the lower spatial entropy bound when the template is square-cross. We then study the structure of traveling wave solutions of 1-d CNN of the advanced type. We show the existence of monotone traveling waves, oscillating wave and eventually periodic waves by using shooting method and comparison principle.

1. Introduction

The types of systems we are interested in are Cellular Neural Networks (CNNs). Such a system has been proposed by Chua and Young [3]. The CNNs without input terms are of the form

$$\frac{dx_{i,j}}{dt} = -x_{i,j} + z + \sum_{|k|\leq 1.|\ell|\leq 1} a_{k,\ell} f(x_{i+k,j+\ell}), i,j \in \mathbb{Z}^2, \tag{1.1a}$$

$$x_{i,j}(0) = x^0_{i,j}. \tag{1.1b}$$

Here the nonlinearity f is a piecewise-linear function of the form

$$f(x) = \frac{1}{2}(|x+1| - |x-1|). \tag{1.2}$$

$A = (a_{kj})$ is called a space-invariant template. A is called symmetric if $a_{-k,-1} = a_{k,\ell}$ for all $|k| \leq 1$ and $|\ell| \leq 1$. The quantites $x_{i,j}$ denote the state of a cell $C_{i,j}$. $y_{i,j} = f(x_{i,j})$ is output and z is biased term.

Stationary solutions $\bar{x} = (\bar{x}_{i,j})$ of (1,1a) are very important in studying CNN systems; their outputs $\bar{y} = (f(\bar{x}_{i,j}))$ are called patterns. It was shown by Chua and Yang [3], and later by and Lin and Shih [9] that (1.1a) behaves like a gradient system when template A is symmetric and the lattice is finite. In this case, every trajectory will tend to a stable stationary solution as time goes up. For certain asymmetric templates, similar results hold [5].

A mosaic solution $\bar{x}$ satisfies $|\bar{x}_{i,j}| > 1$ for all $(i,j) \in \mathbb{Z}^2$. Its corresponding pattern $\bar{y}$ is called a mosaic pattern. Some basic problems in CNN theory can be stated as follows:

1. Direct Problem: Given any $\mathcal{P} \subset \mathcal{P}_{10} = \{(z,A) : A :\ 3\times 3 \text{ real matrix},\ z \in \mathbb{R}\}$, determine $\mathcal{M}(\mathcal{P})$, the set of all stable mosaic patterns of (1.1).

2. Inverse problem: Given a set of stationary patterns $\mathcal{U}$, determine a set of parameters $\mathcal{P} \subset \mathcal{P}_{10}$, such that $\mathcal{U} = \mathcal{M}(\mathcal{P})$.
3. Study the complexity of $\mathcal{M}(\mathcal{P})$ for each subset $\mathcal{P}$ of $\mathcal{P}_{10}$.

Given $(z, A) \in \mathcal{P}_{10}$, the complexity of the mosaic patterns $\mathcal{M}(z, A)$ can be studied as follows: on finite lattice $\mathbb{Z}_{mn}$, define the number of mosaic patterns by

$$\Gamma(m, n; z, A).$$

Then spatial entropy of $\mathcal{M}(z, A)$ is defined by

$$h(\mathcal{M}(z, A)) \equiv \lim_{\mathrm{m,n} \to \infty} \frac{\log \Gamma(m, n; z, A)}{mn}.$$

Now $\mathcal{M}(z, A)$ is called spatial chaos if $h(\mathcal{M}(z, A)) > 0$ and pattern formation if $h(\mathcal{M}(z, A)) = 0$, see [1]. Note that, we have $\Gamma(m, n) \sim e^{hmn}$ as $m, n \to \infty$.

2. Patterns

For simplicity, we consider that the template is square-cross i.e.,

$$A = \begin{bmatrix} 0 & b & 0 \\ b & a & b \\ 0 & b & 0 \end{bmatrix}, \quad 0 < b < \frac{a}{4}. \tag{2.1}$$

Then the parameter space can be partitioned as follows:

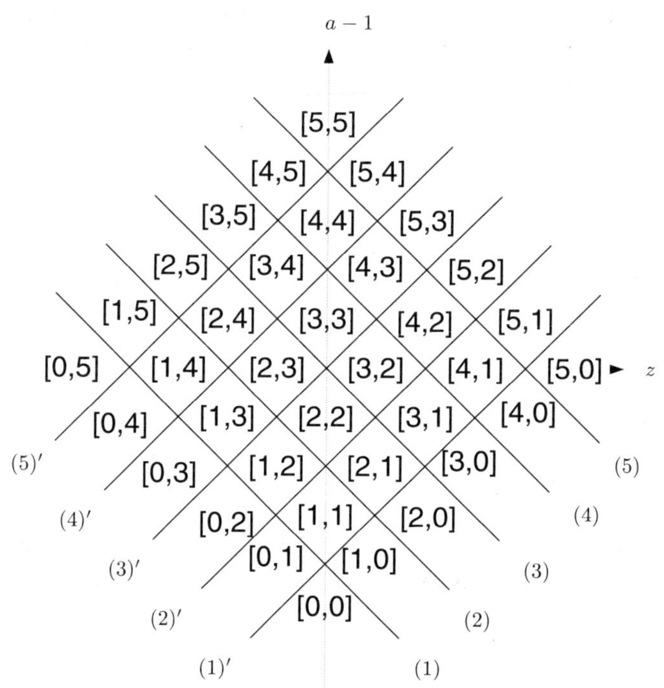

Note that in region $[m, n]$, only local patterns (1), …, (m), (1)′, …, (n)′ are allowed, which can be explained as follows:

It is clear that $\sum_{i=1}^{4} v_i \in \{-4, -2, 0, 2, 4\}$ when $v_i \in \{-1, 1\}$. Therefore, we have 5 parallel lines when the center cell is +1: $b\sum_{i=1}^{4} v_i + a + z - 1 = 0$. The associated local patterns are given by:

$$\underbrace{\begin{array}{ccc} \bullet & - & \bullet \\ - & \oplus & - \\ \bullet & - & \bullet \end{array}}_{(5)}, \quad \underbrace{\begin{array}{ccc} \bullet & + & \bullet \\ - & \oplus & - \\ \bullet & - & \bullet \end{array}}_{(4)}, \quad \underbrace{\begin{array}{ccc} \bullet & + & \bullet \\ - & \oplus & + \\ \bullet & - & \bullet \end{array}}_{(3)}, \quad \underbrace{\begin{array}{ccc} \bullet & + & \bullet \\ + & \oplus & + \\ \bullet & - & \bullet \end{array}}_{(2)}, \quad \underbrace{\begin{array}{ccc} \bullet & + & \bullet \\ + & \oplus & + \\ \bullet & + & \bullet \end{array}}_{(1)}$$

Similarly, when the center cell is -1 the lines and patterns are given by $b\sum_{i=1}^{4} v_i - a + z + 1 = 0$ and

$$\underbrace{\begin{array}{ccc} \bullet & + & \bullet \\ + & \ominus & + \\ \bullet & + & \bullet \end{array}}_{(5)'}, \quad \underbrace{\begin{array}{ccc} \bullet & - & \bullet \\ + & \ominus & + \\ \bullet & + & \bullet \end{array}}_{(4)'}, \quad \underbrace{\begin{array}{ccc} \bullet & - & \bullet \\ + & \ominus & - \\ \bullet & + & \bullet \end{array}}_{(3)'}, \quad \underbrace{\begin{array}{ccc} \bullet & - & \bullet \\ - & \ominus & - \\ \bullet & + & \bullet \end{array}}_{(2)'}, \quad \underbrace{\begin{array}{ccc} \bullet & - & \bullet \\ - & \ominus & - \\ \bullet & - & \bullet \end{array}}_{(1)'},$$

here $\bullet \in \{+,-\}$. Now, we have results of spatial entropy on each $[m,n]$.

THEOREM 1 (Juang & Lin [8]). $\mathcal{M}([m,n])$ is spatial chaos iff

$$\alpha \equiv \max(m,n) \geq 3,$$

and

$$\beta \equiv \min(m,n) \geq 2.$$

Furthermore, we have

$$h([m,n]) = \begin{cases} = \log 2 & \text{if } \beta = 5, \\ \geq \frac{\log 10}{4} & \text{if } \beta = 4, \\ \geq \frac{\log 4}{4} & \text{if } \beta = 3, \\ \geq \frac{\log 4}{9} & \text{if } \beta = 2, \alpha = 5, \\ \geq \frac{\log 4}{12} & \text{if } \beta = 2, \alpha = 4, \\ \geq \frac{\log 2}{16} & \text{if } \beta = 2, \alpha = 3. \end{cases}$$

The main idea of proof is by introducing a building block and compatibility of local patterns. For example, consider the region $[3,2]$. On each $4m \times 4n$ rectangle, we take following 4×4 local patterns

$$\begin{array}{cccccccc} - & - & - & - & - & - & - & - \\ - & - & - & - & - & - & - & - \\ + & + & - & - & - & - & - & - \\ + & + & - & - & - & - & - & - \end{array}$$

as building block, then it is easy to see

$$h([3,2]) \geq \lim_{m,n\to\infty} \frac{\log 2^{mn}}{4m \cdot 4n} = \frac{\log 2}{16}.$$

REMARK 1. The main problem is to compute the exact spatial entropy.

3. Traveling Waves

For simplicity, we only consider the one-dimensional traveling wave problem with template $A = (\alpha, a, \beta)$. Assume $x_i(t) = \phi(i - ct)$, then φ satisfies:

$$\text{(TW)} \qquad -c\varphi'(s) = -\varphi(s) + z + \alpha f(\varphi(s-1)) + af(\varphi(s)) + \beta f(\varphi(s+1)).$$

There are three homogeneous stationary solutions: $x^0 = \frac{-z}{-1+a+\alpha+\beta}$ which is unstable and $x^{\pm} = z \pm (\alpha + a + \beta)$ which are stable. The boundary conditions at $\pm\infty$ are given by $(BC)^- :\ \lim_{s\to-\infty}\varphi(s) = x^0$, and $(BC)^+ :\ \lim_{s\to+\infty}\varphi(s) = x^+$. We have the existence of traveling solutions:

THEOREM 2 (Hsu & Lin[6]). If $\alpha + a + \beta > 1$, then $\exists c^* < 0$ such that $\forall c \in (-\infty, c^*)$, $\exists\,!$ non-decreasing traveling wave solution satisfies $(BC)^+$ and $(BC)^-$.

Moreover, on advanced case, we have the following global result:

THEOREM 3 (Hsu, Lin & Shen[7]). If $\alpha = 0$, $a + \beta > 1$, $a > 0$ and $\beta > 0$. Let $\phi(s,c)$ be the solution of (TW) and satisfies $(BC)^+$ with $\phi(0,c) = 1$.

1. when $a \geq 1 + \beta$, $\exists c_* < 0$ such that
   (a) if $c \leq c_*$, then $\phi(s,c)$ non-decreasing and satisfies $(BC)^-$.
   (b) if $c_* < c < 0$, then $\phi(s,c)$ oscillating and $|\phi(s,c)| < 1$ for $s < 0$.
2. when $a < 1 + \beta$, $\exists c_*$, c_p and c^* with $c_* \leq c_p \leq c^* < 0$ such that
   (a) if $c \leq c_*$, then $\phi(s,c)$ non-decreasing and satisfies $(BC)^-$.
   (b) if $c_* < c < c_p$, then $\phi(s,c)$ oscillating and $|\phi(s,c)| < 1$ for $s < 0$.
   (c) if $c_p \leq c < c^*$, then $\phi(s,c)$ eventually periodic and $\sup_{s<0} |\phi(s,c)| > 1$.
   (d) if $c = c^*$, then $\phi(s,c)$ non-decreasing and $\phi(s,c^*) = x^- \ \forall\ s \leq s^* < 0$.
   (e) if $c^* < c < 0$, then $\phi(s,c)$ non-decreasing and unbounded.

REMARK 2. For traveling waves of lattice dynamical system, see [2,10,11].

References

[1] S.N. Chow, J. Mallet-Paret and E.S. Van Vleck(1996), Pattern formation and spatial chaos in spatially discrete evolution equation, Random Comp. Dyn. 4, 109-178.

[2] S.N. Chow, J. Mallet-Paret and W. Shen(1998), Traveling waves in lattice dynamical system, J.D.E., 149, 249-291.

[3] L.O. Chua and L. Yang (1988), Cellular neural networks: theory & application, IEEE Trans. Circuit Syst. 35, 1257-90.

[4] L.O. Chua (1998), CNN: A paradigm for complexity, World Scientific Series on Nonlinear Science, Series A, vol. 31.

[5] B. Fiedler and T. Gedeon(1998), A class of convergent neural network dynamics, Physica D, 111, 288-294.

[6] C.H. Hsu and S.S. Lin (1998), Traveling waves in lattice dynamical system with application to cellular neural networks.

[7] C.H. Hsu, S.S. Lin and W. Shen (1999), Traveling waves in cellular neural networks, Int. J. Bifurcation and Chaos, vol.9, no.7.

[8] J. Juang and S.S. Lin(1997), Cellular neural network: mosaic pattern and spatial chaos, SIAM. J. Appl. Math.(to appear).

[9] S.S. Lin and C.W. Shih(1999), Complete stability for standard CNN, Int. J. Bifurcation and Chaos, Vol 9, No. 6.

[10] J. Mallet-Paret(1999), The Fredholm alternative for functional differential equation of mixed type, J. Dyn. Diff. Eqns., 11, 1-48.

[11] J. Mallet-Paret(1999), The global structure of traveling waves in spatial discrete dynamical system, J. Dyn. Diff. Eqns., 11, 49-127.

DEPARTMENT OF APPLIED MATHEMATICS, NATIONAL CHIAO TUNG UNIVERSITY, HSIN CHU, 30050, TAIWAN

AMS/IP Studies in Advanced Mathematics
Volume 20, 2001

Long Time Behavior of Curves Evolving by Curvature

Xi-Ping Zhu

1. Introduction

In this paper we study the motion of curves on a surface whose normal velocity is a given function of its position and its tangent direction, as well as its curvature. One of motivation of this study comes from the theory of phase transition. Modeling the dynamics of melting solids or similar phenomena has been a topic of study for long time. For sharp interface the motion of the interface, this is the boundary of the solid, is usually related to its curvature in a certain way. One of the early models was proposed by Mullins for grain boundaries. This two dimensional model is given by the equation

$$V = k \tag{1}$$

where V and k are respectively the normal velocity and curvature of the interface. Equation (1) is sometimes called the "curve shortening flow" because it is the negative L^2-gradient flow of the length of the interface. By drawing pictures one can be easily convinced that an embedded closed curve stays embedded and smooth along (1), and shrinks to a point in finite time, with limiting shape a round circle. However, a mathematical treatment of (1) turned out to be rather delicate. In fact, a rigorous study did not exist until early eighties, when differential geometers considered (1) as a tool in the search for embedded closed geodesic on surfaces. It was also regarded a model case for more general curvature flows, which are believed to be important in the topological classification of low dimensional manifolds. As a first attempt Gage [12] proved that the isoperimetric radio decreases along plane convex curves. Then in Gage and Hamilton [14] it was shown that a plane convex curve stays convex and shrinks to a point in finite time. Moreover, if one normalized the flow by dilating it so that the enclosing area is constant, the normalized flow converges smoothly to a circle. Finally Grayson [16] completed this line of investigation for plane curves. He showed that an embedded curve evolves into a convex one before it shrinks to a point. For alternative approaches to this result one may consult Hamilton [20] and Huisken [21].

Recently Mullins' theory was generalized by Gurtin [18] and by Angenent and Gurtin [3] to include anisotropy and the possibility of a difference in bulk energies between phases. Anisotropic is indispensable in dealing with crystalline materials. For perfect conductors the temperatures in the both phases are constant. The equation becomes

$$\beta(\theta) = g(\theta)k + F \tag{2}$$

where θ is the tangent angle of the interface, $\beta > 0$ is a kinetic coefficient, g is of the form $g = \frac{d^2 f}{d\theta^2} + f$, $f > 0$ the interfacial energy, and the constant F is the energy of solid relative to its surrounding. The presence of θ reflects anisotropy, and the particular form in which f appears in g is a consequence of thermodynamics.

In this note we describe some recent developments for (1), (2) and their generalizations on surfaces. Section 2 discusses the global behaviors of (2) for convex, closed plane curves. In section 3 the formations of shrinking point for the evolving embedded curves are considered. Finally in section 4 we discuss the curve shortening flow (1) for complete, noncompact curves.

2. Anisotropic Flows for Convex Plane Curves

In the theory of phase transition, the interface is described as an embedded closed curve $\gamma(\cdot, t)$ in the plan, of which evolution is governed by the equation (2). We shall consider regular interfaces and always assume that the interfacial energy is strictly stable, this is, g is positive. This means that (2) is essentially an equation of parabolic type. Using the standard theory of parabolic PDE's, it is a fairly straightforward matter to prove short time existence of solution to (2) for initial curves which are C^2. With the presence of F, apart from shrinking the flow may expand depending on the "size" of the initial interface as well. Denote by $L(t)$ and $A(t)$ the length of , and the area enclosed by the interface at time t respectively. It was known in [3] that when $F \geq 0$, $A(t)$ tends to zero in finite time; and when $F < 0$, $A(t)$ tends to zero in finite time provided the length of the initial interface is sufficiently small, and it tends to infinity provided the initial interface contains a sufficiently large disk. Both the expanding and collapsing cases have been taken up by several authors. Soner [24] considered as well as its higher dimensional generalization by means of the concept of viscosity solutions. He proved that in the expanding case, the solution is asymptotic to a dilation of the Wulff region of $\frac{1}{\beta}$, answering a conjecture raised in [3]. This result was further strengthened by Angenent and Gurtin [4], where, in particular, the assumption required in [24], which states that the polar diagram of β is convex,was removed.

The study of the collapsing case is more difficult. When F vanishes, the equation becomes

$$\beta(\theta) = g(\theta)k. \tag{3}$$

Equation (3) arises from geometry. When $\left(\frac{g}{\beta}\right)(\theta) = \left(\frac{g}{\beta}\right)(\theta + \pi)$, it has a natural interpretation as the curve shortening problem for a Minkowski geometry. Using this point of view Gage [13] showed that a convex curve shrinks to a point as in the isotropic case. Moreover, the normalized curve converges smoothly a self-similar solution of (3), which is unique up to dilation and translation. Without assuming any symmetry on $\frac{g}{\beta}$, Gage and Li [15] showed that the normalized curve "subconverges" to self-similar solution of (3), of which uniqueness is not known.

One can also observe that when F is negative, (2) has a stationary solution of which curvature satisfies

$$g(\theta)k + F = 0. \tag{4}$$

The specified form $g = \frac{d^2 f}{d\theta^2} + f$ guarantees that this equation is uniquely solvable (up to translation) for a convex curve. Linearized stability of this stationary solution was studied and, in fact, employed to prove the above mentioned asymptotic result in the expanding case by Angenent and Gurtin in [4]. They showed that the stationary point is a hyperbolic fixed point for the associated linearized semi-flow. The unstable manifold is of dimension one and consists of two trajectories $v^{\pm}(\theta, t)$, both decay to zero as $t \to -\infty$. v^+ is positive and corresponds to the expanding case; v^- is negative and corresponding to the collapsing case (for convex curves, since the stationary solution is uniformly convex). Angenent and Gurtin studied the expanding case and posed the asymptotic behavior of the collapsing case as an open problem in [4] (see also [19]). In this context the results of Gage[13] and Gage and Li [15] may be regarded as a partial answer to their question.

In recent joint work of Chou and the author [6] the evolution of (2) for convex interfaces was studied and the following rather complete results were obtained.

Theorem 1. *Consider (2) where β, g are positive, smooth 2π-periodic functions and F is a constant. For any given smooth and convex closed initial curve γ_0, there exists a negative F^* such that the maximal solution $\gamma(\cdot, t)$ of (2) starting at γ_0 is uniformly convex for $t \in (0, t_{max})$, and*
(a) when $F > F^*$, t_{max} is finite and $\gamma(\cdot, t)$ shrinks to a point as $t \to t_{max}$. Furthermore, if we normalized γ so that its enclosing area is constant, the normalized flows subconverges smoothly to a self-similar solution of the flow (3);
(b) when $F = F^*$, $t_{max} = \infty$ and the flow converges to a stationary solution of (4) smoothly; and
(c) when $F < F^*$, $t_{max} = \infty$ and the interface $\gamma(\cdot, t)$ expands to infinity as t approaches ∞. If, in addition, it is assumed that the polar diagram of β is uniformly convex, i.e. $\frac{d^2\beta^{-1}}{d\theta^2} + \beta^{-1} > 0$, for all θ, $\frac{\gamma(\cdot,t)}{t}$ converges smoothly to the boundary of the Wulff region of $\frac{-F}{\beta}$.

In the shrinking case this theorem completes the answer to the above mentioned question raised by Angenent and Gurtin in [4] (see also [19]). In fact, we proved that the curvature of the normalized flow has a uniform (positive) lower bound and an upper bound. By the construction of a Liapnov function it was shown that any limit point of $\gamma(\cdot, t)$ must be a self-similar solution of (3). In the case self-similar solution is unique (up to dilations and translations) as, for instance, when $\left(\frac{g}{\beta}\right)(\theta) = \left(\frac{g}{\beta}\right)(\theta + \pi)$, it implies that the normalized flow actually converges smoothly to a self-similar solution. Since the uniqueness of self-similar solution of (3) is in general not known, we could only make a weaker assertion: the normalized flow "subconverges" to a self-similar solution. It remains an open problem whether the normalized flow converges smoothly to a self-similar solution of (3) in the general case. In view of the results of Simon [23], we conjectured that this is true.

In the expanding case this theorem improves previous works by establishing smooth rather than uniform convergence of the corresponding support functions when the polar diagram of β is uniformly convex. Moreover it was shown in [6] that this polar diagram convex condition is also necessary for the smooth convergence.

3. Asymptotic Behavior of Evolving Embedded Curves

The curve shortening flow (1) for a curve embedded in the plane has been studied in a great detail by Gage, Hamilton and Grayson etc. It was well-known that the embedded planar curve becomes convex before its curvature can blow up, and then shrinks to a point with round limiting shape. Less was known for a curve embedded in a Riemannian surface. Grayson [17] proved that the evolving curve of (1) on a surface either shrinks to a point in finite time, or exists for infinite time.

In a sequence of two papers [1],[2], Angenent developed a theory of an arbitrary uniformly parabolic equation

$$V = V(T, k) \tag{5}$$

for a curve $\gamma(\cdot, t)$ on a surface, where T is the unit tangent vector to $\gamma(\cdot, t)$, and k is the geodesic curvature of $\gamma(\cdot, t)$ on the surface and $V(T, k)$ is some suitable function which is uniformly Lipschitz in k. Under an additional symmetry assumption, $V(-T, -k) = -V(T, k)$, he proved that whenever an immersed curve $\gamma(\cdot, t)$ evolving by (5) develops a singularity it either loses at least one self-intersection, or else its total curvature drops at least π. Oaks [22] improved the Angenent's argument and showed the latter case never occurs. As a consequence, if at the start the curve is embedded, the evolving curve either shrinks to a point in finite time or exists for infinite time. So they extended the Grayson's theorem from the curve shortening flow (1) to the general flow (5).

Naturally the further development leads to investigating the formation of the shrinking point. In fact, even in the simplest case of the curve shortening flow (1) for an embedded curve on a surface, it had already been raised as a conjecture by Grayson [16] that the limiting shape of the shrinking point would be a circle. In [13], Gage also gave a similar conjecture for anistropic flow (3). In [7], [25], we had studied the formation for the following cases,

$$V = \Phi(T)k + \Psi(T) \tag{6}$$

where Φ, Ψ are bounded smooth functions satisfies $\Phi \geq \lambda$ for some positive constant λ, and the symmetry condition $\Phi(-T) = \Phi(T)$, $\Psi(-T) = -\Psi(T)$. To analysis the formation of the shrinking point, without loss of generality, we may always assume that the whole family $\gamma(\cdot, t)$, $t \in (0, t_{max})$ is contained in a small neighborhood of the shrinking point on the surface. Then it is not hard to see that the problem (6) is equivalent to the following problem on the plane $\mathbf{R}^2$,

$$V = \Phi(x, y, \theta)k + \Psi(x, y, \theta), \tag{7}$$

where we denote by (x, y) the position vector of the curve $\gamma(\cdot, t)$, and θ the angle between the tangent vector and the x-axis.

For any fixed $(x_0, y_0) \in \mathbf{R}^2$, $\Phi(x_0, y_0, \theta)$ is a 2π-periodic function satisfying the symmetry condition: $\Phi(x_0, y_0, \theta + \pi) = \Phi(x_0, y_0, \theta)$, for all θ. From [13], one can write $\Phi(x_0, y_0, \theta)$ as

$$\Phi(x_0, y_0, \theta) = \frac{\overline{h}(\theta)}{\overline{k}(\theta)} \tag{8}$$

where $\overline{h}$, $\overline{k}$ are the support function and the curvature of some smooth, symmetry, strictly convex curve f and J is uniquely determined up to a dilation. We call the curve J a Minkowski isoperimetrix associated $\Phi(x_0, y_0, \theta)$. The definition is a generalization of the concept of circle. In fact, if $\Phi(x_0, y_0, \theta)$ is a positive constant, the corresponding Minkowski isoperimetrix is just a circle.

The following theorem was gotten by Chou and the author [7] in the case $M = \mathbf{R^2}$ and by the author [25] in the general surface cases.

Theorem 2. *Let $\gamma(\cdot,t) : [0,t_{max}) \times S^1 \to \mathbf{R^2}$ be an evolving embedded curve of (7) with $t_{max} < +\infty$. Then as $t \to t_{max}$, the evolving curve $\gamma(\cdot,t)$ shrinks to a point $(\overline{x},\overline{y}) \in \mathbf{R^2}$ and the rescaled evolving curve $(t_{max}-t)^{-\frac{1}{2}}(\gamma(\cdot,t)-(\overline{x},\overline{y}))$ converges to a Minkowski isoperimetrix associated $\Phi(\overline{x},\overline{y},\theta)$ in C^∞ topology.*

As consequences, this confirms the above mentioned both Grayson's conjecture and Gage's conjecture and the C^∞ convergence shows that the evolving embedded curve becomes convex before it shrinks to a point.

4. Shortening Complete Plane Curves

In this section we discuss the curve shortening flow (1) for complete, non-compact embedded curves. Very few results in this direction are known. In [9] and [10] Ecker and Huisken studied the mean curvature flow for complete non-compact hypersurfaces. They proved that the mean curvature flow for entire graphs exists for all time. Problem (1) is the mean curvature flow for curves. Their result asserts that for a complete initial curve which is a graph over a straight line, (1) has a solution for all $t \geq 0$. In general, hypersurfaces in $\mathbf{R}^{n+1}$ $(n \geq 2)$ which are not graphs could develop singularities under the mean curvature flow. The notion of a viscosity solution has been used successfully to study the problem, see, Chen-Giga-Goto[5] and Evans-Spruck[11]. However, classical solutions are possible when $n = 1$. In recent joint work of Chou and the author [8] we established a long time existence result which embraces a large class of complete curves. To state it let us recall that according to a theorem of Jordan a complete properly embedded curve γ_0 divides the plane into two regions Ω_1 and Ω_2.

Theorem 3. *Let $\gamma_0 : \mathbf{R} \to \mathbf{R^2}$ be a locally Lipschitz continuous, complete, properly embedded curve dividing the plane into two regions Ω_1 and Ω_2. Suppose that the areas of Ω_1 and Ω_2 are infinite. Then there exists a solution $\gamma(\cdot,t)$ of (1) for all $t > 0$ such that*
(a) for each t, $\gamma(\cdot,t)$ is a smooth, complete, properly embedded curve which divides the plane into two regions of infinite area; and
(b) $\gamma(\cdot,t)$ converges to γ_0 uniformly on every compact subset of $\mathbf{R}$ as $t \searrow 0$.

One does not expect (1) admits a solution for all time when one of the regions divided by the initial curves has finite area. In fact, it was also shown in [8] that for complete curves with finite total absolute curvature the life span of the solution is precisely equal to the minimum of the areas of Ω_1 and Ω_2 divided by π.

We briefly describe the proof of Theorem 3, which was separated into two parts. First we established the long time existence of a unique solution of (1) when γ_0 has finite total absolute curvature. To prove this one needs to derive a priori estimates of the flow. We observed that in this case the total absolute curvature is non-increasing in time. Outside a bound region, each of the two ends of the curve can be represented as a graph over some semi-infinite axis. As a consequence one may use the localized estimates of Ecker and Huisken [10] to bound the curvature. To control the curvature in the bounded region we employed a method of Hamilton [20], which relies on an isoperimetric quantity and a blow up argument. We pointed out the isoperimetric quantity is not defined for non-compact curves. We introduced a geometric quantity that is no longer isoperimetric. However, just like the isoperimetric quantity it is bounded positively from below, and this is sufficient for our purpose. After proving the long time existence for curves with finite total absolute curvature, we turned to the second half of the proof of the theorem, namely, we used the result in the first half to construct approximating solutions and foliations transversal to a given γ_0 satisfying the hypothesis of the theorem. The construction yielded "gradient estimates" of the flow from which curvature estimates followed. The idea of using foliations to study (1) can be found in Grayson [17]. It was further developed in [2] and [22] to obtain gradient estimates that are local in time. We adapted the idea and applied it to a global setting. After obtaining the curvature estimates, long time existence was deduced by an approximation argument. The paper [8] also established a strong uniqueness property of (1) for a class of complete curves.

Proposition 4. *Let γ_0 be a locally Lipschitz continuous, properly embedded curve whose two ends are representable as graphs over some semi-infinite lines. Then (1) has at most one solution taking γ_0 as its initial value.*

More precisely, we mean that if γ_1 and γ_2 are solutions of (1) which tend to γ_0 uniformly on compact subsets of $\mathbf{R}$ as t goes to 0, then γ_1 and γ_2 must coincide. At first glance this uniqueness property is rather surprising. In fact, when restricted to graphs, (1) is equivalent to

$$u_t = \frac{u_{xx}}{1+u_x^2}. \tag{9}$$

This proposition implies that the Cauchy problem to the quasilinear parabolic equation (9) has a unique solution without any assumption concerning the growth of the initial data. The proof involved a variant of energy estimate.

References

[1] S. Angenent, *Parabolic equations for curves on surfaces (I). Curve with p-intergrable curvature*, Ann. of Math. 132 (1990) 451-483.

[2] S. Angenent, *Parabolic equations for curves on surfaces (II). Intersections, blow up and generalized solutions*, Ann. of Math. 133 (1991) 171-215.

[3] S. Angenent & M. Gurtin, *Multiphase thermomechanics with interfacial structure 2. Evolution of an isothermal interface*, Arch. Rational Mech. Anal. 108 (1989) 323-319.

[4] S. Angenent & M. Gurtin, *Anisotropic motion of a phase interface*, J. Reine Angew. Math. 446 (1994) 1-47.

[5] Y. G. Chen, Y. Giga & S. Goto, *Uniqueness and existence of viscosity solutions of generalized mean curvature flow equations*, J. Differential Geom. 33 (1991), 749-786.

[6] K. S. Chou & X. P. Zhu, *Anisotropic flows for convex plane curves*, Duke Math. J. 97 (1999), 579-619.

[7] K. S. Chou & X. P. Zhu, *A convexity theorem for a class of anisotropic flows for plane curves*, to appear in Indiana Univ. Math. J.

[8] K. S. Chou & X. P. Zhu, *Shortening complete plane curves*, J. Differential Geom. 50 (1998), 471-504.

[9] K. Ecker & G. Huisken, *Mean curvature evolution of entire graphs*, Ann. of Math. 130 (1989) 453-471.

[10] K. Ecker & G. Huisken, *Interior estimates for hypersurfaces moving by mean curvature*, Invent. Math. 105 (1991) 547-569.

[11] L. C. Evans & J. Spruck, *Motion of level sets by mean curvature I*, J. Differential Geom. 33 (1991) 635-681.
[12] M. Gage, *An isoperimetric inequality with applications to curve shortening*, Duke Math. J. 50 (1983) 1225-1229.
[13] M. Gage, *Evolving plane curves by curvature in relative geometries*, Duke Math. J. 72 (1993) 441-466.
[14] M. Gage & R. Hamilton, *The shrinking of convex plane curves by the heat equation*, J. Differential Geom. 23 (1986) 69-96.
[15] M. Gage & Y. Li, *Evolving plane curves by curvature in relative geometries II*. Duke Math. J. 75 (1994) 79-98.
[16] M. Grayson, *The heat equation shrinks embedded plane curves to points*, J. Differential Geom. 26 (1987) 285-314.
[17] M. Grayson, *Shortening embedded curves*, Ann. of Math. 129 (1989) 71-111.
[18] M. E. Gurtin, *Multiphase thermomechanics with interfacial structure. Toward a nonequilibrium thermomechanics of two phase materials*, Arch. Rational Mech. Anal. 100 (1988) 275-312.
[19] M. E. Gurtin, *"Thermomechanics of Evolving Phase Boundaries in the Plane"*, Clarendon Press, Oxford 1993.
[20] R. Hamilton, *Isoperimetric estimates for the curve shrinking flow in the plane, Modern methods in complex analysis, The Princeton conference in honor of Gunning and Kohn, (ed. T. Bloom, et al.)*, Ann. of Math. Stud. 137, Princeton Univ. Press, 1996.
[21] G. Huisken, *A distance comparison principle for evolving curves*, Asian J. Math. 2 (1998) 127-134.
[22] J. Oaks, *Singularities and self intersections of curves evolving on surfaces*, Indiana Univ. Math. J. 43 (1994) 959-981.
[23] L. Simon, *Asymptotics for a class of nonlinear evolution equations, with applications to geometric problems*, Ann of Math. 118 (1983) 525-572.
[24] H. M. Soner, *Motion of a set by the curvature of its boundary*, J. Differential Equations, 101 (1993) 313-372.
[25] X. P. Zhu, *Asymptotic behavior of anisotropic curve flows*, J. Differential Geom. 48 (1998) 225-274.

DEPARTMENT OF MATHEMATICS, ZHONGSHAN UNIVERSITY, GUANGZHOU,P.R.CHINA

AMS/IP Studies in Advanced Mathematics
Volume 20, 2001

Discontinuous Solutions to Nonlinear Evolutionary Partial Differential Equations

Gui-Qiang Chen

ABSTRACT. We analyze some recent developments in studying discontinuous solutions to nonlinear evolutionary partial differential equations. The central problems include the existence, compactness, and large-time behavior of discontinuous solutions. The nonlinear equations we discuss include nonlinear hyperbolic systems of conservation laws (especially the compressible Euler equations) and the compressible Navier-Stokes equations. Some of recent ideas, approaches, and methods are also discussed.

1. Introduction

We are concerned with global solutions of nonlinear evolutionary partial differential equations:

$$\partial_t U + AU = 0, \qquad U \in \mathbb{R}^m, \tag{1.1}$$

with *discontinuous* initial data:

$$U|_{t=0} = U_0(x), \tag{1.2}$$

and certain appropriate boundary conditions provided the domain under consideration is a subset of $\mathbb{R}^n$, where A is a nonlinear differential operator in $x \in \mathbb{R}^n$ and $U_0(x)$ is a discontinuous function of large oscillation.

The discontinuous problems arise from many physical situations involving sharp fronts and singularities as observed in nature, such as shock waves and vortices. Mathematically, such problems are generic in the sense that solutions (even starting from smooth initial data) generally develop singularities in a finite time because of strong nonlinearity of the equations without enough dissipation, and hence only discontinuous solutions may exist in the large. One of the main challenges in handling discontinuous solutions is that many powerful analytical techniques, which predominate in the theory of dissipative partial differential equations, may not be directly applied. Novel mathematical ideas, approaches, and methods need to be developed efficiently to handle discontinuous solutions.

The mathematical study of discontinuous solutions started with the pioneering work of Riemann [**58**], who first solved a special initial value problem, called the Riemann problem, for the one-dimensional isentropic Euler equations. The

Riemann problem plays an essential role in the theory of hyperbolic conservation laws and related numerical analysis and computation of discontinuous solutions (see Glimm [**34**]; also see [**22, 35, 36, 45, 51, 59, 60**] and the references cited therein).

Many Chinese mathematicians have made important contributions to the Riemann problem and related problems. See Gu-Li-Yu-Hou [**38**] and Ding-Chang-Wang-Hsiao-Li [**25**]. Also see recent contributions in Chang-Hsiao [**4**], Ding-Liu [**27**], Li-Zhang-Yang [**49**], Li-Yu [**50**], and the references cited therein.

The problems we are interested for discontinuous initial data include:

(1). *Existence and solution spaces*: For hyperbolic systems of conservation laws with small initial data, the BV space is well-posed for Glimm solutions [**34, 54**] (also see [**2, 3**]). Recent examples show that, for large initial data, the total variation of solutions may blow up in a finite time, even for strictly hyperbolic systems [**41**]. Therefore, one expects at most that the solutions are in L^∞ or weighted L^p in general for large initial data.

(2). *Compactness and regularity as $t > 0$*: If the initial data is a large oscillatory sequence, then the question is whether the corresponding solution sequence is compact as long as $t > 0$. That is, the solution operator is compact.

(3). *Uniqueness and stability*: Given a class of solutions containing an entropy solution, we examine whether the solution is unique and stable in this class.

(4). *Asymptotic behavior*: Given an entropy solution without apriori reference to the method for construction of the solution, the question is how the solution behaves asymptotically as $t \to \infty$.

In general, evolutionary nonlinear partial differential equations can be classified into three types:

(1). *Nondissipative systems*: The prototype of such systems is nonlinear hyperbolic systems of conservation laws including the compressible Euler equations and Hamilton-Jacobi equations.

(2). *Weakly dissipative or nonuniform dissipative systems* such as hyperbolic-parabolic systems: The prototype of such systems is the compressible Navier-Stokes equations.

(3). *Strongly dissipative systems*: Such systems include uniform parabolic equations or systems. In general, solutions of such systems with discontinuous initial data become instantaneously regular as long as $t > 0$.

In this paper, we focus on types (1)-(2), mainly on hyperbolic systems of conservation laws and the compressible Navier-Stokes equations, and discuss some recent developments and advances through these prototypical systems.

2. Hyperbolic Conservation Laws

We are concerned with a hyperbolic system of conservation laws:

$$\partial_t U + \nabla_x \cdot F(U) = 0, \quad U \in \mathbb{R}^m, x \in \mathbb{R}^n, \tag{2.1}$$

where $F : \mathbb{R}^m \to (\mathbb{R}^m)^n$ is a nonlinear mapping. Since the conservation law is a fundamental law in nature, most of partial differential equations arising from physical and engineering sciences can be formulated into form (2.1) or its variants. The hyperbolicity of system (2.1) requires that, for all $\xi \in S^{n-1}$, the matrix $(\sum_{k=1}^n \xi_k \nabla F_k(U))_{m\times m}$ have m real eigenvalues $\lambda_j(U,\xi), j = 1, 2, \cdots, m$, and be diagonalizable.

2.1. Entropy Solutions. Let $(\eta, q) : \mathbb{R}^m \to \mathbb{R} \times \mathbb{R}^n$ be an entropy-entropy flux pair (an entropy pair, for short), determined by the linear hyperbolic systems:

$$\nabla q_k(U) = \nabla\eta(U)\nabla F_k(U), \quad k = 1, 2, \cdots, n. \tag{2.2}$$

An entropy pair is called convex if $\nabla^2\eta(U) \geq 0$. If the inequality is strict, then (η, q) is called a strictly convex entropy pair. Friedrichs-Lax [**32**] observed that most of the conservative systems that result from continuum mechanics are endowed with a globally defined, strictly convex entropy. An L^∞ function $U(x,t)$ is called an entropy solution if, for any convex entropy pair (η, q),

$$\int_0^\infty \int_{\mathbb{R}^n} (\eta(U)\partial_t\varphi + q(U)\cdot\nabla_x\varphi)dxdt + \int_{\mathbb{R}^n} \eta(U_0(x))\varphi(x,0)dx \geq 0, \tag{2.3}$$

for any nonnegative test function $\varphi(x,t) \in C_0^\infty(\mathbb{R}^{n+1}_+)$.

2.2. Difficulties. Two of the main difficulties to solve (2.1) are singularity of the solutions and resonance of the systems.

Singularity: Since $F(U)$ is a nonlinear mapping, solutions of the Cauchy problem (even starting from smooth initial data) generally develop singularities in a finite time, and then the solutions become discontinuous functions. This feature reflects the physical phenomenon of breaking of waves and development of shock waves. For this reason, attention must focus on solutions in the space of discontinuous functions satisfying (2.3).

Resonance: A system is called nonstrictly hyperbolic or resonant if there exist some $\xi_0 \in S^{n-1}$ and $U_0 \in \mathbb{R}^m$ such that $\lambda_i(U_0,\xi_0) = \lambda_j(U_0,\xi_0)$ for some $i \neq j$. Such a degeneracy is quite generic. For example, when $n = 3, m = 2 \,(\mathrm{mod}\, 4)$, Lax [**44**] showed that any system must be nonstrictly hyperbolic. The same situation occurs for $m = \pm 2, \pm 3, \pm 4 \,(mod\, 8)$ (see [**33**]).

The problems we consider include the existence, compactness, and large-time behavior of entropy solutions. The compactness and large-time behavior are more challenging, especially for nonstrictly hyperbolic systems, since the solutions are discontinuous and all the information about the solutions is the entropy inequality (2.3), only in the sense of distributions.

2.3. Decay via Compactness and Entropy. It is well known that, for the linear case with a nonzero propagation speed and periodic initial data, the solution is periodic in time t. However, this is not true for the genuinely nonlinear case. The observation is that the genuine nonlinearity of equations forces the nonlinear waves of each characteristic family to interact vigorously and to cancel each other. The theory of Glimm-Lax [**35**], for 2×2 strictly hyperbolic and genuinely nonlinear systems, has indicated that the resulting mutual cancellation of interacting shock and rarefaction waves of the same family induces the decay of periodic Glimm solutions, provided the initial oscillation is small. Recently, Dafermos [**24**] showed that any periodic solution with small locally bounded variation and certain regularity for the 2×2 systems decays asymptotically. A further problem is whether the decay phenomenon holds for more general cases: (a) any L^∞ large periodic solutions without restrictions of either small oscillation or local bounded variation, and (b) more general nonlinear hyperbolic systems, especially nonstrictly hyperbolic systems and multidimensional scalar equations. The counterexample of Greenberg-Rascle [**37**] indicates that the asymptotic behavior is very sensitive with respect to the smoothness of the flux functions.

2.3.1. Framework. One of our main observations in Chen-Frid [**9**] is that the compactness of L^∞ solution operator in L^1_{loc}, coupling with the weak convergence of periodic initial data to the mean, yields the decay of the L^∞ periodic solution $U(x,t)$ in L^1, provided that (2.1) is endowed with a strictly convex entropy. This is achieved via the compactness of the self-similar scaling sequence $U^T(x,t) = U(Tx, Tt)$.

THEOREM 2.1. [**9**]. *Suppose that system* (2.1) *is endowed with a strictly convex entropy. Assume that* $U(x,t) \in L^\infty(\mathbb{R}^{n+1}_+)$ *is a periodic solution of* (2.1) *and* (1.2) *with period* $P \subset \mathbb{R}^n$, *and its scaling sequence* $\{U^T\}$ *is compact in* $L^1_{loc}(\mathbb{R}^{n+1}_+)$. *Then* $U(x,t)$ *asymptotically decays to* $\bar{U} \equiv \frac{1}{|P|}\int_P U_0(x)dx$ *in* L^1:

$$\text{ess} \lim_{t\to\infty} \int_P |U(x,t) - \bar{U}|dx = 0. \tag{2.4}$$

In Theorem 2.1, we assume that $\{U^T\}$ is compact in $L^1_{loc}(\mathbb{R}^{n+1}_+)$, which is a corollary of the compactness of solution operator. Such a compactness can be achieved by the method of compensated compactness, the averaging method, and other analytical techniques. For example, we have

THEOREM 2.2. [**9, 12**]. *Consider system* (2.1) *with a strictly convex entropy pair* (η_*, q_*). *Assume that the uniformly bounded sequence* $U^T(x,t) \in L^\infty(\mathbb{R}^{n+1}_+)$ *satisfies* (2.3) *for any* convex *entropy pair* $(\eta, q) \in \Lambda$, *where* Λ *is a linear space of entropy pairs of* (2.1) *including* (η_*, q_*). *Then the measure sequence*

$$\partial_t\eta(U^T) + \nabla_x \cdot q(U^T) \qquad \textit{is compact in } H^{-1}_{loc}(\mathbb{R}^{n+1}_+), \tag{2.5}$$

for any *entropy pair* $(\eta, q) \in \Lambda \cap C^2(\mathbb{R}^m; \mathbb{R}^{n+1})$.

Therefore, the method of compensated compactness (e.g. [**61, 57, 7**]) is one of the efficient methods to achieve the compactness with the aid of Theorem 2.2.

2.3.2. Multidimensional Scalar Conservation Laws. For multidimensional scalar conservation laws with C^2 flux function $f : \mathbb{R} \to \mathbb{R}^n$:

$$\partial_t u + \nabla_x \cdot f(u) = 0, \tag{2.6}$$

and initial data $u_0(x) \in L^\infty(\mathbb{R}^n)$ with period P, the existence of global entropy solutions of the Cauchy problem is due to Kruzkov [**43**]. Combining Theorem 2.1 with a theorem by Lions-Perthame-Tadmor [**46**], we conclude

THEOREM 2.3. [**9, 12**]. *Assume that the flux function* $f(u) \in C^2$ *satisfies*

$$\text{meas}\left\{ u \in \mathbb{R} \mid \tau + f'(u)\cdot\vec{k} = 0 \right\} = 0, \quad \textit{for any } (\tau, \vec{k}) \in S^n. \tag{2.7}$$

Let $u(x,t) \in L^\infty(\mathbb{R}^{n+1}_+)$ *be an entropy solution with periodic initial data* $u_0(x)$. *Then* $u(x,t)$ *asymptotically decays to* $\bar{u} = \frac{1}{|P|}\int_P u_0(x)dx$ *in* L^1.

2.3.3. The Isentropic Euler Equations. Consider the isentropic Euler equations for compressible fluids:

$$\partial_t\rho + \partial_x m = 0, \qquad \partial_t m + \partial_x(m^2/\rho + p(\rho)) = 0, \tag{2.8}$$

where ρ, m, and p are the density, the momentum, and the pressure, respectively. In the non-vacuum state ($\rho > 0$), $v = m/\rho$ is the velocity, $p = p(\rho) \in C^4(0,\infty)$ is a given function of ρ depending on compressible fluids under consideration. Strict hyperbolicity and genuine nonlinearity away from the vacuum require that

$$p'(\rho) > 0, \quad \rho p''(\rho) + 2p'(\rho) > 0, \qquad \rho > 0. \tag{2.9}$$

Near the vacuum, $p(\rho)$ is allowed very singular: The principal singular part of $p(\rho)$ coincide with the γ-law pressure for some $\gamma > 1$, but additional singularities are allowed (as real gases). That is, for sufficiently small $\rho \ll 1$,

$$p(\rho) = \kappa\rho^\gamma(1 + P(\rho)), \quad |P^{(k)}(\rho)| \le C\rho^{1-k}, \ 0 \le k \le 4. \tag{2.10}$$

Consider the Cauchy problem for (2.8) with initial data

$$(\rho, m)|_{t=0} = (\rho_0, m_0)(x), \quad 0 \le \rho_0(x) \le C_0, \ |m_0(x)/\rho_0(x)| \le C_0 < \infty. \tag{2.11}$$

The main difficulty of this system is that strict hyperbolicity fails, and the flux function is only Lipschitz continuous at the vacuum state $\rho = 0$, which occurs in fluid mechanics.

THEOREM 2.4. [**18**]. *Consider the isentropic Euler equations* (2.8)-(2.10). *Then*

(i). There exists a global solution $(\rho, m)(x,t)$ *of* (2.8)-(2.11), *satisfying* $0 \le \rho(x,t) \le C, |m(x,t)/\rho(x,t)| \le C$, *for some* C *depending only on* C_0 *and* γ, *and* $\partial_t\eta(\rho, m) + \partial_x q(\rho, m) \le 0$ *in the sense of distributions for any convex weak entropy pairs* (η, q). *If* $(\rho_0, m_0)(x)$ *is periodic with period* P, *then* $(\rho, m)(x,t)$ *is also periodic in* x *with same period* P.

(ii). The solution operator $(\rho, m)(\cdot, t) = S_t(\rho_0, m_0)(\cdot)$, *determined by (a), is compact in* $L^1_{loc}(\mathbb{R}^2_+)$ *for* $t > 0$.

(iii). Let $(\rho, m)(x,t), 0 \le \rho(x,t) \le C, |m(x,t)/\rho(x,t)| \le C$, *be a periodic entropy solution of* (2.8)-(2.11) *with period* P. *Then* $(\rho, m)(x,t)$ *asymptotically decays to* $\frac{1}{|P|}\int_P(\rho_0, m_0)(x)dx$ *in* L^1.

For polytropic gases, the existence result was proved by DiPerna [**29**] for the case $\gamma = 1 + 1/N, N \ge 5$ odd, for $L^2 \cap L^\infty(\mathbb{R})$ initial data, by Ding-Chen-Luo [**26**] and Chen [**6**] for $1 < \gamma \le 5/3$ for usual gases with general L^∞ initial data. The existence result was also obtained by Lions-Perthame-Tadmor [**47**] for $\gamma \ge 3$ and by Lions-Perthame-Souganidis [**48**] for $5/3 < \gamma < 3$. More general pressure law will be handled in [**19**].

REMARK 2.1. The same decay results hold for periodic entropy solutions in L^∞ with large oscillation for some important systems (see [**9**]). The approach described here can be easily generalized to the decay problem for L^p entropy solutions and to the asymptotic problems for entropy solutions of hyperbolic conservation laws with relaxation (see [**9, 20**]).

2.4. Asymptotic Stability via Compactness and Entropy. The ideas and observations in §2.3 can be refined into new approaches for solving the asymptotic stability of entropy solutions $U(x,t) \in L^\infty(\mathbb{R}^{n+1}_+)$ or $L^\infty \cap BV_{loc}(\mathbb{R}^{n+1}_+)$ of (2.1) and (1.2) with

$$U_0(x) = R_0(x/|x|) + P_0(x), \quad P_0(x) \in L^1(\mathbb{R}^n). \tag{2.12}$$

Let $R(x/t)$ be the classical Riemann solution of (2.1) with Riemann data

$$R|_{t=0} = R_0(x/|x|). \tag{2.13}$$

2.4.1. Framework. The first approach is to invoke the compactness of the scaling sequences of the perturbation solutions and the uniqueness of the Riemann solution in a class of solutions which includes all possible limits of the scaling sequences.

THEOREM 2.5. [**10, 12**]. *Let $\mathcal{S}(\mathbb{R}^{n+1}_+)$ denote a class of functions in L^∞ defined on $\mathbb{R}^{n+1}_+$. Assume that the Cauchy problems* (2.1) *and* (2.12)-(2.13) *satisfy the following.*

(i). The Riemann solution is unique in the class $\mathcal{S}(\mathbb{R}^{n+1}_+)$;

(ii). Given an entropy solution of (2.1) *and* (2.12), *$U \in \mathcal{S}(\mathbb{R}^{n+1}_+)$, the self-similar scaling sequence $U^T(x,t)$ is compact in $L^1_{loc}(\mathbb{R}^{n+1}_+)$, and any limit function of its subsequences is still in $L^1_{loc} \cap \mathcal{S}(\mathbb{R}^{n+1}_+)$.*

(iii). The Riemann solution $R(x/t) = R(\xi)$ is piecewise Lipschitz in $\xi \in \mathbb{R}^n$. Then the Riemann solution $R(x/t)$ is asymptotically stable in $\mathcal{S}(\mathbb{R}^{n+1}_+)$ with respect to the corresponding initial perturbation $P_0(x) \in L^1(\mathbb{R}^n)$:

$$\text{ess}\lim_{t\to\infty} \int_{|\xi|\le K} |U(\xi t, t) - R(\xi)|d\xi = 0, \quad \textit{for any} \quad K \in (0,\infty). \tag{2.14}$$

For BV solutions, the compactness of the scaling sequence is obtained through the following observation.

THEOREM 2.6. [**10**]. *Assume that $U(x,t) \in BV_{loc}(\mathbb{R}^{n+1}_+)$ satisfies*

$$TV_{\{|x|\le cT_0\}\times(0,T_0)}(U) \le CT_0^n, \quad \textit{for } c > 0, T_0 > 0, \textit{ and } C > 0 \textit{ indept. of } T_0. \tag{2.15}$$

Then $U^T(x,t)$ also satisfies (2.15) *with the same constant C.*

This condition is satisfied by the entropy solutions possessing total variation in x uniformly bounded for all $t > 0$, which is the case for the solutions constructed by Glimm's method (see [**34, 35**]). Hence, the compactness follows from the uniform boundedness of the sequence in BV. For L^∞ solutions, the method of compensated compactness and the Young measures can be applied to yield the compactness of uniformly bounded sequences of entropy solutions (see [**7, 8, 28, 31, 59**]).

The uniqueness of Riemann solutions in the class of BV solutions for 2×2 systems is due to DiPerna [**30**] and for the 3×3 system of compressible Euler equations with a general pressure law to Chen-Frid [**11**] (also see [**10**]).

The second approach is to estimate the difference between the perturbation solution U and the Riemann solution R in L^p: Assume that $U(x,t), R(x/t) \in L^\infty(\mathbb{R}^{n+1}_+)$ satisfy that there exist $C > 0$, independent of t, and $\gamma \in [0,1)$ such that

$$\|U - R\|^p_{L^p(\mathbb{R}^n)}(t) \le Ct^{n-1+\gamma}, \quad \text{for some } 1 \le p < \infty. \tag{2.16}$$

Then the Riemann solution $R(x/t)$ is asymptotically stable in L^1 with respect to the corresponding initial perturbation $P_0(x)$ in the sense of (2.14).

Another approach is to understand directly the asymptotic behavior of the solution along the rays $\xi = x/t, \xi \in \mathbb{R}$, with the aid of the theory of divergence-measure fields (see [**10, 13**]).

2.4.2. Multidimensional Scalar Conservation Laws. A direct application is the asymptotic stability of Riemann solutions $R(x/t)$ of the multidimensional scalar conservation law (2.6) under initial perturbation (2.12), with the aid of Kruzkov's Theorem [**43**].

THEOREM 2.7. *Any Riemann solution $R(x/t)$ of* (2.6) *with data $R_0(x/|x|)$ is asymptotically stable with respect to large $L^\infty \cap L^1$ perturbation $P_0(x)$ in the sense of* (2.14).

2.4.3. The p-System. Consider the p-system

$$\partial_t v - \partial_x u = 0, \qquad \partial_t u + \partial_x p(v) = 0, \tag{2.17}$$

where $p \in C^2(a,b)$ satisfies $p'(v) < 0$ and $p''(v) > 0$. Combining Theorems 2.5-2.6 with the uniqueness of Riemann solutions by DiPerna [**30**], we have

THEOREM 2.8. [**10**]. *Any Riemann solution $R(x/t)$ of (2.17) with Riemann data $R_0(x/|x|)$ is asymptotically stable with respect to large L^1 perturbation $P_0(x) \in L^\infty \cap BV_{loc}(\mathbb{R})$ in the sense of (2.14). Any Riemann solution consisting of only rarefaction waves is asymptotically stable even with respect to large $L^\infty \cap L^1$ perturbation $P_0(x)$ in the sense of (2.14).*

2.4.4. The Compressible Euler Equations. The Euler system for compressible fluids in Lagrangian coordinates reads

$$\partial_t v - \partial_x u = 0, \quad \partial_t u + \partial_x p = 0, \quad \partial_t(e + \frac{u^2}{2}) + \partial_x(pu) = 0, \tag{2.18}$$

where u, p, v, and e represent the velocity, the pressure, the specific volume ($v = 1/\rho$, ρ the density), and the internal energy of the fluids, respectively. Other important physical variables are the temperature θ and the entropy S. To close system (2.18), one needs the basic law of thermodynamics which translates into the differential equation: $de = \theta\, dS - p\, dv$. We choose (v, S) as the independent variables to get the constitutive equations of state $p = p(v,S)$, $\theta = \theta(v,S)$, and $e = e(v,S)$ satisfying

$$e_v(v,S) = -p(v,S), \quad e_S(v,S) = \theta(v,S), \quad \theta_v(v,S) = -p_S(v,S). \tag{2.19}$$

As usual, we assume that the function $p = p(v,S)$ satisfies

$$p_v(v,S) < 0, \quad p_{vv}(v,S) > 0, \quad \nabla^2_{v,S} p(v,S) \geq 0, \quad \text{in the domain } v > 0. \tag{2.20}$$

For ideal polytropic gases, $pv = R\theta$, $e = c_v\theta$, $p(v,S) = \kappa e^{S/c_v} v^{-\gamma}, \gamma = 1 + R/c_v > 1$, where R and c_v are positive constants.

System (2.18) can be written in the conservation form (2.1) with $U = (v, u, E)$ and $F(U) = (-u, p, pu)$, where $E = \frac{1}{2}u^2 + e$, and p is a function of U. We are interested in the large-time behavior of solutions in $L^\infty \cap BV_{loc}$ of the Cauchy problem for (2.18) with initial data:

$$U|_{t=0} = (v_0, u_0, E_0)(x) \equiv R_0(x) + P_0(x), \quad \text{with } P_0(x) \in L^1(\mathbb{R}). \tag{2.21}$$

The entropy-entropy flux pair $\eta_*(U) = \frac{u^2}{2} - \int_{v_0}^{v} p(\sigma, S)\, d\sigma + K(S), q_*(U) = up(v,S)$ is a strictly convex pair in (v, u, S) in the bounded domain $\mathcal{V} \subset \{v > 0\}$ where $(v, u, S)(x,t)$ in the solutions assume their values, provided that

$$K''(S) > \int_{v_0}^{v} p_{SS}(\sigma, S)\, d\sigma - \frac{p_S^2(v,S)}{p_v(v,S)}, \qquad (v,u,S) \in \mathcal{V}. \tag{2.22}$$

Observe that any entropy solution $U(x,t) \in L^\infty(\mathbb{R}^2_+), v(x,t) > 0$, satisfies the entropy inequality

$$\int_0^\infty \int_{-\infty}^\infty \{\eta_*(U)\partial_t\varphi + q_*(U)\partial_x\varphi\}\, dx\, dt + \int_{-\infty}^\infty \eta_*(U_0(x))\varphi(x,0)\, dx \geq 0, \tag{2.23}$$

for any nonnegative $\varphi \in C_0^1(\mathbb{R}^2)$, provided $K(S), v_0 > 0$ are chosen such that

$$K'(S) \leq \theta(v_0, S). \tag{2.24}$$

We choose $K(S), v_0 > 0$ such that, besides (2.22), both (2.24) and

$$K'(S) \geq \theta(v_0, S) - \theta(v, S) \tag{2.25}$$

hold in $\mathcal{V} \subset \{(v,u,S)\,|\,v>0\} \subset \mathbb{R}^3$. It is easy to check for polytropic gases that there exist $K(S), v_0 > 0$, depending on $\mathcal{V}$, such that (2.22) and (2.24)-(2.25) hold.

By the approach developed in Chen-Frid [**10**], we have

THEOREM 2.9. [**11**]. *Let $U(x,t) = (v,u,E)(x,t) \in BV(\mathbb{R}^2_+;\mathcal{V})$ be an entropy solution of* (2.18) *and* (2.21) *in $\mathbb{R}^2_+$ satisfying $TV_{(-cT,cT)\times(0,T)}(U) \leq CT$, for some $c, C > 0$ and any $T > 0$. Assume* (2.20), (2.22), *and* (2.24)-(2.25) *are satisfied. Then $U(x,t)$ asymptotically tends to the corresponding Riemann solution $R(x/t)$ in the sense of* (2.14), *provided that $\mathcal{V}$ is small. Moreover, if the Riemann solution $R(x/t)$ consists of at most rarefaction waves and contact discontinuity, then any entropy solution $U(x,t) \in L^\infty(\mathbb{R}^2_+;\mathcal{V})$ asymptotically tends to the corresponding Riemann solution $R(x/t)$ in the sense of* (2.14), *for arbitrarily large $\mathcal{V} \subset \{v > 0\}$.*

The uniqueness and asymptotic stability of Riemann solutions in the class of entropy solutions for the Cauchy problem (2.18) and (2.21) can be extended to the Euler equations in Eulerian coordinates (see [**11**]).

We recall recent fundamental results for $m \times m$ systems in Bressan-Crasta-Piccoli [**2**], Bressan-Liu-Yang [**3**], Liu-Yang [**54**], and the references cited therein on the L^1-stability of entropy solutions in $L^\infty \cap BV$ obtained by the Glimm scheme and the wave front-tracking method, or satisfying an additional regularity, with small total variation. We remark that the uniqueness results described here cannot be directly obtained from these stability results, since the results presented here neither impose smallness restrictions on the total variation and additional regularity of the solutions, nor need specific reference to any particular method for constructing the entropy solutions.

Another important example is Hamilton-Jacobi equations, which can be transformed into a system of conservation laws. A general framework to establish global L^∞ solutions of Hamilton-Jacobi equations has been established in Chen-Su [**21**].

3. The Compressible Navier-Stokes Equations

In this section we consider the global existence and large-time behavior of discontinuous solutions of the compressible Navier-Stokes equations in Lagrangian coordinates:

(3.1)

$$\partial_t v - \partial_x u = 0,\ \partial_t u + \partial_x p = \partial_x\left(\frac{\epsilon\partial_x u}{v}\right),\ \partial_t\left(e + \frac{u^2}{2}\right) + \partial_x(up) = \partial_x\left(\frac{\epsilon u\partial_x u + \lambda\partial_x e}{v}\right),$$

with *large* discontinuous initial data, where $p = p(v,e)$ is the pressure as in §2.4.4, ϵ and λ are fixed positive viscosity parameters, and x is the Lagrangian coordinate so that $x =$ constant corresponds to a particle path. For polytropic gases, the pressure p is given via the equation of state: $p = (\gamma-1)e/v = (\gamma-1)c_v\theta/v$, where θ is the temperature, e is the internal energy, and $\gamma > 1, c_v > 0$ are constants. For concreteness, we focus on the following initial-boundary value problem for (3.1):

$$(v,u,e)|_{t=0} = (v_0,u_0,e_0)(x), \qquad 0 \leq x \leq 1, \tag{3.2}$$

$$u(i,t) = \partial_x e(i,t) = 0, \qquad i = 0,1, \quad t \geq 0, \tag{3.3}$$

where the initial data satisfy that, for some constant $C_0 > 0$,

$$v_0(x), e_0(x) \geq C_0^{-1}, \qquad \mathrm{TV}(v_0, u_0, e_0) \leq C_0. \tag{3.4}$$

Such a problem is motivated by the fact that large discontinuous solutions are fundamental in both the physical theory of nonequilibrium thermodynamics and the mathematical theory for compressible fluid flow (also see [**39**]). These considerations require to seek a rigorous theory to accommodate the large discontinuities. It is worthwhile, at this point, to give a heuristic argument of the convection of discontinuities in solutions of (3.1). If v_0 is piecewise smooth, having isolated jumps at points $y_1 < \cdots < y_N$, then the discontinuities in $v, p, \partial_x u$, and $\partial_x e$ convect only along the corresponding particle paths $x = y_d$ and satisfy the jump conditions:

$$\left[p(v,e) - \frac{\epsilon \partial_x u}{v}\right] = 0, \qquad \left[\frac{\partial_x e}{v}\right] = 0, \tag{3.5}$$

where $[w]$ denotes a jump of function w across $x = y_d : [w(y_d, t)] = w(y_d + 0, t) - w(y_d - 0, t)$. We remark that the convective contact discontinuities here should not be confused with convective contact discontinuities present in solutions of the Euler equation (2.18) (i.e. (3.1) with $\epsilon = \lambda = 0$). A solution of (2.18) obtained as a limit (as $\epsilon, \lambda \to 0$) of solutions of (3.1) is not necessarily continuous, which has often unrelated contact discontinuities of magnitude $|v_r - v_l|$ at $x = 0$.

THEOREM 3.1 (Well-Posedness and Regularity). *Given initial data $(v_0, u_0, e_0)(x)$ satisfying (3.4), there exists a unique weak solution $(v, u, e)(x, t)$ of (3.1)-(3.3):*

$$(v, u, e)(x,t) \in C([0,\infty); L^2(0,1)), \quad (u,e)(x,t) \in C((0,\infty); H^1(0,1)),$$

satisfying that there exists $M > 0$, independent of t, such that

$$M^{-1} \leq v(x,t) \leq M, \quad e(x,t) \geq M^{-1}, \tag{3.6}$$

$$TV_{[0,1]}(v(\cdot,t)) \leq M, \quad \|v(\cdot,t') - v(\cdot,t)\|_{L^2} \leq M|t'-t|^{1/2}, \tag{3.7}$$

$$\|\partial_x u(\cdot,t)\|_{L^2} \leq M\sigma^{-1/4}(t), \quad \|\partial_x e(\cdot,t)\|_{L^2} \leq M\sigma^{-1/4}(t) \tag{3.8}$$

hold for all $t, t' > 0$, where $\sigma(s) = \min(s, 1)$. Furthermore, the solution is locally stable with large discontinuous initial data.

The existence results in Theorem 3.1 immediately extend to the Cauchy problem with the aid of estimates (3.6)-(3.8) and the arguments as in [**5, 42**]; and the BV assumption (3.4) on the initial data can be relaxed (see [**15, 16**]). We emphasize that the bounds of estimates (3.6)-(3.8) are independent of time, even with large discontinuous initial data, which are crucial to allow for the determination of the large-time behavior of the discontinuous solutions. The time-independent estimate (3.6) for the specific volume indicates that vacuum and concentration do never occur for all time, even starting from large discontinuous initial data. This phenomenon shows the essential difference between the Navier-Stokes equations and the Euler equations for which vacuum may occur for the solutions of large oscillations, even for the Riemann solutions with large Riemann initial data (cf. [**4, 14, 26**]). With the aid of estimates (3.6)-(3.8), we have

THEOREM 3.2 (Discontinuities and Large-Time Behavior). *Let the initial data $(v_0, u_0, e_0)(x)$ satisfy (3.4). Then the discontinuous solution $(v, u, e)(x, t)$ has the* *following behavior: The functions $v(\cdot,t), p(\cdot,t), \partial_x u(\cdot,t)$, and $\partial_x e(\cdot,t)$ have one-sided limits at each discontinuous point y_d for $t > 0$, and the jump conditions* (3.5) *hold pointwise. Furthermore, when $t \to \infty$,*

$$|[(v,p,\partial_x u,\partial_x e)(y_d,t)]| \le Me^{-M^{-1}t} \to 0, \tag{3.9}$$

$$\|(v - \int_0^1 v_0(x)dx, u, e - \int_0^1 (e_0(x) + \frac{u_0^2(x)}{2})dx)(\cdot,t)\|_{L^\infty(0,1)} \to 0. \tag{3.10}$$

Theorem 3.2 indicates that the large jumps of initial discontinuities decay exponentially, although they persist all the time. Moreover, all physical variables decay asymptotically in L^∞.

Regarding early work on (3.1) with large discontinuous initial data, we refer to Matsumura-Yanagi [**55**] for the isothermal case and to Amosov-Zlotnick [**1**] for the existence of weak solutions and some of their *time-dependent* estimates for certain initial and boundary data. We also refer to Hoff [**39**]-[**40**] for the global well-posedness and large-time behavior for small discontinuous initial data.

The results we obtain for (3.1)-(3.4) can be converted to the ones for the corresponding initial-boundary value problem of the compressible Navier-Stokes equations in Euler coordinates (cf. [**5**]). The approaches developed here have been applied to establishing the global well-posedness and large-time behavior of discontinuous solutions to the Navier-Stokes equations for a compressible reacting flow, which describe dynamic combustion (see [**16**]).

Acknowledgments. Most of results discussed here are based on the joint projects with C. M. Dafermos, H. Frid, D. Hoff, P. LeFloch, and K. Trivisa, respectively. Gui-Qiang Chen's research was supported in part by NSF grants DMS-9971793, DMS-9708261, and INT-9726215.

References

[1] A. A. Amosov and A. A. Zlotnick, A semidiscrete method for solving equations of the one-dimensional motion of a non-homogeneous viscous heat-conducting gas with nonsmooth data, *Izv. Vyssh. Uchebn. Zaved. Mat.* OVYR(**4**) (1997), 3–19.

[2] Bressan, A., Crasta, G., Piccoli, B., *Well posedness of the Cauchy problem for $n \times n$ systems of conservation laws*, *Memoirs Amer. Math. Soc.* 2000 (to appear).

[3] A. Bressan, T.-P. Liu, and T. Yang, L_1 stability estimates for $n \times n$ conservation laws, *Arch. Rational Mech. Anal.* **149** (1999), 1–22.

[4] T. Chang and L. Hsiao, *The Riemann Problem and Interaction of Waves in Gas Dynamics*, Longman Scientific & Technical, Harlow; John Wiley & Sons, Inc.: New York, 1989.

[5] G.-Q. Chen, Global solutions to the compressible Navier-Stokes equations for a reacting mixture, *SIAM J. Math. Anal.* **23** (1992), 609–634.

[6] G.-Q. Chen, Convergence of the Lax-Friedrichs scheme for isentropic gas dynamics (III), *Acta Math. Sci.* **8** (1988), 243–276 (in Chinese); **6** (1986), 75–120.

[7] G.-Q. Chen, *Compactness Methods and Nonlinear Hyperbolic Conservation Laws*, AMS/IP Studies in Advanced Mathematics, vol. **15**, pp. 34–75, AMS: Providence; IP: Cambridge, MA, 2000.

[8] G.-Q, Chen and C. M. Dafermos, The vanishing viscosity method in one-dimensional thermoelasticity, *Trans. Amer. Math. Soc.* **347** (1995), 531–541.

[9] G.-Q. Chen and H. Frid, Decay of entropy solutions of nonlinear conservation laws, *Arch. Rational Mech. Anal.* **146(2)** (1999), 95–127.

[10] G.-Q. Chen and H. Frid, Large time behavior of entropy solutions for conservation laws, *J. Diff. Eqs.* **152** (1999), 308–357.

[11] G.-Q. Chen and H. Frid, Uniqueness and asymptotic stability of Riemann solutions for the compressible Euler equations, Trans. Amer. Math. Soc. 2000.

[12] G.-Q. Chen and H. Frid, Large-time behavior of entropy solutions in L^∞ for multidimensional conservation laws, In: *Advances in Nonlinear Partial Differential Equations and Related Areas*, G.-Q. Chen et al (eds.), pp. 28-44, World Scientific: Singapore, 1998.

[13] G.-Q. Chen and H. Frid, Divergence-measure fields and hyperbolic conservation laws, *Arch. Rational Mech. Anal.* **147** (1999), 89-118.

[14] G.-Q. Chen and J. Glimm, Global solution to the compressible Euler equations with geometrical structure, *Commun. Math. Phys.* **180** (1996), 153-193.

[15] G.-Q. Chen, D. Hoff, and K. Trivisa, Global solutions of the compressible Navier-Stokes equations with large discontinuous initial data, *Commun. Partial Diff. Eqs.* 2000 (to appear).

[16] G.-Q. Chen, D. Hoff, and K. Trivisa, Global discontinuous solutions for viscous compressible reacting flow, Preprint, September 1999.

[17] G.-Q. Chen, and P. LeFloch, Entropies and weak solutions of the compressible isentropic Euler equations, *C. R. Acad. Sci. Paris, Série I*, **324** (1997), 1105-1110.

[18] G.-Q. Chen and P. LeFloch, Compressible Euler equations with general pressure law, *Arch. Rational Mech. Anal.* 2000 (to appear).

[19] G.-Q. Chen and P. LeFloch, In preparation.

[20] G.-Q. Chen, C. D. Levermore, and T.-P. Liu, Hyperbolic systems of conservation laws with stiff relaxation terms and entropy, *Comm. Pure Appl. Math.* **47** (1994), 787–830.

[21] G.-Q. Chen and B. Su, Discontinuous solutions for Hamilton-Jacobi equations, *Chinese Ann. of Math.* **21B** (2000), 165–186.

[22] C. M. Dafermos, *Hyperbolic Systems of Conservation Laws in Continuum Physics*, Springer-Verlag: New York, 2000.

[23] C. M. Dafermos, Hyperbolic systems of conservation laws, In: *Proceedings of the International Congress of Mathematicians*, Vol. **1, 2** (Zürich, 1994), 1096–1107, Birkhäuser, Basel, 1995.

[24] C. M. Dafermos, Large time behavior of periodic solutions of hyperbolic systems of conservation laws, *J. Diff. Eqs.* **121** (1995), 183-202.

[25] X. Ding, T. Chang, C.-H. Wang, L. Hsiao, and T.-C. Li, A study of the global solutions for quasi-linear hyperbolic systems of conservation laws, *Sci. Sinica* **16** (1973), 317–335.

[26] X. Ding, G.-Q. Chen, and P. Luo, Convergence of the Lax-Friedrichs scheme for isentropic gas dynamics (I)–(II), *Acta Math. Sci.* **7** (1987), 467–480, **8** (1988), 61-94 (in Chinese); **5** (1985), 415–432, 433–472.

[27] X. Ding and T.-P. Liu, *Nonlinear Evolutionary Partial Differential Equations*, AMS/IP Studies in Advanced Mathematics **3**, AMS: Providence; IP: Cambridge, MA, 1997.

[28] R. DiPerna, Convergence of approximate solutions to conservation laws, *Arch. Rational Mech. Anal.* **82** (1983), 27–70.

[29] R. DiPerna, Convergence of the viscosity method for isentropic gas dynamics, *Comm. Math. Phys.* **91** (1983), 1–30.

[30] R. DiPerna, Uniqueness of solutions to hyperbolic conservation laws, *Indiana Univ. Math. J.* **28** (1979), 137–188.

[31] R. DiPerna, Measure-valued solutions to conservation laws, *Arch. Rational Mech. Anal.* **88** (1985), 223-270.

[32] K. Friedrichs and P. Lax, Systems of conservation laws with a convex extension, *Proc. Nat. Acad. Sci. U.S.A.* **68** (1971), 1686-1688.

[33] S. Friedlands, J. W. Robbin, and J. Sylvester, On the crossing rule, *Comm. Pure Appl. Math.* **37** (1984), 19-64.

[34] J. Glimm, Solutions in the large for nonlinear hyperbolic systems of equations, *Comm. Pure Appl. Math.* **18** (1965), 95–105.

[35] J. Glimm and P. D. Lax, Decay of solutions of nonlinear hyperbolic conservation laws, *Memoirs Amer. Math. Soc.* **101**, 1970.

[36] E. Godlewski and P.-A. Raviart, *Numerical Approximation of Hyperbolic Systems of Conservation Laws*, Applied Mathematical Sciences, **118**, Springer-Verlag: New York, 1996.

[37] J. Greenberg and M. Rascle, Time-periodic solutions to systems of conservation laws, *Arch. Rational Mech. Anal.* **115** (1991), 395-407.

[38] C.-H. Gu, T.-T. Li, W.-C. Yu, and Z.-Y. Hou, Discontinuous initial value problems for systems of quasilinear hyperbolic equations (I)-(III), *Acta Math. Sinica* **11** (1961), 314-323, 324-327; **12** (1962), 132-143.

[39] D. Hoff, Global well–posedness of the Cauchy problem for nonisentropic gas dynamics with discontinuous initial data, *J. Diff. Eqs.* **95** (1992), 33-74.

[40] D. Hoff, Discontinuous solutions of the Navier-Stokes equations for multidimensional heat-conducting fluids, *Arch. Rational Mech. Anal.* **139** (1997), 303–354.

[41] H. K. Jenssen, Blow up for systems of conservation laws, *SIAM J. Math Anal.* **31** (2000), 894–908.

[42] A. V. Kazhikhov and V. V. Shelukhin, Unique global solution with respect to time of initial-boundary-value problems for one–dimensional equations of a viscous gas, *J. Appl. Math. Mech.* **41** (1977), 273–282.

[43] S. N. Kruzkov, First-order quasilinear equations in several independent variables, *Math. USSR Sb.* **10** (1970), 217–243.

[44] P. D. Lax, The multiplicity of eigenvalues, *Bull. Amer. Math. Soc.* **6** (1982), 213-214.

[45] P. D. Lax, *Hyperbolic Systems of Conservation Laws and the Mathematical Theory of Shock Waves,* CBMS. **11**, SIAM, Philadelphia, 1973.

[46] P. L. Lions, B. Perthame, and E. Tadmor, A kinetic formulation of multidimensional scalar conservation laws and related equations, *J. Amer. Math. Soc.* **7** (1994), 169–192.

[47] P. L. Lions, B. Perthame, and E. Tadmor, Kinetic formulation of the isentropic gas dynamics and p-system, *Comm. Math. Phys.* **163** (1994), 415–431.

[48] P. L. Lions, B. Perthame, and T. Souganidis, Existence of entropy solutions for the hyperbolic systems of isentropic gas dynamics in Eulerian and Lagrangian coordinates, *Comm. Pure Appl. Math.* **49** (1996), 599-638.

[49] J. Li, T. Zhang, and S. Yang, *The Two-Dimensional Riemann Problem in Gas Dynamics,* Longman (Pitman Monographs **98**): Essex, 1998.

[50] T.-T. Li and W.-C. Yu, *Boundary Value Problems for Quasilinear Hyperbolic Systems,* Duke University Mathematics Series **5**, Duke University, Durham, N.C., 1985.

[51] T.-P. Liu, The deterministic version of the Glimm scheme, *Comm. Math. Phys.* **57** (1975), 135-148.

[52] T.-P. Liu, Linear and nonlinear large time behavior of solutions of general systems of conservation laws, *Comm. Pure Appl. Math.* **30** (1977), 767-796.

[53] T.-P. Liu, Z. Xin, and T. Yang, Vacuum states for compressible flow, *Discrete Contin. Dynam. Systems* **4** (1998), 1–32.

[54] T.-P. Liu and T. Yang, L_1 stability for systems of hyperbolic conservation laws, In: *Nonlinear Partial Differential Equations*, G.-Q. Chen and E. DiBenedetto (eds.), Contemporary Mathematics, AMS: Providence, 1999.

[55] A. Matsumura and S. Yanagi, Uniform boundedness of the solutions for a one-dimensional isentropic model system of compressible viscous gas, *Comm. Math. Phys.* **175** (1996), 259–274.

[56] F. Murat, L'injection du cone positif de H^{-1} dans $W^{-1,q}$ est compacte pour tout $q < 2$, *J. Math. Pures Appl.* **60** (1981), 309–322.

[57] F. Murat, Compacité par compensation, *Ann. Scuola Norm. Sup. Pisa Cl. Sci. (4)* **5** (1978), 489-507.

[58] B. Riemann, Uber die Fortpflanzung ebener Luftwellen von endlicher Schwingungsweite, *Abhabdl. Koenig. Gesell. Wiss., Goettingen,* **8** (1860), 43.

[59] D. Serre, *Systémes de Lois de Conservation (I)-(II),* Diderot Editeur, Paris, 1996.

[60] J. Smoller, *Shock Waves and Reaction-Diffusion Equations*, 2nd edition, Springer-Verlag: New York, 1994.

[61] L. Tartar, Compensated compactness and applications to partial differential equations, In: *Research Notes in Mathematics, Nonlinear Analysis and Mechanics,* R. J. Knops (ed.), **4**, Pitman Press, New York, 1979, pp. 136–211.

[62] D. H. Wagner, Equivalence of the Euler and Lagrangian equations of gas dynamics for weak solutions, *J. Diff. Eqs.* **68** (1987), 118-136.

Department of Mathematics, Northwestern University, 2033 Sheridan Road, Evanston, IL 60208-2730, USA

E-mail address: gqchen@math.nwu.edu

AMS/IP Studies in Advanced Mathematics
Volume 20, 2001

Well-Posedness Theory for Hyperbolic conservation laws

Tai-Ping Liu

ABSTRACT. We survey the development of the well-posedness theory for systems of hyperbolic conservation laws.

1. History

Shock theory started with the paper by Riemann, [**24**], where he solves the shock tube problem for isentropic Euler equations for gas dynamics. It is the system,

$$\rho_t + (\rho v)_x = 0 \qquad \text{(conservation of mass)}$$
$$(\rho v)_t + (\rho v^2 + p)_x = 0 \qquad \text{(conservation of momentum)}$$

with $p(\rho) = a\rho^\gamma$.

Actually, Riemann made the mistake of conserving energy, instead of entropy. Nevertheless, the need to include the shock waves in the solution requires the fundamental concept of weak solutions. Riemann's mistake was subsequently corrected by Rankine and Hugoniot (see Courant-Friedrich, [**5**]). It was realized that the second law of thermodynamics is crucially needed when one considers the full Euler equation with the inclusion of,

$$(\rho E)_t + (\rho EV + pV)_x = 0 \qquad \text{(conservation of energy)}.$$

As a consequence of the second law of thermodynamics, the entropy increases across a shock. This is the entropy condition for the full Euler equations.

It has been a long standing open problem to show that the initial value problem for the Euler equations is well-posed if the weak solutions are required to satisfy the entropy condition. Meanwhile, the shock wave theory has been generalized to a general system of hyperbolic conservation laws,

$$u_t + f(u)_x = 0$$

The theory for scalar conservation laws, $u \in \mathbb{R}^1$, is complete due to the efforts of Hopf [**12**], Lax [**14**], and particularly the Russian School, Oleinik [**22**], Volpert [**25**], and Krushkov [**13**]. The entropy condition has been identified and the initial value

1991 *Mathematics Subject Classification.* Primary 37DXX 70SXX; Secondary 35L63, 35A05.

problem is shown to be well-posed in the $L_1(x)$ topology. In fact, the solution operator is an L_1-contraction semi-group:

$$\frac{d}{dt}\int_{-\infty}^{\infty}|u_1(x,t)-u_2(x,t)|dx \le 0$$

for any two weak solutions, u_1 and u_2.

The problem becomes much harder for the system, $u \in \mathbb{R}^n, n \ge 2$. We assume that the system is strictly hyperbolic, that is, the $n \times n$ matrix $f'(u)$ has real and distinct eigenvalues $\lambda_1(u) < \lambda_2(u) < \cdots < \lambda_n(u)$. The dependence of the characteristic speeds $\lambda_i(u)$ on the dependent variables makes the waves of distinct characteristic families to interact nonlinearly and new waves are produced. Interaction between two elementary waves can produce infinitely many shock waves in a finite time. There is the entropy condition of Lax [**14**] generalizing that for the Euler equations. In the important, now classical, paper by Glimm [**10**], the wave interactions are studied locally and globally, which yields the existence theory. Much has been achieved based on the Glimm estimates of wave interactions. There is the deep work of Glimm-Lax [**11**] on regularizing and the decay properties of solutions. There are also works on N-waves, (DiPerna [**7**], Liu [**15**]), on regularity, (DiPerna [**8**], Liu [**18**]) and large-time behavior (Liu [**16, 18**]).

The study of uniqueness of solutions has focused basically on piecewise smooth solutions (Oleinik [**23**], DiPerna [**9**]). These and other works use the Holmgren's approach and thereby are framed in the L_2 space. However, since the solution operator cannot be stable in the L_2 topology due to the presence of compression waves, these results cannot be generalized to general weak solutions. The only possible topology is L_1, which makes the usual functional analytic approach nonfeasible. There are now two approaches on the L_1 well-posedness of the initial value problem. The two approaches are different in their general setups, overall philosophy and also in terms of analytical ideas. These are described in the next two sections. In the final section we make some remarks and raise open problems.

2. Homotopy Approach

In Bressan [**1**] it is shown that the initial value problem is stable in the L_1 topology if two solutions, u_1 and u_2 are infinitesimally close. Hard analysis is then carried out (Bressan-Columbo [**2**], Bressan-Croatia-Piccoli [**3**]), for general two solutions u_1 and u_2 not necessarily close to each other. The idea is to homotopy u_1 to u_2 through a one-parameter family of solutions. The L_1 distance between the two solutions is then estimated using the above result for infinitesimally close solutions. To make sure that errors do not accumulate and becomes unbounded in the process, the variation of wave interactions need to be closely studied for two nearby solutions. This is done with the aid of a specific, explicit front tracking construction of approximate solutions originated in Dafermos [**6**]. With the control of wave interactions, the L_1 distance of two solutions is analyzed and shown to be dominated by the initial distance at $t = o$.

This approach forces the investigation of variation of wave interactions as the initial data vary. While such investigation is useful for the understanding of solution behavior, it is not intrinsic for the L_1 stability in question. The construction of approximate solutions for this purpose is complex and so far has been done basically only for genuinely nonlinear systems.

3. Entropy Functional Approach

The approach of Liu-Yang [**19, 21**] is called the entropy functional because it is based on the construction of a functional of which the entropy functional is the hey component (Liu-Yang [**20**]). The functional is time-decreasing and equivalent to the L_1 distance between solutions. There are three parts to the functional,

$$\begin{aligned} H[u_1(\cdot,t),u_2(\cdot,t)] &= L[u_1(\cdot,t),u_2(\cdot,t)] + KQ_d[u_1(\cdot,t),u_2(\cdot,t)] \\ &= +KE[u_1(\cdot,t),u_2(\cdot,t)]. \end{aligned}$$

The basic part L is equivalent to the L_1 distance between u_1 and u_2, and dominates the other parts. Q_d measures the coupling of waves in the solutions on the L_1 distance. The entropy functional is defined for the scalar conservation law for waves in each characteristic field. The entropy functional makes straightforward and crucial usage of the strong nonlinearity, the genuine nonlinearity, of the system. The entropy functional is new even for the simplest Burgers equation

$$u_t + \left(\frac{u^2}{2}\right)_x = 0, \quad u \in \mathbb{R}'$$

One of the main considerations in the construction of the functional H is the realization that one may regard the solutions as linear superpositions of nonlinear waves. This is due to the following two facts: The first is the boundedness of the total amount of wave interactions as a consequence of the Glimm functional and estimates, [**11**]. The second is a wave tracing device (Liu [**17**]), which implies that the L_1 distance between the solutions is accurately approximated by wave patterns consisting of linear superpositions of nonlinear waves.

The aforementioned simplification is useful only if a robust nonlinear functional can be constructed for such wave patterns. This is where the main effort lies. The difficulty is that, unlike the scalar equation, the solution operator is not an L_1 contraction. That is,

$$\frac{d}{dt}L[u_1(\cdot,t),u_2(\cdot,t)]$$

is in general not non-positive. The reason is not primarily due to nonlinear wave interactions in the solutions, u_1 and u_2. It is because of the coupling of waves of distinct families, which disallows the representation of the general system as a collection of uncoupled scalar conservation laws. On the other hand, nonlinearity also has regularizing effects for the waves pertaining to each characteristic field. The entropy functional, $E[u_1(\cdot,t),u_2(\cdot,t)]$, makes effective use of this regularizing property so that dE/dt contains a negative third order terms. This is used to counterbalance the positive third order terms in dL/dt resulting from the aforementioned coupling of waves.

There is the classical entropy functional directly related to the entropy condition and also yields third order estimates. The classical functional applies to an individual solution and not to the difference of two solutions. The new entropy functional is related, not so much to the entropy condition, but more to the nonlinearity of the equation.

The Liu-Yang approach on L_1 stability focuses on wave patterns without nonlinear interactions. The Glimm existence theory is primarily on nonlinear wave interactions. Thus the Liu-Yang approach is orthogonal to and supplements the Glimm theory in the well-posedness theory for hyperbolic conservation laws.

4. Concluding Remarks

The stability analysis yields an L_1 semi-group for the solution operator. It follows that the solution to the initial value problem for the hyperbolic conservation laws is unique within a space of functions, which can be approximated well, local in time, by the solutions produced by the semi-group [**4**]. There remains the question of regularity of solutions obtained by other means, for instance, those produced by vanishing viscosity or finite differences from the theory of compensated compactness.

This article focuses solely on the on the part of shock wave theory which is of direct relevance to the recent works on the well-posedness problem. For instance, we have not mentioned the approach using the theory of compensated compactness, and the works on multi-dimensional shock waves. The Beijing School and Fudan University have done significant works on these topics.

References

[1] **A. Bressan**, *A locally contractive metric for system of conservation laws*, Estratto dagli Annali della Scuola Normale Superiore di Pisa, Scienze Fisiche e Matematiche-serie IV, **XXII** (1995), Fasc. 1.

[2] **A. Bressan** and **R. M. Colombo**, *The semi-group generated by* 2×2 *conservation laws*, Arch. Rational Mech. Anal., **133** (1995), pp. 1–17.

[3] **A. Bressan, G. Crasta** and **B. Piccoli**, *Well posedness of the Cauchy problem for* $n \times n$ *systems of conservation laws*, Memoir Amer. math. Soc., (to appear).

[4] **A. Bressan**, and **P. LeFloch**, *uniqueness of weak solutions to systems of conservation laws*, preprint S.I.S.S.A., Trieste 1996.

[5] **R. Courant** and **K. O. Friedrich**, *Supersonic Flows and Shock Waves*, Wiley-Interscience; New York, 1945.

[6] **C. M. Dafermos**, *Polygonal approximations of solutions of initial value problems for a conservation law*, J. Math. Anal. Appl., **38** (1992), pp. 33–41.

[7] **R. DiPerna**, *Decay and asymptotic behavior of solutions to nonlinear hyperbolic conservation laws*, Indiana Univ. Math. J., **24** (1975), pp. 1047–1071.

[8] **R. DiPerna**, *Singularities of solutions of nonlinear hyperbolic system of conservation laws*, Arch. Rat. Mech. Anal., **60** (1975), pp. 75–100.

[9] **R. DiPerna**, *uniqueness of solutions to hyperbolic conservation laws*, Indiana Univ. Math. J., **28** (1979), pp. 137–188.

[10] **J. Glimm**, *Solutions in the large for nonlinear hyperbolic systems of equations*, Comm. Pure Appl. Math., **18** (1965), pp. 699–715.

[11] **J. Glimm** and **P. D. Lax**, *Decay of solutions of systems of hyperbolic conservation laws*, Memoirs, Amer. Math. Soc., **101** (1970).

[12] **E. Hopf**, *The partial differential equation* $u_t + uu_x = \mu u_{xx}$. Comm. Pure Appl. Math.**3** (1950), pp. 201–230.

[13] **s. Krushkov**, *First-order quasilinear equations with several space variables*, Mat. Sb. **10** (1970), pp. 228–255; English transl. in Math. USSR Sb., **10** (1970), pp. 217–273.

[14] **P. D. Lax**, *Hyperbolic systems of conservation laws, II*, Comm. Pure Appl. Math., **10** (1957), pp. 537–566.

[15] **T.-P. Liu**, *Pointwise convergence to N-waves for solutions of hyperbolic conservation laws*, Bull. Inst. Math., Academia Sinica, **15** (1987), pp. 1–17.

[16] **T.-P. Liu**, *Linear and nonlinear large-time behaviors of solutions of hyperbolic conservation laws*, Comm. Pure Appl. Math., **30** (1977), pp. 767–796.

[17] **T.-P. Liu**, *The deterministic version of the Glimm scheme*, Comm. Pure Appl. Math., **57** (1995), pp. 135–148.

[18] **T.-P. Liu**, *Admissible solutions of hyperbolic conservation laws*, Memoirs, Amer. Math. Soc., **241** (1981).

[19] **T.-P. Liu** and **T. Yang**, L_1 *stability for* 2×2 *systems of hyperbolic conservation laws*, J. Amer. Math. Soc., **12** (1999), pp. 729–774.

[20] **T.-P. Liu** and **T. Yang**, *A new entropy functional for a scalar conservation law*, Comm. Pure Appl. Math.,(1999), pp. 1427–1442.

[21] **T.-P. Liu** and **T. Yang**, *Well-posedness theory for system of conservation laws*, Comm. pure Appl. Math., (1999).

[22] **O. Oleinik**, *Discontinuous solutions of nonlinear differential equations*, usp. Mat. Nauk., (N.S.) **12** (1957), pp. 3–73; English transl. in Amer. Math. Soc. Transl. Ser. 2, **26**, 95–172.

[23] **O. Oleinik**, *On the uniqueness of the generalized solution of the Cauchy problem for a nonlinear system of equations occuring in mechanics*, usp. Mat. Nauk., (N.S.) **12** (1957), pp. 169–175 (Russian).

[24] **B. Riemann**, *Gesammeltte Werke*, 1896, pp. 1490.

[25] **A. I. Volpert**, *The spaces BV and quasilinear equations*, Math. USSR Sb., **2** (1967).

DEPARTMENT OF MATHEMATICS, STANFORD UNIVERSITY, STANFORD, CA 94305-2125

AMS/IP Studies in Advanced Mathematics
Volume 20, 2001

Two–dimensional Riemann Problems for conservation laws in gas dynamics

Jiequan Li and Tong Zhang

ABSTRACT. This paper is to survey some recent results on two–dimensional Riemann problems for hyperbolic conservation laws in gas dynamics. The focus will be on the global structures of solutions in two dimensions.

1. Introduction

The Riemann problem was initiated and solved for one-dimensional Euler equations of isentropic flow in gas dynamics by Riemann in 1858 in his pioneering work on the mathematical theory of shock waves [23]. Since then a lot of interesting work has been contributed to the one–dimensional Riemann problem for various hyperbolic systems of conservation laws [e.g., 4, 12]. It is fully recognized that the Riemann problem is the most fundamental problem in the entire field of nonlinear hyperbolic consevations laws.

After considering one–dimensional Riemann problem, it is natural for us to investigate two–dimensional Riemann problems. What is a two–dimensional Riemann problem? It should be an initial–value problem with the simplest discontinuous initial data, being scale invariant and piecewise constant. The focusing of the rays of discontinuity of initial data at the origin forms a geometrical singularity. The discontinuity rays emit, in $1+2$ space, planar elementary waves with the edges from the origin. These waves may intersect or they may not. What the two–dimensional Riemann problem should answer is how these elementary waves interact and match together in a cone having the origin as the vertex in $1 + 2$ space. So, the Riemann problem in the two–dimensional case is substantially different from that in the one–dimensional case.

The most familiar examples of two–dimensional Riemann solutions are the regular and Mach reflections of oblique shocks, which were observed experimentally by Mach in the last century. Research interest on it has lasted for over a century. But it is still a great challenge for mathematicians nowadays.

This paper tries to give a survey on some recent results of two–dimensional Riemann problem. First we recall some basic concepts on the one–dimensional Riemann problem so as to understand two–dimensional ones much easily and fully in Section 2. Then in Section 3 the Riemann solutions to scalar conservation laws

Partially supported by NNSF of China.

are constructed, and the emphasis is placed on the global structure of solutions, Mach–reflection–like configuration. In Section 4, we discuss the four–wave Riemann problem for the full compressiblem Euler equations. In Sections 5 and 6, the zero–pressure gas dynamics and pressure–gradient equations, resulting from the flux splitting of the Euler system, are brought into our attention. Especially we will compare the Euler equations and the zero–pressure gas dynamics and illustrate that the latter is the dominant factor of the former.

2. 1–D Riemann Problem

Let us start by investigating the shock tube probem [4, 23]. For a tube that is filled by an invisid, compressible fluid, assume that initially it is separated by a thin membrane into two parts (see Figure 1.1). If this membrane is broken instantaneously, the flowfield will be formed at once in the entire tube. We want to see what will happen by taking the isentropic Euler system as the governing equations

$$\begin{cases} \rho_t + (\rho u)_x = 0, \\ (\rho u)_t + (\rho u^2 + p)_x = 0, \end{cases} \tag{2.1}$$

subject to the initial data

$$(\rho, u)(t=0,x) = (\rho_0, u_0)(x) = \begin{cases} (\rho_1, u_1), & x < 0, \\ (\rho_2, u_2), & x > 0, \end{cases} \tag{2.2}$$

where $\rho(t,x)$ is the density, p is the pressure, and $u(t,x)$ is the velocity. The simple initial data consists of two pieces of constant states but it has a discontinuity at the origin. So we have to consider the solution to this problem beyond the classical sense. Next we give a definition of generalized solutions to this problem.

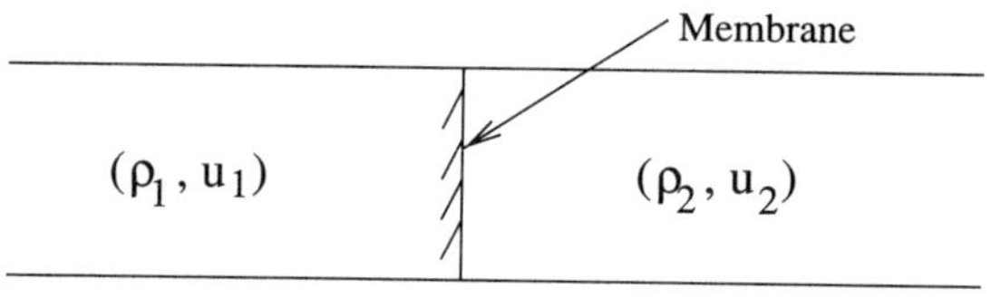

Figure 1.1 Shock tube experiment

Definition 2.1 *A pair $(\rho, u)(t,x) \in L^\infty(R^+ \times R)$ is a solution of (1.1) and (1.2) if and only if*

$$\begin{cases} \displaystyle\int_{R^+\times R} (\rho\phi_t + \rho u \phi_x) dx dt + \int_R \rho_0(x)\phi(0,x)dx = 0, \\ \displaystyle\int_{R^+\times R} (\rho u \psi_t + (\rho u^2 + p)\psi_x) dx dt + \int_R \rho_0(x) u_0(x)\psi(0,x)dx = 0 \end{cases} \tag{2.3}$$

for all $(\phi, \psi)(t,x) \in C_0^\infty(R^+ \times R)$.

This definition allows us to consider the solutions in a broader class of essentially bounded functions. Thus we can construct the solution of (1.1) and (1.2) in the class of piecewise smooth functions with the method of characteristic analysis in phase plane.

The solutions consists of elementary waves, rarefaction waves and shock waves, besides some constant states. The rarefaction waves are continuous and satisfy

$$R_1: \begin{cases} x/t = \lambda_1 = u - \sqrt{p'(\rho)}, \\ \dfrac{du}{d\rho} = -\dfrac{\sqrt{p'(\rho)}}{\rho}, \end{cases} \qquad R_2: \begin{cases} x/t = \lambda_2 = u + \sqrt{p'(\rho)}, \\ \dfrac{du}{d\rho} = \dfrac{\sqrt{p'(\rho)}}{\rho}; \end{cases} \tag{2.4}$$

where λ_1, λ_2 are eigenvalues, R_1 and R_2 are called backward and forward rarefaction waves, respectively. While shock waves are discontinuous and satisfy

$$S_1: \begin{cases} x/t = \sigma_1 = u_1 - \sqrt{\dfrac{\rho_1[p]}{\rho_2[\rho]}}, \\ u = u_1 - \sqrt{\dfrac{[p]}{\rho_1\rho_2[\rho]}}[\rho], \\ p > p_1, \end{cases} \qquad S_2: \begin{cases} x/t = \sigma_2 = u_2 + \sqrt{\dfrac{\rho_2[p]}{\rho_1[\rho]}}, \\ u = u_2 + \sqrt{\dfrac{[p]}{\rho_1\rho_2[\rho]}}[\rho], \\ p > p_2, \end{cases} \tag{2.5}$$

where σ_1, σ_2 are the propagation speeds, [p] is jump of the pressure p etc., S_1 and S_2 are called backward and forward shock waves. These identities are Rankine–Hugoniot Relations, describing the relationship between the propagation speed and the states on two sides.

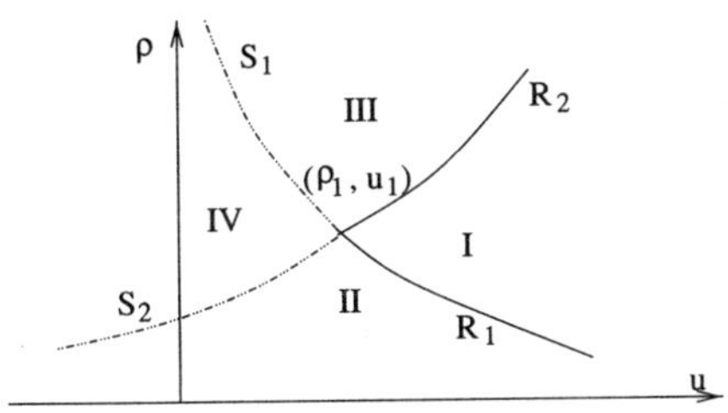

Figure 2.2

It is easily seen from Figure 2.2 that the solution may not be unique. Some additional conditions must be imposed in order to select physically relevant solutions and gurantee the uniqueness of solutions. Here this condition is an entropy condition stating that the entropy increases across a shock or that the pressure on the wave front is larger than that on the wave back. Then the shock wave curves only consists of two half branches. Thus the whole half phase plane ($\rho \geq 0$) is divided into four regions, I, II, III and IV. The solution of (2.1) and (2.2) has four kinds of structures, see Figure 2.3. The intermediate state (ρ_*, u_*) is the intersection point of the respective wave curves drawn from (ρ_1, u_1) and (ρ_2, u_2).

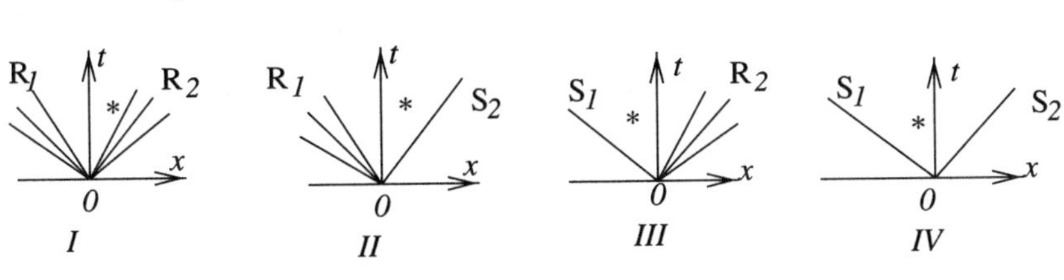

Figure 2.3

3. Scalar Conservation Laws

Now we turn to two–dimensional Riemann problems. Let us first take a look at scalar conservation laws

$$u_t + f(u)_x + g(u)_y = 0, \quad u \in R, \tag{3.1}$$

where the fluxes f and g are supposed to be in C^3 for simplicity. Since Guckenheimer [11] gave his first Riemann solution with the specific fluxes $f(u) = u^2/2$, $g(u) = u^3/3$ subject to the initial data of -1, 0 and 1 in three angular domains having vertexes at the origin, much attention has been paid to this problem, [e.g., 28, 32]. Till recently, an simple and easiler understood method, the generalized characteristic analysis method, is used to constructed the Riemann solutions of (3.1) in [31]. The basic idea is to introduce some singularity point sets associated with characterisics and discontinuities, originating from the singularity of initial data at the origin, and to combine the self–similar plane and phase plane so that the solution can be constructed with the standard method of characteristics. This method not only greatly simplifies the proof in the construction process of solutions before, but derived the phenomenon observed by Guckenheimer for very general fluxes. This is the genuinely multidimensional phenomenon, never appearing in one-dimensional case, which will be described in the following.

Take a ray l with the origin as the vertex such that it is not located in the first quadrant of the plane $\{t = 0\}$. Then this initial plane is divided into three angular domains Ω_i $(i = 1, 2, 3)$, in which u takes u_i respectively. $u_1 < u_3 < u_2$ are ordered so that three shocks emit from the initial discontinuities under the non–degeneracy condition $f''(u) > 0$, $g''(u) > 0$ and $(f''(u)/g''(u))' > 0$ (see Figure 3.1). It can be proved in [31] that the planar wave from l is a shock if and only if $\alpha \in (\pi + \alpha_0, 2\pi)$ where α is the polar angle of l, $\alpha_0 = \arctan \dfrac{g'(u_3) - g'_{32}}{f'(u_3) - f'_{32}}$ and $f'_{32} = (f(u_3) - f(u_2))/(u_3 - u_2)$. Denote the intersection point of $\Gamma(u_1)\Gamma_s(u_1, u_2)$ and $\eta = g'_{13}$ by $P = (\xi_p, \eta_p)$, and $\alpha_1 = \arctan \dfrac{\eta_p - g'_{32}}{\xi_p - f'_{32}}$, where $\Gamma(u)$ and $\Gamma_s(u, u_1)$ stand for the points $(\xi, \eta) = (f'(u_1), g'(u_1))$ and $(\xi, \eta) = (f'_{u,u_1}, g'_{u,u_1})$, $(\xi, \eta) = (x/t, y/t)$ respectively. The solution displays two kinds of configurations.

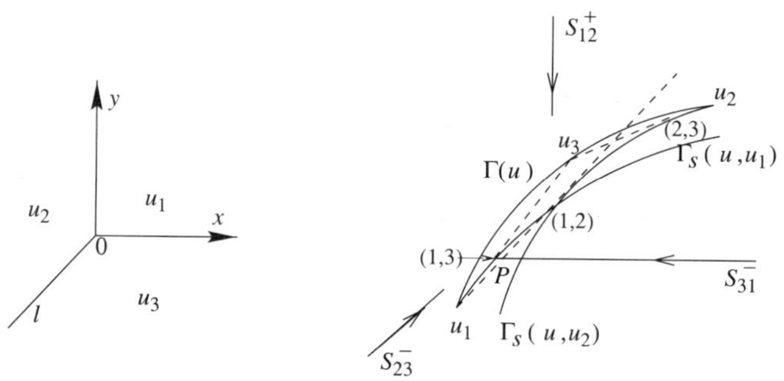

Figure 3.1 The initial data and the exterior waves: the first illustrates the distribution of initial data; the second displays the exterior waves in (ξ, η)–plane.

Case 1 $\pi + \alpha \le \alpha < 2\pi$

At this time, the intersection point Q of S_{23}^- and S_{31}^- is located on the right–hand side of P. We solve the Riemann problem locally at Q to obtain a shock S_{12}^-

connecting the states u_1 and u_2 and finally this new shock matches the exterior wave S_{12}^+ from the infinity at the singularity point $\Gamma_s(u_1, u_2)$ (see Figure 3.2). This solution is the consequence of direct collision of two shocks. We call it the regular–reflection–like configuration, after comparing it with the resemblence of the oblique shock experiment in gas dymanics.

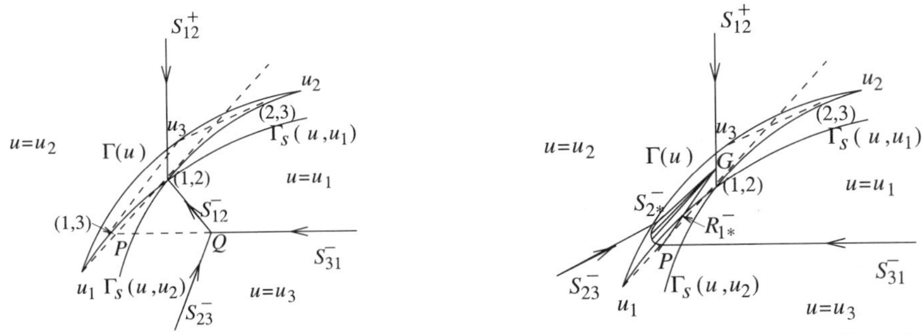

Figure 3.2 The first is regular–reflection–like configuration; and the other is Mach–reflection–like configuration

Case 2 $\pi + \alpha_0 < \alpha < \alpha_1 + \pi$

In contrast to the first case, we cannot get the global solution by solving the Riemann problem locally at the intersection point of S_{23}^- and S_{31}^-, which is on the left–hand side of P. It can be proved that S_{12}^+ from the infinity bifurcates somewhere, before reaching its singularity point $\Gamma_s(u_1, u_2)$, into a half contact discontinuity S_{2*}^- and a rarefaction wave R_{1*}^-. Then S_{31}^+ penetrates R_{1*} and finally matches with S_{23}^- and S_{2*}^-. This phenomenon just resembles to the Mach reflection at the triple point in gas dynamics. Hence it is named the Mach–reflection–like configuration and the consequence of global interaction of waves.

4. Four–wave Riemann Problem for Compressible Euler Equations

In [33], the authors conjectured on the plausible structure of the four wave Riemann problem for the full compressible Euler equations

$$\begin{cases} \rho_t + \nabla \cdot (\rho U) = 0, \\ (\rho U)_t + \nabla \cdot (\rho U \otimes U) + \nabla P = 0, \\ (\rho E)_t + \nabla \cdot (\rho U(E + p/\rho)) = 0, \end{cases} \tag{4.1}$$

where ρ, p and U are the density, pressure and velocity, respectively, $E = |U|^2/2 + p/((\gamma-1)\rho)$ is the total energy with the adiabatic index $\gamma > 1$. The initial data is restricted to be constant in each quadrant and only one elementay waves, a planar rarefaction wave, a planar shock wave or a planar contact discontinuity, emits at each sectorial interface. They classified the combination of four waves into 16 kinds, which was supplemented in [24]. Afterwards, several numerical simulations are added, see [25] by Gudonov scheme, [3] by MmB scheme, [13] by positive scheme and recently [15]. Almost all these results are consistent with the analysis in [33]. They exhibits very complicated flowfield patterns. We select some numerical results as follows.

Case 1 The interaction of rarefaction waves

The result shows that the interaction of rarefaction waves results in the occurence of shocks, which separate the subsonic region from the supersonic regions. This is just because the symmetric axes $\xi = \pm\eta$ are pseudo–stream lines that form compressive corners.

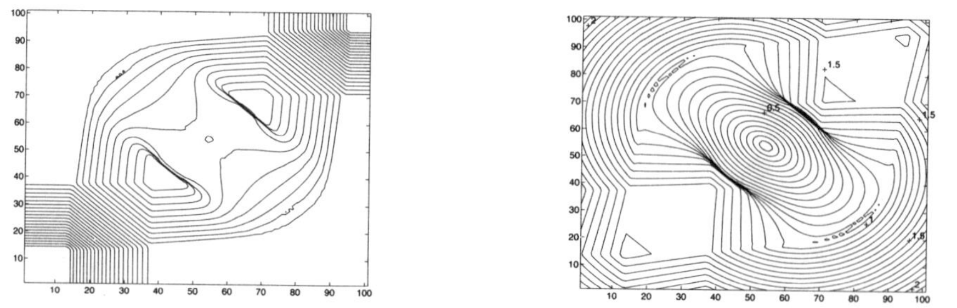

Density contour curves   Pseudo-Mach number contour curves

Figure 4.1 Interaction of rarefaction waves

Case 2 Interaction of shock waves

In this case, the Mach reflections occur. The symmetric axis $\xi = \eta$ can be regarded as the rigid wall in the oblique shock reflection experiment.

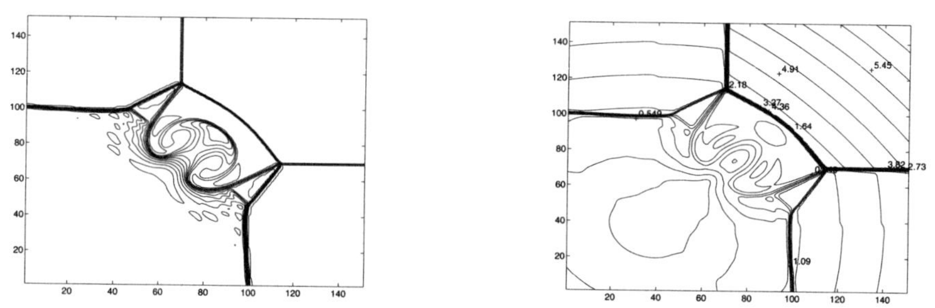

Density contour curves   Pseudo-Mach number contour curves

Figure 4.2 Interaction of shock waves

Case 3 The interaction of contact discontinuities with the same sign

The result displays an interesting spiral in the center of interaction domain when the contact discontinuities have the same sign of curls of velocity on them. This structure is familiar to us in hurricanes and tornadoes.

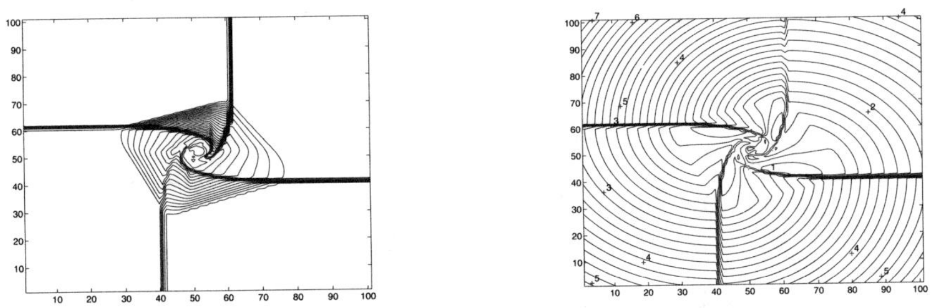

Density contour curves Pseudo-Mach number contour curves

Figure 4.3 Interaction of contact discontinuities with the same sign

Case 4 The interaction of contact discontinuities with different signs

In Figure 4.4, we show the result with a high Mach number. The density concentrates in the subsonic domain. This can be compared to the second example in Section 5 and shows that it may be impossible for us to consider the solution of multidimensional conservation laws in the traditional spaces such as L^p $(1 \leq p \leq \infty)$ or BV–spaces. This new wave is called the smoothed delta–shock wave.

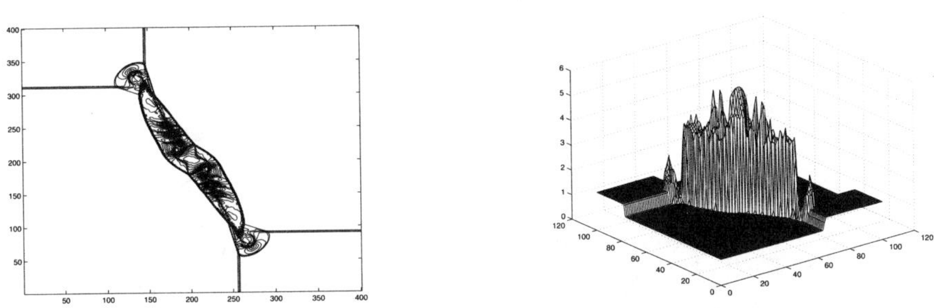

Density contour curves Three–dimensional version of density

Figure 4.4 Interaction of contact discontinuities with different signs

Unfortunately, there is no analytic result to prove the above conjecture and numerical results. So we have to turn to recognizing these structures from the physical point of view.

As is well–known, the motion of an invisid fluid is determined by inertia and pressure gradient effects. When one of them plays a dominant role, we neglect the other effect so that the flowfield patterns can be understood. Thus the Euler equations are splitted into two models [1, 19]: Transport equations or zero-pressure gas dynamics and pressure–gradient equations. In next two sections we will discuss these models.

5. Zero-pressure gas dynamics

The zero–pressure gas dynamics has the form

$$\begin{cases} \rho_t + \nabla \cdot (\rho U) = 0, \\ (\rho U)_t + \nabla \cdot (\rho U \otimes U) = 0. \end{cases} \tag{5.1}$$

It not only originates from the flux splitting scheme of Euler equations mentioned above, but it has other origins, for example, the limit of compressibe Euler equations with zero temperature [16], the adhesion particle dynamics proposed by Zeldovich in astrophysics to describe the motion process of free particles sticking under collision [26], and the first order closure of Boltzman equation [2] etc..

The distinct feature is that two blowup mechanisms occur simultaneously and the density may become a singular measure even though the initial data is smooth enough [e.g., 20]. So we have to consider the solutions beyond the traditional spaces.

Definition 5.1 [2, 14] *Denote by $BM(R^n)$ the space of bounded Borel measure on R^n. Consider a mass distribution $\rho(t,\cdot)$ and a momentum distribution $Q(t,\cdot)$ satisfying*

$$\begin{array}{c} \rho(t,\cdot) \in C([0,T], BM(R^n)), \quad \rho \geq 0, \\ Q(t,\cdot) \in C([0,T], (BM(R^n))^n) \end{array} \tag{5.2}$$

such that

$$|Q(t,\cdot)| \leq C\rho(t,\cdot), \quad \forall t \in [0,T] \tag{5.3}$$

in the sense of measures, where C is bounded. By Radon-Nykodym theorem, there is a mean velocity $U(t,\cdot) \in L^\infty(\rho(t,\cdot))$ for each fixed $t \geq 0$ such that

$$U(t,x) = \frac{dQ}{d\rho}. \tag{5.4}$$

Thus we say a pair $(\rho, U)(t,\cdot)$ is a measure solution of (5.1) if and only if

$$\begin{cases} \int_0^T \int_{R^n} (\phi_t + U \cdot \nabla\phi) d\rho dt = 0, \\ \int_0^T \int_{R^n} U(\psi_t + U \cdot \nabla\psi) d\rho dt = 0, \end{cases} \tag{5.5}$$

for all $\phi, \psi \in C_0^\infty([0,T] \times R^n)$.

Based on this definition, we consider the solution of the form

$$(\rho, U)(t,X) = \begin{cases} (\rho_1, U_1)(t,X), & (t,X) \in \Omega_1, \\ (w(t,s)\delta(X - X(t,s)), U_\delta(t,s)), & (t,X) \in S, \\ (\rho_2, U_2)(t,X), & (t,X) \in \Omega_2, \end{cases} \tag{5.6}$$

where (ρ_i, U_i) $(i = 1,2)$ are smooth in Ω_i respectively, $s = (s_1, \cdot, s_{n-1}) > 0$ is a parameter, δ is a Dirac delta measure with support $X = X(t,s)$ which is called **delta-shock** because the density becomes a Dirac delta measure and the velocity is discontinous there. It is proved in [14] that (5.6) is a measure solution if the generalized Rankine–Hugoniot conditions

$$\begin{cases} \dfrac{\partial X}{\partial t} = U_\delta(t,s), \\ \dfrac{\partial w(t,s)}{\partial t} = ([\rho], [\rho U]) \cdot (N_t, N_X), \\ \dfrac{\partial (wU_\delta)(t,s)}{\partial t} = ([\rho U], [\rho U \otimes U]) \cdot (N_t, N_X), \end{cases} \tag{5.9}$$

is satisfied, where $[\rho] = \rho_1 - \rho_2$ etc., and $N = (N_t, N_X)$ is oriented from Ω_1 to Ω_2. In addition, the entropy condition

$$U_2 \cdot N_X < U_\delta \cdot N_X < U_1 \cdot N_X, \tag{5.10}$$

is imposed in order to select physical relevant solutions and it describes the sticking process of free particle when colliding.

The stability problem has been solved in [27, 29] by taking the viscosity vanishing process of

$$\begin{cases} \rho_t + \nabla \cdot (\rho U) = 0, \\ (\rho U)_t + \nabla \cdot (\rho U \otimes U) = \epsilon t \Delta U, \end{cases} \tag{5.11}$$

where $\epsilon > 0$ is small viscosity, Δ is the Laplacian. The planar delta–shock is verified to be the limit of solutions of (5.11) as the viscosity vanishes.

The one-dimensional Riemann solutions can be founded in [2, 18]. The two–dimensional Riemann Problem was initiated in [26] and solved finally in [8]. It exhibits interesting configurations. We take some examples here and compared them with those of Euler equations.

First we take two examples corresponding to Cases 3 and 4 in Section 5 [26, 8]. When the initial discontinuities are four contact discontinuities with the same sign, the solution consists of four contact discontinuities, which surround the vacuum (see the first case of Figure 5.1). While the delta–shock waves appears in the solution when the initial contact discontinuities have different signs (see the second case of Figure 5.2).

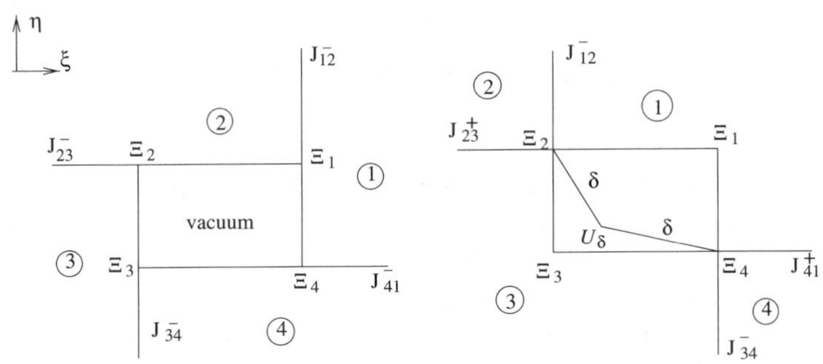

Figure 5.1 The configurations with contact discontinuities in the initial data

Corresponding to the global interaction of waves, we consider the case that the initial discontinuities are just four delta-shocks shown in Figure 5.2. The delta–shocks bifurcates into new delta–shocks surrounding a vacuum region [8].

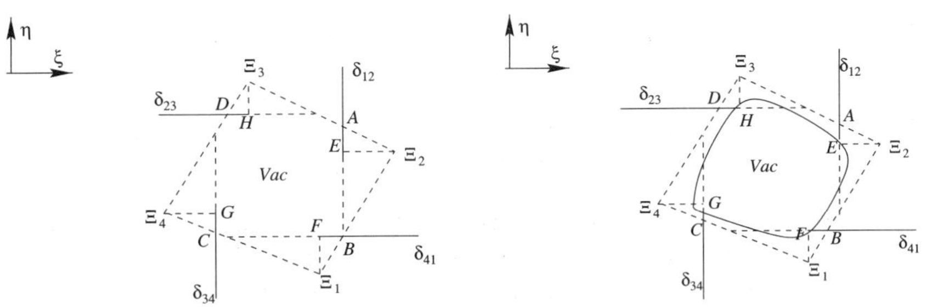

Figure 5.2 The Mach–reflection–like configuration

For the comparison of zero–pressure gas dynamics and Euler equations, we can refer to [21].

Finally we point out in passing that the Cauchy problem has been solved in [10, 7] with different methods independently. The uniqueness problem for one–dimensional case has been proved in [17] and remains open for multidimensional case.

6. Pressure–gradient equations

The pressure–gradient equations take the form

$$\begin{cases} \rho_t = 0, \\ (\rho U)_t + \nabla p = 0, \\ (\rho E)_t + \nabla \cdot (pU) = 0, \end{cases} \tag{6.1}$$

where the notations have the same meanings as those of Euler equations. This system can approximate the Euler system when the Mach number is very large. It is similar to the classical wave equations. Especial interest for us is that this equation can be decoupled into a single mixed–type partial differential equations of second order,

$$(p-\xi^2)p_{\xi\xi} - 2\xi\eta p_{\xi\eta} + (p-\eta^2)p_{\eta\eta} + \frac{1}{p}(\xi p_\xi + \eta p_\eta)^2 - 2(\xi p_\xi + \eta p_\eta) = 0. \tag{6.2}$$

This model sheds some light on the recognization of complicated flowfield pattern of Euler equations.

The two–dimensional Riemann solutions are analyzed in [30] where numerical solutions are accompanied. By the comparison with the Euler equations, we found that the flowfield pattern are surprisingly similar except the cases involving contact discontinuities. And a degenerate elliptic problem for (6.2) with some smooth convex boundary is solved in [36]. Then in [9], the authors prove the existence and uniqueness of smooth solution of the problem of the expansion of gas into vacuum.

7. Axisymmetric solutions of Euler equations

The Euler equations subject to the initial data

$$(\rho, u, v)(t=0, x, y) = (\rho_0, u_0\cos\theta + v_0\sin\theta, u_0\sin\theta - v_0\cos\theta) \tag{7.1}$$

was investigated in [34, 35], where ρ_0, u_0 and v_0 are constants and θ is the polar angle. The problem is transferred into solving the problem

$$\begin{array}{l} (\rho_r, u_r, v_r) = (\frac{\rho}{r}\frac{v^2 - u(r-u)}{p'(\rho) - (u-r)^2}, \frac{1}{r}\frac{v^2(r-u) - up'(\rho)}{p'(\rho) - (u-r)^2}, \frac{uv}{r(r-u)}), \\ \lim_{r\to+\infty}(\rho, u, v) = (\rho_0, u_0\cos\theta + v_0\sin\theta, u_0\sin\theta - v_0\cos\theta). \end{array} \tag{7.2}$$

This is a problem to seek a trajectory connecting the infinity to the origin. The results disclose many interesting patterns, such as spirals, shock waves, expansion waves and the vacuum. The detailed proof can be found in [37]

References

[1] R. K. Agarwal and D. W. Halt , A modified CUSP scheme in wave/particle split form for unstructured grid Euler flow, *Frontiers of Computational Fluid Dynamics,* Eds., D. A. Caughey and M. M. Hafez, 155-163, 1994.

[2] F. Bouchut, On zero pressure gas dynamics, *Advances in kinetic theory and computing, Series on Advances in Mathematics for Applied Sciences,* World Scientific, **Vol. 22**: 171–190, 1994.

[3] Tung Chang, Guiqiang Chen and Shuili Yang, On the 2-D Riemann problem for the compressible Euler equations, I. Interaction of shocks and rarefaction waves, *Discrete continuous dynamical systems*, **Vol. 1**, 4: 555–584, 1995.

[4] R. Courant and K. O. Friedrichs, Supersonic flow and shock waves, *Interscience*, New York. 1948

[5] Tung Chang (Tong Zhang) and Ling Hsiao (Ling Xiao), The Riemann problem and interaction of waves in gas dynamics, *Pitman Monographs and Surveys in Pure and Applied Mathematics 41, Longman Scientific and Technical, Harlow*, 1989.

[6] Guiqiang Chen, Dening Li and Dechun Tan, Structure of Riemann solutions for 2-Dimensional scalar conservation law, *J. Diff. Equat.*, **127**: 124–147, 1996.

[7] Shaozhong Cheng, Jiequan Li and Tong Zhang, Explicit construction of measure solutions of the Cauchy problem for the transportation equations, *Science in China (Series A),* **Vol. 40**, 12: 1287–1299, 1997.

[8] Shaozhong Cheng, Jiequan Li and Tong Zhang, On two dimensional Riemann problem for zero–pressure gas dynamics, *preprint, Institute of Applied Mathematics, Academia Sinica,* 1998.

[9] Zihuan Dai and Tong Zhang, Global smooth solution of degenerate Coursat problem for a quasilinear mixed type equation, *Preprint, Institute of Mathematics, Academia Sinica,* 1999.

[10] Weinan E, Yu. G. Rykov and Ya. G. Sinai, Generalized variational principles, global weak solutions and behavior with random initial data for systems of conservation laws arising in adhesion particle dynamics, *Comm. Phys. Math.*, 177: 349-380, 1996.

[11] J. Guckenheimer, Shocks and rarefactions in two space dimensions, *Arch. Rational Mech. Anal.*, 59: 281-291, 1975.

[12] P. D. Lax, Hyperbolic system of conservation laws II. *Comm. Pure Appl. Math.* 10: 537-66, 1957.

[13] P. D. Lax and X. D. Liu, Solution of two–dimensional Riemann problem of gas dynamics by positive schemes, *SIAM Sci. Comp.*, **Vol. 19**, No. 2, 319–340, 1998.

[14] Jiequan Li, Generalized Rankine–Hugoniot conditions of delta-shocks and Cauchy problem for transportation equations, *Ph.D. dissertation, Institute of Mathematics, Academic Sinica*, 1996.

[15] Jiequan Li, Numerical solutions of two–dimensional Euler equations in gas dynamics, *preprint, Institute of Applied Mathematics, Academic Sinica* , 1998.

[16] Jiequan Li, Note on the compressible Euler equations with zero–temperature, to appear in *Lett. Appl. Math.*, 1999.

[17] Jiequan Li, On the uniqueness of entropy solutions to zero-pressure gas dynamics, *Preprint*, 1999.

[18] Jiequan Li and Wei Li, The Riemann problem for zero–pressure gas dynamics, to appear in *Progress in Nat. Sci.,*, 2000.

[19] Yinfan Li and Yiming Cao, Second order "large particle" difference method, *Sciences in China,* in Chinese (Series A), **8**: 1024–1035, 1985.

[20] Jiequan Li and Tong Zhang, Generalized Rankine–Hugoniot relations of delta-shocks in solutions of transportation equations, *Nonlinear PDE and Related Areas*, eds. G. Q. Chen, Y. Li, X. P. Zhu and D. Cao, World Scientific, Singapore, 219–232, 1998.

[21] Jiequan Li and Tong Zhang, The dominator of compressible Euler equations, *Proceedings of Differential Equations in Macou*, Eds, T. T. Li, 1998.

[22] Jiequan Li, Tong Zhang, and Shuli Yang, The two–dimensional Riemann problem in gas dynamics, *Pitman Monographs and Surveys in Pure and Applied Mathematics 98*, Longman, 1998.

[23] B. Riemann,Uber de forpflanzung ebener luftwellen von endlicher schwingungsweite, *Abhandl. Koenig. Gesell. Wiss, Goettingen,* **Vol. 8**: 43, 1858.

[24] Carsten W. Schulz-Rinne, Classification of the Riemann problem for Two-dimensional Riemann problem, *SIAM J. Math. Anal.*, 24: 76-88, 1993.

[25] Carsten W. Schulz-Rinne, James P. Collins and Harland M. Glaz, Numerical solutions of the Riemann problem for two-dimensional gas dynamics, *SIAM J. Sci. Compt.*, **Vol. 4**, 6: 1394-1414, 1993.

[26] S. F. Shandarin and Ya. B. Zeldovich, The large-scale structure of the universe: Turbulence, intermittency, structures in a self-gravitating medium, *Rev. Mod. Phys.,* 61: 185-220, 1989.

[27] Wancheng Sheng and Tong Zhang, The Riemann problem for transportation equations in gas dynamics. *Memoirs of AMS*, **Vol. 137**, No. 654, 1997.

[28] D. H. Wagner, The Riemann problem in two space dimensions for a single conservation law, *SIAM J. Math. Anal.* 14: 534-559, 1983.

[29] Hanchun Yang and Jiequan Li, Delta-shocks as limits of vanishing viscosity for multidimensional zero–pressure flow in gas dynamics, *Q. Appl. Math.,* 1998.

[30] Peng Zhang, Jiequan Li and Tong Zhang, On two–dimensional Riemann problem for pressure-gradient equations of Euler system, *Discrete and Continuous Dynamical Systems*, **Vol. 4**, No. 4, 609–634, 1998.

[31] Peng Zhang and Tong Zhang, Generalized characteristic analysis and Guckenheimer structure, *J. Diff. Equat.*, **Vol. 52**, 2: 409–430, 1999.

[32] Tong Zhang and Yuxi Zheng, Two-dimensional Riemann problem for a single conservation law, *Trans. AMS*, **312**, 2: 589-619, 1989.

[33] Tong Zhang and Yuxi Zheng, Conjecture on the structure of solution of the Riemann problem for two–dimensional gas dynamics systems, *SIAM J. Math. Anal.*, **Vol. 21** 3: 593-630, 1990.

[34] Tong Zhang and Yuxi Zheng, Exact spiral solutions of the two–dimensional Euler equations, *Discrete and Continuous Dynamical Systems,***Vol. 3**, 1: 117-133, 1997.

[35] Tong Zhang and Yuxi Zheng, Axisymmetric solutions of the Euler equations for polytropic gases, *Arch Rat Mech Anal,* in press, 1998.

[36] Yuxi Zheng, Existence of solutions to the transonic pressure-gradient equations of the compressible Euler equations in elliptic regions, *Comm. P. D. Es*, 22(11&12): 1849-1868, 1998.

[37] Yuxi Zheng, Two-Dimensional Riemann Problems for Systems of Conservation Laws, *Birkhauser, in the series of Progress in Nonlinear Differential Equations,* 1999.

INSTITUTE OF MATHEMATICS, ACADEMIA SINICA, 100080, P. R. CHINA

AMS/IP Studies in Advanced Mathematics
Volume 20, 2001

Boundedness of Cascade Algorithms in Besov Spaces

S. L. Lee and Qiyu Sun

ABSTRACT. Necessary conditions for refinable functions to belong to Sobolev spaces or Besov spaces are given based on the boundedness of the cascade algorithms.

1. Introduction

For a finitely supported real sequence $h = \{h(j)\}_{j\in\mathbb{Z}^d}$, define an operator T_h on the space of all compactly supported distributions by

$$T_h f = \sum_{j\in\mathbb{Z}^d} h(j) f(2\cdot\ -j). \tag{1.1}$$

We shall call T_h the *refinement operator* for the mask h. Starting with a compactly supported nonzero distribution f_0, the sequence $\{f_n\}_{n\in\mathbb{Z}}$ defined iteratively by

$$f_n = T_h f_{n-1}, \quad n \geq 1, \tag{1.2}$$

is called a *cascade sequence*. The iterative scheme (1.2) is known as the *cascade algorithm* for the mask h. By (1.1) and (1.2) each f_n can be expressed as

$$f_n = \sum_{j\in\mathbb{Z}^d} (S_h^n \delta)(j) f_0(2^n\cdot\ -j), \tag{1.3}$$

where S_h is the *stationary subdivision operator* defined on the space $\ell(\mathbb{Z}^d)$ of all sequences by

$$(S_h b)(i) = \sum_{j\in\mathbb{Z}^d} h(i-2j) b(j), \quad i \in \mathbb{Z}^d.$$

It is well known that if $\sum_{j\in\mathbb{Z}^d} h(j) = 2^d$, which we shall assume throughout the paper, there exists a unique compactly supported distribution ϕ that satisfies the refinement equation

$$\phi = \sum_{j\in\mathbb{Z}^d} h(j)\phi(2\cdot\ -j), \tag{1.4}$$

1991 *Mathematics Subject Classification.* 41A15, 41A30, 42C05, 42C15.

Key words and phrases. Cascade algorithm, Sobolev space, Besov space, Strang-Fix condition.

and $\widehat{\phi}(0) = 1$. Further, $T_h^n f_0$ tends to $\hat{f}_0(0)\phi$ in distributional sense as n tends to infinity for any compactly supported distribution f_0. The compactly supported distribution ϕ that satisfies (1.4) is called a *refinable distribution*.

Here, $\widehat{\phi}$ denotes the Fourier transform of ϕ. The Fourier transform $\widehat{f}$ of an integrable function f is defined by

$$\widehat{f}(\xi) = \int_{\mathbb{R}^d} f(x)e^{-ix\xi}dx,$$

and this is extended in the usual way to compactly supported distributions. Taking Fourier transforms, (1.4) is equivalent to

$$\widehat{\phi}(\xi) = H(2^{-1}\xi)\widehat{\phi}(2^{-1}\xi), \quad \xi \in \mathbb{R}^d \tag{1.5}$$

where

$$H(\xi) = 2^{-d}\sum_{j\in\mathbb{Z}^d} h(j)e^{-ij\xi}.$$

The polynomial H is called the *symbol* of the refinement equation (1.4). Iterating (1.5) leads to

$$\widehat{\phi}(\xi) = \prod_{j=1}^{n} H(2^{-j}\xi)\widehat{\phi}(2^{-n}\xi), \quad n = 0, 1, \ldots. \tag{1.6}$$

Similarly, (1.2) is equivalent to

$$\widehat{f_n}(\xi) == \prod_{j=1}^{n} H(2^{-j}\xi)\widehat{f_0}(2^{-n}\xi), \quad n = 0, 1, \ldots. \tag{1.7}$$

The limiting behaviour of a cascade algorithm for a given mask and that of the corresponding stationary subdivision scheme have been extensively studied, very often under some assumptions on the regularity of the corresponding refinable function ϕ. Subdivision schemes have been studied in [**1**], [**3**], [**6**], and [**7**] in connection with geometric modelling. While interest in the convergence of subdivision schemes is mainly confined to the uniform norm, Jia [**7**] gave a characterization of L_p convergence in terms of joint spectral radius of the corresponding two-slanted matrices associated with the mask h. On the other hand, convergence of cascade algorithms in $L_2(\mathbb{R}^d)$ was independently studied in [**2**], [**12**] and [**9**] in connection with wavelet construction. Shen [**11**] extended the results of [**9**] to the matrix setting. Further extensions to nonstationary cascade algorithms can be found in [**4**] and [**5**]. A characterization of convergence of cascade algorithms in Sobolev space W_2^k is given in [**8**].

It is easy to see that if $\hat{f}_0(0) = 1$ and the cascade sequence $\{f_n\}_{n\in\mathbb{Z}}$ is bounded in $L_p(\mathbb{R}^d), 1 < p < \infty$, then it converges weakly to the solution of the corresponding refinement equation. In particular, the scaling function belongs to $L_p(\mathbb{R}^d)$. The object of this paper is to extend this result, and give necessary conditions on the boundedness of the cascade sequence in order that the refinable distribution belongs to a Sobolev space or Besov space.

In the next section we give the preliminaries and state the main theorem. The proof is given in Section 3. The last section is devoted to remarks and more corollaries.

2. Preliminaries and Main Results

We shall deal with the Sobolev space, W_p^α which is defined for all $-\infty < \alpha < \infty$ and $1 \le p < \infty$ as the space of all tempered distributions f for which

$$\|f\|_{W_p^\alpha} := \big\|(\widehat{\Phi}\widehat{f})^\vee\big\|_p + \Big\|\Big(\sum_{n\ge 0} 2^{2n\alpha}\big|(\widehat{\Psi}(2^{-n}\cdot)\widehat{f})^\vee\big|^2\Big)^{1/2}\Big\|_p < \infty, \tag{2.1}$$

where $\|\cdot\|_p$ denotes the L_p-norm, $f^\vee$ denotes the inverse Fourier transform of f, and Ψ and Φ are Schwartz functions that satisfy the following conditions:

(2.2) $\widehat{\Phi}(\xi) \ge 0, \quad \widehat{\Psi}(\xi) \ge 0, \quad \xi \in \mathbb{R}^d,$

(2.3) $\widehat{\Psi}$ is supported in $(-4\pi, 4\pi)^d \backslash [-\pi, \pi]^d$ and $\widehat{\Phi}$ is supported in $(-2\pi, 2\pi)^d$,

(2.4) $\widehat{\Phi}(\xi) + \sum_{n\ge 0} \widehat{\Psi}(2^{-n}\xi) = 1, \quad \xi \in \mathbb{R}^d.$

The Besov space B_p^α is similarly defined as the space of all tempered distributions f that satisfies

$$\|f\|_{B_p^\alpha} := \big\|(\widehat{\Phi}\widehat{f})^\vee\big\|_p + \sup_{n\ge 0} 2^{n\alpha}\big\|(\widehat{\Psi}(2^{-n}\cdot)\widehat{f})^\vee\big\|_p < \infty. \tag{2.5}$$

The topologies of W_p^α and B_p^α are induced by the norms $\|\cdot\|_{W_p^\alpha}$ and $\|\cdot\|_{B_p^\alpha}$ respectively. For convenient reference we state some basic properties of these function spaces. More properties and details can be found in [**13**].

Suppose that $-\infty < \alpha < \infty$ and $1 \le p < \infty$. Then

$$W_p^\alpha \subset B_p^\alpha, \tag{2.6}$$

$$B_p^{\alpha+\delta} \subset W_p^\alpha \text{ for any } \delta > 0, \text{ and} \tag{2.7}$$

$$W_p^0 \subset L_p \subset B_p^0. \tag{2.8}$$

Further, for any compactly supported Schwartz function g with $0 \notin \operatorname{supp} g$, there exists a constant C_g such that

$$\|(g(2^{-n}\cdot)\widehat{f})^\vee\|_p \le C_g 2^{-n\alpha}\|f\|_{B_p^\alpha} \text{ for all } n \ge 1 \text{ and } f \in B_p^\alpha, \tag{2.9}$$

and for any compact set K, there exists a constant C_K such that

$$\|f\|_{B_1^\alpha} \le C_K\|f\|_{B_p^\alpha} \text{ for all } f \in B_p^\alpha \text{ with } \operatorname{supp} f \subset K. \tag{2.10}$$

If k is a nonnegative integer, then for $1 < p < \infty$,

$$W_p^k = \{f:\ D^\nu f \in L_p,\ \forall\, 0 \le |\nu| \le k,\ \nu \in \mathbb{Z}_+^d\}.$$

Here, $\mathbb{Z}_+^d$ denotes the set of all vectors in $\mathbb{Z}^d$ with nonnegative integer components. Note that for $1 < p < \infty$, $W_p^0 = L_p(\mathbb{R}^d)$, but $W_1^0 \ne L_1(\mathbb{R}^d)$.

A nonzero compactly supported distribution f_0 is said to satisfy *Condition F* if for any $\xi \in \mathbb{R}^d\backslash\{0\}$, there exists $k_\xi \in \mathbb{Z}^d$ such that $\widehat{f_0}(t\xi + 2k_\xi\pi) \not\equiv 0$ for $t \in \mathbb{R}$.

We shall prove the following theorem.

THEOREM 1. *Let $1 \le p < \infty$, $-\infty < \alpha, \beta < \infty$, T_h be the refinement operator with mask h, and ϕ be the corresponding refinable distribution. Suppose that $f_0 \in B_p^\alpha$ satisfies Condition F. If $2^{n\beta}\|T_h^n f_0\|_{B_p^\alpha}$ is uniformly bounded for $n \ge 0$, then $\phi \in B_p^{\alpha+\beta}$, and for any $k \in \mathbb{Z}^d\backslash\{0\}$ there exists a constant C_k such that*

$$|\widehat{\phi}(2k\pi + \xi)| \le C_k|\xi|^{\alpha+\beta} \quad \forall\ |\xi| \le 1. \tag{2.11}$$

The second result can be viewed as an extension of the Strang-Fix Condition. Suppose that $\alpha + \beta > 0$. Define κ as the smallest integer larger than $\alpha + \beta$. Then we have the following result as a corollary of Theorem 1.

COROLLARY 2. *Let p, α, β, T_h and ϕ be as in Theorem 1, and suppose that $\alpha+\beta > 0$, and $f_0 \in B_p^\alpha$ satisfies Condition F. If $2^{n\beta}\|T_h^n f_0\|_{B_p^\alpha}$ is uniformly bounded for $n \geq 0$, then $\phi \in B_p^{\alpha+\beta}$ and satisfies the Strang-Fix Condition of order κ, i.e., $\widehat{\phi}(0) = 1$ and $D^\gamma \widehat{\phi}(2\pi k) = 0$ for all $\gamma \in \mathbb{Z}_+$ with $|\gamma| < \kappa$ and for all $k \in \mathbb{Z}^d \setminus \{0\}$.*

Also by (2.7), $B_p^{\alpha+\beta} \subset W_p^{\alpha+\beta-\delta}$, for any $\delta > 0$. Therefore, we have

COROLLARY 3. *Let p, α, β, T_h and ϕ be as in Theorem 1, and suppose that $f_0 \in B_p^\alpha$ satisfies Condition F. If $2^{n\beta}\|T_h^n f_0\|_{B_p^\alpha}$ is uniformly bounded for $n \geq 0$, then $\phi \in W_p^{\alpha+\beta-\delta}$ for any $\delta > 0$.*

3. Proof of Main Theorem

In this section, we give a proof of Theorem 1.

PROOF. Since f_0 is a compactly supported distribution, $\widehat{f_0}(t\xi + 2k\pi)$ is an analytic function of $t \in \mathbb{R}$ for any $\xi \in \mathbb{R}^d \setminus \{0\}$ and $k \in \mathbb{Z}^d$. The assumption on f_0 implies that for any $\xi \in \mathbb{R}^d \setminus \{0\}$ there exists a $k_\xi \in \mathbb{Z}^d$, such that $\widehat{f_0}(t\xi + 2k_\xi\pi) \not\equiv 0$ for $t \in \mathbb{R}$. Take any $\xi \in [-4\pi, 4\pi]^d \setminus (-\pi/2, \pi/2)^d$, and choose $k_\xi \in \mathbb{Z}^d$ such that $\widehat{f_0}(t\xi + 2k_\xi\pi) \not\equiv 0$ for $t \in \mathbb{R}$. Since $\widehat{f_0}(t\xi + 2k_\xi\pi)$ can have at most an isolated zero at 0, it is nonzero for all sufficiently small $t \neq 0$. Hence there exists $j_\xi \in \mathbb{Z}$

$$2^{j_\xi}\xi + 2k_\xi\pi \neq 0 \quad \text{and} \quad \widehat{f_0}(2^{j_\xi}\xi + 2k_\xi\pi) \neq 0. \tag{3.1}$$

Let Ψ be the Schwartz function in the definition of the space B_p^α. By (3.1), the compactness of the set $[-4\pi, 4\pi]^d \setminus (-\pi/2, \pi/2)^d$ and the continuity of $\widehat{f_0}(\xi)$, there exists a finite number of compactly supported Schwartz functions $h_\ell \geq 0$ and $g_\ell \geq 0$, integers $j_\ell \in \mathbb{Z}$, $k_\ell \in \mathbb{Z}^d$ and a constant C_0 such that

$$\sum_\ell h_\ell = 1 \text{ on } \operatorname{supp}(\widehat{\Psi}), \tag{3.2}$$

$$g_\ell = 1 \text{ on } \operatorname{supp}(h_\ell), \tag{3.3}$$

with supports of h_ℓ and g_ℓ in $(-4\pi, 4\pi)^d \setminus [-\pi/2, \pi/2]^d$, and

$$\min(|\xi + 2^{-j_\ell+1}k_\ell\pi|, |\widehat{f_0}(2^{j_\ell}\xi + 2k_\ell\pi)|) \geq C_0 \text{ on } \operatorname{supp}(g_\ell). \tag{3.4}$$

Set

$$m_\ell(\xi) = \widehat{\Psi}(\xi) h_\ell(\xi) \widehat{\phi}(2^{j_\ell}\xi) / \widehat{f_0}(2^{j_\ell}\xi + 2k_\ell\pi). \tag{3.5}$$

Then m_ℓ are compactly supported Schwartz functions and $m_\ell^\vee$ are integrable by (3.4).

By (1.7), (1.6), (3.2), (3.3), (3.4) and (3.5), for sufficiently large n,

$$
\begin{aligned}
&2^{(\alpha+\beta)n}\|(\widehat{\Psi}(2^{-n}\cdot\,)\widehat{\phi})^{\vee}\|_p \\
&\quad\le\; C\,2^{(\alpha+\beta)n}\sum_{\ell}\Big\|\Big(\widehat{\Psi}(2^{-n}\cdot\,)h_\ell(2^{-n}\cdot\,)\widehat{\phi}\Big)^{\vee}\Big\|_p \\
&\quad=\; C\,2^{(\alpha+\beta)n}\sum_{\ell}\Big\|\Big(\widehat{\Psi}(2^{-n}\cdot\,)h_\ell(2^{-n}\cdot\,)\widehat{\phi}(2^{-n+j_\ell}\cdot\,)\prod_{j=1}^{n-j_\ell}H(2^{-j}\cdot\,)\Big)^{\vee}\Big\|_p \\
&\quad=\; C\,2^{(\alpha+\beta)n}\sum_{\ell}\Big\|\Big(m_\ell(2^{-n}\cdot\,)g_\ell(2^{-n}\cdot\,)(T^{n-j_\ell}f_0)^{\wedge}(\cdot\;+2^{n-j_\ell+1}k_\ell\pi)\Big)^{\vee}\Big\|_p \\
&\quad\le\; C\,2^{(\alpha+\beta)n}\sum_{\ell}\Big\|m_\ell^{\vee}\Big\|_1\times\Big\|\Big(g_\ell(2^{-n}\cdot\,)(T^{n-j_\ell}f_0)^{\wedge}(\cdot\;+2^{n-j_\ell+1}k_\ell\pi)\Big)^{\vee}\Big\|_p \\
&\quad\le\; C\sum_{\ell}2^{(\beta+\alpha)(n-j_\ell)}\Big\|\Big(g_\ell(2^{-n}\cdot\;-2^{-j_\ell+1}k_\ell\pi)(T^{n-j_\ell}f_0)^{\wedge}\Big)^{\vee}\Big\|_p.
\end{aligned}
$$

This together with (3.4) and (2.9) lead to

$$
2^{(\alpha+\beta)n}\|(\widehat{\Psi}(2^{-n}\cdot\,)\widehat{\phi})^{\vee}\|_p\le C\sum_{\ell}2^{\beta(n-j_\ell)}\|T^{n-j_\ell}f_0\|_{B_p^{\alpha}} \tag{3.6}
$$

which is uniformly bounded, since the cascade sequence $\{T_h^n f_0\}_{n\ge 0}$ is assumed to be uniformly bounded on B_p^{α}. It follows from (2.5) that $\phi\in B_p^{\alpha+\beta}$.

We now proceed to prove the second part of Theorem 1. Since $\phi\in B_p^{\alpha+\beta}$ and is compactly supported, by (2.10) we have $\phi\in B_1^{\alpha+\beta}$. Take any $k\in\mathbb{Z}^d\backslash\{0\}$. Since

$$
\sum_{j\ge 0}\widehat{\Psi}(2^{-j}\xi)=1\quad\text{for}\quad \xi\in\mathbb{R}^d\backslash[-2\pi,2\pi]^d,
$$

there is a nonnegative integer k_0 such that $\widehat{\Psi}(2^{-k_0+1}k\pi)\neq 0$. Recall that $\widehat{\phi}$ is continuous and $\widehat{\phi}(0)=1$. Therefore we can choose $\delta>0$ such that both $|\widehat{\Psi}(2^{-k_0+1}k\pi+\eta)|$ and $|\widehat{\phi}(\eta)|$ are bounded away from zero for $|\eta|\le\delta$. For sufficiently large n, this together with (1.6) give

$$
\begin{aligned}
&2^{n(\alpha+\beta)}\sup_{|\eta|\le\delta}|\widehat{\phi}(2k\pi+2^{-n}\eta)| \\
&\quad\le\; C_k\,2^{n(\alpha+\beta)}\sup_{|\eta|\le\delta}\Big|\widehat{\Psi}(2^{-k_0+1}k\pi+2^{-n-k_0}\eta)\widehat{\phi}(2k\pi+2^{-n}\eta)\widehat{\phi}(\eta)/\widehat{\phi}(2^{-n}\eta)\Big| \\
&\quad\le\; C_k\,2^{n(\alpha+\beta)}\sup_{\xi\in\mathbb{R}^d}\Big|\widehat{\Psi}(2^{-n-k_0}\xi)\prod_{j=1}^{n}H(2^{-j}\xi)\widehat{\phi}(2^{-n}\xi)\Big| \\
&\quad=\; C_k\,2^{n(\alpha+\beta)}\sup_{\xi\in\mathbb{R}^d}\Big|\widehat{\Psi}(2^{-n-k_0}\xi)\widehat{\phi}(\xi)\Big| \\
&\quad\le\; C_k\,2^{n(\alpha+\beta)}\Big\|\Big(\widehat{\Psi}(2^{-n-k_0}\cdot\,)\widehat{\phi}\Big)^{\vee}\Big\|_1.
\end{aligned}
$$

Applying (2.9) to the last inequality, and taking into account that $\phi\in B_1^{\alpha+\beta}$, we have

$$
2^{n(\alpha+\beta)}\sup_{|\eta|\le\delta}|\widehat{\phi}(2k\pi+2^{-n}\eta)|\le C_k\|\phi\|_{B_1^{\alpha+\beta}}\le C_k, \tag{3.7}
$$

for all sufficiently large n. Take ξ with sufficiently small $|\xi|$. There exists an integer $n \in \mathbb{Z}_+$ such that $2^{-n}|\xi| \le \delta < 2^{-n+1}|\xi|$. By (3.7)

$$|\widehat{\phi}(2k\pi + \xi)| \le C_k 2^{-n(\alpha+\beta)} \le C_k |\xi|^{\alpha+\beta},$$

which proves (2.11). □

A similar argument as in the proof of the second part of Theorem 1 gives

THEOREM 4. *Let p, α, β, T_h and ϕ be as in Theorem 1.1. If $2^{n\beta}\|T_h^n f_0\|_{B_p^\alpha}$ is uniformly bounded for $n \ge 0$, then for any $k \in \mathbb{Z}^d \backslash \{0\}$ there exists a constant C_k such that*

$$|\widehat{f_0}(2k\pi + \xi)| \le C_k |\xi|^{\alpha+\beta} \quad \forall\ |\xi| \le 1. \tag{3.8}$$

We remark that in Theorem 4 we do not assume that f_0 satisfies Condition F.

4. Remarks and Corollaries

In this section, we shall give some remarks about Condition F and some consequences of the main theorem. In particular, the connection with some known results on the convergence of cascade algorithm is given.

Remark 1. In one dimension ($d = 1$), $\widehat{f_0}$ is a nonzero analytic function if f_0 is a nontrivial compactly supported distribution. Thus for any $\xi \in \mathbb{R}\backslash\{0\}$, $\widehat{f_0}(t\xi) \not\equiv 0$ on $t \in \mathbb{R}$. Therefore any nonzero initial function f_0 *always* satisfies Condition F in one dimension. Hence we have

COROLLARY 5. *Let p, α, β, T_h and ϕ be as in Theorem 1 with $d = 1$, and assume that f_0 is a nonzero compactly supported distribution in $B_p^\alpha(\mathbb{R})$. If $2^{n\beta}\|T_h^n f_0\|_{B_p^\alpha}$ is uniformly bounded for $n \ge 1$, then $\phi \in B_p^{\alpha+\beta}$ and there exists a constant C_k for any $k \in \mathbb{Z}^d\backslash\{0\}$ such that*

$$|\widehat{f_0}(2k\pi + \xi)| + |\widehat{\phi}(2k\pi + \xi)| \le C_k |\xi|^{\alpha+\beta} \quad \forall\ |\xi| \le 1.$$

Remark 2. Condition F imposed on f_0 in Theorem 1 cannot be removed in higher dimensions. For instance, for dimension $d = 2$, let $\alpha = 0$,

$$\phi(x, y) = \chi_{[0,1]}(x)\delta(y)$$

be the tensor product of the characteristic function on the unit interval $[0, 1]$ and the delta distribution at zero, and let

$$f_0(x, y) = (\chi_{[0,1]}(x) - \chi_{[0,1]}(x-1))\chi_{[0,1]}(y).$$

Then ϕ satisfies the refinement equation

$$\phi = 2\phi(2\cdot\ ) + 2\phi(2\cdot\ -(1, 0)).$$

and $\phi \notin B_1^\beta$ for any $\beta > 0$, since $\phi \notin L_1(\mathbb{R}^2)$. Further, ϕ does not satisfies (2.11) with $\beta > 0$ because $\widehat{\phi}(0, 2k\pi) = 1$ for all $k \in \mathbb{Z}$. Also f_0 does not satisfy condition F since $\widehat{f_0}(2k\pi, \eta) = 0$ for all $k \in \mathbb{Z}$ and $\eta \in \mathbb{R}$. A straightforward computation gives

$$f_n(x, y) = 2^n(\chi_{[0,2^{-n}]}(x) - \chi_{[0,2^{-n}]}(x-1))\chi_{[0,2^{-n}]}(y)$$

and

$$2^{n\beta}\|f_n\|_{B_1^0} \le C2^{n\beta}\|f_n\|_1 = C2^{-n+1+n\beta}$$

which tends to 0 as n tends to infinity for $0 < \beta < 1$.

Remark 3. If f_0 satisfies Condition F, the uniform boundedness of $2^{n\alpha}\|f_n\|_{B_p^0}$ does not imply $\phi \in W_p^\alpha$ and hence does not imply $\phi \in B_p^{\alpha+\delta}$ for all $\delta > 0$. Here is an example. Let $\phi = \chi_{[0,1]}$ be the characteristic function on the unit interval $[0,1]$ and $f_0 := \phi - \phi(\cdot\ - 1)$. Then $\phi \notin W_p^{1/p}$. A straightforward computation gives

$$f_n := T_h^n f_0 = \phi - \phi(\cdot\ - 2^{-n}) = \chi_{[0,2^{-n}]} - \chi_{[1,1+2^{-n}]}, \quad n \geq 1,$$

and

$$\|f_n\|_{B_p^0} \leq C\|f_n\|_p \leq C2^{-n/p}.$$

Hence $2^{n/p}\|f_n\|_{B_p^0}$ is uniformly bounded for $n \geq 0$.

Remark 4. It is shown in [**10**] that if $\lim_{n\to\infty} 2^{n\alpha}\|T_h^n f_0\|_p = 0$, there exists a constant $\delta > 0$ such that $2^{n(\alpha+\delta)}\|T_h^n f_0\|_{B_p^0}$ is bounded for $n \geq 0$. The following Corollary then follows from Theorems 1 and 4 and the relation (2.7).

COROLLARY 6. *Let p, β, T_h and ϕ be as in Theorem 1, and suppose that $f_0 \in L_p$ satisfies Condition F. If $\lim_{n\to\infty} 2^{n\beta}\|T_h^n f_0\|_p = 0$, then $\phi \in W_p^\beta$ and*

$$\lim_{\xi\to 0} |\xi|^{-\beta}\Big(|\widehat{f_0}(2k\pi + \xi)| + |\widehat{\phi}(2k\pi + \xi)|\Big) = 0, \quad \forall\ k \in \mathbb{Z}^d\backslash\{0\}.$$

References

[1] A. Cavaretta, W. Dahmen and C. A. Micchelli, *Stationary subdivision,* Memoir Amer. Math. Soc., 453(1991), 1-186.

[2] I. Daubechies and J. C. Lagaria, *Two-scale difference equations I. Existence and global regularity of solutions,* SIAM J. Math. Anal., 22(1991), 1388 - 1410.

[3] N. Dyn and D. Levin, *Stationary and nonstationary subdivision schemes,* in "Mathematical Methods on Computer Aided Geometric Design II" (T. Lyche and L. L. Schumaker, eds.), Birkhauser, Basel, 1992, pp. 209 - 216.

[4] T. N. T. Goodman and S. L. Lee, *Convergence of cascade algorithms,* In " Mathematical Methods for Curves and Surfaces II", (M. Dahlen, T. Lyche and L. L. Schumaker eds.), Vanderbilt University Press, Nashville, 1998, pp. 191-212.

[5] T. N. T. Goodman and S. L. Lee, *Convergence of nonstationary cascade algorithms,* Preprint, National University of Singapore, 1997.

[6] T. N. T. Goodman, C. A. Micchelli and J. Ward, *Spectral radius formulas for subdivision operators,* In "Recent Advances in Wavelet Analysis", (L. L. Schumaker and G. Webb eds.), Academic Press, New York, 1995, pp. 335-360.

[7] R. Q. Jia, *Subdivision schemes in L_p spaces,* Adv. Comput. Math., 3(1995), 309-341.

[8] R. Q. Jia, Q. Jiang and S. L. Lee, *Convergence of cascade algorithm in Sobolev spaces and integrals of wavelets,* Preprint, National University of Singapore, 1998.

[9] W. Lawton, S. L. Lee and Z. Shen, *Convergence of multidimensional cascade algorithms,* Numer. Math., 78(1998), 427-438.

[10] S. L. Lee and Q. Sun, *Stability lemma, cascade algorithm and smoothness of refinable distributions,* Preprint, National University of Singapore, 1998.

[11] Z. Shen, *Refinable function vectors,* SIAM J. Math. Anal., 29(1998), 239-250.

[12] G. Strang, *Eigenvalues of $(\downarrow 2)H$ and convergence of cascade algorithm,* IEEE Trans. Sig Proc. 44(1996), 233-238.

[13] H. Triebel, *Theory of Function Spaces,* Monographs in Math. No. 78, Birkhäuser, 1983.

DEPARTMENT OF MATHEMATICS, NATIONAL UNIVERSITY OF SINGAPORE, 10 KENT RIDGE CRESCENT, SINGAPORE 119260

AMS/IP Studies in Advanced Mathematics
Volume 20, 2001

Circulant Preconditioners for Ill-Conditioned Hermitian Toeplitz Matrices

Raymond H. Chan, Ming-ham Yip, Chi-wah Leung, and Michael K. Ng

ABSTRACT. In this paper, we propose a new family of circulant preconditioners for solving ill-conditioned Hermitian Toeplitz systems $A\mathbf{x} = \mathbf{b}$. The eigenvalues of the preconditioners are given by the convolution products of the generating function f of A with some summation kernels. When f is a nonnegative 2π-periodic continuous function defined on $[-\pi, \pi]$ with a zero of order $2p$, we show that the circulant preconditioners are positive definite and the spectrum of the preconditioned matrix is uniformly bounded in $[a, b]$ with at most $2p+2$ outliers where $0 < a \leq b < \infty$. Hence the linear system can be solved by the preconditioned conjugate gradient method efficiently. We emphasize that the construction of the circulant preconditioner does not require the explicit knowledge of the generating function. Numerical results are included.

1. Introduction

An n-by-n matrix $A_n = [a_{ij}]$ is said to be Toeplitz if $a_{ij} = a_{i-j}$. Toeplitz systems of the form $A_n\mathbf{x} = \mathbf{b}$ occur in a variety of applications in mathematics and engineering [**7**]. There are a number of specialized fast direct methods for solving Toeplitz systems that require $O(n^2)$ operations, see for instance [**17**]. Faster methods requiring $O(n \log^2 n)$ operations have also been developed, see [**4, 3**].

Strang in [**16**] proposed using the preconditioned conjugate gradient method with circulant matrices as preconditioners for solving Toeplitz systems. The number of operations per iteration is of order $O(n \log n)$ as circulant systems can be solved efficiently by the Fast Fourier Transform (FFT). Several successful circulant preconditoiners have been introduced and analyzed; see for instance [**10, 5**]. In these papers, the Toeplitz matrix A_n is assumed to be generated by a generating function f, i.e., the diagonals of A_n are given by the Fourier coefficients of f. It has been shown that if f is a positive function in the Wiener class (i.e., the Fourier coefficients of f are absolutely summable), then these circulant preconditioned systems converge superlinearly [**5**]. However, if f has zeros, the corresponding Toeplitz system will be ill-conditioned. Tyrtyshnikov [**18**] has proved that the Strang [**16**] and T. Chan [**10**] preconditioners will fail in this case.

1991 *Mathematics Subject Classification.* Primary 65F10, 65F15, 65T10.

Raymond Chan was supported in part by Hong Kong Research Grants Council Grant No. CUHK 4207/97 and CUHK DAG Grant No. 2060143.

Michael Ng was supported in part by Hong Kong University CRCG grant no. 10201939.

To tackle this problem, non-circulant type preconditioners have been proposed, see [**6, 8, 2, 15, 13**]. The basic idea behind these preconditioners is to find a function g that matches the zeros of f. Then the preconditioner is constructed based on the function g. These approaches work when the generating function f is given explicitly. However, in some applications, we are given only a Toeplitz system where the underlying generating function is unknown and therefore these preconditioners cannot be constructed. In contrast, circulant preconditioners only use the entries of the given Toeplitz matrix. Recently, Di Benedetto [**1**] proved that the condition numbers of the preconditioned matrices by sine transform preconditioners are uniformly bounded. However, the preconditioners themselves may be singular or indefinite in general. Our aim in this paper is to propose a family of positive definite circulant preconditioners that do not require the explicit knowledge of f and that work for ill-conditioned systems.

Chan and Yeung [**9**] showed that circulant preconditioners can be derived in general by considering convolution products of the generating function f with some kernels. For instance, the Dirichlet kernel $\mathcal{D}_{\lfloor \frac{n}{2} \rfloor}$ gives to the Strang preconditioner. They proved that for any positive 2π-periodic continuous function f, if $\mathcal{C}_n$ is a kernel such that $\mathcal{C}_n * f$ tends to f uniformly on $[-\pi, \pi]$, then the corresponding preconditioned system $C_n^{-1} A_n$ will have clustered spectrum. This result turns the problem of finding a good preconditioner to the problem of approximating f with $\mathcal{C}_n * f$. We note that $\mathcal{D}_{\lfloor \frac{n}{2} \rfloor} * f$ depends solely on the first $n-1$ Fourier coefficients $\{a_j\}_{|j|<n}$ of f. Thus the Strang preconditioner does not require explicitly knowledge of f.

Our idea here is to construct positive definite preconditioners by approximating f by $\mathcal{K}_{n,r} * f$ (defined in §2) that matches the zeros of f and depends only on $\{a_j\}_{|j|<n}$. In this proceeding paper, we mainly consider nonnegative 2π-periodic continuous function with a single zero. We will show that if $f(\theta)$ has a zero of order $2p$ at $\theta = \gamma$, then $\mathcal{K}_{n,2r} * f$ matches the zero of f for $r > p$. We will also show that the spectrum of the preconditioned system is uniformly bounded with at most $2p + 2$ outliers. The general case for generating functions with multiple zeros will be considered in a future paper.

The outline of the paper is as follows. In §2, we give an efficient method for computing the eigenvalues of the preconditioners. In §3 we show that $\mathcal{K}_{n,r} * f$ matches the zero of f. We then estimate in §4, the spectrum of the preconditioned matrices. Numerical results are given in §5 to illustrate the effectiveness of our preconditioner in solving ill-conditioned Toeplitz systems.

REMARK: This paper was first presented at the International Congress of Chinese Mathematicians in Beijing in December, 1998. Recently, similar results have also been obtained independently by Potts and Steidl [**14**].

2. Construction of Circulant Preconditioners

Denote the space of all 2π-periodic continuous real-valued functions by $\mathbf{C}_{2\pi}$. The Fourier coefficients of a function f in $\mathbf{C}_{2\pi}$ are given by

$$a_k = \frac{1}{2\pi} \int_{-\pi}^{\pi} f(\theta) e^{-ik\theta} \, d\theta, \quad k = 0, \pm 1, \pm 2, \cdots .$$

The Hermitian Toeplitz matrix generated by f is then denoted by $A_n[f] = [a_{i-j}]$. We remark that $A_n[f]$ is positive definite when f is a nonnegative function that takes positive values on a nondegenerate interval, see [**6**]. Suppose θ_0 is a zero of f. We say that θ_0 is of *order* p if p is the smallest positive integer such that $f^{(p+1)}(\theta)$ is continuous in a neighborhood of θ_0 and $f^{(p)}(\theta_0) \neq 0$. We note that when f is nonnegative, $f^{(p)}(\theta_0) > 0$ and p must be even.

For the Toeplitz systems considered in this paper, we will assume that f is a nonnegative function with a single zero. The systems will be solved by using preconditioned conjugate gradient (PCG) methods with circulant preconditioners. It is well known that circulant matrices can be diagonalized by the Fourier matrix F_n, see [**7**]. Therefore, a circulant matrix is uniquely determined by its set of eigenvalues. For a given function f, we define the circulant preconditioner $C_n[f]$ to be an n-by-n matrix with its j-th eigenvalue given by

$$\lambda_j(C_n[f]) = f(\frac{2\pi j}{n}), \quad 0 \le j < n. \tag{2.1}$$

We note that $C_n[f] = F_n^* \mathrm{diag}(\lambda_0, \lambda_1, \cdots, \lambda_{n-1}) F_n$, see [**7**].

In the following, we will use the kernel function

$$\mathcal{K}_{n,2r}(\theta) = \frac{k_{n,2r}}{n^{r-1}} \left(\frac{\sin(\frac{n\theta}{2})}{\sin(\frac{\theta}{2})} \right)^{2r}, \quad r = 1, 2, \ldots \tag{2.2}$$

to construct our circulant preconditioners. Here $k_{n,2r} (= O(1))$ is a normalization constant such that the integral of $\mathcal{K}_{n,2r}$ over $[-\pi, \pi]$ is 1. We note that $\mathcal{K}_{n,2}$ is the Fejér kernel and $\mathcal{K}_{n,4}$ is the Jackson kernel [**12**, p.57].

For any N, the Fejér kernel $\mathcal{K}_{N,2}$ can be expressed as

$$\mathcal{K}_{N,2}(\theta) = \sum_{k=-N+1}^{N-1} b_k^{(N,2)} e^{ik\theta},$$

where

$$b_k^{(N,2)} = \frac{N - |k|}{2\pi N}, \quad k = 0, \pm 1, \pm 2, \cdots, \pm(N-1).$$

Hence by (2.2),

$$\mathcal{K}_{N,2r}(\theta) = \sum_{k=-r(N-1)}^{r(N-1)} b_k^{(N,2r)} e^{ik\theta} \tag{2.3}$$

where the coefficients $b_k^{(N,2r)}$ can be obtained by convolving the vector $(b_k^{(N,2)})_{k=0}^{N-1}$ with itself for $r-1$ times using FFTs. Hence the cost of constructing the coefficients is of $O(rN \log N)$ operations.

The convolution product of $\mathcal{K}_{n,2r}$ with a function f is defined as

$$(\mathcal{K}_{n,2r} * f)(\theta) \equiv \int_{-\pi}^{\pi} \mathcal{K}_{n,2r}(t) f(t - \theta)\, dt. \tag{2.4}$$

Since $f(\theta) = \sum_{k=-\infty}^{\infty} a_k e^{ik\theta}$, by (2.3), the convolution product in (2.4) becomes

$$(\mathcal{K}_{N,2r} * f)(\theta) = 2\pi \sum_{k=-r(N-1)}^{r(N-1)} a_k b_k e^{ik\theta}.$$

Hence our circulant preconditioners $C_n[\mathcal{K}_{N,2r} * f]$ can be constructed by using only the first $r(N-1)+1$ coefficients of f. Thus when we are given a Toeplitz system $A_n \mathbf{x} = \mathbf{b}$, we can choose the largest N such that $r(N-1) < n$. Then $\mathcal{K}_{N,2r} * f$

only requires a_k for $|k| < n$. This means that $C_n[\mathcal{K}_{N,2r} * f]$ can be constructed efficiently without the explicit knowledge of the underlying generating function f.

3. Properties of the Kernel $\mathcal{K}_{n,2r}$

In this section, we study some properties of $\mathcal{K}_{n,2r}$ in order to see how good the approximation of f by $\mathcal{K}_{n,2r} * f$ will be. We first note the following Lemma.

LEMMA 3.1. [**12**, p.55] *Let p and r be positive integers with $r > p$. Then*

$$\left(\mathcal{K}_{n,2r} * \theta^{2p}\right)(0) = O\left(\frac{1}{n^{2p}}\right).$$

Next we simplify $(\mathcal{K}_{n,2r} * \theta_{2\pi}^{2p})(\phi)$ where $\theta_{2\pi}$ is the periodic extension of θ on $[-\pi, \pi]$.

LEMMA 3.2. *Let p be a positive integer. Then there exist positive α_i and β_i for $i = 1, 2, 3$ such that*

$$\alpha_1 \le \frac{\left(\mathcal{K}_{n,2r} * \theta_{2\pi}^{2p}\right)(\phi)}{\left[\mathcal{K}_{n,2r} * \theta^{2p}(\theta+2\pi)^{2p}\right](\phi)} \le \beta_1, \quad \forall \phi \in [-\pi, -\pi/2],$$

$$\alpha_2 \le \frac{\left(\mathcal{K}_{n,2r} * \theta_{2\pi}^{2p}\right)(\phi)}{\left(\mathcal{K}_{n,2r} * \theta^{2p}\right)(\phi)} \le \beta_2, \quad \forall \phi \in [-\pi/2, \pi/2],$$

and

$$\alpha_3 \le \frac{\left(\mathcal{K}_{n,2r} * \theta_{2\pi}^{2p}\right)(\phi)}{\left(\mathcal{K}_{n,2r} * \theta^{2p}(\theta-2\pi)^{2p}\right)(\phi)} \le \beta_3, \quad \forall \phi \in [\pi/2, \pi].$$

PROOF. We just note that there exist positive α_i and β_i for $i = 1, 2, 3$ such that

$$\alpha_1 \le \frac{(\phi - t)_{2\pi}^{2p}}{(\phi - t)^{2p}(\phi + 2\pi - t)^{2p}} \le \beta_1, \quad \forall t \in [-\pi, \pi], \forall \phi \in [-\pi, -\pi/2],$$

$$\alpha_2 \le \frac{(\phi - t)_{2\pi}^{2p}}{(\phi - t)^{2p}} \le \beta_2, \quad \forall t \in [-\pi, \pi], \forall \phi \in [-\pi/2, \pi/2], \tag{3.1}$$

and

$$\alpha_3 \le \frac{(\phi - t)_{2\pi}^{2p}}{(\phi - t)^{2p}(\phi - 2\pi + t)^{2p}} \le \beta_3, \quad \forall t \in [-\pi, \pi], \forall \phi \in [\pi/2, \pi]. \tag{3.2}$$

□

With Lemmas 3.1 and 3.2, we have the following theorem.

THEOREM 3.3. *Let p and r be positive integers with $r > p$. Then there exist positive numbers α and β and a natural number M such that*

$$\alpha \le \frac{\left(\mathcal{K}_{n,2r} * \theta_{2\pi}^{2p}\right)(\phi)}{\phi_{2\pi}^{2p}} \le \beta \tag{3.3}$$

for all $n > M$ and $\pi/n < |\phi| \le \pi$.

PROOF. By Lemma 3.2, for different values of ϕ, $(\mathcal{K}_{n,2r}*\theta_{2\pi}^{2p})(\phi)$ can be replaced by different functions. Hence, we proceed the proof for different ranges of values of ϕ.

We note from (3.1) that $\phi_{2\pi}^{2p} = \phi^{2p}g(\phi)$ for some bounded positive function g. For $\phi \in [\pi/n, \pi/2]$, $(\mathcal{K}_{n,2r} * \theta_{2\pi}^{2p})(\phi) = [\mathcal{K}_{n,2r} * \theta^{2p}g(\theta)](\phi)$. By the Mean Value Theorem, there exists $\zeta \in [-\pi, \pi]$ such that

$$[\mathcal{K}_{n,2r} * \theta^{2p}g(\theta)](\phi) = g(\zeta)(\mathcal{K}_{n,2r} * \theta^{2p})(\phi).$$

It follows that

$$\begin{aligned}\frac{\left(\mathcal{K}_{n,2r} * \theta_{2\pi}^{2p}\right)(\phi)}{\phi_{2\pi}^{2p}} &= \frac{g(\zeta)\left(\mathcal{K}_{n,2r} * \theta^{2p}\right)(\phi)}{g(\phi)\phi^{2p}} \\ &= \frac{g(\zeta)}{g(\phi)}\sum_{k=0}^{p}\binom{2p}{2k}\phi^{-2k}\int_{-\pi}^{\pi}\mathcal{K}_{n,2r}(t)t^{2k}\,dt.\end{aligned}$$

By applying Lemma 3.1, $\int_{-\pi}^{\pi}\mathcal{K}_{n,2r}(t)t^{2k}\,dt = c_k/n^{2k}$, for $k = 0, \cdots, p$ where $c_k = O(1) > 0$. Then

$$\frac{\left(\mathcal{K}_{n,2r} * \theta_{2\pi}^{2p}\right)(\phi)}{\phi_{2\pi}^{2p}} = \frac{g(\zeta)}{g(\phi)}\sum_{k=0}^{p}\frac{d_k}{\phi^{2k}n^{2k}},$$

where $d_k = c_k\binom{2p}{2k}$. Let $g_{\max} = \max_{\phi\in[\pi/n,\pi/2]} g(\phi)$ and $g_{\min} = \min_{\phi\in[\pi/n,\pi/2]} g(\phi)$. Since

$$0 < 1 = d_0 \le \sum_{k=0}^{p}\frac{d_k}{\phi^{2k}n^{2k}} \le \sum_{k=0}^{p}\frac{d_k}{\pi^{2k}},$$

we have

$$\frac{g_{\min}}{g_{\max}} \le \frac{\left(\mathcal{K}_{n,2r} * \theta_{2\pi}^{2p}\right)(\phi)}{\phi_{2\pi}^{2p}} \le \frac{g_{\max}}{g_{\min}}\sum_{k=0}^{p}\frac{d_k}{\pi^{2k}}.$$

For $\phi \in [\pi/2, \pi]$, we note that $(\pi/2)^{2p} \le \phi_{2\pi}^{2p} \le \pi^{2p}$. By (3.2), we have $\phi_{2\pi}^{2p} = \phi^{2p}(\phi - 2\pi)^{2p}h(\phi)$ for some bounded positive function function h. Hence, for $\phi \in [\pi/2, \pi]$

$$\left(\mathcal{K}_{n,2r} * \theta_{2\pi}^{2p}\right)(\phi) = \left[\mathcal{K}_{n,2r} * \theta^{2p}(\theta - 2\pi)^{2p}h(\theta)\right](\phi) = h(\zeta)\left[\mathcal{K}_{n,2r} * \theta^{2p}(\theta - 2\pi)^{2p}\right](\phi)$$

for some $\zeta \in [-\pi, \pi]$. On the other hand,

$$\begin{aligned}\left[\mathcal{K}_{n,2r} * \theta^{2p}(\theta - 2\pi)^{2p}\right](\phi) &= \int_{-\pi}^{\pi}\mathcal{K}_{n,2r}(t)(\phi - t)^{2p}(\phi - t - 2\pi)^{2p}\,dt \\ &= \int_{-\pi}^{\pi}\mathcal{K}_{n,2r}(t)\left(\phi^{2p}(\phi - 2\pi)^{2p} + s(t)\right)\,dt,\end{aligned}$$

where $s(t)$ is a degree $4p$ polynomial such that $s(0) = s'(0) = 0$. By Lemma 3.1, $\int_{-\pi}^{\pi}\mathcal{K}_{n,2r}(t)s(t)\,dt = O(1/n^2)$. Thus,

$$\frac{\left[\mathcal{K}_{n,2r} * \theta^{2p}(\theta - 2\pi)^{2p}\right](\phi)}{\phi_{2\pi}^{2p}} = (\phi - 2\pi)^{2p} + O(\frac{1}{n^{2j}}).$$

It follows that for sufficiently large n, the above ratio is bounded below and above by some positive constants.

For the case $\phi \in [-\pi, -\pi/2]$, it can be proved in a similar way as above that for sufficiently large n, (3.3) holds. The last case $\phi \in [-\pi/2, -\pi/n]$ is similar to the case where $\phi \in [\pi/n, \pi/2]$. □

Using the fact that $[\mathcal{K}_{n,2r} * (\theta-\gamma)_{2\pi}^{2p}](\phi) = (\mathcal{K}_{n,2r} * \theta_{2\pi}^{2p})(\phi-\gamma)$, we obtain the following corollary which deals with functions having a zero at $\gamma \neq 0$.

COROLLARY 3.4. *Let $\gamma \in [-\pi, \pi]$, p and r be positive integers with $r > p$. Then there exist positive numbers α and β and a natural number M such that*

$$\alpha \le \frac{\left[\mathcal{K}_{n,2r} * (\theta-\gamma)_{2\pi}^{2p}\right](\phi)}{(\phi-\gamma)_{2\pi}^{2p}} \le \beta$$

for all $n > M$ and $\pi/n < |\phi - \gamma| \le \pi$.

Now we are in the position of proving a main result for functions with a single zero of order $2p$.

THEOREM 3.5. *Let $f \in \mathbf{C}_{2\pi}$ be a nonnegative function with a zero of order $2p$ at $\gamma \in [-\pi, \pi]$ and let $r > p$ be any integer. Then there exist positive numbers α and β and a natural number M such that*

$$\alpha \le \frac{(\mathcal{K}_{n,2r} * f)(\phi)}{f(\phi)} \le \beta$$

for all $n > M$ and $\pi/n < |\phi - \gamma| \le \pi$.

PROOF. The result can be obtained by writing

$$\frac{(\mathcal{K}_{n,2r} * f)(\phi)}{f} = \frac{(\mathcal{K}_{n,2r} * f)(\phi)}{\left[\mathcal{K}_{n,2r} * (\theta-\gamma)_{2\pi}^{2p}\right](\phi)} \cdot \frac{\left[\mathcal{K}_{n,2r} * (\theta-\gamma)_{2\pi}^{2p}\right](\phi)}{(\phi-\gamma)_{2\pi}^{2p}} \cdot \frac{(\phi-\gamma)_{2\pi}^{2p}}{f},$$

and then applying the Mean Value Theorem to the first factor, Corollary 3.4 to the second factor, and the definition of the zero (cf. §2) to the third factor. □

So far we have considered only the interval $\pi/n < |\phi-\gamma| \le \pi$. For $|\phi-\gamma| \le \pi/n$, we have the following theorem.

THEOREM 3.6. *Let $f \in \mathbf{C}_{2\pi}$ be a nonnegative function with a zero of order $2p$ at $\gamma \in [-\pi, \pi]$ and let $r > p$ be any integer. Then for any $|\phi - \gamma| \le \pi/n$, we have*

$$(\mathcal{K}_{n,2r} * f)(\phi) = O(\frac{1}{n^{2p}}).$$

PROOF. We first prove the theorem for the function $f(\phi) = \phi_{2\pi}^{2p}$. In view of (3.1), there exists a positive continuous function $g(t)$ such that $(\phi - t)_{2\pi}^{2p} = g(t)(\phi - t)^{2p}$. Hence,

$$\begin{aligned}
\left(\mathcal{K}_{n,2r} * \theta_{2\pi}^{2p}\right)(\phi) &= \int_{-\pi}^{\pi} \mathcal{K}_{n,2r}(t) g(t)(\phi-t)^{2p}\, dt \\
&= g(\zeta) \int_{-\pi}^{\pi} \mathcal{K}_{n,2r}(t)(\phi-t)^{2p}\, dt \\
&= g(\zeta) \int_{-\pi}^{\pi} \mathcal{K}_{n,2r}(t) \sum_{j=0}^{2p} \binom{2p}{j} \phi^{2p-j}(-t)^j\, dt,
\end{aligned}$$

where the second equality is obtained from the Mean Value Theorem for some $\zeta \in [-\pi, \pi]$. For odd j, $\int_{-\pi}^{\pi} \mathcal{K}_{n,2r}(t) t^j \, dt = 0$. Thus

$$
\begin{aligned}
(3.4) \qquad \left(\mathcal{K}_{n,2r} * \theta^{2p}\right)(\phi) &= g(\zeta) \int_{-\pi}^{\pi} \mathcal{K}_{n,2r}(t) \sum_{j=0}^{p} \binom{2p}{2j} \phi^{2p-2j} t^{2j} \, dt \\
&\le g(\zeta) \int_{-\pi}^{\pi} \mathcal{K}_{n,2r}(t) \sum_{j=0}^{p} \binom{2p}{2j} \left(\frac{\pi}{n}\right)^{2p-2j} t^{2j} \, dt \\
&= g(\zeta) \sum_{j=0}^{p} \binom{2p}{2j} \left(\frac{\pi}{n}\right)^{2p-2j} \int_{-\pi}^{\pi} \mathcal{K}_{n,2r}(t) t^{2j} \, dt.
\end{aligned}
$$

By Lemma 3.1, $\int_{-\pi}^{\pi} \mathcal{K}_{n,2r}(t) t^{2j} \, dt = O(1/n^{2j})$ for each j. Hence $\left(\mathcal{K}_{n,2r} * \theta^{2p}\right)(\phi) \le O(1/n^{2p})$. On the other hand, from (3.4),

$$
(\mathcal{K}_{n,2r} * \theta^{2p})(\phi) \ge g(\zeta) \int_{-\pi}^{\pi} \mathcal{K}_{n,2r}(t) t^{2p} \, dt = g(\zeta) O(\frac{1}{n^{2p}}).
$$

Thus the theorem holds for $f(\phi) = \phi_{2\pi}^{2p}$.

If $f(\phi) = (\phi - \gamma)_{2\pi}^{2p} g(\phi)$ for some positive function $g \in \mathbf{C}_{2\pi}$, then by the Mean Value Theorem, there exists a $\zeta \in [-\pi, \pi]$ such that

$$
\begin{aligned}
(\mathcal{K}_{n,2r} * f)(\phi) &= [\mathcal{K}_{n,2r} * (\theta - \gamma)_{2\pi}^{2p} g(\theta)](\phi) \\
&= g(\zeta)[\mathcal{K}_{n,2r} * (\theta - \gamma)_{2\pi}^{2p}](\phi) = g(\zeta)(\mathcal{K}_{n,2r} * \theta_{2\pi}^{2p})(\phi - \gamma).
\end{aligned}
$$

Hence the theorem follows. □

4. Spectral Properties of the Preconditioned Matrices

In this section, we analyze the spectra of the preconditioned matrices. We will need the following lemma.

LEMMA 4.1. [**15, 13**, Lemma 3.2] *Let $f, g \in \mathbf{C}_{2\pi}$ be nonnegative such that $0 < a \le f/g \le b$. Then*

$$
a \le \frac{\mathbf{x}^* A_n[f] \mathbf{x}}{\mathbf{x}^* A_n[g] \mathbf{x}} \le b, \qquad \forall \mathbf{x} \ne \mathbf{0}.
$$

Finally, we have our main theorem.

THEOREM 4.2. *Let $f \in \mathbf{C}_{2\pi}$ be a nonnegative function with a zero of order $2p$ at γ. Then for all $r > p$, there exists an integer M such that for $n > \max\{M, 2p\}$, at most $2p+2$ eigenvalues of $C_n^{-1}[\mathcal{K}_{n,2r} * f] A_n[f]$ are not contained in the interval $[a, b]$ where $0 < a < b < \infty$.*

PROOF. Throughout this proof, we denote by $R_n(m)$ an arbitrary n-by-n matrix of rank at most m. For any function g, we let $\tilde{C}_n[g]$ to be the circulant matrix whose j-th eigenvalue is given by

$$
(4.1) \qquad \lambda_j(\tilde{C}_n[g]) = \begin{cases} \dfrac{1}{n^{2p}}, & \text{if } \left|\dfrac{2\pi j}{n} - \gamma\right| \le \dfrac{\pi}{n} \\ g(\dfrac{2\pi j}{n}), & \text{otherwise.} \end{cases}
$$

Note that $\tilde{C}_n[g] = C_n[g] + F_n^* \Phi_n F_n$ where $C_n[g]$ is defined in (2.1) and Φ_n is a diagonal matrix of rank not more than 2.

We use the following decomposition to prove the theorem:

$$
\begin{aligned}
(4.2) \qquad & \frac{\mathbf{x}^* A_n[f]\mathbf{x}}{\mathbf{x}^* C_n[\mathcal{K}_{n,2r} * f]\mathbf{x}} \\
&= \frac{\mathbf{x}^* A_n[f]\mathbf{x}}{\mathbf{x}^* A_n\left[\sin^{2p}\left(\frac{\phi-\gamma}{2}\right)\right]\mathbf{x}} \cdot \frac{\mathbf{x}^* A_n\left[\sin^{2p}\left(\frac{\phi-\gamma}{2}\right)\right]\mathbf{x}}{\mathbf{x}^* \tilde{C}_n\left[\sin^{2p}\left(\frac{\phi-\gamma}{2}\right)\right]\mathbf{x}} \cdot \\
&\quad \cdot \frac{\mathbf{x}^* \tilde{C}_n\left[\sin^{2p}\left(\frac{\phi-\gamma}{2}\right)\right]\mathbf{x}}{\mathbf{x}^* \tilde{C}_n[f]\mathbf{x}} \cdot \frac{\mathbf{x}^* \tilde{C}_n[f]\mathbf{x}}{\mathbf{x}^* \tilde{C}_n[\mathcal{K}_{n,2r} * f]\mathbf{x}} \cdot \frac{\mathbf{x}^* \tilde{C}_n[\mathcal{K}_{n,2r} * f]\mathbf{x}}{\mathbf{x}^* C_n[\mathcal{K}_{n,2r} * f]\mathbf{x}}.
\end{aligned}
$$

By Lemma 4.1, the first factor in the right hand side of (4.2) is uniformly bounded. For the second factor, since $C_n[\sin^{2p}\theta]$ is just the Strang preconditioner for $A_n[\sin^{2p}\theta]$ when $n > 2p$, we have

$$
\begin{aligned}
A_n\left[\sin^{2p}\left(\frac{\phi-\gamma}{2}\right)\right] &= C_n\left[\sin^{2p}\left(\frac{\phi-\gamma}{2}\right)\right] + R_n(2p) \\
&= \tilde{C}_n\left[\sin^{2p}\left(\frac{\phi-\gamma}{2}\right)\right] + R_n(2p+2).
\end{aligned}
$$

By the definition of zeros, the third factor in (4.2) is uniformly bounded. By Theorem 3.5 and the definition of $\tilde{C}_n$ in (4.1), the fourth factor in (4.2) is uniformly bounded. Finally, by Theorem 3.6 and (4.1) again, we see that the last factor of (4.2) is also uniformly bounded.

The theorem now follows by using the properties of Rayleigh quotient and the Weyl's theorem [**11**, p.184]. □

We emphasize that we do not need the explicit knowledge of where the zero γ is in order to construct the preconditioners $C_n[\mathcal{K}_{n,2r} * f]$. Moreover, since $\mathcal{K}_{n,2r}$ are positive kernels, the preconditioners $C_n[\mathcal{K}_{n,2r} * f]$ are positive definite, see [**9**].

By Theorem 4.2, the number of PCG iterations required for convergence is of $O(\log n)$, see [**1**]. Since each PCG iteration and also the construction of the preconditioner both require $O(n \log n)$ operations, the total complexity of our PCG method is at most $O(n \log^2 n)$.

5. Numerical Experiments

In this section, we illustrate by numerical examples the effectiveness of the preconditioner $C_n[\mathcal{K}_{n,2r} * f]$ in solving ill-conditioned Toeplitz systems. As for comparisons, we also test the Strang and the T. Chan circulant preconditioners, see [**16, 10**]. We solve Toeplitz systems $A_n[f]\mathbf{x} = \mathbf{b}$ by the preconditioned conjugate gradient method for eight test functions: θ^2, $(\theta^2-1)^2$, $\theta^2(\pi^2-\theta^2)$, $\theta^2(\pi^4-\theta^4)$, θ^4, $\theta^4(\pi^2-\theta^2)$,

$$
\sum_{|k|\le 1024} \frac{1}{|k|+1} e^{ik\theta} - 0.3853, \quad \text{and} \quad \sum_{|k|\le 1024} \frac{1}{|k|^{0.5}+1} e^{ik\theta} - 0.4134.
$$

Regarding the last two functions, we found numerically that the minimum values of $\sum_{|k|\le 1024} \frac{1}{|k|+1} e^{ik\theta}$ and $\sum_{|k|\le 1024} \frac{1}{|k|^{0.5}+1} e^{ik\theta}$ are approximately equal to 0.3853 and 0.4134 respectively. Hence the two test functions are approximately zero at some points in $[-\pi, \pi]$.

	θ^2							$(\theta^2-1)^2$						
n	16	32	64	128	256	512	1024	16	32	64	128	256	512	1024
I	16	36	79	170	362	753	†	21	53	141	293	547	†	†
S	6	7	7	7	7	7	8	11	10	12	9	10	8	12
T	11	12	16	19	23	29	39	15	18	24	30	27	36	46
$K_{N,4}$	9	8	9	10	9	9	9	13	13	13	14	12	13	11
$K_{N,6}$	9	10	10	10	10	9	9	13	13	13	14	14	13	13
$K_{N,8}$	10	9	10	10	10	10	10	14	13	13	15	15	14	13

	$\theta^2(\pi^2-\theta^2)$							$\theta^2(\pi^4-\theta^4)$						
n	16	32	64	128	256	512	1024	16	32	64	128	256	512	1024
I	16	32	61	116	220	428	835	16	32	64	128	256	510	†
S	9	9	9	9	10	11	12	8	9	9	9	10	11	12
T	11	12	14	17	20	26	33	11	12	15	17	22	27	38
$K_{N,4}$	10	10	11	11	11	11	11	10	10	11	11	11	11	11
$K_{N,6}$	10	10	11	11	11	11	12	10	10	11	11	11	11	13
$K_{N,8}$	11	11	12	12	11	12	13	11	11	12	12	12	12	12

	θ^4							$\theta^4(\pi^2-\theta^2)$						
n	16	32	64	128	256	512	1024	16	32	64	128	256	512	1024
I	22	63	209	790	†	†	†	16	46	131	410	†	†	†
S	12	13	18	21	24	31	41	12	13	16	20	22	31	38
T	17	26	42	71	161	167	247	16	24	35	58	106	144	196
$K_{N,4}$	13	15	17	20	24	26	26	14	15	16	20	22	27	26
$K_{N,6}$	13	15	16	18	18	17	18	14	15	16	18	18	18	21
$K_{N,8}$	14	16	17	19	19	19	20	16	16	18	19	20	21	23

	$\sum_{\lvert k\rvert<1024} 1/(\lvert k\rvert+1)e^{ik\theta}-0.3853$							$\sum_{\lvert k\rvert<1024} 1/(\lvert k\rvert^{0.5}+1)e^{ik\theta}-0.4134$						
n	16	32	64	128	256	512	1024	16	32	64	128	256	512	1024
I	19	45	112	184	240	296	343	18	41	92	238	715	†	†
S	10	10	10	7	8	8	8	9	11	10	9	10	9	8
T	12	15	15	14	12	10	8	11	14	13	12	17	15	13
$K_{N,4}$	7	8	8	7	8	8	7	10	11	10	9	9	9	8
$K_{N,6}$	7	10	7	8	8	8	8	10	11	10	9	10	9	9
$K_{N,8}$	9	9	8	8	8	8	8	11	11	11	10	10	10	9

TABLE 1. Numbers of iterations for different preconditioners.

In our test, we use $\mathcal{K}_{N,2r}$ as kernel functions where N is the largest interger such that $N(r-1)<n$. The right-hand side vector $\mathbf{b}$ is formed by multiplying a random vector to $A_n[f]$. The initial guess is the zero vector and the stopping criteria is $||\mathbf{r}_q||_2/||\mathbf{r}_0||_2 \le 10^{-7}$ where $\mathbf{r}_q$ is the residual vector after q iterations. All computations are done by Matlab on a Sun Ultra SPARC 1 workstation.

Table 1 shows the numbers of iterations required for convergence for different choices of preconditioners. In the table, I denotes no preconditioner, S is the Strang preconditioner, T is the T. Chan preconditioner, and $K_{N,2r}$ are the preconditioners from $\mathcal{K}_{N,2r}$ with $r=2$, 3 and 4. We recall that T is just $K_{N,2}$. For those iterations more than 1000, we denote them by †.

The first four test functions have single or multiple zeros of order 2 on $[-\pi,\pi]$. We see that T. Chan preconditioner does not work. This can be explained by the fact that the order of $\mathcal{K}_{N,2} * f$ does not match that of f at the zeros of f. The Strang and T. Chan preconditioners do not work for the next two test functions which are functions with a zero of order 4. As expected, $K_{N,6}$ and $K_{N,8}$ still work very well. The last two functions are functions with slowly decaying Fourier coefficients. It is nontrivial to determine the order of their zeros. However, the results show that $K_{N,r}$ perform better than the Strang and T. Chan preconditioners.

References

[1] F. Di Benedetto, *Analysis of preconditioning techniques for ill-conditioned Toeplitz matrices*, SIAM J. Sci. Comput., **16** (1995), pp. 682–697.

[2] F. Di Benedetto, G. Fiorentino and S. Serra, *C.G. preconditioning for Toeplitz Matrices*, Comput. Math. Appl., **25** (1993), pp. 35–45.

[3] R. Bitmead and B. Anderson, *Asymptotically Fast Solution of Toeplitz and Related Systems of Equations*, Lin. Alg. Appl., **34** (1980), pp. 103–116.

[4] R. Brent, F. Gustavson and D. Yun, *Fast Solution of Toeplitz Systems of Equations and Computations of Padé Approximations*, J. Algorithms, **1** (1980), pp. 259–295.

[5] R. Chan, *Circulant Preconditioners for Hermitian Toeplitz Systems*, SIAM J. Matrix Anal. Appl., **10** (1989), pp. 542–550.

[6] R. Chan, *Toeplitz Preconditioners for Toeplitz Systems with Nonnegative Generating Functions*, IMA J.Numer. Anal., **11** (1991), pp. 333–345.

[7] R. Chan and M. Ng, *Conjugate Gradient Method for Toeplitz Systems*, SIAM Review, **38** (1996), pp. 427–482.

[8] R. Chan and P. Tang, *Fast Band-Toeplitz Preconditioners for Hermitian Toeplitz Systems*, SIAM J. Sci. Comput., **12** (1994), pp. 164–171.

[9] R. Chan and M. Yeung, *Circulant Preconditioners Constructed from Kernels*, SIAM J. Numer. Anal., **29** (1992), pp. 1093–1103.

[10] T. Chan, *An Optimal Circulant Preconditioner for Toeplitz Systems*, SIAM J. Sci. Statist. Comput., **9** (1988), pp. 766–771.

[11] R. Horn and C. Johnson, *Matrix Analysis*, Cambridge University Press, Cambridge, 1985.

[12] G. Lorentz, *Approximation of Functions*, Holt, Rinehart and Winston, 1966.

[13] D. Potts and G. Steidl, *Preconditioners for Ill-Conditioned Toeplitz Matrices*, BIT, to appear.

[14] D. Potts and G. Steidl, *Preconditioners for Ill-Conditioned Toeplitz Systems Constructed from Positive Kernels*, preprint, 1999.

[15] S. Serra, *Preconditioning Strategies for Hermitian Toeplitz Systems with Nondefinite Generating Functions*, SIAM J. Matrix Anal. Appl., **17** (1996), pp. 1007–1019.

[16] G. Strang, *A Proposal for Toeplitz Matrix Calculations*, Stud. Appl. Math., **74** (1986), pp. 171–176.

[17] W. Trench, *An Algorithm for the Inversion of Finite Toeplitz Matrices*, SIAM J. Appl. Math., **12** (1964), pp. 515–522.

[18] E. Tyrtyshnikov, *Circulant Preconditioners with Unbounded Inverses*, Lin. Alg. Appl., **216** (1995), pp. 1–24.

DEPARTMENT OF MATHEMATICS, THE CHINESE UNIVERSITY OF HONG KONG, SHATIN, HONG KONG. EMAIL: `rchan@math.cuhk.edu.`

DEPARTMENT OF MATHEMATICS, THE UNIVERSITY OF HONG KONG, POKFULAM ROAD, HONG KONG. EMAIL: `mhyipa@hkusua.hku.hk.`

SCHOOL OF BUSINESS AND ADMINISTRATION, THE OPEN UNIVERSITY OF HONG KONG, HO MAN TIN, HONG KONG. EMAIL: `cwleung@ouhk.edu.hk.`

DEPARTMENT OF MATHEMATICS, THE UNIVERSITY OF HONG KONG, POKFULAM ROAD, HONG KONG. EMAIL: `mng@maths.hku.hk.`

AMS/IP Studies in Advanced Mathematics
Volume 20, 2001

What is Computational Geometry

Ren-Hong Wang

As human kind entered the end of the 20th century, the computer led us into an information era. Information technology is playing more and more important roles in various aspects such as society, science, life, and economy, while in many cases, computer exhibits a colorful world by means of geometries.

In this survey, we will be concerned with the relationship between multivariate splines and geometries including computational geometry, differential geometry, and algebraic geometry.

It is well known that analytic geometry is a subject to deal with geometric objects by using the system of coordinates established by R. Descartes.

Concerning the computational geometry, there have been several definitions. Yet I would like to suggest a new one to it:

Computational geometry is a subject for studying the theory, method, generation and realization of geometric objects under the environment concerning computers, networks, and other information tools.

We can say that the computers, networks and other information tools, play important roles in computational geometry similar to Descartes' coordinates in analytical geometry.

Computational geometry is concerned with computer aided design, computer aided manufacture, computer aided engineering, image processing, shape from X, computer graphics, scientific visualization, shape reconstruction, and all the other fields related to geometric objects.

The computational geometry is an important borderline subject related to computational mathematics, differential geometry, algebraic geometry, approximation theory, and so on.

It is well known that the polynomial is an important approximation tool of functions as well as curves and surfaces. In fact, according to classical Weierstrass approximation theorem, any continuous function on a bounded closed domain can be uniformly approximated by polynomials on this domain.

Unfortunately, polynomials have such strong global properties that a polynomial can be determined only by its properties on a neighbourhood of a given point in the domain. This property is not satisfied by a number of geometric objects, for

---

Project was supported by the National Natural Science Foundation of China, and Morningside Center of Mathematics, Chinese Academy of Sciences .

instance, the surfaces appeared in aircrafts, cars, ships, satellites, and most other practical geometric objects.

Splines being piecewise polynomials instead can be used to approximate any continuous, smooth, and even discontinuous function within any given tolerence.

Moreover a spline is easy to store, to evaluate, and to manipulate on a digital computer; a myraid of applications in scientific and engineering computing have been found. Spline has become a kind of fundamental tools for computational geometry, numerical analysis, approximation, and optimization, etc..

Now let us turn to multivariate splines. Multivariate splines are piecewise polynomials defined on domains of more than one variable. In what follows, we will only consider bivariate splines, for it is easy to understand, while the treatment on higher dimensional splines with more details can be found in [11] [5] and [13].

Let D be a domain in R^2 and Δ a partition of D consisting of finite irreducible algebraic curves $\Gamma_i : \ell_i(x, y) = 0$, $i = 1, \cdots, N$, where the coefficients of $\ell_i(x, y)$ are real numbers. Denote by Δ_i, $i = 1, \cdots, T$, all the cells of Δ. For integer $k > \mu \geq 0$, we say that

$$S_k^\mu := \{s \in C^\mu(D) | s|_{\Delta_i} \in P_k, \forall i\},$$

is a multivariate spline space with degree k and smoothness μ, where P_k denotes the collection of polynomials

$$P_k := \left\{ \sum_{i=0}^{k} \sum_{j=0}^{k-i} c_{ij} x^i y^j | c_{ij} \text{ real} \right\}$$

if $k \geq 0$, and $P_k = \{0\}$ if $k < 0$.

According to the definition above, multivariate splines will be very useful in solving problems in computational geometry and at the same time, we will encounter and have to deal with a number of problems in differential geometry, such as tangent, curvature, offset, and minimal surface problems. However, most people have not mentioned the relationship between algebraic geometry and multivariate splines.

First, by using Bezout's theorem in algebraic geometry, the author discovered the following fundamental theorem on multivariate splines [10]

THEOREM 1. *$s(x, y) \in S_k^\mu(\Delta)$ if and only if the following conditions are satisfied:*

1. For each interior edge of Δ, which is defined by $\Gamma_i : \ell_i(x, y) = 0$, there exists a so-called smoothing cofactor $q_i(x, y)$, such that

$$p_{i1} - p_{i2} = \ell_i^{\mu+1} q_i, \tag{1}$$

where the polynomials p_{i1} and p_{i2} are determined by the restriction of $s(x, y)$ on the two cells Δ_{i1} and Δ_{i2} with Γ_i as the common edge and $q_i \in P_{\alpha-(n+1)n_i}$, $\alpha =$ max(degrees of p_{i1} and p_{i2}), $n_i =$ degree of ℓ_i.

2. For any interior vertex v_j of Δ, the following conformality conditions are satisfied

$$\sum [\ell_i^{(j)}(x, y)]^{\mu+1} q_i^{(j)}(x, y) \equiv 0 \tag{2}$$

where the summation is taken on all the interior edges $\Gamma_i^{(j)}$ passing through v_j, and the signs of the smoothing cofactors $q_i^{(j)}$ are refixed in such a way that when a point crosses $\Gamma_i^{(j)}$ from Δ_{i2} to Δ_{i1}, it goes around v_j counter-clockwisely.

Theorem 1 shows that the multivariate spline, in principle, is equivalent to an algebraic subject defined by Theorem 1. Furthermore, Theorem 1 has also shown a general method than any others for studying the multivariate splines over any given partition. It was called the smoothing cofactor-conformality method.

On the other hand, multivariate splines have strong background of algebraic geometry. In fact, from Theorem 1, the dimension of the space $S_k^\mu(\Delta)$ can be expressed as

$$\dim S_k^\mu(\Delta) = \binom{k+2}{2} + \Theta$$

where Θ is the dimention of the linear space defined by the conformality conditions (2).

However the second term of the right-side is dependent on the topological and geometric properties of the partition Δ.

In fact, a key problem on $\dim S_k^\mu(\Delta)$ is to calculate the rank of matrix A derived from the conformality conditions (2). However, the rank of A is dependent on the topological and geometric propeties of Δ.

Let us consider in detail the special partition Δ consisting of a finite straight lines. By using map $\varphi : (x, y) \to [x, y, 1]$, the partition Δ is embedded in CP^2, and any edge Γ_i can be represented in $\alpha_i x + \beta_i y + \gamma_i z = 0$. It is clear that the matrix M_v determined by the conformality condition (2) at any given interior vertex v is a $\binom{k+2}{2} \times \binom{k-\mu+1}{2} N_v$ matrix.

It is clear that the dimension of solution space defінded by $M_v\zeta = 0$ is $\binom{k-\mu+1}{2}N_v - \binom{k+2}{2} + \tau$, where τ is the dimension of the solution space of the following linear system

$$M_v^T \xi = 0. \tag{3}$$

However, (3) shows that the points $\{(\alpha_i, \beta_i, \gamma_i)\}_{i=1}^{N_v}$ lie on algebraic curves as follows

$$\sum_{|\lambda|=\mu+1} C_\lambda \alpha_i^{\lambda_1} \beta_i^{\lambda_2} \gamma_i^{\lambda_3} = 0, \quad i = 1, \cdots, N_v,$$

where N_v is the number of edges $\Gamma_i : \alpha_i x + \beta_i y + \gamma_i z = 0$ passing v, $\lambda = (\lambda_1, \lambda_2, \lambda_3)$, $\lambda_1, \lambda_2, \lambda_3 \geq 0$, $|\lambda| = \lambda_1 + \lambda_2 + \lambda_3$, and the unknown $\binom{\mu+3}{2}$-vectors C_λ are taken from a $\binom{k+2}{2}$-vector in different ways.

Hence the problems on $\dim S_k^\mu(\Delta)$ should be also the problems in algebraic geometry.

As we show that for any given triangulation Δ, the problems of finding the exact formulae on $\dim S_2^1(\Delta)$ and $\dim S_3^1(\Delta)$ are still open[13].

They seem to be heavily dependent on the geometric properties of the triangulation Δ.

From the algebraic geometric point of view, for instance, by Pascal's Theorem, H. Du [2] shows that for Morgan-Scott triangulation Δ_{MS} ([3])

$$\dim S_2^1(\Delta_{MS}) = \begin{cases} 7, & \text{if and only if } p_1, p_2, p_3, p_4, p_5, \text{ and } p_6 \\ & \text{are on an irreducible conic curve} \\ 6, & \text{otherwise} \end{cases}$$

The close relationship between bivariate C^1-quadratic spline based on Morgan-Scott partition and the classical Pascal's Theorem and some results in multivariate

C^1-quadratic splines on Morgan-Scott partition in R^n(cf. [6]) stimulate us to generalize Pascal's Theorem to n-dimensional cases.

To do this, one first takes Pascal's Theorem as the problem that when the hexagon obtained from a triangle by cutting out its three corners inscribes a conic (In fact, any hexagon can be obtained in this way). A natural problem is when the polyhedron obtained by cutting out the corners of an n-simplex (this polyhedron is called Pascal's polyhedron) inscribes an n-dimensional quadratic hypersurface. We have ([7])

THEOREM 2. *An n-Pascal's polyhedron inscribes an n-dimensional quadratic hypersurface if and only if each of its* $(n-1)$*-faces inscribes an* $(n-1)$*-dimensional quadratic hypersurface.*

A basic problem on interpolation by the multivariate spline in $S_k^\mu(\Delta)$ is how to choose a set of so-called suitable knots $\{(x_i, y_i)\}_{i=1}^d$, $d = \dim S_k^\mu(\Delta)$, such that for any given $\{z_i\}_{i=1}^d$, there exists a unique $s \in S_k^\mu(\Delta)$ satisfying

$$s(x_i, y_i) = z_i, \quad i = 1, \cdots, d.$$

It is clear that the above interpolation problem has a unique solution if and only if the knot-set $\{(x_i, y_i)\}_{i=1}^d$ does not lie on the following curve

$$\Gamma := \{(x, y) | s(x, y) = 0\}, \quad s \in S_k^\mu(\Delta). \tag{4}$$

The curve Γ is called a piecewise algebraic curve ([9]).

It is obvious that the piecewise algebraic curve is a kind of generalization of the classical algebraic curve.

Because of the possibility $\{(x, y) \in D | s|_{\delta_i} = p(x, y) = 0\} \cap \bar{\Delta}_i = \emptyset$, (empty), it is more difficult to study the piecewise algebraic curve.

Bezout's theorem is an important and classical result in algebraic geometry, its weak form says that two algebraic curves of degree m and n will have infinitely many intersection points if they have more than mn (this threshold is called Bezout's number) intersection points. For $S_m^r(\Delta)$ and $S_n^t(\Delta)$, we denote by $BN = BN(m, r; n, ; \Delta)$ the Bezout's number, i.e. the threshold number that for any two piecewise algebraic curves $f \in S_m^r(\Delta)$, $g \in S_n^t(\Delta)$, they will have infinitely many number intersection points if they have more than BN intersection points.

For piecewise algebraic curves, a fundmental problem is how to find the Bezout's number. In general, this problem is very difficult. For this problem, we have[8]

THEOREM 3. $BN(m, 0; n, 0; \Delta) = mnT$ *if* Δ *is a 2-triangle-signed triangulation or if* mn *is even, and* $BN(m, 0; n, 0; \Delta) \le mnT - \left[\frac{V_{\rm odd}+2}{3}\right]$ *in general, where* T *is the number of triangles in* Δ *and* $V_{\rm odd}$ *is the number of odd vertices in* Δ*. Here by a triangulation being 2-triangle-signed, we mean that each triangle in* Δ *can be marked by* -1 *and* 1 *so that any two adjacent triangles in* Δ *are marked by different signs.*

References

[1] Böhm, W., Algebraic and differential geometric methods in C.A.G.D., in *Computation of Curves and Surfaces*, eds. W. Dahmen, etc., Kluwer Acad. Press, 1990, pp. 425–455.
[2] Du, H, A geometric approach to $\dim S_2^1(\Delta_{MS})$, Proc. Comp. Geo., Intern. Press, 1999.
[3] Morgan, J., and R. Scott, The dimension of piecewise polynomials, manuscript, 1975.
[4] Shafarevich, I.R., *Basic Algebraic Geometry* 1 & 2, Springer-Verlag, 1977, 1974, Berlin, Heiderberg.

[5] Shi, X.Q., *Higher-dimensional Splines,* Ph D Thesis, 1988, Jilin University, Changchun China.
[6] Shi, X.Q., *The singularity of Morgan-Scott partition,* CAGD, 8(1991), 201-206.
[7] Shi, X.Q., and R.-H. Wang, The generalization of Pascal's Theorem and Morgan-Scott's partition, Proc. Comp. Geo., Intern. Press, 1999.
[8] Shi, X.Q., and R.-H. Wang, Bezout number for piecewise algebraic curves, BIT, No. 2(1999).
[9] Wang, R.-H., Multivariate spline and its application in science and techonlogy, Rend. del Seminario Mat. e Fis. di Milano, vol. LXIII (1993), 213–229.
[10] Wang, R.-H., The structural characterization and interpolation for multivariate splines, Acta Math. Sinica, **18**(1975), 91–106.
[11] Wang, R.-H., On the analysis of multivariate splines in the case of arbitrary partition II: The case of k-dimension, Numer. Math. (A Jour. of Chinese Univ.) **1**(1980), 78–81.
[12] Wang, R.-H., On piecewise algebraic curves, in *Advanced Topics in Multivariate Approximation,* eds., F. Fontanella, etc., World Scientific, 1996, pp 355–361.
[13] Wang, R.-H., X.Q. Shi, Z.X. Luo, and Z.X. Su, *Multivariate Spline and its Applications,* Science Press, 1994, Beijing.

INSTITUTE OF MATHEMATICAL SCIENCES, DALIAN UNIVERSITY OF TECHNOLOGY, DALIAN 116024, CHINA

AMS/IP Studies in Advanced Mathematics
Volume 20, 2001

A survey on circular chromatic numbers of graphs

Gerard J. Chang

ABSTRACT. The circular chromatic number of a graph, introduced by Vince in 1988 under the name star chromatic number, is a natural generalization of the chromatic number. In this paper, we survey results on this topic, concentrating on the relations between the circular chromatic number and the chromatic number of a graph.

1. Introduction

The concept of circular chromatic number is a good model for the problem of designing a traffic control system at a road intersection in such a way that each traffic flow is assigned a unit interval of time during which it faces a green light. The object is to minimize the total length of a complete traffic period.

We may model the problem as follows. Each traffic flow is represented by a vertex, and two vertices are adjacent if the green lights intervals of the corresponding traffic flows do not overlap. We may view a complete traffic period as a circle C, and each vertex is assigned a unit-length arc which is the time interval in which the corresponding traffic flow has the green light. Thus adjacent vertices of the graphs are assigned with disjoint arcs of C. And the object is to minimize the circumstance of the circle C. More formally, we have the following definition.

Suppose C is a circle of (circumstance) length r. An *r-circular coloring* of a graph G is a mapping f which assigns to each vertex x of G a closed-open arc $f(x)$ of C such that $f(x) \cap f(y) = \emptyset$ for each edge xy of G. The *circular chromatic number* of G is

$$\chi_c(G) = \inf\{r : G \text{ has a } r\text{-circular coloring}\}.$$

The circular chromatic number is closely related to the chromatic number. The *chromatic number* of a graph G is the minimum positive integer k such that G has a k-coloring which is a mapping $f : V(G) \to \{0, 1, \ldots, k-1\}$ with $f(x) \neq f(y)$ for each edge xy in G.

THEOREM 1. *For any graph G, we have $\chi(G) - 1 < \chi_c(G) \leq \chi(G)$ or equivalently $\chi(G) = \lceil \chi_c(G) \rceil$.*

Proof. Suppose f_c is a $\chi_c(G)$-circular coloring of G, which is defined on a circle C. The circle C can be viewed as a closed-open interval $[0, \chi_c(G))$. For each vertex x, the arc $f_c(x)$ contains at least one integer $f(x)$ with $0 \le f(x) < \chi_c(G)$. For each edge xy in G, since $f_c(x) \cap f_c(y) = \emptyset$, it is clear that $f(x) \ne f(y)$. This gives a $\lceil \chi_c(G) \rceil$-coloring. Hence $\chi(G) \le \lceil \chi_c(G) \rceil$.

On the other hand, suppose f is a $\chi(G)$-coloring of G. Consider a circle C of length $\chi(G)$. Define $f_c(x)$ as the arc $[f(x), f(x)+1)$. This gives a $\chi(G)$-circular coloring. Thus $\chi_c(G) \le \chi(G)$ and so $\lceil \chi_c(G) \rceil \le \chi(G)$. □

Theorem 1 implies that two graphs having the same circular chromatic number have the same chromatic number. However, two graphs with the same chromatic number may have different circular chromatic numbers. In some sense, $\chi_c(G)$ is a refinement of $\chi(G)$ and it contains more information about the graph.

Theorem 1 was first proved by Vince [**23**] who introduced the concept of circular chromatic number under the name star chromatic number and the notation $\chi^*(G)$. The name circular chromatic number was concluded after several discussions by Zhu, Hahn and others.

Vince [**23**] actually define the term in a equivalent way, which can be viewed as a discrete version of the above definition. For two integers $1 \le d \le k$, a (k,d)-coloring of a graph G is a mapping f which maps each vertex x an integer $f(x)$ in $\{0, 1, \ldots, k-1\}$ such that $d \le |f(x) - f(y)| \le k - d$ for each edge xy in G. It is then equivalent that

$$\chi_c(G) = \inf\{k/d : G \text{ has a } (k,d)\text{-coloring}\}. \tag{1}$$

As a $(k,1)$-coloring is a k-coloring, one may also see that $\chi_c(G) \le \chi(G)$. Vince [**23**] and Bondy and Hell [**4**] proved that the infimum in Eq. (1) is actually attained by some $k \le |V(G)|$. Hence $\chi_c(G)$ is always a rational number.

The circular chromatic number has been extensively studied during the past decade. Zhu [**34**] gave a detail survey of the topics. This article serves as a compact survey with some more new results discussed.

2. Which value can be $\chi_c(G)$?

As the circular chromatic number of a graph is a rational number, A natural question is to determine which rational number can be the circular chromatic number of a graph. It is clear that $\chi_c(G) = 1$ if G has no edge and $\chi_c(G) \ge 2$ otherwise. In fact, for any rational number $k/d \ge 2$, where k and d are integers with $1 \le d \le k$, we may define the graph G_k^d whose

$$V(G_k^d) = \{0, 1, \ldots, k-1\} \text{ and } E(G_k^d) = \{xy : d \le |x - y| \le k - d\}.$$

It is clear that $f(x) = x$ define a (k,d)-coloring of G_k^d and so $\chi_c(G_k^d) \le k/d$. Vince [**23**] proved that in fact $\chi_c(G_k^d) = k/d$. Hence we have

THEOREM 2. *A rational number $r = \chi_c(G)$ for some graph G if and only if $r = 1$ or $r \ge 2$.*

The following stronger result is proved in [**27**]:

THEOREM 3. *For any rational number $r \ge 2$ and any integer $g \ge 3$, there is a graph G of girth at least g such that $\chi_c(G) = r$.*

The proof of Theorem 3 in [**27**] is by using probabilistic method. A constructive proof can be found in [**34**]. Also, by using the categorical product, Kirsch [**19**] gave a construction of graphs with arbitrary large odd girth, and with $\chi_c(G) = r$ for an arbitrary rational $r \geq 2$. This result was generalized in [**31**] where uniquely H-colorable graphs of arbitrarily large odd girth were constructed.

It follows from the Four Color Theorem that a necessary condition for a rational number r to be the circular chromatic number of a planar graph is that $r = 1$ or $2 \leq r \leq 4$. It turned out that this condition is sufficient as well. It was proved by Moser [**20**] that every rational number r between 2 and 3 is the circular chromatic number of a planar graph. The method was simplified by Zhu [**29**] and then generalized in [**30**] to prove that every rational number r between 3 and 4 is the circular chromatic number of a planar graph.

THEOREM 4. *For any rational number r between 2 and 4, there is a planar graph G with $\chi_c(G) = r$.*

An interesting question is to prove that $\chi_c(G) < 5$ for every planar graph G, without using the Four Color Theorem. It was proved by Hilton, Rado and Scott [**16**] that every planar graph has fractional chromatic number $\chi_f(G)$ strictly less than 5, without using the Four Color Theorem. However, as $\chi_c(G) \geq \chi_f(G)$ for any graph G, it would be a stronger result if it were proved that every planar graph has circular chromatic number strictly less than 5.

We close this section by discussing two questions concern other classes of graphs.

QUESTION 1. *Is it true that for any $n \geq 5$ and any rational number $2 \leq r \leq n-1$ there exists a K_n-minor free graph whose circular chromatic number is equal to r?*

The results on planar graphs shows that the answer is "yes" if $n = 5$, because planar graphs are K_5-minor free. However, a recent result of Hell and Zhu [**15**] that if a graph G is K_4-minor free, then $\chi_c(G)$ is either equal to 3 or at most $8/3$. For $n \geq 6$, it was proved in [**33**] that for any rational $2 \leq r \leq n-2$, there exists a K_n-minor free graph G with $\chi_c(G) = r$. The problem remains open for $n \geq 6$ and for $n-2 < r < n-1$.

QUESTION 2. *For which rational number r, there is a graph G whose line graph $L(G)$ has circular chromatic number r?*

3. Graphs G for which $\chi_c(G) = \chi(G)$

The question that for which graphs G we have $\chi_c(G) = \chi(G)$ was raised by Vince [**23**], and investigated in [**1, 10, 13, 22, 23, 24, 26, 27, 31, 32**]. It was shown by Guichard [**13**] that it is NP-hard to determine whether or not an arbitrary graph G satisfies $\chi_c(G) = \chi(G)$. Indeed, using an oracle which determines whether or not an arbitrary graph G satisfies $\chi_c(G) = \chi(G)$, we can easily determine the chromatic number of a graph G as follows: Let $G \cup H$ denote the disjoint union of graphs G and H. Since $\chi_c(G_k^d) = k/d$, it is straightforward to verify that $\chi(G) = n$ if and only if

$$\chi_c(G \cup K_n) = \chi(G \cup K_n) \text{ and } \chi_c(G \cup G_2^{2n+1}) \neq \chi(G \cup G_2^{2n+1}).$$

A similar arguments also proves that it is NP-hard to determine whether or not a planar (or line) graph G satisfies $\chi_c(G) = \chi(G)$.

The complexity of the following decision problem seems more interesting, and the authors have not seen any discussion on this problem: Given a graph G of chromatic number n, determine whether or not $\chi_c(G) = n$? It is likely that this problem is also NP-hard for general graphs. The problem maybe more interesting if restricted some special classes of graphs, say for planar graphs, line graphs, etc.

In light of the above results, it is unlikely that there is a simple characterization of graphs G with $\chi_c(G) = \chi(G)$. Nevertheless, some sufficient conditions are found in [**1, 10, 13, 22, 23, 24, 26**] under which a graph G satisfies $\chi_c(G) = \chi(G)$. The following result, which was proved in [**22**], is the strongest result along this line:

THEOREM 5. *Suppose $\chi(G) = n$. If there is a non-trivial subset A of V (i.e., $A \neq V$ and $A \neq \emptyset$) such that for any n-coloring f of G, each color class X of f is either contained in A or disjoint from A, then $\chi_c(G) = \chi(G)$.*

The Petersen graph P is an example which does not satisfy the condition of this theorem, and yet $\chi_c(P) = \chi(P) = 3$. So the condition of this theorem is not a necessary condition. There are other sufficient conditions for a graph G to satisfy $\chi_c(G) = \chi(G)$, and it turns out that all the other known sufficient conditions can be easily derived from this one:

1. The complement of G is disconnected. (Take A a component of $\overline{G}$, see [**1, 10**].)
2. G has a universal vertex x. (Take $A = \{x\}$, see [**13, 26**].)
3. G is uniquely n-colorable. (Take A any color class, see [**22**].)

It was proved by Bollobás and Sauer [**3**] that for any integers $n \geq 1$ and $g \geq 3$, there is a graph G of girth at lest g and that G is uniquely n-colorable. Therefore we have the following corollary [**22**]:

COROLLARY 6. *For any integers $n \geq 1$ and $g \geq 3$, there is a graph G of girth at lest g and $\chi_c(G) = \chi(G) = n$.*

In a search of triangle-free graphs with arbitrary large chromatic numbers, Mycielski [**21**] developed an interesting graph translation as follows. For a graph $G = (V, E)$ with vertex set V and edge set E, the *Mycielskian* of G is the graph $\mu(G)$ with vertex set $V \cup V' \cup \{u\}$, where $V' = \{x' : x \in V\}$, and edge set $E \cup \{xy' : xy \in E\} \cup \{y'u : y' \in V'\}$. For $k \geq 2$, let $\mu^k(G) = \mu(\mu^{k-1}(G))$. Myceilski showed that $\chi(\mu(G)) = \chi(G)$ for any graph G, and $\omega(\mu(G)) = \omega(G)$ for any graph G with at least one edge. Hence $\mu^k(K_2)$ is a triangle-free graph of chromatic number $k + 2$. The following results are from [**5**] (for (1), (2) and (3)) and [**17**] (for (4)).

THEOREM 7. (1) *If $\chi(G) = 3$, then $\chi_c(\mu(G)) = \chi(\mu(G)) = 4$.*
(2) *If $n \geq 3$, then $\chi_c(\mu(K_n)) = \chi(\mu(K_n)) = n + 1$.*
(3) *If $n \geq 4$, then $\chi_c(\mu^2(K_n)) = \chi(\mu^2(K_n)) = n + 2$.*
(4) *$\chi_c(\mu(G_k^d)) = \chi(\mu(G_k^d)) = \lceil k/d \rceil + 1$.*

The following questions remain open.

QUESTION 3. *Is it true that if $n \geq m + 2$, then $\chi_c(\mu^m(K_n)) = \chi(\mu^m(K_n)) = n + m$ for $m \geq 3$?*

QUESTION 4. *What is $\chi_c(\mu^n(K_n))$ for $n \geq 3$?*

4. Graphs G with $\chi_c(G)$ close to $\chi(G) - 1$

We have shown that $\chi(G) - 1 < \chi_c(G) \leq \chi(G)$ for every graph G. Section 3 discusses the extremal case that $\chi_c(G) = \chi(G)$. The other extremal case is that $\chi_c(G)$ is close to the lower bound $\chi(G) - 1$.

It was shown by Guichard [**13**] that if a graph G is n-critical, and has girth at least $n+1$, then $\chi_c(G) < \chi(G)$. The following result of [**22**] reveals a quantitative relation between the girth and the circular chromatic number of an n-critical graph.

THEOREM 8. *Let $m \geq 1$ and $t \geq 1$ be integers. If G has a vertex x such that $G - x$ is m-colorable, and every cycle of G containing x has length at least $m(t-1)+2$, then $\chi_c(G) \leq m + 1/t$.*

COROLLARY 9. *If G is $(n+1)$-critical and has girth at least $n(t-1)+2$, then $\chi_c(G) \leq n + 1/t$.*

In some sense the bound in Corollary 9 is the best bound for the circular chromatic number of a critical graph in terms of its girth. For example, when $n = 3$, then 3-critical graphs are just odd cycles, and $\chi_c(C_{2t+1}) = 2 + 1/t$. On the other hand, the following problem remains open:

For any integer n, let $g(n)$ be the least integer such that any n-critical graph G with girth greater than $g(n)$ satisfies $\chi_c(G) < \chi(G)$. What is the value of $g(n)$?

Certainly $g(n) \geq 3$. On the other hand, Corollary 6 implies that $g(n) \leq n$. Indeed, if an n-critical graph G has girth greater than n, then it follows from Corollary 9 that $\chi_c(G) \leq n - 1/2$. Therefore $g(3) = 3$. Another value of $g(n)$ we know is that $g(4) = 4$ [**5**]. For all other n, we know nothing more than the bounds $3 \leq g(n) \leq n$. In particular, we do not know whether or not the function $g(n)$ is non-decreasing.

THEOREM 10. *If G is a graph for which $\chi_c(G) \leq \chi(G) - 1/d$ with $d = 2$ or 3, then $\chi_c(\mu^2(G)) \leq \chi(\mu^2(G)) - 1/d$.*

It is unknown whether or not Theorem 10 remains true when d is an arbitrary number. We are attempted to conjecture that for any graph G, $\chi(G) - \chi_c(G) \leq \chi(\mu^2(G)) - \chi_c(\mu^2(G))$. If this conjecture is true, then we would still wonder if the sequence $\chi(\mu^{2k}(G)) - \chi_c(\mu^{2k}(G))$ $(k = 0, 1, \cdots)$ would decrease, and approaches 0, as k goes to infinity.

THEOREM 11. *If G is a graph of chromatic number n, then $\chi_c(\mu^{n-1}(G)) \leq \chi(\mu^{n-1}(G)) - 1/2$.*

We do not know if Theorem 11 is sharp, in the sense that for any $m < n-1$ there would be an n-chromatic graph G such that $\chi_c(\mu^m(G)) = \chi(\mu^m(G))$.

References

[1] H. L. Abbott and B. Zhou, *The star chromatic number of a graph*, J. Graph Theory **17** (1993), 349-360.

[2] B. Bauslaugh and X. Zhu, *Circular coloring of infinite graphs*, Bulletin of the ICA **24** (1998), 79-80.

[3] B. Bollobás and N. Sauer, *Uniquely colorable graphs with large girth*, Canad. J. Math. **28** (1976), 1340-1344.

[4] J. A. Bondy and P. Hell, *A note on the star chromatic number*, J. Graph Theory **14** (1990), 479-482.

[5] G. J. Chang, L. Huang and X. Zhu, *Circular chromatic numbers of Mycielski's graphs*, Discrete Math. **205** (1999), 23-37.

[6] G. J. Chang, L. Huang and X. Zhu, *The circular chromatic numbers and the fractional chromatic numbers of distance graphs*, European J. Combin. **19** (1998), 423-431.

[7] G. J. Chang, D. Liu and X. Zhu, Distance graphs and T-coloring, J. Combin. Theory, Series B **75** (1999), 259-269.

[8] W. Deuber and X. Zhu, *Circular coloring of weighted graphs*, J. Graph Theory **23** (1996), 365-376.

[9] G. Gao, G. Hahn and H. Zhou, *Star-chromatic number of flower graphs*, preprint, 1992.

[10] G. Gao, E. Mendelsohn, and H. Zhou, *Computing star chromatic number from related graph invariants*, J. Combin. Math. Combin. Comput. **16** (1994), 87-95.

[11] G. Gao and X. Zhu, *Star extremal graphs and the lexicographic product*, Discrete Math. **152** (1996), 147-156.

[12] L. A. Goddyn, M. Tarsi and C. Q. Zhang, *On* (k,d)*-colorings and fractional nowhere-zero flows*, J. Graph Theory, **28** (1998), 155-161

[13] D. R. Guichard, *Acyclic graph coloring and the complexity of the star chromatic number*, J. Graph Theory **17** (1993), 129-134.

[14] L. Haddad and H. Zhou, *Star chromatic numbers of hypergraphs and partial steiner triple systems*, Discrete Math. **146** (1995), 45-58.

[15] P. Hell and X. Zhu, *The circular chromatic numbers of series-parallel graphs*, J. Graph Theory, **33** (2000), 14-24.

[16] A. J. W. Hilton, R. Rado and S. H. Scott, *A* (< 5)*-colour theorem for planar graphs*, Bull. London Math. Soc. **5** (1973), 302-306.

[17] L. Huang and G. J. Chang, *The circular chromatic number of the Mycielskian of* G_k^d, J. Graph Theory **32** (1999), 63-71.

[18] L. Huang and G. J. Chang, *Circular chromatic numbers of distance graphs with distance sets missing multiples*, Europ. J. Combin. **20** (1999), 1-8.

[19] Y. Kirsch, Personal communication via Zhu.

[20] D. Moser, *The star-chromatic number of planar graphs*, J. Graph Theory **24** (1997), 33-43.

[21] J. Mycielski, *Sur le chromatic numbers*, Colloq Math. **3** (1955), 161-162.

[22] E. Steffen and X. Zhu, *On the star chromatic numbers of graphs*, Combinatorica **16** (1996), 439-448.

[23] A. Vince, *Star chromatic number*, J. Graph Theory **12** (1988), 551-559.

[24] B. Zhou, *Some theorems concerning the star chromatic number of a graph*, J. Combin. Theory, Series B **70** (1997), 245-258.

[25] X. Zhu, *On the bounds for the ultimate independence ratio of a graph*, Discrete Math. **156** (1996), 229-236.

[26] X. Zhu, *Star chromatic numbers and products of graphs*, J. Graph Theory **16** (1992), 557-569.

[27] X. Zhu, *Uniquely* H*-colorable graphs with large girth*, J. Graph Theory **23** (1996), 33-41.

[28] X. Zhu, *The circular chromatic number and the fractional chromatic number of a graph*, preprint, 1996.

[29] X. Zhu, *A simple proof of Moser's theorem*, J. Graph Theory **30** (1999), 19-26.

[30] X. Zhu, *Planar graphs with circular chromatic numbers between* 3 *and* 4, J. Combin. Theory, Series B **76** (1999), 170-200.

[31] X. Zhu, *Construction of uniquely* H*-colorable graphs*, J. Graph Theory **30** (1999), 1-6.

[32] X. Zhu, *Graphs whose circular chromatic number equal the chromatic number*, Combinatorica **19** (1999), 139-149.

[33] X. Zhu, *Circular chromatic number and graph minor*, preprint, 1997.

[34] X. Zhu, *Circular chromatic number: a survey*, Discrete Math., to appear.

DEPARTMENT OF APPLIED MATHEMATICS, NATIONAL CHIAO TUNG UNIVERSITY, HSINCHU 30050, TAIWAN

E-mail address: gjchang@math.nctu.edu.tw

AMS/IP Studies in Advanced Mathematics
Volume 20, 2001

Approximations for Color-Covering Problems

Lu Ruan , Ding-Zhu Du , Xiaodong Hu , Xiaohua Jia , and Deying Li

ABSTRACT. Consider a connected graph G. Suppose each edge of G is in a set of colors. The problem is to find the minimum number of colors such that all edges in those colors contains a spanning tree of G. In this note, we show that the problem has a polynomial-time approximation with performance ratio $\ln n + O(1)$ and this performance ratio is the best possible unless $NP \subseteq DTIME(n^{poly \log n})$ where n is the number of vertices in G. We also show an application in optical networks.

Key Words: Network optimization, optical networks.

1. Introduction

Let $G = (V, E)$ be a connected graph. Suppose each edge of G has a set of colors, $c : E \rightarrow 2^N$. A set of colors is called a *color-covering* if all edges in those colors contains a spanning tree of G. In this paper, we study three problems on the color-covering.

Problem 1: Given a graph $G = (V, E)$ and edge-color-sets $c : E \rightarrow 2^N$, compute a color-covering of minimum cardinality.

The optimal solution of Problem 1 is called a *minimum color-covering.* For each vertex x, define the set of colors of x, $c(x)$ to be the union of color sets of edges incident to x.

Problem 2: Given a graph $G = (V, E)$ and edge-color-sets $c : E \rightarrow 2^N$, find the minimum number of vertices such that all colors of those vertices form a color-covering.

Both Problems 1 and 2 will be showed to have polynomial-time approximations with performance ratio $\ln n + 1$ and not to have a polynomial-time approximation with performance ratio $\rho \ln n$ for $\rho < 1$ unless $NP \subset DTIME(n^{poly \log n})$ where n is the number of vertices in G.

In this paper, a subset of vetices is said to be *color-connected* if connecting every pair of vertices in the same color gives a connected graph.

Problem 3: Given a graph G and color-sets for each edge of G such that for every color all edges in the color form a connected subgraph, find the minimum color-connected subset of vertices such that all colors of those vertices form a color-covering.

We will also show that Problem 3 has a polynomial-time approximation with performance ratio $2(\ln n + 1)$ and no polynomial-time approximation with performance ratio $\rho \ln n$ for $\rho < 1$ unless $NP \subset DTIME(n^{poly \log n})$ where n is the number of vertices in G.

Moreover, we discuss the application background of these problems in optical networks.

2. Minimum Color-Covering

In this section, we study Problem 1, the minimum color-covering problem. Before present our approximation, let us first show an easy negative result.

THEOREM 1. *Problem 1 has no polynomial-time approximation with performance ratio $\rho \ln n$ for $\rho < 1$ unless $NP \subset DTIME(n^{poly \log n})$ where n is the number of vertices in input graph.*

Proof. The proof is based on a recent result of Raz and Safra [**6**] which is obtained by improving a result of Lund and Yannakakis [**3**]. Consider the following problem.

Set-Covering: Given a (finite) collection $\mathcal{S}$ of subsets of a set U of n elements, find a minimum cardinality subcollection of $\mathcal{S}$ such that the union of all subsets in the subcollection covers U.

They proved the following.

LEMMA 2. *For any $0 < \rho < 1$, there is no polynomial-time approximation algorithm with performance ratio $\rho \ln n$ for the set-covering problem unless $NP \subset$ DTIME $(n^{poly \log n})$.*

Now, suppose the color-covering problem has a polynomial-time approximation with performance ratio $\rho \ln n$. We then show that the set-covering problem also has a polynomial-time approximation with performance ratio $\rho \ln n$.

For each input collection $\mathcal{S}$ of subsets of a set U of n elements, construct a graph G with edge-colors as follows: G has $n+1$ vertices consisting of all elements in U and a special vertex o not in U, and n edges (o, a) for $a \in U$. Suppose $\mathcal{S} = \{S_1, S_2, \cdots, S_m\}$. Then for each $a \in S_i$, assign a color i to edge (o, a). Clearly, $\{S_{i_1}, S_{i_2}, \cdots, S_{i_k}\}$ is a set-covering of U if and only if $\{i_1, i_2, \cdots, i_k\}$ is a color-covering of G. It follows immediately that if $\{i_1, i_2, \cdots, i_k\}$ is a color-covering of G with a factor of $\rho \ln n$ from optimal, then $\{S_{i_1}, S_{i_2}, \cdots, S_{i_k}\}$ is a set-covering of U with a factor of $\rho \ln n$ from optimal. By Lemma 2, Theorem 1 is proved. □

COROLLARY 3. *Problem 1 is NP-hard.*

Next, we show that a greedy algorithm reaches the best possible approximation performance ratio.

Let E_i denote the set of all edges with color i. Let $r(E_i, H)$ be the number of components reduced by adding E_i to graph H.

Greedy Algorithm 1

```
1.  input G = (V, E) and c : E → 2^N;
2.  C ← ∅;
3.  H ← the empty graph with vertex set V;
4.  E_i ← {e ∈ E | i ∈ c(e)};
5.  while H is not connected
6.  do select a color i that maximizes r(E_i, H);
7.      C ← C ∪ {i};
8.      H ← H ∪ E_i;
9.  return C.
```

THEOREM 4. *Greedy Algorithm 1 produces an approximation solution within a factor of* $\ln n + 1$ *from optimal for Problem 1 where* n *is the number of vertices in input graph.*

Proof. Suppose C is the color-set obtained by Greedy Algorithm 1 and $H = \cup_{i\in C} E_i$. Let $1, 2, \cdots, k$ be elements of C in the order of their appearance. Denote by H_i the graph with vertex set V and edge set $E_1 \cup \cdots \cup E_i$. Denote $r_i = r(E_i, H_{i-1})$.

Suppose C^* be an optimal solution and $H^* = \cup_{e\in C^*} E_e$. For $e \in C^*$, denote $c_{ei} = r(E_e, H_{i-1})$ and

$$w(e) = \sum_{i=1}^{k-1}(c_{ei} - c_{e,i+1})\frac{1}{r_i} + c_{ek}\frac{1}{r_k} = \frac{1}{r_1}c_{e1} + \sum_{i=2}^{k}(\frac{1}{r_i} - \frac{1}{r_{i-1}})c_{ei}.$$

Our first claim is that

$$|C| \le \sum_{e\in C^*} w(e).$$

To show the first claim, note that

$$|C| = \sum_{i=1}^{k}\frac{r_i}{r_i} = \frac{1}{r_1}\sum_{j=1}^{k} r_j + \sum_{i=2}^{k}(\frac{1}{r_i} - \frac{1}{r_{i-1}})\sum_{j=i}^{k} r_j.$$

Moreover, from the line 6 of Greedy Algorithm 1, we know

$$r_1 \ge r_2 \ge \cdots \ge r_k.$$

Thus, it suffices to prove that for any $i = 1, 2, \cdots, k$,

$$\sum_{j=i}^{k} r_j \le \sum_{e\in C^*} c_{ei}.$$

This inequality holds since

$$\sum_{j=i}^{k} r_j = r(H, H_{i-1}) = r(H^*, H_{i-1}) \le \sum_{e\in C^*} r(E_e, H_{i-1}) = \sum_{e\in C^*} c_{ei}.$$

The first claim is proved.

Our second claim is that for every $e \in C^*$,

$$w(e) \le \sum_{i=1}^{k}\frac{1}{i}.$$

To show it, note that from the line 6 of Greedy Algorithm 1, we know $c_{ei} \leq r_i$ for all $i = 1, 2, \cdots, k$. Thus,

$$\begin{aligned} w(e) &= \frac{1}{r_1} c_{e1} + \sum_{i=2}^{k} \left(\frac{1}{r_i} - \frac{1}{r_{i-1}}\right) c_{ei} \\ &\leq \frac{1}{r_1} r_1 + \sum_{i=2}^{k} \left(\frac{1}{r_i} - \frac{1}{r_{i-1}}\right) r_i \\ &= \sum_{i=1}^{k-1} (r_i - r_{i+1}) \frac{1}{r_i} + r_k \frac{1}{r_k} \\ &\leq \sum_{i=1}^{k} \frac{1}{i}. \end{aligned}$$

By the two claims,

$$|C| \leq \sum_{e \in C^*} w(e) \leq |C^*| \sum_{i=1}^{k} \frac{1}{i} \leq |C^*|(1 + \ln n).$$

□

3. Converter Placement

When wavelength configuration is done, an optical network can be considered as a graph $G = (V, E)$ that each edge has a set of wavelengths [**1, 2, 4, 5, 7**]. The multicast routing requires a spanning subgraph of G. If a message from an edge to another edge uses different wavelengths, then a converter is required at the common endpoint of the two edges.

Let each color represent a connected component in a subgraph with vertex set V and all edges with a certain wavelength. Then every color-covering corresponds a routing for multicast in the optical network. An optimization problem about it is to minimize the number of converters required by multicast routing. The mathematical model is Problem 3.

To solve Problem 3, let us first study Problem 2.

THEOREM 5. *Problem 2 has no polynomial-time approximation with performance ratio* $\rho \ln n$ *for* $\rho < 1$ *unless* $NP \subset DTIME(n^{poly \log n})$ *where* n *is the number of vertices in input graph. Moreover, Problem 2 has a polynomial-time approximation with performance ratio* $\ln n + 1$.

Proof. Consider the set-covering problem. For each input collection $\mathcal{S}$ of subsets of a set U of m elements, construct a graph $G = (V, E)$ as follows: The vertex set V consists of all elements in U, all subsets in $\mathcal{S}$, and a special vertex o not in U and $\mathcal{S}$. The edge set E consists of (o, S) for all $S \in \mathcal{S}$ and (a, S) for all $a \in S \in \mathcal{S}$. Let $U = \{a_1, a_2, \cdots, a_m\}$. Define $c : E \to 2^N$ by

$$\begin{aligned} c((o, S)) &= \{m + 1\} \text{ for all } S \in \mathcal{S}, \\ c((a_i, S)) &= \{i\} \text{ for all } a_i \in S \in \mathcal{S}. \end{aligned}$$

For contradiction, suppose Problem 2 on input $< G, c >$ has a polynomial-time approximation $V' \subseteq V$ with performance ratio $\rho \ln n$. Without loss of generality we may assume $V' \subseteq \mathcal{S}$ since, otherwise, we can easily find a solution in $\mathcal{S}$ with the same (possibly small) cardinality by replacing each vertex not in $\mathcal{S}$ with one of its adjacent vertices. It is easy to verify that V' is also a set-covering for $\mathcal{S}$. Since an optimal solution $V^* \subseteq \mathcal{S}$ for the set-covering problem is an optimal solution for Problem 2, V' is an approximation for Problem 2 with performance ratio $\rho \ln n = \rho \ln(m + |\mathcal{S}| + 1)$. When $|\mathcal{S}| \le n$, $\rho \ln(m + |\mathcal{S}| + 1) \le \rho' \ln m$ for $\rho < \rho' < 1$ and sufficiently large m. Therefore, V' is a polynomial-time approximation with performance ratio $\rho' \ln /m$ for the set-covering problem in the special case $|\mathcal{S}| \le n$. From **[6]**, we know that Lemma 2 holds in this special case, a contradiction is found.

Consider an input $< G, c >$ of Problem 2. Let $v_1, v_2, \cdots, v_n$ be all vertices of G. , we define $c' : E \to 2^N$ by

$$c'(e) = \{i' \mid e \text{ is in a color of } v_i\}.$$

Now, $\{v_{i_1}, v_{i_2}, \cdots, v_{i_k}\}$ is a solution for Problem 2 on input $< G, c >$ if and only if $\{i'_1, i'_2, \cdots, i'_k\}$ is a solution for Problem 1 on input $< G, c' >$. By Theorem 4, the second half of the theorem is proved. □

Now, we present the following result about Problem 3.

THEOREM 6. *Problem 3 has no polynomial-time approximation with performance ratio* $\rho \ln n$ *for* $\rho < 1/2$ *unless* $NP \subset DTIME(n^{poly \log n})$ *where* n *is the number of vertices in input graph. Moreover, Problem 3 has a polynomial-time approximation with performance ratio* $2(\ln n + 1)$.

Proof. It follows immediately from Theorem 5 and a fact that for any solution of k vertices for Problem 2 there exists a solution of at most $2k - 1$ vertices for Problem 3. □

Aknowledgement

The authors wish to thank Professor Manuel Blum for helpful discussion and comments.

References

[1] D.-Y. Li, X.-F. Du, X.-D. Hu, and X.-H. Jia, Minimizing number of wavelengths in multicast routing trees in WDM networks, manuscript.

[2] P.E. Green, *Fiber-Optic Networks*, (Prentical-Hall, Cambrige, MA, 1992).

[3] C. Lund and M. Yannakakis, On the hardness of approximating minimization problems, *J. ACM* 41 (1994) 960-981.

[4] B. Mukherjee, WDM-based local lightwave networks Part I: single-hop system, *IEEE Networks* 3 (1992) 12-26.

[5] B. Mukherjee, WDM-based local lightwave networks Part II: multihop system, *IEEE Networks* 4 (1992) 22-32.

[6] R. Raz and S. Safra, A sub-constant error-probability low-degree test, and a sub-constant error-probability PCP characterization of NP, *Proc. 28th STOCS* (1997).

[7] R. Ramaswami and G. Sasaki, Multiwavelength optical networks with limited wavelength coversion, *IEEE/ACM Trans. on Networking* 6 (1998) 744-754.

ICCM98 出席会议者名单

(一九九八年十二月十二日)

总人数： 465 人 总共 9 页

No: 1 **Jun Ai** (Chinese U HK)
No: 2 **R. Borcherds** (Cambridge Univ, UK)
No: 3 **Ching-Li Chai** (U Penn, USA)
No: 4 **Raymond Chan** (Chinese U HK)
No: 5 **Tony Chan** (UCLA, USA)
No: 6 **Gerard J. Chang** (Nat'l Chiao Tung U)
No: 7 **Mei-Chu Chang** (UC Riverside, USA)
No: 8 **S. C. Chang** (Nat'l Tsing-Hua U)
No: 9 **S. Y. A. Chang** (UCLA & Princeton)
No: 10 **Beifang Chen** (HK U Sci Tech)
No: 11 **C. C. Chen** (Nat'l Taiwan U)
No: 12 **Gui-Qiang Chen** (Northwestern U, USA)
No: 13 **Jungkai Chen** (Nat'l Chung Cheng U)
No: 14 **Roger Chen** (Nat'l Cheng Kung U)
No: 15 **Yuqun Chen** (Chinese U HK)
No: 16 **Chong Q. Cheng** (Nanjing Univ.)
No: 17 **Jih-Hsin Cheng** (Inst Math,AS,TW)
No: 18 **Nen-Fon Cheng** (Nat'l Chengchi U)
No: 19 **Sui Sun Cheng** (Nat'l Tsing-Hua U)
No: 20 **Sun-Jen Cheng** (Nat'l Cheng-Kung U)
No: 21 **Hua-Huai Chern** (Nat'l Chung Cheng U)
No: 22 **I- Liang Chern** (Nat'l Taiwan U)
No: 23 **Chin P. Cheung** (Chinese U HK)
No: 24 **Man-Duen Choi** (U Toronto, Canada)
No: 25 **Kai Seng Chou** (Chinese U HK)
No: 26 **Shui-Nee Chow** (Georgia Tech & NUS)
No: 27 **Cho-Ho Chu** (London U, UK)
No: 28 **C.K. Chu** (Columbia U, USA)
No: 29 **Fan Chung** (U Penn, USA)
No: 30 **John Coates** (Cambridge U, UK)
No: 31 **Bo Cui** (Harvard U, USA)
No: 32 **Jinqiao Duan** (Clemson U, USA)
No: 33 **Ding-Zhu Du** (U Minnesota, USA)
No: 34 **Keqin Feng** (Chinese U Sci Tech)
No: 35 **Xinchu Fu** (Heriot-Watt U, UK)
No: 36 **Shen Gong** (Chinese U Sci Tech)
No: 37 **Ron Graham** (AT&T Bell Lab, USA)
No: 38 **V. N. Grebenev** (Inst Comp Tech, RU)
No: 39 **Chaohao Gu** (Fudan Univ.)
No: 40 **Lei Guo** (Inst System Sci, CAS)
No: 41 **Yongsheng Han** (Auburn U, USA)
No: 42 **Jiaxing Hong** (Fudan Univ.)
No: 43 **Tom Hou** (Caltech, USA)
No: 44 **Po-Hsun Hsieh** (Nat'l Chung Cheng U)
No: 45 **C. H. Hsu** (Nat'l Dong-Hwa U)
No: 46 **Shu-Yu Hsu** (Chinese U HK)
No: 47 **Sze-Bi Hsu** (Nat'l Tsing-Hua U)
No: 48 **Xiaojun Huang** (Rutgers U, USA)
No: 49 **Ching Nam Hung** (Chinese U HK)
No: 50 **Frank Hwang** (Nat'l Chiao Tung U)
No: 51 **Lizhen Ji** (U Michigan, USA)
No: 52 **Xinhua Jiang** (City Univ. HK)
No: 53 **Xuan Jiang** (Univ. NSW, AU)
No: 54 **Xiao-Qing Jin** (Univ.of Macau)

No: 55 **Naihuan Jing** (N Carolina State,USA)
No: 56 **Ning Ju** (Indiana Univ., USA)
No: 57 **Ming-chang Kang** (Nat'l Taiwan U)
No: 58 **Dexing Kong** (Kyoto Sangyo U)
No: 59 **Yin To Kong** (Chinese U HK)
No: 60 **Kit-Ian Kou** (Univ. of Macau)
No: 61 **Chun Kit Kwan** (Univ.of HK)
No: 62 **Hang-Chin Lai** (I-Shou Univ.)
No: 63 **Ming-Jun Lai** (U Georgia, USA)
No: 64 **Peter C. B. Lam** (HK Baptist U)
No: 65 **Ka-Sing Lau** (Chinese U HK)
No: 66 **Shui-kong Lau** (Chinese U HK)
No: 67 **Wayne Lawton** (Nat'l U Singapore)
No: 68 **Seng Luan Lee** (Nat'l U Singapore)
No: 69 **Ka Keung Lei** (Chinese U HK)
No: 70 **Siu-Long Lei** (Univ. of Macau)
No: 71 **Daqian Li** (Fudan Univ.)
No: 72 **Guanxin Li** (Chinese U HK)
No: 73 **Hua-Chieh Li** (Nat'l Tsing-Hua U)
No: 74 **Jianhua Li** (Lawrence Berkeley Lab)
No: 75 **Jian-Shu Li** (U Maryland & HKUST)
No: 76 **Jiequan Li** (Inst App Math, CAS)
No: 77 **Jun Li** (Stanford U, USA)
No: 78 **Peter Li** (UC Irvine, USA)
No: 79 **Song-Ying Li** (UC Irvine, USA)
No: 80 **Wen-Ching W Li** (Penn State, USA)
No: 81 **Yanyan Li** (Rutgers U, USA)
No: 82 **Bong Lian** (Brandeis U, USA)
No: 83 **Chang-Shou Lin** (Nat'l Chung-Cheng U)
No: 84 **Chin-Cheng Lin** (Nat'l Central U)
No: 85 **Fang-Hua Lin** (New York U, USA)
No: 86 **Ke-Pao Lin** (Univ Ill Chicago, USA)
No: 87 **Song Sun Lin** (Nat'l Chiao Tung U)
No: 88 **Tai-Chia Lin** (Nat'l Chung-Cheng U)
No: 89 **Wen-Hsiung Lin** (Nat'l Tsing-Hua U)
No: 90 **Xiao-Song Lin** (UC Riverside, USA)
No: 91 **Jian-Guo Liu** (U Maryland, USA)
No: 92 **Kefeng Liu** (Stanford U, USA)
No: 93 **Tai-Ping Liu** (Stanford U, USA)
No: 94 **Tong Liu** (Tsinghua Univ)
No: 95 **Xu-Dong Liu** (UC Santa Barbara)
No: 96 **Kening Lu** (Brigham Young U, USA)
No: 97 **Ya Yan Lu** (City U HK)
No: 98 **Wei Luo** (MIT, USA)
No: 99 **Duy-Minh Nhieu** (Inst Math, AS, TW)
No: 100 **Wen-Xiu Ma** (City U HK)
No: 101 **Zhiming Ma** (Inst App Math, CAS)
No: 102 **Kin-wai Mak** (HK U Sci Tech)
No: 103 **Guo-Wu Meng** (HK U Sci Tech)
No: 104 **Ngaiming Mok** (Hong Kong Univ.)
No: 105 **Siu-Ah Ng** (U Natal, S Africa)
No: 106 **Ronghua Pan** (SISSA, Italy)
No: 107 **Duong Phong** (Columbia U, USA)
No: 108 **Bob Plemmons** (Wake Forest U, USA)
No: 109 **Kin-Keung Poon** (HK Baptist U)
No: 110 **Zhenbo Qin** (Oklahoma State, USA)
No: 111 **Haobo Ren** (HK Poly Univ)
No: 112 **Bernard Russo** (UC Irvine, USA)
No: 113 **Wei-Chang Shann** (Nat'l Central U)
No: 114 **Ian J. Shelley** (Cambridge U Press, UK)

No: 115 **Jie Shen** (Penn State, USA)
No: 116 **Zengmin Song** (HK Baptist U)
No: 117 **Daniel Stroock** (MIT, USA)
No: 118 **Hsuan-Wen Su** (Nat'l Chong-Cheng U)
No: 119 **Hugo S. H. Sun** (CSU Fresno, USA)
No: 120 **Xiaofeng Sun** (Stanford U, USA)
No: 121 **Weiwei Sun** (City U HK)
No: 122 **Luen-fai Tam** (Chinese U HK)
No: 123 **Eng Chye Tan** (Nat'l U Singapore)
No: 124 **Sheng-Li Tan** (East China N U)
No: 125 **Tai-man Tang** (Hong Kong Univ.)
No: 126 **Tao Tang** (HK Baptist U)
No: 127 **Luo Tao** (CUHK & Inst Math)
No: 128 **Gang Tian** (MIT, USA)
No: 129 **Dong-Ho Tsai** (Nat'l Chung Cheng U)
No: 130 **Man Tsoi** (Chinese U HK)
No: 131 **Jengnan Tzeng** (Nat'l Taiwan U)
No: 132 **Frederic Wan** (UC Irvine, USA)
No: 133 **Tom Y. H. Wan** (Chinese U HK)
No: 134 **Bai-Ling Wang** (U Adelaide, AU)
No: 135 **Chin-Lung Wang** (Nat'l Tsing-Hua U)
No: 136 **Jenn-Nan Wang** (Nat'l Cheng Kung U)
No: 137 **Mu-Tao Wang** (Stanford U, USA)
No: 138 **Ren-Hong Wang** (Dalian U Tech)
No: 139 **Wei-cheng Wang** (Nat'l Tsing-Hua U)
No: 140 **Xiaowei Wang** (U Minnesota, USA)
No: 141 **Zhen Wang** (City U HK)
No: 142 **Zhifeng Wang** (HK U Sci Tech)
No: 143 **Roderick Wong** (City U HK)
No: 144 **Berlin Wu** (Nat'l Chengchi U)
No: 145 **Jyh-Yang Wu** (Nat'l Chung-Cheng U)
No: 146 **Jun Wu** (Vancouver, Canada)
No: 147 **Sijue Wu** (U Maryland, USA)
No: 148 **Zhijian Wu** (U Alabama, USA)
No: 149 **Zongmin Wu** (Fudan U & City U HK)
No: 150 **Hasi Wulan** (U Joensuu, Finland)
No: 151 **Jeff Zhihong Xia** (Northwestern U, USA)
No: 152 **Ganquan Xie** (Lawrence Berkeley Lab)
No: 153 **Zhouping Xin** (NYU & HK Chinese U)
No: 154 **Haipeng Xing** (HK U Sci Tech)
No: 155 **Jinchao Xu** (Penn State, USA)
No: 156 **Zhenhua Xu** (Wuhan Univ.)
No: 157 **Larry Xue** (Bradley U, USA)
No: 158 **Tong Yang** (City U HK)
No: 159 **Horng-Tzer Yau** (New York U, USA)
No: 160 **Stephen S T Yau** (U Illinois Chicago, USA)
No: 161 **S.-T. Yau** (Harvard Univ, USA)
No: 162 **Yangbo Ye** (U Iowa, USA)
No: 163 **Ming Ham Yip** (City U HK)
No: 164 **Jing Yu** (Inst Math, TW)
No: 165 **Ya-xiang Yuan** (Chinese Acad Sci)
No: 166 **Qiang Zhang** (City U HK)
No: 167 **Shouwu Zhang** (Columbia U, USA)
No: 168 **Tong Zhang** (Inst Math, CAS)
No: 169 **Yu-Qiu Zhao** (City U HK)
No: 170 **Sifen Zhong** (Princeton, USA)
No: 171 **Ding-Xuan Zhou** (City U HK)
No: 172 **Jiazu Zhou** (U West Alabama, USA)
No: 173 **Xiangyu Zhou** (Inst Math, CAS)
No: 174 **Changjiang Zhu** (City U HK)

No: 175 **Chengbo Zhu** (Nat'l U Singapore)
No: 176 **Kehe Zhu** (SUNY Albany, USA)
No: 177 **Xiping Zhu** (Zhongshan Univ)
No: 178 **Yongchang Zhu** (HK U Sci Tech)
No: 179 **Zuonong Zhu** (HK Baptist U)
No: 180 **Jun Zou** (Chinese U HK)
No: 181 羊丹平 (山东大学)
No: 182 程爱杰 (山东大学)
No: 183 江成顺 (山东大学)
No: 184 仪洪勋 (山东大学)
No: 185 杨连中 (山东大学)
No: 186 扈培础 (山东大学)
No: 187 袁益让 (山东大学)
No: 188 刘兆理 (山东大学)
No: 189 娄本东 (山东大学)
No: 190 伍鹏程 (贵州民族学院)
No: 191 王万恒 (贵州民族学院)
No: 192 王　林 (贵州民族学院)
No: 193 索洪敏 (贵州民族学院)
No: 194 吴有富 (贵州民族学院)
No: 195 李翊神 (中国科技大学)
No: 196 梅加强 (中国科技大学)
No: 197 史济怀 (中国科技大学)
No: 198 任广斌 (中国科技大学)
No: 199 叶向东 (中国科技大学)
No: 200 刘太顺 (中国科技大学)
No: 201 刘　浩 (中国科技大学)
No: 202 章　璞 (中国科技大学)
No: 203 李志斌 (兰州大学)
No: 204 钟承奎 (兰州大学)
No: 205 罗彦锋 (兰州大学)
No: 206 张志强 (兰州大学)
No: 207 陈特为 (华南师大)
No: 208 郑　权 (华中理工大学)
No: 209 张显文 (华中理工大学)
No: 210 王志成 (湖南大学)
No: 211 唐先华 (湖南大学)
No: 212 李庆国 (湖南大学)
No: 213 李开泰 (西安交大)
No: 214 侯延仁 (西安交大)
No: 215 梅立泉 (西安交大)
No: 216 周小林 (西安交大)
No: 217 张　斌 (西安交大)
No: 218 何银年 (西安交大)
No: 219 陈口斌 (西安交大)
No: 220 李功胜 (西安交大)
No: 221 赵春山 (西安交大)
No: 222 李东升 (西安交大)

No: 223 李洪军 (西安交大)
No: 224 李明军 (西安交大)
No: 225 王立周 (西安交大)
No: 226 王巨平 (复旦大学)
No: 227 陈纪修 (复旦大学)
No: 228 邱维元 (复旦大学)
No: 229 王亚光 (复旦大学)
No: 230 丁 青 (复旦大学)
No: 231 蒋尔雄 (复旦大学)
No: 232 薛军工 (复旦大学)
No: 233 王兴华 (浙江大学西溪校区)
No: 234 代雄平 (浙江大学西溪校区)
No: 235 沈一兵 (浙江大学西溪校区)
No: 236 东瑜昕 (浙江大学西溪校区)
No: 237 程晓良 (浙江大学西溪校区)
No: 238 李 方 (浙江大学西溪校区)
No: 239 陈志国 (浙江大学西溪校区)
No: 240 刘宗光 (浙江大学西溪校区)
No: 241 孙利明 (浙江大学西溪校区)
No: 242 王 伟 (浙江大学西溪校区)
No: 243 唐绍祥 (宁波大学)
No: 244 黄土森 (宁波大学)
No: 245 陶祥兴 (宁波大学)
No: 246 杜先能 (安徽大学)
No: 247 顾荣宝 (安徽大学)
No: 248 李世雄 (安徽大学)
No: 249 汪继文 (安徽大学)
No: 250 陈兰祥 (同济大学)
No: 251 陈志华 (同济大学)
No: 252 陆洪文 (同济大学)
No: 253 苏维宜 (南京大学)
No: 254 程崇庆 (南京大学)
No: 255 魏宝社 (南京大学)
No: 256 王绍立 (南京大学)
No: 257 秦厚荣 (南京大学)
No: 258 丁德成 (南京大学)
No: 259 陆 宏 (南京大学)
No: 260 黄震宇 (南京大学)
No: 261 马吉溥 (南京大学)
No: 262 刘士强 (宁夏大学)
No: 263 田振夫 (宁夏大学)
No: 264 张卫国 (宁夏大学)
No: 265 王明新 (东南大学)
No: 266 孙志忠 (东南大学)
No: 267 潮小李 (东南大学)
No: 268 尹永成 (浙江大学)
No: 269 梁金荣 (浙江大学)
No: 270 施锡泉 (大连理工大学)

No: 271 苏志勋 (大连理工大学)
No: 272 徐定华 (上海大学)
No: 273 陆鸣皋 (上海大学)
No: 274 蔡迎春 (上海大学)
No: 275 王汉兴 (上海大学)
No: 276 王天泽 (河南大学)
No: 277 王　术 (河南大学)
No: 278 郑驻军 (河南大学)
No: 279 卢克平 (河南大学)
No: 280 李起升 (河南大学)
No: 281 定光桂 (南开大学)
No: 282 王日生 (南开大学)
No: 283 何　华 (南开大学)
No: 284 陈万义 (南开大学)
No: 285 刘明学 (福建师大)
No: 286 李永青 (福建师大)
No: 287 林火南 (福建师大)
No: 288 王维克 (武汉大学)
No: 289 杜金元 (武汉大学)
No: 290 刘培德 (武汉大学)
No: 291 文志英 (武汉大学)
No: 292 饶　辉 (武汉大学)
No: 293 吴　军 (武汉大学)
No: 294 姚家燕 (武汉大学)
No: 295 奚李峰 (武汉大学)
No: 296 齐民友 (武汉大学)
No: 297 陈　华 (武汉大学)
No: 298 陈文义 (武汉大学)
No: 299 叶瑞松 (汕头大学)
No: 300 战荫伟 (汕头大学)
No: 301 殷慰萍 (首都师大)
No: 302 张增喜 (首都师大)
No: 303 刘晓玫 (首都师大)
No: 304 袁超伟 (北京邮电大学)
No: 305 胡泽军 (郑州大学)
No: 306 陈绍春 (郑州大学)
No: 307 慕小武 (郑州大学)
No: 308 刘祖汉 (扬州大学)
No: 309 林支桂 (扬州大学)
No: 310 项筱玲 (贵州大学)
No: 311 宋永忠 (南京师范大学)
No: 312 孙文瑜 (南京师范大学)
No: 313 陈才二 (南京师范大学)
No: 314 汪明义 (西南交大)
No: 315 侯振挺 (长沙铁道学院)
No: 316 刘再明 (长沙铁道学院)
No: 317 张汉君 (长沙铁道学院)
No: 318 邵儒林 (云南师大)

No: 319 吴　鲜 (云南师大)
No: 320 郭聿琦 (云南大学)
No: 321 郑喜印 (云南大学)
No: 322 周作领 (中山大学)
No: 323 黄　煜 (中山大学)
No: 324 胡国权 (中山大学)
No: 325 施武杰 (西南师大)
No: 326 陈贵云 (西南师大)
No: 327 杨　雷 (上海交大)
No: 328 许德良 (上海交大)
No: 329 余军扬 (上海交大)
No: 330 周春琴 (上海交大)
No: 331 王国荣 (上海师大)
No: 332 许庆祥 (上海师大)
No: 333 彭国华 (四川大学)
No: 334 张起帆 (四川大学)
No: 335 于秀源 (杭州师院)
No: 336 阚家海 (南京邮电学院)
No: 337 孙国正 (安徽师大)
No: 338 严　平 (安徽师大)
No: 339 丁万鼎 (安徽师大)
No: 340 祝东进 (安徽师大)
No: 341 董新汉 (湖南师大)
No: 342 郭晋云 (湖南师大)
No: 343 黄元秋 (湖南师大)
No: 344 陈传淼 (湖南师大)
No: 345 张福基 (厦门大学)
No: 346 赵俊宁 (厦门大学)
No: 347 谭　忠 (厦门大学)
No: 348 郭铁信 (厦门大学)
No: 349 徐宪民 (浙江师大)
No: 350 何国龙 (浙江师大)
No: 351 陈芳跃 (浙江师大)
No: 352 李金全 (浙江师大)
No: 353 陈胜敏 (浙江师大)
No: 354 叶慈南 (上海理工大学)
No: 355 向淑文 (贵州省科委)
No: 356 韩　波 (哈尔滨工业大学)
No: 357 马昕东 (哈尔滨工业大学)
No: 358 高荣兴 (湖南财经学院)
No: 359 付希林 (山东师大)
No: 360 李师正 (山东师大)
No: 361 张立琴 (山东师大)
No: 362 瞿文广 (山东师大)
No: 363 樊太和 (陕西师大)
No: 364 赵　彬 (陕西师大)
No: 365 畅大为 (陕西师大)
No: 366 张利军 (陕西师大)

No: 367 张京良 (陕西师大)
No: 368 陈 露 (陕西师大)
No: 369 霍海峰 (陕西师大)
No: 370 罗建书 (国防科大)
No: 371 吴振德 (河北师大)
No: 372 王彦英 (河北师大)
No: 373 陈述涛 (哈师大)
No: 374 赖万才 (华侨大学)
No: 375 欧阳才衡 (武汉物理与数学所)
No: 376 周焕松 (武汉物理与数学所)
No: 377 李工宝 (武汉物理与数学所)
No: 378 钱黎文 (武汉物理与数学所)
No: 379 丁义明 (武汉物理与数学所)
No: 380 汤铭瑞 (航天工业总公司 204 所)
No: 381 刘德贵 (航天工业总公司 204 所)
No: 382 宋晓秋 (航天工业总公司 204 所)
No: 383 陈丽容 (航天工业总公司 204 所)
No: 384 陈翰麟 (中科院数学所)
No: 385 陆启铿 (中科院数学所)
No: 386 张立群 (中科院数学所)
No: 387 王跃飞 (中科院数学所)
No: 388 张 景 (中科院数学所)
No: 389 崔景安 (中科院数学所)
No: 390 吉新华 (中科院数学所)
No: 391 林 勇 (人民大学)
No: 392 肖爱国 (湘潭大学)
No: 393 王 键 (湘潭大学)
No: 394 陈仲沪 (湘潭大学)
No: 395 龚志明 (湘潭大学)
No: 396 舒 适 (湘潭大学)
No: 397 高随祥 (中科院研究生院)
No: 398 杨庆之 (中科院研究生院)
No: 399 杨大春 (北师大)
No: 400 林润亮 (清华大学)
No: 401 曾云波 (清华大学)
No: 402 马仁义 (清华大学)
No: 403 王晓峰 (清华大学)
No: 404 简怀玉 (清华大学)
No: 405 郑建华 (清华大学)
No: 406 马 力 (清华大学)
No: 407 陆志奇 (河南师大)
No: 408 袁平之 (郑州信息工程学院)
No: 409 黄乘规 (天津师院)
No: 410 关颖男 (东北大学)
No: 411 徐 涛 (吉林工业大学)
No: 412 罗学波 (西北工大)
No: 413 韩 东 (新疆大学)
No: 414 焦晓祥 (南昌大学)

No: 415 杨 一 都 (贵州师大)
No: 416 欧 业 林 (广西民族学院)
No: 417 张 学 知 (河北大学)
No: 418 王 晓 敏 (河北大学)
No: 419 揣 建 军 (河北大学)
No: 420 陈 纪 阳 (福州师范高等专科学校)
No: 421 **J P Bourguignon** (IHES, France)
No: 422 **Felix E Browder** (Rutgers, USA)
No: 423 **Ronnie Chan** (Morningside Group, HK)
No: 424 **Kun Soo Chang** (Korean U & KMS)
No: 425 **Eugene Durenard** (USA)
No: 426 **Happy Harun** (Morningside Group, HK)
No: 427 **Arthur M. Jaffe** (Harvard, USA)
No: 428 **Juergen Jost** (Max Planck Inst, Germany)
No: 429 **Mitsuo Morimoto** (Int Christ U, Japan)
No: 430 **Toshigazu Sunada** (Tohoku U, Japan)
No: 431 **Martin Taylor** (London Math Soc, UK)
No: 432 **Abraham Zaks** (Israel Math Soc)
No: 433 **Huai-Dong Cao** (Texas A&M U, USA)
No: 434 **Yeneng Sun** (Nat'l U Singapore)
No: 435 **Shih-Tsai Feng** (Nat'l Tsing-Hua U)
No: 436 **Hsungrow Chan** (Nat'l Ping-Tung T C)
No: 437 **Kwok-Keung Au** (Chinese U HK)
No: 438 **Jann-Long Chern** (Nat'l Central U)
No: 439 **Chin-Man Fok** (HK Baptist U)
No: 440 **Louis H Y Chen** (Nat'l U Singapore)
No: 441 忻 元 龙 (复旦大学)
No: 442 王 元 (中科院数学所)
No: 443 冯 琦 (中科院数学所)
No: 444 陆 柱 家 (中科院数学所)
No: 445 杨 乐 (晨兴数学中心)
No: 446 黄 且 圆 (中科院软件所)
No: 447 韩 锋 (中科院数学所)
No: 448 张 洪 云 (中科院数学所)
No: 449 任 南 衡 (中科院数学所)
No: 450 范 同 春 (中科院数学所)
No: 451 李 炳 仁 (中科院数学所)
No: 452 陈 翰 馥 (中科院系统所)
No: 453 章 祥 荪 (中科院应用数学所)
No: 454 袁 亚 湘 (中科院计算数学所)
No: 455 蔡 大 用 (清华大学)
No: 456 袁 向 东 (中科院数学所)
No: 457 冯绪宁 (中科院数学所)
No: 458 李福安 (中科院数学所)
No: 459 孟 实 华 (湖南教育出版社)
No: 460 郑 丽 华 (广东教育出版社)
No: 461 慕 鹏 (山东教育出版社)
No: 462 刘 贵 廷 (河北教育出版社)
No: 463 黄 明 雨 (江西教育出版社)
No: 464 刘 嘉 善 (科学出版社)
No: 465 孙 建 华 (南京大学)
No: 466 安 丰 稳 (南京大学)

No: 467 瞿 成 勤 (南京大学)
No: 468 ()